Charles Seale-Hayne Library
University of Plymouth
(01752) 588 588
LibraryandITenquiries@plymouth.ac.uk

LANDSLIDES
GLISSEMENTS DE TERRAIN

PROCEEDINGS OF THE SIXTH INTERNATIONAL SYMPOSIUM
COMPTES RENDUS DU SIXIEME SYMPOSIUM INTERNATIONAL
10-14 FEBRUARY 1992/CHRISTCHURCH/10-14 FEVRIER 1992

Landslides
Glissements de terrain

Editor/Rédacteur

DAVID H. BELL

Geology Department, University of Canterbury, Christchurch
New Zealand/Nouvelle Zélande

VOLUME 3

Additional papers/Contributions additionelles
Reports/Rapports
Discussions

A.A.BALKEMA/ROTTERDAM/BROOKFIELD/1995

Cover photograph: Courtesy of Otago Daily Times, Dunedin

The texts of the various papers in this volume were set individually by typists under the supervision of each of the authors concerned. / Les textes des divers articles dans ce volume ont été dactylographiés individuellement sous la supervision de chacun des auteurs concernés.

Published by / Publié par
A.A. Balkema, Postbus 1675, 3000 BR Rotterdam, Netherlands (Fax: +31.10.4135947)
A.A. Balkema Publishers, Old Post Road, Brookfield, VT 05036, USA (Fax: 802.2763837)

Complete set of three volumes: / Collection complète de trois volumes: ISBN 90 5410 032 X
Volume 1: ISBN 90 5410 033 8 / Volume 2: ISBN 90 5410 034 6 / Volume 3: ISBN 90 5410 035 4

Landslides, Bell (ed.) © 1995 Balkema, Rotterdam, ISBN 90 5410 032 X

Editor's note

Volume 3 includes the Theme Addresses and Theme Reports that were formally presented at the Sixth International Symposium on Landslides held at Christchurch, New Zealand, in February 1992. In addition those papers that were received and accepted for publication after the deadline for inclusion in either Volume 1 or Volume 2 are printed here, together with discussion notes and questions posed at various stages of the Symposium. Following the format of Volume 1 and 2, papers are arranged by Theme commencing with G1 and concluding with S5, whilst discussion is recorded by session.

It had been my intention to include in Volume 3 reprints of the field trip guides that were issued to participants at ISL92, but that proved to be impossible because of the large amount of editing and figure redrafting required. Nevertheless, some 64 new papers and reports are included here, and represent a significant additional data source on landslides which will undoubtedly be of interest to the geotechnical profession. I note in particular the Theme Addresses which include state-of-the-art information relating to the various Symposium Themes.

David H. Bell
Symposium Convenor

Landslides, Bell (ed.) © 1995 Balkema, Rotterdam, ISBN 90 5410 032 X

Table of contents

G2   *Stability analysis techniques*
       Techniques des analyses de stabilité

G1 Landslide investigations
Investigations sur les glissements de terrain

Landslides, Bell (ed.) © 1995 Balkema, Rotterdam, ISBN 90 5410 032 X

Keynote paper: Geological modelling in landslide investigation

David Stapledon
Adelaide, S.A., Australia

ABSTRACT: Several post-failure studies of Vaiont Landslide, and a few studies of subsequent slides appear to have been conducted without consideration of all relevant issues or questions, and without proper definition of the slide geological model. With these deficiencies in mind this paper first lists questions which should usually be addressed during studies aimed at the recognition of and investigation of landslides, including dormant, active and potential first-time slides. It then addresses the answers to the questions, with emphasis on methods which contribute most to geological modelling. It concludes with a brief account of geological studies of the downstream right abutment ridge at Thomson Dam in Victoria, Australia. The studies were conducted because at one stage it appeared that with the reservoir filled part of the ridge had potential for first-time failure.

RESUME: Plusieurs études d'après rupture du glissement Vaiont et quelques études de glissements subséquents semblent avoir été conduites sans considération aucune de toutes les issues ou questions qui s'y rapportent, et sans définition propre du modèle géologique de glissement. En tenant compte de ces carences ce document fait d'abord la liste des questions qui devraient normalement être posées au cours d'études dont le but est la reconnaissance et également les recherches sur les glissements de terrain, y compris les glissements en repos, en activité et latents pour la première fois. Il aborde ensuite l'application et les limites des méthodes qui peuvent être utilisees pour donner des réponses aux questions, en insistant sur les méthodes qui contribuent le plus à la création de modèles géologiques. Il conclue avec un bref rapport sur des études géologiques du contrefort droit en aval de la crête à Thomson Dam, dans le Victoria, Australie. Les études ont été faites car il est apparu à un certain moment qu'une fois le reservoir plein une partie de la crête présentait pour la première fois une possibilité de rupture.

1. AFTER VAIONT

One of the world's greatest landslide disasters occurred in 1963 at Vaiont. During the following 15 years numerous investigations of the slide were made by eminent specialists, but their reports showed no general agreement on important issues such as the presence of clays along the failure surface, water pressures, the failure mechanism, and whether the slide was a first time event or reactivation of an old slide.

Twenty-two years after the event and after a comprehensive study of the slide carried out over several years, Hendron and Patton (1985) provided the first assessment to use 3-dimensional analyses together with a reliable geological model. The latter was based largely on extensive field work done both before and after the slide, by Italian geologists Professors Edoardo Semenza and Daniele Rossi.

Although there may be further debate about some aspects of the Hendron and Patton conclusions there is no doubt that their report has been accepted as, and is likely to remain, the authoritative account of the slide.

As well as showing serious deficiencies in the pre-slide investigations, the report concludes that the earlier post-slide studies mentioned above varied from "useful accounts to misleading fiction," and that "The most misleading accounts.... have generally been given by those who have not visited the site or who are not familiar with the geology."

Hendron and Patton also provided a more fundamental conclusion:

"Because of the great diversity in geologic and hydrogeologic environments among projects, it is difficult and perhaps misleading to attempt to set rules for analyses and field exploration programs which would cover all landslide studies."

I would be strongly opposed to "rules". However, the first two conclusions quoted above, and a few papers describing studies of more recent landslides, suggest that even post-Vaiont the need to define the geological model is not understood fully by some who are involved seriously in landslide investigations. With this in mind the present paper will discuss the scope of landslide studies and methods commonly used for them, with particular emphasis on definition of geological models.

2. QUESTIONS FOR LANDSLIDE INVESTIGATORS

Geotechnical investigations are about answering questions. The questions fall naturally into two categories :
  A. Questions associated with types of engineering structures or projects, and
  B. Questions arising from the geological environments at particular sites.
  In many cases of failures, delayed projects and contractual disputes resulting from "changed geological conditions" it has been found (Stapledon, 1976, 1983a), that a major contributing factor was the failure of the site investigators to define (and in some cases to understand) all of the questions which needed to be answered. The following observation, attributed to the famous French detective Bertillion, is considered to apply:

" We only observe that which we can see, but we can only see that which is already in the mind."

Recognising this as a deficiency in dam engineering, Fell et al (in press) have compiled check lists of questions in the above two categories, based on experience in the planning and construction of many embankment dams in eleven different geological environments.
  I suggest that comparable check lists could be compiled for the guidance of investigators of landslides, and lists in Category A, considering landslides as a type of "project", are set out below in 2.1 and 2.2 and in Appendices 1 and 2. I have not listed questions in Category B

but some should become evident from the discussions of case histories in Section 3 of the paper.
  At any particular site not all questions listed may be highly relevant, but it is important that they all be listed so that the relevance of each can be judged. Some questions may be answered adequately in an hour or less from surface inspection or from published data, while others may take weeks or months and involve expensive subsurface exploration.
  The questions in 2.1 apply to slope movements of all types. Those in 2.2 and Appendices 1 and 2 relate only to rotational and translational slides, as defined by Varnes (1978) and Hutchinson (1988). In compiling these lists it has been assumed that the slides occur, or are suspected, at the sites of proposed civil works. When slides are discovered during construction or after development, the decision options (broad questions) will usually differ from those stated, but most of the detailed questions will be the same.

2.1 Questions relating to recognition of landslides

From the practical engineering viewpoint, the most important aspect of landslide investigation is early recognition, as very often the best approach to a slide is to avoid it, by relocation of the project or by placing restrictions on development of the affected area.
  Fortunately, active landslides and even the scars of very old, dormant slides can be recognised usually without great difficulty from geomorphological and other features, observed and plotted by air photo interpretation followed by direct examination of the ground. These methods are discussed in Section 3 below.
  The questions to be asked by investigators relate mainly to the features referred to above, and include:
  1) Anomalies in topography?
  2) Anomalies in vegetation?
  3) Anomalies in surface soil profile?
  4) Evidence of diverted or temporarily dammed stream or streams?
  5) Disruption of man-made features, eg. pipelines, pavements, kerbs, etc?
  6) Anomalies in the geological situation?
  These questions are addressed in detail by Rib and Liang (1978), Sowers and Royster (1978) and Hunt (1984).

2.2 Questions relating to a dormant slide, or scar of an old landslide

Recognition of a dormant slide at the site of proposed construction works usually raises the following broad questions:
A. Probability of reactivation?
B. Damage potential if reactivated?
C. Relocate proposed works or design them to stabilise or accommodate slide?
Investigations to provide answers to these will usually be aimed at all or most of the following detailed questions:

1) Geological setting, regional and local?
2) Type of slide?
3) History of past movements?
4) Any current creep movements?
5) Nature of any climate change since past movements?
6) Nature of any site change since past movements?
7) Boundaries, dimensions and volume?
8) Is the slide part of a larger one?
9) Configuration and properties of basal failure surface or zone?
10) Configuration and properties of other bounding surfaces or zones?
11) Internal structure of the slide?
12) Relationships between 7) to 11) and the structure of the parent mass?
13) Regional groundwater picture?
14) Groundwaters; distribution and pressures in slide and parent mass?
15) Relationships between groundwater picture and structure of the slide and parent mass?
16) Effect of rainfall/streamflows on groundwater pressures?
17) Slide mechanism; cause of past movements?
18) Reason for cessation of movements?
19) Allowances for seismic and/or extreme rainfall events?
20) Annual Exceedance Probabilities (AEPs) of these events?
21) Current factor of safety and/or probability of reactivation?
22) Effect of proposed works on factor of safety?
23) Amount of displacement if activated?
24) Speed of displacement if activated?

3. ANSWERING THE GEOLOGICAL QUESTIONS

Some methods commonly used to determine the geological and groundwater models during landslide investigations are listed below and discussed in 3.1 to 3.13 which follow. As the "basics" of most of these methods are well known, the discussions concentrate on the questions to which each method may be applicable, and the limitations of each method. The question numbers referred to are those listed in 2.2.

1. Assessment of local records
2. Assessment of regional geology
3. Interpretation of photographs
4. Geomorphological/geological mapping
5. Pedological studies
6. Dating past movements
7. Estimation of slope retreat rates
8. Monitoring slide movements
9. Geophysical methods
10. Trenches and pits
11. Drilling
12. Shafts and drives
13. Groundwater studies

3.1 Assessment of local records

Old newspapers and books on history can provide useful accounts of past movements of dormant or active landslides. Anecdotal accounts and photographs can often be obtained from elderly residents. Such accounts are often lacking in precision but can help the investigator to establish the cause of initial movements, or to locate important field evidence. Thus contributions can be made to Questions 1 to 8.
The classic example of use of anecdotal evidence was in Switzerland in 1881 where Albert Heim leaned heavily on eyewitness accounts in his study of the rockfall and sturzstrom which killed 115 persons in the town of Elm (Hsu, 1978).
Moon et al (1992) describe how the need for debris flow risk zoning in a well established housing area was appreciated from an historical account of a debris flow which occurred through the area in 1891, when it was sparsely inhabited.

3.2 Assessment of regional geology

A regional geological study of the area surrounding a site can usually provide an initial indication of its "geotechnical environment", as defined by Patton and Hendron (1972). In particular it should provide general information about:
a) the age and history of development of the slope in question,
b) the stress history and seismicity of the area,
c) the material in which the slide developed, ie. its parent or source, and
d) the regional groundwater system.

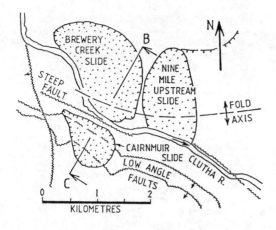

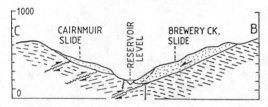

Figure 1. Part of regional geology plan, and section across valley, Cromwell Gorge. (From Gillon and Hancox, 1992).

The regional study should therefore generally contribute to answering Questions 1, 3, 5, 6, 13 and 20. It can also provide information about possible slide boundaries (Questions 7 to 10) by indicating the likely positions of bedrock faults and/or folds concealed beneath a landslide but not exposed near it. Failure surfaces of many rockslides have developed wholly or partly along such features.

This is illustrated by Figure 1 which shows part of the regional plan prepared specially for the study of the Cromwell Gorge slides (see Gillon and Hancox, 1992). It shows two low angle faults (probably thrusts) on the south bank of the river. In some places seepages were indicated from these faults. It was postulated that similar faults were present beneath the north side of the valley, and that Brewery Creek and Nine Mile slides had developed along them. Exploration including drives showed this to be true for Brewery Creek and Nine Mile (Upstream) Slides.

The seepages from the fault exposures on the south side of the valley suggested that these faults formed barriers to the regional groundwater flow. This was confirmed by drilling and groundwater monit-

oring at the Cairnmuir Slide (Gillon et al, 1992a).

The regional map thus provided a logical base for the planning and interpretation of more detailed geological mapping and subsurface exploration at and around the landslides.

There is one aspect of this regional map which calls for attention in future studies. Classification of the schist bedrock types was based on metamorphic textural zones (Turnbull, 1987) and did not distinguish between rocks which were quite variable from the engineering point of view. During the detailed landslide studies, rocks within one textural zone (Zone IV) were found to range from very strong, almost gneissic types (eg at Clyde and Jacksons Slides) to much weaker, more fissile and ductile types (eg. at Nine Mile Downstream).

This large difference in rock substance properties has had a major influence on the types of slope movements which have occurred in the different areas. For example, the slides in the more gneissic types show basically wedge or slab type mechanisms with failure along pre-existing defects of tectonic origin (Section C-B on Figure 1), while the detailed studies at Nine Mile (Downstream) have indicated that in much of this area the very ductile schist has been deformed pseudo-plastically by "sagging" and toe buckling as described by Beetham et al (1992) and McSaveney et al (1992). It is therefore considered important that "classical" maps to be used in engineering studies be checked carefully for any such aspects caused by differences in outlook during geological mapping.

3.3 Interpretation of photographs

The boundaries and much of the structure of many landslides (Questions 6 to 12) can be identified by stereoscopic examination of vertical air photographs. The method is described by Rib and Liang (1978), Sowers and Royster (1978), Hunt (1984) and Lillesand (1987). The ease and reliability of the method depend on the degree of development of the morphological features, the angle of lighting by the sun, and the height and density of the vegetation.

Figure 2 shows an extremely large slide in the Southern Highlands of Papua New Guinea. The slide is in folded karstic limestone and appears to have blocked a former valley of the Mubi River, when the river was displaced 400m horizontally by a fault (Figures 3 and 4). Despite the dense

Figure 2. Air photo showing landslipped mass blocking old river valley, Papua New Guinea.

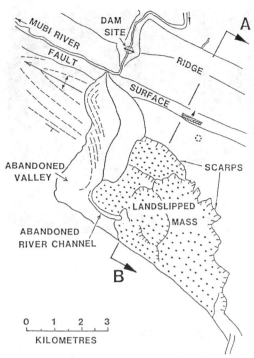

Figure 3. Sketch plan showing geology at landslide in Figure 2.

rain forest cover, the large sharply defined scarps and oblique lighting make this slide easy to recognise.

Figures 5 and 6 show the scar of a relatively shallow slump and mudflow, at the site of a portal for a tunnel in Victoria, Australia.

The geomorphic features are more subtle (Figure 7) but oblique lighting and lack of dense vegetation make this slide also quite easy to recognise under a stereoscope. Air photos were not examined, however, until a dispute arose about poor portalling conditions (Stapledon, 1983b). The slide is in extremely weathered dacite (high plasticity silt, MH) and the portal works were in a lateral shear zone in which the material was wet and soft due to remoulding.

Some slide-developed features on the very old, creeping slides of the Cromwell Gorge (described by Gillon and Hancox, 1992) are difficult to recognise on air photos despite the lack of vegetation on them. Most scarps are relatively small and well rounded due to subsequent and ~ntinuing erosion.

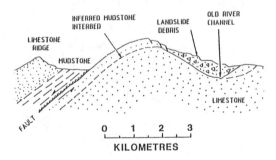

Figure 4. Sketch section showing geology at landslide in Figure 2.

Oblique photographs taken either from the air or from suitable ground locations are useful both for the recognition of slides and for delineation of subtle features such as minor, rounded scarps. The latter are usually difficult to see in direct frontal views.

The difficulty of interpretation in densely vegetated areas and the importance of checking on the ground have been illustrated by Fookes et al (1991), by comparing the results of interpretations

Figure 5. Air photo showing scar of old landslide and mudflow, near Emerald, Victoria, Australia.

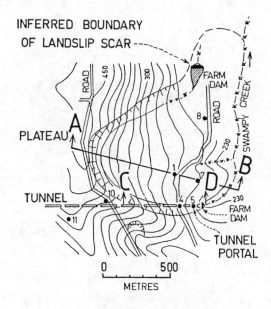

Figure 6. Plan of landslide shown on Figure 5 and cross section C-D along tunnel.

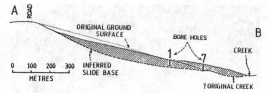

Figure 7. Section through the landslide shown on Figure 5.

by different professionals at the same site. Five interpretations were made of the site of Ok Ma Tailings Dam in Papua New Guinea, prior to the landslide which demolished the site, in January, 1984. Two people found no landslips at the site. The other three found evidence of slips but each in different configurations.

3.4 Geomorphological/geological mapping

It is clear from the the Ok Ma experience that when dense, tall vegetation limits the effectiveness of air photo interpretation, ground surface mapping of the types of features listed in 2.1 above is vital, just for the recognition of some slides.

For any landslide, systematic mapping of these features in the slide area and of the geology of the surrounding area is essential for a cost-effective and sucessful investigation, in my opinion. Usually it contributes much to answering Questions 1 to 15.

Where contours are not available, slope angles and all morphological features are mapped using symbols such as those of Savigear (1965) and Gardiner and Dackombe (1983). Examples of landslide maps made in this way are in Walker et al (1987) and Moon (1984).

Where an accurate contour plan exists or can be prepared it is preferable in my opinion to use this as a base plan onto which geological, geomorphological and groundwater data are plotted mainly by means of symbols. Slope angles and symbols for some morphological items are not required if they are shown adequately by the contours.

"Geotechnical mapping" of this kind was carried out at each of the Cromwell Gorge landslides. Figure 8 shows a small part of the geotechnical plan of the Clyde slide, the original of which covered the whole slide and its surrounds on 1:2000 scale. By comparison with Figure 9 which shows most of the slide boundaries in broad outline, it can be seen that the mapping delineated the upper boundaries of the

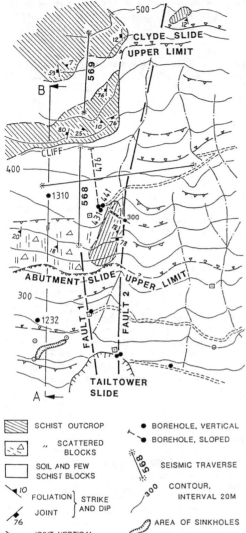

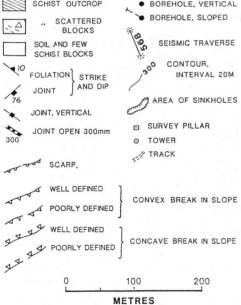

Figure 8. Clyde Slide, Cromwell Gorge. Part of detailed geotechnical plan, with legend.

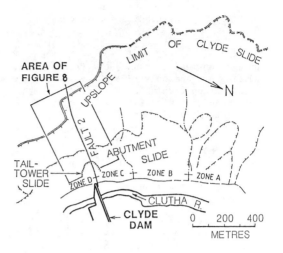

Figure 9. Plan showing outline of Clyde Slide and inferred zones within it.

slide and of the more active Abutment Slide which occurs within it. Also indicated was the upslope-downslope trending Fault 2, which forms the boundary between Zones C and D of the slide.

Similar results were achieved at the other Cromwell Gorge slides. In this way the mapping provided targets or questions to be answered by ongoing subsurface exploration. Without reasonably well defined questions to answer, these expensive exploratory works (boreholes, trenches, etc) are "wildcats" and as such are not likely to be cost effective.

It can be seen also on Figure 8 that as well as natural features, the positions of cross sections and of all exploratory works and monitoring points are shown. These are added to the plan as they are constructed, and enable the user of the plan (and sections) to have an uptodate, integrated picture of the results of the investigation. Also shown are tracks and all features which assist users to locate themselves on the ground.

To fit all of the above types of data on a plan while maintaining clarity, and also contrast where necessary, requires careful choice of symbols and line thicknesses, and limitation of the amount of detail to that appropriate to the scale of the map. Suggested "standard" symbols eg. those of Matula (1981) and Dearman and Fookes (1974) do not always give a clear picture. Where necessary to achieve clarity, some symbols should be designed specifically for the project.

1505

3.5 Pedological studies (Questions 3 to 8)

Surface or "pedological" soils which may have taken many thousands of years to develop are commonly removed by, buried by or caught up within landslides. Also, when an exposed scarp or landslipped mass of "engineering soil" becomes stabilised, a new surface soil profile starts to develop on it. Hence in areas where the distribution of the established surface soils is known, past landsliding may be indicated by profiles which are absent, thinner than or different to the norm, or buried under obviously younger soil. Landsliding is not always indicated, because soil profiles can be modified by other (eg. creek) erosional and depositional processes.

Moon et al (1992) describe how careful mapping of pedological soil profiles at Montrose, Victoria, led to recognition of a younger soil deposited by a debris flow which occurred in 1891. This led to the location of the scar of the precursor landslide, hidden in dense forest near the crest of Mt. Dandenong.

3.6 Dating past movements (Questions 3 to 6)

Methods commonly used for determining the approximate age of landslide movements include radiocarbon dating, stratigraphic relationships and tree-ring dating. A less common archaeological approach used at Mam Tor landslide is described by Johnson (1987). Only the three common methods are discussed here.

3.6.1 Radiocarbon dating

Where a landslipped mass contains the remains of trees which died as a result of burial at the time of sliding, the approximate age of the slide event can be found by radiocarbon dating of samples of the timber (wood or charcoal). Worsley (1981) describes the principles of the method, requirements for sampling and preservation of samples, and errors which may be introduced by natural contamination (ie. in the ground, prior to sampling).

Recent landslide studies in which this method was used are described by Schlemon and Wright (1987), Osterkamp and Hupp (1987) and Hutchinson and Chandler (1991).

3.6.2 Stratigraphic relationships.

The stratigraphic relationship between slide debris and other deposits of known age can provide a general indication of the history of slide activity. For example the landslides in the Cromwell Gorge abut or occur immediately upslope of terraces formed by glaciofluvial gravels believed to range in age from 16,000 to 35,000 yrs. As reported by Gillon and Hancox (1992) there is no evidence that the very large slides have overridden or failed onto these terraces, indicating no major slide movements during the last 35,000 yrs. Thin layers of slide material intertongued with the gravels at some slides are believed to result from minor slide movements during river erosion and gravel deposition.

Hutchinson and Gostelow (1976) carried out detailed sedimentological and geomorphological studies on colluvial materials downslope from an abandoned sea cliff in London Clay in Essex, UK. They concluded that four main stages of landsliding had occurred in the last 10,000 years, each correlating with known periods of wetter climate.

3.6.3 Tree-ring dating

Trees which are tilted during landslide movements and then become bent due to upward growth, record the year of the tilting by developing assymmetric growth rings. Schroder (1978) and Osterkamp and Hupp (1987) describe the application of this phenomenon to landslide dating.

3.7 Estimation of slope retreat rates (Questions 3 to 6).

Published maps have been used to show rates of recession of cliffs due to landsliding and erosion, in areas where the rates are high enough to be discernable and greater than any errors in the maps. Holmes (1972) used map evidence to show that the London Clay cliffs at Warden Point on the Isle of Sheppey (UK) had retreated at about 2.2m/yr between 1819 and 1953. Also Brunsden and Jones (1976) showed an average rate of recession of 0.71m/yr between 1887 and 1964, with variable intermediate rates, for Stonebarrow Cliff, in Dorset, UK. They used 1:2500 Ordnance Survey plans.

The studies of Hutchinson and Gostelow (1976) mentioned above in 3.6 showed that the London Clay cliff had retreated about 50m during the last 10,000 years.

The coastline near Sydney, Australia has been formed by the "drowning" of relatively steep rocky slopes during the rapid rises in sea level which occurred during late Pleistocene and Recent times. The rise has been shown to have stopped about 6000 years ago (Fairbridge, 1961, Bowen,

1978, Lambeck and Nakada, 1990 and Emery and Aubrey, 1991). Since that time a wave cut terrace has developed (Figure 10). The width of this terrace gives an indication of the average recession rate of the mainly sandstone cliffs, during the last 6000 years. The average recession rate found in this way was used recently in studies of the potential for a first-time slide at the site shown.

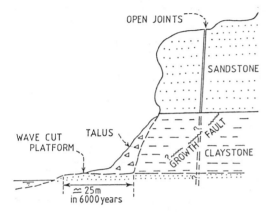

Figure 10. Projecting coastal cliff near Sydney showing wave cut platform believed to have formed during the last 6000 years.

3.8 Monitoring slide movements (Questions 3 to 9 and 16 to 18)

Monitoring methods are being dealt with in detail under Theme 5.

3.9 Geophysical methods (Questions 7 to 12 and 14 and 15)

In my experience geophysical methods have yielded best results where
a) the work is planned, conducted and interpreted with well defined objectives and good understanding of the geological situation
b) the ground studied contains bodies or layers with contrasting properties and which are separated by sharp, regular boundaries, and
c) the groundwater picture is simple

Criterion a) is simply a matter of good management, but b) and c) are seldom met in landslides, and this limits the value of geophysical work in many landslide studies.

3.9.1 Soil slides

Slides on and derived from low strength soils rarely show enough boundary contrast for seismic methods to be useful. Even in the deep weathering mantle present over much of Australia, it is not always possible to distinguish between variably weathered soil-rock masses in situ, and similar materials which have been displaced or disrupted by landsliding. Also, in deep soils the size of explosive charge necessary to achieve depth penetration may initiate or accelerate slide movements. This happened at a slide in wet extremely weathered granite in the Snowy Mountains (Australia) as described by Hosking (1960).

Electrical resistivity soundings from surface electrodes is also of limited value in soil slides due to the usual variability in slide materials, irregular shape of basal surface, and complex groundwater picture (Hutchinson, 1983).

Notwithstanding the above, Palmer and Weisgarber (1988) claim that seismic refraction and confirmatory resistivity studies have delineated the slip surface (up to 12m deep) at the Stumpy Basin slide near the Cuyahaga River in Ohio. This slide is in "heavily overconsolidated" silts and clays - lake deposits. Typical geophysical properties from 7 traverses around the fairly irregular slope were indicated to be as follows:

Table 1. Summary of geophysical results, Stumpy Basin Landslide.

	Longitudinal velocity Vp km/s	Resistivity (ohm-m)
Slide material	0.35	30
Undisturbed material	1.6	60

Very complex travel-time curves were interpreted assuming that non-linearity within them was due to changes in configuration of the refracting interface rather than to gradients in the velocities of the layers. Good correlations are claimed between the seismic and resistivity profiles and with the mapped surface scarp of the slide. However I suggest that it would be prudent to check the interpretations against borehole or other direct subsurface data on the soils and groundwater,

Table 2. Summary of seismic refraction results, Downie and Cromwell Gorge Slides (data from Piteau et. al, 1978, Bryant et. al, 1992 and Macfarlane et. al, 1992).

| | DOWNIE SLIDE | | CROMWELL GORGE SLIDES | | |
Zone	Description	Vp km/s	Zone	Description	Vp km/s
A	Rock-soil mixture, dry, local large voids	0.4 - 0.8	Chaotic debris (dry)	Disoriented blocks in matrix of sheared, crushed and gouge materials	0.6 - 0.8
B	Moderately to highly fractured rock	1.25 - 2.8	Dis-placed schist	Large intact blocks touching or separated by infill or sheared or crushed material	1.2 - 3.0
C	Rock slightly fractured by slide movement	2.5 - 3.5			
— — — — — — — — — — — Assumed base of slide movements — — — — — — — — — —					
D	Undisturbed bedrock	4.0 - 6.2	Insitu rock		2.8 - 4.6

before relying heavily on them.

Basal failure surfaces of soil slides have been located successfully using downhole or jacked probe electrical resistivity profiling. Hutchinson (1961, 1965 and 1983) describes the use of this method at several sites, including the Furre and Vibstad quick clay slides in Norway. At Furre the slip surface was defined by a sharp change in resistivity, due to large displacements placing leached soil directly over more saline soil. The resistivity probe was jacked to more than 30m without preboring. At Vibstad the slip surface was defined by a contrast in void ratio across it.

3.9.2 Rock slides

The seismic refraction method has proved useful in rock slide investigations as a supplement to direct exploration methods such as drilling, shafts and drives.

Piteau et al (1978) describe results of about 12km of traverses at the Downie (schist and gneiss) slide in British Columbia, and Bryant et al (1992) report on about 62km of traverses (including both refraction and reflection) at the Cromwell Gorge slides, which are also in schistose and gneissic rocks. Table 2 summarises and compares seismic results at Downie and Cromwell Gorge slides.

The text of Piteau et al (1978) is vague in its interpretation of the base of the slide at Downie, but it seems likely that the base of Zone C is roughly equivalent to the base of Displaced Schist at Cromwell Gorge slides.

Bryant et al (1992) have not found precise correlations between boundaries identified by refraction and reflection methods, and by downhole seismic profiles, drill cores and drives. However they found "general trends" to be reliable and useful for checking and extending data from other exploratory methods.

Irregular, steep surface topography can limit the lengths of seismic traverses and hence the depth to which data can be obtained. Short seismic lines 568 and 569 near the upper boundary of Clyde Slide (Figures 8, 11 and 12) did not penetrate to undisturbed bedrock but showed distinct steps in the 2.5/3.2 km/s refractor, across and just above the cliff face which is believed to mark the slide boundary in this area.

The seismic results therefore support the stepped rather than the deep curved failure surface shown on Figure 12. See also 3.11 for discussion of drilling results in this area.

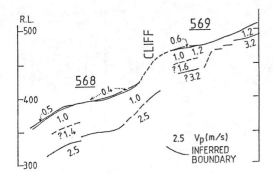

Figure 11. Section along Seismic Lines 568 and 569 near the top of Clyde Slide. For position of lines see Figure 8.

3.9.3 Rock-noise monitoring to locate active failure surface.

Rock-noise or microseismic monitoring was developed initially to give warning of imminent collapse in underground mines (Obert and Duvall, 1957). Piteau et al (1978) report on its use at Downie Slide, to indicate areas of slide activity and to locate active failure surfaces. For this latter purpose three boreholes were monitored by geophone lowered by cable. This method would be unsuitable for use in more active slides due to likely loss of the probe down the hole.

Novosad et al (1977) describe the use of this method at creeping soil and rock slides. Polyethylene casing is grouted into holes and filled with water which acts as an acoustic conductor between the tube and probe. In the soil slides the noise events originate from fracturing of the grout and plastic at failure surfaces.

3.10 Trenches and pits (Questions 1 to 3, 6 to 12, and 17)

Trenches cut either by ripper and bull-dozer or by large backhoe type excavator are considered to be a very cost-effective method of obtaining data on mechanisms and boundaries of rock-slides, if carefully located and oriented to answer questions on these aspects. Hence trenching is best carried out after completion of the surface mapping described in 3.4.

Bulldozer trenches were used extensively at Sugarloaf Dam project in Victoria, Australia, where thinly interbedded silt-stones and sandstones contain numerous crushed seams mostly 1 to 20mm thick formed by interbed displacements during

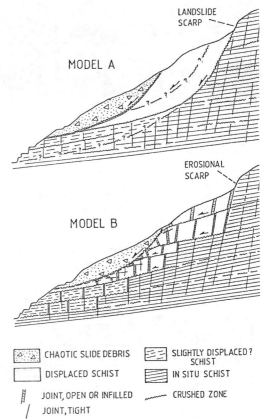

Figure 12. Diagrams showing alternative models for Clyde Slide.

folding (Casinader and Stapledon, 1979). The rocks are weathered within 10 to 20m of the surface, and form dip-slopes and near dip-slopes ranging from 20 to 30 degrees at the sites of both the main 85m high dam and pumping station.

The slopes generally showed few rock exposures and most showed no topographic evidence of past instability. However long trenches cut upslope-downslope showed much evidence of past downslope movements, including clay-infilled near-vertical joints and major extension and collapse features such as those shown on Figure 13.

The downfolded collapse feature on Figure 13 is believed to have formed due to upper beds falling into a widely gaping joint formed by concentration of strain in lower beds. Because of the weathered and broken nature of the rock these near-vertical features were rarely recovered or recognisable in diamond drill cores.

The trenches were used also to study the

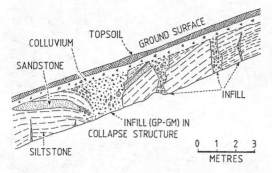

Figure 13. Elevation view of trench wall showing extension features in dipslope at Yering Gorge, Victoria. (Regan, 1980)

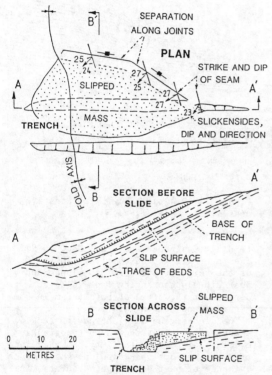

Figure 14. Plan and sections through trench cut down local folded section of right bank dipslope at Sugarloaf Dam, Victoria.

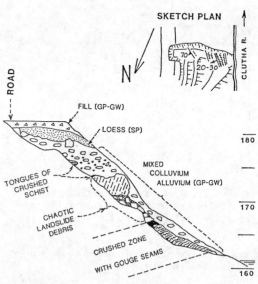

Figure 15. Section along wall of trench cut into the breakout zone at the toe of Brewery Creek Slide, Cromwell Gorge.

waviness of the seams and to get samples for direct shear and other tests.

The trenches were cut directly down-dip in order to avoid risk of collapse, as the direct shear tests on undisturbed seams had shown residual strengths of about 10 degrees. Despite this precaution, the side of one trench cut slightly oblique to dip direction failed across-dip as shown on Figure 14.

This failure occurred in dry weather more than 30m above the water table. It illustrates the need for great care during the excavation and logging of trenches and pits, specially those cut into landslipped materials. Also, local regulations on ground support must be followed.

The understanding of past slope movements provided by the trenching, plus the almost zero waviness on the seams, led to adoption of the laboratory effective residual strength of 10 degrees for the seams, in design of the foundation treatment for the dam. The wisdom of this was confirmed by a failure of part of one dipslope by sliding and buckling during construction.

Trenches were not used effectively until a relatively late stage in the studies of some Cromwell Gorge slides, partly because of concern that the depth of "chaotic debris" at the surface of slides would prevent trenches reaching displaced schist in which meaningful geological evidence might be found. Also at some sites there was concern about rocks rolling downslope onto the road or works areas. However trenches were eventually used with good results at the toes of Brewery and Nine Mile Upstream slides, where they exposed basal and other failure zones breaking out near river level. Figure 15 is a section

1510

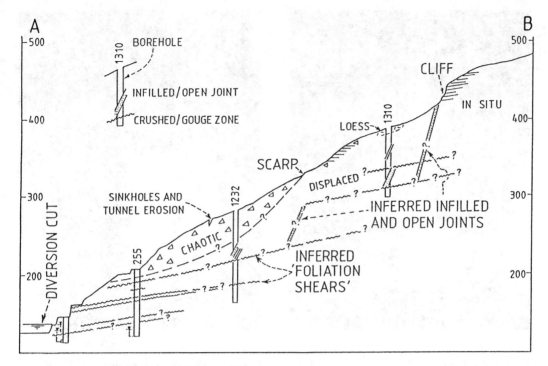

Figure 16. Geological section through part of Clyde Landslide, based on results of surface mapping, seismic refraction and PQ size diamond drilling.

along a toe trench at Brewery Creek, showing 8.5m of crushed rock and silty gouge, forming part of the basal defect zone, beneath slide debris and mixed colluvial/alluvial sands and gravels.

Deep trenching by heavy duty tractor and ripper with cleanup by backhoe would probably have been successful at higher levels at the Cromwwell Gorge slides. For example, a deep trench down from the cliff at the top of Clyde Slide (Figures 12 and 16) would have been an effective way of confirming the stepped failure model.

Trenches are sometimes incorporated into slide drainage systems. Disadvantages of trenches include the danger of collapse if the sides are not battered or supported, and the devastation they cause to the land. However these are usually minor compared with the damage potential should a landslide fail unexpectedly because of an inadequate (but tidy) investigation.

Pits are also useful for exposing slide materials and structure, and for sampling, but because of their small lateral extent they can be ineffective unless located precisely on specific targets. They don't give the valuable continuity of exposure provided by trenches.

3.11 Boreholes (Questions 1 to 15)

Boreholes cause little environmental damage and enable sampling, testing and monitoring through a landslide and into the insitu material. In soil slides and commonly in rock slides, exploratory holes are vertical because of excessive caving and jamming of equipment in angled holes.

Being a small, expensive one dimensional "probe" a borehole needs to be located and designed carefully if it is to be cost-effective. It should be aimed at answering questions arising from mapping, trenching and in many cases geophysical work completed previously.

The holes near the top of Clyde Slide (Figures 9 and 16-18) were aimed at showing which of the two "models" on Figure 12 is valid, in this critical area above the dam. These holes illustrated the benefits of large diameter cores in broken ground. The first holes, 431, 441 and 476 recorded "cavities" and major zones of broken rock with soil properties, but due to the degree of disturbance in the 50mm core it was not possible to determine how much of this material was crushed rock (fault or failure surface) and how much was infill ie. had migrated into steeply dipping joints (Figure 12). The more recent Holes

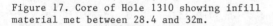

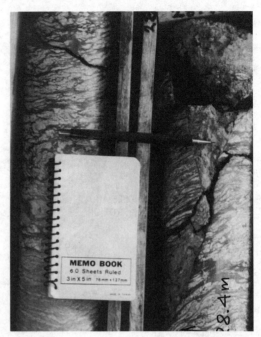

Figure 17. Core of Hole 1310 showing infill material met between 28.4 and 32m.

Figure 18. Closeup view of gravelly silt infill in irregular joint; core of Hole 1310 between 28.4 and 28.7m. Note that the adjacent schist has coarse gneissic texture

1310 and 1232 were drilled PQ size for the installation of inclinometer casing. The 85mm cores (Figures 17 and 18) were much less disturbed and so infilled zones could be distinguished easily from crushed zones, and the stepped model (B on Figure 12) was confirmed.

In contrast, exploration of the portal area on Figures 6 and 7 by 5 boreholes was unsuccessful because the slump scar was not recognised and so the holes drilled (diamond core) were not appropriate for sampling landslipped soil. Actually some of the material was recovered and showed clear slickensides, but these were not recognised (because the question was not asked?).

3.11.1 Logging of boreholes

I have purposely said "boreholes" here rather than "samples" because the record of the borehole needs to be much more than a description of samples. Columns A, B and C on Table 3 summarise the data usually recorded.

Drill cores and soil samples are simply subsurface geological exposures and they are best treated as such by preparation of

a graphic log or columnar section down the hole. If well drawn this part of the log assists the logger and later users by reducing the number of descriptive words required. It is important that the symbols chosen for soils or rocks are able to show the apparent or true dips of their structural fabrics; for example see the schist foliation and fault fabric on Figure 19. Also, boundaries and defects are shown graphically at their apparent or true dips.

When too much detail is recorded the geological picture becomes obscured. 1:100 scale is usually adequate for the basic detailed log. Important features, the details of which cannot be shown at this scale can be described separately by a small number of detailed sketches or photographs.

3.11.2 Orientation of samples

Core orientation devices and impression packers are of limited use in many landslides due to instability of boreholes. Where the orientation of bedding or foliation at the collar is known and can be assumed to be constant along the depth of

Table 3. Principles involved in the logging and interpretation of borehole data.

A	B	C	D
WHAT WE DID	WHAT HAPPENED	WHAT WE GOT UP	WHAT IS DOWN THERE
Penetration method, type of fluid	Penetration rates and pressures, machine behaviour	SAMPLES: Geological description	MUST BE INFERRED from A, B & C plus understanding of the geological situation
Downhole Tests	Behaviour of hole	Engineering description	
	Fluid loss	TEST RESULTS (Properties depend on WHAT WE DID to WHAT IS DOWN THERE)	
	Groundwater behaviour		
	Downhole test results		

the hole, this can be used to orient defects and boundaries met in core samples. However this method can be unreliable, particularly in schistose rocks. Figure 19 shows a steep fault which has developed from a "kink zone" in schist. The fault zone has retained the schistose fabric but in a sheared and crushed condition and in a different orientation. If, as commonly happens, the crushed boundaries of the fault zone are disturbed during coring, the core logger may assume that this fault is parallel to the foliation which dips at 30 degrees. Thus a steeply dipping fault could be inferred to be a gently dipping one, or a basal failure surface of a slide.

An impression packer would have shown the orientation of the fault boundaries on Figure 19, if the hole was not caving. However, boundaries of crushed zones in schist commonly show up to 30 degrees of local waviness, and this also limits the confidence which can be placed on correlations or extrapolations based on oriented intersections in drill cores.

3.11.3 Boxing of drill cores

For rock cores, which generally do not need to be wrapped in plastic (except for samples to be tested) the systematic boxing method shown on Figure 20 is recommended most strongly. This method "reconstructs" the core in the box, to scale, packed with all joint or broken faces fitting, and without any spacers except in sections where core losses have occurred.

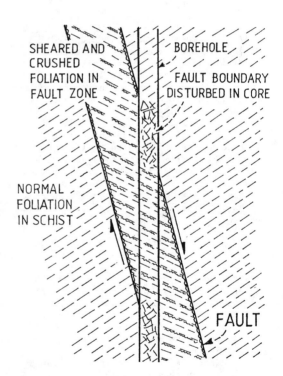

SHEARED AND CRUSHED FOLIATION IN FAULT ZONE

BOREHOLE

FAULT BOUNDARY DISTURBED IN CORE

NORMAL FOLIATION IN SCHIST

FAULT

Figure 19. Diagram showing one of several situations which make it difficult to determine the dip angle of defect zones in schistose rocks, from core samples.

In the example shown, three runs have been placed in the box, from the collar to 3.65m, and the core from the next run (3.65 to 4.91m) is about to be placed in the box.

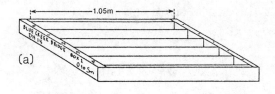

(a)

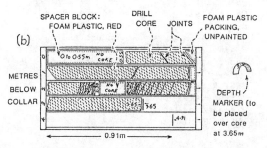

(b)

Figure 20. Systematic boxing of core
samples.

This method is used by most dam-building
authorities and some consultants in
Australia. It is not at all difficult for
the driller. It makes core logging and all
subsequent inspections of the core much
easier, less prone to error and less time
consuming, as little effort is needed to
locate depths of core runs, boundaries,
defects etc. Core photographs are easier
to read and details can be scaled from
them. Cores from adjacent holes at the
same depths can be compared directly when
the cores are laid out side by side.
Further details of the method are in Fell
et al (in press).

3.12 Shafts and drives (Questions 1 to
18)

Shafts and drives can contribute to answering all of these questions because they
allow one to step inside the landslide and
get a 3-dimensional view and record of a
relatively large cross section or sample
of it. They also have the following
advantages:

1. Undisturbed, oriented samples can be
taken for laboratory testing.
2. Large scale tests, eg. direct shear,
can be conducted in them.
3. Monitoring systems can be installed
in them.
4. Drives can be used as drains, and as
access for drilling holes for exploratory
and drainage purposes.

Bored shafts have been used to locate
basal failure surfaces of slides in soils
and very weak rocks (Cummings and Clark,

1988, Hutchinson, 1973, 1983 and Depman et
al 1972. A thick remoulded zone can make
logging difficult in clayey materials.
This is not a problem in hand dug shafts
which have been used to explore slides in
marl (Hutchinson et al, 1973) and in basalt which is extremely weathered to high
plasticity clay (Macgregor et al, 1990).

Hand dug shafts are likely to be of
greatest value where the slide materials
are mixtures of soil and rock which are
impossible to excavate by drilling, and
are too deep or unstable to be explored by
trenching, eg. deep bouldery colluvium.

Where they extend below the water table
shafts require pumping and can be very
difficult to excavate and support. Shafts
in poor ground are also difficult to log;
often the geologist needs to be full-time
at the site so as to log the floor at the
end of mucking and any exposed wall before
it is obscured by lagging.

As noted above drives have great value
as drains and for providing access for
drainage drilling in landslides. Hence
they are used widely for these purposes,
examples being Sierra Slide in Brazil
(Fox, 1957), Fisher Slide in Tasmania
(Paterson et al, 1975), Tablachaca Slide
in Peru (Arnao et al, 1984), Downie and
Dutchman's Ridge slides in British
Columbia (Lewis and Moore, 1988) and the
Cromwell Gorge slides (Gillon et al
1992b).

In some of these cases the main drives
were located mostly below the slides, and
exploration was achieved both by short
drives and exploratory and drainage boreholes generally sloped upwards into them.
This can be necessary because of the very
difficult tunnelling conditions often
encountered in slides. However, for dormant slides which are often difficult to
prove adequately by drilling or other
methods it is considered that highest
priority should be given to the exploratory drives into them to confirm their
status and boundaries.

Drives are used commonly also to explore
and drain dam abutments and slopes just
downstream from them, where some potential
is seen for water pressures to initiate
first-time sliding (see Appendix 2, and
Section 5 below).

3.13 Groundwater studies (Questions 13 to
18, and 21 and 22)

The required understanding of the broad
or regional groundwater picture can often
be obtained by relating observations on
the climate and surface hydrology, to the
regional geological picture. In areas

where groundwater is used for domestic or other purposes there should be records of aquifers, and water qualities, levels and extraction rates, and (hopefully) some published data on the regional model.

Plotting of relevant surface features (eg. drainage pattern, vegetation, sink-holes, swamps, springs, etc.) during the geotechnical mapping of the slide and surrounding area gives an initial indication of the site groundwater picture and how it relates to the regional one, and to the structure of the slide and parent soil or rock mass (ie. Questions 14 and 15). For example, high infiltration rates were suspected from sparse vegetation, surface cracks and sinkholes mapped at Cairnmuir Slide in Cromwell Gorge. Subsequent infiltration and monitoring studies have shown that this slide accelerates after periods of heavier than usual rainfall cause raising of the level in an internal aquifer overlying the active failure surface (Gillon et al (1992a).

Much general groundwater information can be obtained during the drilling of bore-holes eg. from fluid depth versus hole depth (DVD) plots and permeability values from downhole tests. This information, when assessed against a) the other drilling results eg. borehole and machine behaviour, samples and logged defects and b) the evolving geotechnical model, has generally been used to select zones for installation of standpipe piezometers.

Pressure levels from these single or multiple piezometer installations in boreholes form the basis for the slide's groundwater model, and thus for values adopted in stability analyses and planning of drainage works.

Piezometers (especially multiple) are not easy to instal in landslide materials (Patton, 1989) and even if properly inst-talled may not provide a reliable ground-water picture in complex slides. For example, all water bearing zones may not have been located, and long completion zones can give misleading resulting from intersection of more than one aquifer. These and other limitations are illustrat-ed by Patton (1984) using data from 5 boreholes at Downie Slide, and from a 6th hole in which the Westbay CPI modular system was installed. Such modular systems allow monitoring of an almost unlimited number of short zones in a borehole. Other systems which allow monitoring of short zones are the Westbay MP (Black et al, 1986, Patton, 1988) the Piezodex (Kovari and Koeppel, 1985, 1987) and the Piezofor (Groupe de Travail, CFGB, 1970, Debreuille et al 1979).

To obtain good sample recovery in the very poor ground in most rockslides it is necessary to use either mud or foam (Brand and Phillipson, 1984) as drilling fluid. Polymer muds which revert in a few hours to the viscosity of water are used so as to avoid clogging of pores and defects in the rock or soil mass, and of piezometer installations. Experience shows that some clogging and lowering of permeability occurs even when these muds, or simply water, are used (Howsam and Hollamby, 1990, and Fell et al, in press).

Temperature measurements, chemical anal-yses and tritium dating of groundwaters are sometimes used to assist in under-standing of their distribution, sources and flow patterns. Paterson et al (1975) used tritium dating at a slide near the Fisher pressure tunnel in Tasmania. The differences in tritium contents of tunnel water and that from below the slide showed that although the slide resulted from water pressure transmitted rapidly from the leaking tunnel to the base of a talus slope, no tunnel water had actually reached the base of the talus.

4. ANALYSIS AND PRESENTATION OF GEOLOGICAL AND GROUNDWATER DATA

All of the traditional geological methods of analysis and presentation are used, including correlation on plans, cross sections, fence diagrams and isometric projections, and analysis by stereographic methods, structure contouring and ground-water contouring. Computers are now used commonly to carry out or assist with all of these activities. Predictably, the transition from manual to computer methods has not been without teething problems. These will be solved when competent and experienced engineering geologists become more closely involved in production of the software. In my view the worst deficiency in available programmes is their inability to produce graphic sections of borehole logs, or the graphics on geological plans and sections, to the standard needed for investigations in folded and faulted rock masses. This needs to be addressed.

When analysing geological data in a complex area, construction of 3-dimension-al models can assist greatly in the correlation, and is also useful for ex-plaining the geological picture to others. Such models commonly represent boreholes by rods with the geological detail painted on them, and show correlations between holes on sheets of transparent plastic.

Computer generated 3-D images which allow viewing in all directions are likely

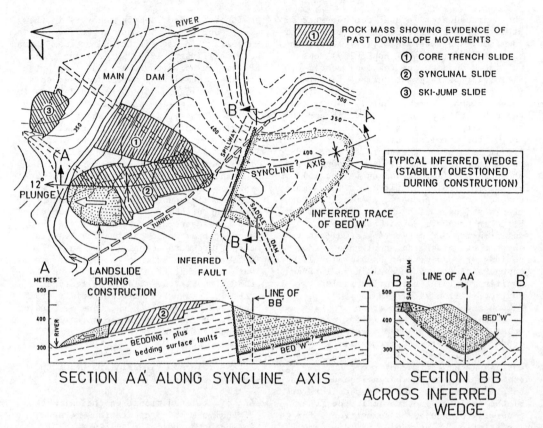

Figure 21. Thomson Dam, plan and sections showing the overall geological structure and landslides met during construction.

to be widely used for this purpose in the future.

5. CLASSICAL GEOLOGY ANSWERS FIRST-TIME QUESTION AT THOMSON DAM

The following brief account of studies of a potential first-time slide area is included here because it illustrates the value of careful and disciplined classical geology, carried out with engineering questions in mind. The studies have been described in detail in Marshall (1985).

Thomson Dam is a 166m high earth and rockfill structure, located 120km east of Melbourne, Australia. It was built between 1977 and 1984.

Figure 21 shows a plan and geological sections at the site. The area is underlain by a "monotonous" sequence of thinly interbedded siltstones and sandstones, which contains no distinct marker beds. Flexural slip folding has produced bedding surface faults (crushed seams) generally

less than 20mm thick and spaced less than 2m apart. There are several folds, and many steeply dipping faults in two sets, one striking roughly north-south and the other roughly east-west. The main fold is the Thomson Syncline which plunges about 12 degrees upstream and towards the river, and underlies a ridge near and just under the upstream shoulder of the dam on the right bank (see also Section A on Figure 21)

Three dormant landslides numbered 1 to 3 on Figure 21 were uncovered during construction of the dam. Slides 1 and 3 had been anticipated from the site investigations, but Slide 2, down the plunge of the syncline, was not found until it was partly reactivated by haul road construction.

Detailed studies were put in hand to determine the structure and mechanism of Slide 2, for use in the design of temporary stabilising works and modifications to the dam design. It was found that the active part of the slide was moving along a crushed seam about 150mm thick, a thrust fault which coincided locally with bedding

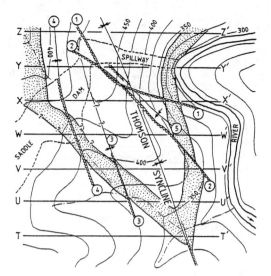

Figure 22. Downstream right abutment ridge at Thomson Dam. Plan showing actual shape of synclinal fold, and major structural features, as found using the detailed stratigraphic picture.

and had been folded within the syncline. Back analyses suggested that this seam, and the other bedding surface faults, had effective residual shear strengths of about 10 degrees, with zero cohesion.

Early in these studies it became clear that the pattern of folds and faults was so complex that the structure of the site and landslides could not be understood until the stratigraphy was properly defined. However in all prior mapping and core logging no marker horizons had been recognised and correlation was generally impossible.

Acting on advice from Dr. Barry McMahon, all drill cores were re-logged, and all surface exposures re-mapped to produce true thickness strip sections on 1:100 scale. The logging and mapping was done following these standards:

a) Beds thicker than 100mm were plotted with boundaries.

b) Individual beds between 50 and 100mm thick were recorded only when they were within thicker beds, and were plotted as line symbols.

c) Individual beds less than 50mm thick were ignored.

d) Sequences of beds less than 100mm thick were recorded as interbedded units.

On these strip logs, siltstone beds were coloured blue and sandstones yellow. The strips were then hung on the office wall and adjusted up and down until distinctly similar sequences of beds were indicated,

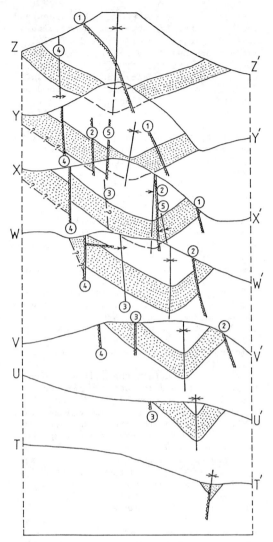

Figure 23. Cross sections through the downstream right abutment ridge.

by more or less continuous bands of blue, yellow and green.

The degree of correlation was remarkable considering that the measured sections were up to a kilometre apart. Individual beds varied in thickness, and some lensed out, but groups of beds and sequences of mainly siltstone and mainly sandstone could be traced across the site. Four separate sequences were found, separated by major thrust faults. In the Main Dam Sequence, 33 mappable units were recognised and numbered from 81 Sandstone (the lowest) to 113 Siltstone at the top.

With the rock sequence now well defined,

the structure at the known slide areas and the dam foundation began to become clear. However, projection of the geological picture southwards raised a serious question about the stability of the ridge on which the spillway and part of the saddle dam were located (see Sections A and B on Figure 21). Using residual strengths from the back-analyses of Slide 2, this ridge appeared to have safety factors of less than unity against sliding along any of the numerous bedding-surface seams which daylighted above river level. This was extremely serious, because if the ridge .was shown to be really unsafe, options for remedial works were very limited at this stage of construction of the project.

Major new site studies were carried out over a period of 16 months, including 1.6km of trenches, adits 250 and 460m long diamond drilling and field and laboratory testing. The primary aim was to determine the detailed structure of the ridge.

Figures 22 and 23 summarise the most important features found. On these figures the actual shape of the synclinal fold and bed displacements across faults are shown diagrammatically by the hatched "layer" which is bounded by the mapped bases of Units 83 and 92.

It can be seen that the radius of the hinge of the fold is much smaller than assumed previously on Figure 21, and that the fold becomes tighter then changes into a fault at the southern end. On Sections W to Z the fold axis dips steeply to the west rather than vertically as on Figure 21. Each of these changes to the assumed model clearly improves the stability.

Other structural features found which also improve the stability of the ridge are numbered 1 to 5 on Figures 22 and 23. Features 1 and 2 are major faults. It had been assumed that these faults, and many smaller faults of the same set, would show normal displacements as on Figure 24. It can be seen on this diagram that if sliding was initiated along Seam A, the normal fault displacement shown would simply result in the failure surface stepping down to Seam B. The same would happen at reverse displacements.

However, Faults 1 and 2 were found to be of the rotational or scissor type shown on Figure 25. Clockwise east side rotations of between 1 and 14 degrees were proven by measured offsets (up to 40m) of the stratigraphic succession, across these faults. These rotations were confirmed by a detailed survey of bedding attitudes which showed values ranging from 3 to 13 degrees.

Features 3 to 5 on Figures 22 and 23 are of the type shown on Figure 26. At one end

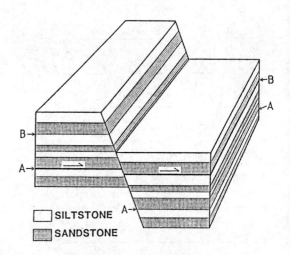

SILTSTONE
SANDSTONE

Figure 24. Diagram showing that a normal fault displacement would simply cause a sliding failure surface along a bedding crushed seam to step down to the closest bedding seam on the foot wall.

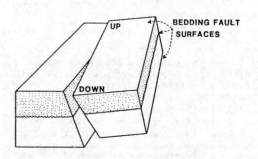

Figure 25. Diagram showing how a rotational or scissors fault disrupts the continuity of bedding, and bedding surface faults.

(A) they are simple flexures in the beds, but along strike they change into fold couples (B) and eventually into faults (C).

The rotations and minor folds of Features 1 to 5 have caused major disruptions to the continuity of the bedding faults. For movement to occur on the bedding faults, either
  a) shearing of rock substance would have to occur, or
  b) large steps downwards of the failure surface would have to occur at the faults, effectively increasing the size of the passive side of the wedge.

Other results of the study which showed the ridge to be much more stable than indicated previously included measured roughness angles of bedding surfaces averaging 3.6 degrees, and discovery that the east-west faults had been rotated slightly during the folding of the syncline and would therefore not provide continuous lateral release surfaces through the ridge, as had been assumed previously.

It was concluded that except for a shallow, weathered zone at its crest, the ridge was kinematically stable.

The Thomson study appears to have disproved a well known fundamental law of site investigation!

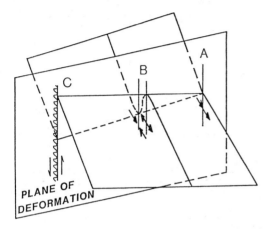

Figure 26. Diagram showing the changes in character of Features 3 to 5, along strike.

CONCLUSIONS

Effective landslide investigations start with definition of general questions such as those listed in the paper. If the geological environment of the slide area is already known, then all known questions relevant to that environment are listed.

Answers to these questions, and further questions which arise as the study proceeds, are then found progressively by applying various investigation techniques in a staged, iterative manner, ie. in a series of loops as suggested by Stevenson and Moore (1976) and Stapledon (1983a).

The methods discussed in this paper are usually appropriate for answering the questions relating to the geological and groundwater models. For best results, the geological work should include classical methods applied in a disciplined manner and always with the engineering questions understood and in mind.

ACKNOWLEDGEMENTS

Permission to publish details of slide studies at the following projects by the organisations listed is gratefully acknowledged:

Clyde Reservoir Project, Cromwell Gorge: Electricity Corporation of New Zealand Ltd WORKS Consultancy Services Ltd and DSIR Geology and Geophysics.

Sugarloaf and Thomson Dam Projects: Melbourne Water.

The geological work relating to Cromwell Gorge slides was carried out by DSIR and specialist subconsultants.

The Thomson Dam studies were carried out by geologists of Melbourne and Metropolitan Board of Works and Snowy Mountains Engineering Corporation, but mostly by the late Allan Marshall, with guidance from Dr. Barry McMahon.

Thanks are due also to Alan Moon for his helpful suggestions on the layout of this paper.

APPENDIX 1.

QUESTIONS RELATING TO AN ACTIVE SLIDE.

When an active slide has been recognised at the site of proposed works, the broad questions will usually be as follow:

A. Avoid slide by relocation?

B. Damage potential if not stabilised?

C. Design works to accommodate or stabilise the slide?

The following detailed questions for the investigator vary only slightly from those listed in 2.2 for a dormant slide:

1) Geological setting, regional and local?

2) Type of slide?

3) Rates and directions of movements?

4) History of past movements?

5) Boundaries, dimensions and volume?

6) Is the slide part of a larger one?

7) Configuration and properties of basal failure surface or zone?

8) Configuration and properties of other bounding surfaces or zones?

9) Internal structure of the slide?

10) Relationships between 5) to 9) and the structure of the parent mass?

11) Regional groundwater picture?

12) Groundwaters; distribution and pressures in slide and parent mass?

13) Effect of rainfall/streamflows on groundwater pressures?

14) Relationships between groundwater picture and structure of the slide and parent mass?

15) Slide mechanism; cause of movements?

16) Relationships between rainfall/ streamflows and slide movement rates?

17) Current factor of safety?

18) Allowances for seismic and/or extreme rainfall events?
19) AEPs of these events?
20) Likely effects on works if not stabilised?
21) How to stabilise?

APPENDIX 2

QUESTIONS RELATING TO A POTENTIAL FIRST-TIME SLIDE.

The slope to be considered here shows no evidence of past or present instability but the possibility of first-time slope failure is to be examined because such failure may cause serious damage to the proposed works. Two alternative situations will be considered, namely
a) where construction or operation of the works will tend to lower the factor of safety of the slope, eg. by the load of structures on it, excavation into it, or by storing water against or behind it, and
b) where the slope will be unaffected by the works but the possibility of its failure due to the actions of natural processes needs to be assessed.

1. Potential first-time slide induced by filling a reservoir

For this common example of a) above, suggested questions to be examined by the site investigator are as follow:

1) Geological setting, regional and local?
2) History of development of slope?
3) Evidence of active or dormant slides nearby?
4) Is geological situation at any nearby slides similar to that at site?
5) Are there active or dormant slides in similar geological, topographic and climatic situations elsewhere?
6) Regional groundwater picture?
7) Groundwaters; distribution and pressures in the slope?
8) Relationships between groundwater picture and geological structure?
9) Effect of rainfall/streamflows on groundwater pressures?
10) Effect of reservoir operation on groundwater picture?
11) Presence of soluble or swelling soils or rocks, known to weaken on inundation?
12) For rock slopes, does geological structure indicate kinematic stability?
13) For all slopes, is there a feasible failure model?

14) Boundaries, dimensions and volume of any potentially unstable mass?
15) Configuration and properties of potential basal & bounding failure surfaces?
16) Factors of safety with ranges of properties, and assumptions for seismic loading, rainfall and reservoir operation?
17) AEPs of assumed seismic and rainfall events?
18) Probabilities of failure of the mass under these various conditions?
19) Effect on facilities, of failure of this mass?

2. Possible first-time slide induced by natural processes

First time slides occur commonly in mountainous areas in response to extreme rainfall events or earthquakes, and also quite often in actively retreating coastal cliffs. The suggested questions which follow are for an investigation of a possible first-time cliff failure which would affect an important facility located at its top.

1) Geological situation, regional and local?
2) Evidence that the cliff has been, and is still, retreating due to coastal erosion processes?
3) Assuming yes to both, which processes have been and are active?
4) Mechanism(s) by which retreat occurs?
5) Average or other known rates of retreat in Recent and Modern times?
6) Does Modern average rate endanger the facility during its economic life?
7) Regional groundwater picture?
8) Groundwater distribution and pressures near the cliff?
9) Relationships between groundwater picture and local geological structure?
10) Effects of rainfall/streamflows/sea levels on groundwater pressures?
11) For rock cliffs, does geological structure indicate kinematic stability?
12) For all cliffs, is there a feasible model for failure which would affect the facility during its economic life?
13) Configuration and properties of potential basal/bounding failure surfaces?
14) Does the potential failure model require erosion/undercutting at the cliff toe during a major storm? If so what is the AEP of this event?
15) Factors of safety with ranges of groundwater pressures and earthquake loadings?
16) AEPs of rainfall and earthquake events capable of causing failure?

REFERENCES

Arnao, B.M., Garca, V.K., Wright, R.S. and Peres, J.Y. 1984. The Tablachaca Slide No 5, Peru, and its stabilization. Proc 4th Intl. Conf. on Landslides, 597-604.

Black, W.H., Smith, H.R. and Patton, F.D. 1986. Multiple-level groundwater monitoring with the MP System. Proc. MWWA-AGU Conf. on surface and borehole geophysical methods and groundwater instrumentation: 46-61.

Barton, M.E. and Coles, B.J. 1984. The characteristics and rates of the various slope degradation processes in the Barton Clay Cliffs of Hampshire. Quart. J. Engg. Geol., London. 17: 117-136.

Beetham, R.D., Moody, K.E., Fergusson, D.A., Jennings, D.N. and Waugh, P. 1992. Toe buckling as a mechanism for landslide development in schist terrain. Proc. 6th Intl. Symp. on Landslides.

Bowen, D.Q. 1978. Quaternary geology. Pergamon.

Brand, E.W. and Phillipson, H.B. 1984. Site investigation and geotechnical engineering practice in HongKong. Geotech. Engg. 15: 97-153.

Brunsden, D. and Jones, D.K.C. 1976. The evolution of landslide slopes in Dorset Phil. Trans. Roy. Soc. A283, 605-631.

Bryant, J.M., Woodward, D.E., Beetham, R.D. and Logan, T.C. 1992. The usefulness of seismic methods in defining landslide structure. Proc. 6th. Intl Symp. on Landslides. Balkema.

Chandler, J.H. and Moore, R. 1989. Analytical photogrammetry: a method for monitoring slope instability. Quart. J. Engg. Geol., London. 22: 97-110.

Cummings, D., and Clark, B.R. 1988. Use of seismic refraction and electrical resistivity surveys in landslide investigations. Bull. Assoc. Engg. Geol. XXV. 4: 459-464.

Dearman, W.R. and Fookes, P.G. 1974. Engineering geological mapping for civil engineering practice in the United Kingdom. Quart. J. Engg Geol. London. 7: 223-256.

Debreuille, P.J., Franq, J.B. and Londe, P. 1979. Auscultation fine d'un glissement de terrain a l'aide d'instruments nouveaux. Proc. 4th. Congress Int. Soc. for Rock Mechanics, 2: 77-87. Balkema.

Depman, A., Dodds, K. and Parrillio, D. 1972. Tocks Island project spillway rock mechanics studies. In Cording (Ed) Stability of Rock Slopes: Proc. 13th Symp. on Rock Mechanics, Urbana: 443-486.

Emery, K.O. and Aubrey, D.G. 1991. Sea Levels, Land Levels, and Tide Gauges. Springer-Verlag.

Fairbridge, R.W. 1961. Eustatic changes in sea level. Physics and chemistry of the earth. 4: 99-185.

Fell, R., MacGregor, J.P. and Stapledon, D.H. Geotechnical engineering of embankment dams. In press, Balkema, Rotterdam.

Fookes, P.G., Dale, S.G. and Land, J.M. 1987. Some observations on a comparative interpretation of a landslipped area. Quart J. Engg Geol., London. 24: 249-265.

Fox, P.P. 1957. Geology exploration and drainage of the Sierra slide, Santos, Brazil. Geol. Soc. America Engg. Geol. Case Histories 1: 17-23.

Gardiner, V. and Dackombe, R. 1983. Geomorphological Field Manual. London, George Allen and Unwin.

Gillon, M.D. and Hancox, G.T. 1992. Cromwell Gorge Landslides - a general overview. Trans. 6th. Intl. Symp. on Landslides. Balkema.

Gillon, M.D., Riley, P.B., Halliday, G.S. and Lilley, P. B. 1992a. Movement history and infiltration, Cairnmuir Landslide, New Zealand. Proc. 6th. Intl. Symp. on Landslides. Balkema.

Gillon, M.D., Denton, B.N. and Macfarlane. 1992b. Field investigation of the Cromwell Gorge landslides. Proc. 6th. Intl. Symp. on Landslides. Balkema.

Hendron, A.J. Jr. and Patton, F.D. 1985. The Vaiont slide - A geotechnical analysis based on new geologic observations of the failure surface. U.S. Army Corps of Engineers Technical Report GL-85-5.

Holmes, S.C.A. 1972. Geological applications of early large scale cartography. Proc. Geol. Assn. 83: 121-138.

Hosking, A.D., 1960. Slope stability studies in the valley of the Geehi River. Proc. 3rd. Australia-New Zealand Conf. on Soil Mech. and Foundn. Engg. 169-176.

Howsam, P. and Hollamby, R. 1990. Drilling fluid invasion and permeability impairment in granular formations. Quart. J. Engg. Geol. 23: 161-168.

Hunt, R.E., 1984. Geotechnical Engineering Investigation Manual. McGraw-Hill, New York.

Hutchinson, J.N., 1961. A landslide on a thin layer of quick clay at Furre, Central Norway. Geotechnique 11: 69-94.

Hutchinson, J.N., 1965. The landslide of February, 1959 at Vibstad in Namdalen. Norwegian Geotech. Inst. Pub. 61: 1-16.

Hutchinson, J.N., Somerville, S.H. and Petley, D.J. 1973. A landslide in periglacially disturbed Etruria Marl at Bury Hill, Staffordshire. Quart. J. Engg Geol. London, 6: 377-404.

Hutchinson, J.N. and Gostelow, T.P., 1976. The development of an abandoned cliff in London Clay at Hadleigh, Essex. Phil. Trans. Roy. Soc. A 283:557-604.

Hutchinson, J.N., 1988. Morphological and geotechnical parameters of landslides in relation to geology and hydrogeology: State of the Art Report. Proc. 6th. Intl Symp. on Landslides, Vol.1, 1-35.

Hutchinson, J.N. and Chandler, M.P. 1991. Investigations of landslides at St. Catherine's Point, Isle of Wight. Slope stability engineering. Thomas Telford, London: 151-161.

Hsu, K.J. 1978. Albert Heim: Observations on landslides and relevance to modern interpretations. Ch.1 in Voight (Ed) Rockslides and Avalanches, 1. Elsevier.

Johnson, R.H. 1987. Dating of ancient, deep-seated landslides in temperate regions. Chapter 18 in M.G. Anderson and K.S. Richards (Eds), Slope stability. Wiley.

Kirkby, M. J. 1987. General models of long-term slope evolution through mass movement. Chapter 11 in M.G. Anderson and K.S. Richards (Eds), Slope stability. Wiley.

Kovari, K. and Koeppel, J. 1987. Head distribution monitoring with the sliding piezometer system Piezodex". Proc. 2nd. Int. Symp. on Field Measurements in Geomechanics. Kobe, Japan.

Lambeck, K. and Nakada, M. 1990. Late Pleistocene and Holocene sea-level change along the Australian coast. Palaeogeog., Palaeoclim., Palaeoecol. Global & Planetory Change Section. 89: 143-176 Elsevier.

Lewis, M.R., and Moore, D.P. 1988. Construction of the Downie Slide and Dutchman's Ridge drainage adits. Proc. 7th Ann. Canadian Tunnelling Conf., Tunnelling Assoc. of Canada, Edmonton, Alberta: 238-247.

Lillesand, T.M. 1987. Remote sensing and image interpretation. Wiley.

MacGregor, J.P. Olds, R. and Fell, R. 1990. Landsliding in extremely weathered basalt, Plantes Hill, Victoria. In The Engineering Geology of Weak Rock, Engg. Group of the Geological Society, Leeds.

Marshall, A.J. 1985. The stratigraphy and structure of the Thomson damsite and their influence on the foundation stability. South Australian Institute of Technology M.App.Sc. Thesis, unpublished.

Matula, M. 1981. Recommended symbols for engineering geological mapping. Report by the IAEG Commission on Engineering geological mapping. Bull. Intl. Assoc. Engg. Geol. 24: 227-234.

McSaveney, M.J., Thomson, R. and Turnbull, I.M. 1992. Timing of relief and landslides in Central Otago, New Zealand. Proc. 6th Intl. Symp. on Landslides. Balkema.

Moon, A.T., 1984. Investigation of Bovills Landslip, near Devonport, Tasmania. Uni. of Tasmania, MSc Thesis, Unpub.

Moon, A.T., Olds, R.J., Wilson, R.A. and Burman, B.C. 1992. Debris flow risk zoning at Montrose, Victoria. Trans. 6th. Intl. Symp. on Landslides.

Novosad, S., Blaha, P. and Kneijzlik, J. 1977. Geoacoustic methods in the slope stability investigation. Bull. Intl. Assoc. Engg. Geol. 16: 129-131.

Obert, L. and Duvall, W.I. 1957. Microseismic method of determining stability of underground openings. US Bur. Mines Bull. 573.

Osterkamp, W.R. and Hupp, C.R. 1987. Dating and interpretation of debris flows by geologic and botanic methods at Whitney Creek Gorge, Mount Shasta, California. Geol. Soc. Am. Reviews in Eng. Geol. VIII, 157-163.

Palmer, D.F. and Weisgarber, S.L. 1988. Geophysical survey of the Stumpy Basin Landslide, Ohio. Bull. Assoc. Engg Geol. XXV.3: 363-370.

Paterson, S.J., Hale, G.E.A. and Ikin, D.B. 1975. Stabilizing a landslide above Fisher Penstock, Tasmania. Proc. 2nd Aust.- New Zealand Conf. on Geomechanics : 314-318.

Patton, F.D., and Hendron, A.J. Jr. 1972. General report on mass movements. Proc. 2nd. Int. Congress of I.A.E.G., V-GR1 - V-GR57.

Patton, F.D. 1988. The concept of quality in geologic and hydrogeologic investigations. In: C. Bonnard (Ed) Landslides. Proc. 5th Intl. Conf. on Landslides 3: 1405-1411. Balkema.

Patton, F.D. 1989. Groundwater instrumentation for determining the effect of minor geologic details on engineering projects. Ch.5 in Cording et al (Eds) The art and science of geotechnical engineering, 73-95. Prentice-Hall.

Piteau, D.R., Mylrea, F.H. and Blown, I.G. 1978. Downie Slide, Columbia River, British Columbia, Canada. Chapter 10 in Voigt (Ed) Rockslides and avalanches. Elsevier.

Regan, W.M. 1980. Engineering geology of Sugarloaf Dam. South Australian

Institute of Technology M.App.Sc. Thesis, Unpub.

Rib, H.T. and Liang T. 1978. Recognition and identification. Ch.3 in Schuster, R.L. and Krizek, R.J.(Eds) Landslides, analysis and control. National Academy of Sciences, Washington, D.C.

Schlemon, R.J., Wright, R.H., and Montgomery, D.R. 1987. Anatomy of a debris flow, Pacifica, California. Geol. Soc. Am. Reviews in Eng. Geol. VIII, 181-199.

Schroder, J.F. 1978. Dendrogeomorphological analysis of mass movement on Table Cliffs Plateau, Utah. Quaternary Research 9: 168-185.

Selby, M.J. 1982. Hillslope materials and processes. Oxford University Press.

Sowers, G.F. and Royster, D.L. 1978. Field investigation. Ch.4 in Schuster, R.L. and Krizek, R.J. eds. Landslides, analysis and control. National Academy of Sciences, Washington, D.C.

Stapledon, D.H. 1976. Geological hazards and water storage. Bull. Int. Assoc. Engg. Geol. 14: 249-262.

Stapledon, D.H. 1983a. Towards successful waterworks. Proc. Symp. on Engineering for Dams and Canals, Alexandra. Instn. of Prof. Engineers, New Zealand: 1.3-1.15.

Stapledon, D.H. 1983b. The geotechnical specialist and contractual disputes. In: Minty and Smith (Eds) Collected case histories in engineering geology. Geol. Soc. Aust. Special Publication 11.

Stevenson, P.C. and Moore, W.R. 1976. A logical loop for the geological investigation of dam sites. Quart. J. Eng. Geol. London 9: 65-71.

Turnbull, I.M. 1987. Sheet S138 Cromwell. Geological Map of New Zealand, 1:63360. DSIR, Wellington, New Zealand.

Varnes, D.J. 1978. Slope movement types and processes. Ch.2 in Schuster, R.L. and Krizek, R. J. (Eds) Landslides, analysis and control. National Academy of Sciences, Washington, D.C.

Walker, B.F., Blong, R.J. and MacGregor, J.P. 1987. Landslide classification, geomorphology, and site investigations. In: Walker and Fell (Eds) Soil slope instability and stabilisation. Balkema.

Worsley, P. 1981. Radiocarbon dating: principles, application and sample collection. Ch.5.1 in Goudie et al.(Eds) Geomorphological Techniques. George Allen and Unwin.

Landslides, Bell (ed.)© 1995 Balkema, Rotterdam, ISBN 90 5410 032 X

Theme report

D. M. Cruden
University of Alberta, Edmonton, Alb., Canada

ABSTRACT: Of the 39 papers submitted 20 discuss individual landslides, 9 are regional surveys of landslide activity and 3 are regional surveys of landslides in particular materials. Public policy concerns 3 papers, investigation methods are reviewed in 3 papers and a paper proposes a new investigation method.

RESUMÉ: 20 des articles parmi trente neuf articles presentés discutent des glissements de terrain individuelles. 9 sont des études regionales de l'activite des glissements de terrain et trois autres descrirent des études regionales des materiaux particulieres. Trois articles discutent d'affaires d'etat, trois autres articles sont des revue de methodes d'investigation, une contribution propose une methode nouvelle de recherche.

1 INTRODUCTION

This report first classifies and then summarizes the range of topics covered by the papers submitted to the theme-Landslide Investigations. It then briefly discusses points from some of the papers in order to identify significant trends and developments. Finally it identifies highlights and some challenges.

2 CLASSIFICATION

Most of the papers discuss the investigation of individual landslides. The characteristics of these landslides form Table 1, the major methods used to investigate the landslides are summarized in Table 2. Paper 34 reviews 6 similar landslides and these are recorded as 1 movement in the Tables. Brief discussions of the other papers submitted to this theme are gathered in Section 4; among them there are 12 regional surveys of landslides, 3 papers on public policy for landslide investigations, 3 reviews of investigation methods and 1 proposal of a new investigation tool.

3 SUMMARY

The investigations of individual landslides can be summarized by histograms compiled from Tables 1 and 2. Figures 1-5 show the location, state of activity, rate of movement, material and type of landslide. Figures 6 and 7 show subsurface and surface investigation techniques.

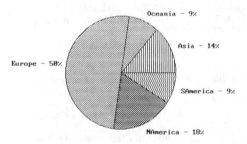

Figure 1 Location

4 REVIEW OF OTHER PAPERS

Regional surveys of landslides have come from Poland (Bazynski et al.,) Algeria (Chermouti and Gribici), New Zealand (Beetham, Moody et al., Gillon and Hancox, Gillon, Denton and Macfarlane) South Africa (Mortimer and Bell), Italy (Polloni et al.,) Austria (Scheidegger) and Sri Lanka (Sithamparapillai, Senanayake and de Silva). All the authors are from the countries they discuss. The papers from Algeria, Italy, Austria and Sri Lanka are particularly welcome as they fill geographic gaps in Brabb and Harrod's (1989) compilation.

It is interesting to notice how rapidly focus of attention may shift, the New Zealand papers describe movements in materials mentioned only briefly by Blong and Eyles (1989). This

Table 1. Characteristics of the landslides

1	2	3	4	5	6
1	4	1	2	2	3
4	3	1	2	2	5
7	4	3	2	2	3
12	4	1	2	1	3
14	3	1	2	2	3
16	2	6	-	1	3
17	2	5	-	1	3
18	2	4	7	1	3
20	6	1	3	2	3
21	4	1	2	2	5
23	4	4	-	1	1
25	4	1	-	1	3
	4	1	-	1	3
27	6	3	2	2	3
28	4	1	4	1	2
33	5	1	-	1	3
	5	1	-	1	4
34	4	5	-	2	3
35	4	5	1	2	3
36	5	4	7	1	3
37	4	5	1	2	3
39	5	5	-	1	3

Column 1, paper number in alphabetical order of first author;
Column 2, Location, 1) Africa, 2) Asia, 3) Australia, 4) Europe, 5) North America, 6) South America;
Column 3, State of Activity 1) Active 2) Suspended 3) Reactivated 4) Dormant 5) Stabilized 6) Relict;
Column 4, Rate of Movement 1) extremely slow 2) very slow 3) slow 4) moderate 5) rapid 6) very rapid 7) extremely rapid;
Column 5, Material 1) Rock 2) Debris 3) Earth;
Column 6, Type of landslide 1) Fall 2) Topple 3) Slide 4) Spread 5) Flow.

Table 2. Landslide investigation techniques

1	1,2,3,4,6,7,11,12
4	1,3,4,5,6,7,11,12,14,16,17
7	1,3,4,5,11,19
12	1,3,4,5,6,7,11,12,14,15,16,19
14	1, 3, 4, 6, 7, 11, 12, 14, 15, 16 19
16	11, 12
17	6, 7, 13
18	11, 12, 19
20	1, 3, 4, 6, 7, 11, 12, 14, 19
1	1, 6, 7, 11, 12, 14, 19
23	12, 13, 14
25	11, 13
	11, 13
27	1, 3, 4, 11, 12, 13, 14, 16, 19
28	11, 12, 13, 14, 16, 19
33	11, 12
	11, 12
34	1, 6, 7, 11, 19
35	1, 3, 4, 6, 7, 11, 12, 14, 15, 19
36	11, 13
37	1, 3, 5, 6, 7, 11, 17, 18, 19
39	11, 13

Subsurface Investigation Techniques

1) Drilling, 2) Geophysical logging, 3) Piezometers, 4) Inclinometers, 5) Permeability tests

Laboratory Tests

6) Classification tests, 7) Shear strength

Surface Investigation Techniques

11) Geological Mapping, 12) Geomorphological Mapping, 13) Fabric Mapping, 14) Geodetic Monitoring, 15) Extensometers, 16) Photogrammetric monitoring, 17) Seismic Exploration, 18) Resistivity Mapping, 19) Hydrological Observations.

emphasizes the importance of maintaining comprehensive inventories of landslides extending over the full period for which records are available. Information gathered at considerable expense during rare climatic events, such as the storms described from Durban (Mortimer and Bell) and Valtellina (Polloni et al.), or in the course of construction activities such as those in New Zealand (Gillon and Hancox) may soon be obliterated by equally dramatic events elsewhere unless it is formally recorded. Three papers limit themselves to particular geological materials in particular areas. Farkas describes Hungarian clays, Kovacik describes the flysch strata of the western Carpathians and Wang discusses landslides in Chinese loess.   Each paper demonstrates how the common properties of the materials lead to similar performance in slope movements.

Three papers, Cruden and Brown, Ottosson, Slosson and Slosson, address public policy.

Cruden and Brown describe the success of the International Geotechnical Societies UNESCO Working Party on World Landslide Inventory in promoting international communication about landslides. Ottosson documents the work of the Swedish commission on Slope Stability, initiating and coordinating research both on the causes of landslides and their prevention. Slosson and Slosson make an emphatic case for the economic benefits to communities of "diligent pre-development analysis and realistic design criteria".

Three other papers are reviews. Schoeneich describes the use of carbon-dating, pollen analysis and dendrochronology in landslide studies, supporting his argument that dating is an efficient method of generating important scientific information with a number of Swiss examples.

Duffaut finds that the stability of very high, steep rock slopes in the Alps and Norway cannot be described satisfactorily by any existing theory. His comment "Afin d'échapper á l'influence souvent determinante de structures geologiques diversifiés, les exemples sont choisis dans des massifs cristallins" invites controversy.

The paper by Bryant et al. describes the use of seismic methods in exploring the Cromwell Gorge landslides, highlighting the strengths and weaknesses of various types of surveys and the contributions they can make to comprehensive landslide investigations.

Crosta et al. show how a base friction table can be used to model the flow of plastic materials, a technique useful for predicting features in earth and debris flows.

5 SIGNIFICANT TRENDS AND DEVELOPMENTS

The continued success of the Symposium in attracting papers on landslides from around the world is worth noting. African landslides are the subject of two regional studies not shown on the diagram (Figure 1).

Slow moving landslides have attracted much attention. (Figure 3) There is clearly an urgent need for investigative techniques that will allow us to distinguish chronically slow landslides from those which have the potential to acelerate to more dangerous velocities.

The absence of any papers describing landslides in fine grained soils might be taken as an indication that landsliding in these materials is now comparatively well understood (Figure 4).

6 HIGHLIGHTS AND CHALLENGES

The Cromwell Gorge landslides are highlighted in this theme, not only by the magnitude of their contribution to the theme, 3 reviews (Bryant et al., Gillon, Denton and Macfarlane; Gillon and Hancock) and 3 more focused papers (Beetham, Smith, Jennings and Newton; Beetham, Moody, Ferguson, Jennings and Waugh; Gillon, Riley, Haliday and Lilley) but also by the intensity of their investigation during the stabilization works. They may also be illuminated by comparison, Beetham et al. on landslide development by toe buckling is particularly helpful in suggesting a slide development model. They draw attention to the work of Zischinsky (1966) which figured in the landslide classifications of Nemcok, Pasek and Rybar (1972) and Varnes (1978). Zischinsky's examples were sketch sections but fortunately more detailed description of similar movements are available; Malatrait (1975) mapped deeply eroded stream sections across the strike of a schist ridge. Sliding caused toppling below the slide rupture surface (Antoine and Fabre, 1980, Fig. 93), a process imprecisely called slide-toe toppling by

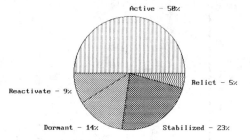

Figure 2 State of Activity

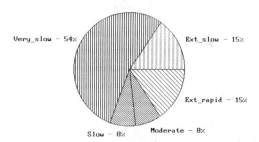

Figure 3 Rate of Movement

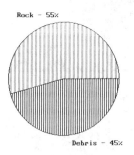

Figure 4 Material

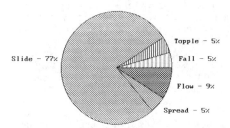

Figure 5 Type of Landslide

Goodman and Bray (1976) and "fauchage" by Alpine engineering geologists. This latter analogy with scythed grass is evocative but hardly mechanically helpful. Similarly the description by

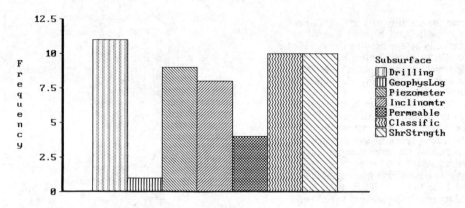

Figure 6 Subsurface Investigation Techniques

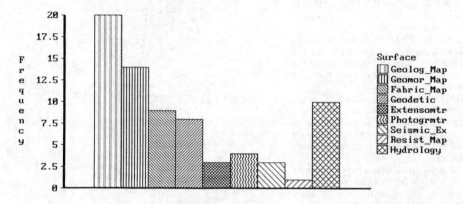

Figure 7 Surface Investigation Techniques

Harrison and Falcon (1934) of perfectly exposed "roof and wall structures" on the long dip slopes of Asmari limestone in S.W. Iran only added another example and another metaphor. Clearly the model described a complex rock slide (Varnes, 1978), a rock-slide rock-topple in the terminology of the Working Party on World Landslide Inventory (1990). Varnes has recently suggested to the Party's Working Group on Activity that complex landslides where different types of movement occur not in sequence but simultaneously in separate volumes of the displacing mass be called composite landslides. The landslide illustrated in Beetham et al. (Fig. 4) may thus be a composite rock-slide rock-topple, without buckling or bedrock flow. Perhaps further exploration has established whether the slope movement followed their suggested model or the slope really buckled and Beetham et al. have documented a new type of slope movement.

Technology challenges us to make full use of the new opportunities it provides. The total station, electronic, distance-measuring, survey instruments provide us with relatively rapid methods of determining vectors of movement on the surfaces of slow-moving landslides. With dense networks of points related to surface features of the landslides and frequent surveys, it is possible to determine much more precisely when the landslide moves and which portions of the displacing masses are moving by how much in which particular directions. In other words, comprehensive real-time kinematic analyses of slow landslides are attainable now. If changes in driving forces on the landslide can be measured which adequately explain the displacements, then confidence in predictions of the behaviour of the landslide under similar future changes in driving forces is considerably enhanced.

Another challenge this Symposium provides is communication in an efficient manner. Fortunately, through the papers of previous Symposia (Hutchinson, 1988, for instance) and the efforts of Commissions and Working Parties (IAEG Commission, 1990, WP/WLI, 1990, 1991) standard technical nomenclatures are becoming available. Many contributors use standard terminology, though not everyone quotes its source. Those who are still writing in private languages should consider how much more difficult it is for their readers to understand their important messages. In this decade of international cooperation to reduce natural disasters, we can all make a personal contribution by considering our terminology. It is particularly gratifying that many contributors have learned the languages of the Symposia in order to be able to contribute to them. Those of us who are fortunate enough to be able to speak our first languages at these meetings will recognize a particular obligation to choose our terminology with care, never using a new word when an old word, and there are so many of them (Cruden, 1991), will do.

ACKNOWLEDGEMENTS

I am grateful to the Natural Sciences and Engineering Research Council of Canada for supporting my travel to deliver this Report. J. de Lugt prepared the illustrations.

REFERENCES

Antoine, P., & D. Fabre 1980. Géologie appliquée au génie civil, Masson, Paris, 291.

Brabb, E.E. & B.L. Harrod 1989. Landslides: Extent and Economic Significance. Balkema, Rotterdam, 385.

Blong, R.J. & G.O. Eyles 1989. Landslides: Extent and Economic Significance in Australia, New Zealand and Papua, New Guinea. In E.E. Brabb and B.L. Harrod. Editors, Landslides: Extent and Economic Significance. Rotterdam: Balkema, 343-355.

Cruden, D.M. 1991. A simple definition of a landslide. Bulletin International Association for Engineering Geology, 43: 27-29.

Goodman, R.E. & J.W. Bray 1976. Toppling of rock slopes. Proceedings, Speciality Conference on Rock Engineering for Foundations and Slopes. American Society of Civil Engineers. Boulder, Colorado, 201-234.

Harrison, J.V. & N.L. Falcon 1934. Collapse Structures, Geological Magazine, 71: 529-539.

Hutchinson, J.N. 1988. General report: Morphological and geotechnical parameters of landslides in relation to geology and hydrogeoloy. Proc. 5th International Symposium on Landslides. Rotterdam: Balkema, 1: 3-35.

IAEG Commission on Landslides 1990. Suggested nomenclature for landslides. Bulletin. International Association for Engineering Geology, 41:13-16.

Malatrait, A.M. 1975. Analyse et classement des mouvements gavitaires (feuille Saint Jean de Maurienne á 1:50 000) These, Universite de Grenoble, Grenoble, 219.

Nemcok, A., J. Pasek & J. Rybar 1972. Classification of landslides and other mass movements. Rock Mechanics, 4: 71-78.

Varnes, D.J. 1978. Slope movement types and processes, in Schuster, R.L., Krizek, R.J., Landslides, analysis and control. Special report 176, Transportation Research Board, National Research Council, 11-33.

WP/WLI (International Geotechnical Societies' UNESCO Working Party on World Landslide Inventory) 1991. A suggested method for a landslide summary. Bulletin International Association for Engineering Geology, 43: 101-110.

WP/WLI (International Geotechnical Societies' UNESCO Working Party on World Landslide Inventory) 1990. A suggested method for reporting a landslide. Bulletin International Association for Engineering Geology, 41: 5-12.

Zischinsky, U. 1966. On the deformation of high slopes. Proc. 1st International Conference, International Society for Rock Mechanics. Lisbon, 2: 179-185.

Landslides, Bell (ed.) © 1995 Balkema, Rotterdam, ISBN 90 5410 032 X

Le mouvement de versant de Bonvillard, Savoie, France

P. Antoine & A. Giraud
Institut de Recherches Interdisciplinaires de Géologie et de Mécanique, Université de Grenoble, France

P. Desvarreux
Association pour le Développement des Recherches sur les Glissements de Terrains, Gières, France

J. Villain
Centre d'Études Techniques de l'Équipement de Lyon, France

ABSTRACT : In the western french Alps, the river Arc cuts its valley, heading E-W, through a very thick series of strongly folded coal measures made up with sandstones, shales, conglomerates and coal seams, carboniferous in age, which are very sensitive to slope instabilities. As a consequence of the late glacial history, huge slope movements have developed in some occurrences from the crests down to the bottom of the valley. The Bonvillard slide is one of the most famous but, up to a recent period, it has never been investigated nor monitored. Civil engineering projects for a next future (highways and high speed railways) need tunnels due to the lack of room in the narrow bottom of the valley. They have recently allowed some investigations at depth as well as some surface monitoring, the results of wich are presented. The natural evolution of the Bonvillard post-glacial slides gives way to a progressive reworking of unstable rock masses in such a manner that the most recent of the movements look like mud flows. The deepest failure surface is located below the river bed itself wich was once covered by the displaced masses (post-Würm period). The average speed of the faster of these slides is only about 2 cm/yr near the toe but this could be enough to damage concrete works.

1. LOCALISATION

La vallée de la Maurienne (SE de la France) est une voie naturelle de pénétration vers l'Italie du Nord (liaison Paris - Turin). Elle est empruntée par des voies de communication existantes très importantes via les tunnels ferroviaires et routiers du Fréjus. A la fin du siècle, ces voies seront complétées par un tracé autoroutier et par une voie ferrée à grande vitesse (T.G.V.).

L'étroitesse de la vallée de l'Arc entre Saint-Michel de Maurienne et Modane impose des tracés en souterrain. Un tunnel est envisagé en rive droite de l'Arc, où d'importants mouvements de terrain post-glaciaires se développent jusqu'à une profondeur importante, mais pour le moment inconnue, ce qui laisse présager des difficultés pour le choix du tracé et l'exécution des travaux.

2. CONTEXTE GEOLOGIQUE.

La vallée de l'Arc, entre Saint-Michel de Maurienne et Modane, recoupe une des zones structurales majeures des Alpes Occidentales : la zone houillère briançonnaise. Celle-ci est constituée par une puissante formation montrant une alternance de schistes et de grès, avec des horizons conglomératiques et des veines de charbon.

Cet ensemble est affecté de plis N-S, déversés symétriquement vers l'Est et vers l'Ouest.

La vallée étant orientée sensiblement E-W, cette structure est à priori plutôt favorable sur le plan de la stabilité du massif rocheux. Cependant, les terrains carbonifères, lorsque leur cohésion a été fortement amoindrie par une fracturation dense, sont bien connus dans les Alpes pour être le siège d'importants glissements de terrain (Martin-Cocher, 1984).

La rive droite de l'Arc en offre de bons exemples, probablement déclenchés au moment du retrait des glaciers würmiens.

Parmi ces mouvements de versant de grande ampleur, dont les surfaces de rupture sont susceptibles de se développer jusqu'à une profondeur importante, se trouve le glissement de Bonvillard, objet de cette communication.

3. LE GLISSEMENT DE BONVILLARD

3.1. *Description*

Le glissement de Bonvillard se développe sur une dénivellation maximale de 1700 m entre 2 550 m et 850 m d'altitude (soit 50 m environ sous le fond actuel de la vallée). Sa pente moyenne est de 26° et sa longueur maximale de 3 450 m (Fig.1).

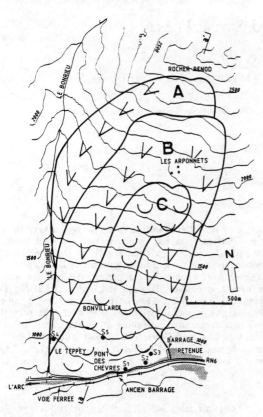

FIG. 1. Le mouvement de versant de Bonvillard.

Trois grandes familles de fractures affectent le versant :
- des fractures E-W, proches de la verticale, parallèles à la vallée de l'Arc,
- des fractures subverticales de direction N-S,
- des fractures subhorizontales.

Sur ce versant orienté au Sud, les fractures E-W découpent la masse rocheuse en tranches verticales, ce qui tend à favoriser le fauchage et à augmenter la désorganisation de la partie supérieure du versant.

Les fractures N-S participent au découpage du versant en blocs de taille variable. Quant aux fractures proches de l'horizontale et à faible pendage Sud (20°), elles créent des zones de faiblesse défavorablement orientées par rapport à la vallée.

Le glissement initial s'est produit dans un substratum rocheux très fracturé lequel, par remaniements successifs, a fini par se comporter globalement comme un matériau pulvérulent, ce qui nous amène à considérer que l'hypothèse d'une surface de rupture circulaire est représentative pour certains des glissements étudiés (Antoine et al., 1988).

3.2. *Types de mouvements* (Fig.1)

Ce glissement est en fait constitué de deux glissements très importants, emboîtés l'un dans l'autre. Le plus ancien (glissement A) a affecté une ligne de crête à 2 550 m d'altitude (Rocher Rénod) probablement dès la fin du retrait des glaciers würmiens. Il couvre une superficie de 5 km^2 pour un volume estimé de 250 à 300 millions de m^3.

Le glissement B, plus récent, reprend une partie de la masse précédemment glissée. Cette zone B, formé par une accumulation de blocs, est active, et des déplacements concentrés dans certains secteurs ont pu y être décelés et mesurés. Ce glissement couvre une superficie de 4 km^2 pour un volume estimé de 200 millions de m^3.

Cette succession de mouvements entraîne une diminution progressive de la dimension des éléments constituants, qui a permis l'apparition d'une coulée locale (C). Cette coulée, naissant à 1 950 m d'altitude (secteur des Arponnets), est formée d'une argile silteuse à blocs qui représente le terme ultime du remaniement de la zone B.

La superficie couverte est de 1,6 km^2 pour un volume estimé de 50 millions de m^3. Le front de cette coulée s'évase vers le bas et repousse l'Arc vers la rive opposée.

3.3. *Indices d'activité.*

Ces glissements, et tout particulièrement la partie basale de la coulée (au niveau de la vallée) montrent un certain nombre d'indices d'activité (murs fissurés, radier d'un ancien barrage déformé). Des mesures y sont pratiquées depuis maintenant une trentaine d'années par Electricité de France et, depuis 1974, par le Centre d'Etudes Techniques de l'Equipement de Lyon dans le cadre d'aménagements routiers et autoroutiers.

De 1964 à 1972, Electricité de France a mesuré le déplacement de la culée rive droite d'un petit barrage au fil de l'eau, qui existait alors au lieu dit Le Pont des Chèvres. Un déplacement de 19 cm en 9 années de mesures nous donne un déplacement moyen annuel de 2 cm. Ceci est concordant avec les mesures inclinométriques d'un sondage (S1) implanté en 1974 en bordure de la route nationale (RN 6). De plus, ce sondage montre qu'un mouvement se produit à 11 m sous le niveau de l'Arc actuel, et cela est confirmé par la mise en compression et le soulèvement d'un radier au droit du barrage actuellement détruit.

Des sondages récents et des mesures inclinométriques effectuées depuis 1989 ont bien confirmé que des mouvements se produisent à une dizaine de mètres au dessous du lit actuel de l'Arc (sondage S2). Enfin le sondage S3 décèle des mouvements à 37 m de profondeur, c'est-à-dire au niveau de la vallée actuelle (Fig.1) .

Cette auscultation confirme que la base du versant est instable, avec des vitesses de déplacement mesurées de 2 à 4 cm par an. Ces mouvements

FIG. 2. Vue aérienne de la partie basse du versant.

s'accompagnent d'une tendance au bombement de la partie inférieure, bien mise en évidence par la vue aérienne (Fig.2).

Par ailleurs, des mesures topographiques effectuées en 1989 et 1990 ont détecté, près de Bonvillard, plus haut dans le versant, des mouvements de l'ordre de 10 cm/an. Ceux-ci semblent affecter aussi bien les matériaux de la zone B que ceux de la zone C. Enfin, ces mêmes mesures effectuées en deux points situés dans la partie ouest de la zone A, ont montré que les déplacement y étaient nuls.

Par conséquent, l'ensemble des données acquises nous incite à penser que les mouvements actuels affectent un volume de l'ordre de 200 millions de m3.

4. ESTIMATION DE L'EPAISSEUR DE LA MASSE GLISSEE.

La disposition de la langue terminale de la coulée centrale (C) suggère que l'actuelle vallée de l'Arc est épigénique. Un ancien lit doit donc se trouver enfoui sous le glissement.

Les sondages profonds (S4 et S5), ainsi que la campagne sismique réalisée à partir de 1989, avaient pour but d'estimer la profondeur du substratum rocheux stable.

4.1. *Résultats apportés par les sondages profonds.*

Le sondage S4, implanté au Teppey, en bordure ouest du glissement, a recoupé sur 40 m d'épaisseur des blocs rocheux altérés dans une matrice argilo-silteuse, avant de rencontrer le rocher sain.

Le sondage S5, implanté au centre de la coulée (cote 1100) a été bloqué à 185 m de profondeur par la poussée des terrains et par des éboulements, sans avoir rencontré le substratum rocheux stable.

4.2. *Résultats apportés par la prospection sismique.*

La prospection sismique réalisée au niveau du sondage S4 montre que le rocher sain, à 40 m de profondeur, est caractérisé par des vitesses de 4 000 à 4 500 m/s. Les matériaux situés au dessus, qui correspondent au rocher fracturé et glissé de la zone B, sont caractérisés par des vitesses de 3000 à 3200 m/s.

La prospection sismique réalisée au niveau du sondage S5 fait apparaître des vitesses de 3100 m/s à partir de 80 m de profondeur, et cela pourrait correspondre aux matériaux rocheux fracturés de la zone B. Dans la tranche de 0 à 80 m, les vitesses sismiques varient de 1 500 à 2 500 m/s et doivent correspondre aux matériaux de la coulée C.

4.3. *Synthèse (Fig.3).*

Sur la coupe longitudinale interprétative passant par

S   N

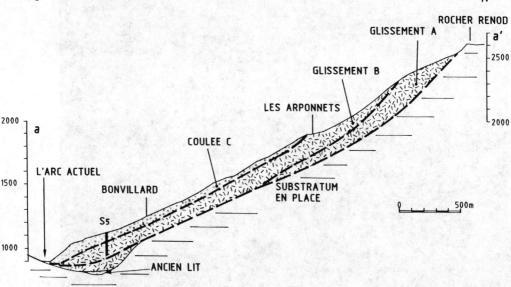

FIG. 3. Coupe longitudinale interprétative a a'.

le sondage S5, nous avons indiqué les hypothèses les plus probables, dans l'état actuel de nos connaissances :
- l'ancienne vallée de l'Arc est remplie, sur une épaisseur qui peut atteindre 300 m, par des matériaux provenant de glissements et coulées ;
- les incidents de foration de S5, à 185 m de profondeur, peuvent correspondre au fait que ce sondage a atteint une surface de glissement ;
- des mouvements affectant la zone B et la zone C se produisent au-dessous du lit de l'Arc actuel, ce qui est cohérent avec l'observation précédente. Les glissements, qui viennent buter contre la rive gauche rocheuse et stable, sont responsables des mouvements de compression constatés au niveau des alluvions de la vallée actuelle ;
- la base du glissement actuel le plus profond doit se situer en moyenne à 200 m au dessous de la surface topographique.

5. CONCLUSIONS

Le mouvement de versant de Bonvillard fournit un bon exemple des grands glissements qui affectent la zone houillère briançonnaise.
Il permet de suivre tous les stades de remaniements successifs qui transforment la roche en place fracturée en un matériau meuble susceptible d'évoluer en coulée.Le sondage S5 montre que les épaisseurs concernées par de tels mouvements sont considérables .Par contre, à l'échelle du versant pris dans son ensemble, cette épaisseur reste relativement faible.

Les quelques mesures topographiques montrent que les vitesses, en surface, peuvent ne pas être négligeables (10 cm par an), bien que les perturbations liées aux mouvements soient peu visibles, que ce soit au niveau des constructions ou du tracé routier.
En ce qui concerne les ouvrages souterrains en projet dans ce secteur, leur tracé doit éviter de recouper la surface basale la plus profonde. Bien qu'il soit maintenant établi que les masses glissées atteignent une cote inférieure à celle du fond de la vallée, il reste à préciser la géométrie profonde du contact entre le glissement et la roche stable. A moins d'admettre à priori un déplacement du tracé vers la rive opposée, ce n'est qu'au prix de nouvelles reconnaissances détaillées que le meilleur tracé pourra être déterminé sur cette rive droite.

REFERENCES.

P. Antoine, D. Fabre, A. Giraud, M. Al Hayari (1988), Propriétés géotechniques de quelques ensembles géologiques propices aux glissements de terrains. Cinquième symposium International sur les Glissements de terrain, Lausanne, P. 1301-1306.

Martin-Cocher (1984). Etude géologique de la stabilité des versants de la rive droite de l'Arc entre Saint-Michel de Maurienne et La Praz (Savoie). Thèse Université de Grenoble.

Landslides, Bell (ed.) © 1995 Balkema, Rotterdam, ISBN 90 5410 032 X

The European 'RIVET' research project: Degradation of mountain slopes

J.P.Asté
Bureau de Recherches Géologiques et Minières, France

ABSTRACT : The European Economic Community launched the EPOCH program (European Program for Climatic hazards); among the proposals received, that of a French-Italian group directed by the B.R.G.M. in France and the ERSAL in Italy was accepted.

The main objective of the two-year program accepted is to make available within the EEC a proven methodology for monitoring the degradation of mountain slopes.

The main features of this methodology are:
– Constitution of a database for past events,
– Development of territorial information systems covering all data useful for understanding degradation mecanisms.
– Recommending of methods of monitoring slope evolution.

The present article describes the status of this recent project at the end of 1991.

1 INTRODUCTION

The degradation of land on mountain slopes is a complex, quasi-continuous process whose rapidity has varied during different geological periods; even minor climatic changes may cause it to accelerate, with very damaging consequences.

In the mountainous parts of Europe, and those of developing countries, land resources exposed to this risk must be protected: the long-term development of these regions depends on effective protective strategies. Such measures require scientific knowledge and techniques covering many disciplines, from earth sciences and meteorology to social sciences.

For this reason, in the framework of the European program EPOCH (European program on climatology and natural hazards), the EEC has decided to support a program presented jointly by Italian and French scientific teams.

* Coordinator of a working group, with :
– P. ANTOINE and A. GIRAUD, professors at IRIGM of Grenoble,
– P. DESVARREUX and C. AZIMI, from ADRGT - Grenoble.
– J.P. ROTHEVAL, L. COUDERCY, M.C. LEBORGNE, from CETE - Lyon,
– E. LEROI, M. TERRIER, from BRGM - Orléans.

The program, named RIVET (acronym of "Recherche Intégrée sur la dégradation des VErsants en Territoire de montagne"), has a dual goal:
– to survey preventive methods developed on both sides of the Alps,
– to develop these methods by use of modern tools for data acquisition, formalism and transfer of know-how.

In order to make the best use of the skills of the members of the participating teams, the work was broken down as follows:
– Evaluation of degradation in terms of mass movements and engineering works on the French side;
– Evaluation of degradation in terms of erosion and agricultural practices on the Italian side.

PROGRAM OF THE FRENCH TEAMS

Four teams are participating in the program:
– Interdisciplinary Research Institute for Geology and Mechanics (IRIGM), Joseph Fourier University, Saint Martin d'Hères,
– Association for Landslip Research (ADRGT), Grenoble,
– Technical Studies Center of the Ministère d'Equipement (CETE), Lyon,

– Geological and Mineral Research Board (BRGM), Orléans, who assures coordination of the teams.

The work of these French teams is supported through a contract involving the State and the Rhône-Alpes region for research into natural hazards in mountainous areas.

General strategy

The general strategy is designed to respond to the two requirements of the program: Survey of preventive methods and development with modern tools.

As regards preventive methods: two preliminary tasks have been carried out:

– Preparation of an inventory, as complete as possible, of all unstable slopes in the region under study. This work, known as the "INVI" project (Asté, 1992) provides not only technical but also economic and administrative data on the observed events. This work was considered an indispensable preliminary to the establishment of any preventive policy.

– Establishment of a realistic, regionalized typology of localized phenomena of instability and degradation. Typologies traditionally used throughout the world, although of some interest, are not always appropriate for the evaluation of a phenomenon and its consequences in a particular locality or region. Such typologies must be based not only on phenomenologic criteria, but also on local factors such as the speed of evolution, the associated damage, and the characterization of associated critical situations.

This preliminary work facilitates the survey of preventive methods themselves:

– Preparation of maps of hazard maps

A number of cartographic surveys of zones exposed to landslip in the Alps have been performed in the last twenty years. Determination of their extent is at present a highly subjective task given that the spatial distribution of most of the factors involved is extremely variable. As regards the frequency of events and the reactivation of the phenomena concerned, these depend on various aggravating circumstances sometimes difficult to quantify. The objective is therefore to refine and harmonize earlier analyses for selected test zones, then to propose new methods based on modern data processing techniques, such as database management systems, geographic information

systems and other analytical tools used by thematicians of slope instability.

– Surveillance of situations considered dangerous and population alerts.

In many cases the goal of preventive action is to live with an active phenomenon too large to be controlled. Such phenomena and their aggravating factors must be monitored permanently. The objective here is to survey the many events experienced in the region and, in order to consolidate the data , to identify in each case the positive features, and to give guidelines and recommendations to mitigate weaknesses which the analyses have revealed.

– Palliative or curative work.

Much expensive work has been carried out in an attempt to stabilize potential or active landslips. Unfortunately there is very little data showing the real effectiveness of these measures, or even their initial and maintenance costs. A survey will therefore be carried out among the various bodies responsible for this work: Ministries of "Equipement", "Agriculture", local authorities and private engineering companies.

As regards the use of modern data input and processing tools and the analysis of new data, the strategy adopted is to use :

– satellite or aerial photographs for the determination of surface permanent factors, with, if necessary, a 3D localisation,

– geostatistic methods

– specific geographic information systems and associated databases (Bonnefoy, 1988)

– specific models for runoff, infiltration and stability analysis.

Finally, some expert rules will be established or consolidated in order to build new methodologies for hazard assessment and that coud be a most decisive contribution for future prevention expert systems.

Choice of study sites

For this research program, two zones in the northern French Alps had to be selected: areas with significant practical experience of landslip prevention and prone to significant socio-economic risks.

To satisfy these criteria the zones needed to be sufficiently extensive, of the order of several hundred square kilometers (accepting that for certain phenomena it would be necessary to study smaller zones of only a few tens of square kilometers). Consequently the work is carried out on a regional scale with the objective of preparing maps at a scale of 1/25,000. The

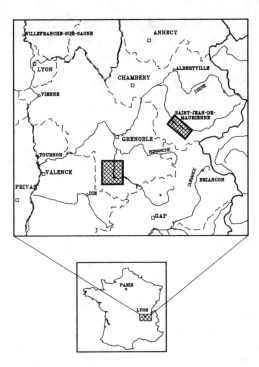

Figure 1 : Localisation of study sites

main sources of data are existing maps and satellite photos at a scale of 1/60,000 (SPOT). For the localized studies data are obtained by ground or aerial photography and by detailed field reconnaissance.

The zones, chosen in liaison with regional authorities and in particular with the Ministry of Agriculture's mountain land restoration services, are shown in Figure 1.

The first is the Maurienne Valley between Saint Jean de Maurienne and Modane. Landslips on the slopes of the valley are very numerous and varied; the geologic context is essentially sandstone and carboniferous shales. The socio-economic stakes are high: in addition to agricultural and industrial activities, and the important development of winter sports facilities, the valley is an essential communications route between France and Italy. A new motorway and railway are currently under study.

The second zone is the region of Trièves, between Grenoble and Sisteron. Slopes here are commonly unstable, the geologic context being dominated by glacial clays. The main activities of the region are agriculture and tourism. A large motorway construction project is in progress and facing extremely difficult geological conditions.

Program schedule

The program has three milestones:
- 1991: selection and preparation of data
- 1992: establishment of data processing procedures and validation of preventive methods
- 1993: synthesis and final report.

Sharing of tasks

Each team, in addition to its general participation in the group, has a specific responsibility:
- The ADRGT steers the study of the methodologic guide for monitoring works.
- The BRGM coordinates the French teams, the development of inventory procedures and of methods of planning of preventive action.
- The CETE studies the methods of analysing of the effectiveness of the preventive works.
- The IRIGM defines the regional typology.

PROGRAM OF THE ITALIAN TEAMS

The Italian program, under the control of the ERSAL (Ente Regionale Sviluppo Agricolo della Lombardia), will not be described in detail here. It corresponds more specifically to the pedologic and agronomic aspects of soil degradation in mountains and will be presented in its various phases at specialized colloquia.

The same strategy is used, however, and the same knowledge engineering tools, in particular for all hydro-meteorologic data.

The French and Italian programs are designed to be as complementary as possible and their results will be combined for the final reports.

STATE OF ADVANCEMENT OF THE PROJECT AT THE END OF 1991

At the end of 1991, the various tasks have been started as scheduled.

The inventory work has led to the identification in the region under study of a hundred known events of different nature. The INVI project is described in a separate report at the same congress. In addition, new information has been obtained on the study zones through research funded by this project.

An initial typology has been prepared based on the identified events; for the moment it is based only on phenomenologic factors (Antoine, 1990), but it will be developed and complemented by an analysis of circumstances created by the various phenomena, and then by an analysis of socio-economic consequences.

The work on planning of preventive measures was started by analyzing and collating knowledge of factors associated with instability and aggravating factors.

The collection of available information and the assessment of its quality and pertinence relative to hazard assessment have been started. Special SPOT surveys were ordered over the test zones and the results were processed by computer to obtain digital elevation models (DEM) of these areas. Their accuracy is obviously limited, yet they allow three-dimensional geographic location of the main factors identifiable at the scale used. Some of these factors can even be identified directly from the SPOT images. All available aerial photographic surveys have also been recorded and localized areas of the more interesting ones have been modelled numerically. Finally, all existing maps of use to the project have been digitalised and the contours projected on the DEMs.

In parallel a general collection of all available data on aggravating factors has been started, notably of hydro- meteorologic data, and suitable processing is in progress.

Finally, for zones for which hazard maps had already been produced, the contours of ground believed to be unstable have been digitalised so as to determine whether this judgement, based on traditional methods of analysis, is the same using new methods which will be tested in 1992.

As regards monitoring and alerts, all sites within the region which have been under surveillance have been analysed and their characteristics described. This has already enabled some rules to be defined for designing monitoring systems adapted to specific objectives. Work is now concentrated on the harmonization of requirements for sensors and data transmission and processing.

Concerning the survey of palliative or curative works, a survey has been undertake bodies responsible for such works of in order to analyze the nature, method of programming and financing of the various types of work. We must then define standard cases for which the situation, the technology used, the effectiveness, the range of costs, and the contractual framework can be described. In addition, the constitution of a databasespecific to this work, and connectable to the INVI, has been started.

CURRENT SITUATION AND PERSPECTIVES

After only six months of operation, the work of the teams of the RIVET project is progressing well. The ultimate goals of the project will be attained: objective methods for estimation of the probability of occurrence of earth movements in mountainous areas, which are highly random events, will be proposed after tests in representative areas of the northern Alps. The work carried out in parallel by the Italian team on erosion and degradation of agricultural land will provide a useful complement. The structure of the data necessary for both team's work will be harmonized. New research will then be possible with the objective of more detailed explanation of the more important phenomena, using consolidated data, and, if the same methods are adopted worldwide, a harmonized global survey of supposedly dangerous zones.

REFERENCES

ANTOINE P., ASTÉ J.P., AZIMI C., DESVARREUX P. et ROTHEVAL J.P. - 1990 - Les mouvements de terrain dans les Alpes du Nord. Typologie des phénomènes, méthodologie de la prévention. DATAR. Commissariat à l'Aménagement des Alpes du Nord.

ASTÉ J.P., GOUISSET Y., LEROI E. - 1992 - The French "INVI" project: national inventory of instables slopes. Proc. 6th International Symposium on Landslides. Christchurch Rotterdam - Balkema.

BONNEFOY D., GUILLEN A. - 1988 - Mappable data integration techniques in mineral exploitation. International Symposium on Computer and Geosciences, Helsinki.

Landslides, Bell (ed.) © 1995 Balkema, Rotterdam, ISBN 90 5410 032 X

GIS, SPOT, DEM and morphology of major land movements

J.P.Asté & F.Girault
Bureau de Recherches Géologiques et Minières, France

ABSTRACT : The increasing number of techniques for processing remote sensing data is opening up new avenues for the reconnaissance of land movements. Geographic information systems and digital image processing are used to characterize, sometimes with old data, the morphological evolution of large unstable slopes, such as those of La Clapière (Alpes Maritimes Department) and Friolin (Savoie Department) in France. The results, presented in map form, enable qualitative and quantitative analysis of events, and shed particular light on the dynamics of slope evolution.

INTRODUCTION

The development of new territorial information means, notably geographic information systems (GIS), has increased our knowledge of major land movements occurring over the last fifty years. The creation of digital elevation models at grids of 2 to 5 meters, using aerial or ground photographs taken before and after the event at scales of between 1:30,000 and 1:5,000, makes it possible to create topographic databases which can be compared to each other, and to draw maps of vertical deformation.

These maps give a continuous image of the deformation field at the surface. Selective processing of data makes possible new methods of interpretation of morphology and dynamics.

Combination of these analyses with stereo restitutions obtained by processing SPOT images further enriches the procedure, enabling the deformed zone to be situated in its geographic and geological context.

GEOGRAPHIC INFORMATION SYSTEMS (GIS)

Geographic information systems are digital systems for storage, processing and restitution of cartographic information. The data may be processed in either vector or raster mode. The advantage of GISs is that they enable the combination of information from different sources, with a view to updating the mapping of a region. They therefore constitute real geo-referenced databases.

GIS, REMOTE SENSING, DEM AND LAND MOVEMENTS

Remote sensing obviously plays a major role in these systems; this is true in particular as regards satellite data, which, by their nature, are digital, and which for this reason can easily be projected onto a reference cartographic system and stored on magnetic media. This is not the case for aerial photographs, which must be digitized before they can be input into the GIS.

Satellite images have a ground resolution of about ten to a few tens of meters (depending on the sensors), which restricts their uses. In the field of civil engineering, for example, aerial photos are used, whose higher resolution is more suitable. However, the synoptic effect of satellite data, which show the geographic, morphological and structural context on a regional scale, combined with their multispectral aspect, which enables many differentiations and notably identification of several different plants species, make it a valuable complement to aerial photography.

In addition, the frequency of satellite observations - generally about every 2 to 3 weeks, but higher for SPOT thanks to its lateral viewing - enables monitoring of the evolution of rapid phenomena (e.g. forest fires, floods, etc.), when weather conditions are suitable, and provides permanent updating of territorial information.

Thus, satellite remote sensing - which should develop further with the 3rd-generation satellites with

Figure 1: Image SPOT XS 052/261 du 06/01/1987, canal 1

a future resolution of 5m - is increasingly used in GIS, along with many more traditional tools such as aerial photos, research into ancient movements, land use, hydrogeology, etc.

At the same time, the production and exploitation of digital elevation models (DEM) - altimetric files integrated into GISs - is becoming general.

Two types of DEM exist:

- those deriving from classical photogrammetric analyses, based on stereoscopic pairs of aerial photographs (the data are in this case given in the form of a scattered data set - raw data - or level curves - smoothed data),

- those stemming from autocorrelative analysis by autocorrelation of digital files (SPOT images or digitized aerial photographs), for example ISTAR-type DEMs.

Thus, the development of GISs and image-processing software has already, within the framework of a study carried out by BRGM on behalf of the Ministry of the Environment, (Délégation aux Risques Majeurs) enabled processing of topographic data relative to several large unstable slopes, including that of La Clapière, in the Alpes Maritimes Department of France and Le Friolin in Savoie Department.

La Clapière

The La Clapière landslide has already been discussed in numerous papers. For the sake of convenience, however, we shall briefly recall the context.

The Duminières slope, situated in the Tinée valley (Alpes-Maritimes Department) and constituted essentially of migmatites, is part of the crystallophyllian basement of the Argentera-Mercantour massif. For several decades at least, it has been affected by movement(s) of great extent, whose evolution has accelerated during the last few years. Given the threat of disastrous evolution of these instabilities and the socio-economic interests at risk, the slope has been placed under surveillance. In view of the absence of means of reinforcement, civil engineering works have been carried out to change the Road layout, and divert the Tirée river through a tunnel.

The major fractures of the fracture set at Colle di Stau, oriented N120E, appear very clearly in the northern slope of the Mercantour chain; the same thing can be seen in the Auron region, where the Col du Ciavalet fracture set, parallel to the Upper Tinée valley, has the same orientation.

La Clapière is easily identified when its characteristic features are known: first the large NW glacis, then the upper landslide zones and the SE scree slope which are visible, but show up differently on the different images (Fig. 1).

The panchromatic image, although "hazy", shows clearly the landslide and scree zones, while on channel 1 of the XS image the two lobes of the upper landslide cannot be distinguished, and the SE scree can hardly be seen. On the other hand, this image shows the influence of the lithology on the morphology of the slopes, and their terraced structure can inferred.

The snow line is seen at an altitude of approximately 2000m, slightly above the permanent (winter) snow line at 1600m.

Directional statistical analysis of the fractures observed on the SPOT stereo pair (fig.2) brings out the extent of NW-SE fracturing on a regional scale: 37% of the fractures are oriented N130-150E, while only 17% are N70-90E, the rest being distributed in the other directions. These observations can be compared with those made in the field, where the most common fracture direction, N10- 30E (59%), divides the zone affected into compartments. The apparent difference between local and regional fracturing can be explained by the different scales of observation; the regional fracturing is shown by satellite or aerial remote sensing, while the "small" fracturing is noticeable only in the field.

1 dérochoir NW
2 dérochoir SE
3 barre d'Iglière
4 formation d'Iglière atypique
5 migmatites d'Anelle (série sup.)
6 migmatites d'Anelle (série inf.)
7 fluvio-glaciaire

▲▲▲ les 3 granges

Figure 2: La Clapière schéma lithologique (d'apres LRPC)

The unstable zone, 1km long at its base, affects La Clapière over a height of 650m and covers a surface area of some 85ha (Fig.2). Its upper part is delimited by a landslide zone continuous over 751m (in 1988), characterized by the presence of 2 lobes:
– a NW lobe which reaches an altitude of 1730m,
– a SE lobe, separated from the other one by a thalweg (La Gardiole), and whose summit is at 1700m. In this lobe, the scarp slope had in 1988 an average height of 25 to 30m (50 to 60m max.) and a slope of 50 to 55.

The NW border of the slope is represented by a narrow, very active scree corridor, which feeds a cone with a very extensive base. From the left side of this corridor a large scree zone has developed, which extends as far as Les Trois Granges.

Half-way up the slope is a band of quartz diorite some 80m thick, which runs through the whole length of the slope and is known as the "Barre d'Iglière".

Analysis of the measurements supplied by the monitoring network have shown the existence of:

Figure 3: MNT 1970 - MNT 1989

- lateral compartmenting of the slope, controlled by N10E fracturing;
- deformations which affect the whole of the slope, but in different ways depending on their date of occurrence and their rapidity of evolution (the disorders are more rapid in the SE part);
- transverse movements occurring towards the SE in the upper part (above the Barre d'Iglière);
- the major role played by the Barre d'Iglière, in the general behavior of the slope - its top, overlaid with mica, favours tangential movements of the overlying rocks;
- hydraulic control of the unstable zones.

To improve knowledge of the site, on which conventional photogrammetric data were available, stored in the form of scattered data sets (data derived from the analysis of aerial photos from 1970, 1987 and 1989 at a scale of 1:10000), 3 DEMs, coded on 16 bits, were prepared, with a definition of 5m x 5m.

The grids of the DEMs are defined in the horizontal plane by the x and y coordinates of the middle of the grid sections; along the vertical axis, the altitude z is given by the average altitude of the points situated in each grid section.

Calculation of differentials between the three DEMs (1989-1970, 1989-1987 and 1987-1970), and transformation of these differentials into images coded on 8 bits makes it possible to produce maps showing vertical deformation. Analysis of these maps has produced the following information (Fig.3):

- all the lower part of the slope, situated below the Barre d'Iglière, is affected by instability; the differentials here are high, showing large-scale displacements,
- the disorders in the upper part are more localized, and the movements of greatest extent correspond to the retreat of the scarp of the landslide,
- in the base of the slope, the displacement maxima are aligned along N10E fractures, which confirms the role played by this fracturing in the distribution of the disorders,
- the most marked "positive" deformations are at the intersection of two fractures, at the base of the SE compartment, near the La Gardiole valley.

The figures obtained by quantitative analysis have established the order of magnitude of the deformations:

total surface area affected by the disorders:
791600m^2 (79ha)
volumes affected:
period from 1970 to 1989: 6976675 m^3
period from 1987 to 1989: 2014775 m^3

Le Friolin

The Le Friolin site has so far been little studied, in spite of the fact that it seems to be exceptional in many ways.

Le Friolin (2678m) is situated in the Bellecôte massif, belonging to the internal part of the Briançon zone of the Alps. This massif unconformably overlies the Briançon Carboniferous coalfields, which crop out to the north in the Peisey-Nancroix valley.

Stratigraphically, above the Carboniferous we find quartzites remaining from the old Mesozoic cover (Lower Triassic), then Triassic gypsums and dolomites; the morphology of this last level is very rounded since glaciers have left many moraines and several cirques.

Le Friolin is a small triangular plateau, greatly fractured, formed of greenstone thrust over the Triassic.

While the southern flank is regular, the eastern slope, constituted of a rocky projection almost 250m high and approximately 600m long, underwent a sudden change in 1982. The phenomenon is certainly associated with progressive evolution: Professor J.Goguel had already noticed signs of instability in the 1960's, which disappeared in the years after that. The events of 1982, however, which occurred after a very rainy autumn, were the result of the generalized subsidence, of about 50m, of the whole rock shield of the face. The magnitude of the surface events, however - essentially rock-falls - is out of all proportion to that of this subsidence, so that its occurrence could have passed unnoticed, and in fact almost did.

The eastern face of the summit of Le Friolin has the following morphological characteristics:

– since 1982, a landslide 50 to 55m high in the summit part,
– under this landslide, a series of monolithic rocky spurs, more or less separated, whose base is at an altitude of about 2400m. These spurs, which are extremely fissured, show scaling, particularly in their lower part,they are bounded to the north by a large corridor (the "NE corridor" taken by scree), and to the south by the "south corridor", situated directly below the southern summit of the Le Friolin. Some fractures running diagonally across a fair width of the slope, affecting several pillars.

Finally, the base of the slope, from north to south, has the following characteristics:

– bulging at the mouth of the large NE corridor, at about 2200/2300m;

– a landslide zone which in 1982 took the place of a small escarpment, visible on photographs taken in 1970;
– a fold situated under the cone of blocks from the south corridor, between the S-shaped thalweg into which the corridor runs and the landslide zone.

It should be noted that neither the other slopes nor the summit plateau have been affected by this evolution; there is simply an old landslip below the northern slope.

If the 1982 event was the result of a landslide along a more-or-less circular slip surface, this would imply the possibility of a disaster situation where a mass of rock (estimated at 10million/m^3) could sweep down the slopes of the Bellecôte mountain into the Ponturin valley. Les Lanches, a village situated above Nancroix, would certainly be destroyed with grave consequences for human activity in the valley.

If the 1982 event resulted from a subsidence into voids left by dissolution of gypsum, this threat would not exist; the slope will have recovered a certain stability, probably for many years to come.

There are plenty of arguments for the first hypothesis (morphology of the upper landslide, bulging at the base of the large NE Corridor, the fold under the south corridor), especially as events of this type have already occurred in the vicinity of this site.

In an attempt to learn about the transformations having occurred in this face, and to characterize them as far as possible, BRGM contracted the ISTAR company to make two DEMs from the two photographic pairs (IGN missions in 1970 and 1986). These DEMs have a horizontal definition of 4m x 4m. They are coded using 16 bits, and altitude measurements are to a precision of 3m.

The Montagne de l'Arc has been examined on both DEMs, at about the altitude of 2200m, along the path leading from the Lac de l'Etroit to the Grand Plan. In this sector there are three gypsum dissolution sinkholes, which are clearly visible on the aerial photographs; their morphology can be clearly under the stereoscope. The smallest of these sink-holes is approximately 20m in diameter, and the largest 30m, however neither of them appear on the DEMs.

Some remarks should be made concerning the difference between the two DEMs. The imprecision inherent in the determination of the altitude of each point (RMS = 3m) gives a non-zero differential in areas which are clearly stable. The differential values must therefore be thresholded to "erase" these artificial movements. Visual interpretation of the corresponding image during interactive processing shows that this threshold is +/- 5m.

Figure 4: Carte d'iso-déformation verticale de la pointe du Friolin
(d'après MNTAERO ISTAR)

After filtering, the unstable zone shows up clearly. It is of course difficult to quantify the vertical component of the movement with any degree of precision, but the image obtained (Fig.4) appears to be in agreement with the description of the disorders given by J.Goguel. The legend, reproduced here for information, refers to the raw digital results of the processing.

On the global image, we first notice some zones in which the differential has significant values: for example, the region of the dissolution sink-holes. Other anomalies can be observed either in very steep slopes (NW cliff of the Friolin top), on the top edge of escarpments, or in forested areas. These are very probably artificial, caused by the problems in calibrating altitude measurements (or occasionally x,y measurements on ridges) for the DEMs.

The 1986 orthophoto has been "projected", using the coloring technique known as "Hue, Saturation, Brightness", on this differential, enabling the transformations of the slope to be marked precisely.

The first remarkable peculiarity of the disordered zone is a stable zone, slightly oblique to the line of greatest slope, which separates the subsided zone from the bulging zone.

The second particularity is that the upper, subsided part of the slope shows two sectors of maximum subsidence, one in the southern part of the face (south spurs), and the other near the NE spur, while the bulging zone has only one maximum.

Finally, a certain symmetry can be seen in the morphology of the disordered zone, the stable sector appearing as an axis of symmetry.

Despite all the reservations already mentioned on the remaining uncertainty, quantitative analysis of the disorders (Fig.5) shows that the subsidences are on a larger scale than the bulges. The surface ratio is 3:5 and the volume ratio 2:4. This observation suggests that the origin of the disorders is in the subsidence of part of the slope in holes due to dissolution of the underlying gypsum. Numerous arguments can be found to support this hypothesis: extensive fracturing of the plateau - and therefore high permeability -, abundant infiltration and circulation of water, etc.

On the other hand, the morphology of the disturbed zone, and in particular its apparent symmetry, would indicate a rotational type of landslide.

There is insufficient information to choose between these two hypotheses. The observations made by J.

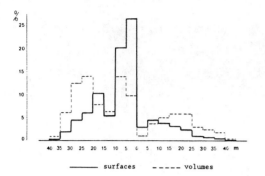

——— surfaces   - - - - - volumes

Figure 5: Distribution statistique de la composante verticale des désordres du Friolin par pas de 5m.

and B.Goguel, and the photogeological analysis, backed up by the use of DEMs, presented here, would require complementary follow-up work, especially since the two hypotheses would have very different consequences: probable stabilization by blocking of the movement, if subsidence is the cause, but a high risk of continuing landslide hazard if not.

Conclusions

The two examples presented here show the possibilities offered by digital processing of geographic information, and their application to the study of land movements.

The use of DEMs based on photogrammetric data, or calculated from pairs of aerial photographs (e.g. of IGN type), and their restitution in the form of images, sheds new light on the morphological evolution of unstable slopes. At the same time, DEMs constitute an effective tool for understanding the dynamics of surface movements, making possible maps of iso-deformation (vertical), continuous images of surface deformation fields.

REFERENCES

FOLLACI J.P. - Rapport de synthèse du Laboratoire des Ponts et Chaussées de Nice : Mouvements du versant de La Clapière, Saint Etienne de Tinée. Périodique trimestriel.

GIRAULT F., ASTÉ - 1991 - Caractérisation morphologique numérique de versants instables ; Rapport BRGM R 33096 ENV 4S 91.

GOGUEL B. - 1989 - Le glissement du Frioilin, en Savoie : un mouvement majeur révélé par les photographies; Rev. Franç. de Géotech. n° 48.

LEYMARIE P., DARDEL J., PISOT N. - 1990 - Les applications géologiques de la cartographie numérique en relief, Géologues n° 93, p. 31-34.

Landslides, Bell (ed.) © 1995 Balkema, Rotterdam, ISBN 90 5410 032 X

The French 'INVI' project: National inventory of unstable slopes

J.P.Asté, Y.Gouisset & E.Leroi
Bureau de Recherches Géologiques et Minières, France

ABSTRACT : In order to develop methods of landslide prevention it is indispensable to collect, analyze and highlight data concerning events which have already taken place. This requires the constitution of databases, a task undertaken by many groups throughout the world.

All are concerned with the same problems: data acquisition is difficult and the users of these data have a wide range of objectives.

INVI was structured to overcome these problems:
- Data collection is ordered as a function of their availability; it is assisted by the computerized process and the information fields are classified in order of increasing complexity.
- The data collected are not only technical, but also concern the socio-economic impact of events and administrative aspects.

Moreover, after a prototype phase, INVI was developed using dBASE and ORACLE, taking into account related experiences in other European countries, notably England and Italy, and also the specifications of the World Landslide Inventory.

INVI interfaces with the geographic information systems and the SISYPHE expert system developed by the B.R.G.M.

1 INTRODUCTION

The instability of natural slopes, notably in mountainous areas, is the cause of large-scale damage to possessions, activities and populations. The first requirement for a preventive strategy is to inventory the zones where instability phenomena have already been noted.

Such an inventory enables:
- location in space and time of an event and its consequences,
- storage of this information,
- comparison of the information collected.

It therefore constitutes a very important working tool for the community of decision-makers, scientists and technicians responsible for protection against hazards generated by slope instability.

The development and operation of such a tool is too laborious and difficult a task to be undertaken before obtaining a general consensus of the potential users. So far, it appears that few significant experiments have been performed in any country. Those described in scientific papers are often rather limited, as

regards their geographic coverage (Foietta, 1990), their purpose (Durville, 1989), and their content or structure.

It is only recently, through national and international programmes, that sufficient encouragement has been expressed for this work to be undertaken:
- In France, the policy of major hazard prevention, inaugurated by the State in 1982, has gradually led to the creation of inventories of instability phenomena and associated damage.
- In Europe, the launching of the EPOCH project by the EEC has enabled methodological research to be developed concerning the territorial information systems in mountainous areas (Asté, 1992).
- The United Nations declaration of the International Decade of Prevention of Natural Disasters has also provided inventory support. A specialized commission from the International Association for Engineering Geology (IAEG) has also been working for the last few years on a World Inventory of major earth movements (Crudden, 1988).

2 NEEDS, INITIAL SPECIFICATIONS AND DATA SOURCES

2.1 Needs

The INVI project was initiated by BRGM with the assistance of the Délégation aux Risques Majeurs (DRM) of the French Ministry of the Interior, for the production of a prototype and its application in the Isère Department of France.

The first task accomplished was the identification of the various potential users, i.e. decision-makers, managers, insurers, populations exposed to risk, scientists and technicians involved, and the analysis of their needs.

This showed that the information required varies greatly depending on the user. The scientific and technical information, which is essential for the understanding of landslip-related phenomena, is not sufficient for risk assessment, any more than for information and communication with the population or for management of the technical methods to be used, in particular in a crisis.

The decision-makers and their needs can be split into three main categories:
- Those responsible for prevention require the most exhaustive information possible on events having occurred in the past, to help them to detect potentially dangerous situations. They must therefore collect and make available this information.
- Those responsible for emergency services in a crisis require references to situations experienced elsewhere, to provide them with factors on which to base a short-term decision.
- Those responsible for planning need to be able to measure the factors at stake in their various long-term planning proposals.

Managers need as much information as possible on the facts and the associated damage.

Insurers need an evaluation of the damage caused to the various insured items (vulnerability), with which they can associate an overall cost.

The populations concerned require communication and education; the inventory facilitates both of these, and renovates and maintains collective memory of the events.

Scientists and technicians in this field need tools to manage their knowledge - the inventory is only the most basic of these tools.

Compared to the information required, the data actually available are often minimal, particularly when they concern events that occurred a long time ago. These data are also very disparate, which makes them very difficult to input and organize.

It was therefore necessary to find a compromise between the needs expressed, the information available, the ease of use of the tool, its capacity to interface with other data sources and finally its future development.

2.2 Initial specifications

The information was structured into several subsets, taking into account its degree of availability and its ease of access:
- location in space and time,
- brief description of the phenomenon,
- identification of information sources,
- evaluation of damage suffered,
- detailed technical description.

The first two points are almost always available; access to the last two depends on the third and often makes a phase of additional research necessary, which may be deferred once it is known where the necessary information can be found.

In the longer term, other categories will be added, for example administrative and legal information. However, given the disparate nature of the information in these fields when it exists, and although the structure of the system would make it possible, it has been decided to wait and install these subsets in a future version of the inventory.

To process this information, a relational database management system was naturally used. At first ORACLE was the system chosen. One of the essential reasons for this choice, when it was made in 1988, apart from the recognized performance of this software, was the mode of management intended for the INVI base: it was to operate not only on PC in regional mode, in each Department of France, in various services and authorities, but also on workstations at a central site, with much greater capacity. We shall see below how this initial choice was gradually extended and enriched.

Another initial specification was to be able to obtain graphic and geographic representations of the information. This implied a link with geographic information systems. Here again, several solutions were possible, as a function of the needs and the means of the services using the information.

Finally, on a scientific level, the use of the database by expert systems (Asté, 1988 ; Ke, 1990) had to be possible. This required the installation of a "knowledge of the world" conforming to that of the expert system envisaged. The glossaries declared for each data field were therefore drawn up in accordance with the semantic analysis and the methods of representation of knowledge used within the framework of the research on expert systems.

2.3 Data sources

Data sources are very numerous but of widely differing interest. Efforts to gather data have been led mainly in France by three organizations:
- two public organizations: the network of civil engineering laboratories (LRPC) of the Ministère de l'Equipement, and the Restoration of Mountainous Lands services of the Ministry of Agriculture (RTM),
- one semi-public organization, the Bureau de Recherches Géologiques et Minières (BRGM: Geological and Mining Research Board).

Research has also been carried out in universities.

Access to these data, however, is very difficult. Every time it becomes necessary to investigate known past events, for the technical procedure of drawing up risk exposure plans (PER - plans d'exposition aux risques), this means searching through the archives of the various services concerned, the archiving for each organization with information being structured in a different way. What is more, if the search is performed with no particular precautions, it simply increases the disorder in the archives.

Moreover, the data are very heterogeneous. Files may be very brief, containing a vague indication of a type of event with its approximate data of occurrence, especially when events occurred long ago. On the other hand, in the case of studies for reinforcement projects, the information is too abundant to be used in its entirety, and filtering is necessary.

3 THE INVI 1 PROTOTYPE: TESTS IN THE ISERE AND SAVOIE DEPARTMENTS

Initially planned only for the Isère Department, the prototyping work has been extended to the Savoie Department thanks to financial assistance obtained through a contract between the State and the Rhône-Alpes region for research on natural hazards in mountainous areas.

3.1 Technical results

In these two Departments in the French Alps, approximately 3500 events have been recorded, the earliest of them being major events dating back to the 12th century, and the more recent ones being fed into the database as they occur, regardless of the scale of the event. Figure 1 shows an example of information processing on a typical community:
- location of known events,
  - type of listing, depending on the type of request. As in any relational database system, the operator can program the type of request as he pleases. Nevertheless, to make the system more user-friendly the most common requests have been defined in advance.

The table below gives the percentage of each category filled in for each set of information.
- location in space and time   100 %
- brief description of the phenomenon 100 %
- identification of information sources 70 %
- evaluation of damage suffered   30 %
- detailed technical description   10 %

These data were collected from the three organizations mentioned above, from the Geological and Mechanical Research Institute (IRIGM) of the University of Grenoble and from the Association for the Development of Research on Landslides (ADRGT), in Grenoble. The files available from the various organizations often referred to the same events. The mode of input enables double entry of information to be minimized.

Identification of the compiler of the file and the sources of information makes possible rapid access to the base data in case it is necessary to verify or complete some of the knowledge of an event.

3.2 The lessons learned from the prototype phase

Among the many things learned are some important points which should be mentioned.

The first point concerns the reluctance of organizations possessing data to communicate the information they hold. The reason behind this may be an unwillingness to share what they have, or more simply a reluctance to carry out the work required to make their own archives available.

The first reason is the most difficult to deal with and initially caused some problems for INVI. However, when the project came to be seen as having national interest, supported by the relevant Ministries, the organizations approached did cooperate, particularly once the first results had been made available. In addition, since these first results appeared, the contributing bodies have taken part in the critical analysis which enabled the operational version, INVI2, to be specified.

As regards the second reason, it is true that reorganizing one's own archives is a costly operation,

INVI

FRENCH NATIONAL UNSTABLE SLOPES INVENTORY

1 - Printed outputs

N REF.	COMMUNE	N INSEE	LOCALISATION	DATE	TYPE DE MOUVEMENT	Nb MORTS	Nb BLES.	BIENS ENDOMMAGES
38 0774	MEAUDRE	225		NON CONNUE	GLISSEMENT	0	0	
38 0775	MEAUDRE	225		NON CONNUE	CHUTE DE PIERRES	0	0	
38 0776	MEAUDRE	225		NON CONNUE	CRUE TORRENTIEL.	0	0	
38 0575	MENS	226	ROUTE DE CLELLES A MENS	OCTOBRE 1889	ECROULEMENT	0	0	ROUTE
38 0574	MENS	226	TUNNEL DES ROCHERS	16/05/1954	CHUTE DE BLOCS	0	0	ROUTE
38 0573	MENS	226	LES ROCHERS	28/02/1957	ECROULEMENT	0	0	ROUTE
38 0778	MENS	226		NON CONNUE	COULEE	0	0	
38 0777	MENS	226		NON CONNUE	CRUE TORRENTIEL.	0	0	
38 0075	MENS	226	CD254 MOULIN DE CHARDAYRE.AU	NON CONNUE.	GLISSEMENT	0	0	ROUTE
38 0055	MENS	226	A CHARDAYRE. TOUS LES FONDS	NON CONNUE	INCON./INDETERM.	0	0	CHAMP
38 0960	MENS	226	CD526 PK 11,2 a 12,2	NON CONNUE	GLIS. DE TALUS	0	0	
38 1055	MENS	226	"LE FOREYRE"	NON CONNUE	Mvt LENT DE VERS.	0	0	
38 1056	MENS	226	RD DU RUISSEAU DE MENS	NON CONNUE	GLIS. DE TALUS	0	0	
38 1058	MENS	226	"LE PONTILLARD"	NON CONNUE	GLIS. DE TALUS	0	0	
38 0572	MEYLAN	229	VILLAGE DE LA BATIE	18/12/1886	ECROULEMENT	0	0	
38 0056	MEYLAN	229	LA BATIE MEYLAN.ROCHER DE ST-	NON CONNUE	CHUTE DE PIERRES	0	0	

2 - Graphical outputs

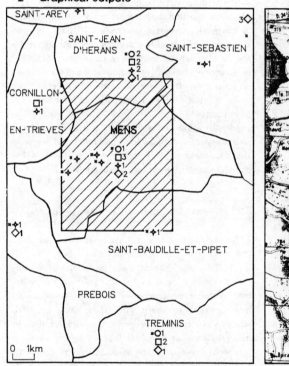

Fig.-1a - Localisation of recorded events

Fig.-1b- I.G.N. 1/50 000 map

Commentaire : Within the MENS commune , 11 events have been recorded

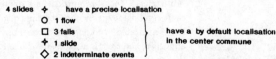

4 slides ✢ have a precise localisation

O 1 flow
☐ 3 falls
✢ 1 slide
◇ 2 indeterminate events
} have a by default localisation in the center commune

Figure 1

1550

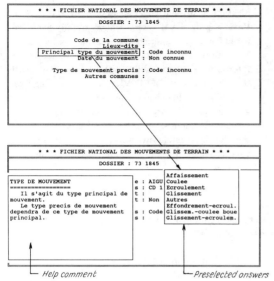

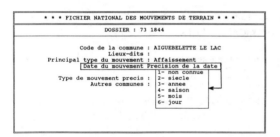

| | Help comment | | Preselected answers |

Fig. 2 : Input screens

which does not always seem beneficial in the short term. Here again, the first results helped clarify the advantages.

The second point concerns the ease of use necessary for input operations, without which the work becomes very laborious.

During development of the prototype, ORACLE did not give full satisfaction in this respect. An input module was therefore developed at the same time, using the CLIPPER programming language, which is very rich in functions and is compatible with dBASE files.

This solution has enabled real progress to be made in input operations, in particular the installation of pull-down menus. Figure 2 shows an input screen.

4 THE SPECIFICATIONS FOR INVI 2

From the experience acquired with the prototype INVI1, it was decided to develop an operational version, INVI2, responding to more complete specifications. This project obtained the assistance of the Delegation for Major Hazards (DRM : Délégation aux Risques Majeurs) of the Ministry of the Environment, and of the Directorate of Civil Safety (DSC : Direction de la Sécurité Civile) of the Ministry of the Interior.

The main new specifications for INVI2 concern:
- The organization and structure of the files:
  - alphabetic and not organized, such as the dictionary of communities within a given Department,
  - thematic and structured, such as the dictionary of types of landslips.
- The creation of detailed functions:
  - input of new records,
  - selective consultation, modification and deletion of records,
  - use of maps and selective display on the screen,
  - statistical processing,
  - complete or selective listing, either preprogrammed or specified by the user,
  - database management.
- New information storage possibilities:
  - in particular, scanner images.

In addition, INVI2 will be easily transposable into foreign languages, notably into English, Spanish and Italian.

Beyond INVI2, thinking on INVI3 has already begun. It concerns:
- interfacing with the other relevant environmental databases: geotechnical (Faure, 1991), hydro-meteorological and socio-economic databases,
- interfacing with complex systems of territorial information (ref. 5), and with expert systems (Asté, 1988, 1992 ; Ke, 1990),
- the use of hypertexts and hyperdocuments.

5 CONCLUSION

At the present stage of the project, INVI seems to provide a response to the requirements of the various potential users:
- for decision-makers, it provides a tool for storage, listing and graphic or geographic representation,
- for regional planners it provides the storage tool necessary to maintain awareness of the risks specific to the area,
- for insurers, it constitutes an excellent source of information on the nature and the distribution of damage,
- it improves communication with and information of populations at risk,
- and finally, on a scientific and technical level, it

is the first basis for information and exchange, and will therefore be a very important factor in the development of knowledge.

Like any inventory, however, it is useful only if it is exhaustive and complete, and if the effort made to create it continues.

It is for this reason that it was set up as a national "activity-generating" project, and that it has been proposed to use it within the framework of the European project RIVET (Asté, 1992). The procedure for data input management and information communication is presently being studied by the French. Various provisional versions have already been proposed for tests to a number of international partners. This effort of cooperation in the development of an international system will continue more easily with the development of INVI2, with its linguistic adaptations. This should enable consolidation of the very important work done in recent years by the specialized commission of the IAEG in the context of the World Landslide Inventory, which deals only with very large landslides, of a volume of over 1 million/m^3.

BIBLIOGRAPHY :

ASTÉ J.P. - 1992 - The European "RIVET" research project : degradation of mountain slopes. Proc. 6th International Symposium on Landslides - Christchurch. Rotterdam. Balkema.

ASTÉ J.P. - 1988 - Some reflexions on methodologies of landslide prevention in France. International Symposium on natural disasters in European Mediterranean Countries, Perugia. Italy.

CRUDDEN D.M. and IAEG Commission on Landslides - 1990 - Suggested Nomenclature for Landslides. Bulletin of International Association for Engineering Geology n° 41, pp. 13-16.

DURVILLE J.L., LACUBE J. - 1989 - Un essai de fichier informatique sur les mouvements de terrain. Bulletin de Liaison des LPC. Paris. n° 161. Mai-Juin 1989.

FAVRE J.L., HICHER P.Y., KENLIS J.M., TOUATI K. - 1991 - "Modelisol : database for reliability analysis in geotechnics". Sixth International Conference on Applications of Statistic and Probabilities in Civil Engineering. Mexico.

FOIETTA P. et FORLATTI F. - 1990 - La Banca Dati Geologica. A cura del settore prevenzione del rischio geologico, meteorologico e sismico - Regione Piemonte - CNR - Istituto di ricerca per la protezione idrogeologica del bacino Padano Torino.

KE C. - 1990 - Un système de représentation des connaissances et d'aide à la décision pour la prévention des mouvements de terrain. Doctorat de spécialité en Informatique Appliquée. Université Claude Bernard - Lyon I.

Landslides, Bell (ed.) © 1995 Balkema, Rotterdam, ISBN 90 5410 032 X

Analysis of gravitative deformation of Mt. Fana (Italian Alps)

S. Chiesa
C.N.R., Milan, Italy

A. Frassoni & A. Zanchi
ISMES, Bergamo, Italy

I. Fornero
ENEL DPT SOIC, Torino, Italy

G. Mazza & A. Zaninetti
ENEL DSR CRIS, Milan, Italy

ABSTRACT: The results of a multidisciplinary analysis of the mechanisms responsible for the deformation of the tunnels of the E.N.E.L. hydro-electric power plant of Quart (Aosta, Italy) are presented. Repeated in situ stress measurements have shown a progressive increase of loads in the concrete line also associated to progressive severe cracking. The geological and structural analysis coupled with photo-interpretation have revealed the existence of a wide "paleo-landslide" which has affected the entire southern slope of Mt. Croce di Fana (2211 m). The areal extent of the slide is about 12 km^2, and the genesis of the slide appears to be connected with widespread post-Wurm gravitative collapse triggered by the withdrawal of the Balteo glacier. The main paleo-landslide body is composed of several smaller secondary slides, some of which are still active. The structural analysis indicates a possible genetic connection between the Aosta Valley Line and the shear planes reponsible for the observed landslides. The reconstructed geologic scenery provides a thorough understanding of the displacements observed in the tunnels.

1 INTRODUCTION

Repeated cracking along the tunnels of the E.N.E.L (National Electricity Board) hydro-electric power plant of Quart have suggested the activation of a permanent instrumentation network with the purpose of understanding the cause of the damage of the concrete lining. In situ measurements also included extenso-inclinometric instrumentation in the main chamber of the plant in order to recognize possible tilting of the structure.

The observed deformational phenomena were so severe and widespread that a geological and geomorphological study of the area was necessary to understand such a situation. The study revealed that the whole southern slope of Mt. Croce di Fana (2211 m) is affected by an important gravitational sliding process which was ignored by the previous specific literature. An effort has been made to relate the deformations observed in the tunnels to the deformation and instability processes observed on the slopes of Mt. Croce di Fana.

In this paper we present and discuss the results of the instrumental tests conducted in the tunnels and of the geological and photo-geological study of the southern slope of Mt. Croce di Fana.

2 GEOLOGY

The study area is located in the axial portion of the Western Alps, on structural units belonging to the Austroalpine and Pennidic domains (Figure 1). The main metamorphic and structural imprinting is related to the Eo-Alpine phase of deformation under HP/LT conditions, resulting from the build-up of the accretionary complex which accompained the collision between the African and European lithospheric plates (Polino et al., 1990). A partial back-metamorphism to greenschists facies was the result of the later evolution of the Alpine chain (Dal Piaz, 1976).

The studied rock mass includes an upper unit composed of ortogneiss, paragneiss, micaschists and phyllades belonging to the Austroalpine "Arolla Series" (Dal Piaz et al., 1972) forming the Mont Mary klippe, which have been thrusted onto a lower unit composed of calc-schists and ophiolites of the Pennidic Zermat-Saas zone. The thrust plane, which is characterized by the presence of a thick band of mylonitic and cataclastic rocks, outcrops along the lower part of the studied slope and is generally inclined southward at 10°-20°, following the main metamorphic foliation.

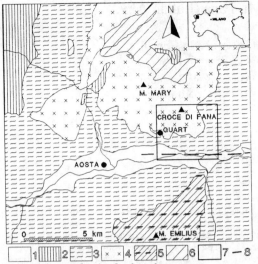

Figure 1. Geological sketch map.
1=Alluvial deposits; 2=Brianzonese zone;
3=Calc-schists and ophiolites; 4=Mt Mary
klippe; 5=Mt. Emilius klippe; 6=Val
Pelline series; 7=Investigated area;
8=Main fault.

The Plio-Quaternary neotectonic evolution
of the area is characterized by brittle
deformation. Strike-slip and extensional
faults are the main active structures. A
fault with regional relevance, called
Aosta-Chatillon-Colle Ranzola (Carraro et
al., 1982) cuts the study area with E-W
orientation. According to the authors the
fault shows no evidence of post-Wurm
movement; nevertheless, its possible
passive contribution to the recent morpho-
genesis will be discussed.

3 GEOMORPHOLOGICAL OUTLINE

Mt Croce di Fana is located on the left
hydrographic side of Aosta Valley, a few
kilometers to the East of the town of
Aosta. The East and the West borders of
the southern slope of the mountain are
sharply defined by the Vallone di Saint
Barthelemy and the Valle Castello di Quart
respectively. The entire slope is
imprinted with glacial morphology
(terraces and mounds). Wide glacial
terraces, related to the maximum expansion
of the Balteo glacier during the Wurm, are
present between elevation 1000 m and 1500
m. To a lesser extent morain deposits and
terraces related to the effluent glaciers
from Vallone di Saint Barthelemy are also
present on the upper eastern part of the
slope (approximate elevation 1700 m). On
the lower portion of the slope, where

elevation is lower than 800 m a.s.l.,
different orders of glacial, fluvio-
glacial, and fluvial terraces connected to
post-Wurm phases are present. Moreover,
recent geophysical surveys carried out
along the Aosta valley have shown at least
200 hundred metres of glacial and
fluvioglacial deposits laying on the
bedrock.
At the foot of the slope, fluvio-glacial
conglomerates unconformably overlying
morain deposits and overlain by less
tilted gravel, outcrop along the old road
"Provinciale" west of the village of
Plantaz, indicating southward tilting of
at least 50°. These post-glacial deposits
allow the identification and dating of the
recent movements, connected with the
failure of the overlying mountain slope.
Other evidences of very recent movements
are present in the upper part of the
slope, where the main niche of the slide
cuts the stadial morains of the Vallone di
Saint Barthelemy.
The morphologic evidence of the
gravitative processes is well defined at
the scale of the whole mountain slope and
of smaller portions. A 3-D picture (Figure
2) with view point from the SSE provides a
qualitative analysis of the morphologic
patterns of the slope, in which the large
trench which separates the main collapsed
body from the bedrock is particularly
evident. Furthermore, several escarpments
and concentric counterslopes can be
recognized also within the main collapsed
body, expecially in the eastern portion of
the slide.

4 PHOTO-INTERPRETATION

The analysis through the aereal
photographs has produced a morpho-tectonic
map (Figure 3), which outlines landforms
derived from processes with structural
relevance (morpho-structures) in addition
to those derived from erosional and

Figure 2. Perspective view of the southern
slope of Croce di Fana Mountain.

depositional processes. The entire collapsed sector is characterized by several slide surfaces developed at different depth and often with complex geometries. The major morphologic element of the whole studied area is the large main scarp developed in the bedrock of the crown area of the southern slope of Mt Croce di Fana. The slip of the scarp is about 100 m and an area of slope failure of about 12 km^2 with width at the toe of about 5 km is defined. An overall symmetry of the geometry and the measurable slip of the main detachment surface centered on a N-S/NNW-SSE axis can be recognized. This observation suggests that the movement of the collapsed body happened in a N-S direction, thus orthogonal to the orientation of the Aosta Valley.

The upper part of the main scarp, which surrounds the entire Mt. Croce di Fana southern slope, is composed of a family of shear planes developed along the main detachment surface, is semi-circular in shape, and widens downslope until an helliptic shape is achieved. Several secondary planes, also with arcuate shape, some of which are concentric, have developed in the upper section and give the slope a characteristic step-like profile. Along the whole eastern side of the main scarp, typical doubling of the mountain crest, counter slopes, and elongated mounds in E-W direction, clearly influenced by E-W trending fractures of regional relevance, are present.

Three additional minor slope failures have been identified inside the area defined by the main scarp and seem to represent still active portions of the major landslide. These slidden blocks are defined by minor scarps with markedly arcuate shape, counterslopes, and collapse-trenches. The largest one consists in a minor scarp, concentric to the main scarp, present between elevation 1950 m and 1450 m. The second one is located on the SW portion of the slope, along the ridge of the left hydrographic flank of the Castello di Quart Valley. The third active area is located in the corresponding south-eastern sector of the rockslide, where the

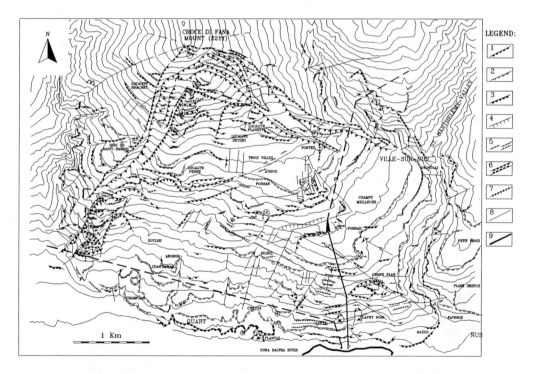

Figure 3. Photogeological map of the morphotectonic evidences and of the main tecton structures and gravity induced collapses of the southern slope of Croce di Fana mountain.
1=Active and inactive scarps; 2=Glacial terrace; 3=Fluvial or glacio-fluvial terrace; 4=Slide surface; 5=Trench or couterslope; 6=Double ridge; 7=Rocky ridge in erosion; 8=Faults and fractures; 9=Hydroelectric plant

underground power plant was excavated.
Lineaments such as faults and fractures
and gravity-induced structures have been
separately represented in fig.4 in order
to evaluate the link with the regional
structures.

Two rose diagrams of fig.5, one for the
strike vs percentage of occurrence and one
for strike vs cumulative lengths, have
been obtained relatively to the identified
structural elements. Four main peaks are
evident in both the diagrams: at 60°-90°,
at 100°-120°, at 30°-40°, and at 160°-
190°. The strongest peak corresponds to
the Aosta-Chatillon-Ranzola fault line.
The higher cumulative length of this trend
can be explained with their younger age
with respect to the other lineaments,
since older structures show a
discontinuous evidence at the surface.

5 GEO-STRUCTURAL ANALYSIS

The geo-structural field analysis was
planned with the purpose of investigating
in further detail some of the structural
trends and fracture systems outlined with
the photo-interpretation.

Along the main scarp wide open fractures
were observed in highly fractured quartz-
ortogneiss with sub-horizontal
schistosity. The main discontinuities are
here oriented in ENE-WSW or NE-SW
direction and the dip is higher than 70°
toward the south. Inside the slidden mass,
attention has been paid to the central
spur below the inner scarp, the
counterslopes SE of the top of the slope,
and the eastern portion of the main scarp
near Fonteil. Outstanding structures have
been observed on the central spur, which
is separated from the main slope by large
trenches whose aperture is on the order of
100 m. The couterslopes associated with

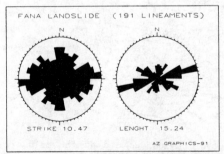

Figure 4. Rose diagrams relative to
tectonic lineaments structures represented
in Figure 3. The radius of the rose
corresponds to the largest percentage; its
value is reported under each rose.

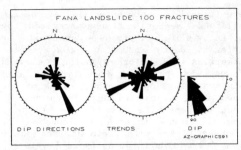

Figure 5. Rose diagrams relative to the
main sets of discontinuities measured in
the field along the Croce di Fana southern
slope. The diagrams report the percentual
frequency of each class as in Figure 4.

the spur have formed large depressions in
the bedrock filled with blocks and coarse
detritus.

Small E-W and ESE-WNW trending normal
faults associated to NO-SW joints have
been observed in the calc-schists
outcropping along the eastern portion of
the slope. On the western side of the
middle portion of the slope the surface of
detachment is outlined by a 10 m high
scarp in the bedrock surrounded by open
fractures with E-W trends and subvertical
dip. Large collapsed wedges of bedrock,
have also been isolated by N-S trending
fractures.

On the eastern side of the middle portion
of the slope, where roches moutonees of
Wurmian age outcrop, wide N-S collapse
trenches filled with recent colluvium were
observed in continuation of a regional
fault system with a similar trend,
separating serpentinites and calc-schists
upslope at 1500 m elevation. Along the
whole central part of the rockslide, along
the edge of the Condomine terrace and in
the southeastern part of the slope, the
passive role of the E-W and ENE-WSW
tectonic structures can be observed. Open
fractures associated with dangerously
instable rock masses, grabens, trenches,
and counterslopes have formed along
subvertical joints with these trends.

The deformations present in the lower
sector of the slope in the fluvio-glacial
conglomerates provide a mean to date the
movements occurred in the study area.
These deposit, overlying moraine deposits,
are younger than 10,000 years and show at
least 50° dip toward the main valley
floor. They are covered by gravel with
much lower dip, and at the top of the
section subhorizontal layers indicate
progressive diminution of tilting. NE-SW
trending normal faults and small folds due
to water drag and water escape processes

are here observable. The geometrical characteristics (dip direction, strike and dip) of the main observed fractures are summarized in the rose diagrams of fig.5. The main peak correspond to an ENE-WSW set of fractures, showing an outstanding asymetrical plunge toward the valley in the same direction as the slope failure. The dip values have a preferential concentration between 60° and 80°.

6 DEFORMATION IN TUNNELS AND GEOMECHANICAL INVESTIGATIONS

Damages of the concrete linining of the main tunnel of the Quart plant have been observed since 1970. The cracking pattern, denoting a strong non symmetry, consist in a series of crackings due to compression in the upper part of the lining and in almost continuos tensile cracking along the left side of the tunnel. First reconstruction works were started during the second half of 80'S.

The consolidation of the rock around the tunnel lining was carried out by means of injections using tie rods having the double function to stabilize the rock mass and to relieve the loadings on the concrete lining. During the works two sections were instrumented by installing rod extensometers, vibrating wire pressure trasducers, two inclinometer-extensometer, vibrating wire load cells, pore pressure trasducers, and termometers. Almost all measuraments were automatically recorded

(Figure 6).

The evolution of the crackings along not repaired stretches of the tunnel and the trend of the measuraments recorded (mainly estensometers) showed that the phenomena were in progress, hence further in-situ testings were decided. Boreholes placement, down-hole video survey, flat-jack measurements, hydraulic fracturation and doorstopper tests have been carried out in 1989 and 1990 (Figure 7).

Moreover, a survey using photo-aerial techniques was carried out from which the general situation already shown in the previous chapters was discovered. The geological situation pointed out by the new investigations has allowed a qualitative interpretation of the phenomena in progress and has given suggestions about the potential residual life of the tunnel and on possible design modifications of the hydroelectric plant. Quantitative information on the loadings acting on the tunnel lining, headed to design a further restoration of the most damaged stretches of concrete, were obtained by means of a numerical model based on a back-analysis approach starting from the stress measuraments obtained with flat-jack tests.

7 DISCUSSION AND CONCLUSION

Fotointerpretation and field structural analysis allowed to identify a deep gravitative slope failure on the whole

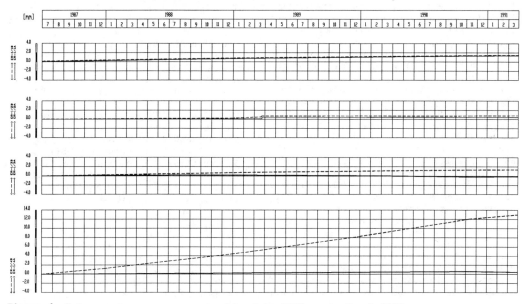

Figure 6. Extensometers measurements from July 1987 up to March 1991.

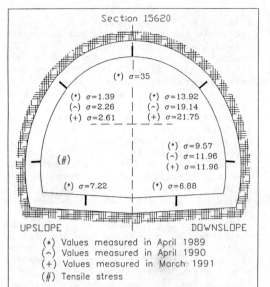

Section 15620

(*) σ=35

(*) σ=1.39   (*) σ=13.92
(^) σ=2.26   (^) σ=19.14
(+) σ=2.61   (+) σ=21.75

(#)   (*) σ=9.57
               (^) σ=11.96
               (+) σ=11.96

(*) σ=7.22   (*) σ=6.88

UPSLOPE   DOWNSLOPE

(∗) Values measured in April 1989
(^) Values measured in April 1990
(+) Values measured in March 1991
(#) Tensile stress

Figure 7. Flat-jack maesurements.

southern slope of Mt. Croce di Fana. The areal extent of the phenomenon is 12 km^2.
The detachment surface of the slide is outlined by a continuous main scarp from the mountain top (2211 m elevation) to the altitude of 100 m, while from here to the Aosta Valley floor the scarp is not clearly exposed. The scarp shows a vertical slip around 100 m and an outstanding morphological evidence.
The slide has developed on a slope that was deglaciated about 10,000 years ago and it is suggested that the gravitative instability was originated by the slope de-compression that followed the glacier withdrawal. The deformation of post-Wurm terraced deposits at the base of the slope near Plantaz indicate the young age of the phenomenon.
All the recognized morphologic elements are typical of most of the major rock slides observed in the Alps, indicating deep gravitational collapse (see Giraud et al., 1990 for an updated review).
Regional tectonic structures like the Aosta Valley Line (known in the literature) influence the development of the gravitative instability, at least as passive control on the movements. This hypothesis is supported by the prevailing ENE-WSW strike, with southward dip higher than 70°, observed on the collapsed body, which is parallel to both the strike of the main detachment surface and the Aosta Valley line. Similar trend of the structural lineaments has been observed on the aereal photographs. In particular, the main families of lineaments show

identical peak orientations of the cumulative length measured on aereal photos and of the percentage occurrence measured in the field (rose diagrams of figure 4 and 5).
The sliding slope can be divided into sectors with different structure and evolution of the movement. In addition to the main scarp, three minor semi-circular scarps, collapse trenches, and counterslopes have been commonly identified. These surface manifestations of slope failures are related to detachment surfaces at different depth.
Most of the damage to the tunnel lining and the other structures of the power plant of Quart has occurred in areas where the geological investigation has revealed the most recent evidence of movement. Our study does not allow the precise areal definition and the possible evolution with time of the gravitative process.

REFERENCES

Carraro F., Ferrero E., Forno M.G., Ricci R., 1979. Dati preliminari sull'evoluzione neotettonica dell'arco delle Alpi Occidentali. C.N.R., Contr. Prel. alla realizzazione della Carta Neotettonica d'Italia, Pubbl. 231 Prog. Fin. Geodin., 239-249.
Dal Piaz G.V., Hunziker J.C. & Martinotti G., 1972. La zona Sesia-Lanzo e l'evoluzione tettonico-metamorfica delle Alpi nordoccidentali interne. Mem. Soc. Geol. It., 11, 433-466.
Dal Piaz G.V., 1976. Il lembo di ricoprimento del Pillonet: la Falda della Dent Blanche nelle Alpi Occidentali. Mem. Ist. Geol. Mineral., Univ. di Padova, 31, 63 pp.
Elter G., 1960. Carta geologica della Val d'Aosta, scala 1:100.000. Consiglio Nazionale delle Ricerche, anno 1987.
Fornero I., Frassoni A., Giuseppetti G., Mazza' G., 1988. The restoration of a concrete tunnel lining: the role of field measurements in the safety and control of the structural stability. 1st International Conference on Computer Methods and Water Resourches, Marocco.
Giraud A., Rochet L. and Antoine P., 1990. Processes of slope failure in crystallophillian formations. Eng. Geol., 29, 241-253.
Polino R., Dal Piaz G.V. & Gosso G., 1990. Tectonic erosion at the Adria margin and accretionary processes for the cretaceous orogeny of the Alps. Deep structure in the Alps. Roure F., Heizmann R. & Polino R. Editors.

Landslides, Bell (ed.) © 1995 Balkema, Rotterdam, ISBN 90 5410 032 X

The importance of historical observations in the study of climatically controlled mass movement on natural slopes, with examples from Italy, Poland and UK

M. Del Prete
University of Basilicata, Potenza, Italy

T. P. Gostelow
British Geological Survey, Nottingham, UK

J. Pininska
University of Warsaw, Poland

ABSTRACT: The importance of long-term historical observations in mass movement studies on natural slopes is briefly discussed followed by examples from Italy, Poland and UK which consider the effects of climate and climatic change on slopes of different ages and geotechnical properties. Most historical studies consider individual slides or slopes and it is suggested that more regional investigations are needed to explore the incidences of mass movement with climatic events so that the controls of their future susceptibility to reactivation are better understood.

1 INTRODUCTION

The need to understand the relative importance of natural and man-made changes in the environment has prompted the publication of several articles which draw attention to the value of long term historical observations. For example NERC (Anon,1991) in a general statement, suggests historical changes can take place at global, regional and local levels and that observational scale is also important. The document suggests that long time series data must be collected if their relative impacts are to be understood. It summarises their role in science as

1. A study of long-term processes
2. Early warning of man-made environmental change
3. Detection of rare events and episodic phenomena
4. A study of processes showing high variability
5. Measurement of subtle or complex processes
6. Testing hypotheses and environmental models

Mass movement, in its various forms, is usually triggered by a combination of natural, and man-made influences and is ideally suited for sudies of long term processes and historic events especially when related to climatic records. In this paper, we first, briefly consider the kinds of long term data which need to be collected, and secondly continue with some examples which illustrate their use. We conclude by attempting, through these examples to summarise the main factors which control the susceptibility of soil slopes to mass movement following climatic changes.

2. LONG TERM OBSERVATIONS OF ENVIRONMENTAL DATA AND MASS MOVEMENT

2.1 Timescales

Many of the natural landslipped slopes which cause engineering problems were degraded by processes associated with the climatic changes of the Quaternary glaciations. The collection of data regarding the slope age and subsequent reactivations can therefore be extended back from recent decades through the Holocene period to at least the last glaciation. Gerard (1991) suggests four broad timescale divisions for the climatic conditions which might be responsible for triggering past mass movement events,ie

1. Minor fluctuations   - operating over 25-100 years
2. Post-Glacial, Historic   "   " 250-1000 years
3. Glacial- Interglacial   "   " 50-100000 years
4. Minor Geological   "   " 1-10 million years

Baker (1989) in a study of past flood events makes the distinction between paleofloods and historic floods, the latter referring to those which have been documented by human records. However, although broad divisions give an impression of data quality and quantity it seems likely that increasing numbers of time series plots will emerge, and therefore absolute time scales, Before Present (BP) which can be expanded in scale during historic time are probably the most useful. The examples reviewed here include landslides and degraded slopes of Quaternary age which have been or could be reactivated in the future. An assessment of risk requires careful observations of trends in the factors responsible for mass movement rather than a staightforward engineering analysis for a factor of safety, at a particular time.

2.2 Historical Observations

The historical factors responsible for mass movement can perhaps be summarised under five interelated headings, ie

1. Climatic change (long-term) of trends in both precipitation and temperature

2. Short term climatic fluctuations, frequency and magnitude

3. Seismicity, frequency and magnitude

4. Rates of fluvial and marine erosion associated with relative sea level changes

5. Man-made factors, such as past and present mining activity, agricultural practices, construction, groundwater abstraction and recharge. Greenhouse effect.

Historical observations, and past evidence for them may be made for each group and related to i) the morphology of slopes and landslides of different ages, ii) the distribution of first time slides, and iii) mass movement activity or reactivation events. Most papers on landsliding include some historical discussion, but as suggested below there is a need for a regional approach to the subject, especially when linked to climatic studies. In this paper we consider climatic influences and how they might interact with soil slopes of different age and plasticity.

3. LANDSLIDES AND CLIMATIC CHANGE

3.1 General

There are many examples, worldwide, of large, relict currently stable landslides on comparatively shallow slopes which date from previous climates associated with the Quaternary. These are comparatively easy to identify and date by conventional geomorphological and geological mapping, but it is more difficult to determine their subsequent history of movement, if any, or to rank them in terms of hazard or risk from future rising groundwater levels associated with climatic fluctuations and/ or man-made change. There are also other slopes, especially in mountainous areas which are geologically susceptible to shallow first-time debris slides following high intensity rainfalls. The triggering intensities and trends in climate or rainfall patterns which might increase the frequency of these also need to be established with the help of historical studies.

The type of climatically triggered slope response (reactivation) or extent of first-time post failure displacement is largely controlled by the brittleness of the materials involved and the age of the slope; older slopes generally being of lower angle. It is perhaps convenient for the purposes of this discussion to consider two material groups and two ages of slope in which they occur i) brittle soils of medium to high plasticity which develop shear surfaces, and ii) non-brittle soils of low to medium plasticity which do not, both in slopes of Glacial/Late-Glacial and Post-Glacial (Holocene) age.

3.2 Soils of Medium to High Plasticity

3.2.1. Late-Glacial Slopes

The development of solifluction mantles and lobes as a result of glacial and late glacial freeze-thaw processes on slopes underlain by plastic clays in the UK has been well documented in the literature (Skempton and Hutchinson, 1976). Most, if not all natural slopes above 2° or 3° in those parts of the clayey geological formations of the UK which lie beyond the edge of the last ice sheet contain shear surfaces which have been caused by downslope movements, or differential volume change associated with ice segregation. However, where these formations are overlain by stiffer materials (often relatively permeable and of low plasticity) the natural slopes are steeper reaching up to 8° or 10° in places sometimes with relict deep-seated slides at the slope crest. The overall form of such slopes is concave compared to the generally convex, flatter forms of lower angle which are found on clays only. The presence of the different overlying lithologies encourages higher groundwater tables, discharge areas, and springlines, and, where they are associated with landslides and solifluction deposits the slopes are susceptible to reactivation. Although there are many areas in S. England where late glacial slopes, may in the future cause problems during construction, there are other parts of Europe more distant from the edge of the last ice sheet where the same geological conditions are found, but the geotechnical problems are more severe, perticularly when they have extreme, seasonal rainfall patterns. For example, there are extensive areas in southern Italy where relict slides dating from Quaternary pluvial climatic conditions are found. Two typical examples can be found below the town of Grassano in the Basento valley, Basilicata, which adjoins river terraces thought to be of Wurm or Riss age, and in the Appennino near Avigliano (fig 1).

3.2.2 Grassano Landslide

The solid geology of the slide area consists of marine sedimentary rocks, clays, sands and

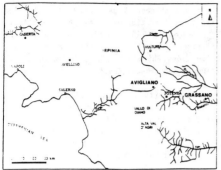

Fig 1 Location of landslide studied in Italy.

conglomerates of Pliocene and Calabrian age. Following the Calabrian, a period of uplift initiated fluvial downcutting, which has left the conglomerates and sands at the summits of isolated hills along the interfluves. The majority of towns in the Basilicata region occupy the tops of these hills which are flat-topped and plateau like. the adjoining valley slopes thus now consist of the Calabrian clays, overlain by about 70m of cohesionless sediments. The slopes average 12° and continue down to the valley floor, but are heavily dissected in places by gulley erosion which has led to steep, post glacial slopes with 50° to 60° angles in places. These are subjected to soil erosion, and exhibit the characteristics of badland topography , known locally as calanchi. The river terrace remnants above the present valley floor are well developed below Grassano (Boenzi et al, 1978) suggesting that the slopes have undergone several periods of degradation related to changing valley base levels. The development of the slopes has involved mass movement of both clays and overlying conglomerates which have become detached from the hilltop summits, being completely removed in places. The landslip types seem to be mainly rotational with some translational movements with shear surfaces in the clays. These are blue, stiff, sometimes fissured with a clay fraction varying between 40% and 50% and plasticity index of 20% to 30% . In the reversal shear box a $\phi'p$ of 20° to 26° and $c'p$ of 10 to 15 kPa and residual, $\phi'r$ of 13° to 16° with c' equal to zero has been obtained. The overlying sands and conglomerates are soft and weakly cemented.

Grassano is a good example of a typical S. Italian hilltop town, being built largely on the conglomerates at an elevation of circa 575m with a mean annual rainfall of 550mm, but also at the edge of an ancient landslide (figs 2 and 3). This slide doesn't seem to be have been reactivated during the history of the town which dates back to the middle ages. However in the last major earthquake of the area there may have been some small movements because several houses were damaged (Cotecchia and Del Prete, 1986). Although the slide area is very ancient recent

reactivations have developed at the edges and it seems likely that the equilibrium of the entire mass and safety of the town will be controlled by this renewed degradation. These reactivations, or minor landslides are known as the Frana Calvario and Frana Cimitero slides (fig) which are gradually undermining the edges of the old (Frana Centro Storico) slide.

The first mention of these renewed movements in the Frana Calvario was in 1956, when after heavy rainfall from 4a.m. to 6 a.m. on the 14th of February over 30 houses were severely damaged and the area evacuated. Since these first movements there have been small deformations each winter, but the most noticeable were in 1957, 1959, 1960, 1973 and 1976 (fig 4). Reactivations take place during periods of extreme rainfall, as shown in fig 5. The most dangerous movements for the equilibrium of the entire slope are from the Frana Cimitero which is removing the toe of the Centro storica. The Via Appia passes across this slide and an attempt has been made to stabilise it by continuous bored piles and drainage which have, up to present, been successful. The dates of the reactivated movements of this slide agree only in part with the Frana Calvario and the last deformations were recorded in December 1991.

3.2.3 Avigliano Landslides

The second example in Italy lies in the Appennino mountain chain at an average elevation of 900m between the rivers Avigliano, Braida and Valle Boni (fig 6). The solid geology consists of scaley, agille varicolori with a capping of some 200m of Pliocene conglomerates and sands which include beds of clays. The slope form is concave, varying from 50° at the crest to circa 15° in the landslipped areas. Eight large landslides (debris slides, consisting of plastic detritus, P.I., 20% to 35% and large conglomerate blocks) have affected the town during the last century. Following an historical study of local archives it was found that the oldest recorded movements were registered at La Vanga in 1883. Other landslide movements (fig 6) were first recorded as follows: Franas Pantenello, 1903, Lagariello, 1915, Gianturco, February 1915, S. Vito February 1915, S. Lucia September 1929, Italia, September 1929, and Fontana January 1940.
An analysis of historical factors which may have caused the movements suggested there was a small relation with human interference or seismicity and a direct link with extreme rainfall as shown on fig 7. The amount of precipitation required for each reactivation is not the same in each case, some slides being more sensitive than others. For example, Frana Gianturco was reactivated in 1929, 1933, 1940, 1945, 1976, 1978, 1980, and 1983, in contrast to Frana La Vanga which reactivated in 1947 and 1956 only. This suggests local differences in geology, topography and the sequences of antecedent precipitation are

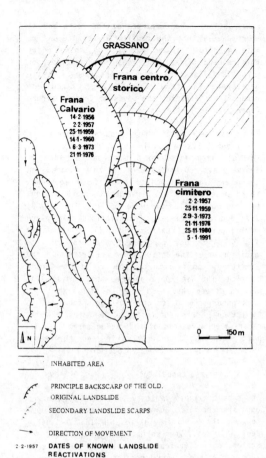

INHABITED AREA

PRINCIPLE BACKSCARP OF THE OLD.
ORIGINAL LANDSLIDE

SECONDARY LANDSLIDE SCARPS

DIRECTION OF MOVEMENT

2·2·1957 DATES OF KNOWN LANDSLIDE
REACTIVATIONS

FIG. 2   GRASSANO LANDSLIDE

important in each case. Figs 8 and 9 show detailed cumulative precipitation plots for March and April 1978 and April 1985 which resulted in reactivation. They illustrate the typical rising intensity limbs which are associated with storms resulting in landslides (Campbell, 1975).

The susceptibility to reactivation following rainfall is greater at Avigliano than Grassano. This is most likely due to the higher mean annual rainfall, although the difference in geology which have resulted in a debris layer of greater hydraulic conductivity may also be a contributory factor.

There are similar examples of ancient landslides throughout Basilicata, often close to inhabited areas and towns. The question arises as to whether these observed changes in slope behaviour at Grassano and Avigliano are a local phenomena, associated partly with man made activity, (through construction and expansion of the inhabited areas) or whether it is a response to a regional change in climate or rainfall pattern which increases erosion (calanchi) and/or groundwater levels in the upper parts of valley slopes. A collection and analysis of historic data and rainfall records throughout the area is curently being carried out as part of an EC funded investigation, to try and determine which of these possibilities is most likely.

3.2.4 Post-Glacial Slopes

Many post glacial slopes have been recently eroded and are thus generally in a more critical state of

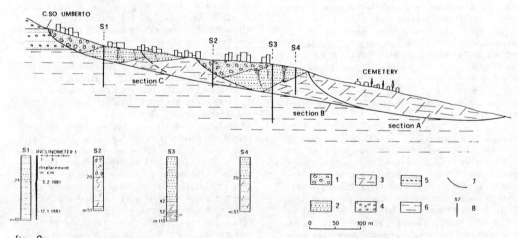

fig. 3

Geological profile of Historic Centre slide: 1| Detrital – deposits:
2| sandy – conglomeratic slipped mass: 3| Clayey slipped mass:
4| Conglomerates: 5| Sands: 6| Blue Clays: 7| Slip surfaces: 8| Boreholes.
(after Cotecchia and Del Prete. 1986).

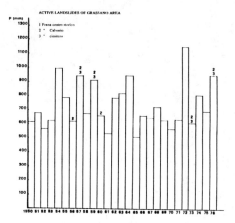

Fig.4. Histogram showing annual precipitation and landslide reactivations in the period 1950–76.

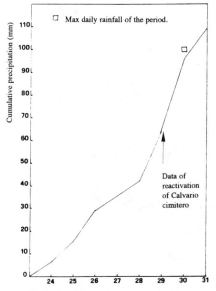

Fig. 5 - Cumulative precipitations at Grassano in march

1973 responsable for landslide reactivation.

equilibrium with regard to pre-existing landslides and colluvial accumulations at the slope toe. However, the same questions regarding future climatic reactivation arise, especially when considering slopes of early post-glacial age.

Good examples of historical analyses in Quebec, Canada have recently been published by Begin and Filion (1988) and Filion (1991) who used radiocarbon dating and dendrochronology to date landslides next to river terraces which date back 6000 years. The tree ring characteristics were also used for the interpretation of climate. For example the rings suggested that landslides dating from 1818 may have been associated with a general

lowering of temperature and increased precipitation associated with an 1815 volcanic eruption in Indonesia. They also found evidence to suggest that the greatest frequency of mass movements were connected to the 'Little Ice Age'.

Similar investigations using radiocarbon and archaeological dating methods were used at Hadleigh Castle in Essex, UK by Hutchinson and Gostelow (1976) who were able to show that the degradation of an abandoned marine cliff in London Clay was related to long term climatic change rather than local extreme isolated events, and that the onset of degradation periods was followed by substantial slope movements and accumulation of colluvium. There, following an initial phase of solifluction, the main periods of first time sliding and downslope accumulation of colluvium were dated at between 6500 and 7000 years BP and 2000 and 2100 years BP, corresponding to the beginning of the wetter Atlantic and Sub-Atlantic Holocene periods respectively. The most recent phase of movement occurred at the slope crest in the 19th century which may have been influenced by man-made activity connected with the Castle. The colluvium toe at the base of the slope has not changed greatly since that last single event. This quiescent behaviour is connected, in part, to a comparatively low mean annual rainfall of c. 546mm. However, in other parts of the UK where the mean annual rainfall is greater there is evidence to suggest that slides are more active and degradation of some early post glacial slides has been virtually continuous. Skempton et al (1989) have shown by careful historical analysis that an earthflow below a springline consisting of weathered britttle shales at Mam Tor in Derbyshire dating from 8000 years BP (mean annual rainfall 1250mm) moves when the winter monthly rainfall exceeds about 200mms. They suggested that as the slide moved downslope the factor of safety increased and hence a greater winter rainfall or longer return period would eventually be required to initiate movement.

Recent reactivations in old Post Glacial slopes occur elsewhwere in Europe. For example in southern Poland large landslides in weathered Carpathian flysch have been dated back to between 5300 and 3100 years BP when the climate was generally wetter. However they are still unstable under present conditions when the mean annual rainfall exceeds 1000mm per year, with the highest monthly totals taking place between April and August following snowmelt (fig 10).

These selected examples of dated slopes and reactivation activity are geographically widely, spaced, but are all related to climate. Collaborative investigations are required in similar physical settings, to establish whether a) there are regional connections between the onset of landsliding, or b) they are local events.

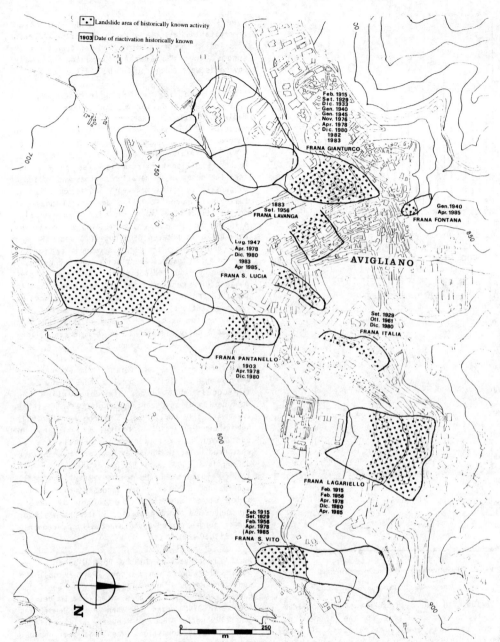

Feb. 1915
Set. 1929
Dic. 1933
Gen. 1940
Gen. 1945
Nov. 1976
Apr. 1978
Dic. 1980
1982
1983
FRANA GIANTURCO

1883
Set. 1956
FRANA LAVANGA

Gen. 1940
Apr. 1985
FRANA FONTANA

Lug. 1947
Apr. 1978
Dic. 1980
1983
Apr 1985.
FRANA S. LUCIA

AVIGLIANO

Set. 1929
Ott. 1961
Dic. 1980
FRANA ITALIA

FRANA PANTANELLO
1903
Apr. 1978
Dic. 1980

FRANA LAGARIELLO
Feb. 1915
Set. 1956
Apr. 1978
Dic. 1980
Apr. 1985

Feb 1915
Set. 1929
Feb. 1956
Apr. 1978
(Apr. 1985
FRANA S. VITO

N

0   250
m

FIG. 6 PLAN SHOWING LANDSLIDE AREAS OF AVIGLIANO

3.3 Soils of low to medium plasticity

3.3.1 General

The non-plastic to low plasticity geological formations of low brittleness do not develop shears with a substantially lower frictional strength than the surrounding matrix. This characteristic influences the form of natural slopes and overall angle. In a study of abandoned marine and fluvial slopes in glacial till deposits in the UK (fig 11) a very clear distinction between those of Late glacial and post glacial age could be made. Fig 11 from Gostelow and Hutchinson (in preparation) also illustrates the influence of friction angle on the slope angle and form. The most obvious contrast to the more plastic materials is the lack of complex landslide deposits and hummocky terrain. Colluvium accumulates at the slope toe leaving an intact,

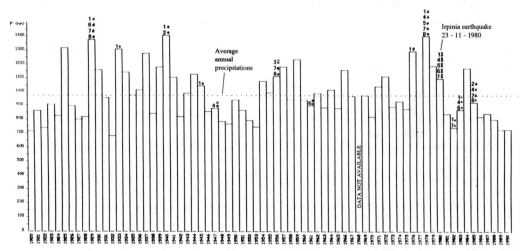

ACTIVE LANDSLIDES OF AVIGLIANO AREA

1 Frana Gianturco	5 Frana Pantanello
2 " Fontana	6 " Via Italia
3 " Lavanga	7 " Lagariello
4 " S. Lucia	8 " S. Vito

POSSIBLE CAUSES OF REACTIVATION

* Precipitations
□ Earthquake
• Human

FIG. 7 HISTOGRAM SHOWING ANNUAL PRECIPITATION AND LANDSLIDE REACTIVATIONS IN THE PERIOD 1920-90

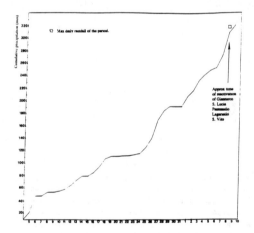

Fig.8. Cummulative precipitations at Avigliano in march and april 1978 responsible for landslide reactivation.

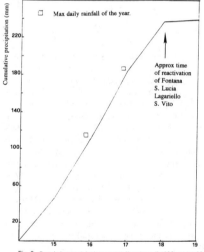

Fig. 9 - Cumulative precipitations at Avigliano in april 1985 responsable for landslide reactivation.

degrading slope segment at the slope crest. Climatic control of mass movement is thus largely confined to first-time slides (mainly from the slope crest) and is similar to that found on residual soil slopes and weathered rocks.

3.3.2 Late glacial slopes

Fig 11 shows that the slopes of late-glacial age in northern England and Scotland all tend to be of low plasticity. Unlike their more plastic counterparts the slope profiles show strongly convex crests with no evidence of instability under present climatic conditions. Because of

1565

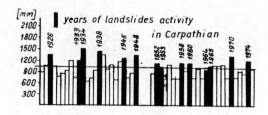

FIG. 10 - LANDSLIDE ACTIVITY IN CARPATHIANS, POLAND.

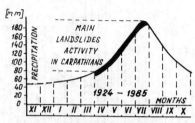

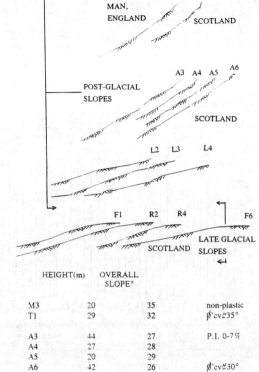

	HEIGHT(m)	OVERALL SLOPE°	
M3	20	35	non-plastic
T1	29	32	$\phi'cv \angle 35°$
A3	44	27	P.I. 0-7%
A4	27	28	
A5	20	29	
A6	42	26	$\phi'cv \angle 30°$
L2	12	15	P.I. 7-20%
L3	16	18	
L4	15	16	$\phi'r \angle 24°$
F1	21	14	P.I. 0-7%
R2	25	13	
R4	20	15	
F6	19	11	$\phi'cv \angle 30°$

Fig. 11 Comparison of late and post glacial slopes in low plasticity materials, UK.

their low primary brittleness these late glacial slopes are perhaps the least susceptible to long term climatic, changes although weathering may produce loose soil structures at shallow depth, which on the steeper slope segments, (greater than 24˚) may be susceptible to first time failure following heavy rainstorms although no evidence was found for these during the survey.

3.3.3 Post Glacial Slopes

The post glacial slopes in fig 11 are not precisely dated, but have been abandoned from erosion by isostatic movements relative to sea level following the last glaciation. Radiocarbon dating from adjacent raised beach or estuarine/alluvial sediments suggests that most have been freely degrading for at least 5000 years. The slopes are concave to rectilinear in profile and overall angles depend on the soil plasticity, varying from 28˚ to 35˚ in non plastic soils to 14˚ to 18˚ in soils of intermediate plasticity (PI, circa 20%). These slopes are still susceptible to first time movements, following extreme rainfall events, but reactivations are less common, in contrast to the steeper actively eroding coastal and fluvial examples in the same materials.

4. CONCLUSIONS

Conventional engineering analyses of natural slopes normally only take into acount the environmental conditions which prevail at a particular time. However, there are many circumstances where there is a need to predict the possibility of future instability from long-term climatic change. We have discussed here, some of the factors which may influence the response of slopes to climatic change and presented some selected examples from Italy, Poland and UK which have explored the relationships between the age of the slope, broad geotechnical characteristics, the stage of development, and subsequent reactivation in historic time following precipation events. Although these studies are useful at the local level, more long-term historical observations at the regional level are required, for example in the mediterranean area, central Europe, and northern Europe, so that the links between climatic fluctuations and dates of mass movement can be made. The examples reviewed here suggest that ancient landslipped slopes in plastic soils in the European area are still susceptible to natural reactivation, with the susceptibility decreasing as the plasticity index and material brittleness also decreases.

5. ACKNOWLEDGMENTS

This paper was supported by Italian CNR. The authors also acnowledge the help given by Mario Bentivenga, Vito Summa and Anna Grignetti with data collection.

6. REFERENCES

Anon (1991) The environment in time: Long term observations of the natural environment. Natural Environment Research Council, Swindon, UK.

Baker VR (1989) The magnitude and frequency of palaeofloods, in Floods, Hydrological, Sedimentological and Geomorphological Implications ed by K Beven and P Carling. J Wiley

Begin C, Filion L (1988) Age of landslides along the Grande Riviere de la Baleine estuary, eastern coast of Hudson Bay, Quebec, Canada. Boreas, 17, 289-29

Boenzi F, Digennaro MA, Pennetta L (1978) I terrazzi della valle del Basento (Basilicata) Rivista Geografica Italiana Annata LXXXV, 4, 397-418

Campbell RH (1975) Soil slips, debris flows and rainstorms in the Santa Monica mountains and vicinity, southern California: US Geological Survey Prof. Pap. 851, 51 p

Cotecchia V Del Prete M (1986) Considerations on stability of landslides in the historic centre of Grassano after the earthquake of 23rd November 1980. Symp. Int. sui problemi in aree sismiche, IAEG Bari

Filion L Quinty F Begin C (1991) A chronology of landslide activity in the valley of Riviere du Gouffre, Charlevoix, Quebec. Can. Jnl. Earth. Sci. 28, 250-256

Gerrard J (1991) The status of temperate hillslopes in the Holocene. The Holocene, 1, 1, 86-90

Hutchinson JN Gostelow TP (1976) The development of an abandoned cliff in London Clay at Hadleigh, Essex. Phil. Trans. R. Soc. Lond. A. 283, 557-604

Skempton AW Hutchinson JN (1976) Phil. Trans. R. Soc. Lond. A. 283

Skempton AW Leadbetter AD Chandler RJ (1989) The Mam Tor landslide, North Derbyshire Phil. Trans. R. Soc. Lond. A 329, 503-547

Landslides, Bell (ed.) © 1995 Balkema, Rotterdam, ISBN 90 5410 032 X

Slope stability – The influence of granular particle chemistry

A. B. Hawkins & C. McDonald
Department of Geology, University of Bristol, UK

ABSTRACT

The peak and residual shear strengths of a soil are affected by the grain size, grading and chemistry. The granular fraction of soils, which commonly consists of quartz, calcite or gypsum, is subject to both long term diagenesis and short term chemical alteration which leads to marked changes in shear strength. While in most situations quartz remains stable, in dark calcareous mudstones near-surface weathering may remove the calcite and chemcial reactions result in the formation of gypsum crystals. Decalcification processes involve not only downward leaching but also rounding of the calcite grains, resulting in a reduction of shear strength. In contrast the sucrose selenite crystals which typically occur between 1.5 and 3 m enhance the shear strength. The combination of these factors has a major influence on the development of shear surfaces at shallow depths.

1 INTRODUCTION

The shear strength of a soil depends on the cohesive forces between the contained particles, the soil grading, the shape of the grains and the nature of the contained water. Quartz, calcite and gypsum are three minerals commonly found in clay-rich soils or mudrocks and are all effectively electrically inert. However, their granular nature exerts a strong influence over both peak and residual shear strength. In addition, all three minerals are subject to diagenesis which may modify their grain size, grain angularity and intergranular fabric and hence influence the shear strength of the soil material. This is of particular importance in the case of calcite and gypsum as the changes may occur within the engineering life of a project.

2 NON-COHESIVE (GRANULAR) SOILS

Sowers and Sowers (1951) stated that the three main factors influencing the shear strength of granular soils are particle shape, particle size and relative density. The highest angles of peak shearing resistance are obtained for dense, well-graded soils with angular particles that have a high degree of particle interlocking during shear. Table 1 shows that for a sand there is a 6 to 8° difference in peak friction angle between the loose and the dense states and a 5 to 6° difference between soils with

Table 1 Effect of angularity and grading on peak friction angle (Sowers & Sowers, 1951).

Shape/ Grading	Peak Friction Angle	
	Loose	Dense
Rounded, uniform	30°	37°
Rounded, well-graded	34 °	40°
Angular, uniform	35°	43°
Angular, well-graded	39°	45°

angular and rounded grains. Sowers and Sowers suggested that for granular soils the Mohr failure envelope is straight and passes through the origin. However, Holtz and Gibbs (1956) show that for a sand/gravel mixture the envelope is curved.

Hough (1957), drawing attention to the significance of grain size, grading and density, quoted a series of data summarised in Table 2. The lower values within each range refer to well rounded particles and the higher values to angular particles. Hough also stated that when the interlocking between soil particles decreases such that continuous shear deformation can continue without further volume change, an "ultimate" strength is achieved. It is unfortunate that the term "ultimate strength" has a number of usages; in this paper it refers to the straight line portion of a curved failure envelope, replacing the expression "lowest residual strength" of Hawkins and Privett, 1985.

Table 2 Peak and residual shear angles for a number of granular soils (after Hough 1957).

	Friction Angles		
	Peak Str		"Ultimate" Str
	ϕ		ϕ_{cv}
	Med Dense	Dense	
Non-plastic silt	28-32°	30-34°	26-30°
Uniform fine to medium sand	30-34°	32-36°	26-30°
Well-graded sand	34-40°	38-46°	30-34°
Sand and gravel	36-42°	40-48°	32-36°

Dusseault and Morgenstern (1978, 1979) introduced the term locked sand to explain the 65 m high valley sides with slopes averaging 55° which occur in the Athabasca Oil Sands in Canada. Examination of the sand grains and undisturbed intergranular fabric showed a relatively high incidence of long and interpenetrative grain contacts and a surface rugosity resulting from diagenetic grain solution. As a result the sands have both very high peak strengths due to the interlocking nature of the particles and very high residual shear strengths, in the order of 30 to 35°. The robustness of the quartz particles prevents the shearing of grains and grain surface asperities, resulting in considerable dilatancy during shear and hence a pronounced curvature of the Mohr envelope. In summarising they considered the important characteristics of locked sands to be a porosity less than the minimum obtainable in the laboratory, a quartzose composition, a diagenetic fabric and associated grain surface texture, an age generally older than the Quaternary and little or no intergranular cement.

The role of water in granular soils is variable depending on whether the material is moist or fully saturated, whether the water is static or moving and the direction of flow. In moist granular soils water may move by vapour rise or capillary tension and form a meniscus at the grain contact which actually results in an increase in the soils shear strength by the process of soil suction. In a similar manner water passing through a sufficiently permeable soil may create a reduction in static head. When the water flow is rising, however, an almost static state occurs above a positive (artesian) pressure. In this manner the non-cohesive soils experience a major rise in pore pressure, reducing the effective shear strength which creates a "blowing sand" or quick-sand condition.

3 COHESIVE SOILS

Bishop, Webb & Llewellyn (1956), when undertaking detailed testing on London Clay, showed clearly that for most cohesive soils the Mohr failure envelope was curved. Surprisingly, however, the curved nature of the failure envelope is still frequently ignored and the shear strength established from an approximated straight line. The envelopes are frequently drawn to pass through the origin both in the literature and in commercial stability analyses, thus both the cohesion intercepts, c' and c'r are assumed to be low or zero. Particularly with residual shear strength analyses, the ease of testing at high effective normal stresses, the drawing of straight lines and the ignoring of the cohesion intercept have led many practitioners to use artificially low residual shear angles especially for shallow slips.

The concept of residual shear strength was first explained in detail by Skempton (1964). As shear box tests and drained triaxial tests to establish the peak strength of intact samples are somewhat more difficult to perform than ring shear tests on remoulded soil, the influence of silt/sand particles within a silty clay on the peak strength is less well appreciated than on the residual shear strength. Lupini, Vaughan and Skinner (1981) demonstrated that during failure in a mixed subangular quartz sand/mica sample, only when more than 40% clay is present is a distinct shear surface developed through the finer fraction. When less than 40% clay fraction is present, failure through the material involves significant dilation and particle rolling, hence with progressively coarser material no pronounced shear surface is developed but rather a shear zone in which the grains are disturbed by rolling.

Although in both granular and cohesive soils quartz frequently forms a high percentage of the non-clay mineral part of the soil mass, in cohesive soils calcite typically forms up to 50% of the mass and in near-surface situations gypsum may account for over 20% by volume (Russell and Parker, 1979; Hawkins and Wilson, 1990). For all three minerals there is little ionic bonding between the cohesive material and the coarser fraction. When quartz occurs it has usually undergone sufficient abrasion during transport for the grains to be relatively well-rounded. Calcite, however, is present either as a diagenetic cement infill within the pore matrix of the sediment, as comminuted shell debris or as euhedral authigenic grains which frequently have a rhombic structure. Gypsum is present at near-surface levels having developed as a consequence of the weathering of iron sulphides; the resultant sulphuric acid reacting with

any calcium carbonate present in the ground to form calcium sulphate. This is observed as euhedral crystals of selenite, the crystal form of gypsum. Calcite and gypsum differ from quartz in that they are liable to chemical alteration over relatively short periods of time which may be significantly less than the engineering life of a project.

It has been shown by Hawkins, Lawrence and Privett (1988) that for samples taken from a core through the Fuller's Earth Formation north of Bath the index limits have an inverse relationship to the calcite content (Figure 1). Analysis of samples taken from the 2.5 m thick Fuller's Earth Bed, characterised by the high percentage of contained montmorillonite, revealed that the residual shear strength changed dramatically with calcite content, reducing from 27° to 11° at low effective normal stresses (Hawkins, 1985). When the sample with the highest calcium carbonate content was decalcified and re-tested by Lawrence (1985) it was seen that the straight line part of the curved failure envelope (ultimate residual shear angle - ϕ'_r ult) dropped from 21° to 6° (Hawkins, Lawrence and Privett, 1988). These authors noted that calcareous silty particles elevated the residual shear strength above the sliding mode value of the clay mineral particles alone.

McDonald (1990) and Hawkins and McDonald (1992a) have shown that, at high effective normal stress, there is a threshold of 28% calcite content which separates the sliding and transitional modes of failure for natural soils taken through the Fuller's Earth succession. Above this threshold the soils failed predominantly in the transitional mode. To confirm the threshold of 28% calcite content for high effective normal stress, one sample of Upper Fuller's Earth clay was progressively decalcified in the chemistry laboratory and its geotechnical behaviour charted (Figure 2). Clearly a calcareous silty clay shearing in the transitional mode may decalcify resulting in a marked reduction in residual shear strength. This is observed over both

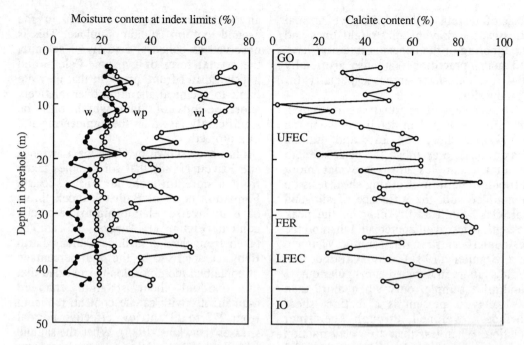

Figure 1: The inverse relationhip between index limits and calcite content for the Fuller's Earth formation at Soper's Wood, near Bath, UK. GO - Great Oolite; UFEC - Upper Fuller's Earth clay; FER - Fuller's Earth Rock; LFEC - Lower Fuller's Earth clay; IO - Inferior Oolite (after Hawkins Lawrence & Privett, 1980).

the curved and straight portions of the failure envelope although the phenomena is increasingly pronounced at high effective normal stress. The process of decalcification also involves rounding of the calcite grains which, compared with the angular nature of much fresh calcite, results in a reduction of the shear strength (Hawkins and McDonald, 1992b). The rounding and leaching of the calcite are likely to be significant factors in the development of shear surfaces at about 0.5 to 1.5 m below ground level.

Hawkins and McDonald (1992a) draw attention to shear surfaces acting as natural zones of increased permeability which facilitate enhanced decalcification along the shear surface encouraging failed slopes in calcareous mudrocks to decrease in strength progressively with time. When designing for long term stability in calcareous mudrocks therefore, engineers should ensure that potential shear strength

values after decalcification are taken into account. Care must also be taken to avoid over-zealous installation of counterfort drains to reduce pore pressure build-up. Where drainage is enhanced, the rounding and dissolution of the calcareous silty particles by carbonic and humic acids (Fookes and Hawkins 1988) may be faster than under normal ground weathering conditions. In addition to lowering the shear strength of calcareous mudrocks decalcification may result in precipitation within outlet pipes, restricting the effectiveness of the drainage. In this way a short term remedy may create a long term problem.

The formation and consequent draining of slopes exposes the pyrite which is present in many dark mudstones to chemical oxidation, frequently exacerbated by bacterial activity. It is well known (Quigley, Zajic, McKeyes and Yong, 1973; Penner, Eden and Gillott 1973; Hawkins and Pinches 1987) that the bacteria

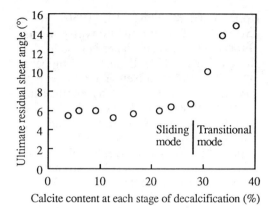

Figure 2: The change in ultimate residual shear angle (the straight line section of the curved failure envelope) for one sample of Upper Fuller's Earth clay progressively decalcified in the laboratory.

Thiobacillus catalyse the oxidation of iron pyrite which produces sulphuric acid. This sulphuric acid will preferentially combine with calcium carbonate to create selenite (calcium sulphate) which, due to its high solubility, may be removed or redistributed depending on the groundwater regime. Typically there are few selenite crystals in the upper 0.75 m of a soil profile as a consequence of the downward migration of vadose water. As a result the main sulphate-bearing horizon is at about the groundwater level or at an evaporation front where the proportions may be very high (Chandler, 1972). The crystallising sulphates may also be disseminated throughout the soil mass, but in a laminated soil, crystals will naturally develop in the areas of least stress, such as along bedding partings or in stress relief fissures; as a consequence ground heave can occur as described by Hawkins and Pinches (1987). The presence of sucrose gypsum will undoubtedly enhance the shear strength of the soil mass and hence be an additional factor in controlling the depth at which shallow shear surfaces typically form, ie 0.5 to 1.5 m.

SUMMARY

1.   Although quartz, calcite and gypsum are effectively electrically inert, their presence in mudrocks has been shown to have an important influence on the shear strength of the soil.

2.   If the silt-sized fragments consist of quartz, its inherent resistance to weathering will ensure that any measured shear strength is likely to remain constant in engineering time.

3.   In calcareous soils, decalcification involves the rounding and leaching of grains which has been shown to reduce the residual shear strength of silty clays such that the mode of shear changes from the transitional to the sliding. This is likely to be one reason why slopes frequently fail in the leached zone, about 0.5 to 1.5 m below ground level.

4.   Carbonates may also be dissolved along the more permeable shear surfaces, such that the residual shear strength decreases progressively with time.

5.   The rate at which calcite is removed is likely to be accelerated by the use of counterfort drains.

6.   The chemical reactions leading to the development of sucrose selenite crystals can result in an enhancement of the shear strength at depths of 1.5 to 3 m, typically related to past or present groundwater regimes. Such zones of enhanced shear strength may also influence the depth to which large scale translation surfaces develop.

7.   Slope stability analyses must consider not only the dissolution of carbonates within the soil by natural and acid rain but in addition, the precipitation, dissolution or redistribution of sulphates which can also change the shear strength parameters from those operating at the time of investigation.

ACKNOWLEDGEMENTS

The authors thank Marian Trott for help in preparing the manuscript.

REFERENCES

Chandler, R.J. (1972). Lias clay: weathering processes and their effect on shear strength. Géotechnique, 22, 403-431.

Dusseault, M.B. & Morgenstern, N.R. (1978). Shear strength of Athabasca Oil Sands. Canadian Geotechnical Journal, 15, 216-238.

Dusseault, M.B .& Morgenstern, N.R. (1979).Locked sands. Quarterly Journal of Engineering Geology, 12, 117-131.

Fookes, P.G. & Hawkins, A.B. (1988). Limestone weathering: its engineering significance and a proposed classification scheme. Quarterly Journal of Engineering Geology, 21, 7-31.

Hawkins, A. B. (1985). Geological Factors Affecting Residual Shear Strength. Proceedings IVth International Conference and Field Workshop on Landslides, Tokyo, 239-244.

Hawkins, A.B., Lawrence, M.S. & Privett, K.D.(1988). Implications of weathering on the engineering properties of the Fuller's Earth formation. Géotechnique, 38, 517-532.

Hawkins, A.B. & McDonald, C. (1992a) Decalcification and residual strength reduction in Fuller's Earth Clay. Géotechnique (in press).

Hawkins, A.B. & McDonald, C. (1992b). The influence of particle shape on residual shear strength. Manuscript submitted.

Hawkins, A.B. & Pinches, G. M. (1987). Cause and significance of heave at Llandough Hospital Cardiff: a case history of ground floor heave due to gypsum growth. Quarterly Journal of Engineering Geology, 20, 41-57.

Hawkins, A B & Privett, K D (1985). Measurement and use of residual shear strength of cohesive soils. Ground Engineering, 18, 22-29.

Hawkins, A.B. & Wilson, S.L.S. (1990). Technical Note: Sulphate increase in laboratory prepared samples. Quarterly Journal of Engineering Geology, 23, 383-385.

Holtz, W.G. & Gibbs, H.J. (1956). Shear Strength of Pervious Gravelly Soils. Proceedings ASCE, Paper No. 867.

Hough, B.K. (1957). Basic Soil Engineering, Ronald Press, New York.

Lawrence, M.S. (1985). The engineering geology of selected slopes on the Jurassic strata of the south Cotswolds. Unpublished Ph.D Thesis, University of Bristol.

Lupini, J.F., Skinner A.E.&Vaughan, P.R. (1981). The drained residual strength of residual soils. Géotechnique, 31, 181-213.

McDonald, C. (1990). Decalcification and residual shear strength reduction in Fuller's Earth clay. Unpublished Ph.D Thesis. University of Bristol.

Penner, E., Eden, W.J. & Gillot, J.E. (1973). Floor heave due to the biochemical weathering of shale. Proceedings of the 8th International Conference on Soil Mechanics and Foundation Engineering, Moscow, II, 151-158.

Privett, K.D. (1980). The engineering geology of slopes in the south Cotswolds. Unpublished Ph.D.Thesis, University of Bristol.

Quigley, R.M., Zajic, J.E., McKeyes, E. & Yong, R.N. (1973). Biochemical alteration and heave of black shale; detailed observations and interpretations. Canadian Journal of Earth Science, 10, 1005-1015.

Russell, D.J.& Parker, A. (1979). Geomechanical, mineralogical and chemical relationships in weathering profiles of an overconsolidated clay. Quarterly Journal of Engineering Geology, 12, 107-116.

Skempton, A.W. (1964). Long-term stability of clay slopes. Géotechnique, 14, 77-102.

Sowers, G.B. & Sowers, B.F. (1951). Introductory Soil Mechanics and Foundations, Macmillan, New York.

Landslides, Bell (ed.) © 1995 Balkema, Rotterdam, ISBN 90 5410 032 X

Data on slope angles of talus formations with special reference to the shape of rock fragments

P. E. Ihalainen & R. P. J. Uusinoka
Tampere University of Technology, Finland

ABSTRACT: Size, shape and fabric of material and slope angles were measured in four talus slopes consisting of shaly, rod-shaped and cubic fragments of sedimentary rocks in N Norway. Elongation index of fragments proved to be straightly proportional to slope stability. Also large blocks on finer fragments play an important role in stabilizing certain talus slopes.

1 INTRODUCTION

Talus in a hillslope profile is formed by the disintegration and fall of debris from the cliff (free face) above. Slope angle in the upper straight part of the slope profile approaches a constant value as the cliff retreats either through discrete particle (one-at-a-time) rockfall or rock avalanches. The slope profile increases its concavity downslope where it passes to the pediment. According to e.g. Kirkby and Statham (1975), Carson (1977) and Kirkby (1987) the concavity is less prominent where the scree is high relative to the cliff height. At the same time the angle of the straight section also grows. According to Kirkby and Statham (1975), the concave profile of the scree base is due to accumulations of debris in the up-slope parts resulting in mass-movements at the talus slope.

According to Ward (1945), the angle of the upper straight parts of talus slopes could be regarded as the angle of repose as well as that of shearing resistance of the slope-forming coarse-grained material. Critical studies have, however, cast doubt on this belief (Chandler 1973, Statham 1976, Carson 1977). According to Chandler (1973), the angle of shearing resistance of most talus-forming material is likely to be 39-40° while the angle of talus slopes is typically 35°. The frequent repetition of the 35° characteristic angle for talus is suggested by him to be due to degradation process (creep, ero-sion) being more active than the supply of rockfall debris. There must be a host of factors affecting the slope angle, such as the height of the cliff above the talus, the rate of supply of rock fragments (discrete particle rockfall versus rock avalanche), size and shape of rock fragments, intensity of degradation of slope material, undercutting of talus slopes, etc.

Many studies (see e.g. Van Burkalow 1945 and Carson 1977) pay attention to experimental studies either by using certain artificial apparatuses, such as tilt boxes or through observations of artificial rockfalls (stockpiling of rock aggregate). Observations in Nature, such as those from Northern Sweden (Rapp 1960), Spitsbergen (Chandler 1973) and Isle of Skye (Statham 1976) are of great importance from the point of view of the present study treating slope angles of coarse-grained (d> 50 mm) rock fragments in subarctic talus slopes in Norway.

According to Melton (1965), two aspects to the surface stability exist: the ease with which fragments can be turned in motion, and the tendency of dislodged fragments to continue moving downhill once set in motion. However, as stated by Statham (1976), the instability has often been exaggerated, since most failures, such as the small avalanches caused by e.g. the persons walking on the scree slope, are confined to small scale motions at the point of disturbance.

The purpose of the present study is to

find out the possible influence of the shape and the joined properties of shape and size of the talus material on the slope angles and slope stability of natural talus formations. Elongate, oblate, cubic and mixed shapes of rock fragments of different lithologies exist in relatively similar conditions of slope height, rate of supply, degradation, etc. within a distance of few kilometers between the two extreme points of study. The slopes studied consist of strata of sedimentary rocks often forming complete series of slope elements, i.e. from convex upper slope through free face and talus to concave pediment. Subarctic conditions leave parts of the talus slopes practically free of vegetation. Freeze-and-thaw phenomena can be regarded as the most important process as regards the disintegration and detachment of material from the cliff as well as the degradation of the debris.

2 STUDY AREA

2.1. *Topography and geological history*

The area, the western slopes of the Blåberget and Rödberget, is located in Tana commune, Finnmark province, N Norway, and forms the eastern side of the valley occupied by the last 10 km course of the Tana river for the Tanafjord. The long axis of the valley is from SSW to NNE, and the cross section is about 5 km wide, asymmetrical and, because of glacial scouring, U-shaped. The western fell tops reach as high as 500-570 m while the eastern tops remain at the altitude of 200-330 m above the river surface.

The talus formations in front of the cliffs are not older than the last phase of glaciation which ended at the present river mouth about 11 000 B.P. (Mansikkaniemi 1970). The glacial movement was directed from the southern highlands toward the Arctic Ocean in the north while the direction of the deglaciation was the reverse. The highest shoreline reaches here about 56 m above sea level. The shoreline gradually lowered because of crustal upwarping.

The last phase of glaciation must have occupied the valleys only. The fell tops and the upper slopes have already been freed of ice. This phase might have lasted several hundreds of years. However, the products of supraglacial frost-wedging as well as subglacial nivation in the hillslopes have surely ended to the prey of glacial and meltwater transport within the domain of the valley glacier. Thus, the accumulation of the talus slopes most likely began together with the last phase of deglaciation some 11 000 B.P. During the next 10 000 years together with crustal upwarping, the lower parts of the talus slopes have suffered from progressively waning forces of undercutting caused by summertime coastal action. The lower parts of some talus slopes clearly became reactivated and unstable through roadcutting in this century. A considerable part of the talus slope of the Blåberget has been reactivated by the opening of a quarry where plates of shale are taken for stabilization material for road ditches.

2.2. *Bedrock*

The bedrock of the area is characterised by slightly metamorphosed allochtonous or parautochtonous sedimentary rocks of late Proterozoic to late Cambrian (see Beynon et.al. 1967). The cliffs studied are called Blåberget (Blue Cliff), because of the blue-green shale it contains and Rödberget (Red Cliff), because of the talus formations of of red-brown marly mudstone on its western slope. The mineral components of the blue-green shale in Blåberget are quartz, plagioclase and muscovite with chlorite as the accessory constituent. The red-brown marly mudstone consists of quartz, plagioclase, carbonates (calcite and dolomite), and muscovite.

In the middle part of the cliff Rödberget the free face is composed of a complex tectonical setting of several different beds with orthoquartzite, tillite and shale. Farther to the north the cliffs consist of quartzite.

3 STRUCTURE OF THE SLOPES

3.1. *Blåberget*

The shale in the free face has a strong slaty cleavage cutting almost perpendicularly the lamination, the latter dipping toward the synclinal axis in SE. The strike and dip of the slaty cleavage determining the morphology of the free face are SSW-NNE (i.e. along the direction of the valley) and 70° NW, respectively. The dip was measured on the convex upper slope and at the rim of the free face. However, the dip becomes more gradual downward being 45° at the top of the talus formation. The slaty cleavage is cut by sparsely (>2 m) spaced irregular cross joints.

The heights of the free face and talus are 80 m and 90 m, respectively.

The slaty cleavage in the upper slope is obviously parallel with the fold axis. An explanation to its more gently dip below the rim is found among the process of glacial erosion: tension fractures have developed in the incompetent shale due to pressure release and dilation after withdrawal of the glacier (cf. Sudgen and John 1976). The cleavage clearly follows the U-shaped profile of the valley.

Discrete particle rockfall mostly due to frost wedging along the slaty cleavage has produced a mighty talus consisting of flat pebbles (100-200 mm wide, 10-30 mm thick) of shale. On both sides of the remaining free face the talus has already reached the upper convex slope. Dense bushes of fell birch overwhelm these old and stable talus slopes.

The angle of the upper unstable talus slope is 33°-36° while angles of 24-29° were measured in the lower stable slope. The talus-forming mechanism can be compared with that in the slopes of schistose fragments in N Sweden (Rapp 1960). The detached material falls from the free face on the upper slope where it begins its slow movement towards the lower parts through the various processes (see Rapp 1960) of talus-creep. In the upper slope the dip of the platy fragments approaches to the slope angle, and on this unstable slope the fragments easily slide downslope past and upon each other. The shaly fragments become stabilized downslope by grouping into heaps of several plates upon each other. These heaps show an imbricated fabric like roof tiles. The dip of the individual plates is almost zero, in certain instances even a few degrees reverse to the slope. The thickness of the layer consisting of the imbricated heaps grows downslope. It is several meters thick at the base of the talus slope, where the dip of the individual plates is reverse thus supporting the lower talus. Imbrication of the platy fragments begins in the lower part of the upper unstable slope as stripes along which the talus creep starts taking place. These stripes become broader downslope and finally join each other to form fabric described above. The fabric on this lower slope thus brakes the downslope movement of the material. The formation of this fabric can be explained through the decreasing speed of the downslope movement of the talus material forcing the fragments to compile above each other in the lower parts of the talus.

3.2. Southern Rödberget

The mudstone does not form any clear free face, except few 0,5-2 m high bluffs above each three talus formations along the 200 m high slope. The rock is more competent than the shale in Blåberget showing strong lineation due to intersection of bedding and fracture cleavage. These planes are very closely spaced whence the rock is disintegrated into rod-shaped debris. The jointing is thus as follows: bedding joints (bedding cleavage) dipping 60° SE cut more or less perpendicularly by fracture cleavage along the strike of the bedding (N 20° E). Both the systems (spacing 1-3 cm) are cut by the cross joints spacing 10-30 cm and thus determining the length of the detached elongate pebbles forming the talus.

The three separate talus slopes on Southern Rödberget consist of stable slopes on the upper part and on the middle part, and an unstable lower talus slope. The angles of the stable upper and middle slopes vary between 31° and 38°, while the angles of unstable lower part are between 36° and 38°. All the three talus slopes act independently. They have developed stairways on the mudstone below the most resistant parts of the rock acting as small free faces. The instability of the lowest slope is due to the river action and the road cutting.

The rod-shaped fragments detach from 1-2 m high walls on the slopes without falling. The talus-creep starts through processes similar to those prevailing in Blåberget. The movement of the individual fragments starts with the long axis of the rods more or less parallel to the slope gradient. However, the rods soon begin to dip steeper than the slope forming a texture resembling a dense cluster of pins in a pincushion in which the rod may point to any direction. The slope angle is steepest on this "pincushion" texture. Downslope this structure is gradually buried below a 5-10 cm thick layer of loose rods sliding on this "pincushion" and mainly dipping more or less towards the slope gradient. On the rims of the unstable lower slope the sliding rods have overwhelmed grassy vegetation growing on the stable "pincushion" fabric.

3.3. Middle Rödberget

The free faces some 500 m to the north from the previous site consist of a complex setting of strata dipping both NW and

SE. The dominant component takes a more or less equidimensional form as detached blocks. Such is also the case with the tillite, a softy and massive rock with a badly sorted texture. It consists of angular pebbles and blocks (here up to 20 cm in diameter or even more) mostly of dolomite, while the matrix contains quartz, feldspars, carbonates and mica. Both the quartzite and the tillite show joint systems consisting of bedding and cross joints cut by diagonal joints although the joints in the tillite are more irregular and sparsely spaced as compared with those in the quartzite. The detached blocks range from less than 10 cm up to 1 m in diameter. Strata consisting of shale yield talus material resembling that of the Blåberget shale. The 80 m high talus is cone-shaped with a funnel-shaped free face of equal height above.

The slope angle on the steep upper part of the talus is 36°, on the lowest undisturbed part 38-39° and on the lowest disturbed parts (because of road cutting) 30°. According to the varying lithology of the free face, the talus material consists of different rocks with a wide range of shape values among the fragments. The smaller (d<10-15 cm) fragments are covered by a 0.5 to 1 m thick layer of greater, mostly rounded, more or less cubic blocks of quartzite. The stability of the lower bed with smaller fragments is relatively weak being easily displaced when the upper layer is disturbed.

3.4. Northern Rödberget

The cliff, some 5 km N of the previous site, consists exclusively of the orthoquartzite with the same joint system and block size as in the cliffs farther south. However, a base fringe of huge blocks, even more than 2 m in diameter, is met with at the talus base. (Such were also the case at the previous site unless the Tana River flows at the talus base swallowing the biggest blocks eventually rolling over the road cut 20 m above the talus base.) The height of the free face is about 100 m, while the talus slope begins only 50 m above its base resting upon the upper parts of the recent Tana River delta flat.

The quartzitic blocks are more or less isomorphic with diameters ranging from 5 cm up to 1,5 m above the rim. A stable surface with slope angle 35°, was met with in places where the largest blocks covered the smaller fragments, while places without these large blocks lower angles 33° were measured.

4 RESULTS OF THE MEASURING PROCEDURES: CONCLUSIONS AND DISCUSSION

In Table 1 the shape indices (flakiness, b/a, and elongation, c/a, with a<b<c as the axial lengths of the talus material), as well as the slope angles are presented. Stability values are estimated by using symbols from ++ (=highly stable slope) to -- (=highly unstable slope). The codes SHALE, ROD, CUBE I and II, as well as AGGR 8-20 and 20-35 indicate (from top to bottom in the column): shaly material in Blåberget, rod-shaped material in Southern Rödberget, cubic material in Middle Rödberget and in Northern Rödberget, on different levels above the river as indicated by the numbers (m) after each code, as well as sorted aggregate material in nearby stockpiles consisting of 8-20 and 20-35 mm, respectively.

Table 1. Data obtained from field measurements (see text).

Code	elong.	flakin.	size	angle	stab.
SHALE 40	13	8	22	24°	++
SHALE 55	14	7	22	29°	++
SHALE 85	11	6	25	33°	+
ROD 185	9	1.5	3	33°	++
ROD 212	7	1.6	5	38°	+
ROD 195	3	1.5	0.5	28°	--
ROD 32	5	1.5	0.9	37°	-
CUBE I25	4	2.5	13	30°	-
CUBEII25	2.5	1.8	78	35°	-
AGGR 8-20	2.3	1.6	3	38°	--
AGGR20-35	2.5	1.5	17	38°	--

Fig.1 shows cumulative curves of flakiness and elongation values of some of the materials in Rödberget and Blåberget. Curves are drawn on the basis of weighing the mass relationships of the different classes of shape indices as shown in Fig 1. The average particle size is obtained from the formula $a*b*c/1000$ (cm3) (see Table 1). The shape indices have been measured by using a Mitutoyo Digimatic caliper connected with Toshiba T 1000 computer.

From the results it is evident that the

slope angle cannot be correlated with shearing resistance although the latter is not herewith determined numerically. If the talus slope is regarded as stable, e.g. in accordance with the principles presented by Melton (1965), its slope angle tends to take the value which is determined by the shape and size of the fragments. This is why the slope angle in an unstable talus may exceed that in a stable one even though both slopes consist of similar material.

The talus creep through processes presented by Rapp (1960) is thus caused by the oversteepening of the slope. The relatively stable upper slopes become oversteepened through the supply of rock frag-

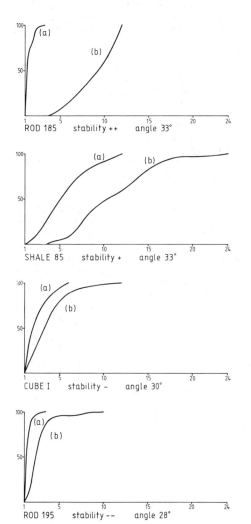

ROD 185   stability ++   angle 33°

SHALE 85   stability +   angle 33°

CUBE I   stability −   angle 30°

ROD 195   stability −−   angle 28°

Fig.1. Cumulative curves of flakiness (a) and elongation (b) values.

ments resulting in downslope movement. This model of behaviour is observed on the slopes with a small free face (such as in Southern Rödberget). The kinetic energy of the fragments detached from the outcrop is extremely weak as compared with that owned by the fragments falling from a high free face. The shape of the upper slope is even convex in a stable state. Concave upper slopes are met with on talus formations below a high free face (M and N Rödberget).

As to the shape and size of the rock fragments, the increase of elongation allows a higher slope angle in an unstable slope as indicated by the codes ROD 32 and ROD 195 in Table 1. The unstable slopes become stabilized through the re-arrangement of the fragments with high values of elongation through the formation of the "pincushion" fabric as described above.

The increase of elongation is positively correlated with the state of slope stability observed (Table 1). Practically no correlation is observed between flakiness and slope angle or slope stability as indicated by all the SHALE-codes in Table 1. Comparing the code SHALE 85 with ROD 185 for example, one observes a less prominent increase of stability through a higher value of elongation in a slope with material also having a high value of flakiness than in a slope consisting of rod-like fragments with elongation as the more prominent shape index, although smaller than among the shaly fragments. The high values of flakiness as such already seem to decrease the slope stability. The slopes with fragments of high values of flakiness become stabilized by grouping into an imbricated fabric as described above. The almost horisontal position of the individual flaky fragments decreases their downslope movement. A flaky fragment has two axial directions to control its orientation while in a rod-shaped fragment there is only one such a direction. The orientation among flaky fragments is more restricted as compared with that among the rod-shaped ones.

The positive correlation between the fragment size and the slope angle is in the present study clearly observed when compared the codes ROD 195 with ROD 212 in Table 1. Furthermore, when compared the codes CUBE I and II between each other one observes a higher slope angle among the fragments with a smaller value of elongation with larger block size as compared with those of higher elongation and smaller size. Especially the larger blocks on the talus surface make the slope more stable than judged from the value of elon-

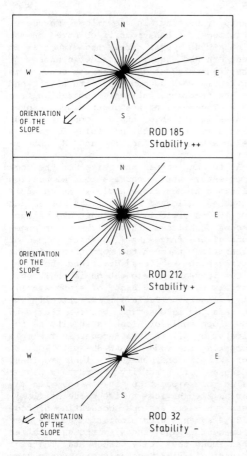

Fig.2. Orientation diagrams of the rod-shaped fragments.

gation among the fragments. When compared the codes CUBE I and ROD 32 in Table 1 one observes the same degree of stability among these slopes with a greater block size and smaller elongation among the material in the former and with a smaller grain size and greater elongation as well as greater flakiness among the material in the latter.

A talus slope consisting of large blocks is more stable because of the great weight of the blocks is evenly distributed (i.e. including an upslope component) in deeper layers. Secondly, rod-shaped fragments, particularly those with a relatively high value of elongation as compared with that of flakiness (see Fig. 1 and the codes in Table 1), have a better ability to become orientated transverse or randomly ("pin-cushion" fabric) against the slope and the downslope movement. (see Fig. 2). Thirdly, when disturbing or removing the coarse-grained surface layer of a talus slope, the deeper-seated finer-grained material looses its stability, which was clearly observed in Middle Rödberget after road-cutting.

REFERENCES

Beynon, D.R.V., Chapman, G.R., Ducharme, R.O. & Roberts, J.D. 1967 The geology of the Leirpollen area. Tanafjord, Finmark. *Norg.Geol.Unders*. 247: 7-17.

Carson, M.A. 1977 Angles of repose, angles of shearing resistance and angles of talus slopes. *Earth Surface Processes* 2: 363-380.

Chandler, R.J. 1973 The inclination of talus, Arctic talus terraces, and other slopes composed of granular materials. *J.Geol* 81: 1-14.

Kirkby M.J. 1987 General models of long-term slope evolution through mass movement. In M.G. Anderson & K.S. Richards (eds), *Slope stability*: 359-379. Chichester: John Wiley & Sons.

Kirkby, M.J. & Statham, I. 1975 Surface stone movement and scree formation. *J.Geol* 83: 349-362.

Mansikkaniemi, H. 1970 Deposits of sorted material in the Inarinjoki-Tana river valley in Lapland. *Ann.Univ. Turkuensis A II* 43.

Melton, M.A. 1965 Debris-covered hill-slopes of the southern Arizona desert - consideration of their stability and sediment contribution. *J.Geol*. 73: 715-729.

Rapp, A. 1960 Recent development of mountain slopes in Kärkevagge and surroundings, northern Scandinavia. *Geogr. Annaler* 42: 67-200.

Statham, I. 1976 A scree slope rockfall model. *Earth Surface Processes* 1: 43-62.

Sudgen, D.E. & John, B.S. 1976 *Glaciers and Landscape*. London: Edward Arnold.

Van Burkalow, A. 1945 Angle of repose and angle of sliding friction: an experimental study. *Bull.Geol.Soc.Am*. 56: 669-708.

Ward, W.H. 1945 The stability of natural slopes. *Geogr.J*. 105: 170-197.

Landslides, Bell (ed.) © 1995 Balkema, Rotterdam, ISBN 90 5410 032 X

Winterslides – A new trend in Norway

N. Janbu
Norwegian Institute of Technology/SINTEF, Trondheim, Norway

J. Nestvold
Kummeneje A/S, Trondheim, Norway

L. Grande
Norwegian Institute of Technology, Trondheim, Norway

ABSTRACT: Rain-induced slides have always been fairly common in Norway, particulary during the spring and the fall seasons, when the heaviest rain periods of long duration usually occur. In the winter of 1981 a unique slide took place in a 10 m high compacted moraine fill, killing two persons. For several years this unusual slide was considered to be a freak incident. However, during the winter of 1988-89, some 100 rain-induced slides took place within 3 months, all in the Trondheim region and its nearest surroundings. Hence, the question: is this a new trend, and in case why? The majority of the cases were smaller slides. The total material damages can be estimated at approx. NOK 20 mill. Only one person was killed. Quick clays were not involved in more than a few cases.

1 INTRODUCTION

In all natural slopes, exposed to climatic changes, the ground water regime will change with time. Hence, along a shear surface the average pore pressure will change between low values in steady dry weather, and higher values during prolonged rainy periods, or shortly thereafter.

When pore pressure increases, the effective stress level decreases, and so does the effective strength of the soil. The consequence is a

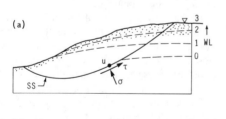
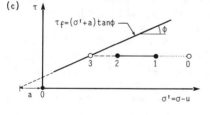
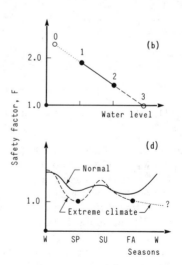

Fig.1 Effect of variable ground water conditions

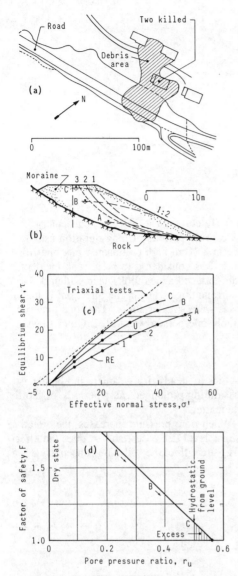

Fig.2 The Skei Slide, Grande, Janbu (1981)

During spring the frozen ground thaws, and at the same time heavy rain may last for days in a row. Therefore, the statistics of landslides may well be topped during springtime. Prolonged periods of moderate to heavy rain may some years occur in the fall, and if so, it is invariably associated with an increasing number of landslides, see Figure 1d.

In tropical parts of the world a majority of landslides occurs during heavy (tropical) rainstorms of much higher intensity than in the temperate to cold regions, Costa Nunes (1969), Brand (1984). These are truly rain-induced slides, Janbu (1989).

2 THE SKEI SLIDE

During the night of February 3., 1981, a compacted road fill sled out, after a very mild period of heavy rain. The sliding mass damaged three houses, and killed two persons in the nearest house, see plan view, Figure 2a. A cross-section of the fill is drawn in Figure 2b, containing three levels of ground water. The fill was a compacted moraine.

Triaxial tests were performed on the fill material for the in situ porosity (n = 33-35%) and for the actual range of normal stresses 10-60 kPa. From these tests an attraction of 3 to 8 kPa and a friction $\tan\varphi$ of 0.77 to 0.85 were found. The dotted line in Figure 2c corresponds to an average set of parameters a = 5 kPa and $\tan\varphi$ = 0.80.

The numerical results of the stability analyses for the three ground water levels are plotted in the form of resistance envelopes Figure 2c. The continuous envelopes are obtained by direct methods using dimensionless diagrams, Janbu (1982). For the (critical) shear surfaces marked 1, 2 and 3 in Figure 2b, control analyses were carried out by the computerized Generalized Procedure of Slices, GPS, Janbu (1957 1973).

The results of the analyses demonstrates clearly the all-important influence of increasing pore pressure on the safety of the slope. The variation in ground water level from A to C reduces the safety factor from roughly 1.7 to 1.0, with very moderate or no increase in equilibrium shear stress.

Hydrostatic pore pressure corresponds to r_u = 0.5, while F = 1 for r_u = 0.56. Hence,

gradual decrease in factor of safety (F) as long as the pore water pressure is increasing, as illustrated in Figure 1, sketches (a), (b) and (c).

In the moderate to cold regions of the world there may be four distinctly different seasons, the winter (W), the spring (SP), the summer (SU) and the fall (FA). During the winters, with steady frost, and during the dry summers, natural slopes seldom fail, because these are seasons with the lowest ground water levels, and the least erosion.

overpressure must have existed prior to the violent failure. Overpressure was possible because severe frost prior to the heavy rain covered the ground with an ice sheet.

Hence, it is only the reduction in effective normal stress that creates the failure, and this is a process that goes on within the soil mass, without any visual or a priori sign, detectable from the outside.

The most frightening experience in this case is that a compated fill of fairly wellgraded material can become so liquid, by internal excess pore pressure build-up, that it can flow on almost horizontal ground with such a speed that it damages well-built houses 20-30 m away so severely that lives are lost.

3 CLIMATIC CONDITIONS IN 1988/89

Parts of Mid-Norway experienced an unusually high frequency of slides during the winter of 1988/89. Particularly the city of Trondheim and its neighbouring community to the south (Melhus) were severely struck. To the north the Helgeland region (Vefsn and Rana) had also shorter period of intense sliding activity. Altogether some 100 single cases are known to have occured in these areas during this winter.

The slides varied in size from small slumps to large slides. One life was lost, and the material damage was appreciable. The major part of the slides occured in slopes of firm clay, silt and even sand and moraine.

The situation of poor stability was caused by extreme precipitation and high temperatures, leading to rise in ground water levels and increased pore pressures, with corresponding loss of effective stress and strength.

From July 1988, and especially through the winter months of November to January, these parts of the country had a precipitation far above the normal. For instance, the central part of the area (Leinstrand) had 340 % in November, 195 % in December and 314% in January, as shown in Figure 3a. During this same period the temperature was well above average, and very often around +10°C.

Most of the precipitation fell as rain. This was combined with melting snow after shorter periods of snowy weather. Since the ground remained unfrozen, the rain and the melted snow accumulated in the ground water regime,

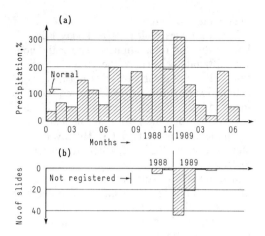

Fig.3 Precipitation and slide frequency in the Trondheim region (Leinstrand/Melhus)

leading to rizing ground water level and increased pore pressure. At the same time the ground surface was vulnerable to erosion activity along brooks and rivers.

It was generally found that by far most slides occured in areas of generally "good subsoil conditions" manifested by steep, high slopes. Many cases encountered firm clays and dense moraines in addition to medium sand-silt strata.

Some slopes had been steepened in connection with building roads and houses. Slides occured also in natural slopes, and in areas where construction works had taken place more than 50 years ago. Quick clays did not exist in most of the cases, but in a few single cases the slides reached into quick clay layers or pockets, without leading to an accellerating retrogressive activity which is otherwise so common in this country. External erosion was barely registered.

A few examples of the sliding activity during 1988/89 will be covered in more detail.

4 THE HOVINÅSEN SLIDE

The only slide causing death happened in Hovin in Gauldalen on January 29., 1989. During a period of heavy rain and snowmelting a slide was released in a 35 to 40 m high, steep slope. A family house at the foot of the slope was severely crushed, killing its woman inhabitant, see Figure 4.

The house was built i 1978. Today it is known that the foundation works for the house was difficult "because of the ground conditions". It is not clear wether or not there is a connection between the 1978 ground-movement and the 1989 slide, because the damage was so total that the failure mechanism could not be reconstructed in detail. However, the slip surfaces of 1978 and 1989 are entirely different.

Medium stiff to stiff clay dominates the ground conditions in Hovinåsen. The depth to rock is small, just a few meters. Extensive stabilizing works were required to secure the neighbouring houses.

5 THE NORDHAUGVEGEN SLIDE

A somewhat unique slide took place in Nordhaugvegen, Trondheim, on February 15, 1989.

A steep slope rizes from Nidelven up to 80 m above the Oslovegen. The lower part of the slope contains only shallow pockets of soil above bedrock, while sand and clay up to 20 m thick was found in the upper one third of the slope. The slide occured in this upper part, Figure 5.

The slide debris covered the Trondheim-Oslo railroadline (NSB) in the middle of the slope, and blocked the Oslo-Trondheim road at the bottom of the slope on the banks of Nidelven. The railroad was opened after a day and a half, but with severe safety restrictions for weeks afterwards.

Most of the works (cuts) for the railway was carried out 60 to 70 years ago. The NSB archives indicate heavy rain periods and frequent slides in 1924-25, and again in 1931-32. This slide, at Nordhaugvegen, is unique because it took place 1 to 2 weeks after the most intensive rainperiod. However, the topography is such that the slide area may have ground water supply from neighbour areas at high

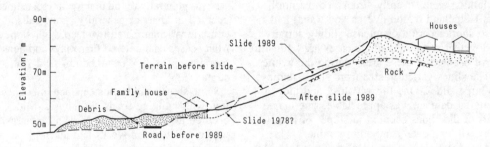

Fig.4 Slide at Hovinåsen, January 29, 1989

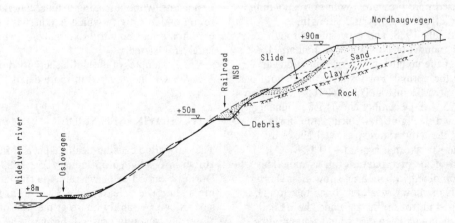

Fig.5 Slide at Nordhaugvegen, Trondheim

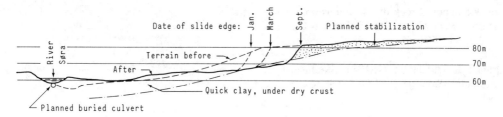

Fig.6 Slide at Lersbakken, Trondheim, Jan. 1989

elevations. The build-up of maximum pore pressure could therefore be delayed since it takes time to infiltrate the ground, and flow towards the slope.

Geotechnical investigations showed that the main seat of the slide was located within the clay layer near bedrock. It is a possibility that an excess pore water front had gradually been built up and concentrated towards the foot of the slope, and that this foot gave away first.

6 THE LERSBAKKEN SLIDE

Only a few slides occured in areas of potential danger for releasing retrogressive quick clay slides. One of those occured around Jan. 1, 1989, in Lersbakken, near Trondheim to the south, Figure 6.

After the first major slide into a quick clay layer around New Years Eve, leaving a steep back scarp 14 m high, several afterslides have taken place within intervals of months.

Once a slump occurs, suction is developed in the remaining bank making it stable for a while. But as suction disappears with time the soil swells and becomes weaker, so a next slump takes place. Each time a new slide occurs the debris flows almost horizontally towards the river Søra, as long as the quick clay mud has not reconsolidated.

Stabilization of the area consist of leading the river Søra in a buried culvert, and flatening the unstable upper part of the profile.

7 CONCLUDING REMARKS

The statistics from the large sliding activity in Mid-Norway during the winter of 1988-89 is clearly correlated with intense rain during the same period, Figure 3. Such a correlation is observed many times earlier. However, the fact that the slides occured during the regular winter months, even above 63° north, is really a unique and novel experience. It is therefore solely attributed to the extreme weather conditions of the season, unusually high temperatures and precipitation, lasting for months. Most of the slides therefore belong to the nature-induced cathegory.

A large number of the slides are characterized by a steep rock slope, overlain by fairly thin sediments, say between 5 and 15 m. The interface between soil and rock (fractured) carries water. The conditions for developing large ground water gradients towards unprotected grasscovered slopes are there.

It is the extreme values of the pore pressure regime that played the predominant role in all these failures. By comparison, uncertainties in effective strength values are of much lesser importance than even moderate seasonal pore pressure changes.

Over the last years there has been many unusually warm and wet winters in the near-coastal areas of Norway. It is, as yet, anybodys guesswork whether this is a warning of a more lasting change of climatic conditions to come. The result of such a change would no doubt be a reoccurance of similar sliding activities as were registered during the winter of 1988-89.

8 REFERENCES

Brand, E.W. 1984. Landslides in Southeast Asia: A state-of-the-art report. *Proc. 4th Int. Symp. Landslides:* Toronto, (1), 17-59.

Da Costa Nunes, A.J. 1969. Landslides in soils

of decomposed rock due to intense rain storms. *Proc. 7th ICSMFE:* Mexico City, Vol.2, 547-554.

Grande, L. and Janbu, N. 1981. Skeiraset i Surnadal. *Rapport O.81.06-1,* Geoteknikk, NTH.

Janbu, N. 1989. Slope stability evaluation. *7th S.E.Asian Geot. Conf.:* Hong Kong, Vol.II, 1-24.

Janbu, N. 1989. Slope failure due to rainfall and loss of suction. *DeMello Volume:* 205-283. E. Blücher, LTDA, Sao Paulo, Brasil.

Nestvold, J. 1989. Vinterskred i Trøndelag 88/89, statistikk og årsaksforhold. Fellessesjonen, Geoteknikkdagen 1989.

Landslides, Bell (ed.) © 1995 Balkema, Rotterdam, ISBN 90 5410 032 X

Microstructures of clayey soils in relation to shear strength

A. Klukanova

Dionyz Stur Institute of Geology, Bratislava, Czechoslovakia

ABSTRACT: We have evaluated the fabric of soils, disturbed by shear on the basis of microstructural analysis. We compared the fabric of soils from shear zones in natural landslides with fabric soils disturbed under laboratory conditions at movements of different characters. A comparison of fabric of these soils showed some similarities indicating possible identification of shear planes. We have studied slope deformations in the area of the town Handlova.

INTRODUCTION

The development of instrumental technic enable still more exact determination of soil strength. Advanced shear apparatus tended towards modelling of the non-linear relation between stress and strain of soil. Electric driving devices and microprocessor control and computer techniques are widely applied in laboratories. It is useful for providing the quality of input data concerning static and stable resolution and for revealing the gaps in our knowledge of the influence of the soil fabric upon their properties, transformation and strength. Detailed knowledge of the arrangement and bonds of the tiniest particles of soil facilitates the determination of interaction between the building and the foundation soil. In our article entitled "Change in soil microstructure caused by shear movements" (Klukanova – Modlitba, 1990) we dealt with the artificial shear planes created under laboratory conditions. The resulting data were testified on natural shear planes. This article deals with the slope deformation near the town Handlova.

2 METHODS

Samples for microstructural analysis were taken from engineering geologi-cal drill holes JH-1, JH-2 and JH-5 in the Handlova landslide. All the three drill holes penetrated the shear zones. The analyzed samples were taken from the drill holes by 0,5 meters. The purpose of micro-structural analysis was to get information about changes in the fabric of cohesive soils in the land-slide body, and to reveal the possi-bility of a more exact identifica-tion of shear planes, their genesis, a.o.

Microstructural analyses were performed by the scanning electron microscope JSM-840 - JEOL (Japan). The mineral composition was examined by X-ray and DTA method. Laboratory tests of samples proceed in the laboratory of soil mechanics accord-ing to methods corresponding to the respective Czechoslovak standards or in accordance with the generally applied procedures.

3 CHARACTERISTICS OF SHEAR ZONES

The research revealed three shear zones differing in age and genesis. In the drill hole JH-5 two separate shear zones above each other were found out. The drill hole JH-2 pene-trated one shear zone associated with subparallel shear planes in the underlying zone and with vertical charriage planes in the overlying zone. The character of microstruc-

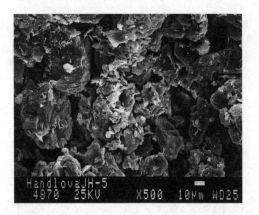

Micrograph 1. Skeletal-matrix microstructure - primary microstructure

Micrograph 2. Matrix microstructure
- primary microstructure

Micrograph 3. Turbulent microstructure with relics of the primary
microstructures. Plane of discontinuity.

tural elements is indicative of
different structural conditions of
the origin of shear zones, due to
mineral composition and type of
microstructure on the shear plane.
Shear planes do not form a continuous plane but consist of many partial shear planes – so we have denoted them as them as the shear
planes zone.

3.1 The first shear zone

It is the oldest shear zone. It was
found in the drill hole JH-5 in the
upper part of the landslide body.
Originally, the soil microstructure
was skeletal or matrix (in the sense
of Grabowska-Olszewska et al. 1984)

The skeletal-matrix microstructure
(micrograph 1) is homogeneous, its
fabric is isotropic.
It consists of aleuritic and psammitic grains of quartz, feldspars,
calcite. The grains are well rounded, isometric. Their mutual connection is provided by clay films and
clay buttresses, consisting of illite and montmorillonite. Kaolinite
forms individual particles. Most
pores are intergranular and microaggregate, scarcely intragranular.

The matrix microstructure (micrograph 2) is homogeneous, slightly
isotropic. Clay aggregates consisting of preferred-oriented sheets of
illite, montmorillonite and kaolinite are the basical structural
elements. Pores represented the
interaggregate type. Contacts between clay microaggregate are predominantly face-to-face.

In the shear zone only relics of
the primary microstructures have
been preserved. The zone alone shows
a considerable structural heterogeneity (micrograph 3) caused by several sets of inhomogeneity planes and
of discontinuity planes.
The primary microstructures were by
slope movements changed into laminar
and turbulent microstructures synge-
netic with the origin of shear
planes. In soils they manifest themselves in disconformable systems of
preferred-oriented primary clay
particles and microaggregates. Among
deformations of structural particles
the brittle deformation prevails. It
is covered with recrystallization
and healing phenomena (micrograph
4).

The course of planar structural elements is usually sigmoid, with different curvature degree. Originally isometric types of pores are rare, elongated pores, often filled with secondary minerals are predominant.

3.2 The second shear zone

The shear zone formed in the soil with a homogeneous primary microstructure i.e. skeletal-matrix to matrix. Slope movements resulted in a mixed laminar-turbulent microstructure (micrograph 5). According to the distribution and proportions of particular fabric elements the soil is heterogeneous. With respect to the structural elements orientation the fabric of the soil is anisotropic, with preferred-oriented planar and linear fabric elements, mainly clay particles. In contrast to the preceding zone, the secondary and authigene particles are practically absent. Moisture, shape and bordering of microaggregates predominate. Contacts between microaggregates represent the-face to-face and face-to-edge types. Broken microaggregates are neither recrystallized, nor healed. They are plastic deformed – bent in the nearness of aleuritic and psammitic grains. Larger aggregates are broken and so the intermicroaggregate longitudinal pores represent the basical type of pores.

3.3 Third shear zone

Basic material of the soil consist of grey to brown plastic clay with dispersed rigid "pseudo-ooids". With respect to petrographic character of the parent rock and mineral composition we suppose that the rigid "pseudo-ooids" are restides of volcanoclastic rock. Microstructure of the basic material is turbulent-laminar, deformed by shear movements. The microstructure was formed from the original matrix microstructure. It is characterized by a high degree of anisotropy due to preferred-oriented clay minerals (illite, montmorillonite, chlorite). Clay particles are intensely plastic-deformed, and brittle-deformed on brims. On surfaces parallel to the shear plane

Micrograph 4. Laminar microstructure with relics of the primary microstructures. Brittle deformation with healing phenomena and recrystallization.

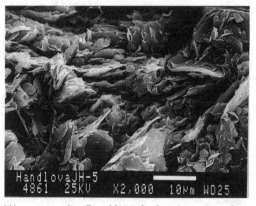

Micrograph 5. Mixed laminar-turbulent microstructure. Face to face contacts between clay particles. Longitudinal pores as basical type of pores.

borders between microaggregates get lost and the soil gets partly "pseudomorphous" and "homogenized" – like the linear movement created over a linear plane under laboratory conditions (Klukanova et al, 1990) – micrograph 6.

The turbulent microstructure of "pseudo-ooids" shows a high degree of homogeneity and isotropy. Structural particles are mostly represented by secondary clay particles, less by primary aleuritic grains.

The "roofing" phenomenon, characteristic of shear planes, resulting from shearing over a linear plane (micrograph 6), was found in a sample taken from the place near the shear plane. This phenomenon occurs only in primary clay particles. It is fading out with the increasing distance from the shear plane.

Linear elements, so frequent on shearing by shear apparatus, are absent. The pores are directly on the plane filled with clay minerals. In the nearest of shear planes the pores are clearly elongated, parallel to the shear plane.

4 CONCLUSIONS

The microstructural analysis of soil samples from the drill hole JH-5 showed that the drilling penetrated two shear zones separated by an interjacent zone. The upper shear zone seems older than the lower one. We also suppose a different mechanism and conditions of their origin. The upper shear zone comprises some predominant preserved relics of disconformable sets of discontinuity planes whereas the lower shear zone only comprises one prominent system.

In the upper shear zone the microstructure, resulting from shearing is covered (with recrystallization and thus formed secondary minerals), in the lower shear zone it is dominant. We have also found differences in the deformation of particular clay particles. It is indicative of a different shearing mechanism. The upper shear zone formed in a semi-liquid state, the lower one – in a more viscose state.

The activation of pore pressure and stress in the upper shear zone is more extensive than in the lower zone. We may suppose that the activation was more intensive in less permeable layer.

The main structure-forming factor in the upper shear zone is represented by the consequent groundwater circulation (intraaggregate pores, weak deformation of aggregates, mineral composition, a.o.), and by the shearing process in the lower shear zone.

On shearing, the matrix and skeletal microstructures change into turbulent and laminar microstructure. It is important to notice that slope movements took place on the boundary between two types of micro-

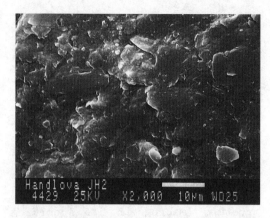

Micrograph 6. Laminar microstructure. Typical anisotropy fabric. "Roofing" phenomenon.

structure, also differing in amount, type and shape of pores.

The soil of the upper lying zone in the drill hole JH-2 shows the matrix microstructure. Our study of aleuritic and pelitic non-clay particles shows that the soil resulted from aerial weathering of neovolcanics of the Kremnicke vrchy mountains in an oxidated environment. This opinion is also supported by the presence of intraaggregate types of pores, high-grade magmatic corrosion of quartz grains. The increasing depth in associated with the orientation of clay microaggregates. We have also found subhorizontal partial foliations conformable with the shear plane.

The basic material shows a turbulent – laminar deformed microstructure with a high degree of anisotropy caused by completely preferred-oriented sheets of clay minerals. They are markedly plastic-deformed, and brittle-deformed on their brims.

Boundaries between particular microaggregates disappear on surfaces parallel to the shear plane, and the soil becomes partly "pseudomorphous" and "homogenized".

Some features, like "roofing", brittle deformation of clay particles and sheeting of clay microaggregates indicate that partial subparallel movements genetically and in time associated with the main slope movements in the shear zone also formed in the underlying zone.

REFERENCES

Grabowska-Olszewska, B., Osipov, V.I., Sokolov, V.N. 1984: Atlas of the microstructure of clay soils. Panstw. wydaw. nauk.,Warszawa, 414.

Klukanova, A., Modlitba, I.,1990: Change in soil microstructure caused by shear movements. Proceedings Sixth International Congress IAEG, Amsterdam, Netherlands, 2211 - 2215.

Investigation, monitoring and emergency remedial works at the La Butte landslide, Mauritius

T.I. Longworth
Building Research Establishment, Garston, Watford, UK

ABSTRACT: A large slow-moving landslide in colluvium has caused serious damage to buildings, important power lines and trunk water mains in Port Louis, Mauritius. An initial risk of a devastating debris flow originating in the 26° sloping head region of the landslide during torrential cyclonic rainfall, compounded by flow from broken water pipes, was averted by emergency works. An investigation of landslide geology, morphology and movement gave understanding of the landslide mechanism and causes, and allowed more accurate risk assessment. The cause of the landslide was natural long-term creep movements in colluvium, aggravated by leakage from a reservoir and water mains, by intense seasonal rainfall and by poor maintenance of drainage.

1 INTRODUCTION

In 1987, the Building Research Establishment, BRE, was asked by the UK Overseas Development Administration to advise the Government of Mauritius on the causes and remedy of serious damage to buildings occurring in the La Butte area of the capital, Port Louis. Local engineers were uncertain as to the cause: their opinions ranged from swelling of clay and bearing capacity failure due to wetting up of clay sub-soil, through to local slope instability.

An initial inspection by the author found the damage to be caused by an extensive slow-moving landslide in colluvium mantling the lower slopes of an eroded volcanic mountain. In following days a more thorough observation was made of the extent and character of the landslide and enquiries made about affected infrastructure. BRE concluded that there was a risk of devastating slope instability during torrential cyclonic rainfall and advised that emergency remedial works should be carried out to reduce landslide risk and preserve infrastructure. Provision should also be made for warning and evacuating the occupants of threatened buildings.

A programme of agreed works was immediately put in hand, and BRE was requested to specify and supervise a limited site investigation and monitoring to confirm the landslide mechanism and facilitate a more accurate risk assessment.

Recommendations were then prepared for a more detailed geotechnical study leading to remedial works.

2 LOCATION AND TOPOGRAPHIC SETTING

The island of Mauritius lies in the southern Indian Ocean at a latitude of 20° south. Its climate is tropical with abundant rainfall being brought by south-east trade winds. Rainfall is much the heaviest in the hottest months of December to March when trade-wind rainfall is added to by occasional tropical cyclones.

The La Butte area is situated in the western suburbs of Port Louis and lies partly on the coastal flat-land and partly on the gentle seaward-facing lower slopes of Signal Mountain. Figure 1. There has been development in the area for about 200 years since the time of the French administration. The major buildings, including offices, shops and a 120 year old mosque, border the Brabant and D'Entrecasteaux Streets. Up slope, buildings mainly comprise a mix of small houses and apartment blocks.

Signal Mountain is a remnant of the basalt lavas of the Older Volcanic Series, which were erupted as a shield volcano to form the island of Mauritius about 7 m years ago, the lavas dipping gently seaward. Subsequently, after deep erosion of the Older Volcanic Series had formed

Signal Mountain, further lava flows of the Intermediate and Younger Volcanic Series spread around its roots forming a shelf near sea level.

Over millions of years the slopes of Signal Mountain and surrounding Younger Volcanic Lava shelf have been covered with deep colluvium, material eroded from exposed rock, transported down hill and partially weathered to silt and clay.

3 INITIAL APPRAISAL OF DAMAGE AND GROUND MOVEMENTS

An initial reconnaissance of the La Butte area was made by the author starting in the area of greatest reported damage, adjacent to the Mosque on Brabant street, Figure 1. It was immediately apparent that the ground under the Mosque and adjacent buildings including a petrol filling station, shops and offices was being thrust forward and heaved by a well defined compression ridge. This could be traced in an arc extending for some 300 m. Buildings in this area were severely structurally damaged by differential vertical movements and several had required demolition. The arcuate shape, size and position of the compression ridge at the foot of a mountain slope immediately suggested that it was the toe of a large landslide.

Walking up hill from the mosque little ground disturbance or building damage was apparent for some 200 m, at which point a severely crack damaged school, comprising four long single-storey concrete buildings, was noted, Figure 1. Up hill of this an empty service reservoir was found with 50 mm wide cracks in its metre thick concrete walls. Above and below the reservoir the presence of trunk water mains was apparent from dozens of pits (some flooded) obviously dug for the purpose of pipe repair. Walking up hill again, on the undeveloped mountain side, a large crevasse-like tension crack was found open to some 500 mm and with vertical displacements of up to 300 mm. This was traced for some 300 m running sidelong down the mountain from a highest point above the reservoir.

On the basis of the evidence gathered in this first brief inspection it was concluded that the ground movement and building damage was being caused by a single large landslide.

4 DETAILED WALK-OVER OBSERVATIONS AND DESK STUDY

Following the initial reconnaissance, which established the existence of the landslide, meetings were held with government

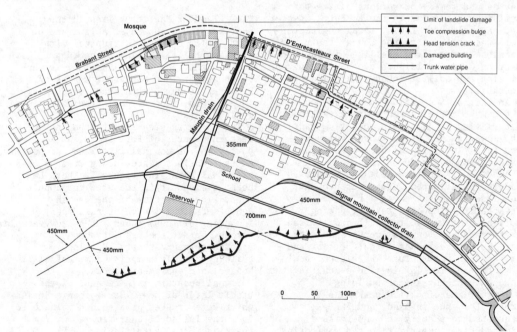

Fig 1. Plan of La Butte area, showing extent of landslide and features observed in walk-over survey

departments, the Municipality of Port Louis (MPL) and the Central Water Authority (CWA) to inform them of the problem and request information which would throw light on the mechanism, causes and likely consequences. Requests were made for large scale town plans, aerial photographs ranging back over a period of years, plans of water services, records of water main repairs, plans of drainage services, rainfall records for the previous 5 years, and damage inspection records.

4.1 Extent of landslide

The first priority was to establish the approximate limits of the landslide and identify the areas where building and infrastructure were most at risk. To do this the author systematically walked the area street-by-street with a 1:2500 map in hand noting existing building damage and visible signs of ground movement. The principal observations are indicated in Figure 1. Further discontinuous sections of toe compression ridge and head tension crack were discovered, the former being traceable in an arc for some 550 m, and the latter 450 m. Both were less prominent towards the extremities.

Between the head crack and the toe ridge, signs of ground movement and crack damage to buildings were found all the way down the slope in a zone extending laterally 700 m. In the mid-slope area, however, damage was generally light with just a few buildings affected, these mostly having minor crack damage. The best indicator of landslide movement in this area was the

buckling of long downslope runs of roadside kerb and of the downhill masonry walls of the Maupin surface water drain, Figure 1. Above and below this central area damage increased towards the head and toe of the landslide, the worst damage being in the vicinity of the previously mentioned school and mosque. From these observations the area of the landslide was concluded to be some 700 m by 400 m in extent. The area was found to be inhabited by 1500 persons in some 500 buildings.

The observations also allowed a provisional section through the landslide to be drawn indicating the vertical extent of the landslide and its likely geometry. Figure 2, shows this section, based on the 1:2,500 map, running up hill from the Mosque and through the location of the reservoir. The lower part of the landslide was found to mainly be in clayey colluvium sloping at an average of 8°. In contrast, the upper part, terminating at the head tension crack, was on undeveloped mountain side comprised largely of basalt boulders and cobbles, sloping at 25°. The landslide was judged to be translation in type with a shear surface approximately parallel to the ground surface over much of its length, this explaining the relatively light damage to buildings in the central part.

4.2 History of landslide

Research of the history of damage obtained from MPL files and discussion with officials and residents indicates that movements in the head area of the landslide date back at least a decade. A resident of

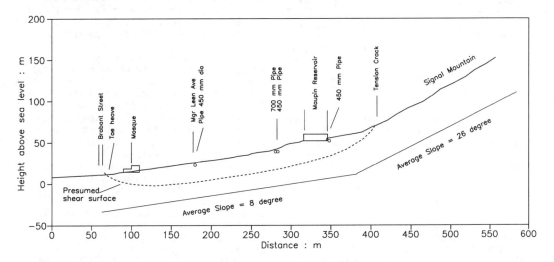

Fig 2. Provisional section through the landslide based on 1:2,500 map and walk-over observations

a terraced house in Mgr. Leen Avenue just below the school (Figure 1), which presently has 10 mm wide cracks in the walls, recalled that cracking started around 1970. The cracks had been repaired many times and each time had reopened. Substantial movements in the toe area were noted in 1985, following heavier than average seasonal cyclonic rainfall. A car showroom, comprising a large reinforced concrete structure, located just to the east of the mosque, Figure 1, near the toe of the landslide showed signs of differential vertical movement. An investigation carried out at that time concluded that columns supporting the front part of the building had settled due to bearing capacity failure following wetting up of soil under the foundations. In reality, the rear of the building was being affected by the landslide toe compression ridge. Ground movement in the vicinity of the Mosque continued slowly until the rainy period of January to April 1987. During this time ground movements accelerated, so that by June 1987 the car showroom was a total wreck and adjacent buildings including the Mosque severely cracked. At the same time large ground movements occurred further up the slope of Signal Mountain, the school was badly damaged and the head tension cracks opened.

4.3 Electricity supply lines

Two major Central Electricity Board (CEB) overhead power lines were observed to cross the head of the landslide above the reservoir. Distress to these lines was apparent, with the insulators supporting the power cables being rotated at angles of up to 45° from the original vertical. It was established that these high voltage power lines carried 65 % of the Port Louis electricity supply.

4.4 Water services and leakage

The La Butte area has long been a corridor for water services to Port Louis, Figures 1 & 2. The first major pipeline (450 mm diameter, cast iron with lead-caulked joints) was constructed running along the 55 m contour just above the reservoir in 1860, and the second (480 mm diameter, cast iron) in 1883. In the 1950's these trunk mains were connected to the Maupin Reservoir, a 4,000 m³ service reservoir, constructed in reinforced concrete at 50 m altitude. Three distribution pipes of diameters up to 355 mm were installed running directly downhill of the reservoir to service the western part of Port Louis.

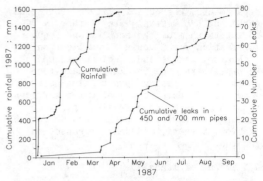

Fig 3. Comparison of occurrence of leaks in trunk water pipes running through the head of the landslide with cumulative rainfall in 1987

Subsequently two further trunk water mains of 700 mm and 450 mm diameter were constructed on the 42 m contour just downhill of the reservoir.

It was soon realised by observation and discussion with officials that all the major pipes, together supplying some 95% of Port Louis's water passed through the landslide area. Moreover, none of the pipelines had shut-off valves closer than 2 km upstream of the landslide boundary.

Figure 3 shows the date of occurrence of 78 leaks in the trunk water mains passing through the head of the landslide, repaired by the CWA during the period January to October 1987. Also shown is the cumulative rainfall from 1 January on. Most of these leaks were due to pulling at joints. One typical major leak is reported to have spilt 1350 litres of water per minute into the ground for two weeks (a total of 27 Ml) before it was located and repaired. It can be seen from Figure 3 that only one leak was reported in the period up to late March 1987. Thereafter leaks occurred at a fairly steady rate. It would therefore appear that landslide movement accelerated greatly in March following exceptionally heavy rainfall (1100 mm) in the preceding two months, including two spells of cyclonic rainfall when 386 mm and 313 mm occurred in 48 hours in January and February. Thereafter landslide movement apparently continued steadily, though rainfall was much reduced after the end of the rainy season in May. A vicious circle was apparent in which landslide movement was leading to pipe breakage which in turn released water to continue driving the landslide.

As well as these major leaks many smaller spillages from water service pipes and sewage drains were noted by the author, particularly in the vicinity of the toe pressure ridge.

Leakage from the Maupin service reservoir was reported by residents to have occurred for many years. The seepage was sufficient to maintain a surface-water stream below the reservoir, until this was taken out of service due to cracking in March 1987.

4.5 Surface water drainage

During heavy rainfall water flows down Signal Mountain in natural gullies. In the 1800's major interception drains were built to collect this natural flow and prevent ponds forming on the lower slopes and coastal flat land in which mosquitoes could breed. One such major drain, the Maupin Drain, runs down hill in the centre of the landslide area. The lower part of this was in good condition except where buckled by compressive landslide movement. However, some of the natural gullies which once led to it have been blocked by construction of water supply pipes and the reservoir. A second important drain, the Signal Mountain Collector, Figure 1, was located on the 30 m contour. This was constructed 4 m wide by 2 m deep with basalt block walls. Weep holes in the uphill side drained the ground as well as catching surface run off. Sadly this drain had not been maintained for some decades, being nearly full of silt and boulders, with the walls breached by landslide movement. This lack of drain maintenance undoubtedly lead to greater rainfall infiltration, contributing to landslide activation.

5 POSSIBLE CONSEQUENCES OF LANDSLIDING

From foregoing observations it was evident that the landslide had been moving for many years at rates dependent on seasonal rainfall and was obviously seated in colluvium (by nature full of relic shears at near residual strength). It was thus concluded that movement would inevitably continue and could well accelerate in the coming rainy season, causing further extensive building damage. Some lateral extension of the landslide might occur, but extension across Brabant and D'Entrecasteaux Streets was unlikely as these were underlain by basalt bedrock at shallow depth.

The worst case scenario was concluded to be a risk, during high intensity rain, of rapid movement of the 26° sloping head of the landslide. It was thought that this could have occurred if large quantities of surface-water flowed into the open tension cracks on Signal Mountain. If these cracks

became filled, very large temporary hydraulic pressures could be generated, triggering rapid landslide movement. Such movement could have severed all the CWA pipes simultaneously, cutting off 95 % of the water supply to Port Louis. It would additionally have allowed an enormous flow of water to escape into broken ground as several kilometres of pipe were emptied over an estimated period of four hours. Such a water flow, combined with cyclonic rainfall could have turned the disrupted ground at the head of the landslide into a fast moving debris flow which would have engulfed buildings downhill. The first casualty would likely have been the school as this lies in a hollow on the centre-line of the landslide, directly below the pipelines. Additionally the head of landslide movements would have severed the high-voltage power cables which carried 65% of the Port Louis electricity supply.

6 EMERGENCY REMEDIAL WORKS

The author advised that emergency remedial works should be undertaken to reduce the risk of a devastating debris slide which would hazard life, property and infrastructure. The aim should be to reduce the possible inflow of water into the colluvium from broken pipelines and high intensity rainfall. Works should include sealing of the head tension crack, clearance and repair of drainage channels, placing of shut-off valves in the pipelines just off the landslide limits, and that a scheme be set up for warning and evacuating the occupants of buildings at risk. In the longer term the author advised that the trunk water mains and power-lines be re-routed around the landslide. These measures were accepted by various authorities responsible. Additionally, a decision was made to close the school for the duration of the rainy season.

6.1 Diversion of water pipes and power lines

The emergency re-routing of the CWA large diameter pipes supplying water to Port Louis commenced in early November 1987: the work being carried out in 3 stages: Stage 1, the installation of a by-pass of the La Butte area in 450 mm diameter pipe, completed in January 1988; Stage II, a further by-pass with 800 mm pipeline started early in 1988, following the delivery to Mauritius of the required pipe sections, and completed in May 1988; Stage III the re-organisation of local water services in the La Butte area, with small diameter distribution mains in the landslide being replaced by flexible plastic pipes, carried out in late 1988.

Prior to completion of the by-pass pipelines it was necessary to keep in operation the existing large diameter water pipes passing through the upper part of the landslide. To prevent the feared major spillage of water emergency shut-off valves were installed just off the western limit of the landslide between October and December 1987. It was agreed that CWA operatives would man the valves when a cyclone or heavy rain was forecast. It was estimated that close-down of the pipes could be achieved within one hour.

The high voltage power lines which crossing the landslide were re-routed by the Central Electricity Board to by-pass the La Butte area in early 1988.

6.2 Clearance and repair of storm water drains and sealing of tension cracks

Clearance of the important Signal Mountain Collector drain was carried out in November 1987. Removal of the many years of accumulation of silt and boulders from the drain revealed the base of the drain to be of compacted clay and in reasonably good condition. A few deep cracks were noted and it was recommended that these be filled by ramming in moist clay. Repairs to the Maupin Drain, comprising a resetting of basalt blocks in the lower section and a new concrete channel adjacent to Mgr. Leen Avenue was also carried out in December 1987.

Natural drainage channels, terminating in the Maupin and Signal Mountain Collector Drain were also cleared of boulders, trees and fill from pipeline and reservoir construction, increasing slope runoff.

The head tension cracks were prepared by removing any surface bridges formed by roots, boulders etc., and then plugged with stone-free silty-clay. Thereafter the cracks were inspected regularly for further opening and re-plugged if necessary.

7 CONTINGENCY MEASURES FOR EVACUATION

The Cyclone and Other Natural Disasters Committee of Mauritius was convened to manage arrangements for warning and evacuating persons from the landslide zone if catastrophic movement appeared likely. This committee included the MPL engineering staff, the fire service, medical services and police, the Ministry of Housing, Land & Environment and the Meteorological Service.

Arrangements were made for staged warnings and evacuation of persons in the landslide area. Provisional criteria adopted for evacuation were: a Class II cyclone warning (70% probability of a cyclone passing over Mauritius); the occurrence of 100 mm rain in 12 hours; a measured increase of the width of the head tension crack of 10 mm in one hour.

8 MONITORING TO DETERMINE LANDSLIDE EXTENT AND RATE-OF-MOVEMENT

Ground movement monitoring was carried out under BRE supervision between December 87 and April 88 to confirm the extent of the landslide, determine the rate-of-movement, get warning of any acceleration, and detect any uphill regression of slope instability. A grid of ground survey points was established and the vertical and downslope movements observed by precision surveying. Additionally, tape measurements were made between pairs of steel stakes, each with a central punch reference mark, spaced downhill across the tension crack.

An initial four pairs of stakes, installed in early October 1987, indicated that the tension crack was continuing to open at a steady rate of between 5 and 10 mm per week. These initial stakes were added to in December '88 by 20 more pairs, spaced 3 m apart, spanning the tension crack at representative points along its length, and by 4 lines of stakes spaced 5 m apart, running for a distance of 20 m above and below the crack. Tape measurements between these stakes were carried out at weekly intervals to an accuracy of 1 mm.

The Survey Department of the Ministry of Housing, Lands and the Environment, undertook to survey the landslide at two week intervals with first-order survey instruments commencing in December '87. Precise levelling was carried out to measure vertical displacements to an expected accuracy of 2 to 3 mm, concentrating on four lines of ground survey stations located on the slope contours, as shown in Figure 4. These stations comprised either masonry nails in concrete kerbs or steel rods set in concrete on open ground.

Electronic distance measurement (EDM) was used to measure down-slope displacements from a stable monument erected on a convenient vantage point high on a road up Signal Mountain. Distance measurements were made from this monument to five downhill lines of survey stations A to E (Figure 4), established on the mountain side above and below the tension crack and in the streets which run uphill through the built-up area.

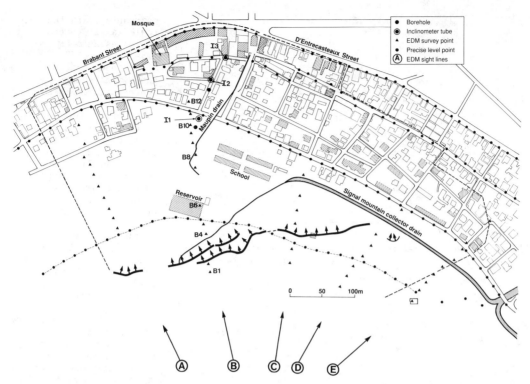

Fig 4. Plan of La Butte area showing location of ground survey station, boreholes and inclinometer tubes

Since repeated measurements were required on a large number of points, only the slope distance was measured. Accuracy of measurements was 5 to 10 mm (depending on location).

The EDM results for line B near the centre-line of the landslide are given in Figure 5, and comparison can be made with rainfall over the same period in Figure 6. The rate of downslope movement was found to be 5 mm a week in December 87, increasing to 11 mm per week following fairly heavy rains in mid February 88. A large area in the centre of the landslide, stations B6 to B12 (see locations in Figure 4), can be seen to be sliding down-slope as a block without significant ground strain, thus explaining the relative lack of damage to buildings in this region. Survey point B1, on the mountain side above the tension crack shows some trend for downhill displacement after March '87, possible indicating up slope regression of the landslide.

9 SUB-SURFACE INVESTIGATION

9.1 Geological and geotechnical investigation

An initial series of three NX (65 mm diameter) cored boreholes B, C and D, located in the more accessible lower part of the landslide (Figures 4 and 7) was drilled in November 1987, to investigate the geology of the landslide and ground water regime. Core recovery was of the order of 95%.

Figure 7 gives the deduced geological section through the lower and central parts of the landslide between Brabant Street and the Maupin reservoir. Borehole D passed through 34.8 m of colluvium before reaching bedrock comprising Old Series Lava. The colluvium mostly consisted of cobbles and boulders of basalt embedded in a plastic clay matrix. However, two seams of highly plastic clay were encountered at depths of 19.9-22.9 m and 24.6-26.9 m which had a very low clast content and contained several slickensided shear surfaces. The upper 20 m or so of the Old Series Lava was mostly highly weathered to friable rock, silt and clay.

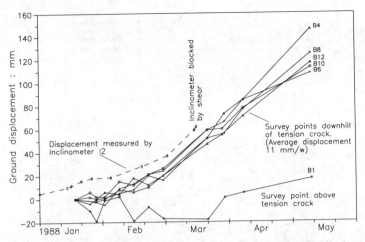

Fig 5. Ground displacements in the centre of the landslide measured by EDM survey on line B and by inclinometer I2

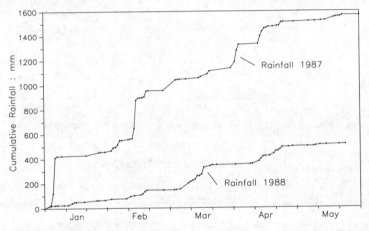

Fig 6. Cumulative rainfall in Port Louis for 1987 and 1988

permeable, with the water table in all three boreholes at near sea-level.

The sequence of geological events represented by these boreholes and interpreted in Figure 7 would appear to be as follows :

(i) Old Series Lava extruded and eroded over several millions of years to produce the deeply weathered basalt core of Signal Mountain, overlain by thick colluvium.

(ii) This old colluvium partly covered (about 1 m years ago) by Intermediate Series Lava, which encroached on Signal Mountain from the seaward side.

(iii) Slight weathering of the Intermediate Series Lava occurred before it was covered by colluvium eroded from the Old Series Lava still exposed in the stump of Signal Mountain.

(iv) At various times during formation of the colluvium adverse climatic and topographical factors combined to produce landslide movement. Shear movement at the base of the sliding colluvial masses locally caused disintegration of the clasts leading to the formation of seams of slickensided highly-plastic clay.

Borehole C passed through 18.3 m of colluvium similar to that encountered in borehole D. A seam of highly plastic clay with few clasts was again present, at the base of the colluvium at 16.0-18.3 m depth. Below the colluvium borehole C entered basalt of the Intermediate Series Lava. Only the top metre or so of this was highly weathered, the lava below being moderately to lightly weathered.

Borehole B had a near identical sequence of strata to borehole C: Colluvium comprising cobbles and boulders of basalt down to 8 m; nearly clast-free highly-plastic clay at 8-11.7 m; and Intermediate Series Lava below 11.7 m.

The borehole water-level data clearly shows that the clayey colluvium supports a perched water table. However, the lava and agglommerate bedrock was found to be highly

A limited series of tests on samples from the highly sheared clay horizons was carried out at the University of Mauritius and at BRE. X-ray diffraction tests showed the clay to mainly comprise smectite (probably a mixed layer montmorillonite /chlorite), with subsidiary amounts of calcium carbonate, plagioclase feldspar and ilmanite, consistent with the clay being weathered from a basalt. Typical index properties were found to be Plastic Limits of 36-43%, Liquid Limits of 84-131% and Plasticity Indices of 40-92%. Residual shear strength parameters determined by ring shear tests in the normal stress range 390-530 kN/m² were found to be $\phi'_r = 10°$ and $c'_r = 26$ kN/m².

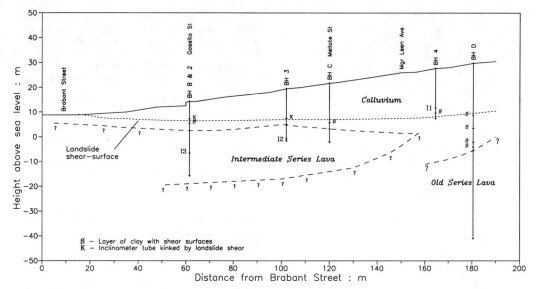

Fig 7. Geological section for the lower-central part of the landslide based on borehole cores, with shear surface deduced from inclinometer tube displacement

9.2 Inclinometers

A further three boreholes (BH 2, 3 & 4) were drilled in December '88 on the same down-slope line as previous boreholes (Figures 4 & 7) for installation of inclinometer tubes. These holes were drilled by open-hole techniques at 225 mm diameter through the colluvium and 150 mm diameter in the bedrock. They were lined on completion with PVC tube. Inclinometer tubes I3 and I2 were installed with their bottom ends firmly grouted into bedrock at depths of 20.5 m and 19.5 m respectively. However, this was not possible for borehole I1 as the PVC lining failed to prevent this borehole closing at a depth of 16 m due to landslide movement. It is significant that closure occurred just above a slickensided highly-plastic clay horizon, seen in the core of adjacent borehole D).

Initial inclinometer observations were made on 10 December '88, after the cement-bentonite installation grout had hardened. The tubes were read to an accuracy of 2 mm with a Geotechnical Instruments biaxial inclinometer probe working upwards from the bottom at intervals of 0.5 m. Thereafter inclinometer observations were taken at weekly intervals until the tubes became blocked by landslide shear movement. Figure 8 shows the successive inclinometer profiles for I2 located (Figure 4) 100 m up hill of the landslide toe. Downslope displacement totalling 60 mm can be seen to be seated in

a shear zone at 10.5-12 m depth. The displacements when plotted against time, Figure 5, show a rate of movement of 5 mm increasing to 11 mm a week, in good agreement with the surface displacements measured by EDM survey instruments.

The observed positions of the inclinometer detected shear movements are indicated on the profile of the lower half of the landslide, Figure 7, and the deduced landslide shear surface drawn taking account of these and of the sheared clay horizons found in the cored boreholes. It can be seen that, except at the toe, the shear surface is near planar but not parallel to ground level as initially presumed.

CONCLUSIONS

Damage to buildings and infrastructure in the La Butte area of Port Louis has been caused by a large slow-moving landslide, some 700 by 400 m in extent, located in weathered basalt colluvium mantling the lower slopes of Signal Mountain. In the lower-central part of the landslide the shear surface was near planar, seated on horizons of highly sheared smectite clay. Ground above this planar part moved as a block without significant deformation, resulting in translocation of many buildings without damage, a fact which led to local misinterpretation of the ground instability.

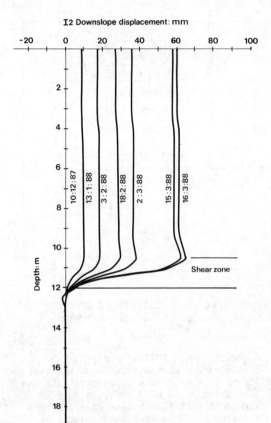

I2 Downslope displacement: mm

Shear zone

Depth: m

Fig 8. Downslope displacement measured by inclinometer tube I2 , located in centre of the landslide area

Before development of the La Butte area the colluvium was probably marginally stable and subject to periodic downhill creep movements in times of heavy cyclonic rainfall. Detrimental action by man in clearing the natural rain forest cover and in large-scale cut and fill works during construction of trunk water mains and reservoirs probably stimulated an increase in creep movement, which in turn resulted in damage to the water services, allowing continued high water leakage. This leakage combined with torrential cyclonic rainfall and aggravated by poorly maintained drainage, raised ground water pressures in the colluvium and eventually caused the major landslide movement of 1987.

An initial risk was identified of a devastating debris flow originating in the 26° sloping head region of the landslide, triggered by torrential cyclonic rainfall and compounded by massive inflow of water from broken trunk mains. This hazard was averted by removing the water services from the head region of the landslide, reducing surface water infiltration by sealing tension cracks, and by improving drainage.

In the longer term, it was considered that these measures would lead to a substantial reduction in groundwater pressures in the colluvium, and it would be likely that landslide movement would decline to levels of a few centimetres per year which could be tolerated by the majority of building in the area.

Full stabilisation of the landslide would undoubtedly be very expensive, and would need a much fuller geotechnical investigation of geology, groundwater pressures and material properties, together with appropriate stability analysis. An outline specification for such a study was drawn up by the author at the request of the Government of Mauritius, and was accompanied by a recommendation that consulting engineers of international standing be engaged to supervise further investigation, a cost-benefit analysis being carried out before proceeding with remedial measures.

ACKNOWLEDGEMENTS

I am indebted to many staff in Government Departments including the Ministry of Economic Planning, the Ministry of Housing, Land & Environment Government, and the Ministry of Works; the Municipality of Port Louis; the University of Mauritius; the Central Water Authority; and private consultants: all of whom who gave priority to the provision of information and facilities required for this investigative study. The BRE study and inclinometer equipment was financed by the UK Overseas Development Administration.

Landslides, Bell (ed.) © 1995 Balkema, Rotterdam, ISBN 90 5410 032 X

A case study on a landslide at the anticline hill in south-east Osaka

K. Nakaseko
Kobe Yamate Junior Women's College, Osaka, Japan

W. Kanamori
Department of Public Works, Osaka Prefectural Government, Japan

Y. Iwasaki & T. Hashimoto
Geo-Research Institute, Osaka, Japan

ABSTRACT: This paper presents a case study on a landslide which occurred in a newly developed residential area in south-eastern Osaka. Development moved from lowlands to the hilly area some 20 years ago. The area of the slide is 70m wide and 60m in length. Geological structures are complex and show reverse faulting which has resulted in the formation of an anticline structure on the upper side block.

  The sliding mechanism was difficult to assess from boreholes due to the disturbed structure so a number of vertical shafts a few meters in diameter were excavated. From these the three-dimensional character of the geology was observed and the slide plane identified. Sliding occurred in sandstones of Tertiary age which were highly fractured and thrust over formations of Quaternary age.

1. INTRODUCTION

  The Osaka area has developed from the lowlands of the Osaka plain and expanded into mountainous regions during the last twenty years. The landslide is located at Nishi-Asahigaoka in Kashiwara City, Osaka Prefecture (Fig. 1).

The slide has developed in the southeastern hilly region on the border between the plains and the mountains where development for residential use began about twenty years ago. The geology of this area is complex and mainly Cenozoic in age.

  Several cracks were found in a vineyard adjacent to the residential area on June 6, 1986 a few weeks after a heavy rainfall of 77mm on May 19-20. The surface area of the slide measured about 70m in width and 60m in length. The uppermost surface cracks were found very near a water supply tank on the top of the hill. The sliding zone was limited to the vineyard but the residential area and the water supply tank are expected to be included if the landslide continues to develop. A photographic view of the area can be seen in Photo 1.

2. GEOLOGICAL SETTING

  The general geology of the area is shown in Fig. 2 and an east-west cross-section in Fig. 3. The hill is 90-100m high and the hill axis runs north-south.

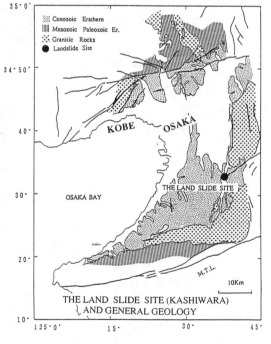

Fig. 1

THE LAND SLIDE SITE (KASHIWARA) AND GENERAL GEOLOGY

- Cenozoic Erathem
- Mesozoic Paleozoic Er.
- Granitic Rocks
- ● Landslide Site

KOBE OSAKA

OSAKA BAY

THE LAND SLIDE SITE

M.T.L.

10Km

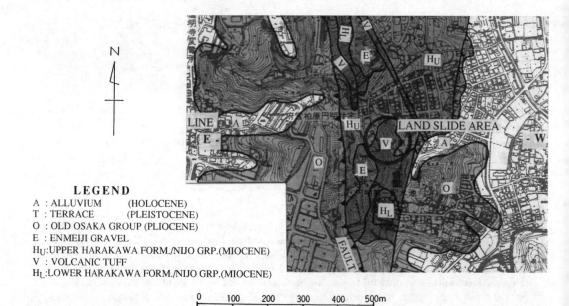

LEGEND

A : ALLUVIUM (HOLOCENE)
T : TERRACE (PLEISTOCENE)
O : OLD OSAKA GROUP (PLIOCENE)
E : ENMEIJI GRAVEL
H_U:UPPER HARAKAWA FORM./NIJO GRP.(MIOCENE)
V : VOLCANIC TUFF
H_L:LOWER HARAKAWA FORM./NIJO GRP.(MIOCENE)

0   100   200   300   400   500m

Fig.2 GEOLOGY OF THE AREA

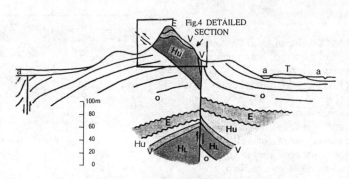

Fig.3 VERTICAL SECTION ALONG E - W LINE

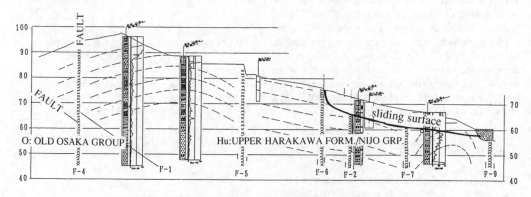

Fig.4 DETAILED GEOLOGICAL E-W SECTION OF LAND SLIDE ZONE

Photo 1 LAND SLIDE AREA VIEWED FROM NORTH EASTERN RESIDENTIAL ZONE

the direction of Photo 1 taken

Photo 2 AERIAL PHOTOGRAPH

The eastern slope, where the sliding occurred, consists of rocks of the Nijo Group formation (gravels, sandstones and mudstones) of Miocene age which have a dip of about 15°. The Nijo Group rocks in this area are termed Harakawa Formation and consist of tuffaceous sands and gravel sands which have been fractured by tectonic movements and also deeply weathered. The western slope consists of sands of the Old Osaka Group of Pliocene age.

The western block is considered to have underthrust the older rocks of the eastern side along a fault striking north-south and dipping 30°E. As the detailed section of the site in Fig. 4 shows, the underthrusting of the western block has created an anticline structure down which sliding is possible.

3. COUNTERMEASURES

After the ground cracks were found on June 6, measurements of ground movements at various points as well as horizontal borings at 71 points (total length = 2825m) to drain water were planned and carried out as shown in Fig. 5. Based on the surface geology, 21 geotechnical borings with a total length of 549.2m were carried out. Study of cored samples from the borings showed the presence of one or two soft and weakened layers. However it was difficult to identify the sliding surface on the basis of cores alone as the geology has been very disturbed by tectonic movements.

4. SHAFT WALL GEOLOGICAL MAPPING

Three vertical shafts were excavated to investigate the geology, in particular the detailed structure and sliding surfaces, as well as for drainage. Each shaft was 20m deep and 3.5m wide. Fig. 6 shows a typical mapping of the inner peripheral surface of the shaft as well as the bore log section at the point of shaft excavation. The boring log shows only the vertical discontinuities of the geology and can give a misleading impression if the geology continues horizontally at depth. On the other hand the vertical shafts allow three dimensional access to the geology. This gives several advantages over boring logs. It is then easier to study the degree of discontinuity of geological formations in situ as well as the groundwater situation in order to understand the geological and hydrological conditions relating to the landslide site. The sliding planes are more easily detected in situ with quantitative as well as qualitative information on the geometry of the sliding surface e.g. inclination angles and plane directions. Finally in situ geotechnical tests can easily be performed if required. Another advantage of vertical shafts is that they can be further used for drainage as well as to drain groundwater into the sliding blocks to stabilize the landslide.

The shaft wall mapping clearly shows the lack of continuity of the formations. The Tertiary formations have been highly deformed during tectonic movement during the Quaternary. It should be noted that the understanding of the geology

1605

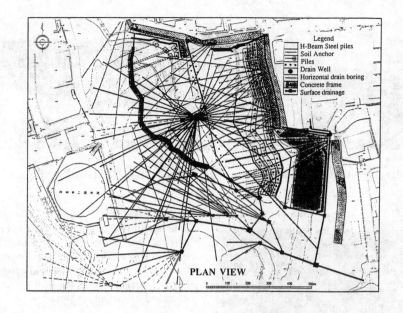

PLAN VIEW

STANDARD SECTION VIEW

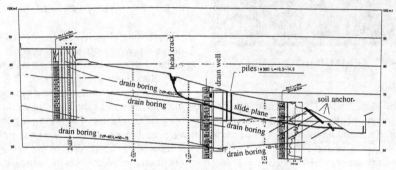

Fig.5 COUNTERMEASURES AGAINST LANDSLIDE

gained even by increasing the number of drill holes would still be limited due to the highly fragmental nature of the formations.

The sliding plane was identified by the presence of clay seams with slickensides in the sandstone formation at a depth of 6m below ground level. The sliding plane dips east at about 10°. Groundwater seepage was identified in a number of locations including the sliding plane. Various discontinuity zones are seen in the shaft map, some of which are nearly vertical in inclination especially near the bottom of the shaft (Fig. 6).

Several in situ tests were carried out in the shafts to determine the strength characteristics in the sliding area. Direct shear tests on the sliding plane and the weakened sandstone gave the following results:

sliding surface $c' = 0$ kg/cm²
                   $\phi' = 13.5°$

weakened sandstone $c' = 0.08\text{-}0.24$kg/cm²
                   $\phi' = 15 - 17°$

sandstone $c' = 0.19 - 0.75$ kg/cm²
                   $\phi' = 29 - 64°$

The weakened sandstones show strengths several times lower than the non-weakened sandstones. The strength parameter obtained for the sliding surface has only a ϕ value for the frictional component. The cohesionless nature of this surface tends to cause easy sliding due to high water pressures. The water contents for the sliding surface ranged from 32 -

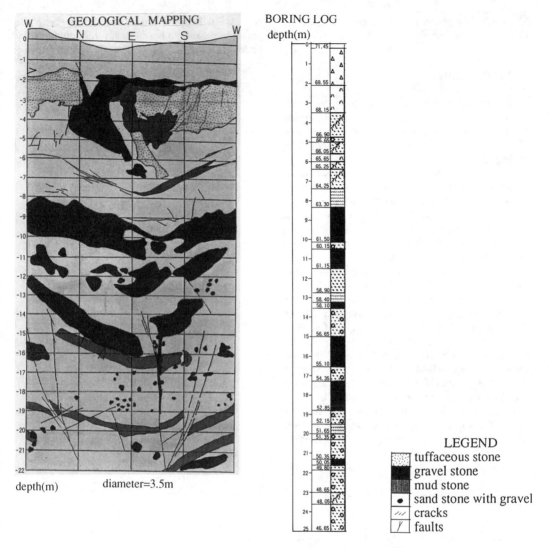

GEOLOGICAL MAPPING

BORING LOG
depth(m)

LEGEND
- (tuffaceous pattern) tuffaceous stone
- (black) gravel stone
- (mud pattern) mud stone
- (dotted) sand stone with gravel
- /// cracks
- Y faults

depth(m)   diameter=3.5m

Fig.6 GEOLOGICAL MAPPING ON VERTICAL SHAFT WALL
WITH BORING LOG AT THE SITE

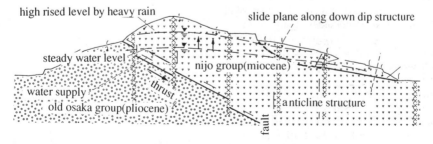

high rised level by heavy rain   slide plane along down dip structure

steady water level

nijo group(miocene)

water supply   thrust

old osaka group(pliocene)   anticline structure

fault

Fig.7 SLIDING MECHANISM AT KASHIWARA LAND SLIDE

90% while the surrounding sandstones contained 20 - 40%.

5. SLIDING SURFACE

Geotechnical characteristics revealed by field tests as well as laboratory studies suggest the following division of the landslide and surrounding area as well as the mechanism of landsliding in this zone (also see Fig. 7).

1) The tectonically weakened zone; where the existence of faults and cracks, predicted by the surface geology, has weakened the sandstone.

2) The landslide block; which may have be active in the past as slickensides on the sliding surface mapped in the shaft wall appear to have been formed by several slide movements.

It is likely that heavy rainfall (77mm) in the Osaka area on the 19 - 20 May, 1986 triggered the landslide. After the heavy rainfall, the groundwater level became higher but not high enough to trigger the complete failure of the area. The total vertical stress and the shear stress acting on the slide plane was more or less constant before and after the increase of groundwater. The effective vertical stress acting on the potential sliding plane decreased due to the increase in pore pressure caused by rising groundwater levels. Accordingly the shear strength of the slide surface dropped to a level where creep was possible, resulting in the apparent displacement with cracks within a few weeks.

6. LESSONS LEARNT FROM THIS CASE STUDY

Areas of residential development in hilly country like Osaka will tend to be affected by slope failures. Examples include Kobe, Hong Kong or Rio de Janeiro. Landslides usually have a long history of movement which will create characteristic topographic features easily recognised as a landslide zone. Thus, had geomorphological studies of the area been carried out prior to residential development, the area would have been identified as a landslide zone and appropriate measures could have been taken.

Also the usefulness of vertical shaft mapping in comparison to drill cores has become apparent in this study. The three dimensional nature of the former method has enabled the distribution of the fragmented formations to be accurately mapped in the problem zone as well as in the landslide itself.

It is especially useful where, as is the case here, the geology of the area has been severely disrupted by tectonic activity.

Landslides, Bell (ed.) © 1995 Balkema, Rotterdam, ISBN 90 5410 032 X

Simulation of landslide behaviour in the time domain

J.D. Nieuwenhuis

Delft Geotechnics & University of Utrecht, Netherlands

The statistical or stochastical analysis of the stability of landslides in the space domain has become standard practice by now. I mean the introduction of statistical variations of the parameters of strength or stiffness or permeability within the potentially sliding mass. Prof. Chowdhury has given us many examples (1980). And if there is a trend in the spatial variations we rather call the analysis a stochastical analysis.

Figure 1. Part of the landslide-prone region over varved clays along the river Drac in the French Alps S. of Grenoble.

As we all know potential landslides sooner or later may turn into real ones often under a specific set of boundary conditions. For instance abundant rain or snow melt or high groundwater levels. These boundary conditions may fluctuate rather in the course of time than from place to place. Hence for certain types of landslides especially small and/or shallow ones a stochastical analysis of the boundary conditions may have more predictive value than the analysis of

Figure 2. View of a particular landslide in the region shown in fig. 1.

Figure 3. Surface topographly of the landslide from fig. 2.

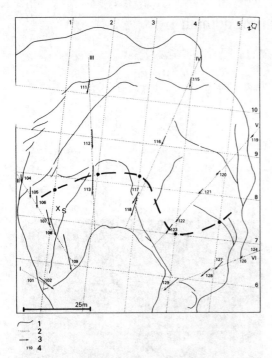

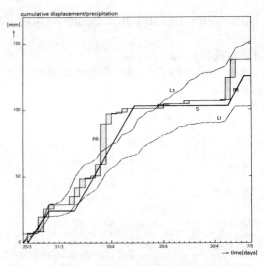

Figure 5. Measured and computed daily
displacements along the slide
planes of the landslide from
fig. 2.
The recording of block
movements and precipitation
(daily totals) during spring
1987.
L1 Relative displacement
monitored by recorder 1
L3 Relative displacement
leading block monitored by
recorder 3
PR Precipitation
S Computed displacements

Figure 4. Surface displacements 1980 –
1987 of the landslide from
fig. 2.
Simplified map of slide 2
showing the lines of cross
sections.
1= Major scarp
2= Line of cross section
3= Displacement vector (at map
scale, 8yr total)
4= Marker number

the spatial variability of the soil
properties. With that idea in mind a
group of students from the University of
Utrecht investigated the behaviour of a
number of small landslides in the French
Alps near Grenoble for a period of 8
Years from 1980 to 1988 (Van Asch et
al., 1984). In fig. 1 a part of the
unstable
area – just above the steep valley
slopes of the Drac river – developed in
outcropping varved clays is shown. Fig.2
focusses on a particular landslide. Its
surface, in light colour in the upper
part of the photograph, measures
approximately 1 ha. Fig. 3 represents
the rugged topography of the landslide's
surface.

The analysis appeared relatively
difficult since the behaviour of those
small landslides was a lot more complex
than expected (van Genuchten, 1989). In

fig. 4 the surface displacements are
depicted.

Of course the surface movements were
larger than the movements at some depth,
especially near the slide planes at 4 to
8 m of depth, but more specifically the
displacements were convergent. Hence
before performing any stochastical
analysis it appeared necessary to
understand the displacement mechanism of
the landslides. I will be brief on this,
but it became clear that some
complications needed introduction into
the stability model. We introduced:
residual strength (through ring-shear
tests), strength regain during periods
of stagnancy, rate-dependent (viscous)
strength increase, increase in stability
by convergent flow and by mass
accumulation at the toe (Nieuwenhuis,
1991). This of course was added to a
classical plane stability analysis
involving residual strength and measured
and predicted porewater pressures.

The comprehensive stability model also

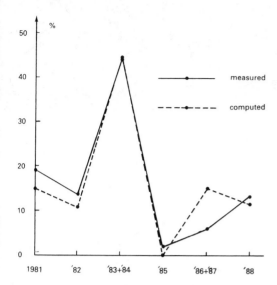

Figure 6. Percentual spread of
         displacements over the years
         of observation, of the
         landslide from fig. 2.

produced displacements due to the
viscosity component. In fig. 5 measured
and predicted daily displacements are
shown. You see that the agreement of
predicted and actual daily displacements
is reasonable.

With a satisfactory relation of
predicted and measured displacements we
could proceed to a long-term
stochastical analysis in which the
amount of rain and snow, together with
the temperature in the case of snow,
were the input parameters.

In this way predictions of the yearly
displacements were made (fig. 6).
As you see the measurements and the
predictions match fairly well.

The displacements of the investigated
landslides are stronger correlated to
the number of raindays than to the
amount of daily rainfall (fig. 5). Since
there is also a clear correlation with
the interruptions between intervals of
consecuctive raindays those two
elements, number of consecutive raindays
in an interval and total number of
raindays in the wet season (Februar thru
May), enter the stochastical analysis.
Drawing more than a 1000 years through
the Monte Carlo simulation method
produced a frequency distribution as
shown in fig. 7.

A probable sequence of events

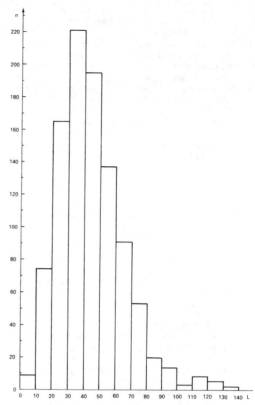

Figure 7. Frequency distribution of the
         number of raindays in the wet
         season obtained by a Monte
         Carlo simulation.

characterized by the sliding of parts of
the landslide rather than the usual
chips sliding over the downslope edge of
the landslide, appears to ask for more
than 110 to 120 raindays in one wet
season (Nieuwenhuis, 1991). According to
fig. 7 such extreme wet conditions could
occur once in 50 to 150 years. Most
fortunately, and one of the reasons to
investigate this particular landslide,
it is known that this landslide, or
rather its predecessor, collapsed in
1910 blocking the local river Drac for
three days. This figure of 80 years ago
fits surprisingly well in the Monte
Carlo simulation of once every 50 - 150
years.

The briefly described investigations
have tought us a number of things which
might be of more general relevance:
1. Active landslides merit scientific
   attention before they collapse.
2. A very close look as done in this
   case asks for much instrumentation

1611

(if possible redundancy) and a
sufficient number of years of
observation.
3. Recording must address displacements
(3D), soil samples and hydrology
(meteorology, surfacewater flow,
groundwater flow)
4. Only a few landslides in a landslide-
prone area ask for a close look. Much
of the behaviour of landslides not or
superficially studied can be inferred
form the extensively investigated
ones. And the same is true for
sanitation methods.

References

- Chowdhury R.N. 1980. Landslides as
natural hazards - mechnisms and
uncertainties Geotech. Eng. 11(2): pp
135-180.
- Nieuwenhuis J.D. 1991. The lifetime
of a landslide. Investigations in the
French Alps. A.A. Balkema, Rotterdam:
148 p.
- Van Asch T.W.J. Brinkhorst W.H.,
Buist H.J., Van Vessem P. 1984. The
development of landslides by
retrogressive failure in varved
clays. Zeitschrift für Geormophologie
4: pp 165-181.
- Van Genuchten P.M.B. 1989. Movement
mechanisms and slide velocity
variations of landslides in varved
clays in the French Alps. Thesis,
University of Utrecht, ISBN 90-6809-
108-5, 157 p.

G2 Stability analysis techniques
Techniques des analyses de stabilité

Landslides, Bell (ed.) © 1995 Balkema, Rotterdam, ISBN 90 5410 032 X

Keynote paper: The role of analysis in the evaluation of slope stability

N. R. Morgenstern
University of Alberta, Edmonton, Alb., Canada

ABSTRACT: A distinction is made between the use of analysis to evaluate the ultimate failure of slopes from analysis to calculate deformations. Failure calculations are commonly based on limit equilibrium methods and recent progress is surveyed with emphasis on three-dimensional methods of analysis. Limitations to limit equilibrium methods as a whole are stressed. Progress in deformation analyses are discussed separately under: 1) empirical evaluation, 2) small deformation studies, 3) pre-failure deformation, 4) slow landslide movements, and 5) fast landslide movements.

1 INTRODUCTION

1.1 *Use of Analysis*

Analysis is used in the evaluation of slope stability to provide a quantitative assessment of slope behaviour. While calculated behaviour should not be the only basis for design and hazard assessment of earth slopes, it is an important, if not dominant, input to these processes. Therefore it is of on-going interest to understand the limitations of current methods of stability analysis and to indicate where progress has been made and further work is necessary. That is the intent of this presentation.

It is convenient to distinguish the use of analysis in the assessment of pre-failure conditions from its application to post-failure conditions or back-analysis. In the former, analysis is used to assess safety in a global sense and to ensure that the slope will perform as intended. That is, it will not fail and it will not deform or crack too much. It is tempting to interpret these applications of analysis within the framework of limit states design. Here one would say that analysis is used to evaluate the ultimate limit state (failure) and the serviceability limit state (deformation). However geotechnical materials are more complex than many other engineering materials and it is not possible to transfer in any simple way the concepts of limit states design as applied to steel or concrete structures and related building codes to geotechnical structures such as slopes. For example, as will be noted, for some soils and rocks the available strength depends upon the pre-failure deformations and hence there is a coupling between the two. Under these conditions, in rigorous terms, the analysis for serviceability or

deformations cannot be separated from the analysis for global stability or ultimate limit state.

Post-failure analyses are used to provide a consistent explanation for landslide events. These back-analyses can then be used as a basis for design of stabilizing measures if engineering works are required. This approach has the appeal that the remedial design is normalized in terms of the post-failure analytical model.

Reliance on post-failure analysis can be misleading. For example, if a slide failed in a retrogressive manner, an analysis of the stability of the overall mass alone can give incorrect results with regard to operational strength and critical mechanisms. Post-failure analysis should be responsive to the totality of processes that led to failure if they are intended to add to our understanding and provide a basis for calculating improvement of stability.

1.2 *Requirements for Analysis*

In the most general terms, the following are required for slope stability analysis:

i) site characterization - this refers to the determination of the geological and hydrogeological conditions acting on the site. Site characterization is a pre-requisite for the other requirements that follow.

ii) mode identification - soils and rocks can fail in a variety of modes such as falls, topples, slides and flows. Hutchinson (1988) provides a comprehensive review of morphological classification. In some instances the appropriate mode is obvious but in others it is not.

Figure 1 is of a rock slope in British Columbia. The linear scarp-like features characteristics of movement are clear. The hazard assessment required the evaluation whether a large volume, high velocity slide might descend into the lake below and generate a wave of sufficient magnitude to endanger a lake-side community. Historical precedent suggests that such hazards are more likely if a slide mechanism along some geological structure exists. If the movement features are a result of toppling, the prospect for a single large volume, high velocity episode is reduced. A considerable field effort was required to determine the likely mode of movement which proved to be toppling.

iii) material characterization - this refers to the process of in-situ testing, sampling, and laboratory testing of the relevant materials. The end products are the strengths, deformation properties and permeability data for these materials.

Figure 1. Mount Breakenridge, British Columbia, Canada

iv) pore pressure characterization - this refers to the actual pore pressure model or models that will be used in the analyses. It is rare that the groundwater conditions interpreted from the site characterization will be exactly the same as those used in the analysis. In general, allowance may have to be made for stress changes, alteration in drainage boundary conditions, seasonal variations and a variety of other factors that might influence the design pore pressure distribution.

v) analytical model - given the above, a specific model is then necessary to conduct the calculations.

In practice the limitations associated with (i)-(iv) above are the most critical. Gross inadequacy in any one of these requirements will dominate the result. However, in the following it will be assumed that the validity of the proposed analysis is not impeded by limitation of input and 'that site and material characterization have been undertaken for pre-failure conditions utilizing best available field and laboratory techniques. The discussion will also be restricted to analysis of sliding and shearing modes.

2 GLOBAL STABILITY ANALYSIS

2.1 Principles

The calculation of global stability is normally expressed in terms of the Factor of Safety calculated by means of limit equilibrium methods (LEA). The principles underlying all of these methods have been clear for some time (Morgenstern and Sangrey, 1978) and are as follows:

i) A slip mechanism is postulated.

ii) The shearing resistance required to equilibrate the assumed slip mechanism is calculated by means of statics.

iii) The calculated shearing resistance required for equilibrium is compared with the available shear strength in terms of the Factor of Safety.

iv) The mechanism with the lowest Factor of Safety is found by iteration.

In order to avoid confusion, it is important to define the Factor of Safety clearly. The definition favoured here is as follows:

The Factor of Safety is that factor by which the shear strength parameters may be reduced in order to bring the slope into a state of limiting equilibrium along a given slip surface.

Morgenstern and Sangrey (op.cit.) expand on the merits of this definition, compare it with others that have been used and note some limitations to it.

The Factor of Safety has several roles. One well-recognized role is to account for uncertainty and to act as a factor of ignorance with regard to the reliability of the inputs to the analysis. These include strength parameters, pore pressure distribution and stratigraphy. However, an additional major role of the Factor of Safety is that it constitutes the empirical tool whereby deformations are limited to tolerable amounts within economic restraints. In this way, the choice of the Factor of Safety is greatly influenced by the accumulated experience with a particular soil or rock mass. Since the degree of risk that can be taken is also much influenced by experience, the actual magnitudes of the Factor of Safety used in design will vary with material type and performance requirements.

2.2 Recent Developments

The various methods of limit equilibrium analysis are well understood and, at least for two-dimensional analyses, have been well-summarized elsewhere (Fredlund, 1984; Nash, 1987). While there are still new studies of these methods emerging, their impact is slight. Powerful, easy-to-use computer programs are widely available that facilitate the adoption in practice of even the most rigourous methods. Numerical complexity or cost of computer time is no longer an excuse for avoiding the more complete methods.

Developments employing optimization techniques

to search for the minimum factor of safety in slope stability analysis have evolved to the stage of practical application. The work Chen and Shao (1988) has discovered an optimization technique that embraces the generalized method of slides and has been applied to several practical problems with impressive results. Additional perspectives on optimization techniques are provided by Sridevi and Deep (1992).

The most important developments in LEA in recent years have related to three-dimensional methods of analysis. The evolution of two-dimensional methods involved analytical solutions for simplified homogenous profiles and numerical methods, primarily based on the method of slices, for more complex sections. The same pattern can be discerned with regard to the development of three-dimensional methods.

Based on the analytical approach, Gens, Hutchinson and Cavounidis (1988) presented a comprehensive treatment of undrained slides in homogeneous cohesive soils. Their work followed on from Azzouz and Baligh (1978) in that the slip surface is assumed to be cylindrical and terminated at its ends by various types of regular surfaces. A variety of end surfaces is chosen and the results are compared with analyses performed assuming cylindrical slides with plane ends. In addition to studying the influence of end shape, valuable stability charts are plotted for the most critical shape. These charts provide stability numbers for slides of length L, height H and slope angle i. The analysis reduces to Taylor's (1937) charts for L/H = ∞. Characteristic geometries for a number of short term failures in soft saturated clays were assembled and interpreted in terms of the new theoretical results. Generally the theory correctly predicted the mode of failure, i.e., slope, toe or base, and this gives confidence in applying the results. The influence of three-dimensional effects for various cases is summarized in Figure 2. The cases reveal that neglecting three-dimensional effects involves errors which range from 3-30% and average 13.9%. As stressed by Gens et al (op.cit.), these errors are comparable in importance with the corrections commonly made with regard to undrained shear strength and, in back-analysis, may be on the unsafe side.

The three-dimensional extension of the method of slices is termed the method of columns. Hovland (1977) appears to have been the first to produce a workable solution. His method is an extension of the ordinary method of slices in that the intercolumn forces were ignored. The normal and shear forces acting on the base of each column are merely the appropriate components of the weight of the column. The method has the same excessive conservatism as its two-dimensional equivalents.

Chen and Chameau (1982) attempted a method of columns analysis that was more general. They found that under certain circumstances the ratio of Factors of Safety, F_3/F_2, was less than 1.0. This cannot be so (Hutchinson and Sarma, 1985) and

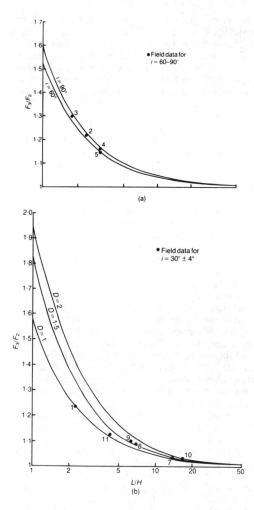

Figure 2. Three-dimensional effects, purely cohesive slopes (Gens et al, 1988)

hence their method must be regarded as suspect.

The most useful solution to date, in terms of clarity of mechanics and utility of available computer programs, is the work of Hungr (1987). This is an extension to three dimensions of Bishop's simplified method of analysis in which the vertical intercolumn shear forces acting on both the longitudinal and the lateral faces of each column are neglected. The Factor of Safety is obtained by summing moments about the axis of rotation for the entire assemblage of columns. Horizontal force equilibrium in both the longitudinal and transverse directions are neglected. The computer program also includes an extension of the Janbu simplified method into three dimensions.

Hungr, Salgado and Byrne (1989) have summarized some test analyses to explore the validity and application of this method. From an analytical perspective, several closed form solutions of simplified three-dimensional mechanisms exist in the literature and initial validation would require replication of these results by any numerical procedure. Gens et al (1988) reported the closed form solution for a spherical sliding surface in a purely cohesive slope. The numerical and analytical solutions are within 1% of each other. Both Leshchinski et al. (1985) and Dennhardt and Forster (1985) have proposed analytical solutions for three-dimensional failure for a slope in a cohesive-frictional material and their results are recovered within 2-3% by assuming ellipsoidal surfaces in the numerical procedure. A comparison is also made between the results of closed form wedge analyses (Hoek and Bray, 1977) and the method of columns. It is found that the Bishop simplified method gives a nearly exact result for symmetric wedges but that the comparison is variable in the cases of wedges that are laterally asymmetric in either geometry, material properties, or both.

At the empirical level, a three-dimensional analysis was also performed on the Lodalen Slide (Sevaldson, 1956) which remains one of the best documented case histories in the literature. The average of two-dimensional slip circle analyses of three sections is a Factor of Safety of about 1.05 and the critical circles depart somewhat from the actual surface determined from borings. A three-dimensional analysis fitting an ellipsoid to coincide closely with the actual surface yielded a Factor of Safety of 1.01. The surface was almost symmetric. It is evident that this method of analysis gives accurate results for symmetric rotational surfaces.

Figures 3 and 4 are taken from Lam (1991) who has recently made additional contributions to three-dimensional analysis. Figure 3 shows a cross-section of a generalized sliding mass in the X-Y direction, with forces on slices prescribed in the usual way. Figure 4 is a longitudinal section in the Z direction showing the additional forces required for equilibrium of columns. With n columns in the X direction and m columns in the Z direction the number of unknowns is $12nm+2$, while the number of equations is $4nm+2$, as summarized in Table 1. The degree of indeterminancy is reduced in the following way:

i) Assume that the point of application of the normal force, N, acts through the centre of the base area. Therefore, a_x, a_y, and a_z are known when the corners of the bottom of the column are defined. The number of unknowns is reduced to $9nm+2$.

ii) Following the approach initially proposed by Morgenstern and Price (1965), it is possible to assume that all the intercolumn shear forces can be related to their respective normal forces. Hence,

$$\frac{X}{E} = \lambda_1 f(1) \qquad (1)$$

$$\frac{H}{E} = \lambda_2 f(2) \qquad (2)$$

$$\frac{V}{P} = \lambda_3 f(3) \qquad (3)$$

$$\frac{Q}{P} = \lambda_4 f(4) \qquad (4)$$

$$\frac{T}{N} = \lambda_5 f(5) \qquad (5)$$

where $f(n)$ is a function that describes the spatial variation of the specific force ratio in a particular direction and $\lambda(n)$ is an undetermined multiplier. This reduces the number of unknowns by 5nm but adds 5 more unknowns, $\lambda(n)$. As a result, the number of unknowns is $4nm+7$ and 5 unknowns need to be defined.

One can anticipate a variety of three-dimensional solutions in the future depending upon how this indeterminancy is resolved. For example, if the influence of H/E, Q/P, and T/N are taken to be insignificant, only λ_1 and λ_3 remain as unknowns.

Lam (1991) observes that the problem is now determinate because the two unknowns can be solved for subject to the restraints that:

i) the Factor of Safety with respect to moments (F_m) must be equal to the Factor of Safety with respect to forces (F_t) when total equilibrium is achieved, and

ii) the most critical Factor of Safety must be the lowest if all other conditions of the slope remain unchanged.

To gain further insight into the force distribution functions Lam (1991) undertook three-dimensional elastic stress analyses of several slope configurations and found that in all instances $f(2)$, $f(4)$ and $f(5)$ were negligible and that in many instances even $f(3)$ could be neglected which would reduce to a modified Spencer (1967) analysis. Test cases were also provided to validate a new computer program. This work clarifies the current situation with respect to three-dimensional analysis.

The adoption of powerful graphical input to PC based computers make three-dimensional analysis a practical reality. Following the collapse of the Kettleman Hills waste landfill (Seed, Mitchell and Seed, 1990), three-dimensional analysis has been made mandatory in the United States by the Environmental Protection Administration. The available programs meet many of these needs.

It should be noted that the three-dimensional elastic stress analyses cited above were all symmetrical configurations and the conclusions that $f(2)$, $f(4)$ and $f(5)$ can be neglected is not likely true in general. Hungr et al. (1989) had previously drawn attention to potential sources of error in their analysis arising from high degrees of mobilization of internal strength and from asymmetry in the

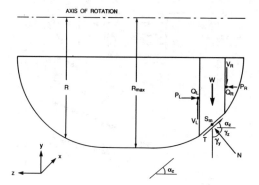

Figure 3. Transverse section of three-dimensional slide (Lam, 1991)

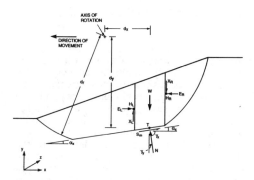

Figure 4. Longitudinal section of three-dimensional slide (Lam, 1991)

longitudinal direction, perpendicular to movement. A numerically derived lateral imbalance index has been developed to correct for this latter form of asymmetry. This index provides a warning with regard to the limitations of the three-dimensional Bishop analysis. The way is clear to explore more-generalized solutions in the future and progress can be anticipated. In the meantime three-dimensional rotational and symmetric sliding surfaces are amenable to solutions that are accurate for practical purposes.

2.3 Limitations to LEA

The most common and significant barriers to reliable LEA arise from practical difficulties in site characterization and material characterization based on sampling and testing. These are well-known and much discussed. However, even if there were no such barriers, there would still be significant limitations that have to be understood if the results of a LEA were to be relied upon.

The two dominant inputs to LEA are pore pressure distribution and the shear strength of the

Table 1. Unknowns and equations in three-dimensional analysis (Lam, 1991)

KNOWNS	DESCRIPTION
nm	$\Sigma F_z = 0$ in Z direction for each column
nm	$\Sigma F_x = 0$ in X direction for each column
nm	$\Sigma F_y = 0$ in Y direction for each column
nm	Mohr-Coulomb failure criterion for each column
1	$\Sigma M = 0$ about the axis of rotation for the whole sliding mass
1	$\Sigma F_x = 0$ in X direction for the whole sliding mass

4nm+2
=====

UNKNOWNS	DESCRIPTION
nm	N, normal force at base of each column
nm	S_m, shear force mobilized at base of each column
3nm	a_x, a_y, a_z, point of application of N
nm	T, shear force in Z direction at base of each column
nm	E, intercolumn normal shear force on YZ plane
nm	X, intercolumn vertical shear force on YZ plane
nm	H, intercolumn horizontal shear force on YZ plane
nm	P, intercolumn normal shear force on XZ plane
nm	V, intercolumn vertical shear force on XZ plane
nm	Q, intercolumn horizontal shear force on XZ plane
1	F_m, factor of safety by moment equilibrium
1	F_f, factor of safety by force equilibrium

12nm+2
=====

material. A site investigation and associated laboratory study are typically a snapshot in time. For many soils and rocks they cannot capture the operational conditions that control stability and the pore pressure distributions and shear strengths that are used in the analysis must be modified in some representative manner.

It is convenient to separate the processes that influence the operational properties into: 1) time-dependent processes and 2) deformation-dependent processes.

One time-dependent process that influences pore pressure distribution is variations in precipitation. The broad interaction between precipitation, infiltration and pore pressure rise is well understood and the literature is growing both in terms of examples and theory. Shallow slides develop pronounced decreases in stability as a function of proximate high rainfall. The proximate, as opposed to cumulative, event appears to trigger shallow slides in Hong Kong (Brand, et al., 1984), but antecedent rainfall is an important factor for deeper slides. Interesting progress is being made in the quantitative forecasting of the rise in pore pressure due to various rainfall projections.

An additional factor that affects saturated low permeability soils is the long duration for equalization of pore pressures following unloading. The interaction between undrained excavation response and long term equilibration by swelling was first characterized by Bishop and Bjerrum (1960) and analyzed by Eigenbrod (1975). Field evidence for the time scale for equilibration in the

London Clay was reported by Vaughan and Walbancke (1975) and Bromhead and Dixon (1984). Tens to hundreds of years are required. Evidence to illustrate the geomorphological significance of pore pressure equilibration has been presented by Fennelli and Picarelli (1990). Their study of pore pressures, erosion and landslides in an area of the Southern Appenines led to the conclusion that pore pressures at depth are depressed due to erosive unloading that started about 300,000 years ago and that the deep clays are still swelling with a swelling coefficient of about $0.03 \, \text{m}^2/\text{year}$.

Pore pressure equilibration exercises considerable control on the delayed failure of specific slopes and evidence is accumulating that reveals its influence on regional instability where unloading is occurring naturally and where the bedrock is composed of stiff argillaceous deposits with low coefficients of swelling.

The strength of both soft and stiff clays are influenced by time-dependent processes. Lefebvre (1981) has shown that the peak strength measured on high quality samples of Eastern Canadian quick clays is much higher than that mobilized in-situ. A strength beyond peak, measured at about 8% strain, is consistent with observed failures. the peak strength measured in the laboratory has been found to be time-dependent and cannot be relied upon in the field. Rationalization of the operational strength to make allowance for this creep rupture process is based on case histories of actual slides.

The long-term reduction in strength of stiff fissured clay was first explained by Terzaghi (1936) and the issue of what strength can be relied upon in the long term has been a difficult problem to be resolved when dealing with these materials. While there has been a broad understanding of the physical processes that lead to softening, the mechanics of softening have not been extensively studied. Based on a review of limited data and some physical reasoning, Yoshida et al. (1990, 1991) have put forward a non-linear failure criterion to embrace time-dependent softening. The major relation is:

$$\sigma_1 = \sigma_3 + A \cdot \sigma_c \left(\frac{\sigma_3}{\sigma_c} - S\right)^{1/B} \qquad (6)$$

where σ_1 and σ_3 are the maximum and minimum principal stresses respectively, σ_c is the uniaxial compressive strength, and A, B and S are material properties that can be determined from triaxial compression tests. By supposing A, B and S to vary with time according to some prescribed function, it is possible to simulate the time-dependent reduction in strength to the fully softened state. This is illustrated in Figure 5. The failure criterion can be used in both LEA and finite element analysis to investigate the influence of time-dependent reduction in strength on stability and deformations.

Many earth materials are strain-weakening.

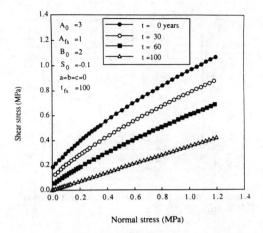

Figure 5. Soil softening (Yoshida et al, 1990)

Because collapse mechanisms rarely develop uniform strains, strain-weakening soil and rock masses will invariably mobilize at failure a strength on average that is less than the peak strength. This phenomenon of progressive failure is not peculiar to stiff clays and mudstones, but occurs in a variety of materials, such as loose undrained granular deposits, quick clays and soft fissured clays.

Progressive failure creates a dilemma for the analyst in that the peak strength as found in the laboratory cannot be operational everywhere when the slope collapses. In general, part of the slope is stressed beyond peak and an additional analysis is required to evaluate this. It cannot be deduced from LEA alone. Deformation analyses are required and for this class of problems, a coupling of both deformation analysis and LEA is necessary to calculate global stability.

Considerable effort is currently being expended on the analysis of progressive failure. Progress has been made in handling strain-weakening behaviour by means of finite element techniques and examples of slope deformations and embankment collapse that displayed progressive failure have been published (Chan and Morgenstern, 1987; Potts et al, 1990).

When a slope is excavated, either by nature or by man, the ground expands laterally. If the ground (soil or rock) is fissured, this expansion induces a mechanical loosening which also leads to reduction in strength. The "talzuschub" of Stini and others (Radbruch-Hall, 1978) has long been evidence of stress relief phenomena acting with gravitational forces on natural slopes. However this form of mechanical loosening has not been much studied and there is no theory of loosening as yet.

A recent instrumented study of an excavation in a coal mine has been revealing (Morgenstern, 1990; Small and Morgenstern, 1992). The layout of instruments is shown in Figure 6. Readings on one inclinometer (S6) as a function of the distance to the highwall are shown in Figure 7. An

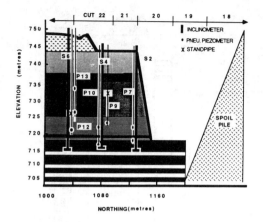

Figure 6. Highvale Mine Instrumentation (Small and Morgenstern, 1992)

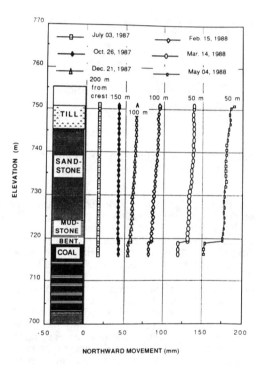

Figure 7. Inclinometer response (Small and Morgenstern, 1992)

additional 100-200 mm of movement typically occurs during removal of the final 50 m. The lateral displacement pattern that emerges from this study linearizes on a plot of distance from highwall crest against logarithm of movement. By differentiation, it can be shown that excavation has induced lateral strains at the highwall of 0.6 - 0.7%,

reducing to zero about 225 m from the highwall crest. The strains near the highwall reduced pore pressures by enhancing drainage. Hence this deformation related process affects both mass strength and pore pressures.

While increased drainage was a stabilizing effect, two small failures developed in a sequential uphill manner. To account for these slides, it was necessary to invoke low mass cohesion for the soft rocks. The lateral expansion results in a reduction of mass strength due to progressive loosening and possibly softening.

3 DEFORMATION ANALYSIS

3.1 *Introduction*

The assessment of deformations enters into the analysis of slopes in a variety of ways. In many instances it is simply sufficient to know that the deformations will not be excessive, without forecasting their magnitude in detail. In other instances, particularly associated with slope excavations, it may be necessary to restrict displacements to small magnitudes and more precise forecasts are required. Stability is implicity assured in both of these situations.

Deformation analyses also enter into slope problems when stability is not assured. The first instance is the requirement to predict failure. Displacements accelerate as failure approaches and it would be of considerable practical value to use the precursory data as a basis for precise forecasting of failure. The second instance is the analysis of post-failure displacements.

Morgenstern (1985) cited a classification of landslide movements that combined the classification of Varnes (1978) with a response scale of Hungr (1981). The velocity terminology is defined in Figure 8. On the basis of case history evidence, six categories have been established according to the manner in which people affected could respond. They are: (1) No response possible; catastrophe of major violence, (2) Some lives lost; velocity too great to permit all persons to escape, (3) Escape evacuation only; structures and equipment destroyed, (4) Temporary and insensitive structures maintained on or in front of the moving mass, others dismantled, (5) Remedial construction undertaken during movement, (6) Permanent structures maintained. The bars in the figure illustrate that specific case histories exist to support the classification. For a simplicity it is convenient to locate a boundary at 1 m/s and describe movements more rapid than this as fast. The other interesting boundary is below about 10 m.yr. and these movements will be grouped as slow.

In the following, the different types of deformation evaluations will be treated separately under 1) empirical evaluation, 2) small deformation studies, 3) pre -failure deformation,

4) slow landslide movements, 5) fast landslide movements.

3.2 Empirical Evaluation

It was emphasized earlier that a major role of the Factor of Safety is to provide an empirical basis to ensure that the soil or rock slope will function in a serviceable manner. That is, if the Factor of Safety is chosen correctly, one knows by experience that the deformations will not be excessive. In many slope problems, it is not necessary to know the magnitude of the deformations. It is often sufficient to know that they will not exceed some cms or tens of cms and this will be acceptable. The reliability of this approach can only be established in an empirical manner and is limited accordingly. Morgenstern (1991) provided several case histories that emphasized that traditional LEA, with sometimes traditional values of Factor of Safety were not always capable of inhibiting undesirable behaviour. It was not that the methods of analysis themselves were wrong, but that their application is bounded by the experience within which they have been developed and applied in the past. The Factor of Safety appropriate for design must reflect this fact. An example of excessive deformations in a clay fill embankment illustrates the point.

Figure 9 shows schematically a long embankment about 15 m high constructed as part of an industrial waste retention scheme. The site was an abandoned clay pit. As the clay pit was developed, weathered clay near the surface, which was unsuitable in brick manufacture, was stripped and deposited on the previously uncovered floor of the pit where it subsequently absorbed water and softened. The economics of the project made it desirable to use this softened clay for part of the fill while freshly excavated hard clay was available for the remainder. This hard clay was used in the upstream portion of the embankment, while the softened clay was used in the downstream section. This softened clay was substantially wet of Proctor optimum moisture content and design of the embankment for stability seemed a relatively straight forward undertaking.

The downstream part of the embankment was to be constructed with a homogeneous remoulded saturated clay on a relatively strong foundation. Slip circle analysis in terms of the undrained strength (C_u, $\phi = 0$) was appropriate and the downstream slope was selected to have a Factor of Safety of about 1.8. This appeared to be a conservative design based on Factors of Safety in conventional use and on the view that the clay fill behaviour was well understood.

When the embankment was completed, substantial bulging developed in the downstream direction and a crack opened between the stiff upstream fill and the ductile downstream material. These features are also illustrated in Figure 9. Although the embankment had not collapsed, it would continue to deteriorate and it was not performing as intended.

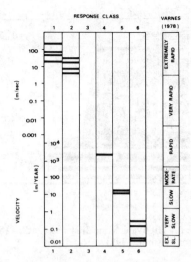

Figure 8. Classification for Landslide Movements (Morgenstern, 1985)

Figure 9. Deformable clay fill embankment (Morgenstern, 1991)

The Factor of Safety was inadequate to ensure the serviceability of the embankment.

What had been ignored in this simple case was the extreme ductility of the clay. Axial strains of possibly 30-40% are mobilized at the peak undrained strength of this material and hence the design Factor of Safety implied strains of 15-20%, far in excess of the levels of strain usually accommodated in conventional earthworks. There had been limited experience in utilizing such wet fill for the shoulders of a berm; it is usually wasted on most jobs. Hence insufficient attention was paid to the level of strain developed at the design Factory of Safety. In order to restrict the deformations to levels commonly accepted, a Factor of Safety of about 4 would be required.

An additional low level berm was constructed in the downstream direction; the cracks were repaired and the embankment subsequently behaved in an acceptable manner. This simple case illustrates the limitations of experience when relying on LEA alone to control deformations.

The empirical method also has some theoretical difficulties for cases where undrained loading can occur. It has been known for some time (Brinch-Hansen,1962) that effective and total stress stability analyses do not coincide except at failure when both theoretically would have a Factor of Safety of unity. For non-failure conditions, they diverge and their relative magnitude is a function of the pore pressures induced at failure under undrained conditions. The magnitude of this divergence has been emphasized by Ladd (1991) who advises that when evaluating stability of stage construction on soft clay, an effective stress analysis can provide a misleading sense of security for a configuration that might fail in an undrained manner. This underscores the need to have a clear understanding of the experience being correlated when the Factor of Safety is being used to avoid overall collapse and to control the deformation.

The example published by Eckersley (1990) is striking in this regard. This was a model study into coal pile collapse. A model 1 m high was constructed with a slope of 36°. The angle of shearing resistance was about 40° and therefore the Factor of Safety was approximately 1.15. The embankment was brought to failure by a slowly rising water table within the model, i.e., under fully drained conditions. Deep-seated failure and flow occurred at a conventionally computed Factor of Safety significantly in excess of unity. Presumably, at this condition sufficient strain was initiated to trigger positive pore pressures which led to the collapse of the slope. Factors of Safety higher than considered routinely are necessary to avert failure in contractive clays and granular deposits.

3.3 Small Deformation Studies

Small deformations of slopes, in the pre-failure range, can be calculated by means of numerical analysis employing finite element or finite difference procedures. The literature is extensive and abounds with examples of both ideal configurations and increasingly realistic cases.

Linearly elastic, non-linear elastic such as the hyperbolic formulation and elasto-plastic formulations such as variants of the Cam-Clay constitutive relation are used routinely in a number of computer codes. Calculations can be performed in terms of both total and effective stress.

Calculations of this kind are undertaken more to forecast deformations associated with the excavation of a slope than for deformations anticipated in a natural slope. It is well understood that for stress changes, and hence deformations, to be modelled accurately it is essential that the loading history of the soil mass be simulated. In principle, this is easier to do for an excavation where the ground starts at some pre-existing in-situ stress, K_0, and is then unloaded; than it is to evaluate the stresses themselves in some natural slope where the unloading process is not so well-defined. Few studies modelling the geomorphological evolution of natural slopes have been undertaken to assess how significant is this limitation. Considerable progress has been made in recent years in

measuring K_0 and understanding the influence that geological processes may have had on pre-existing stresses.

When the Factor of Safety is high, the results from numerical studies can be quite accurate, provided the soil or rock properties are known reliably. As the Factor of Safety decreases, the extent of yielding increases and the ability of many models to represent reality diminishes. Axelson, Yu and Johansson (1992) provide a typical example.

In assessing the ability of these numerical methods to model accurately the increase in deformations as failure is approached, it is important to distinguish between ductile and brittle behaviour. Ductile behaviour implies yielding with no loss of resistance. It may even be work-hardening. The capacity for current numerical methods to simulate accurately this class of behaviour exists to-day, although there may be difficulties in defining the elasto-plastic properties with sufficient accuracy, particularly for a natural soil. However, most natural soils and rocks are strain-weakening and this creates serious limitations for conventional modelling procedures.

As yielding develops in strain-weakening materials localization of the straining occurs with the development of shear bands. Many natural soils and rocks are strain-weakening and therefore this phenomenon is of widespread interest. The phenomenon of localization should be distinguished from progressive failure, although both are characteristic of strain-weakening materials. Localization creates a non-homogeneity that influences the incremental displacements and how they develop. There is little direct field observation of incipient localization although it is often observed in the laboratory. Hence, the data reported by Finno and Harahap (1991) are particularly valuable. Field strain contours behind an excavation in soft clay were calculated from observed horizontal and vertical displacements. Strain contours based on elasto-plastic finite element calculations were compared with the field measurements. There was a good match between computed and observed boundary displacements, but the analytically computed strains failed to capture the observed localization of strains and incipient shear band formation. This discrepancy would become more significant as boundary displacements increased to collapse.

The computation of localized yielding is easier if the location is defined geologically by a thin or otherwise specific zone, as compared to the development of localization during general yielding of the soil or rock mass. In the former, the direction of migration of the shear band is pre-defined, while in the latter, it must be calculated as part of the solution process. Examples of shear band propagation along pre-defined thin zones within slopes have been published (e.g., Chan and Morgenstern, 1987; Wiberg, Koponen and Runesson, 1990). The computation of stress-induced localization with general yielding of a strain-weakening soil remains a difficult problem currently considered at the research level (Wan, 1991).

3.4 Pre-Failure Deformations

There are several reasons to be interested in means of forecasting failure. In many construction and mining operations, large deformations are acceptable provided they can be managed. Management might imply simply avoidance at the time of collapse and therefore there is a need to predict when catastrophic movements will occur. When undertaking construction on moving ground, particularly when the movements are themselves induced by the construction operations, it is not only prudent, but often necessary, to define an allowable velocity in order to control the earthworks. In the case of large natural landslides, the movement signature itself may be the only effective means of assessing risk to public safety should factors like evacuation be under consideration.

Probably the most elaborate and on-going utilization of pre-failure deformations as a means of controlling slope hazards is the monitoring program adopted by Syncrude Canada Ltd, to control its dragline mining operation (Morgenstern, Fair, and McRoberts, 1988). In this operation, which utilizes four large draglines, the machines sit typically within 20 m of the crest of the highwall, about 50 m high. Mining is continuous and the consequences of loss of a dragline are extremely serious. The oil sand ore body displays several instability mechanisms of which two are of particular concern for the safety of the mining operation. The first is block sliding on steeply dipping clay seams. The analysis of such slides is straight forward provided the clay seams can be found, their properties determined and the pore pressures quantified. However, in the complex geology of the Alberta oil sands this is an impossible task and alternate ground management schemes have evolved. The block slide, when it forms, is the most dangerous mechanism since the deformations to failure are relatively small and the time to react is accordingly limited.

The second instability mechanism is associated with gas exsolution in the bitumen saturated dense sand. Gas bubbles form on unloading, which results in expansion and sometimes flow of the highwall. This leads to crest retrogression which can endanger the dragline. Although the deformations associated with this mechanism are larger than those associated with block slides, they happen more slowly and therefore this failure mode is more ductile.

Syncrude have developed comprehensive monitoring techniques to evaluate where hazards exist and if potentially critical block slides are found, continuous monitoring takes place between the dragline and the highwall. If a critical velocity along a clay seam is exceeded, as measured by a borehole inclinometer, mining is stopped, the dragline is moved out of any potential danger and the mining scheme for that location is re-evaluated. For the more ductile bulge process, visual monitoring is adequate to determine critical conditions.

In this way Syncrude have mined through in excess of 500 km of highwall without serious mishap. Confidence in this process does not arise from theoretical analysis. It can only be attained by developing a profound understanding of the specific failure processes involved and by a continual re-evaluation of experience.

The processes involved in a landslide immediately before failure are exceedingly complex and will vary with material, groundwater conditions, and topography. One can imagine some instances where pore pressures will increase with increasing deformation and others where they will decrease. One can imagine some instances where resistance may increase with increasing deformation due to rate-dependent effects and others where, it will decrease due to loosening and loss of three-dimensional support. Some sense of the time-dependent signature of a failing slope may be obtained from creep rupture material studies, but it is necessary to recognize the limitations of oversimplification in developing methods from this perspective. Nevertheless, such studies provide one of the few analytical frameworks for predicting the movements and time to failure.

Saito (1965) appears to have been the first to propose a method making such predictions and since that time there have been several efforts to build on this work. These contributions have been conveniently summarized by Voight (1989) and Fukuzono (1990).

Based on large scale experiments, Fukuzono and others found that the increment of the logarithm of acceleration was proportional to the logarithm of velocity of surface displacement immediately before catastrophic failure. That is,

$$\frac{d^2x}{dt^2} = A \left(\frac{dx}{dt}\right)^m \qquad (7)$$

where x is surface displacement, t is time, and A and m are constants. Applying this relation to available cases reveals that m is larger than 1 for about 80 percent of all measured landslide surface displacements in the past and m is 2 for about 50 percent of available measurements. Simple graphical procedures are available to determine the appropriate constants in a given case. Voight (1989) discusses extensions of these findings.

From their analysis of available data, Hayashi et al (1988) proposed that tertiary creep includes two stages. In the first stage, the relation between velocity and surface displacement is linear, while the increment of logarithm of velocity is proportional to surface displacement in the second stage. They find that the duration of the first stage, following secondary creep, is large compared with the second stage. As a result, they have proposed a new method to predict the time to failure at an early period of the first stage. Preliminary results are promising and the method merits further evaluation.

The value of time to failure forecasts is enhanced if allowable velocity criteria can be established. Fukuzona (op.cit). cites recommendations in this regard that are attributed to Salt (unpublished) and they constitute a useful starting point for further work.

A more comprehensive effort, classifying different soil and rock masses, in terms of their component stabilizing and destabilizing internal mechanisms during failure, is necessary before the limits of application of these various empirical studies can be defined.

3.5 *Slow Landslide Movements*

The mechanics of slow movements were reviewed in some detail by Morgenstern (1985) who observed that most well-recorded case histories occur by sliding within a relatively narrow zone or along a plane. Slides in overconsolidated clays, clay shales and weathered rocks contribute the largest number of examples and case histories are available from many parts of the world.

These cases reveal that slow movements can accumulate over geological time and are generally responsive to precipitation. No complete picture has yet emerged relating local climate, maximum rainfall intensity, antecedent rainfall, depth of sliding and velocity.

In addition to rate-dependence of residual strength of clay, it was noted that the velocity of sliding would also be influenced by the geometrical complexity of the landslide mass. Thickness, water pressure distribution, slope of slip surface, lateral restraint due to channelization and other factors vary from place to place in all but the simplest landslides. Hence not all locations within a slide would be mobilized simultaneously and the velocity would be moderated. Comprehensive field studies were noted as necessary to determine the controlling factors.

Since that time three have appeared that are particularly noteworthy and it is of interest to compare them. They are the observations and analysis of the unstable slope at Salledes (Cartier and Pouget, 1988), the analysis of movements of the Mam Tor Landslide (Skempton, Leadbeater, and Chandler, 1989) and the comprehensive investigations into the La Mure Landslides by Van Genuchten (1989) and Nieuwenhuis (1991). Each involves almost planar slides at residual strength so that the theoretical details of the analyses do not dominate. They are all comparatively shallow, and are in comparable temperate climatic zones.

The analysis of the Mam Tor slide reveals that it is close to equilibrium under normal winter groundwater conditions. At about four year intervals it is reactivated as a result of a rise in water level in response to winter rainstorms. This episodic re-activation converts into an average velocity of 70 mm per year which is to be compared with velocities of 40-150 mm/.year found in similar slides in Cretaceous clay shales in Canada (Morgenstern, 1985). The authors demonstrate that the intermittent displacements can be calculated by means of an analysis based on the transient rise of groundwater level, inferred from observed storm response ratios in clay soils, and laboratory measurements of the increase in residual strength with increasing rates of shear. The limited duration for the reduction of the Factor of Safety and the rate dependence of the residual strength combine to moderate the movements.

This interpretation is to be contrasted with that of the Salledes case. Here direct observations are made between rainfall and fluctuation of groundwater pressure over a number of years. A correlation is also observed between the magnitude of observed pore pressure and surface displacement. Typically a threshold rise is necessary before movements are initiated and, for this case, at a particular value, the velocities increase enormously, by an order of magnitude. There is no evidence of the influence of rate dependence of residual strength in this response, but perhaps it is obscured by the scatter in the data. Cartier and Pouget (1988) prefer to interpret their observations in terms of a relation between Factor of Safety (in excess of unity) and velocity. Velocities vary from about 1 mm/day to 15 mm/day for a computed variation in Factor of Safety of about 10%, see Figure 10. An additional rate-dependent component of the resistance is implied and it is not clear whether this data is consistent with the Mam Tor model.

Vulliet and Hutter (1988) have considered the theoretical basis for slow movements. Central to this study is a correlation between sliding velocity and shear stress level at the base of the slide. The relation implies a rate dependency in the Factor of Safety, but does not have an intercept for zero velocity at a threshold resistance which obscures its utility.

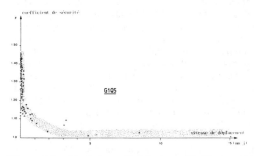

Figure 10. Variation of velocity with Factor of Safety (Cartier and Pouget, 1988)

Nieuwenhuis (1991) has subjected the observations of movements at the La Mure landslide to detailed scrutiny. This slide is on the brink of failure in conventional terms. Slow movements are triggered by rainfall and their analysis requires the following considerations:

i) displacement along existing slide planes

ii) residual shear strength modified by temporary strength regain and rate-dependent resistance

iii) convergent flow

iv) incidental stabilization by accumulation at the toe.

Not all will readily accept these views, but it is a substantial step forward to have them presented in such detail.

Progress in developing an analysis for slow movements can only emerge from comprehensive investigations of the type cited above and a detailed comparative study of these three cases is warranted.

3.6 Fast Landslide Movements

The analysis of fast landslide movements is required in hazard assessment to answer the question, "Given a potential sliding mass, how much will move, how fast will it go, and how far will it go?". This type of analysis is of value in land use zonation, and explaining geomorphological features. A summary of studies into the mechanics of fast movements was provided by Morgenstern (1985). Since that time, several contributions have appeared that enhance our ability to analyze the runout problem.

Hungr (1990) has synthesized empirical correlations concerning the mobility of rock avalanches. He concludes that i) rock avalanche profiles tend to have bi-linear forms consisting of a source/travel segment inclined at 25° to 40° and a near-horizontal deposition segment, ii) there is a rough proportionality between volume of a deposit and the area covered by it and, iii) correlations of the equivalent friction angle and the excessive travel distance with volume do not document a systematic increase in mobility; instead they appear to be a consequence of a fluid behaviour which prevails with volumes beyond a few millions of m³, but is otherwise unrelated to volume. In a critical summary of the mechanisms underlying mobility he concluded that the weight of evidence favours lubrication by the presence of liquefied material. Acoustic fluidization (Melosh, 1979) was also cited as plausible, but convincing experimental confirmation has not been forthcoming. The mobility of wet flows, such as debris flows and submarine landslides in loose sands and silts is much less enigmatic since liquefaction is clearly present.

From an analytical perspective, runout forecasts are influenced by empirical correlations and geological analogues. They can also be made by means of dynamic analyses. Two kinds of analysis are encountered:

i) Centre-of-mass models

ii) Fluid continuum models

Centre-of-mass models employ rigid body dynamics to compute the extent of movement of a landslide. Resistance is provided by friction and a velocity dependent term. Koerner (1976) extended theories of avalanche dynamics to landslide processes and showed how they could be modelled in this way. This approach was adopted by Hardy et al. (1978) in their assessment of landslide hazards in the Garibaldi region of British Columbia. McLellan and Kaiser (1984) provide another example of its use. The method provides no insight into the basis of the resistance terms. However it does enable one to history-match experience and deduce operational parameters from previous slides which can then be used in other settings.

Fluid continuum models based on transient analysis have been used to compute the flow of a fluid under gravity. Jeyapalan et al. (1983) used these concepts to explain the mobility of mine tailings and subsequently Sassa (1988), Hungr (1991) and Sousa and Voight (1991) have developed analyses of different degrees of complexity based on fluid mechanics considerations. In all cases it is still necessary to invoke a history matching technique to determine the soil/rock resistances.

In the latest example of such a method, Sousa and Voight (1991) employ a bi-linear viscous relation to model the rheological behaviour of the flowing material and impressive results are obtained in matching the runout characteristics of several large avalanches. The frictional resistance is embodied in the strength parameter deduced from their analysis and, as with all back-analyses, remarkably low values must be adopted to achieve consistence with observations.

While considerable progress is being made in dealing with the kinematics of fast landslide movements, the acceptance of these analyses and their use in practice is hindered by our limited understanding of the processes involved in reducing the frictional resistance to such low values. The questions are clear: is liquefaction the only process involved?; how do the pore pressures get so high and stay high?; is heat induced pore pressure rise a dominant mechanism?; if other processes are at work, how do we simulate them in the laboratory?

4 CONCLUSIONS

It is useful to distinguish between analysis used to evaluate the ultimate failure of slopes from analysis used to assess the deformation of slopes.

Limit equlibrium analyses (LEA) are commonly adopted to assess ultimate failure in terms of the Factor of Safety. From an analytical perspective, the evolution of two-dimensional theories has reached maturity and powerful, economic computer programs are widely available. In recent years, progress has been made in the theoretical development of three-dimensional methods of analysis. Practical results have been obtained but they are accurate primarily for symmetrical configurations. Extension to asymmetry can be made with more complex assumptions regarding internal force distributions and it is anticipated that progress in extending the capability for three-dimensional analysis will be swift.

Although LEA is a standard tool in geotechnical practice, it has several important limitations. Setting aside the problem of in-situ characterization and laboratory studies, there are additional considerations that affect the selection of the operational properties that enter into LEA. These have been grouped into: 1) time-dependent processes and 2) deformation-dependent processes. The most important processes that affect the operational pore pressure and shear strength have been discussed and illustrated by means of examples.

In most instances of slope stability evaluation, deformations are not calculated explicitly. If the Factor of Safety is chosen correctly, it is assumed that the deformations will be acceptable and that the slope will perform as intended. There are pitfalls to this empirical procedure and illustrations have been provided.

Progress has also been made to calculate small deformations associated with excavations, to forecast the time to failure of slopes, to account for sustained slow landslide movements and to model fast landslide movements. Key case histories have been identified and dominant outstanding issues have been noted.

REFERENCES

Axelson, K., Yu, Y. and Johansson, L., 1992. Finite element analysis of an excavated slope. Proc. 6th Int. Symp. Landslides, Christchurch, in press.

Azzouz, A.S. and Baligh, M.M., 1978. Three-dimensional stability of slopes. Publication R78-8, Massachusetts Institute of Technology.

Bishop, A.W. and Bjerrum, L., 1960. The relevance of the triaxial test to the solution of stability problems. Proc. Res. Conf. Shear Strength of Cohesive Soils, ASCE, Boulder, p. 437-501.

Brand, E.W., Premchitt, J. and Phillipson, H.B., 1984. Relationship between rainfall and landslides in Hong Kong. Proc. 4th Int. Symp. Landslides, Toronto, Vol. 1, p. 377-384.

Brinch-Hansen, J., 1962. Relationship between stability analyses with total and effective stress. Sols-Soils, Vol. 3, p. 28-41.

Bromhead, E.N. and Dixon, N., 1984. Pore water pressure observations in the coastal cliffs at the Isle of Sheppey, England. Proc. 4th Int. Symp. Landslides, Toronto, Vol. 1, p. 385-390.

Cartier, G. and Pouget, P., 1988. Etude du comportement d'un remblai construit sur un versant instable, le remblai de Salledes. (Puy-de-Dome). Rapport de recherche LPC No. 153, Laboratoire Central des Ponts et Chaussées, Paris.

Chan, D.H. and Morgenstern, N.R., 1987. Analysis of progressive deformation of the Edmonton Convention Centre excavation. Canadian Geotechnical Journal, Vol. 24, p. 430-440.

Chen, R.H. and Chameau, J.L., 1983. Three dimensional limit equilibrium of slopes. Geotechnique, Vol. 33, p. 31-40.

Chen, Z.-Y., and Shao, C.M., 1988. Evaluation of minimum factor of safety in slope stability analysis. Canadian Geotechnical Journal, Vol. 25, p. 735-748.

Dennhardt, M. and Forster, W., 1985. Problems of three-dimensional slope stability. Proc. 11th Int. Conf. Soil Mechs. Found. Eng., San Francisco, Vol. 2, p. 427-431.

Eckersley, D., 1990. Instrumented laboratory flowslides. Geotechnique, Vol. 40, p. 489-502.

Eigenbrod, K.D., 1975. Analysis of pore pressure changes following the excavation of a slope. Canadian Geotechnical Journal, Vol. 12, p. 429-440.

Fenelli, G.B. and Picarelli, L., 1990. The pore pressure field built up in a rapidly eroded soil mass. Canadian Geotechnical Journal, Vol. 27, p. 387-392.

Finne, R.J. and Harahap. I.S., 1991. Finite element analyses of HDR-4 excavation. J. Geotechnical Eng., ASCE, Vol. 117, p. 1590-1609.

Fredlund, D.G., 1984. Analytical methods for slope stability analysis. Proc. 4th Int. Symp. on Landslide, Toronto, Vol. 1, p. 229-250.

Fukuzono, T., 1990. Recent studies on time prediction of slope failure. Landslide News, No. 4, p. 9-12.

Gens, A., Hutchinson, J.N. and Cavounidis, S., 1988. Three dimensional analysis of slides in cohesive soils. Geotechnique, Vol. 38, p. 1-23.

Hardy, R.M., Morgenstern, N.R. and Patton, F.D., 1978. Report of the Garibaldi Advisory Panel. British Columbia Department of Highways, Victoria, B.C. (three volumes)

Hayashi, S. et al., 1988. On the forecast of time to failure of slope, II. Approximate forecast in early period of the tertiary creep. J. of Japanese Landslide Society, Vol. 25, p. 11-16.

Hoek, E. and Bray, J., 1977. Rock slope engineering. Revised 2nd ed. Inst. Mining and Metallurgy, London.

Hovland, H.J., 1977. Three-dimensional slope stability analysis method. J. Geotechnical Engineering Division, ASCE, Vol. 103, p. 971-986.

Hungr, O., 1981. Dynamics of rock avalanches and other types of slope movements. Ph.D. Thesis, University of Alberta.

Hungr, O., 1987. An extension of Bishop's simplified method of slope stability analysis to three dimensions. Geotechnique, Vol. 37, p. 113-117.

Hungr., O., 1990. Mobility of rock avalanches. Report of National Research Institute for Earth Science and Disaster Prevention, No. 46, p. 11-20, Japan.

Hungr, O., 1991. Personal communication.

Hungr, O., Salgado, F.M. and Byrne, P.M., 1989. Evaluation of a three-dimensional method of slope stability analysis. Canadian Geotechnical Journal, Vol. 26, p. 679-686.

Hutchinson, J.N., 1988. General Report: Morphological and geotechnical parameters of landslides in relation to geology and hydrogeology. Proc. 5th Int. Symp. Landslides, Lausanne, Vol. 1, p. 3-35.

Hutchinson, J.N. and Sarma, S.K., 1985. Discussion on "Three dimensional limit equilibrium analysis of slopes". Geotechnique, Vol. 35, p. 215-216.

Jeyapalan, J.K., Duncan, J.M. and Seed, H.B., 1983. Analysis of flow failures of mine tailings dams. J. Geotechnical Eng., ASCE, Vol. 109, p. 150-171.

Koerner, H.J., 1976. Reichweite und Geschwindigkeit von Bergstuerzen und Fliesschneelawinen. Rock Mechanics, Vol. 18, p. 225-256.

Ladd, C.C., 1991. Stability evaluation during staged construction. J. Geotechnical Eng., ASCE, Vol. 117, p. 540-615.

Lam, L.W., 1991. Generalized three-dimensional slope stability analysis using method of columns. Ph.D. Thesis, University of Saskatchewan.

Lefebvre, G., 1981. Strength and slope stability in Canadian soft clay deposits. Canadian Geotechnical Journal, Vol. 18, p. 420-442.

Leshchinski, D., Baker, R. and Silver, M.L., 1985. Three-dimensional analysis of slope stability. Int. Journal for Numerical and Analytical Methods in Geomechanics, Vol. 9, p. 199-223.

McLellan, P.J. and Kaiser, P.K., 1984. Application of a two-parameter model to rock avalanches of the Mackenzie Mountains. Proc. 4th Int. Symp. Landslides, Toronto, Vol. 1, p. 559-565.

Melosh, H.J., 1979. Acoustic fluidization: a new geologic process? J. Geophysical Research, Vol. 84, P. 7513-7520.

Morgenstern, N.R., 1985. Geotechnical aspects of environmental control. Proc. 11th Int. Conf. Soil mechs. and Found. Eng., San Francisco, Vol. 1, p. 155-185.

Morgenstern, N.R., 1990. Instability mechanisms in stiff soils and weak rocks. Proc. 10th Southeast Asian Geotechnical Conference, Taipei, Vol. 2, p. 27-36.

Morgenstern, N.R., 1991. Limitations of stability analysis in geotechnical practice. Geotecnia, No. 61, March, p. 5-19.

Morgenstern, N.R., Fair, A.E. and McRoberts, E.C., 1988. Geotechnical engineering beyond soil mechanics - a case study. Canadian Geotechnical Journal, Vol. 24, p. 637-661.

Morgenstern, N.R. and Price, V.E., 1965. An analysis of the stability of general slip surfaces. Geotechnique, Vol. 15, p. 79-93.

Morgenstern, N.R. and Sangrey, D.A., 1978. Methods of stability analysis. In Landslides: Analysis and Control, ed. by R.L. Schuster and R.J. Krizek, Special Report No. 176, Transportation Research Board, National Academy of Sciences, Washington, D.C., p. 155-171.

Nash, D.F.T., 1987. A comparative review of limit equilibrium methods of stability analysis. Chapter 2 in: Slope Stability ed. by M.G. Anderson and K.S. Richards, Wiley & Sons, New York.

Nieuwenhuis, J.D., 1991. The lifetime of a landslide. A.A. Balkema, Rotterdam.

Potts, D.M., Dounias, G.T. and Vaughan, P.R., 1990. Finite element analysis of Carsington embankment. Geotechnique, Vol. 40, p. 79-102.

Radbruch-Hall, D.H., 1978. Gravitational creep of rock masses on slopes. in Rockslides and Avalanches, Vol. 1, ed. by B. Voight, Elsevier, Amsterdam, p. 604-657.

Saito, M., 1965. Forecasting the time of occurrence of slope failure. Proc. 6th Int. Conf. Soil Mechs. Found. Eng., Montreal, Vol. 2, p. 537-541.

Seed, R.B., Mitchell, J.K. and Seed, H.B., 1990. Kettleman Hills waste landfill slope failure. II Stability analyses. J. Geotechnical Eng., ASCE, Vol. 116, p. 669-690.

Sevaldson, R.A., 1956. The slide in Lodalen, October 6th, 1954. Geotechnique, Vol. 6, p. 1-16.

Skempton, A.W., Leadbeater, A.B. and Chandler, R.J., 1989. The Mam Tor Landslide, North Derbyshire. Phil. Trans. Royal Society of London, Vol. 329, p. 503-547.

Small, C.A. and Morgenstern, N.R., 1992. Performance of a highwall in soft rock, Highvale Mine, Alberta. Canadian Geotechnical Journal, Vol. 29, in press.

Sousa, J. and Voight, B., 1991. Continuum simulation of flow failures. Geotechnique, Vol. 41, p. 515-538.

Spencer, E., 1967. A method of analysis of the stability of embankments assuming parallel inter-slice forces. Geotechnique, Vol. 17, p. 11-26.

Sridevi, B. and Deep, K., 1992. Application of global optimisation technique to slope stability analysis. Proc. 6th Int. Symp. Landslide, Christchurch, in press.

Taylor, D.W., 1937. Stability of earth slopes. Journal Boston Society of Civil Engineers, Vol. 24, p. 197-246.

Terzaghi, K., 1936. Stability of slopes in natural clays. Proc. 1st Int. Conf. Soil Mechs. Found. Eng., Vol. 1, p. 161-165.

Van Genuchten, P.M.B., 1989. Movement mechanisms and slide velocity variations of landslides in varved clays in the French Alps. Thesis, Utrecht University.

Varnes, D.J., 1978. Slope movement types and processes, in "Landslides: Analysis and Control" ed. by R.L. Schuster and R.J. Krizek, Transportation Research Board, National Academy of Sciences, Washington, D.C., p. 11-33.

Vaughan, P.R. and Walbancke, H.J., 1973. Pore pressure changes and the delayed failure of cut slopes in overconsolidated clays. Geotechnique, Vol. 23, p. 531-539.

Voight, B., 1989. Material science law applies to time forecasts of slope failure. Landslide News, No. 3, p. 8-11.

Vulliet, L. and Hutter, K., 1988. Continuum model for natural slopes in slow movement. Geotechnique, Vol. 38, p. 199-218.

Wan, R., 1990. The numerical modelling of shear bands in geological materials. Ph.D. Thesis, University of Alberta.

Wiberg, N.E., Koponen, M. and Runesson, K., 1990. Finite element analysis of progressive failure in long slopes. Int. Jnl. for Numerical and Analytical Methods in Geomechanics. Vol. 14, p. 599-612.

Yoshida, N., Morgenstern, N.R. and Chan, D., 1990. A failure criterion for stiff soils and rocks exhibiting softening. Canadian Geotechnical Journal, Vol. 27, p. 195-202.

Yoshida, N., Morgenstern, N.R. and Chan, D., 1991. Finite element analysis of softening effects in fissured, over-consolidated clays and mudstones. Canadian Geotechnical Journal, Vol. 28, p. 51-61.

Landslides, Bell (ed.) © 1995 Balkema, Rotterdam, ISBN 90 5410 032 X

Theme report

G. R. Mostyn

University of New South Wales, Kensington, N.S.W., Australia

ABSTRACT: This report provides a brief review of the papers grouped under the General theme - Stability analysis techniques. It therefore provides some indication of the directions that are currently being actively persued in slope analysis around the world. The methods covered by the papers are extremely broad and include: physical modelling; reliability analysis; numerous case studies; back analysis; and unsaturated slopes.

1 INTRODUCTION

This report presents a brief overview of the fifty six papers that have been grouped into the general theme of stability analysis techniques. This is a large number of papers covering an extremely broad area of research and professional activity; thus the reporter cannot hope to referee all the papers particularly in view of the relatively short time (4 weeks) available to prepare the report. The paper will necessarily reflect the reporter's views and it should be noted that the reporter is more knowledgeable in some areas than in others. The reporter offers his apologies in advance to any authors: (i) that he has not been able to understand (e.g. papers in French or, even, English that he was unable to comprehend); and (ii) for any mistakes that he has made in reporting the various papers.

In view of the number of papers that are being reviewed the reporter has adopted a numbering system in referencing them and the number is shown in [ ] in Section 4 wherein each paper is discussed briefly.

2. GEOGRAPHY

The geographical origin of the papers can be used to provide a clue as to those areas of the world where there is active research on stability analysis techniques. Table 1 provides a summary of the address of the senior author of each paper.

Table 1 Origins of papers

Location	No	Papers
Western Europe		
Italy	10	1 6 7 9 10 15 17 21 27 47
Other	8	4 12 20 23 24 40 46 51
Eastern Europe	5	3 26 33 41 49
India	2	28 48
South East Asia	3	8 38 42
China		
Chendu College	7	14 19 36 45 50 53 57
Other PRC	3	22 29 37
Taiwan	2	13 52
Japan	9	2 5 31 32 43 44 54 55 56
North America	3	16 18 30
Australia & New Zealand	4	25 34 35 39

Table 1 indicates that there is probably considerable research in stability analysis techniques being conducted in Italy, China and Japan. This is not surprising as these countries all have many slope stability problems. What is surprising is that not a single paper was received from South America which equally has large slope stability problems.

3. AREAS OF INTEREST

Table 2 provides an index to the major areas

that are the subject of each paper and therefore allows the reader to select those papers that may be of particular interest.

Table 2 Areas of interest

Interest	Papers
Methods of analysis	10 13 14 20 24 26 29 30 35 48 49 51 52 54 56
Case studies	1 2 4 8 9 10 12 13 14 16 17 18 19 20 21 22 24 25 29 31 32 37 39 40 44 49 50 51 57
Landslides	1 7 10 16 17 18 19 21 22 26 44 45 57
Rock slopes	1 6 7 9 10 14 15 21 29 32 36 38 50 57
Back analysis	2 8 9 12 13 25 27 29 37 45 49 51 54 56
Strenth of materials	2 8 9 12 22 28 37 40 44 54
Time effects, velocity	1 9 10 12 19 20 33 36 40 43 45 53
Progressive failure	3 34 47
Reliability	7 8 15 34 38 49
Physical models	5 6 36 50 53 57
Run out	5 6 44
Stabilization	12 13 24 25 55
Critical surface	8 27 48 52
Janbu method	24 27 48 52 54 55 56
Finite elements	4 32 45 57
Unsaturated soils	4 28 42 43
Groundwater	4 9 17 18 20 30 39 31
Other (see text)	3 6 7 10 11 12 16 18 20 24 28 31 33 39 41 43 51

As with any classification system the allocation of particular areas of interest to a certain paper is somewhat arbitrary but the table does provide a reasonable guide to the content of the papers in this session.

4. THE PAPERS

Following are a few comments on each paper.

Angeli, M.G., Menotti, R.M., Pasuto, A. and Silvano, S. [1] Landslide studies in the Eastern Dolomites Mountains (Italy)

The paper presents general case studies on large (> 1 million m^3) landslides and their relationship to climatic factors. Much development in the Dolomites is on landslide accumulations. Many of the landslides appear to be initiated when pore pressures build up due to freezing of exits.

Aramaki, S., Kitazono, Y., Suzuki A. and Kajiwara, M. [2] Application of direct shear test to analysis of a Tertiary landside in Japan

The authors discuss the choice of shear strength parameters for use in landslide analysis, including many test results. The reporter had difficulty following some of the development. The Swedish method was used in the back analysis of the landslides and the reporter feels this would introduce some bias into the resulting shear strengths. The authors conclude that the softened strength should be adopted in analysis.

Asilbekov, D.A. and Ismagulov, I.K. [3] The evaluation of stability of high slopes based on the determining of the initial failure zone

The reporter found parts of this paper quite difficult to follow and this may be due to the translation. The paper postulates that generally slope failures can only occur if the toe fails and therefore a local factor of safety at the toe may be used as an index of stability. Further the authors claim that the "usual methods" of analysis give too high factors of safety for back analysis of failing slopes; the reporter's experience does not support this contention and thus would suspect the "usual methods" of analysis. Indeed in the 30 slopes discussed, those from "other areas" do not have as high factors of safety. Notwithstanding this the proposed method does give factors of safety closer to unity.

Axelsson, K., Yu, Y. and Johansson, L. [4] Finite element analysis of an excavated slope

The paper provides a comparison of normal total stress finite element analysis and coupled pore pressure and strain effective stress analysis of a trial cut slope in slightly overconsolidated clays. As might be expected the coupled analysis shows superior results. This paper is easy to follow.

Aydan, O., Shimizu, Y. and Kawamoto, T. [5] The reach of slope failures

This paper uses physical block models to investigate the run out of rock like slopes. In their discussion of the various numerical methods of analysing this problem the authors may be a little to dismissive of distinct element methods. An interesting conclusion is shown in Figure 9 in that the centres of mass generally move further than the conventional reach angle, ϕ_s. This work is ongoing and will improve our knowledge of the boundaries between toppling and sliding.

Azzoni, A., Drigo, E., Giani, G.P., Rossi, P.P. and Zaninetti, A. [6] In situ observation of rockfall analysis parameters

The reporter found this a most interesting paper and, in a way, the work reported represents a dream for many youths. The work involved pushing blocks of rock from 0.3 to 3 m^3 off cliffs and videoing their travel. The slopes were up to 1700m long. The records were then analysed and results detemined for coefficients of restitution and rolling friction. This information will be very valuable for those wanting to predict energy dissipation of block falls. The reporter found the presentation of some of the results in Figures 1 and 2 somewhat confusing because there were too many falls included in each figure.

Barisone, G., Bottino, G. and Mandrone G. [7] Correlation among geomechanical characteristics, fracturation degree and the stability of rock slopes

The reporter was somewhat confused as to the purpose of the work reported in this paper or more specifically as to how the results will be useful. As an example the equation in Section 5 of the paper appears to be to predict I_s given J_v. Is this really the intention as the latter is generally the more difficult to measure?

Bergado, D.T., Alfaro, M.C., Patron, B.C. and Chirapuntu, S. [8] Reliability-based analysis of embankment failure on soft ground

This paper develops a probabilistic model for the reliability of a long low road embankment. The method differs from most others in that it includes the effects of spatial correlation on the reliability (see [34]). Indeed a large field investigation was carried to characterise the spatial correlation of the soil strength, these data must rank as some of the best available. The reporter hopes that the data base is available for wider use. The reporter would be wary of adopting the field vane correction factors used in the paper. The reliability analysis is interesting in that the unfailed section had lower mean strength than the failed section but a higher probability of failure. It is also interesting that as the probability of failure is over 50% it is likely that this is one of the rare instances where taking autocorrelation into account will increase the probability of failure. This paper is recommended reading.

Bertini, T., Cugusti, F., D'Elia, B., Lanzo, G. and Rossi-Doria, M. [9] Slow movement investigations in clay slopes

An interesting paper about a medium size failure though softened claystone with good field monitoring. It is interesting the anisotropy and other groundwater flow phenomena reduce the pore pressures to well below those expected in a long slope if flow were parallel to the slope, this provides a good illustration of one of the main points in [30]. Good clear case study.

Borsetto, M., Frassoni, A., La Barbera, G., Fanelli, M., Giuseppeti, G. and Mazza, G. [10] An application of Voight empirical model for the predictions using rate effects on residual shear strength

This paper describes instrumentation installed subsequent to a 35 million m^3 landslide that destroyed several villages. Then an empirical model is applied to predict the time to failure of slopes from monitoring data. The model is demonstrated with synthetic data. It appears that the propsed method is able to predict the time of failure from early monitoring, the prediction method becomes better as the time of the completion of monitoring draws nearer the time of failure. The reporter is concerned that the method relies on invariant external actions (e.g. pore pressures) the variation of which is normally the actual trigger for failures. Nevertheless the method does appear to hold promise.

Bracegirdle, A., Vaughan, P.R. and Hight, D.W. [12] Displacement predictions using rate effects

on residual shear strength

This is an excellent description of the history of a failing site for over one hundred years. The paper sets out a method of analysing the contribution that piles make in stabilizing the slope and on the effect of periodic low tides on the failure rate. The paper reports work indicating that the strength of clay increases about 1-2% per logarithm increase in strain rate which is very similar to that in [40]. The development of the analysis is easy to follow.

Chen, H. [13] Appropriate model for hazard analysis in slope engineering

A case study of an unstable slope in Taiwan is presented. The reporter is interested in the apparent accuracy quoted for the angle of friction for some of the slope materials (i.e. 28.2°).

Chen, M., Wang, L.S. and Li, T. [14] On the deformation starting criterion of sliding-bending model in dip slope

This is an interesting paper that, unfortunately, in translation is difficult to understand. An expression is derived for the buckling of a dip slope in bedded strata and then the results of back analysis of 19 failures on the Yangtze River are presented. It is interesting that this buckling is a problem if the calculated factor of safety is less than 3!

Cherubini, C., Giasi, C.I. and Cucchiararo, L. [15] Probabilistic analysis of slope stability in rocks

This paper presents a comparison of methods for probabilistic analysis of rock slopes but the basic model adopted is what Li [34] calls the single random variable approach, even though c, ϕ and the anchor force are random variables. The reporter also is interested in the models where coefficients of variation are selected apparently with no real support (e.g. as 10 to 30%). Correlation between the parameters is included in one of the analyses investigated. The reporter is not aware of anything new in this paper.

Cornforth, D.H. and Vessely, D.A. [16] Factors of safety during landslide movements

This paper presents the analysis into a slope failure and its stabilisation. A slope above a park failed in 1975 and has moved a total of over 5.2m at up to 90mm/wk but this rate was increased to 245mm/wk in 1990 due to reexcavation. The logic in locating the failure plane is detailed. The reporter is interested in the material that apparently is 11% clay with ϕ_r of 8.5°. The reporter has a number of questions regarding his paper. The monitored average ground water levels appear to have been adopted in the analysis, but did the slope move mainly as a result of perods of high ground water level? In fact the assumption of relatively low GWLs implies a low ϕ from any back analysis, assuming a higher GWL would change the resulting ϕ and also the most appropriate remedial works. Was this considered? The three dimensional analysis is not detailed.

Deganutti, A.M. and Gasparetto, P. [17] Some aspects of a mudslide in Cortina (Italy)

This provides some interesting information about a 1700m long mudslide in Italy including the relationship between the pore pressure and rate of movement (although this appears to be somewhat arbitrarily derived from a small section of the total record). The reporter is interested in the fact that the movement monitors need to be multiplied by a factor of 3.7 to predict the actual movement, why is this?

Diyaljee, V.A. [18] Seepage induced roadway landslide

This presents a case study of a relatively small landslide in South Trinidad. The reporter is not convinced that liquefaction needs to be invoked to explain the failure. In fact very little actually appears to be known with any confidence about this failure.

Dong, X. and Wang, L.S. [19] Some kinematic features of landslide

This paper presents a sensible classification scheme for the velocity of a landslide as it is the velocity that has a large control on the appropriate action to be taken. A formula is then developed for the velocity of a landslide based on geometric parameters. This is then checked against some field data. The reporter wonders why some of the slides ploted on Figure 5 have factors of safety greater than one.

Faure, R.M., Seve, G., Farhat, H., Virollet, M. and Delmas, Ph. [20] A new methodology for evaluation of landslide displacements

This paper looks at a landslide and develops a model for displacement versus time due to a viscous soil, this is very similar to [31]. The method appears to predict displacement versus time after some data fitting has been completed.

Favaretti, M., Previatello, P. and Soranzo, M. [21] Stability analysis of landslides occurred close to a marl and limestone quarry

The authors present a case study of a failure through a relatively thick layer of colluvial soils and establishing if there was any relationship between the failure and nearby quarrying operations.

Feng, L-C. [22] Modification and selection of strength parameters in calculating stabilities of loessal landslide

Unfortunately this paper is difficult to follow in translation. It appears to deal with the many values that c and ϕ may take on for a single soil depending on the failure conditions. Based on their experience the authors suggest the the laboratory values of c and ϕ should be reduced by 7 to 10%.

Gervreau, E., Durville, J.L. and Seve, G. [23] Relations entre precipitations et cinematiqu des glissements de terrain, deduites du suivi de sites

This paper is in French and so the reporter cannot comment on its contents.

Ghosh, N. and Ferguson, P.A.S. [24] Three-dimensional stability analysis of a cone-shaped tip of colliery spoil

This presents a case study and analysis of a colliery spoil tip failure. The authors present a three dimensional analysis which appears an extension of the Swedish method but involving only force equilibrium; they also used Janbu's method (not stated whether "rigorous" or simplified) for the two dimensional analysis. In contrast to most other researchers the authors found that their three dimensional factors of safety were less than those obtained from their two dimensional analysis. The reporter suggests that this may be due to the simplifying assumptions in the three dimensional analysis.

Gillon, M.D., Foster, P.F., Jennings, D.N. and Graham, C.J. [25] Stability analysis applications: Cromwell Gorge landslides

A very general paper about stability analysis of some New Zealand landslides. The reporter considers that a detailed report on a single failure would have been been of more value.

Ginzburg, L.K. [26] Calculation of landslide pressure on the most dangerous sliding surface

This paper presents a method of determing the landslide pressure (i.e. horizontal force on vertical slices) for a landslide. The exact derivation of the method is confused because the authors appear to have assumed zero shear forces on the sides of the slices and E_i is a resultant as shown on Figure 2. This assumption is not stated but given that it has been made there is not much new in the rest of the paper.

Greco, V. [27] Back-analysis procedure for failed slope

The authors present a method of back analysis for homogeneous soil slopes with known pore pressures. The reporter feels that many of the limitations that the authors ascribe to laboratory determination of shear strength are also limitations on back analysis, but also in the laboratory the effective stresses are much better known than in almost all back analyses. The authors state correctly the fact that back analysis must involve not only setting the factor of safety to unity but also ensuring that it is a minimum. The method proposed involves minimizing the area between the actual (often not well known) and critical slip surfaces. The simplified Janbu method is adopted in the analysis, even Janbu himself suggested that this method should only be used for hand calculations, which surely now means it is obsolete. Notwithstanding this the authors do point out an interesting variation in the strength derived, as the water table is lowered the angle of friction also reduces but the cohesion can rise or fall.

Haroon, M. and Shah, S.S. [28] Anisotropy of compacted clay for stability analysis

The authors present some undrained strength test results on anisotropy of a soil compacted in a laboratory. The reporter would have liked to see some discussion on: (i) the relevance of laboratory compacted samples to field conditions, especially as the actual compaction method is likely to greatly influence the anisotropy; (ii) the variability of the results as no experimental data is given only envelopes; and (iii) the relevance of testing at confining pressures of 100 and 200 kPa when undrained strength envelopes on partially saturated soils are generally quite curved and normal stresses on most embankment failures are less than this.

He, D.H. [29]   Back evaluation of shear strength along bilinear slip surfaces

The paper presents a method of back analysing a two wedge failure mode utilising only force equilibrium. Stability charts are presented but unfortunately appear to be only for the case of zero pore pressures which is not common in rock mechanics. Two case studies are presented.

Iverson, R.M. [30]   Sensitivity of stability analyses to groundwater data

The reporter admires the analysis completed by the authors to demonstrate the effects of various groundwater conditions on stability analyses but considers that those who can follow the paper are likely to already know the lessions. Those who need the knowledge may have difficulty in following the presentation here and the effects of anisotropy and inclined equipotentials, etc, can generally be explained in simpler terms. Nevertheless it is important for all professionals involved in slope analysis to understand that simple ground water models can have significant errors (generally overestimation) in estimating pore pressures and these can result in large errors (not always conservative) in stability analysis.

Kawabe, H. [31]   On the influence of pore water pressure on land deformation of a landslide

This paper presents some good data on deformations in an artificial landslide which is really a full scale model of 30m by 20m in plan. They develop a model of a Bingham fluid to account for the strain rate versus pore pressure. The reporter found this an interesting paper.

Kawamoto, T., Aydan, O., Shimizu, Y. and Kiyama, H. [32]   An investigation into the failure of a natural rock slope

The authors have used a variety of methods to back analyse a rock slope failure that had killed fifteen people in 1989. This is an interesting case study and compares LEM, FEM, DEM and physical models. The reporter would like to know what water pressures were adopted in the FEM and DEM?

Kazarnovsky, V.D. and Pavlova, L.N. [33] Consideration of the creep of clay soils of technogenic structure in analysing stabilities of slopes of high embankments

The paper presents charts showing embankment slopes that may suffer from creep problems. The reporter found the translation difficult to follow and unfortunately the references are hard to find.

Li, K.S. [34]   Some common mistakes in probabilistic analysis of slopes

A paper that should be read by those new to probabilistic methods of slope analysis as it outlines some serious limitations, not always stated, of many of the commonly used methods of analysis.

Li, K.S. [35]   A unified solution scheme for slope stability analysis

As with the previous paper a clear exposition of a unified solution scheme for slope stability analysis including an outline of how it can be used to complete many earlier forms of analysis, generally modifying these so they become rigorous. This is a refinement of work similar to that which others have completed over the last fifteen years.

Li, T.B., Xu, J. and Wang, L.S. [36] Ways and methods for the physical simulation of landslides

A very general paper on physical modelling of landslides with some brief case studies.

Liu, G. and Liao, X. [37]   Selection of shear strength parameters of soil and rock in the slip-

zone of landslides

The authors discuss the various ways - testing, experience and back analysis - by which the shear strength of landslides can be determined. Some details of back analysis are provided. The reporter was interested in the back analysis results quoting the angle of friction to the nearest second!

Low, B.K. and Einstein, H.H. [38] Simplified reliability analysis for wedge mechanisms in rock slopes

The paper presents an analysis of a tetrahedral wedge to determine the Hasofer-Lind reliability index. The early part of the paper is a restatement of wedge stability and the reporter found some of the key figures confusing. The method adopted is a single random variable method in terms of [34] and appears to present work completed over a decade ago.

MacFarlane, D.F., Pattle, A.D. and Salt, G. [39] Nature and identification of Cromwell Gorge landslide groundwater systems

A general presentation of the methods used to identify the various groundwater sytems in some New Zealand landslides. The reporter found the formalisation of drilling fluid records as DVD (depth versus depth) plots interesting.

Nieuwenhuis, J.D. [40] Shear resistance along intermediate-depth slide planes in overconsolidated clayey glacial sediments

This is an interesting paper on the effect of the strain rate on the shear strength of varved clay landslides. An interesting method of cutting a 100mm diameter sample to suit a 150mm diameter ring shear device is shown. The results show a small increase (here equivalent to 2% per logarithm) in strength with strain rate, this is similar to the results obtained in [12]. The results also indicate a curved strength envelope.

Pula, O. and Pula, W. [41] On the stability analysis of anchored retaining walls by use of variational method

The paper presents a variational method for determining the stability of anchored walls, but the method appears to not work for the general example given. The reporter feels the method is likely to be of very limited use.

Rahardjo, H., Fredlund, D.G. and Vanapalli, S.K. Use of linear and non-linear shear strength versus matric suction relations in slope stability analyses

The paper presents a method of overcoming some of the limitations in the use of ϕ^b (i.e. angle indicating the rate of change in shear strength relative to changes in matric suction) by using two values of it. This approach appears valid for limited ranges of suction. More work is required in this area.

Sammori, T. and Tsuboyama, Y. [43] Parametric study on slope stability with numerical simulation in consideration of seepage process

The paper looks at the effects of seepage, including rainfall and partial saturation, on stability. It adopts a fairly simple model for the pore pressures in the partially saturated zone. The size of Table 1 indicates how many parameters are required for the analysis of a very simple material and the reporter wonders how often these could be determined in practice.

Sassa, K., Fukuoka, H., Lee, J.H. and Zhang, D.X. [44] Measurement of the apparent friction angle during rapid loading by the high-speed high-stress ring shear apparatus

An interesting paper that presents the results from a specially constructed high speed ring shear device and uses them to derive a relationship between apparent friction angle and landslide volume. The work in [5] is also of interest in this regard.

Shang, Y.Q., Li, Y.G., Li, T.B. and Shu, J.C. [45] Analysis on deformation mechanism in deep zone of the Huanglashi Landslide in the Three Gorges of the Yangtze River

The reporter found this paper difficult to understand but it appears to presents the results of a dynamic FEM program to model the strain rates in a one kilometre long landslide.

Soubra, A.H., Regenas, P. and Kastner, R. [46] Calcul de la butee en presence par une approche variationnelle

This paper is in French so the reporter cannot comment further on it.

Spilotro, G., Fidelibus, C. and Lenti, V. [47] A model for evaluation of progressive failure in earth slopes

An interesting paper presenting a very general simplified procedure for analysing progressive failure. The method appears to be the progressive failure analogue of an infinite slope.

Sridevi, B. and Deep, K. [48] Application of Global Optimisation Technique to slope stability analysis

The paper presents an optimization technique to locate the critical non circular failure surface. The method appears to have merit but evaluates the factor of safety by the simplified Janbu method and the reporter wonders about using a relatively inaccurate method in such calculations.

Tejchman, A., Subotowicz, W. and Gwizdala, K. [49] Calculation and forecast of coastal cliff stability

The paper presents an analysis of a 5 km long cliff that is retreating at about 1 m per year. A probabilistic analysis is included but appears to be a single random variable method in terms of [34].

Wang, L.S., Chen, M. and Li, T. [50] On the turning sliding-cracking slope deformation and failure

The authors present an interesting investigation into a failure mode involving sliding and bending failure of beds on a dip slope. They have undertaken physical modelling and then develop a simple analytical model. The presentation would have been greatly assisted by the inclusion of a key diagram.

West, L.J., Hencher, S.R. and Cousens, T.W. [51] Assessing the stability of slopes in heterogeneous soils

The authors discuss the problem of analysing a slope failure which passes through typical colluvial materials consisting of overconsolidated clay matrix around large cobbles and boulders. They argue that the actual friction angle of the material may be lower than often adopted but that there may be a roughness angle, i, familiar to all rock engineers. They do not point out the obvious corrollary that the higher friction angle generally adopted may unwittingly conpensate for these errors. The paper provides good food for thought even though the concept was not useful in their own back analysis.

Wu, C-S. and Jaung, T-J. [52] Variational approach applied to stability of slopes

The paper presents the application of the calculus of variations to locating the critical surface and minimum factor of safety using Janbu's simplified method. The reporter is not familiar with the concept of critical depth as applied in this paper and sonders if it is being correctly applied here. But it should be noted that it has a severe effect on the conclusions that would generally be obtained by others ignoring the effect.

Xu, J., Chen, M., Li, T. and Wang, L.S. [53] Geomechanical simulation of rock mass deformation and failure on a high slope

The authors present details of a physical model of a dip slope in rock like materials which took over 4 years to fail. The model volume was about one cubic metre and the authors estimate from similitude that a prototype slope 90m thick and 700m long would take 6000y to fail under gravity loading assuming that the materials follow a Kelvin model for creep.

Yagi, N., Yatabe, R. and Enoki M. [54] Stability analysis for landslides using ring shear result

The paper presents details of two ring shear apparatus and presents results with quite large differences in shear strength for the one material depending on the apparatus (including direct shear). The authors' results suggest that there is no reduction in residual friction angle with increasing clay content. Then they use the simplified Janbu analysis to back analyse many failures and demonstrate that the mobilised friction angle is between the residual and peak friction angles.

Yamagami, T., Yawakawa, O. and Suzuki, H. [55] A simplified estimation of the stabilising effect of piles in a landslide slope applying the Janbu method

The authors present the application of a Janbu type method to determining the forces required by a pile to stabilise a slope and claim that the method gives lower forces than conventional methods.

Yamagami, T. and Ueta, Y. [56] Back analysis of strength parameters for landslide control works

The authors develop two methods of back analysis using Janbu's method (it is not stated whether the simplified or "rigorous" method is used). The reporter had difficulty in following the development but is impressed with the answers presented.

Zhang, Z-Y., Wang, S-T., Wang, L-S., Liu, H-C. and Huang, R-Q. [57] Recent researches on the mechanism and geomechanic model of huge landslides in China

This is an interesting paper that presents a tabulation of large (> 1E6 m^3) Chinese landslides from the last decade and notes that in the last century there have been four landslides greater than 100 million m^3. The authors then look in more detail at some of these. The reporter is sorry that no references are given so that more details of these landslides could be obtained.

5 DISCUSSION AND CONCLUSIONS

Having read the fifty seven papers in this session, what can the reporter say about slope stability analysis? There is not very much that could be considered totally new but many of the papers present interesting extensions or modifications of previous developments and some of the remainder document interesting practical investigations.

A feature of the papers is that most contain some information from case studies used to verify the work undertaken, thus academics appear to be moving closer to practising engineers and geologists than was probably the case a decade ago. The reporter was disappointed with the standard of presentation of many of these case studies - often the figures and tables contained either no information or were poorly reproduced and were therefore of less use than they would otherwise have been.

Many of the papers deal with back analysis as a means of determining the strength parameters of a slope but few actually address the effects that uncertainty in the input parameters (e.g. groundwater conditions) can have on the results of these analyses.

It is pleasing to see that so many papers address in some manner the deformation, or deformation rate, of slopes. This is an area that is obviously attracting considerable attention around the world. It was interesting that only about 10% of the papers address reliability or probabilistic methods of slope analysis, the reporter is of the view that this is one of the most natural ways of considering a slope. Further the reporter would like to see further advances in the realistic analysis of three dimensional slope stability as this is an important factor in most slopes but is considered only a little in this symposium.

As might be expected several of the papers deal with groundwater or partially saturated soils and some quite interesting ones deal with the run out or reach of landslides.

Finally the reporter was somewhat disappointed to see so much research in slope analysis based on Janbu's simplified method, which even Janbu recommended in 1973 was only suitable for hand calculations. While the reporter often advocates simplified models or simple methods, he believes that in this day of computers the simplified Janbu method should not be adopted in any serious slope analysis.

6 REFERENCES

All the papers are listed in Section 4 of this report and are included in the Proceedings of the Sixth International Symposium on Landslides, Christchurch, N.Z., February, 1992, published by A.A. Balkema. The number in brackets (e.g. [31]) is used elsewhere in the report to refer to the various papers.

Landslides, Bell (ed.) © 1995 Balkema, Rotterdam, ISBN 90 5410 032 X

Geodetic methods for deformation measurement and analysis at a landslide area in Istanbul

M.O. Altan, T. Ayan & B. Ozuer
Istanbul Technical University, Turkey

ABSTRACT: To design and realize the most effective and appropriate precautions against lands-lide is dependent, besides knowing the geological structure of the soil, on predicting the earth movements in the landslide district with the exact location. In this paper the results of a study in an area near Istanbul which is to be opened for settling will be explained. In this area a 12-point main net was established and condensed by 24 object point. The ground signs of net points are measuring columns except one point. The side measurements were carried out by electronic distance measuring instruments; while the angle measurements were made by Wild T2 Teodolite and the height measurements by Wild NAK2 levelling instruments with parallel sided plate. The plane and height differences of the points have been treated separately; and after a free network adjustment of the plane and height net the root mean square errors of x, y coordinates is found to be max $\pm$ 4 mm and for height max $\pm$ 1 mm. The coordinate differences of points have been tested by means of relative confidense elipse method and displacements of (2–58 cm) have been determined.

1 INTRODUCTION

Landslides are massive soil movements which can be deemed as natural disasters. The information reasons, development conditions and precautions of such movements are within the interest area of soil engineering and related engineering disciplines. Due to concentrated urbanism arising from the need for new settlement areas, the alterations of topographic structures and soil conditions caused the distortion of area stabilities which were in equilibrium gradually and consequently and soil movements were resulted.

The area opened for settling at Pekmez location in Büyükçekmece region of ıstanbul is an old landslide field. The initial massive movements of large size were balanced gradually. Secondary movements were arisen within the creeped mass following the distontion of the existing equilibrium due to various reasons. That the land was preferred as a settling location increased the importance of the zone and the soil movements at excessively consolidated clays were investigated by other researchers from time to time also (Durgunoğlu et.al.1979). In this paper the landslide mechanism was investigated and the validity of the obtained results and drainage precautions taken were evaluated by means of geodetical measurements performed at certain intervals of time.

2 GEOLOGICAL ACTIVITIES

2.1 Geology of research area

The area of investigation covers a sedimented deposition lasting from easone upto now within basin subject to deepening in the direction North–South. A thick sediment unit formed during oligomiocene is embedded over the deeply observed eocene. This unit, known as Gürpınar formation, was deposited principally at lagoinal conditions. The quaternay land is observed in the form of grael and sand deposits and alluviumm which covers the Gürpınar formation irregularly and repeatedly. No signs were observed in the field showing widespread tectonical movements at all. Stratification is in approwimately horizontal extension at areas which are not influenced by landslides. Gravitational faults of small–size were observed at the western part of the field not causing important alterations at geological structure. Apparent units at this formation are green clay series containing limestone concretions and excessive fissures, silted fine sands and sand strata and tuffites consisting of glassy material of sand–silt size containing volcanian parts. Those strata do not display a regular storage at the area of landslides due to soil movements. Local sand strata of about 20 m thickness appear in form of sand focus at at different points. Such strata are water bearing formations (aquifers).

2.2 Landslide mechanism

The area subject to landslide begings with an arched frontier in East and depletes towards sea shore in West. Inclination of horizontal strata in the order of 5°-20° to East at the area where the landslide have started, distortion of surfacial strata in a small-size and dip distribution at landslide facing demonstrate that the movement was developed along a cylindrical sliding surface. This consists of the formation during the initial appearance of the landslide within the area. Later on secondary rupture surfaces and internal sliding slices were developed within the primary landslide mass and these movements resulted also in cylindirical slides. A mass decrease depending upon the shore erosion by waves appeared at shelf and consequently slide wedges were formed in approximately vertical direction at locations near the shore area. The land subjected to an uninterrupted deformation process which can be considered as large-sized creep caused by the secondary movements which showed a development in the direction of primary landslide.

It was observed that the first movements were started within the green clays of Gürpınar formation at the defined landslide mass. Excessively consolidated, fissured clays recover upon decrease of the applied geological loads and the inter-fissural clearance become widened. Also the drying at surfaces subject to weathering causes the cracks and fissures to become distant.

Another characteristic of such clays consists of existence of volume expansion at surfaces subject to strain arising from shearing deformations after soil movements and decrease in shearing strength due the increase of water contents. The influent water at fissures distorted the natural structures of clays as per although they are impervious. Decreases in the strength of clay stratum gradually due to distortion facilitates the occurrence of new movements in case the anti-shearing forces are less than the shearing forces. Consequently secondary movements within the pre-slide mass are observed at clay zones just below the surface. On the other side, sand bands and sand creaters within excessively consolidated clays causes the recuperation and sperading of influent water within the clay strata, and hence strength losses arise in deeper clay strata. Water recuperated within clay craters apply hydrostatic pressure on clay strata simultaneously and facilitates the water movement within clay layers, too.

Building activities commenced without any regulations regarding the natural conditions and topographical formation has been effective for increasing such movements and the unstabil equilibrium is still going on in landslide mass.

3 GEOTECHNICAL ACTIVITIES

22 ground drilling operations with depths varying between 12-20 m were performed in the subject area and destructed and non-destructed soil samples from drillings were evaluated by laboratory tests. Also the consistency and compactness of soil layers were determined basing upon the results of SPT tests performed at drilling depths.

3.1 Soil structure

General soil structure of land is in form of compact and very compact sand-fine gravel series and clay strata containing silts of solid and hard consistency and high plasticity. Any homogen storage from the point of stratification was not observed within the landslide area at all. But soil movements have distorted the stratification of the geological structure. For this reason drained water at sand-fine gravel layers was recuperated within clay strata due to distorted storage and transformed into surface water in form of springs locally.

3.2 Soil parameters

The following values for angle of internal friction and cohesion were found out with regard to total stresses by triaxial shear tests performed on non-destructed soil samples from excessively consolidated and fissured clay strata subject to potential slip surfaces:

$$\emptyset_u = 20^\circ \quad \text{and} \quad C_u = 30 \text{ kN/m}^2$$

And the values for residual slip strength parameters at slip surfaces arising from landslide obtained from repeated shearing box tests (slow tests) are given down below:

$$\emptyset_r = 15^\circ \quad \text{and} \quad C_r = 10 \text{ kN/m}^2$$

Consistency limits and natural water content of the clay layer subject to investigation are as follows respectively:

$$W_l = 68 \% \text{-}74 \%, \quad W_p = 29 \% \text{-}34 \%,$$

$$I_p = 31 \% \text{-}36 \% \text{ and } W_n = 25 \% \text{-}36 \%$$

3.3 Stability analysis

It has been foreseen that the slip circles shall be within a depth of 10 m below soil surface for the analysis performed at sections of Figure 1 and the starting of soil movements were

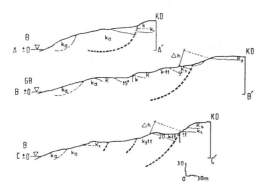

Fig. 1

based on cracks determined at land. Sand bands and interrupted clay strata or clay–sand zones of transtion are available within the said depth of 10.00 m. This zone forms a suitable formation for sliding. Determination of sliding strength parameters which may be valid for stability analyis at excessively consolidated and fissured clays forms the starting point of the problem.

Non–destructive sampling in the capacity of representing the natural structure of the land is rather difficult in case of such clays. The most realistic approach consists of the stability investigation with regard to parameters of residual sliding strength by considering the influence of fissured structure of soil layers upon sliding occurence (Noble, 1973). For this reason the safety coefficients found by analysis performed with regard to the values of C_u and $\emptyset_u$ are larger than those found by the values C_r and $\emptyset_r$.

The issued slope charts show that the sliding movements of land occur at side slopes of approximately $20°$ but any soil movements were not observed at some side slopes exceeding $30°$. This situation depends upon soil sort and stratification types.

Safety factors for slidable fields are approximately $G_s =1$ as per analysis performed by using slice method or simplified Bishop method considering soil conditions of land and topographical conditions (Bishop, 1955). This figure is $G_s > 1.5$ for all other regions. Sliding planes are considered as cylindirical surfaces. Sliding surface and bank regarding the landslide basing upon the performed geological activities were considered for this model and analyis regalized for various sliding surfaces at such areas where landslide is not apparent. Also stability analysis were performed by accepting some of sliding surfaces as lineer upon considering the softening at sand–clay contact surfaces. Thus the safety factors found there from are smaller.

4 GEODETIC MEASUREMENTS, ANALYSIS

Geodetic measurements were performed at regular time intervals in order to determine the displacements of soil masses forming the fundementals of stability analysis and the movements which are observable on land generally and also to investigate the validity of drainage precautions taken.

Measerement columns allowing measurements were formed by the way of forced centering at such locations without potential movements and with high movement probability depending upon the geological surveys at the subject area or by other words at characteristic points capable for representing the soil geology. Two geodetic control nets were formed by connection of those points with geodetic measurements. These are horizontal and vertical control nets for determination of horizontal and vertical components of probable movements respectively. All of soil points within the horizontal control net are also vertical control points.

Angles of horizontal control net which comprises 42 points were measured by Wild T2 Teodelits in four sets of specially manufactured $\emptyset$ 2 mm and 6 cm long signs and an accuracy of $M_r=(0,8-1)$ mgon was reached at a single direction measurement. The side measurements of this net were performed by using electronic distance measuring intstruments as Uniranger and Kern Mekometer ME 3000. Distance measuring instruments used at each measurement period were calibrated at İTü (Technical University of Istanbul) Control Base. A posteriori root mean square error of side measurements is $m_s = \pm$ (2–4) mm. Height measurements of points within horizontal control net located at the top of buildings were also realized by means of zenith angles the same measurement equipment.

The vertical control net based on the concentration of the points within the horizontal control net comprises 34 points principally. Height measurements for the vertical control net were realized by Wild NAK2 levelling instrument having a compensator and parallel plan plates and by a pair of 3 m long levelling rods with cm readings. Root mean square errors of height difference specified for one instrument setting at double way levelling are $\pm$ 0.4 mm.

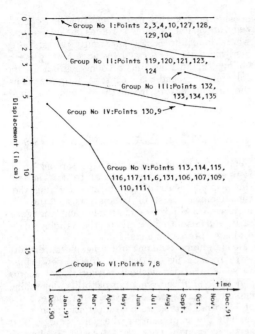

Fig. 2

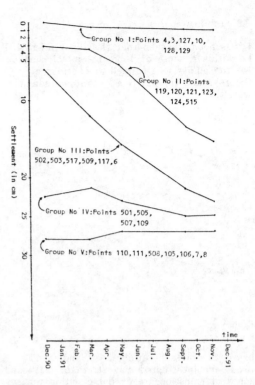

Fig. 3

The horizontal and vertical control nets were independently adjusted before deformation analysis and the nets were subjected to inconsistent measurements test of Baarda and Pope and the inconsistent measurements were eliminated from the measurement plan.

The root mean square horizontal errors of the net points were found to be $M_p = +(1-6)$ mm based upon the free net adjustment of the horizontal control net and maximum value of the large axis length of Helmert mean error ellipse was observed at point number 12 located outer side of the net, and it was due to the net geometry and reached the value of $A = (4.5-6.5)$ mm during different periods.

The horizontal and vertical control nets were measured five times in a year i.e. on February 25, 1991, April 28, 1991, September 10, 1991 and November 8, 1991 as of December 2, 1991. The significant point movements were specified by investigating the coordinate differences of points within the time period between two measurement intervals by integrating the measurements taken at different times as pairs.

As an example of the horizontal and vertical components of point displacements the horizontal movements are given in Figure 4. Size, movement amount and movement direction of the sliding masses are determined precisely by considering equal features like direction and amount of the point movements upon studying these figures. Movement-time diagrams obtained by grouping such points which

have similar movement features within the area of investigation are shown in Figure 2 and Figure 3.

From a combined analysis of the Figures 2, 3 and 4, one can conclude useful information about the landslide mechanism and its development. Therefore it is recommended that beside obtaining tables of displacements and seflements, their graphical representations on mags, the presentation of movements in typical cross sections showing horizontal and vertical displacements (settlements and displacements) is a veriy useful. The interpretation of these figures by civil engineers, geologists and other related non-technician groups are very easy.

5 CONCLUSIONS

Mean deformation rates in horizontal direction are appropriate in areas containing grouped measurement. Also the movements of points belonging to groups No.I and IV are extremely low (in the range of mm). Consequently these zones can be considered as static. However the movements of groups No.II–III–IV and V between those two borders are of permanant nature. And it has been also observed that movements within zone

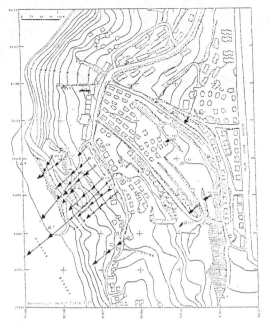

Fig. 4

No.V are in excess with regard to other zones. There is decrease in the rate of horizontal movements measured after September 1991 Figure 2 and Figure 3.

Settlement which arise at soil under building loads is effective in the vertical deformations in addition to the movements of soil masses. Thus the vertical deformations obtained have to be evaluated by considering those two factors mentioned above. Vertical deformations are observed to be as settlements generally, however blastings are also seen, too. This situation is the result of the landslide mechanism. But the assumption that the movements are in form of surface yields of less depth becomes effective because the deformations like blastings are of smaller size with regard to settlements Figure 2 and Figure 3. The movement geometry can be approached on a more realistic basis by increasing the number of measurement points.

It was observed that the directions and courses of horizontal movements reamined unchanged. This situation shows that the movement starting within group No.V is continued without any deformation and couses new movements in group No.II. Changes in directions and courses at some points can be interpreted as an explanation of local movements. And those movements are not a nex landslide formation at all Figure 2 and Figure 3.

The rates of movements within time can be determined by geodetic measurements to be performed for providing an appropriate distribution on the land in general. But the sliding geometry which was forecasted in advance as per geological, topographical and soil conditions can also be checked-up. The validity of precautions taken can also be discussed simultaneusly.

REFERENCES

Ayan, T. 1991. Project Regarding the Determination of Soil Movements within the Landslide Area of Büyükçekmece, Technical Report 1,2,3,4,5-Istanbul, 1990-91 (in Turkish).

Bishop, A.W. 1955. The use of slip circle in the stability of slopes, Geotechnique 10.

Altan, O. Ayan, T. Deniz, R. Özüer B. Tekin, E. 1991. Determination of Soil Movements at Landslide Zone by Geodetic methods and an Application. National Landslide Symposium November 1991, Trabzon (in Turkish).

Durgunoğlu, H.T. & Vardar, A.T. 1979. The natural slope stability of the area between Çekmece Lakes, Turk. National Commitee of Soil Mechanics and Foundation Engineering.

Niemeier, W. 1985. Deformationsanlyse Geodätische Nehe in Landes und Ingenieurvermessung in Vortrage des Kontaktstudiums Februar 1984 in Hannover (Herausgegeben von H.Pelzer).

Noble, H.Z. 1973. Residual strength and landslides in clay and shale, Jour. of the Soil Mechanics and Foundations Divisions, ASCE, Vol.99.

Schmitt, g. u.a. 1990. Deformation Analysis of a Local Terrestrial Network in Romania with Respect to the Vroncea Earthquake of August 30, 1986.

Landslides, Bell (ed.) © 1995 Balkema, Rotterdam, ISBN 90 5410 032 X

The 'SISYPHE' and 'XPENT' projects: Expert systems for slope instability

J.P.Asté
Bureau de Recherches Géologiques et Minières, France

C.Ke
Ilog, France

R.M.Faure
École Nationale des Travaux Publics de l'État, France

D.Mascarelli
Institut National des Sciences Appliquées de Lyon, France

ABSTRACT : Two expert system projects have been developed in parallel by the B.R.G.M. and the ENTPE using common tools, in particular SMECI, a multi-expert engineering design system generator from the company ILOG.

This report shows the status of each of the projects and the development perspectives in the context of close collaboration between the two teams involved.

1 ORIGIN AND JUSTIFICATION OF THE PROJECTS

In 1986 and 1987, two projects for computer-assisted slope stability assessment were launched simultaneously, one by the Bureau de Recherches Géologiques et Minières (BRGM: Geological and Mining Research Board), and the other by the Ecole Nationale des Travaux Publics de l'Etat (ENTPE: National School for Public Works), with whose project Laval University in Montréal very soon became associated.

For the general problems of slope instability and associated hazards, it appeared very useful to benefit from openings created in other disciplines, and particularly in the medical sphere, by the gradual development of expert systems.

The approach used by experts in slope instability has in fact many points in common with that of medical doctors. They have to:

– detect zones suspected of instability as early as possible,
– diagnose the mechanisms of the instability detected,
– forecast the type, intensity, frequency, spatial extent and the moment of occurrence of the event expected,
– assess the levels of direct and indirect damage caused to the environment and the communities affected,
– and finally, recommend and justify measures for consolidation or rehabilitation.

This requires the assistance of many experts: geologists, geotechnicians, hydraulicists, hydro-meteorologists, as well as geographers and civil engineers. It also concerns administrative bodies, sociologists, educationalists, the media and of course the decision-makers.

Some of these specialists are working to design models capable of providing reliable predictions of landslip probability, in terms of safety factors, or evaluation of distance and speed of displacement and of infiltration or drainage rate.

In order to collect together these skills, the ENTPE decided to develop XPENT as an expert system to handle these digital tools, and also as a computer-assisted instruction system (Faure, 1990).

It very quickly appeared interesting to enrich the procedure by adding a knowledge engineering dimension based on the Canadian experience with sensitive clays. Laval University provided this experience and its own contribution to the modelling part of the problem.

BRGM, for its part, was more interested in the geological aspects of instability problems, and launched the project known as SISYPHE (Slope Instability System - Prevention and Hazard Evaluation) with the aim of appraising the three- dimensional framework of the unstable zone and its physical, geological and hydro-meteorological environment (Asté, 1988 ; Ké, 1990).

2 - THE CHOICE OF A SINGLE RANGE OF TOOLS

During the first few months of the projects, experts

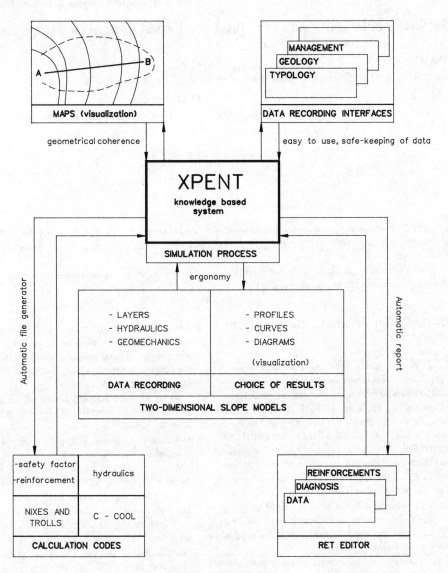

Figure 1 : The XPENT Project

in both teams concentrated on formulating, explaining and justifying their knowledge and experience. In this phase of the project great confidence was placed in the capacity of inference engines to manage rules laid down on paper. Various engines available on the market were thus tested, but with such limited success that the projects were on the point of being terminated (Asté, 1987).

The next phase for both projects was one of in-depth cognitive analysis; it was necessary to structure a knowledge of the world corresponding to the field of expertise, and declare a number of basic objects and concepts, and actions or transformations which these objects should undergo during performance of the expert assessment. We shall see below the directions of development chosen for each of the projects, with reference to the neighbouring article, (Ké, 1992), to illustrate the data- processing and knowledge engineering solutions chosen.

This phase was heavily based on systems analysis techniques (Sadt, 1982). In addition, and very importantly, thanks to the company ILOG making available an engineering design multi- expert system generator, SMECI (Haren, 1985), the two teams were able to work using slightly different approaches on the same system, which should later be very useful for combination of their results.

3 - THE PRESENT STATE OF DEVELOPMENT

3.1 - XPENT Project

XPENT was first developed at the ENTPE on the basis of the experience of specialists from the regional civil engineering laboratories of the French Ministère de l'Equipement, responsible for public civil engineering works. This experience bears essentially on the stability of large slopes of excavated material or embankments formed during roadworks. The analysis and calculation methods used in it are derived from conventional Soil Mechanics techniques and dam-building experience. There are in fact many similarities between a large embankment slope and an earth dam. For this reason, the concept of two-dimensional calculation was given priority in XPENT, especially as this type of approach has been much used at the ENTPE, where was developed an efficient and simple-to- use calculation chain. (Faure, 1985, 1992) To this chain has been added an expert system to enable these programs to be better used. This layer is rich in very user-friendly and ergonomically- designed graphic and textual interfaces.

With Laval University, a computer-assisted instruction dimension (with explanatory modules) was envisaged very early, which led to very meticulous definition, not yet finished, of the vocabulary employed. During this work a first model of an expert system specialised in diagnosis of stability in Canadian soils was produced.

Another article from the same conference (Ké, 1992) describes the data-processing functions of XPENT are described (fig.1):
- a two-dimensional slope editor,
- coupling with high-performance calculation software (Nixes and Trolls, C-Cool) to analyse slope stability in terms of a safety coefficient,
- an interface for input of factual data and its connection with a data base of real cases to produce analogical-type reasoning,
- a knowledge base for qualitative stability diagnosis,
- a knowledge base for the design of stabilizing structures. For modelling of knowledge, the use of customized tools (such as the K-Station) will enable the process to be further accelerated.

The expert system XPENT will develop around this core, by integrating other modules (for CAI, calculation, etc.), since XPENT is intended to be an open, modular and portable system.

3.2 - SISYPHE Project

As we saw above, SISYPHE was developed essentially on the basis of the experience of specialists accustomed to assessing the instability of a slope and proposing solutions for its stabilization ... or justifying the abandoning of the zone concerned.

It was therefore attempted to simulate the actions of the expert in their usual order. In practice, as soon as a request is referred to the expert, he tries to collect all the information available in the form of:
- maps, plans or aerial photos,
- information relative to the event and the place studied,
- references to other events of the same type experienced there or elsewhere in similar geological and geographic contexts, whether personally or by colleagues.

To enable this first phase of actions to be simulated, SISYPHE had to be able to manage geographic information

Such systems enable rapid display of a topography, for example, transcription of a geological map, and even scanned photos.

It was also necessary for SISYPHE to be interfaceable with factual data bases such as the French National Inventory of Unstable Slopes (Asté, 1992).

The expert must then attempt to refine the knowledge acquired on the site by complementary investigations: detailed surveys of the area, geophysical drillholes, measurements of deformation and water level, etc.

SISYPHE therefore had to enable step-by-step updating of the knowledge acquired by the expert. In order to make this possible, the notion of successive surfaces was used, surfaces delimiting a "ground layer" object, a "piezometric surface" object and a "potential landslip surface" object (Fig. 2).

A number of processes enable automatic continuous modification of these surfaces each time a new item of information is input to the system, so that little by little, as in the expert's usual mode of work, an up-to-date integration of the geometric and three-dimensional knowledge of the zone being studied becomes available.

At every step of this updating, the expert must have means of evaluating the mechanical and hydraulic equilibrium of the site. For this purpose, he may run stability calculations using three-dimensional analysis, and selecting a most likely direction of slip.

SLOPE INSTABILITY SYSTEM - PREVENTION AND HAZARD EVALUATION
SISYPHE

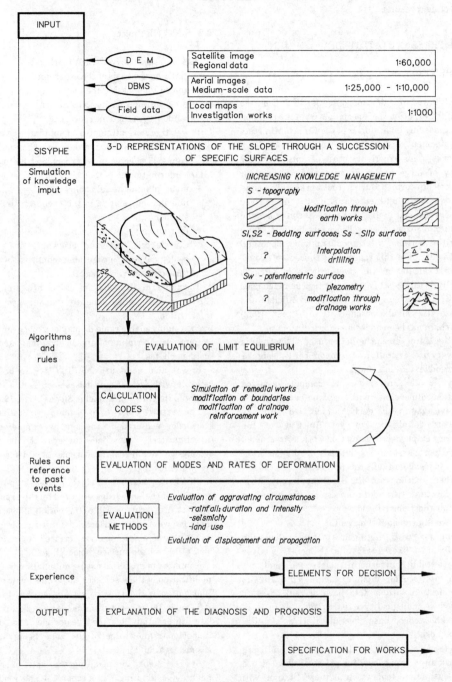

Figure 2 : The SISYPHE Project

SISYPHE is at present coupled with O.Hungr's model CLARA (Hungr, 1988), and will in the future be coupled with hydraulic models enabling the non- saturated medium to be taken into account in equilibrium. SISYPHE must of course make use of mechanical characteristics in these calculations. These characteristics form part of the declared attributes of the "ground layer" objects. If laboratory tests have been performed, the corresponding values are automatically called up. If not, default values (average and standard deviation) are associated in the object base with each category of ground in the catalogue. At a later date, a link will be set up between SISYPHE and MODELISOL, a French data base of mechanical characteristics of soils (Faure, 1991).

This is the present state of SISYPHE, available in prototype form on SUN or VMS stations.

4 - DEVELOPMENT PERSPECTIVES

As we saw above, the two projects, launched at the same time, are at comparable and complementary levels of development. Their combination poses no problems, neither in their substance, thanks to the unicity of the system used and to the partnership of ILOG, nor in their form, thanks to the spirit of partnership reigning between the teams concerned.

Much remains to be done, however, to progress from a prototype to an operational model which could if necessary be distributed in the form of separate modules. There is also much exploratory work to be done in two fields situated before and after the projects with which we are concerned here:

- Before: the capacity to develop such systems for the production of hazard maps. The methodology used today throughout the world to produce this type of document remains insufficient. It must however be developed and rationalized, in order to give a good evaluation of the hazards generated by instability phenomena, with a view to durable development (Asté, 1990).
- After: the capacity to develop such systems integrating the latest advances in computer technology and thus providing design engineers with increasingly productive tools.

The two project teams will come together to begin work on this new phase of development, and they hope to be able to give an account of their work at the next international symposium on landslides.

BIBLIOGRAPHY

ASTÉ J.P. - 1988 - Some reflexions on methodologis of landslide prevention in France. International workshop on natural disaster in European Mediterranean Country, Perugia, Italy.

ASTÉ J.P. - 1990 - Faisabilité d'un système d'aide à l'élaboration d'un diagnostic relatif aux risques générés par des mouvements de terrain. Rapport BRGM. R30703, Décembre 1990, Délégation aux Risques Majeurs du Ministère de l'Environnement.

ASTÉ J.P. - 1992 - RIVET : projet européen de recherche sur la dégradation des versants en montagne, 6th International Symposium on Landslide, Christchurch, New Zealand, February 1992.

ASTÉ J.P. et GIRAULT F. - 1992 - GIS, SPOT, DEM and Morphology of large landslides, 6th International Symposium on Landslide. Christchurch. New Zealand.

ASTÉ J.P., GOUISSET Y., LEROI E. - 1992 - INVI, Projet français d'Inventaire National des Versants Instables, 6th International Symposium on Landslides. Christchurch. New Zealand.

ASTÉ J.P., LOESENER C., KE C. - 1987 - Prevention of earth movements and expert system. Proceeding of the International Symposium on engineering geological environment in mountainous areas. Beijing, China.

FAURE R.M. - 1985 - Stress Analysis in a slope using a perturbation method. Revue Française de géotechnique n° 33, pp. 49-59.

FAURE R.M., LEROUEIL S., RAJOT J.P.., LAROCHELLE P., SEVE G., TAVENAS F. - 1988 - XPENT. Expert system in slope stability, Fifth internaitonal symposium on Landslides Lausanne, pp. 625-629.

FAURE R.M., SEVE G., FARHAT H., VIROLLET M., DELMAS Ph. - 1992 - A new methodology for evaluation of landslides displacements. Sixth Internaitonal Symposium on Landslides, Christchurch, New Zealand.

FAVRE J.L., HICHER P.Y., KERILIS J.M., TOUATI K. - 1991 - "MODELSOL" : a database for reliability analyses in geotechnics", Sixth International Conference on Applications of statistics and Probability in Civil Engineering. Mexico.

HAREN P., NEUVEU B., GIACOMETTI J.P., MONTALBAN M., CORBY O. - 1985 - SMECI : cooperating expert system for Civil Engeenering, SIGART new letter, pp. 67-69.

HUNGR O. - 1988 - Clara, Slope Stability Analysis in two or three dimensions for IBM Compatible Micro-computers, O. Hungr Geotechnical Research Inc. Vancouver, Dec. 88.

KE C. - 1990 - Un système de représentation des connaissances et d'aide à la décision pour la prévention des mouvements de terrain. Doctorat de spécialité en Informatique Appliquée. Université Claude bernard Lyon I. N° 213/90.

KE C., MASCARELLI D., VAUNAT J. - 1992 - Sixth International Symposium on Landslides, Christchurch, New Zealand.

Manuel de référence de la méthode SADT 82, IGL France, Introduction à SADT, Juillet 82, Paris.

Landslides, Bell (ed.) © 1995 Balkema, Rotterdam, ISBN 90 5410 032 X

Charts for the rapid assessment of slope stability

G. E. Barnes
School of Civil Engineering, Bolton Institute, UK

ABSTRACT: Simple charts have been prepared to enable the rapid determination of the minimum factor of safety of a homogeneous slope assuming effective stress conditions. The charts have been produced for slopes of 1 : 1, 2 : 1, 3 : 1 and 4 : 1, and relate all of the relevant parameters, c' and ϕ' of the soil, height of slope H and height of a water table, h_w. The main advantage of the method is that a more realistic water table can be assumed instead of the pore pressure condition being represented by an arbitrarily determined average pore pressure coefficient, r_u. Linear interpolation between F values may be used for intermediate slope inclinations with minimal loss of accuracy.

1 INTRODUCTION

The minimum factor of safety, F, of a homogeneous slope is determined by several variables (c', ϕ', γ, H, slope inclination, water table condition) and some have a greater effect on F than others. The most commonly used chart method to determine the factor of safety of a homogeneous slope using effective stress conditions is that of Bishop and Morgenstern (1960) who published stability coefficients, m and n, to give F in the form

$$F = m - n.r_u \qquad (1)$$

However, this method suffers from the difficulty of representing known water table levels as a single value of the pore pressure coefficient, r_u and can only be determined arbitrarily (Bromhead, 1986). It also has the limitation of fixed depth factors, D, which does not always produce the lowest factor of safety (Barnes, 1991).

The author has produced a method (Barnes, 1992) of obtaining the factor of safety of a slope with a known water table level in the

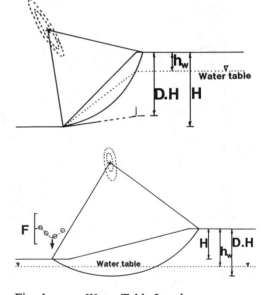

Fig. 1   Water Table Levels

form

$$F = a + b \tan \phi \qquad (2)$$

For a given slope inclination the coefficient a

has been found to be solely a function of $c'/\gamma H$ and the coefficient **b** depends on $c'/\gamma H$ and the level of the water table. The latter has been represented as a line from the toe of the slope at various inclinations within the slope denoted by the parameter, h_w/H, measured below the crest. When the water table lies entirely below toe level it has been represented as a single horizontal surface at a depth h_w below crest level, see Fig.1.

This approach has enabled charts to be prepared which permit the rapid determination of the following:

1. the factor of safety for a particular slope, given the soil parameters and the water table level
2. critical soil parameters and/or critical water table level for a given factor of safety
3. the sensitivity of the factor of safety to the variability of the soil parameters or water table level.

2 USE OF CHARTS

The most straightforward use of the charts is to determine the factor of safety for a given set of parameters in the order c', H, h_w and ϕ', as shown on Fig. 2. Charts for slopes of 1:1, 2:1, 3:1 and 4:1 are given on Figs. 3 - 6, respectively. For intermediate slope inclinations linear interpolation between F values obtained

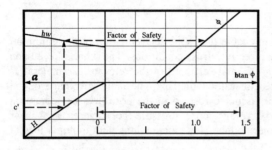

Fig. 2   Use of the Charts

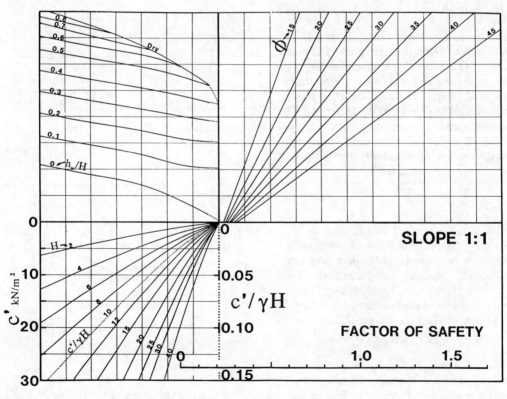

Fig. 3   Slope 1:1

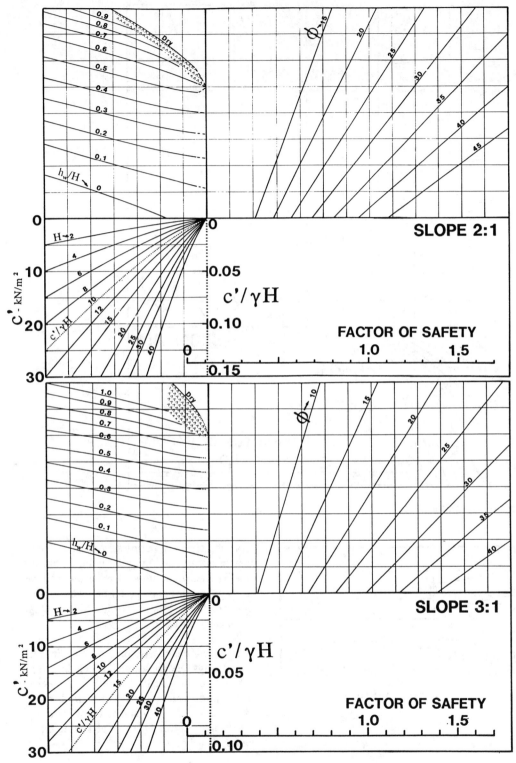

Figs. 4 and 5   Slopes 2:1 and 3:1

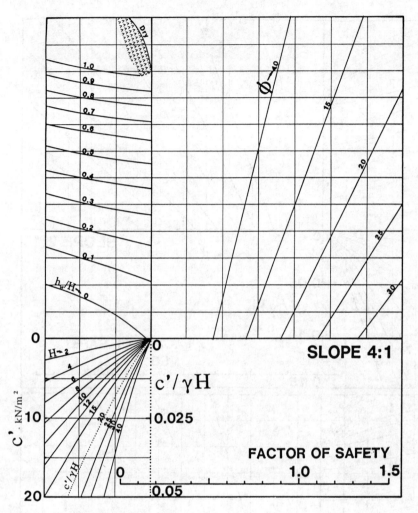

Fig. 6   Slope 4:1

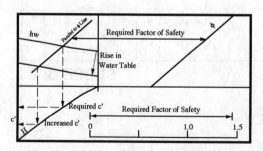

Fig. 7   Estimation of Required c'

from the charts is permissible with minimal loss of accuracy. Any point lying within the shaded area in the upper left portion represents 'dry' conditions and should be taken vertically upwards to the 'dry' curve in order to obtain the factor of safety.

However, the charts could be used in other useful ways. Examples could include using the charts to determine the value of the cohesion intercept c' which would be required for a given factor of safety and to investigate the effect of a rise (or fall) in water table on this value, see Fig. 7.

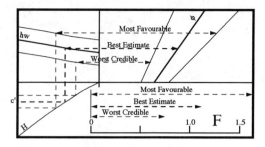

Fig. 8   Sensitivity Analysis

Accurate measurement of the c' intercept is often dubious with standard laboratory apparatus and the value which could be adopted and relied on in practice is often uncertain. It can readily be seen from the charts the significant effect even a small value of c' or small change in c' has on the factor of safety. For example, for slopes in London Clay a cohesion intercept as low as 1 kN/m² is recommended by Chandler and Skempton (1974). Adopting the assumption of c' = 0 for the fully softened case can, therefore, lead to potentially excessive conservatism, particularly for the smaller slope heights. Further research is required into the values of this parameter which could be relied on in practice.

Strength parameters of natural soils show variability, even for apparently homogeneous strata and water tables are likely to fluctuate during the design life of a slope. The charts can be used to rapidly assess the effects of such variability on the factor of safety by carrying out a sensitivity analysis to obtain most favourable, best estimate and worst credible values of F, as shown on Fig. 8.

3 CONSTRUCTION OF CHARTS

The coefficient, a has been found to depend solely on the parameter c'/γH and their relationship has been accurately represented using power regression techniques and is given in the lower left portion of the charts. The coefficient b is dependent on the water table

level and c'/γH so the upper left portion of the charts is a plot of this coefficient for each value of h_w/H and a. The upper right portion is a plot of values of btanϕ. This then enables the factor of safety to be scaled off as the horizontal distance between chosen h_w/H and ϕ' values such that to the left of the vertical axis of the chart is the value a and to the right is the value btanϕ. An explanation of the use of the charts is given on Fig. 2.

The coefficients a and b have been determined for a value of bulk unit weight γ = 20 kN/m³ and the lower left portion of the chart has been plotted for this unit weight so that the slope height H can be read off directly. For other values of unit weight the parameter c'/γH must be determined and the single curve (dotted) used for this parameter. It has been found (Barnes, 1992) that unit weight can affect the results obtained but this effect is minimal for unit weights in the range γ = 18-22 kN/m³ which would apply for most soils.

The values of a for the 'wet' condition, i.e. when the water table lies within the critical circle, were found to be virtually independent of h_w/H. However, for the 'dry' condition when the water table was below the critical circle different a values were obtained and these have been plotted separately. This has meant that an adjustment has been required to the upper left portion of the chart so that the curve for the 'dry' condition is somewhat higher. Any point lying within the shaded area represents 'dry' conditions and should be taken vertically upwards to the 'dry' curve in order to obtain the factor of safety.

4 CONCLUSIONS

It is anticipated that the charts presented will provide busy, practising engineers with a rapid means of assessing the stability of a given slope or designing a slope for given soil parameters without recourse to, or perhaps before using, more detailed computer analysis.

REFERENCES

Barnes, G. E. (1991) A simplified version of the Bishop and Morgenstern slope stability charts. Canadian Geotechnical Journal, August, 1991

Barnes, G. E. (1992) Stability coefficients for highway cutting slope design. To be published in Ground Engineering.

Bishop, A. W. and Morgenstern, N. R. (1960) Stability coefficients for earth slopes. Geotechnique, vol.10, p.129-150

Bromhead, E. N. (1986) The stability of slopes. Surrey University Press, Blackie and sons Ltd., U. K.

Chandler, R.J. and Skempton, A. W. (1974) The design of permanent cutting slopes in stiff fissured clays. Geotechnique, vol.24, No.4, p.457-466

Landslides, Bell (ed.) © 1995 Balkema, Rotterdam, ISBN 90 5410 032 X

Progressive failure considerations in slope stability analysis

R.N.Chowdhury
University of Wollongong, N.S.W., Australia

ABSTRACT: Limit equilibrium methods of slope stability analysis do not generally take progressive action into consideration. However, any of these methods may be used in an innovative way to understand the role of different types of progressive action. For example, one may wish to study the effect of gradual strain-softening on the factor of safety of a slope. In this paper, general considerations for the development of such approaches are discussed. Reference is made to analyses of particular cases of slope failures including earth dams, saturated clay slopes under undrained conditions and long-term stability of clay slopes. Moreover, the decrease in factor of safety due to progressive action is significant except for shallow slopes with planar or subplanar slip surfaces. Back-analyses of failures, based on this approach, are more successful than conventional back-analyses.

1 INTRODUCTION

Every landslide involves some kind of progressive action and there are often typical warning signs before a slope fails. These signs can not always be observed with the naked eye and, therefore, special monitoring systems are installed at selected locations in important slopes and earth structures. Certainly, there are failures which occur 'suddenly' and others which occur 'very slowly'. Nevertheless, in each case progressive action is important although this may be so in different ways and to different extents.

Many slides involve the movement of a mass of soil or rock over a well-defined surface called the slip surface. In most methods of analysis it is assumed that failure occurs simultaneously over the whole of the slip surface. Yet, observations have shown repeatedly that failure of some parts of a slip surface occurs before that of other parts. The importance of this spatial progressive action has been recognised. Thus failures may either initiate at the toe of a slope and progress upwards or at the crest of a slope and progress downwards. In many cases, failure may occur in the interior first and then spread to some proportion of the slip surface. In any particular case, the progression of such 'localised' failure may not lead to a slip involving complete failure over the whole of the slip surface. Yet, it is important to estimate the extent of progressive action or the extent to which the factor of safety has been reduced by it.

The time associated with progressive action may be relatively short or long depending on the failure mechanisms involved. The extent of delay associated with the development of progressive failure is difficult to estimate without an observational approach. Some natural slopes move very slowly and these creep like movements may not lead to a complete failure or a landslide for many years. In such cases, performance monitoring is the only option available to the geotechnical engineer for updating his information and for making some predictions. Modelling the time-dependent slope stability problem is not easy because of the number and complexity of factors which influence natural slope performance.

Quite apart from time-dependent problems, geotechnical engineers have also recognised that predicting the locations where failure will initiate in a slope is also not easy. Limit equilibrium approaches must, in important cases, be supplemented by stress-deformation studies using versatile techniques such as the finite element method. Using such techniques, locations of stress concentration, large strains and large relative deformations along discontinuities can be identified. In the case of earth dams, the potential for hydraulic fracturing can also be recognised based on adequate stress analysis approaches.

In most practical cases of slope stability analysis, however, the geotechnical engineer uses a limit equilibrium approach as his primary tool. The extent to which progressive action can be considered with such a tool is important. Therefore, the discussion in this paper is limited to a consideration of progressive failure in a spatial context. Moreover, only one kind of progressive action, that associated with localised strain-softening, is considered.

2 MOBILISATION OF SHEAR STRENGTH

It is customary, in slope stability problems to consider the values of the factor of safety associated either with the 'peak' shear strength or the 'residual' shear strength. One set of shear strength parameters leads to an upper bound to the factor of safety and the other to a lower bound. These bounds do not take into consideration the variable mobilisation of shear strength along a slip surface. For example the shear strength mobilised may correspond to 'peak' shear strength parameters near the crest of a slope and 'residual' parameters near the toe. The mobilised shear strength parameters between these two points would, in general, be variable and may follow any distribution. However, studies based on simplified and rigorous limit equilibrium methods have shown that the factor of safety depends only on the 'average' mobilised shear strength over the slip surface, regardless of the distribution of mobilised cohesion and mobilised friction angle (Bertoldi, 1988).

The use of limit equilibrium approaches in studying spatial progressive action requires a consideration of changes in the average mobilised shear strength. The decrease in average mobilised shear strength occurs because of shear stress exceeding shear strength over some sections of the slip surface. If this occurs, strain-softening effects are the likely consequence. It is recognised that significant relative deformations are required for shear strength to reduce from a 'peak' to a 'residual' value.

However, strains and deformations can not be estimated, even approximately, within a limit equilibrium framework for analysis. Therefore, it is useful to make the assumption that the material of the slope is brittle so that the shear strength falls to a residual value as soon as the shear stress exceeds the shear strength. Such an assumption is useful for analysis although it may be somewhat conservative.

3 SIMULATION PROCESS

3.1 The steps involved

The approaches for simple analysis of progressive action within a limit equilibrium framework involve the following steps.

(1) An initial limit equilibrium analysis is carried out considering the slip surface to be divided into a number of segments or slices. An initial factor of safety is thus calculated based on 'peak' shear strength parameters.

(2) The shear stress along each slice is calculated using either a simple gravitational stress field or a stress analysis or any other appropriate stress field.

(3) The shear stress along each slice is compared to the shear strength. If the shear strength is exceeded, 'residual' shear strength parameters are assigned to that slice, and it may be referred to as a failed slice.

(4) All such strain-softened slices are identified.

(5) A new limit equilibrium analysis is carried out using 'peak' shear strength parameters for all the slices except those identified as strain-softened for which 'residual' shear strength parameters are used. A new factor of safety is obtained.

(6) Based on the two successive factors of safety, a redistribution of 'excess' shear stresses is simulated. Alternatively, the 'excess' shear stress is redistributed in an arbitrary way along some or all of the slices which have not failed.

(7) The new shear stress on each unfailed slice including any redistributed 'excess' is compared to the shear strength. If the shear strength is exceeded, residual shear strength parameters are assigned to that slice.

(8) Steps (4), (5), (6) and (7) are repeated.

(9) The whole process is completed when no more slices fail and therefore, two successive factors of safety remain the same.

3.2 General Comments

The simulation process involves iterative limit equilibrium calculations. Generally, convergence is always obtained within the first few iterations. In other words, all 'failed' slices are identified within the first few iterations and, therefore, the factor of safety does not decrease further after that.

Any limit equilibrium method may be used to carry out the simulation process and the basic assumptions of a particular limit equilibrium method (whether a simplified or rigorous one) are fully retained. In other words, the concept of limit equilibrium is followed in the same way which has gained wide acceptance.

Each effective stress analysis is performed for a particular distribution of pore water pressure along the slip surface. Short-term stability under "$\phi = 0$" conditions may, however, be considered using a total stress stability approach. This would be relevant for saturated slopes under undrained conditions.

The approach discussed above has not yet been used to study the effect of changing environmental conditions. For example, the effect of rising or falling of reservoir level for slopes adjacent to dams has not been considered. Similarly, seepage pattern changes due to rainstorms and consequent changes in pore water pressure distribution have not been considered in any individual illustrative example or case study.

Seasonal fluctuations in water table also need to be given consideration. Understanding of progressive action due to such repetitive changes requires comprehensive studies at a fundamental level. As stated earlier, time-dependent changes involve complex processes and can not be simulated using simple limit equilibrium approaches unless comprehensive information concerning the

rheological behaviour of the slope material is available.

Studies which take into consideration the incremental process of construction of slopes can, of course, be carried out. However, within a limit equilibrium framework, only the changes in the overall factor of safety can be simulated realistically.

Moreover, progressive action can be incorporated in the sense discussed above. However, strains and deformations associated with the incremental process of construction can not be predicted without a sophisticated stress-deformation approach requiring a knowledge of the deformation parameters.

3.3 Calculation and Redistribution of 'Excess' Shear Stress

The main feature of the proposed approach is the calculation of 'excess' shear stress resulting from local strain-softening along a part of the slip surface and its redistribution along the remaining part. Several approaches can be used to calculate the initial normal and shear stresses along a slip surface. For example, a simple gravitational stress field could be considered. If there is knowledge of high lateral stresses in the region, a modified initial stress field would have to be considered. The main advantage of using these types of stress fields is that calculations are conveniently performed from the same basic data which is used in the limit equilibrium approach. Only, some additional data concerning the ratio of initial horizontal to vertical stresses may be necessary where relevant. The main disadvantage of using these stress fields is that (a) stress concentrations can not be located accurately and (b) the influence of the process of slope formation or construction can not be considered.

From the above, it is obvious that stress-distributions based on a stress-deformation analyses would be desirable. For example, locations of stress and strain concentrations could be more realistically visualised using finite element analysis even if the slope material is modelled as a linear-elastic material. However, non-linear analyses will be necessary for accurate stress-deformation studies and especially if potential slip surface location is to be identified on this basis. The disadvantage in opting for stress-deformation studies is that the simplicity and convenience of working within the limit equilibrium framework is no longer retained. One must have reliable data concerning deformation parameters for linear-elastic or non-linear analyses as the case may be. A reasonably sophisticated computer program which has been proven should be available. A knowledge of the history of slope formation or construction is necessary and must be used in the stress-deformation studies. In other words, the complete approach will have to change. The determination of 'excess' shear stress along a potential slip surface and its redistribution would require the performance of several, iterative stress-deformation studies. The process would, therefore, be complex with no assurance of success in modelling the development of failure. However, in principle, such an alternative simulation process, although complex, should work.

The redistribution of 'excess' shear stress, as envisaged in the proposed studies within a limit equilibrium framework, follows relatively simple procedures. However, some assumptions may have to be made concerning the manner in which redistribution occurs from 'failed' to 'unfailed' segments of a slip surface.

The work which has been done so far, in respect of individual case studies, assumes knowledge of a potential or real slip surface. This is either the observed slip surface based on investigation of a failure or a theoretical critical slip surface within a limit equilibrium framework or some arbitrary slip surface used for illustration purposes. Little attention has so far been given to the identification of potential critical slip surface on the basis of the potential for progressive action and the associated minimum factor of safety reached in an analysis.

4   EXPERIENCE OF APPLICATION OF THE PROPOSED METHOD

The method discussed above has been applied to a number of illustrative and case studies, some of which have been published in the literature. Some studies have been made for hypothetical examples for illustrative purposes only. The main aim here is to consider the overall experience of these analyses in perspective and draw some meaningful conclusions.

4.1 Earth Dam Analysis

This stability analysis was concerned with the upstream face of an earth dam 6.1m high for the steady seepage condition at full reservoir level (Chowdhury, 1986). The dam foundation was considered to be firm (rock) and impermeable. The factor of safety F was first found from the conventional Bishop method considering the peak shear strength parameters as follows:

$$c' = 4.31 \text{ kN/m}^2, \quad \phi' = 32^o,$$

$$F = 1.29$$

The residual shear strenth parameters were

$$c_r' = 0; \quad \phi_r' = 32^o$$

The final factor of safety after applying the proposed method incorporating progressive action was found to be

$$F = 1.14$$

During the iterative simulation process leading up to this value, five of the nine segments of the slip surface were identified as overstressed and, therefore, strain-softened.

It can be seen that, although the soil is only moderately strain-softening in its behaviour, there is a significant reduction in the calculated factor of safety as a result of consideration of potential for progressive action.

One interesting conclusion was that the value of brittleness index alone, which varies from slice to slice along the slip surface, does not determine the potential for progressive action for each slice. Other factors such as shear stress and pore water pressure are also important.

4.2 Short-term Stability of Saturated Clay Slope Under Undrained Conditions; "$\phi = 0$" Analysis

These analyses were carried out for the Bradwell slip case study (Skempton and La Rochelle, 1965). The complete study of development of progressive action is to be presented as a separate paper and,

due to space limitations, the tables of calculation are not included here. However, some significant results are presented here.

For the slip designated as Bradwell Slip 1, two slip surface were considered. These are shown in the appropriate reference (Skempton and La Rochelle, 1965) as Slip Surface 1 and Slip Surface 2. The soil consists of Brown London Clay with a thickness of about 8.5 metres with a layer of Marsh Clay on top about 3 metres thick. There is about 3 metres of fill over the Marsh Clay. The slip

surfaces in both cases were interpreted to be circular in shape. For Slip 1 the central angle was 51° and for Slip 2 its value was 59.5^o. The peak and residual undrained shear strength values were:

| London Clay | Peak C_u | = 71.74 kN/m^2 |
| | Residual C_u | = 38.26 kN/m^2 |

| Marsh Clay | Peak C_u | = 14.35 kN/m^2 |
| | Residual C_u | = 9.61 kN/m^2 |

The analyses were carried out using subdivision of the potential sliding mass into unequal vertical slices, eleven slices for Slip Surface 1 and 10 slices for Slip Surface 2.

Some results are summarised below:

Slip Surface 1

Factor of safety based on peak shear strength $\quad F_p = 1.35$

Factor of safety based on residual shear strength $\quad F_r = 0.67$

Factor of safety after accounting for progressive action $\quad F = 0.90$

No. of slices failed, $\quad N_f = 4$

No. of Iterations required, $\quad i = 4$

Slip Surface 2

$$F_p = 1.37 \quad \text{(Peak shear strength)}$$
$$F_r = 0.73 \quad \text{(Residual shear strength)}$$
$$F = 0.99 \quad \text{(Progressive action considered}$$
$$\simeq 1.00$$
$$N_f = 4$$
$$i = 4$$

It can be seen that 'failure' can be predicted very accurately for Slip Surface 2 and less accurately for Slip Surface 1. The value of F = 0.90 is somewhat low for this surface and may indicate that true "$\phi = 0$" conditions did not prevail throughout the soil mass. Moreover, in traditional limit equilibrium approaches, it is difficult to account for system effects and the retrogression of failure. Nevertheless, the predictions based on either peak or residual shear strength parameters are quite wrong. The proposed simulation of progressive action is largely successful in comparison to these analyses.

The Bradwell slip was examined in more detail again (Chowdhury, 1989) and the results will be published in the comprehensive publication to which reference was made earlier.

4.3 Several 'Effective Stress' Analyses of Slope Stability

Several published case histories (see, for instance, Skempton, 1964) concerning long-term failures of slopes in London Clay were re-analysed and the results are too voluminous to include here (Bertoldi, 1988). For these case histories, the following general conclusions were reached after performing many analyses using both simplified and rigorous methods of analysis:-

(1) Analyses based on peak shear strength parameters overestimated the factors of safety. In other words, factors of safety were significantly higher than one in each case.

(2) Analyses based on residual shear strength parameters generally underestimated the factor of safety. In other words, the factors of safety were smaller than one and, in most cases, significantly so.

(3) Analyses based on simulation of progressive action showed that factors of safety were somewhat lower than values based on the peak shear strength. However, in most cases, the factors of safety were still not close to one.

The general conclusion was reached that, for shallow slip surfaces, progressive action, in the context considered in the proposed approach, is limited in its extent and effect. This is particularly so if the slip surfaces are planar.

CONCLUSIONS

A straightforward approach for simulation of progressive action which may occur in slopes and embankments due to local overstressing of soil is discussed in this paper. Attention is focussed on soils which exhibit strain-softening characteristics either during drained deformation or during undrained deformation or both. The approach works strictly within the framework of conventional limit equilibrium analysis. It is necessary to perform iterative analyses and any simplified or rigorous limit equilibrium method may be used. The concept of limit equilibrium is not violated.

The method proposed here has been extensively studied with reference to a number of applications. Case studies include those concerned with short-term stability of saturated clay slopes, stability of embankment dams and long-term, effective stress analyses. In most cases, involving relatively medium to deep-seated slip surfaces, the method works well and analyses show significant progressive action. Consequently, values of actual factors of safety are significantly lower than those based on peak shear strength parameters and significantly higher than those based on residual shear strength parameters. Actual failures can be 'predicted' with improved accuracy. In other

words, back-analyses of slides are more successful than similar analyses which do not take progressive action into consideration.

The method does not give significantly encouraging results for back-analyses involving shallow slides. This is particularly noticeable with slides in which the failure surfaces are planar.

Recently, an analysis of progressive failure of the Carsington embankment (United Kingdom) was presented by Potts et al (1990). A very sophisticated numerical analysis package was specially developed to simulate the failure process adequately. However, the simple method presented here achieved almost similar success entirely within a limit equilibrium framework. The results are to be published separately.

ACKNOWLEDGEMENTS

The writer acknowledges the encouragement and support of the University of Wollongong for research and scholarly activities.

REFERENCES

Bertoldi, C (1988). Computer Simulation of Progressive Failure in Soil Slopes, PhD Thesis, University of Wollongong.

Chowdhury, R N (1986). Improving Earth Dam Stability Analysis, Proceedings IGC 1986, I.I.T, Delhi, Vol. 2, 109-112.

Chowdhury, R N (1989). Unpublished Work.

Potts, D M, Dounias, G T and Vaughan, P (1990). Finite element analysis of progressive failure of Carsington embankment, Geotechnique, Vol. 40, No. 1, 79-101.

Skempton, A W (1964). Long-term stability of clay slopes, Geotechnique, Vol. 14, No. 1, 77-101.

Skempton, A W and La Rochelle, P (1965). The Bradwell Slip - A short-term failure in London Clay, Geotechnique, Vol. 15, No. 3, 221-242.

Landslides, Bell (ed.) © 1995 Balkema, Rotterdam, ISBN 90 5410 032 X

The ACADS slope stability programs review

I.B. Donald
Department of Civil Engineering, Monash University, Clayton, Vic., Australia

P.S.K. Giam (dec.)
Formerly: Monash University, and Golder Associates Pty. Ltd, Victoria, Australia

ABSTRACT: The paper describes a review of computer programs for slope stability analysis currently in use in Australia. Example problems covering the range of current practice were sent to the profession for analysis and the results are presented in tabular and histogram form. The methodology behind the production of the report of the review is described briefly. Available programs appear to cope with straightforward problems reasonably well, but more complex situations produced an alarmingly wide range of results, highlighting the urgent need for improvements in general practice.

1 INTRODUCTION

In 1988 a project was initiated by ACADS (an Australian/New Zealand Association of technical computer users) to review the state-of-the-art in Australasia of computer analysis of slope stability problems. The authors were contracted to carry out this review which resulted in the publication of the ACADS Report No. U255, Soil Slope Stability Programs Review, 1989. This paper summarises the methodology behind the preparation of the report, the results obtained and the conclusions reached, though a full understanding may be obtained only from the complete report (Donald and Giam, 1989).

1.1 Aims

The aims of the project are summarised as follows:

1. To compile a list of slope stability programs currently in use in Australia, together with tabulated information on their full capabilities.
2. To test these programs on a set of examples devised to cover most situations arising in normal and relatively advanced practice.
3. To encourage the use of more advanced computer methods.
4. To provide an opportunity for users to validate their own software against a set of referee programs.
5. To prepare a state-of-the-art report to assist users in choosing a relevant program for a particular task, having particular regard to its accuracy, capabilities, computer requirements and user friendliness.

1.2 General methodology

The methodology used to satisfy these aims may be divided into five main activities:

1. The preparation of a questionnaire on programs currently in use and the collation of the information received into a user-friendly table.
2. The design and presentation of a set of standard examples covering all important aspects of and problems in stability analyses likely to be met in regular practice.
3. The concept and relevance of correct or referee answers to all problems and the selection and invitation of a panel of experts to provide them.
4. The collection and collation of the solutions to the examples provided by 28 practitioners and the presentation of this material in a readily assimilable form.
5. Analysis and discussion of the data received, together with conclusions on the state-of-the-art and recommendations for future improvements.

2. OTHER SURVEYS

Similar reviews have been reported previously, including Fredlund (1978), Lumsdaine and Tang (1982), Mostyn and Small (1987) and an earlier ACADS Report, document UDC 681.32 - Slope Stability Programs (1977). Neither Fredlund nor the earlier ACADS report attempted any significant testing of program capabilities when applied to a range of example problems, but in the Hong Kong Geotechnical Central Office study described by

Table 1:   Example problems requirements

Prob.	Fig.	Soil Properties	Additional data	Analyses ♠ Required	Comments
Ex. 1(a)	Fig. 1	Table 1.1		Crit. failure surface and corresponding F_{min}	A simple homogeneous slope.
Ex. 1(b)	Fig. 1	Table 1.2		Crit. failure surface and corresponding F_{min}	Tension crack to be estimated by user
Ex. 1(c)	Fig. 1	Table 1.3		Crit. failure surface and corresponding F_{min}	A simplified non-homogeneous slope
Ex. 1(d)	Fig. 1	Table 1.3	Horizontal seismic acceleration = $0.15g$	Crit. failure surface and corresponding F_{min}	(as above)
Ex. 2(a)	Fig. 2	Table 2.1	Tab. 2.3 – Geometrical data	Crit. failure surface and corresponding F_{min}	Talbingo Dam
Ex. 2(b)	Fig. 2	Table 2.1	Tab. 2.3 – Geometrical data Tab. 2.2 – Slip circle	F for the given failure surface	Talbingo Dam
Ex. 3(a)	Fig. 3	Table 3.1		Crit. failure surface and corresponding F_{min}	Homogeneous slope with weak layer
Ex. 3(b)	Fig. 3	Table 3.1	Tab. 3.2 – Failure surface	F for the given failure surface	(as above)
Ex. 4	Fig. 4	Table 4.1	Tab. 4.2 – Ext. Loadings Tab. 4.3 – Piezometric Data	Crit. failure surface and corresponding F_{min}	(as above)
Ex. 5	Fig. 5	Table 5.1	Fig. 6 – Approx. Flow Net	Crit. failure surface and corresponding F_{min}	Excavation with water table

♠ For each analysis it is essential to provide brief details of the methods of analysis used, including program names. (e.g. by reference to questionnaire)

Lumsdaine and Tang, 29 clearly defined test problems based on real-life local situations were offered and respondents required to calculate the safety factor, F, for a specified failure surface. Fifty four sets of solutions were submitted, but, because of local soil conditions and established practice, 75% were based on the Janbu simplified method of analysis, thus limiting the study's general applicability. A wide scatter of results was obtained and 55% of all submissions were returned for recalculation of at least one problem.

The Sydney survey reported by Mostyn and Small (1987) used a single problem geometry, with variations of soil parameters creating five sub-problems, most of the submitted analyses for which were based on Bishop's (1955) simplified method. The results again indicated an unsatisfactory situation with a wide range of answers even for simple, well-defined problems with the failure surface generally specified.

In the current project it was therefore decided to encourage the application of a comprehensive set of programs to a sizeable group of problems including most of the conditions and difficulties likely to be met with in daily practice.

3 QUESTIONNAIRE

A detailed document was prepared for mailing to potential respondents. This included a 3 page questionnaire on which could be entered, for each program in use in a person or firm's practice,

information on the method of analysis used, the handling of 2D or 3D analyses, seismic effects, strength anisotropy, work softening, soil models, pore pressure input, failure surface selection, probabilistic analyses, mechanical slope stabilisation and computer information such as hardware requirements, pre - and post - processing, program user friendliness (self rated) and original source of program.

In all, details were received for 42 programs and the data was collated into a 15 page Appendix to the report

4 EXAMPLE PROBLEMS

A set of 5 examples, some with variants giving 10 in total, was designed to test most aspects of regular practice. The examples were totally defined with regard to geometry, soil properties, ground-water conditions and surcharge loadings and the data was presented in a simple, clear format to minimise the possibility of input errors and provide a constant basis for meaningful comparison of results. The only scope for personalised input lay in the selection of tension crack depth and the choice of critical failure circle or surface. Some consideration was given to specifying a tension crack depth in example 1(b), but it was finally decided that its determination was part of the analysis procedure and should therefore be left to the user. As some of the examples presented significant difficulties in finding the global minimum F corresponding to the critical failure surface, a subsidiary example was provided with a specified failure surface so that answers could be more readily

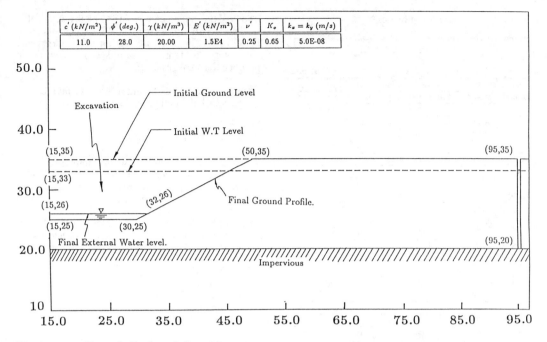

c' (kN/m^2)	ϕ' $(deg.)$	γ (kN/m^3)	E' (kN/m^2)	ν'	K_o	$k_x = k_y$ (m/s)
11.0	28.0	20.00	1.5E4	0.25	0.65	5.0E-08

Fig. 1:   Example 5 - data sheet

compared. Elastic parameters were given for all examples to encourage use of methods such as finite and boundary elements.

A summary of the 10 problems is given in Table 1. (Note that the Figure and Table numbers in Table 1 refer to the ACADS Report U255.) The data page for Example No. 5 is reproduced as Fig. 1. The examples were designed to cover a range of geometries from simple slopes to complex cross sections such as Talbingo Dam, single and multiple soil layers, stable and unstable slopes, steady state pore water pressures or total stress analyses without pore pressures, thin weak layers, tension cracks and both circular and highly non-circular failure surfaces. Critical failure surfaces had to be found by respondents for eight of the problems, while failure surfaces to be investigated were specified for the other two.

5. REFEREE VALUES

5.1 Concept of referee value

The presence of a well defined modal value in a histogram of results is no guarantee that the correct answer has been found. For most geotechnical stability problems it could be argued that there is no such thing as a "correct" safety factor, unless it happens to equal 1.00. For any given problem a range of answers may be found, depending on how safety factor has been defined and what assumptions

are necessary for the method of analysis. However, if a survey such as the present one is to have any meaning, there must be a reference or benchmark value of F for each problem, based on accepted state-of-the-art technology and subjected to all available tests of its validity. Even when a variety of analytical methods has been used to solve a given problem, the mean value of F can still be significantly different from the value which would be accepted as correct by recognized experts in the discipline. With many programs available, often giving different results on the one problem when used by the same operator, the teasing question arises of "who referees the referee values?". There is no simple answer to the question, but in the next section referee values will be discussed in detail and justified as rationally as possible.

5.2 Referee programs

Five significantly different programs were used in the referee calculations. The invited contributors were R. Baker, The Technion, Haifa, with a program based on Spencer's method (Baker, 1980); Z.Chen, Institute of Water Conservancy, Beijing, using Chen and Shao's (1988) improvement of the Morgenstern-Price method and D.G. Fredlund using the most recent version of the General Limit Equilibrium Method (similar to Morgenstern-Price) as described by Fan, Fredlund and Wilson (1986). The authors also contributed two referee solutions with their general wedge analysis programs GWEDGEM and EMU.

Table 2:   Solution statistics

Example No.	Method of Analysis	Factor of safety submitted				Number of submissions
		Mean	Standard deviation	Min. F	Max. F	
1(a)	(ALL)	0.991	0.030	0.940	1.080	33
(1.00)†	BISHOP	0.993	0.015	0.960	1.030	18
	JANBU	0.978	0.041	0.940	1.043	7
1(c)	(ALL)	1.381	0.065	1.230	1.520	31
(1.39)†	BISHOP	1.406	0.039	1.350	1.500	16
	JANBU	1.325	0.057	1.264	1.400	6
1(d)	(ALL)	0.973	0.052	0.900	1.100	15
(1.00)†						
2(a)	(ALL)	2.008	0.103	1.830	2.201	24
(1.95)†	BISHOP	2.049	0.116	1.830	2.201	14
2(b)	(ALL)	2.239	0.110	1.948	2.390	24
(2.29)†	BISHOP	2.204	0.013	2.180	2.220	11
3(a)	(ALL)	1.365	0.166	1.191	1.820	26
(1.26?)†	BISHOP	1.559	0.146	1.400	1.820	7
	JANBU	1.307	0.138	1.191	1.564	8
	NON-CIRCULAR	1.293	0.105	1.191	1.564	19
3(b)	(ALL)	1.291	0.063	1.197	1.450	30
(1.34)†	JANBU	1.252	0.060	1.197	1.333	10
4	(ALL)	0.887	0.255	0.410	1.360	26
(0.78)†	BISHOP	1.151	0.175	0.850	1.360	6
	JANBU	0.984	0.198	0.712	1.287	7
	NON-CIRCULAR	0.808	0.221	0.410	1.287	20
5	(ALL)	1.464	0.139	0.910	1.600	23
(1.53)†	BISHOP	1.431	0.192	0.910	1.600	10

† Adopted Referee Value

It was gratifying to find that in almost all cases the referee solutions were so close that it was possible to state the referee value of safety factor confidently to within ± 0.02: In the small number of cases where a slightly aberrant value was submitted, closer inspection invariably revealed the reason for the discrepancy, e.g. a different assumption used, and again it was possible to give a definitive solution.

6 THE PROGRAMS REVIEW DOCUMENT

It was decided to prepare a detailed document for mailing to all known interested persons and organizations. This document was to include:

1. A statement of the purpose and aims of the review and a request for cooperation.
2. General instructions.
3. Test examples including geometrical and soil property data in readily useable form, to limit the work required for preparing input.
4. Forms for systematic return of results, to ensure that all necessary information was provided by respondents.
5. Several copies of the questionnaire on aspects of program capabilities not revealed by the test examples. These were used in the compilation of Appendix C, Document U255.

The final document, comprising 47 pages, was mailed or handed to approximately 120 recipients. The initial returns were promising but below target and two rounds of reminder letters were required to produce the final result of:-

replies received (excluding referee results) 42
replies including all or some calculations   28

7 DATA PRESENTATION

All calculations submitted were summarised in a large Table as Appendix D in Report U255. The 42 programs or program variants used were each given a name and a clearly recognizable colour symbol in the first column of the Table, with the calculated values of Safety Factor in the remaining 10 columns. Small alphabetic symbols were also attached to the values to indicate the method of analysis used, viz:-

f	Fellenius
b	Bishop (simplified)
bm	Bishop ("rigorous")
j	Janbu (simplified)
jg	Janbu (generalised)
s	Spencer
sr	Sarma
mp	Morgenstern-Price
gw	Generalised Wedge
e	Energy method
wf	Wedge (conventional)
n	Numerical Stress Analysis

The results were also presented in the main body of the text in two ways - firstly as histograms of Factors of Safety, with the x-axis as both numerical value and % difference from the referee value and secondly as plots of critical failure surfaces found from the analyses. Both sets of figures were presented in colour in ACADS Document U255, using the same colours and symbols as described above. Typical examples, regrettably only in black and white, are given in Figs. 2 and 3. It should be emphasised that all results were presented anonymously, as initially promised.

8 ANALYSIS OF RESULTS

A summary of the methods of solution used indicated that Bishop's Simplified Method was still by far the most popular, with Janbu's Generalised Method in a distant second place. For non-circular slips Janbu's Method - in either its simplified or generalised form - was most popular, with the inappropriate Bishop's Simplified Method taking next place ahead of Spencer's, Morgenstern-Price's or Sarma's more relevant methods. A disappointing feature was the limited use made of the more powerful modern numerical techniques. The highest number of returns was 33 for the simplest example, No. 1(a), and lowest

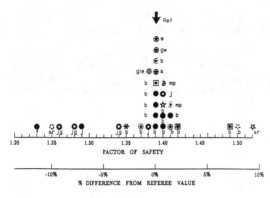

(a)   Example 1(c)

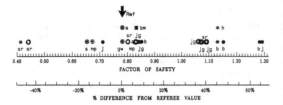

(b)   Example 4

Fig. 2:   Histograms of safety factor

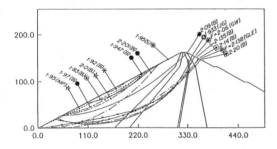

Fig. 3:   Failure surfaces, Example 2(a)

was 15 for example 1(d), which required pseudo-static seismic analysis. The next lowest return was 23 for example 5, which was probably the most conceptually difficult example.

All solutions are summarised in Table 2, taking into account the prominent roles played by the Bishop and Janbu solutions. A glance at the values in this Table show that while for some problems the spread of results is within acceptable limits, for others the limiting variations from the mean are alarmingly large and cause for concern about some current practices.

9 DISCUSSION

It was not possible in this review to isolate the cause of any errors in submitted calculations. The project had the resources to test only the combination of program-operator-computer and where apparent errors or inconsistencies arose it was impossible to attribute them specifically to a single member of this trio. However, it was considered that the form of testing used was representative of actual practice and fully justified.

The histograms presented in Fig. 2 illustrate two of the major findings. For relatively straightforward problems a well defined modal value of F was obtained, in close agreement with the referee value. However for problems such as Example 4, where a weak layer invalidates circular failure surfaces, the spread of F values was alarmingly broad, with no more than two values in agreement anywhere across the range. This behaviour is also visible in fig. 2(a), which shows both well clustered values and a significant spread.

In many cases a value of F submitted reflected a poor choice of critical failure surface. Fig. 3 shows the failure surfaces for Talbingo Dam and they are seen to cover most of the left hand slope. The global minimum safety factor,1.95, is obtained for a planar surface coincident with the outer rockfill surface, while failure mechanisms passing through the core lead to a local minimum of 2.06. In general, failure surfaces were well chosen for problems where circular analyses were valid and often poorly chosen where wedge failures were more appropriate.

10 CONCLUSIONS

The inescapable overall conclusion is that the local state-of-the-art is not as good as it could or should be, especially for problems with complex details or requiring non-circular failure surfaces. For almost all examples the submitted safety factors cover a wide range, as a glance at Table 2 will confirm. The criteria for judging whether or not a result is "good" or "bad" are not clearly defined. Although results have been quoted to two decimal places many would claim that, in practice, only the first decimal is significant. This is undoubtedly often true, particularly when the safety factor is high, but never-the-less modern computer programs have the potential for highly accurate calculations within the constraints of the assumptions involved in any method. This review was looking at the performance of programs in isolation from other factors which might influence the significance of a calculated F value and hence agreement to something better than ± 0.05 was expected for "good" performance. As an extreme case, consider example 4 which is clearly an unsafe slope, yet ten solutions out of twenty three gave F > 1.00, often by appreciable amounts.

A relatively limited range of methods of solution was used. Bishop's simplified Method remains by far the most popular and it performed well for relatively uncomplicated problems with reasonably homogeneous soils. However it was also incorrectly used for problems requiring highly non-circular failure surfaces and, as it was not designed for such cases, naturally its performance was less satisfactory.

A limiting aspect of many programs is the lack of a critical surface search routine or at least of an effective one. Many programs based on circular surfaces still rely on the conventional grid search, star search, line search and tangent search while most programs for generalised surfaces are only designed to analyse a given failure surface. Few programs incorporated Mathematical Programming search routines which have been found to be more efficient (Giam and Donald 1989b) for the selection of critical circular on non-circular failure surfaces. The results show that for many of the examples it is not easy to pick the critical surface without some guidance from the program and some of the largest errors in calculations are caused by this difficulty.

Not many respondents used programs based on the more advanced conventional methods such as Spencer's, Morgenstern-Price's and Sarma's. When properly used these methods gave reasonable to excellent results in the survey, although sometimes hampered by their search routines and with Sarma often misused through poor failure surface choice. A really disappointing feature was the near-total lack of interest shown in advance numerical methods such as Finite Element and Boundary Element, or methods based on energy considerations.

Many parameters were poorly handled by most programs. Pore water pressures were usually input as a piezometric surface coinciding with the phreatic surface. While this did not result in any detectable error in example 5, it is hardly state-of-the-art and for problems where the water table is steeply dipping in the region of the failure surface the errors could become noticeable. Surcharges can usually be handled only as uniformly distributed normal stress and approximations have to be made for complex loadings.

The referee programs stood the test of close scrutiny and the generally excellent agreement between their calculations confirmed, at least in part, the accuracy and relevance of their solutions. The place of their answers in the histograms, relative to the better performed contributor's programs, lends extra weight to their confirmation.

REFERENCES

Baker, R., 1980. Determination of Critical Slip Surface in Slope Stability Computations. Int. J. Num. and Analyt. Methods in Geomech., V11, 333-359.

Bishop, A.W., 1955. The use of the slip circle in the stability analysis of slopes. Geotechnique, V5, 1, 7-17

Chen, Z. and Shao, C., 1988. The use of the method of optimization for minimizing safety factors in Slope stability stability analysis. Chinese Journal of Geotechnical Engineering, V10, No.4, (Total No.45), 1-13

Donald, I.B. and Giam, P.S.K., 1989(a). Soil slope stability programs review. ACADS Publication No. U255, Melbourne.

Donald, I.B. and Giam, P.S.K., 1989(b). Improved comprehensive limit equilibrium stability analysis. Civil Engineering Research Report No.1/1989, Monash University.

Fan, K., Fredlund, D.G. and Wilson, G.W., 1986. An interslice force function for limit equilibrium slope stability analysis. Can. Geotech. J., V23, 287-296.

Fredlund, D.G., 1987. Usage, requirements and features of slope stability computer software (Canada, 1977). Can. Geotech. J., V15, 83-95.

Lumsdaine, R.W. and Tang, K.Y., 1982. A comparison of slope stability calculations. Proc 7th South East Asian Geotech. Conf., Hong Kong, 31-38.

Mostyn, G.R. and Small, J.C., 1987. Methods of stability analysis. Soil Slope Instability and Stabilisation, Walker and Fell (Ed.), Balkema, 71-120.

Landslides, Bell (ed.) © 1995 Balkema, Rotterdam, ISBN 90 5410 032 X

Present state of development of XPENT, expert system for slope stability problems

R.M.Faure
École Nationale des Travaux Public de l'État, France

D. Mascarelli
INSA, Lyon, France

J.Vaunat & S.Leroueil
Université Laval, Que., Canada

F.Tavenas
McGill University, Montréal, Que., Canada

ABSTRACT: Originally created as a geotechnical tool in 1988, XPENT has now been developed into a comprehensive system including modules for the analysis of regional geomorphological information, for the validation of geotechnical data and for the design of slope rehabilitation measures. This paper describes the present state of XPENT with particular reference to the different knowledge bases and the interfaces which comprise this expert system.

1 INTRODUCTION

Expert systems are particularly well-suited to the analysis of slope stability problems since these involve the combination of engineering data, qualitative information of a more or less fuzzy nature, experience and intuition of the person conducting the analysis. Many developments in such expert systems have been reported in recent years (Brown & Singh, 1987; Grivas & Reagan, 1988). In 1988, Faure et al. have presented the preliminary outline of XPENT, an expert system for the analysis of slope stability problems. In its original form, XPENT was intended to serve as a decision support tool for geotechnical engineers confronted with a slope stability problem.

In the last four years, XPENT has been the object of a major development so as to be adaptable to the kind of slope stability problems encountered in different regions of the world; to this end, cooperation has been initiated with researchers in Italy and Morocco. In its new form, presented here, XPENT is characterized by a more complex structure reflecting the enlarged scope and the growing multidisciplinarity of the system, by the application of advanced artificial intelligence (AI) techniques and by the implementation of powerful graphical interfaces.

2 GENERAL STRUCTURE OF XPENT

The solution of a slope stability problem must integrate data relating to the geology, geomorphology, hydrogeology and geotechnique of the site as well as the region.

The representation of the geotechnical problem at the level of the site has been defined by Faure et al (1988). It included a qualitative diagnosis of the stability based on some regional information and on the stratigraphy of the slope, as this determines the possible failure mode. Depending on the complexity of the case, a quantitative diagnosis could be developed using the strength characteristics of the soils and the local pore pressures.

In the last four years, XPENT has been significantly enlarged to allow for the systematic use of local and regional geological and geomorphological information, with the objective to make the expert system adaptable to a wide range of geological settings as they occur in different parts of the world.

2.1 Methodology for the analysis of an unstable slope

Referring to the Structural Analysis and Design Technique (IGL, 1982), Ke (1990) proposed to decompose any slope stability analysis into four steps: detection of the problem, diagnosis, monitoring and correction.

The **detection** of the problem consists in the gathering and logical ordering of mostly qualitative information which must be referenced in a spatial framework. Ideally at this stage, calls should be made to Geographical Information Systems (GIS), in particular to maps of landslide risks as have been produced in many countries (Carrara & Merenda, 1976; Viberg, 1984; Champetier de Ribes, 1987; Brand, 1988) and to various databases. Artificial intelligence techniques are useful to ensure the coherence and the ordering of the information. Advanced interfaces are necessary to make this stage as easy as possible to the user.

The **diagnosis** allows to locate the problem at hand within a series of classes of stability problems. The wide ranges of geological setting as well as soil type and properties make it necessary to use a methodology which takes into account both qualitative and quantitative data at the regional and local levels, and which integrates the use of common engineering analysis methods (Tavenas, 1984; Hartlen & Viberg, 1988). The AI tools to achieve this comprise a knowledge base connected to regional databases and to specific computing codes.

The **monitoring** implies the design of the monitoring system as well as the gathering and the interpretation of the field observations.

Finally, the **correction** of the instability involves a complex design subject to a large quantity of constraints of economic, geotechnical, or organizational nature (Cartier, 1985). It requires a decision support system which integrates a knowledge base with computing codes to check the efficiency and effectiveness of various design solutions.

2.2 XPENT's structure

The global structure of XPENT, which reflects the above described methodology, is shown in figure 1. The expert system will eventually comprise four distinct knowledge bases:
- DOGE, which supports the geotechnical data analysis and ensures its coherence;
- RISQUES, which supports the linkage with regional databases;
- DIAG, which allows the diagnosis of the slope under consideration;
- PACOS, which helps in the design of monitoring systems and corrective measures.

The central system can call on computing modules at various stages of the analysis and design. The whole system is equipped with sophisticated interfaces allowing the user to input text as well as graphical information and to make calls to various databases. At

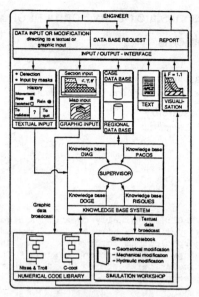

Fig. 1 - XPENT general structure

the end of the session, the system produces a comprehensive report.

2.3 The KOD method

The Knowledge Oriented Design (KOD) method (Vogel, 1988) suggests to describe knowledge relating to a given slope stability problem with three distinct types of information:
- descriptive information on the slope in terms of geometry, stratigraphy, mechanical properties of the soil, hydrogeology, regional characteristics, and on the problem to be solved in terms of risk assessment, diagnosis of failure, type of rehabilitation measures, etc.; in terms of computer formulation, this information is represented by a series of objects.
- actions to be imposed on these objects as part of the analysis, either to simulate a natural process or to examine the sensitivity of the problem to a given parameter, or still to evaluate the effectiveness of a design measure. Within the system, this information is represented by functions associated to the objects.
- empirical rules governing the interaction between the various objects and actions, expressed in the form of logical "production rules" in the expert system.

The KOD method was used to develop the DIAG and PACOS knowledge bases in a homogeneous fashion. The analytical

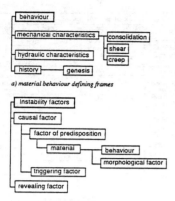

a) material behaviour defining frames

b) instability factors defining frames

*Fig. 2 - Examples of frames in
DIAG knowledge base*

features of diagnosis place greater emphasis on the descriptive information. The DIAG knowledge base is therefore focused on the definition of objects which characterize an unstable slope. Risk analysis was used to make this definition independent of the local context of the slope. Rules have been applied to check the coherence of the data, to estimate the magnitude of possible slope movements and to determine the probability of occurrence.

The design of corrective measures is, in contrast, more a problem of actions and production rules. The KOD method has made it possible to derive these actions and rules from the interview of various classes of experts: generalists (expert responsible for the supervision of this type of project); specialists (expert who has detailed knowledge of a specific technique); validators (experts who were asked to validate the results produced by the expert system).

2.4 Computing hardware and software

XPENT has been developed on SUN and APOLLO and SONY workstations using the expert system shell SMECI (Système Multi-Experts de Conception en Ingéniérie), which provides for both objects/prototypes and rules/strategies (Haren & Neveu, 1987). The development of the user interface was facilitated by the use of the object-oriented graphics language AIDA and of the interface generator MASAI. The SMECI, AIDA and MASAI softwares have been developed by ILOG in France, using the Le-Lisp programming language.

3 DEVELOPMENT OF THE KNOWLEDGE BASES

The DIAG and PACOS knowledge bases are in an advanced state of development; they include both static information in the form of objects and dynamic knowledge in the form of methods and rules as described hereunder.

3.1 Specification of objects

In SMECI, the objects defined are associated with attributes which can be either numerical or symbolic (chain of characters). Prototypes can be defined for specific objects, imposing defined constraints on some attributes of these objects (such as allowable ranges of variations for the mechanical properties of certain soils in a certain region).

In the DIAG knowledge base, two distinct groups of objects and prototypes are defined. The first group describes physically the slope and its environment. It includes a description of the soils, of the slope, of the region and of the geological and climatic context, in the form of coherent frames. Figure 2-a presents a typical frame describing a soil behaviour. The second group of objects relates to the problem to be solved, i.e. the evaluation of the slope stability. This group has been developed using the method for risk analysis proposed by Varnes (1984) and Einstein (1988). This method considers separately the evaluation of the instability factors, of the danger (type and size of the possible event), of the hazard (probability of occurrence of a event) and of the risk (damages created by the event). An example of frames for the definition of instability factors is given in figure 2-b.

Depending on the soils constituting the slope, different prototypes must be considered for the instability factors and the danger. Six classes of soils have been considered:
- hard rock,
- overconsolidated clays, weak rocks, and structurally complex formations
  - weathered rocks and residual soils,
  - loess,
  - cohesionless soils,
  - soft clays.

At the present time, prototypes are fully defined for slope stability problems usually encountered in Quebec, (Lebuis et al., 1983; Tavenas, 1984) i.e. for soft clays; prototypes pertaining to cohesionless soils and the other classes of soils are presently being defined.

The PACOS knowledge base includes, in addition to objects describing the risk

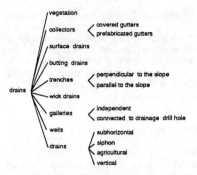

Fig. 3 - *Example of drainage system defining frames in PACOS knowledge base*

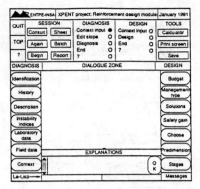

Fig. 5 - *XPENT general board.*

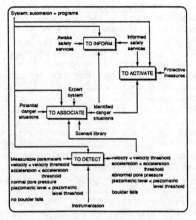

Fig. 4 - *Example of PACOS actinomy: watching actions on slope movement.*

to be mitigated, objects describing the type of work to be carried out to stabilize the slope. Figure 3 illustrates one of the frames relating to the object "drainage systems".

3.2 Specification of rules and methods

In the DIAG knowledge base, rules are used to verify the coherence of the data within each frame. In addition, rules allow the development of new knowledge from the interpretation of raw data. In particular, rules allow the qualitative description of the soil behaviour from its physical identification, e.g. a marine clay with a natural water content in excess of its liquid limit is a brittle material prone to flow slides. Rules also allow the identification of factors favourable to, or factors directly causing failure, from the description of the slope. A

classification system has been developed to determine the different failure mechanisms possibly associated with a given causal factor from the knowledge of the soil characteristics and the predisposing factors; the so-called "simple case" and "stratified case" described by Faure et al (1988) are part of this classification system. Special rules allow the definition of the size of event, the associated hazard and risk in cases of simultaneous occurrence of many causal factors.

In the PACOS knowledge base, methods and rules are defined from actinomies. Figure 4 illustrates the actinomy relating to the monitoring of an unstable slope. The knowledge base includes a module to validate the data and a module suggesting corrective measures. Rules are first used to define a range of solutions depending on the factors of instability and the type of movement. This range is then narrowed by considering rules expressing economic and technical constraints. The remaining solutions are then subjected to numerical analyses to determine the gain in stability. Rules are finally applied to select the solution providing the most favourable combination of gain in stability, ease of implementation and economy.

4 DEVELOPMENT OF THE INTERFACES

XPENT involves three different types of interfaces: interface between the system and the user; interface between the system and data bases; interfaces between the system and numerical computer programs. All interfaces have been designed to be independent of the type of knowledge base to access and to allow for an easy expansion of the system.

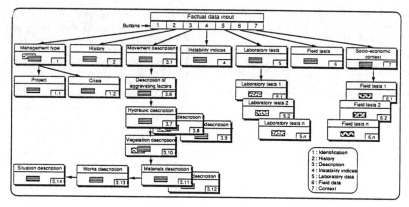

Fig. 6 - Textual input mask tree.

4.1 User interface

This is the most important interface since it conditions the ease of use and thus the effectiveness of XPENT. The user interface is structured as a global entry panel from which the user can proceed to text or graphic input.

Text input. The input of text or numeric information is made easy by the use of sequential masks, organized in a treè structure which facilitates the navigation towards the data to be entered. Data input is subdivided into the geotechnical description of the slope (type of soils, laboratory test data, field test data), the geomorphology of the site (history, evidence of movements, current movement if any, aggravating factors, hydrology of the region, vegetation), the type of intervention (crisis management, slope stability evaluation in connection with a civil engineering project, correction of a long term creep movement), the type of work previously done in the area, and the socio-economic context. It is worth noting that the expert system can operate even if some of the information is lacking. Each mask for data input takes the form of a series of checkboxes, string selectors, radio buttons which force a unique choice, and a line editor to input text. Figure 5 presents the general board of XPENT; figure 6 outlines the tree structure for the text input masks; figure 7 gives an example of input mask.

Graphical input. The input of graphical data is done through the "slope editor". The location map of the slope to be investigated is developed on the basis of a topographical map on which the significant features may then be defined: specific terrain features, outline of the

DESCRIPTION OF AGGRAVATING FACTORS		
Climatic	**Anthropic**	
☐ Freeze thaw	☐ Presence	
☐ Heavy rains	☐ Excavations	
☐ Snow cover	☐ Underground works	
☐ Wind	☐ Reduction of vegetation cover	
☐ Periglacial phenomena	☐ Vibrations	
☐ Snow melt	☐ Modification of mass balance	
	☐ Modification of hydraulic conditions	
Hydrogeological		
☐ Springs		
☐ Water flow	**Geological**	
☐ No flow outlet	☐ Sensitive zone	
☐ Water infiltrations	☐ Sensitive materials	
☐ Impermeable zone at the toe	☐ Seismic zone	
☐ Toe erosion	☐ Volcanic zone	
VALIDATION	?	QUIT

Fig. 7 - Example of mask: description of aggravating factors.

slope movement, location of endangered buildings, location of borings, etc.). The usual techniques for handling graphical objects are used: coincidence of two objects, inclusion of one object in another, intersection, distance between two objects, contacts, projections, cuts and interpolation. The terrain features need not to be precisely referred in the space and they may be qualified in a more descriptive manner (small, large, elongated, compact, distant, etc.) (Buisson, 1989).

The cross-section of the slope is built from graphs representing the geometric profile, the stratigraphy, and the hydraulic regime. The profile and the soil layers are generated automatically from a limited number of input points. The soil strata are described by polygonal surfaces to which are assigned the geotechnical properties of the soils. The data input is done by specifying points and segments or connection between points. Micrometric gauges are available to precisely adjust the location of input points; points may also be specified by entering their coordinates. The automatic generation, from a limited number of data points, of the slope profile or of the soil layers makes it easy to modify these: adding or eliminating a layer or modifying the slope geometry occurs simply by adding, eliminating or displacing a limited number of entry points and segments. The data files corresponding to graphical information are mainly used as input to numerical analyses and they are organized to facilitate the interface with the computer codes used for such analyses.

4.2 Interface with codes for numerical analyses

In order to support both the diagnosis of a slope stability or the design of slope rehabilitation measures, the user of DIAG or PACOS must be able, at some stage in the expert system session, to carry out common engineering analyses of the problem. Two computer codes are available to this end within XPENT: NIXES & TROLLS is the code for the slope stability analysis using a perturbation method for either circular or non-circular failure surfaces (Faure, 1985); C-COOL is a finite element code to determine the hydraulic regime in incompressible porous media in a slope.

Linkage with the slope stability code. To feed NIXES & TROLLS slope stability code, the graphical objects are interpreted into a series of numbers formatted as required by the input data files of the code. In this way, the user may generate from the same set of graphics data any number of input files as required by the analyses to be carried out. Moreover, modifying the input file has no consequence per se on the interface.

The results of the slope stability computation are presented on the screen in the form of the series of failure surfaces investigated with their corresponding factor of safety. It is possible to isolate the most critical failure surface or to present the range of the 10 most critical surfaces.

For use in connection with PACOS for the design of slope rehabilitation measures, the "Simulation Workshop" allows for the interactive modification of the slope geometry or for the introduction of confortative features (anchors, drains) and for the numerical analysis of their effects on the factor of safety of the slope.

Linkage with the groundwater regime code. The usual data entry requirements in the finite element method, the frequently large duration of the

computation and the separation of the code in three modules (mesh generation, FE computation, post-processing), have led to organize the linkage around the use of objects and methods independent of the graphics interface.

An automatic mesh generator is applied to the geometry and stratigraphy defined in the slope cross-section and it is effected whenever the geometry of the problem is modified. This generator creates a series of objects corresponding to nodes, elements and boundary conditions; all these are shown on screen where they can be modified interactively (one can for example easily eliminate a node in the middle of the mesh to simulate a drain). With each series of objects is associated a procedure which automatically calls relevant subroutines of the finite element code, so that the user need only call for assembly and resolution of the problem.

The results of the C-COOL code may be used for two distinct purposes. In a first instance, C-COOL may be used to check the coherence of a series of piezometer readings in a cross-section. Rules have been defined to facilitate the definition of boundary conditions when piezometric values at specified points in the slope are known. If it proves impossible to meet these conditions, the user has tools to examine if this is due to erroneous piezometer readings or to natural features of a complex groundwater regime. In a second instance, C-COOL may be used to generate a mesh of pore pressure values to be used in a stability analysis or to examine the effects of changes to the slope geometry or of the introduction of drainage systems. In this case the results are formatted for direct input in the NIXES & TROLLS program for stability analysis.

4.3 Linkage between the expert system and a database

Two types of databases may need to be linked with XPENT. The first type relates to regional geotechnical databases, allowing for the definition of regional default values for typical soil properties or for the validation of soils investigation results by reference to regional values or to regionally relevant correlations. The linkage with this type of database is presently investigated as part of the development of the DOGE module of XPENT.

The second type of database consists of a registry of slope stability problems in a given region. The objective is to use this registry to compare the features of the problem under investigation to common regional characteristics, thus allowing a reasoning by analogy. Reciprocally, the link between XPENT and the database should facilitate feeding the database with the information on the case under investigation when the XPENT session is completed. Such an interface is presently operational to link XPENT with the INVI database which includes 3,500 case histories of slope problems in France (Aste et al, 1992). The software code DRIVER (Lebastard, 1990) has been developed to this end; this object-oriented code facilitates the interaction between the user and the database.

5 CONCLUSION

As a result of the complexity of any slope stability problem and of the desire to make the expert system applicable world-wide, XPENT has experienced a significant growth in size and scope in the last four years. It has evolved into a system federating different approaches: geographical information systems, regional databases, geomorphological analysis, mechanical analysis. The structure of XPENT has been adapted to this evolution and to support easy linkages between various types of codes. The use of Object Oriented Programming techniques has facilitated this evolution.

In its present state, XPENT is already functional for certain types of problems: diagnosis of slopes in canadian sensitive clays; design of slope rehabilitation measures for common slope instability problems encountered in France.

The next steps of development concern the systematic use of regional mapping information, including the linkage with GIS, the development of the DOGE module for the validation of geotechnical data and the development of RISQUES. It is further hoped that new partners in the XPENT project will enlarge the knowledge bases included in DIAG and PACOS to make the expert system applicable in regional and geological settings other than those considered up to now.

6 ACKNOWLEDGEMENTS

The development of XPENT has been supported by a wide range of sources. NSERC and FCAR have provided funds to support the research activity of the canadian team. The France-Québec cooperative agreement and the Centre Jacques-Cartier in Lyon, France, have provided financial support for meetings, stages for students involved in the project, and for exchange of information and software tools. The authors are indebted to the ILOG company for its continued support in making software development tools available and in providing technical advice.

REFERENCES

Aste, J.P., Gouisset, Y. & Leroi, E. 1992. Projet français d'Inventaire National des Versants Instables. Proc. 6th Int. symp. on landslides, Christchurch (New-Zealand).

Brand, E.W. 1988. Special lecture: landslide risk evaluation in Hong-Kong. Proc. of the 5th int. on landslides, Lausanne (Switzerland), pp. 1050-1074.

Brown, D.J. & Singh, R.N. 1987. An expert system for slope stability assessment: Part I. Int. Journal of Surface Mining, 1: 173-178.

Buisson, L. 1989. Reasoning on space with knowledge object centered representation. Symp. on design and implementation of large spatial database, Santa Barbara (California).

Carrara, A. & Merenda, L. 1976. Landslide Inventory in Northern Calabria, Southern Italy. Bulletin of American Geological Society, 87(8): 1153-1162.

Cartier, G. 1985. Guide pour les études et les confortements de glissements de terrain. Internal Report of Laboratoire Central des Ponts et Chaussées, Paris (France).

Champetier de Ribes, G. 1987. La cartographie des mouvements de terr.: des ZERMOS aux PER. Bulletin de Liaison du Laboratoire des Ponts et Chaussées, Paris (France), 150-151: 9-19.

Einstein, H.H. 1988. Special Lecture: Landslide Risk Assessment Procedure. Proc., 5th Int. symp. on landslides, Lausanne (Switzerland), vol. 2: 1075-1091.

Esu, F. 1977. Behaviour of slopes in structurally complex formations. International Symposium on the Geotechnics of Structurally Complex Formations, Capri (Italie).

Faure, R.M. 1985. Analyse des contraintes dans un talus par la méthode des pertubations. Revue française de Géotechnique, Paris (France), 33: 49-59.

Faure, R.M., Leroueil, S., Larochelle, P., Sève, G. & Tavenas, F. 1988. Xpent, système expert en stabilité des pentes. Proc. 5th int. symposium on landslides, Lausanne (Switzerland), vol. 1: 625-629.

Grivas, D.A. & Reagan, J.C. 1988. An expert system for the evaluation and treatment of earth slope stability. Proc. 5th int. symp. on landslides, Lausanne (Switzerland), vol. 1: 649-654.

Haren, P. & Neveu, B. 1987. SMECI: an system expert for civil engineering design. 1st Int. Conf. on Applications of AI to Engineering Problems; Southampton (UK).

Hartlén, J. & Viberg, L. 1988. General Report: evaluation of landslide hazard. Proc. 5th int. symp. on landslides, Lausanne (Switzerland), vol. 2: 1037-1058.

IGL. 1982. Introduction à SADT. Paris (France).

Ke, C. 1990. Un système de représentation des connaissances et d'aide à la décision pour la prévention des mouvements de terrains. PhD Thesis, 130p., University Claude Bernard, Lyon (France).

Lebastard, F. 1990. DRIVER: Couche Objet Virtuelle Persistante et Raisonnement. Research Report 1155, Sophia Antipolis, Nice (France).

Lebuis, J., Robert, J.-M. & Rissman, P. 1983. Regional Mapping of land-slide hazard in Québec. Proc. of the symp. on slope on soft clay, Linkoping (Sweden), pp. 205-262.

Tavenas, F. 1984. Landslides in Canadian sensitive clays: a State-of-the-Art. Proc. 4th int. symp. on landslides, Toronto (Canada), vol. 1: 141-153.

Varnes, D.J. and the International Association of Engineering Geology Commission on Landslides and Other Mass Movements on Slopes. 1984. Landslides hazard zonation: a review of principles and practice. UNESCO publication, Paris (France).

Viberg, L. 1984. Landslides risk mapping in soft clays in Scandinavia and Canada. Proc. of 4th int. symp. on landslides, Toronto (Canada), vol. 1: 325-348.

Vogel, C. 1988. Génie cognitif. Masson, Paris (France).

Landslides, Bell (ed.) © 1995 Balkema, Rotterdam, ISBN 90 5410 032 X

Failure mechanisms in 3-D limit equilibrium analysis of soil slopes

U. F. Karim
Amsterdam, Netherlands

H. H. Obaid
University of Technology, Baghdad, Iraq

ABSTRACT: This paper presents a three-dimensional (3-D) limiting equilibrium analysis of cohesive and frictional soil slopes using planar (wedge) and rotational failure mechanisms. An explicit expression of the safety factor is derived in this paper for a new wedge mechanism composed of a central prism with back-slope and two tapered end wedges. A parametric analysis shows that the influence of geometric and physical parameters depends on the type of mechanism used. The potential of the new 3-D wedge is demonstrated by comparing it with the rotational mechanisms. Close agreement with results reported by Gens et al. (1988), using more complex analysis based on slip surfaces generated by power curves is obtained.

1. INTRODUCTION

Slope stability analysis is commonly based on well established two-dimensional (2-D) limiting equilibrium procedures, using plane strain assumptions. Limitations of the plane strain approach has been realized in rock slope engineering (Sherard et al. 1963, Paulding 1970, Hoek and Bray 1977) as well as in soil slopes (Baligh and Azzouz 1975, Giger and Krize 1976, Hovland 1977). Simple 3-D planar mechanisms consisting of two end wedges (Hovland 1977), and rotational cylindrical surfaces with conical or ellipsoidal ends (Baligh and Azzouz 1975) were among the first to be investigated. Other three-dimensional procedures were reported by Chen and Chameau (1982), Ugai (1985), Cavounidis (1987), Hunger (1987a; 1987b), Gens et al. (1988), and Xing (1988), using rotational slip geometries.

The most comprehensive 3-D failure analysis of cohesive soil slopes is given by Gens et al. (1988). In this analysis a family of power curves is used to optimize the end shapes of critical failure surfaces to arrive at minimum 3-D and 2-D safety factors. They contend that 3-D analysis will always yield safety factors when the analysis is based on critical slip geometries (see also Cavounidis 1987). This is in contrast with some of the results given previously by Baligh and Azzouz (1975), Chen and Chameau (1982), Hovland (1977), Hunger (1987a), and Michalowski (1989). More research is needed to resolve this, particularly for frictional-cohesive soil slopes. The influence of the physical and geometric parameters as well as the method of analysis on the calculated safety factors need further investigation.

In this paper the influence of soil strength and geometric parameters on the ratio of 3-D to 2-D safety factors is first investigated for a wide range of slip mechanisms. The stability of homogeneous cohesive (c), frictional (∅), and c-∅ soil slopes are then examined using a new wedge mechanism consisting of a back-sloped central prism and two end wedges. A closed form solution is reported for the new 3-D wedge and the results compared with other solutions.

2. THREE-DIMENSIONAL ANALYSIS

The failure mechanisms used in this study include sliding wedge mechanisms I and II and rotational mechanisms III-VI (Fig. 1). Mechanisms II, IV and VI can be produced from the general surfaces I, III and V respectively by setting the length of the central surface 1c to zero. All the surfaces are generated and analysed by a computer program using a random function as proposed by Siegle et al. (1981).

Limiting equilibrium is formulated first for mechanism I (Fig. 2), to obtain a closed-form solution. A simpler version of mechanism II with backfill slope angle β = 0 was first applied to c-∅ soil slopes by Hovland (1977). It is shown later that this version and other limiting cases can be obtained from the general equation derived for mechanism I. Symmetry is imposed and therefore only the volume extending between the plane of symmetry ABG and the end point E in Fig. 2 is considered.

The normal forces N_1 and N_2, acting on the sliding surfaces ECD and ABDC with areas A_1 and A_2 respectively, can be computed from the dot product of the unit weight W_i and unit normal n_i (i = 1, 2).

$$N_i = W_i \cdot n_i \tag{1}$$

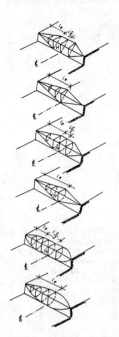

General wedge mechanism I
(planar surfaces)

Wedge mechansim II
(planar surfaces)

Cylinder with end cones
(mechanism III)

End cones (mechanism IV)

Cylinder with elliptic ends
(mechanism V)

Elliptic mechanism (VI)

Fig. (1) Slip mechanism I — VI

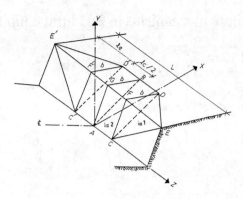

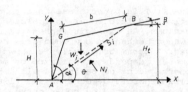

Fig. (2) The geometric parameters of slip
mechanisms I and II

Resolving all vector quantities leads to the following equations:

$$W_1 = \frac{b\ H_t\ l_e\ \sin(\Theta - \beta)}{6\ \sin\Theta} \qquad (2)$$

$$N_1 = \frac{W_1\ \cos\Theta}{B} \qquad (3)$$

$$W_2 = \frac{3\ l_c\ W_1}{2\ l_e} \qquad (4)$$

$$N_2 = W_2\ \cos\Theta \qquad (5)$$

$$B = \left[1 + \left(\frac{b}{l_e}\right)^2 \sin^2(\Theta - \beta) \right]^{1/2} \qquad (6)$$

All the variables appearing in Eqs. 2 to 6 are defined in Fig. 2 and in the notation list. Once these variables are available, safety factors can be derived from:

$$F = \frac{T}{S} = \frac{\text{total available resistance}}{\text{total mobilized resistance}}$$

$$= \frac{c\ A + N\ \tan\phi}{S} \qquad (7)$$

The quantities A, N and T are $A = A_1 + A_2$, $N = N_1 + N_2$ and $T = (W_1 + W_2)\sin\Theta$ respectively. In plane strain problems, $A_1 = N_1 = W1 = 0$. Areas A1 and A2 for sliding wedges 1 and 2 can be obtained from:

$$A_1 = \frac{H_t\ l_e\ B}{2\ \sin\Theta} \qquad (8)$$

$$A_2 = \frac{l_c}{l_e}\ \frac{A_1}{B} \qquad (9)$$

Substituting A_1 and A_2 into Eq. 7 leads to

$$F_3 = \frac{C}{\gamma\ H}\ G_{c3} + \frac{\tan\phi}{\tan\Theta} \cdot G_{\phi 3} \qquad (10)$$

where

$$G_{c3} = \frac{H\ B\ (l_c/l_e B + 1)}{b(l_c/2l_e\ B + 1/3)\ \sin\Theta\ \sin(\Theta - \beta)} \qquad (11)$$

$$G_{\phi 3} = \frac{(l_c/2l_e + 1/3B)}{(2l_c/2l_e + 1/3)} \qquad (12)$$

and γ is the soil's unit weight. For 2-D slopes, $G_{\phi 2} = 1$, while

$$G_{c2} = \frac{H}{b\ \sin\Theta\ \sin(\Theta - \beta)} \qquad (13)$$

Eq. 13 can be obtained from Eq. 7 using A_2, N_2, W_2, S_2 and noting that end resistance (end effect) is not included in 2-D analysis:

$$F_2 = \frac{C}{\gamma H}\ G_{c2} + \frac{\tan\phi}{\tan\Theta}\ G_{\phi 2} \qquad (14)$$

The ratio F3/F2 for the general problem can be found from Eqs. 10 and 14. In the limiting condition of cohesionless soil (c=0), this ratio becomes:

1680

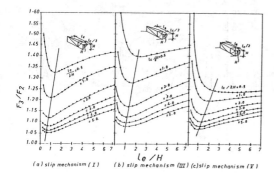

Fig. (3) Effect of shear surface geometry on F_3/F_2 in vertical clay cuts
($\phi = 0$, $C = 49 kN/m^2$, $\gamma = 16 kN/m^3$, $H = 15.3 m$)

(a) slip mechanism (I)   (b) slip mechanism (III) (c) slip mechanism (V)

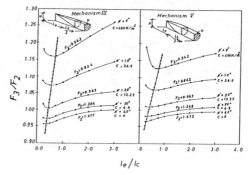

Fig.(5) Effect of strength parameters on F_3/F_2 in vertical cuts
$\gamma = 16 kN/m$, $l_c = 30.5 m$, $H = 15.3 m$ (Mechanisms III, V).

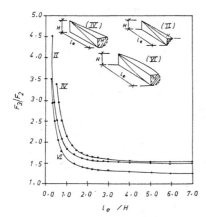

Fig.(4) Effect of end shear surface geometry

on F_3/F_2 in vertical clay cuts.($\phi = 0$,

$c = 49 kN/m^2$, $\gamma = 16 kN/m^3$, $H = 15.3 m$

$l_c = 0$)

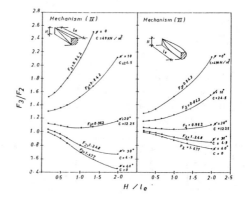

Fig.(6) Effect of strength parameters on F_3/F_2 in vertical cuts
$\gamma = 16 kN/m^3$, $l_c = 0.0$, $H = 15.3 m$ (Mechanisms IV, VI)

$$\frac{F_3}{F_2} = \frac{G_{\emptyset 3}}{G_{\emptyset 2}} = \frac{(1_c/2l_e + 1/3B)}{(1_c/2l_e + 1/3)} \quad (15)$$

And, for frictionless soils ($\emptyset = 0$),

$$\frac{F_3}{F_2} = \frac{G_{c3}}{G_{c2}} = \frac{(1_c/1_e + 1B)}{(1_c/1_e + 2/3)} \quad (16)$$

When the overall length of the failure surface L (L = lc + 2 le) is large relative to the height of the slope H and width b, then B approaches unity and, lc/le approaches zero. Hence, for these limits F3/F2 = 1.5 when ∅ = 0 and F3/F2 = 1.0 for c = 0 (and β = 0). These two limiting situations are exactly equivalent to those considered by Hovland (1977). It should be noted that in the case of mechanism II F3/F2 is at most 1.0 in frictional soil slopes and at least 1.5 in cohesive soil slopes.

Features of the analysis demonstrated earlier for the wedge problem are still valid when curved surfaces (mechanisms III to VI) are considered. The definition of F3 as given by Eq. 7, for instance, is still applicable. But now, the failure surface and center line location of the axes of rotation complicate the computations. It is not possible to obtain analytical solutions to N and S, except for the simple 2-D cylindrical surface, extended to 3-D by including plane end's resistance as is demonstrated by Gens et al. (1988). In this analysis the sliding mass is first subdivided into 3-D slices and forces acting on a slice are evaluated using the procedure given by Hovland (1977).

3. RESULTS OF THE ANALYSIS

The case of homogeneous vertical clay cut (∅ = 0) is used to investigate the influence of geometric parameters for I to VI mechanisms on F3/F2 values (Figs. 3, 4). These plots show that when lc/H is increased relative to le/H, a plane strain condition is reached, indicated by

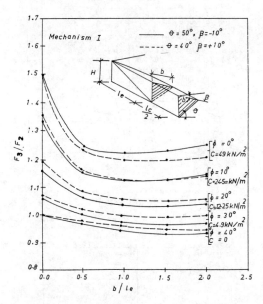

Fig. (7) Effects of strength parameters and backfill angle
(β) on F_3/F_2 in vertical cuts.
(γ = 16 kN/m³, H = 15·3 m, lc = 30·5 m (Mechanism I)

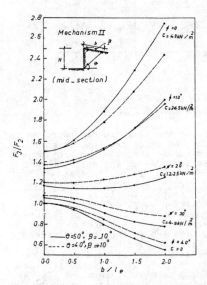

Fig.(8) Effects of strength parameters and backfill

angle(β) on F_3/F_2 in vertical cuts,
γ = 16 kN/m³, H=15·3m, lc=0 (Mechanism II)

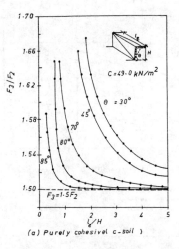

(a) Purely cohesive(c-soil)

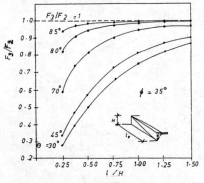

(b) Purely frictional (ϕ - soil)

Fig. (9) Changes in F_3/F_2 with angle of slip plane
(θ)in vertical cuts for slip mechanism(II)
(γ = 16 kN/m³, H=15·3m)

Figures 5 and 6 demonstrate typical
results of the parametric study on the
effect of c and ø on F3/F2 for simple
rotational failure mechanisms in vertical
cuts. The problems defined in these
figures are solved by varying le/lc ratios
(Fig. 5), and H/le ratios (Fig. 6).

The mechanisms in Fig. 5 show more
clearly defined minimum F3/F2 ratios (connected by straight lines) than mechanisms
IV and VI in Fig. 6. Fig. 5 shows that 3-D
effects become more significant with
increasing le/lc and increasing cohesion
relative to the shear angle. It is also
noted that mechanism V gives lower F3/F2
ratios than mechanism III for all values
of le/lc. Also, increasing ø values are
seen to lead to decreasing F3/F2 values
which eventually become less than unity.

Results of the general wedge mechanism
(I) and the simpler version (II) are shown
in Figs. 7 to 10. In Fig. 7 the results
are given for two backfill angles, β = -
10° and +10.0°. With the
dimensions b, H, le, lc fixed, the curves
in Fig. 7 become flatter with increasing ø
and decreasing c values indicating

reduction in F3/F2 ratios. For a slope
with some L/H ratio, there must be one
le/lc which yields a minimum F3 value, as
can be seen from the straight lines
connecting minimum values. These results
also indicate that mechanism I gives
comparable results with the other more
complex mechanisms.

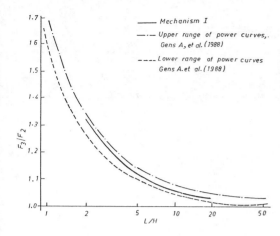

Fig. (**10**) Comparison between wedge mechanism (I) and results

given by (Gens A. et.al)

diminishing 3-D effects. As in the case of rotational mechanisms, wedge mechanism I generally gives decreasing F3/F2 ratios with increasing ø and decreasing c values.

Values of F3/F2 fall slightly below unity for the higher ø values in mechanism I. Mechanism II on the other hand shows a peculiar behavior with F3/F2 falling well below unity for the higher ø values. This behavior is attributed to the mechanism itself and different effects c and ø value have on F3/F2 values as seen in Figs 8,9. As is seen in Fig. 9a, the ratio F3/F2 tends to 1.5 in cohesive vertical cuts when le becomes large relative to H. A similar set of plots is given in Fig. 9b for the purely frictional case, but now all the curves approach unity from below the F3/F2 = 1.0 line. Results by Hovland (1977) show similar trends when β = 0. Mechanism I is investigated further by optimizing the ratio le/lc and changing the slip plane angle Θ to obtain minimum F3/F2 ratios in purely cohesive vertical cuts with flat backfill (β = 0). These results are plotted in Fig. 10 together with those given by Gens et al. (1988) using rotational surfaces defined by a family of power curves and very close agreement is noted.

It is also of interest to note from Fig. 10 that for mechanism I, F3/F2 approaches unity. Mechanism II, on the other hand, used also by Hovland (1977), always leads to F3/F2 ratios other than unity as the 2-D limits are approached. Mechanism II therefore is not a valid failure model since a 2-D limit can never be reached.

5. CONCLUSIONS

Several 3-D rotational and wedge shaped failure surfaces are compared and the effects of several variables on the safety factors of soil slopes investigated. A closed form solution is also given for a new wedge mechanism consisting of a

central prism with a back-slope and two tapered end wedges. The parametric studies show that the influence of geometric and physical parameters depends on the type of mechanism used. It is also noted that in 3-D analysis, increased shear angle may lead to safety factors lower than those obtained from 2-D analysis. Comparison between the general wedge mechanism I and more complex analysis, given by Gens et al. (1988), show very close agreement for the cases studied in this paper. The closed form solution given for the general wedge, due to its simplicity, can be useful for preliminary hand calculations and in back analysis of slope failures.

REFERENCES

Baligh, M.M., and Azzouz, A.S. 1975. End effects on stability of cohesive slopes. J. Geotech. Engrg. Div., ASCE, 101(11): 1105-1117.

Cavounidis, S. 1987. On the ratio of factors of safety in slope stability analysis. Geotechnique, 37(2): 207-210.

Chen, R.H., and Chameau, J.L. 1982. Three-dimensional limit equilibrium analysis of slopes. Geotechnique, 32(1): 31-40.

Gens, A., Hutchinson, J.N., and Cavounidis, S.1988. Three-dimensional analysis of slides in cohesive soils. Geotechnique, 38(1): 1-23.

Hoek, E., and Bray, W. 1977. Rock slope engineering. 2nd ed., The Institute of Mining and Metallurgy, London, U.K.

Hovland, J.H. 1977. Three-dimensional slope stability analysis method. J. Geotech. Engrg. Div., ASCE, 103(9): 971 -986.

Hunger, O. 1987a. An extension of Bishop's simplified method of slope stability analysis to three dimensions. Geotechnique, 37(1): 113-117.

Hunger, O. 1987b. Three-dimensional slope stability analysis by the method of columns, using a micro-computer. Proc. of First Canadian Symposium, Microcomputer Applications to Geotechnique, Regina, Canada.

Hutchinson, J.N., and Sarma, S.K. 1985. Discussion on three-dimensional limit equilibrium analysis of slopes. Geotechnique, (35)2: 215-216.

Michalowski, R.L. 1989. Three-dimensional analysis of locally loaded slopes. Geotechnique, 39(1): 27-38.

Paulding, B.W. 1970. Coefficient of friction of natural rock surfaces. J. of the Soil Mechanics and Foundation div., ASCE, 96(2): 385-394.

Sherard, J.L., Woodward, S.J., Gizienski, S.G., and Clevenger, W.A. 1963. Earth

and earth rock dams. John Wiley and
Sons, Inc., New York, N.Y.

Siegle, R.M., Kovacs, W.D., and Lovell,
C.W. 1981. Random surface generation in
stability analysis. J. Geotech. Engrg.
Div., ASCE, 107: 996-1002.

Ugai, K. 1985. Three-dimensional stability
analysis of vertical cohesive slopes.
Soils and Foundations 25(3): 41-48.

Xing, Z. 1988. Three-dimensional stability
analysis of concave slopes in plane
view. J. Geotech. Engrg. Div., ASCE,
114(6): 658-671.

NOTATION

A_1, A_2	Base areas for sliding prism and wedge as shown in Figure 2
b	Width of wedge
c	Cohesion
F2, F3	2-D and 3-D factors of safety
G_{c2}, G_{c3}	2-D and 3-D factors
$G_{\emptyset2}$, $G_{\emptyset3}$	2-D and 3-D factors
H	Slope height
H_t	Height as defined in Figure 2
L	Length of a sliding mass as defined in Figure 2
l_c, l_e	Central and ends lengths as defined in Figure 2
N_i	Normal force vector (i=1,2)
N_1, N_2	Normal forces on plane 1 and 2
n_i	Unit normal vector components (i=1,2)
S	Total mobilized shear resistance
T	Total available shear resistance
W_i	Weight vector (i=1,2)
W_1, W_2	Weights as defined in Figure 2
α	Slope angle
β	Angle of backfill with the horizontal
γ	Unit weight
Θ	Angle of failure plane with the horizontal
$\emptyset$	Angle of frictional resistance

Landslides, Bell (ed.) © 1995 Balkema, Rotterdam, ISBN 90 5410 032 X

Application of repeated photogrammetric measurements at shaping geotechnical models of multi-layer landslides

Ž. Ortolan, Z. Mihalinec & B. Stanić
Civil Engineering Institute, Zagreb, Yugoslavia

J. Pleško
Faculty of Geodesy, Zagreb, Yugoslavia

ABSTRACT: Repeated photogrammetric measurements are a valuable source of information during final shaping of geotechnical models of complex multi-layer landslides. Photogrammetric measurements of a sequence of detailed unchanged points enable establishing the clear limit between the moved and stable parts of the terrain. The paper provides a number of practical and theoretical findings resulting from an adequate interpretation of photogrammetric measurements made on a thoroughly analyzed landslide where a three-level slide was observed.

1 INTRODUCTION

Stability analyses are preceded by shaping of geotechnical models. An adequate geotechnical model is the basic precondition for the successful stability analysis and for the efficient landslide improvement. Deep landslides are particularly complex since clear discontinuities of minimum shear resistance parameters can not be determined in advance. The problem becomes even more complex in the case of landslides composed of several layers (multi-layer landslides). In such situations, geotechnical model solutions may be so imaginative that their reliability may become questionable. It is in such cases that the old and the repeated aerial survey prior and after the landslide activation may prove valuable. The possibilities of photogrammetric measurement of a number of detailed points, whose configuration remained unchanged in the studied period, will be presented on an example of landslide situated in Podsused near Zagreb.

2 BRIEF LANDSLIDE DESCRIPTION

The cement plant "Croatia" (later called "Sloboda") was first opened in 1908 in the locality of Podsused. This plant was closed in 1988 after the authors of this paper proved that the plant is located on a large landslide (Ortolan et al., 1987).

An area of approx. 1.2 km^2, where roughly 500 residential and industrial structures are located, is influenced by sliding of the sides and north part of the Kostanjek open-cast mine. It was established that the landslide is of three-layer type. The maximum depth of the deepest sliding surface is about 90 m, the depth of the intermediate sliding surface is 65 m, while the superficial sliding surface is about 50 m deep. The total rock mass in movement amounts to approx. 32.6×10^6 m^3 at the deepest sliding surface, while it is about 12.8×10^6 m^3 at the intermediate sliding surface and about 7×10^6 m^3 at the superficial sliding surface. Sliding surfaces are subparallel, they coincide with the position of the layer discontinuities and follow the structural-tectonic elements.

Year 1963 was adopted as the year in which the landslide first started, since it is then that some greater damages of structures at the old part of the plant at the foot of the landslide were observed.

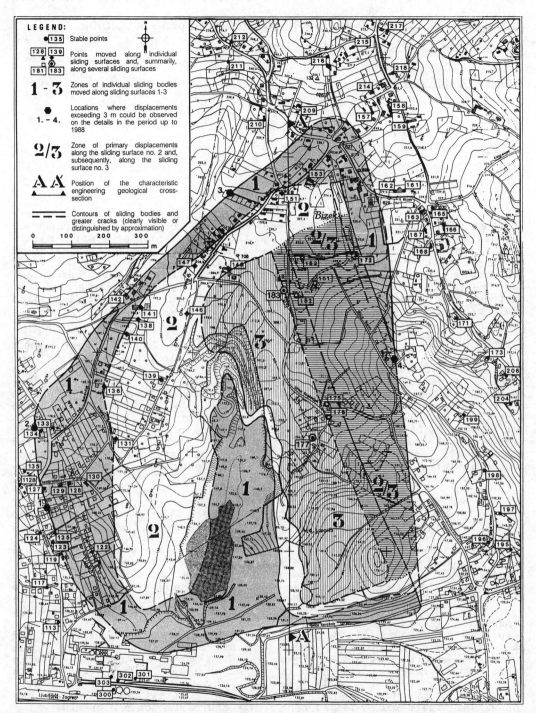

Figure 1. T. C. "Sloboda" landslide with the location of detailed photogrammetric points

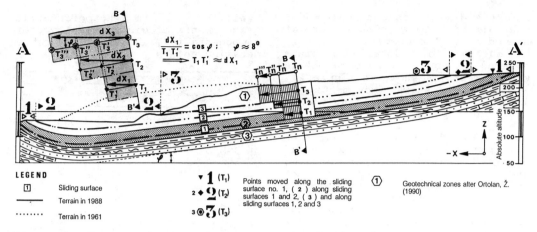

Figure 2. Characteristic engineering geological cross-section A-A'

Approx. 2.1×10^6 m^3 of marl was excavated at the foot of the landslide from the time the plant exploitation started to the year of 1963. The volume of the foot part of the landslide was further reduced by 3.2×10^6 m^3 until 1988 when the plant was closed.

The landslide geometry, the state of pore pressures on sliding surfaces and shear resistance parameters were determined with the high level of accuracy. A detailed view of the landslide is presented in figure 1. The characteristic engineering geological cross-section A-A' is presented in figure 2. A detailed explanation of geotechnical zones including the procedure for determining zones of minimum shear resistance parameters is given by Ortolan, Ž. (1990).

Due to the vastness and complexity of the approach to the complete geotechnical model forming, this review will only be limited to the contribution of photogrammetric measurements to such problem solving.

3 APPROACH TO PROBLEM SOLVING

Horizontal landslide displacement amounting to approx. 3 m (fig. 1) was discovered in 1985 on a number of observed points. Due to the size of displacement and to the fact that aerial surveys from 1963 and 1985 were available, an attempt was made to make a distinction between the landslide and the unmoved part of the terrain by photogrammetric measurement on a number of detailed points. It was established through survey and photo-sketches made on a large scale that the aspect of points remained unchanged. Displacement values for detailed points (details of structures - house ridges, piers, foundations etc.) were derived from the differences of coordinates of the same points obtained by block aerial triangulation of aerial photographs from the above mentioned years. It was established that the values of displacement vectors from different parts of the sliding body do not coincide, while the more detailed geological investigations pointed to the presence of several sliding surfaces, as shown by Ortolan,Ž (1990). The scope of photogrammetric measurements was then widened so that the same and a number of new points from 1979, 1981 and 1988 were measured.

In the first step, the depth of all three sliding surfaces was determined and the structural tectonic structure of the terrain was established. Theoretical sectional lines of the planes of sliding surfaces and of the ground surface were established. Field investigations have shown that collapsed or badly damaged structures are located precisely on these sectional lines or very close to such lines. In other parts of sliding bodies, the structures were displaced without visible damages.

At that time, geotechnical model was already quite clear, but it seemed to be too complex.

Location of detailed points covered by photogrammetric measurements is shown in figure 1.

Figure 2 shows movements of individual points in sliding bodies 1-3. Three points (T_1, T_2, T_3) in the B-B' cross-section were observed. If we assume that the sliding bodies move as rigid blocks, then these points also indicate the itinerary of individual points on the ground surface. Plane displacement vectors in the X-Z vertical plane are presented in the picture. Their horizontal and vertical components were measured photogrammetrically. Horizontal components were determined with greater accuracy. They were used in the following stage of the analysis. Figure 2 shows relationship between displacements that occurred, for instance, in 1985 and those from the "zero survey" made in 1963.

$$T_1 T_1'('85) = dX_1('85)$$

$$T_2'T_2''('85) = dX_2('85) - dX_1('85)$$

$$T_3''T_3'''('85) = dX_3('85) - dX_2('85)$$

Here, dX_{1-3} are photogrammetrically measured summary horizontal components of the displacement vector of points located in the X-Z plane. Corresponding relations are valid for any other year observed in relation with the "zero year" of 1963. The difference

$$T_2'T_2'' = T_2'T_2''('88) - T_2'T_2''('85),$$

represents the partly realized displacement of field points along the sliding surface no.2, while the difference

$$dX = dX_2('88) - dX_2('85)$$

represents the summarily realized displacement of field points along sliding surfaces 1 and 2. Similar relations are also valid for the summarily or partly realized displacements along the sliding surface no.3, while the summary displacement for the sliding surface 1 is equal to the partial displacement.

Table 1. Photogrammetric measurements of displacement vector components (dX, dY) made at stable detailed points

POINT NO.	1979. dY	1979. dX	1981. dY	1981. dX	1985. dY	1985. dX	1988. dY	1988. dX
113	-0,157	-0,343	-0,097	-0,313	-0,082	-0,076	-0,137	-0,303
117	-0,125	-0,713	-0,035	-0,583	-0,029	-0,436	-0,025	-0,543
119	-0,949	-0,859	-0,279	-0,389	-0,258	-0,124	-0,319	-0,339
123	-0,315	-0,733	-0,295	-0,303	-0,223	-0,218	-0,405	-0,053
124	-0,278	-0,629	-0,228	-0,309	-0,132	-0,299	-	-
125	-0,344	-0,503	-0,254	-0,163	-0,056	+0,107	-0,294	-0,073
135	-0,380	-0,654	-0,220	+0,236	-0,169	+0,050	-0,310	-0,244
157	-0,262	-0,211	-	-	-0,315	-0,436	-	-
158	-0,455	+0,160	-	-	+0,248	-0,209	-	-
159	-0,505	+0,327	-	-	+0,235	-0,060	-	-
161	-0,403	-0,062	-	-	+0,018	-0,077	-0,123	+0,068
162	-0,273	-0,414	-	-	-0,034	-0,304	-0,123	-0,214
163	-0,307	-0,099	-	-	-0,064	+0,242	-0,047	+0,091
165	-0,303	-0,100	-	-	-0,230	+0,026	-0,143	+0,590
166	-0,332	-0,293	-	-	+0,044	-0,093	-0,132	+0,595
167	-0,316	-0,606	-	-	-0,031	-0,136	-0,276	-0,466
168	-0,378	-0,589	-	-	-0,175	-0,360	-0,298	-0,459
171	-0,738	-0,472	-	-	-0,300	-0,559	-	-
173	-0,878	-0,740	-0,978	-0,660	-0,546	-0,414	-0,648	-0,370
195	-0,799	-0,496	-0,979	-0,416	-0,671	+0,295	-0,669	-0,356
196	-0,675	-0,357	-0,715	-0,157	-0,584	+0,033	-0,535	-0,047
197	-0,736	-0,168	-0,546	-0,058	-0,668	+0,011	-0,356	-0,088
198	-0,601	-0,289	-0,591	-0,259	+0,153	-0,395	-0,461	-0,019
199	-0,437	-0,343	-0,477	-0,393	-0,399	-0,537	-0,167	-0,563
204	-0,892	-0,450	-0,892	-0,320	-0,016	-0,408	-0,692	-0,110
205	-0,760	-0,200	-0,720	-0,240	-0,271	-0,320	-0,730	-0,140
206	-0,325	+0,106	-0,315	-0,146	+0,633	+0,136	-0,305	+0,206
209	-	-	-	-	-0,027	+0,274	-	-
210	-	-	-	-	+0,109	-0,211	-	-
211	-	-	-	-	-0,360	-0,292	-	-
212	-	-	-	-	-0,062	-0,197	-	-
213	-	-	-	-	-0,238	-0,244	-	-
214	-	-	-	-	-0,030	-0,101	-	-
215	-	-	-	-	+0,100	+0,407	-	-
216	-	-	-	-	-0,062	+0,270	-	-
217	-	-	-	-	-0,207	+0,239	-	-
218	-	-	-	-	+0,040	+0,082	-	-
300	-0,371	-0,499	-0,291	-0,129	-0,536	-0,102	-0,381	-0,449
301	-0,294	-0,942	-0,184	-0,512	-0,067	-0,551	-0,314	-0,712
302	-0,526	-0,517	-0,416	-0,237	-0,342	-0,255	-0,606	-0,517
303	-0,243	-0,843	-0,143	-0,483	+0,179	-0,536	-0,053	-0,663
308	-0,410	+0,340	-	-	-0,240	+0,180	-0,035	+0,042
309	-0,230	+0,110	-	-	-0,030	+0,680	-0,067	+0,216
310	-0,190	+0,380	-	-	-	-	-0,181	+0,075
1128	-0,376	-0,440	-0,286	-0,230	-0,269	+0,112	-0,246	+0,030
AVER. VALUE	-0,44	-0,34	-0,42	-0,27	-0,07	-0,10	-0,30	-0,16

Table 2. Average values of the summary horizontal displacement vector components (dX) at detailed points of the landslide

POINT NO.	dX₁ 1963.-1985.	dX₁ 1963.-1988.	dX₂ 1963.-1985.	dX₂ 1963.-1988.	dX₂/dX₃ 1963.-1985.	dX₂/dX₃ 1963.-1988.	dX₃ 1963.-1985.	dX₃ 1963.-1988.
122	2,752	2,968						
127	2,556	3,144						
128	3,027	3,629						
129	2,482	3,017						
130	3,047	3,732						
133	3,385	3,891						
134	2,607	3,344						
142	3,055	3,823						
153	3,289	3,688						
131			3,498	4,343				
136			3,563	4,441				
138			3,658	4,453				
139			3,491	4,429				
140			3,325	4,128				
141			3,626	4,070				
146			3,489	4,008				
147			3,557	4,124				
148			3,876	4,542				
151			4,038	4,663				
152					4,433	4,959		
175					4,531	5,049		
176					4,572	5,177		
181					4,539	5,212		
182					4,588	5,214		
177							5,078	DESTROYED
183							5,206	6,087
AVER. VALUE (dX)	2,911	3,471	3,612	4,320	4,533	5,122	5,142	6,087

1688

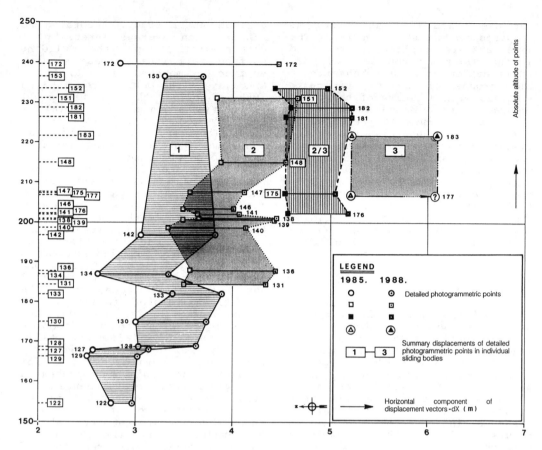

Figure 3. Diagram of summary horizontal components of displacement (dX) of detailed photogrammetric points for 1985 and 1988.

4 MEASUREMENT RESULTS AND THEIR INTERPRETATION

The main data base contains more than 100 points. Average errors in displacement measurement were calculated using standard geodetic procedures (Klak, 1982). In order to obtain a more realistic estimate, 45 points situated outside of the sliding body (where there is no displacement) were defined (table 1). Since their location is fixed for all annual aerial surveys, it is possible to conclude from the average values of components of plane displacement vectors in horizontal plane that the most accurate measurements were made in 1985 and 1988. It was established that the measured displacements on the moved points were practically from north to

south. The above facts (fig.2), indicate that the horizontal component of the displacement vector dX well represents the value of spatial displacement vector .

Table 2 shows the calculation of average values of the measured summary horizontal components of the displacement vector (dX) on a number of detailed landslide points

Table 3. Activities of individual sliding bodies determined on the basis of average values from tab. 2

YEAR	AVERAGE SUMMARY DISPLACEMENT VALUES (m)			PARTIAL DISPLACEMENTS ALONG INDIVIDUAL SLIDING SURFACES (m)		
				1	2	3
	dX_1	dX_2	dX_3	$\eta_1 \eta_1' = dX_1$	$\eta_2' \eta_2' = dX_2 - dX_1$	$\eta_3' \eta_3' = dX_3 - dX_2$
1963.–1985.	2,911	3,612	5,142	2,911	0,701	1,530
1963.–1988.	3,471	4,320	6,087	3,471	0,849	1,767
DIFFERENCE 1988.–1985.	0,560	0,708	0,945	0,560	0,148	0,237

for the years 1985 and 1988, as compared to the state in 1963. The same data are graphically presented in the diagram given in figure 3. It is evident that the behavior of the groups of points is quite similar, within limits of the accuracy of measurement. The location of points is presented in figure 1. Unfortunately, the entire landslide surface is not uniformly covered by an adequate number of points since the configuration of many points changed or they were destroyed due to the great time gap (1963 - 1985/88).

The activity of individual sliding bodies, based on average values from table 2, is presented in table 3.

5 CONCLUSION

The authors indicate how a number of conclusions about the geotechnical model and the landslide activity can be made on the basis of the repeated photogrammetric measurements in an area having three sliding surface.

Three different displacement groups, all made of uniform elements, indicate that the sliding occurs along three sliding surfaces.

Points close to one another, having different displacement values and situated in different parts of the landslide, confirm the thesis that the sliding surfaces followed the bedding discontinuities characterized by minimum shear resistance parameters.

Behavior of the group of points (152-182, table 2) indicates that individual blocks (figure 1) reached their position with a delay. This can also be established from the behavior of the point no.172 (figs. 1 and 3), which obviously moved only along the sliding surface no.1 in the first stage (until the year 1985), while a part of the sliding body to which this points belong also moved in 1988 along the sliding surface no.2.

Average calculated displacements along individual sliding surfaces point to the difference in the displacement intensity in the time periods under study. According to table 3, an average intensity of displacement along the sliding surface no.1 (the deepest sliding surface) amounted to 0.132 m annually in the period from 1963 to 1985, while the displacement intensity amounted to 0.187 m annually in the period from 1985 to 1988. The landslide activity for the remaining two sliding surfaces may also be observed.

It can be seen from figure 3 that the hypsometrically highest points in the frontal part of the landslide as well as the lowest points (in alluvial fan) point to the uniformity of displacement values. It can therefore be concluded that the sliding bodies moved as rigid blocks. This is one of significant theoretical assumptions for proceeding to the spatial analysis of stability.

Abundant information on the behavior of even the most complex sliding bodies and the acceptable accuracy of the method are sufficient reasons for popularization of the repeated photogrammetric measurements and for their more frequent use in the auscultation of great landslides. An average measurement error should be evaluated on the basis of photogrammetric measurements of detailed points that have not been moved.

REFERENCES

Klak, S. (1982): Teorija pogrešaka i račun izjednačenja /Theory of errors and calculation of balanced state/, Geodetski fakultet Sveučilišta u Zagrebu. Sveučilišna naklada Liber: 345 p., Zagreb, Yugoslavia (in Croatian).

Ortolan, Ž. (1990): Le rôle de la méthode de corrélation dans la détermination des zones de paramètres minimaux de résistance au cisaillement. Proceedings Sixth International Congress IAEG: 1675-1679, Amsterdam, the Netherlands.

Ortolan,Ž. et al.(1987): Posljedice rudarenja u laporolomu "Kostanjak" - Podsused. Zbornik referata IX jugoslovenskog simpozija o hidrogeologiji i

inženjersoj geologiji (2): 117-
128. Priština, Yugoslavia (in
Croatian). /Consequences of
mining in the marl deposit
"Kostanjak"-Podsused Proceedings
of the 9th Yugoslav symposium on
hydrogeology and engineering
geology/

Landslides, Bell (ed.) © 1995 Balkema, Rotterdam, ISBN 90 5410 032 X

Laboratory and field measurements of residual shear strength

D.J. Petley
University of Warwick, Coventry, UK

ABSTRACT: The accurate and reliable assessment of residual shear strength is relevant to a large number of stability problems, including cases involving the reactivation of old landslides, areas subjected to tectonic folding, areas affected by solifluction, and also to the design of many stabilisation or remedial schemes. Laboratory measurements of residual shear strength can be obtained by the use of the direct shear apparatus, triaxial apparatus and ring-shear apparatus. Field assessments of residual shear strength can be obtained by appropriate back-analysis of case histories provided that reliable data on the geometry of the moving mass and the relevant pore water pressures are known.

This paper reviews some laboratory methods for the determination of residual shear strength, and compares the results obtained from the tests with values of residual shear strength obtained by the back analysis of actual slope failures. Two cases are considered, (i) an example from a coastal landslide, and (ii) an example involving the reactivation of a previous failure.

1. INTRODUCTION

When a specimen of clay is subjected to relatively slow shearing strain, it builds up an increasing resistance until a maximum or 'peak' shear strength is attained. Frequently shear tests are stopped soon after the peak strength has been defined, but, if the test is continued, the clay exhibits the phenomenon of strain-softening (Skempton, 1964), and the shear strength decreases until, at some large strain, it reaches a minimum or 'residual' value. The decrease in the drained shear strength of a clay is associated with changes in water content in dilatant materials, and with reorientation of clay particles parallel to the direction of shearing. There is considerable evidence to show that the decreases in strength associated with particle reorientation are significant only in clays containing platy clay minerals and having a clay fraction (taken as the percentage by weight of particles smaller than 0.002 mm) exceeding about 20-25%, (Skempton, 1985; Lupini, Skinner and Vaughan, 1981). Examples of the reductions in shear strength from peak to residual conditions are widely available to confirm this point; for example (Skempton, 1964) the boulder clay at Selsea possesses a clay fraction of 17% and exhibits only a small decrease in strength, whilst at Jari, where the clay is heavily over-consolidated due to the erosion of at least 600m of sediments, the clay fraction of this Upper Siwalik clay is approximately 47% and there is a substantial decrease in strength from peak to residual. A summary of the index properties and shear strengths of some clays is given in Table 1.

Substantial displacements are required to cause the drop in strength to the residual condition, and in previously unsheared clays, displacements of 100-500mm may be required before the strength falls to a final steady residual value (Skempton 1985). Clearly, unless significant progressive failure occurs, the residual strength is generally not relevant to first-time slides in previously unsheared clays or fills. If however pre-existing shear surfaces exist, then the relevant strength along these surfaces will be at, or close to, the residual strength. Pre-

Table 1: Index properties and shear strengths of clays

Site (clay)	Liquid limit %	Plastic limit %	Clay fraction %	Peak c' kN/m²	Peak φ' degrees	Residual c'_r kN/m²	Residual $φ'_r$ degrees
[1]Selset (boulder clay)	26	13	17	c.9	32	0	30
[1]Jari (claystone)	70	29	55	c.37	22	0	18
[2]Walton's Wood (colluvium)	57	27	70	15	21	0	13
Peterborough (Oxford clay)	51	25	38	0	23	0	14
Avonmouth (alluvium)	64	28	40	0	26	0	25
Fiddlers Ferry (estuarine)	43	22	30	10	25	0	24
Hendon (brown London clay)	78	30	60	20	23	0	14
Leigh-on-Sea (blue London clay)	69	26	49	17	24	0	14.5
Selborne (Gault clay)	70	22	53	15	25	0	14

Notes: 1 - Skempton 1964   All other data from author's records
          2 - Skempton 1985

existing shear surfaces will exist in old landslides, soliflucted slopes, in bedding shears in folded strata, in sheared joints or faults and in general slope failures. Wherever shear surfaces exist, the accurate and reliable determinations or estimations of the shear strength along them is of fundamental importance in engineering design.

In general, high quality samples containing the shear surface should be obtained and tests performed so that failure occurs along the existing shear surface. The best sampling of this type is obtained from pits, trenches, shafts or adits, from which block samples can be taken. In most engineering projects, it may be unduly expensive to obtain samples in this way from deep shear surfaces. It may be possible to obtain samples using shell and auger or rotary coring techniques, but the author's experience has been that such methods can be unreliable, and the quality of sample obtained may be poor. It must be emphasized however that where pre-existing shear surfaces exist, it is of paramount importance to locate them accurately. If it is possible also to obtain samples containing the shear surfaces, then this should be done.

If, for whatever reason, samples containing the shear surfaces cannot be obtained, then it will be necessary to perform tests which will enable the strength on the shear surface to be estimated. These tests may be performed in the direct shear, triaxial or ring shear apparatus using previously unsheared material which possesses similar characteristics to the material in the sheared specimens.

2.   LABORATORY METHODS FOR RESIDUAL STRENGTH

In the laboratory, the residual strength can be obtained from tests on samples containing the pre-existing shear surface (hereafter termed shear surface tests) or from tests on initially unsheared material. In addition, tests can be performed on samples containing an artificially produced shear surface, in which a previously unsheared clay is carefully cut using a wire-saw and the two surfaces produced are polished by using a glass plate to smooth the surface (and orientate the particles) in the direction of subsequent shearing.

2.1 Shear box tests

The direct shear or shear box apparatus can be used for a variety of tests to measure the residual strength. If samples containing the shear surface are available, the tests can be performed by locating the shear surface as exactly as possible in the plane of the box and arranging the sample so that shearing follows the original direction of movement. It may be necessary when preparing samples of some materials (for example, soft clays) to make allowance for the effects of consolidation before the shearing stage. Tests on existing shear surfaces attain the residual strength at very low displacements (Fig. 1). If, after reaching the maximum displacement in the apparatus, the sample is returned to its original configuration, and then re-sheared, the shear strength in the second and subsequent runs returns almost precisely to the first run values.

When it is not possible to perform tests on existing shear surfaces, the residual strength can be obtained from tests on initially unsheared material using the 'reversal' technique. Following consolidation, the specimen is subjected to an initial shearing stage, following which the apparatus is returned to its original position and then re-sheared. This process is repeated until a steady minimum value of shear strength can be identified. The test is lengthy, since each of the stages must be performed at rates of strain sufficiently slow as to ensure drained conditions. In addition, the reversal in the direction of the applied shear stress may affect the degree of orientation achieved previously, resulting in the development of a small peak during the re-shear (Fig. 2). Material may also be lost between the two halves of the apparatus by slurrying. In some tests, it might not be easy to interpret the results of reversal shear box tests (Skempton and Petley, 1967).

A less time-consuming method for the determination of the residual shear strength using the shear box apparatus can be achieved by the use of the 'cut-plane' technique. In this method, the shear surface is produced artificially by cutting the sample with a wire-saw and smoothing the surface, so produced, to orientate the platy clay particles. Shearing along the surface then produces results similar to those obtained from shear surface tests (Fig. 3). In some very stiff materials, it may be necessary to use the reversal technique in cut-plane tests (see, for example, Skempton 1980).

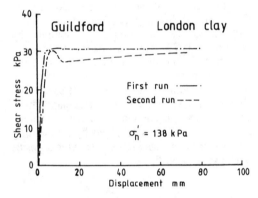

Figure 1: Shear box test on shear surface

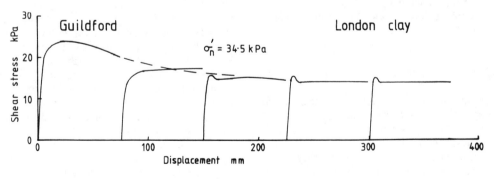

Figure 2: Reversal shear box test

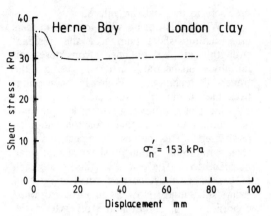

Figure 3: Cut plane shear box test

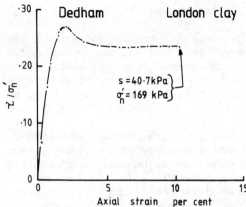

Figure 4: Triaxial test on shear surface

2.2 Triaxial tests

As a general rule, it is not possible to obtain reliable estimates of the residual strength from tests on unsheared material in the triaxial apparatus since the displacement required on a shear plane is limited by the constraints of the apparatus. Both shear surface and cut-plane tests can be performed, in which the sample is set up so that the surface of shearing is inclined at approximately $(45+\phi_r^1/2)$ to the horizontal and care is taken to ensure that it is well clear of the end caps. A double top cap must be used with ball bearings between them to allow for the horizontal components of movement of the upper half of the sample during the test. Corrections for area change and the strength of the rubber membrane and filter strips must be applied (Chandler, 1966) and the results are presented by calculating the normal and tangential stresses on the failure plane. A typical test result is given in Fig. 4.

2.3 Ring shear tests

The ring shear test enables displacement to be applied continuously in one direction, and provided there is no undue loss of material, the test can be continued to large displacements. Early forms of ring shear apparatus were developed by Hvorslev and others in approximately 1934, and some test results were presented by Hvorslev, 1939.

More recently two designs of ring shear apparatus have been developed. Firstly a sophisticated ring shear machine was designed and constructed as a joint venture, by Imperial College and the Norwegian Geotechnical Institute (Bishop et al., 1971), and secondly a simpler machine by Bromhead 1979. Results presented for Gault clay from Folkestone Warren tested in these two machines have given almost identical results (Bromhead and Curtis, 1983). The ring shear tests presented in this paper have been obtained using the Bromhead machine, in which a sample is prepared by remoulding directly into the apparatus. A typical result is shown in Fig. 5.

3. CASE RECORDS

In a reactivated landslide where movement is occurring slowly on an existing shear surface, the factor of safety will be 1.0. Provided that the shape of the failure surface and the relevant piezometric levels are known, back analysis will enable the average normal effective pressure and the average shear stress acting on the slip surface to be calculated from a two-dimensional analysis using an appropriate method of analysis (e.g. Morgenstern and Price, 1965). Since slides are three-dimensional, a correction should be applied, and according to Skempton 1985 this correction, which is a reduction in the shear stress, is given by the factor:

$$\frac{1}{1 + KD/B}$$

where, D and B are the average depth and width of the sliding mass, and K is an earth pressure coefficient (generally assumed to be 0.5).

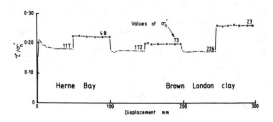

Figure 5: Ring Shear test

3.1 Peterborough

During the early 1960's, extensive restoration work was performed in an area to the south of Peterborough where Oxford clay had been removed for brick-making leaving a number of old pits.

It was the general practice for the clay unsuitable for brick making to be dumped back in the sides of the pits, where, with time weathering occurred, and the clay became brown and firm. Locally the weathered Oxford clay is known as "callow". In order to facilitate the proposed scheme, some re-profiling of the existing bank slopes was carried out, and the angle of slope was steepened from approximately 1:6 to 1:4.

In the late summer of 1963, cracks were observed at the top of part of the bank in one pit. The pit at this point was approximately 12m deep. The cracks widened and deepened and movement at the toe became apparent. The slip was approximately 135m long, and the total movement of the failed section was about 2m.

A number of piezometers, consisting of standpipes connected to mercury manometers were installed in the slip area and in the adjacent stable areas and reasonably accurate measurements of piezometric levels were obtained. In addition, four trenches were excavated through the failed material to define the position of the failure surface, and block samples were taken in the callow at various depths, and in particular some were taken containing the actual failure surface. It was found that the movement had taken place just above the interface between the callow and the in-situ clay. The slip surface was highly polished and slickensided. A section through the bank is given in Fig. 6.

Shear box tests on the unsheared callow indicated peak values of shear strength corresponding to c'=0 and $\phi' = 24°$ (Fig. 7). Since the precise position of the failure surface was known and the

piezometric levels had been measured with reasonable accuracy, stability analyses could be performed with confidence. Based on the peak values of shear strength, the factor of safety was calculated to be 1.71. Assuming a factor of safety of 1.0, the back analysis yields a field residual value of $\phi_r^1 = 14.5°$ (assuming $c_r^1 = 0$). Tests on the slip surface indicate ϕ_r^1 values of 15.0° (Fig. 7). The results of a test performed in the Bromhead ring shear apparatus is also included in Fig. 7, and it corresponds to $\phi_r^1 = 14.0°$, i.e. slightly below the results obtained from slip surface tests. Tests using the reversal technique and cut plane appear to give somewhat higher results (Fig. 7).

3.2 Herne Bay

The London clay cliffs of North Kent are subject to erosion by the sea, and mudslides and deep-seated failures have occurred extensively. The area around Herne Bay has been the focus of considerable research effort (Hutchinson 1970, Hutchinson and Bhandari 1971). Around 1970, a 39m deep borehole was made through the London clay to the rear of the Miramar landslip, and a large number of tests were performed to determine peak and residual shear strength parameters in the London clay. Most of the residual values were obtained from cut-plane tests. The range of values obtained for the brown and blue London clay from shear box cut-plane tests are indicated in Fig. 8. In addition, a number of slip surface tests have also been performed, and are included in Fig. 8. It can be seen that the slip surface tests generally fall near or below the lower limit obtained from cut-plane tests. A few reversal tests were also performed (see Fig. 8).

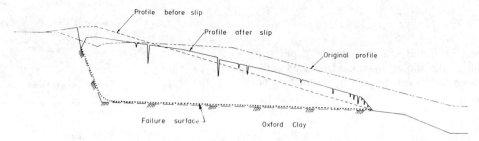

Figure 6: Section through failure at Peterborough

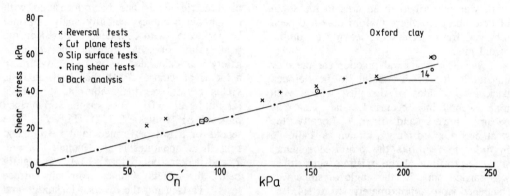

Figure 7: Shear strength tests on Oxford clay at Peterborough

The results obtained from back analyses of a number of failures (taken from Bromhead 1978) are also included in Fig. 8; there appears to be good agreement between these values and the results of the slip surface tests.

Finally, the results of two ring shear tests using the Bromhead apparatus are shown in Fig. 8. One of these tests was performed on blue London clay from a depth of 20m. The second test was performed on brown mudflow material derived from the London clay at Beltinge. These two tests show close agreement, and both fall significantly below the results obtained from the back analyses (Fig. 8).

4. DISCUSSION AND CONCLUSIONS

From the two cases considered, four points can be noted:

(i) There appears to be close agreement between the results of tests measuring the shear strength on a pre-existing shear surface, and the results obtained from back-analysis.

(ii) Cut plane tests generally give results slightly higher than slip surface tests. The difference is small and typically of the order of 1 or 2°.

(iii) Results from multi-reversal shear box tests indicate residual strengths which are higher than the values from slip surface tests and cut plane tests. The difference is most marked at low effective normal stresses. The tests reported here involved five or six reversals, and with a larger number or reversals a further decrease in shear strength may occur, reducing the disparity with the slip surface tests.

(iv) For the two cases considered, ring shear tests are lower than those obtained from other forms of test or from back analysis.

The reasons for these differences are uncertain. It is likely that the residual strength values interpreted from these tests are dependent upon the degree of orientation and relative roughness of the failure surface.

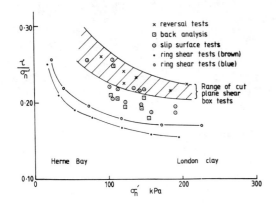

Figure 8: Shear strength tests on London
clay at Herne Bay

In cases where extensive movement has already occurred (i.e. slip surface tests) the degree of orientation may be nearly perfect, and close agreement with back-analysis values may be expected. In cut plane tests, the artificial production of a failure surface may not produce the same degree of orientation as that produced in the field by displacements of some meters. Reversal shear box tests involve only small displacements during each traverse of the box, followed by reversals of the direction of the applied shear stress and possible re-arrangement of particles. Most reversal tests involved only five or six reversals, and the total displacement along the failure surface is small.

In ring shear tests the lowest values of shear strength are developed, possibly as a result of intensive shearing along a continuous failure surface which generally appears to be smooth when the apparatus is dismantled.

REFERENCES

Bishop, A.W., Green, G.E., Garga, V.K., Andresen, A. and Brown, J.D. 1971. A new ring shear apparatus and its application to the measurement of residual strength. Geotechnique, 21, 4, 273-328.

Bromhead, E.N. 1978. Large landslides in London Clay at Herne Bay, Kent. Q.J. of Eng. Geol. 11, 291-304.

Bromhead, E.N. 1979. A simple ring shear apparatus. Ground Engineering, 12, 40-44.

Bromhead, E.N. and Curtis, R.D. 1983. A comparison of alternative methods of measuring the residual strength of London Clay. Ground Engineering, 16, 39-41.

Chandler, R.J. 1966. The measurement of residual strength in triaxial compression. Geotechnique 16, 3, 181-186.

Chandler, R.J. and Hardie, T.N. 1989. Thin sample techniques of residual strength measurement. Geotechnique 39, 3, 527-531.

Hutchinson, J.N. 1970. A coastal mudflow on the London clay cliffs at Beltinge, North Kent. Geotechnique 4, 412-438.

Hutchinson, J.N. and Bhandari, R.K. 1971. Undrained loading: a fundamental mechanism of mudflow and other movement. Geotechnique 21, 4, 353-358.

Hutchinson, J.N., Somerville, S.H. and Petley, D.J. 1973. A landslide in periglacially disturbed Etruria Marl at Bury Hill, Staffordshire. Q.J. of Eng. Geol. 6, 377-404.

Hvorslev, M.J., 1979. Torsion shear tests and their place in the determination of the shearing resistance of soils. Proc. Am. Soc. Test Mator. 39, 999-1022.

Lupini, J.F., Skinner, A.E. and Vaughan, P.R., 1981. The drained residual strength of cohesive soils. Geotechnique, 31, 2, 181-213.

Morgenstern, N.R. and Price, V.E., 1965. The analysis of the stability of general slip surfaces. Geotechnique 15, 1, 79-93.

Skempton, A.W., 1964. Long-term stability of clay slopes. Geotechnique, 14, 2, 75-101.

Skempton, A.W. and Petley, D.J., 1967. The strength along structural discontinuities in stiff clay. Proc. Geotechnical Conf., Oslo 2, 29-46.

Skempton, A.W., 1985. Residual strength of clays in landslides, folded strata and the laboratory. Geotechnique 35, 1, 3-1.

Layered slope profile analysis with random soil parameters

W. Pula & E. Stilger-Szydlo
Technical University of Wrocław, Poland

ABSTRACT: The complete solution which comprises both statics and kinematics of a layered slope in the states of limiting equilibrium is presented. The obtained statically admissible stress field has allowed us to get a lower evaluation of the safety factor and to analyse the influence of the slope stratification on its shape in the state of limiting equilibrium. By the use of kinematic approach an upper bound of safety factor was obtained The statistical analysis of the shape, based on the static solution of the problem is being carried out. The random fluctuations of soil strength parameters are taken into account. The random range of the slope and the random average inclination are analysed.

1 INTRODUCTION

There are many methods basing on the theory of plasticity available for slope designing. Among them the limit states method is of great theoretical significance. It allows us to determine the relationship between the shape of the slope, the limit loading of the overburden and the stress state within the whole slope body. The static and kinematic solutions reported in the literature concern solely homogeneous slopes. A complete solution comprising both statics and kinematics of a layered slope was given by Stilger-Szydlo (1989). The statically admissible stress field obtained in this study made it possible to analyse the influence of the stratification on the limiting shape of slope profile and to obtain the lower estimate of the factor of safety.

The kinematic approach to the problem, gives the upper estimate of the actual factor.

In previous works of this kind, the influence of randomness of soil properties on the shape of slope at the state of limiting equilibrium was not examined. In this paper the random fluctuations of soil strength parameters were taken into account in the static solution. Cohesion c and angle of internal friction Φ in each geotechnical layer were treated as random variables.

2 STATICS AND KINEMATICS OF LIMIT STATE

2.1 Equations of statics

Using the Coulomb-Mohr condition and the coordinate system as in Figure 1, the two-dimensional state of strain is described by a system of hyperbolic equations. These take the form:

$$\frac{\partial \sigma}{\partial x} (1 + \sin\Phi\cos2\varphi) - 2\sigma\sin\Phi\sin2\varphi \frac{\partial \phi}{\partial x} +$$
$$+ \sin\Phi\sin2\varphi \frac{\partial \sigma}{\partial y} + 2\sigma\sin\Phi\cos2\varphi \frac{\partial \phi}{\partial y} = X_s ,$$
(1)

$$\frac{\partial \sigma}{\partial x} \sin\Phi\sin2\varphi + 2\sigma\sin\Phi\cos2\varphi \frac{\partial \phi}{\partial x} +$$
$$+ \frac{\partial \sigma}{\partial y} (1 - \sin\Phi\cos2\varphi) + 2\sigma\sin\Phi\sin2\varphi \frac{\partial \phi}{\partial y} = Y_s$$
(2)

where

$$\sigma = \frac{1}{2} (\sigma_1 + \sigma_2) + c\cdot\cot\Phi = \frac{1}{2} (\sigma_x + \sigma_y) + c\cdot\cot\Phi$$
(3)

is the equivalent stress, σ_x, σ_y and τ_{xy} are components of symmetric stress tensor at the given point of soil medium, σ_1, σ_2

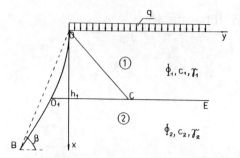

Fig.1 Layered slope scheme

are components of main stress, φ is the direction of the greater main stress inclination to the x-axis, $X_s = \gamma$, $Y_s = 0$ are components of the vector of mass forces, and Φ is the angle of internal friction.

The most convenient way to solve the system of equations (1) and (2) is applying the method of characteristics. The net of characteristics depends on the form of the static boundary conditions. For details see Derski & Izbicki (1988) and Stilger-Szydlo (1989).

2.2 Kinematics of limit state

The statically admissible stress field allows us to obtain the lower estimate of the factor of safety. Now, let us consider the kinematic approach.

Assuming the flow associated with the limit state condition, the equations of kinematics which interrelate the velocity components v_x and v_y have the form:

$$(4)$$

$$(\sin\Phi + \cos2\varphi)\frac{\partial v_x}{\partial x} - (\sin\Phi - \cos2\varphi)\frac{\partial v_y}{\partial y} = 0$$

$$2\sin2\varphi\frac{\partial v_x}{\partial x} + (\sin\Phi - \cos2\varphi)(\frac{\partial v_x}{\partial y} + \frac{\partial v_y}{\partial x}) = 0$$

$$(5)$$

Similarly as the system of equations (1) and (2) the system (4) and (5) can be solved by the method of characteristics.

The field of velocity characteristics is not influenced by the kinematic boundary conditions. It depends on the form of the field of stress characteristics.
A kinematic solution should ensure the non-negativeness of the power being dissipated. The kinematically admissible mechanism of plastic flow is determined by the factor of safety which is the upper estimate of the actual safety factor. These

mechanisms are constructed assuming that a system of rigid elements is formed in the medium and that the elements are sliding on one another along the limiting discontinuity surfaces of velocity dislocations slide surface. If the associated flow principle is assumed, the vector of velocity discontinuity Δv, which propagates along the slide surface, deviates at internal friction angle Φ. Having given a kinematically admissible mechanism of slope failure, the upper estimate of the safety factor F_k can be conveniently determined applying the energetic approach

$$A + L = D(c,\Phi) \quad , \qquad (6)$$

where A is the power of external loading, L the power of mass forces and D the total internal dissipation. The upper estimate of the safety factor F_k is computed by successive approximations, in which the parameters $\tan\Phi_o = \tan\Phi/F$ and $c_o = c/F$ are chosen to satisfy the equation (6). For details see Derski & Izbicki (1988) and Stilger- Szydlo (1989).

3 SLOPE IN THE STATE OF LIMITING EQUILIBRIUM

3.1 Computational model of the slope

The slope body is a two-layered medium characterised by an angle of internal friction Φ, cohesion c and unit weight γ.

The safety of the slope is measured by the factor of safety

$$F = \frac{\tau_f}{\tau_o} = \frac{\sigma\tan\Phi + c}{\sigma\tan\Phi_o + c_o} \quad , \qquad (7)$$

where τ_f is the ultimate shearing resistance and τ_o is the induced resistance. Basing on the limit theorems (see Derski & Izbicki 1988) the upper F_k and the lower F_s estimates of the actual safety factor F have been determined

$$F_s \leq F \leq F_k \quad , \qquad (8)$$

where F_s results from the statically admissible state of stress and F_k from kinematically admissible mechanism of failure.

3.2 Solution of statics

A diagram illustrating the slope in the limiting equilibrium is shown in Figure 1. The loading q of the slope is:

$$q = \frac{2c_1 \cos\Phi_1}{1 - \sin\Phi_1} \quad . \qquad (9)$$

Assumption that $\sigma_x = q$ and $\tau_{xy} = 0$ along the axis $y \geq 0$, results in the following conditions:

1. In the interior of the first layer along the $y \geq 0$ axis

$$\sigma = \frac{q + c_1 \cot\Phi_1}{1 + \sin\Phi_1} \quad , \quad \varphi = 0 \ , \quad (10)$$

inside the area OAC

$$\sigma = \frac{\gamma_1 x + q + c_1 \cot\Phi_1}{1 + \sin\Phi_1} \quad , \quad \varphi = 0 \ , \quad (11)$$

along the boundary OB of the slope

$$\sigma_n = \tau_{nt} = 0 \ , \qquad \sigma = \frac{c_1 \cot\Phi_1}{1 - \sin\Phi_1} \ ,$$

$$\qquad\qquad\qquad\qquad\qquad (12)$$

$$\frac{dy}{dx} = \tan\varphi$$

2. Along the boundary $O_1 E$ between two layers $x = h_1$

$$|\tau_{xy}| \leq \tan\Phi_2 + c_2 \ , \qquad (13)$$

the equivalent stress affecting the lower layer is:

$$\sigma_d = \frac{\sigma_x + c_2 \cot\Phi_2}{\cos^2\Phi_2 \cos\delta}\left(\cos\delta - \sqrt{\sin^2\Phi_2 - \sin^2\delta}\ \right),$$

$$\qquad\qquad\qquad\qquad\qquad (14)$$

$$\varphi_d = -\frac{\delta}{2} - \frac{1}{2}\sin^{-1}\left(\frac{\sin\delta}{\sin\Phi_2}\right) \ , \quad (15)$$

where

$$\tan\delta = \frac{\tau_{xy}}{\sigma_x + c_2 \cot\Phi_2} \ , \qquad (16)$$

and at the point O_1 of the slope shape in the lower layer it is:

$$\sigma_{P_{01}} = \frac{c_2 \cot\Phi_2}{1 - \sin\Phi_2} \quad , \qquad (17)$$

$$\varphi_P = \frac{\cot\Phi_2}{2}\ln\frac{\sigma_P}{\sigma_d} + \varphi_{d_{01}} \ . \qquad (18)$$

The problem of finding the stable shape of a layered slope was solved using the method of characteristics (cf Figure 2 in section section 5).

Statically admissible stress fields for the layered slope have allowed us to obtain the lower estimate of the safety factor $F_s = 1$.

4 PROBABILISTIC APPROACH

In order to consider random fluctuations of soil strength parameters c and Φ it was assumed that (c,Φ) constituted a two-dimensional random variable (random vector). As a consequence of the presence of two geotechnical layers two random vectors

(c_1,Φ_1) and (c_2,Φ_2) were considered and assumed as stochastically independent. Random fluctuations of unit weight, as small, were neglected. The probability distributions of random vectors (c_1,Φ_1) and (c_2,Φ_2) were assumed in the following way. Let

$$X = \frac{c - c_{min}}{c_{max} - c_{min}} \ ; \quad Y = \frac{\Phi - \Phi_{min}}{\Phi_{max} - \Phi_{min}} \ , (19)$$

where the values c_{min}, c_{max} and Φ_{min}, Φ_{max} determine the variability intervals of cohesion and angle of internal friction, respectively. It was assumed that the random vector (X,Y) had bivariate Dirichlet distribution with the probability density function

$$f(x,y) = \qquad\qquad\qquad\qquad (20)$$

$$= \begin{cases} \dfrac{\Gamma(\alpha_1 + \alpha_2 + \alpha_3)}{\Gamma(\alpha_1)\Gamma(\alpha_2)\Gamma(\alpha_3)} \cdot x^{\alpha_1 - 1} \cdot y^{\alpha_2 - 1}(1-x-y)^{\alpha_3 - 1} \\ \\ \text{for } x,y \geq 0 \text{ and } 0 \leq x + y \leq 1 \\ \\ 0 \quad \text{elsewhere} \qquad\qquad , \end{cases}$$

where $\alpha_i > 0$, $i=1,2,3$ denote parameters of the distribution and Γ is the Euler function. This kind of description of soil strength parameters was suggested by

Athanasiou-Grivas & Harr (1979). In such an approach marginal distributions of cohesion and angle of internal friction are beta distributions (see Wilks 1962). This kind of marginal distributions corresponds to the suggestion given by Lumb (1970), who proposed these distributions in a separate description of random fluctuations of cohesion and friction angle (see also Athanasiou-Grivas & Harr 1979). The parameters $\alpha_1, \alpha_2, \alpha_3$ of Dirichlet distribution can be determined by the use of the following relationships (cf Wilks 1962 and Pula & Stilger-Szydlo 1990):

$$\bar{X} = \frac{\alpha_1}{\alpha_1 + \alpha_2 + \alpha_3} \quad , \quad \bar{Y} = \frac{\alpha_2}{\alpha_1 + \alpha_2 + \alpha_3} \quad , \quad (21)$$

$$d_x^2 = \frac{\alpha_1(\alpha_2 + \alpha_3)}{(\alpha_1 + \alpha_2 + \alpha_3)^2(\alpha_1 + \alpha_2 + \alpha_3 + 1)} \quad , \quad (22)$$

$$d_y^2 = \frac{\alpha_2(\alpha_1 + \alpha_3)}{(\alpha_1 + \alpha_2 + \alpha_3)^2(\alpha_1 + \alpha_2 + \alpha_3 + 1)} \quad , \quad (23)$$

$$\rho_{x,y} = -\left(\frac{\alpha_1 \alpha_2}{(\alpha_1 + \alpha_3)(\alpha_2 + \alpha_3)}\right)^{1/2} \quad (24)$$

where $\bar{X}$, $\bar{Y}$, d_x, d_y, $\rho_{x,y}$ denote mean values of random variables X and Y (cf eq. (19)), standard deviations of X and Y, correlation coefficient of X and Y, respectively. Using transformation (19), analogical quantities $\bar{c}$, $\bar{\Phi}$, d_c, d_Φ, $\rho_{c,\Phi}$ for the random variables c and Φ can be easily evaluated (see Pula & Stilger-Szydlo 1990). It is assumed that values $\bar{c}$, $\bar{\Phi}$, d_c, d_Φ, $\rho_{c,\Phi}$ are given from laboratory testing.

The use of Dirichlet distribution here has the following three advantages:
1. We are able to take into account that the parameters c and Φ have bounded intervals of variability. So we can avoid negative or senselessly large values of these parameters, which often happens when a normal distribution is used.
2. We are able to assume a correlation between c and Φ. The presence of this correlation and its negative value was pointed out by many authors, (e.g. Lumb 1970).
3. Marginal distributions of c and Φ are beta distributions. This fact corresponds to proposals made by the authors cited above.

The consideration of random variability of soil strength parameters in finding profiles of a slope in limiting equilibrium creates some computational difficulties, all the more that even in deterministic case the computations must be done numerically. In this paper the random variability was analysed using simulation method. To use this method, a new pseudorandom number generator of Dirichlet distribution was proposed (see Pula & Stilger-Szydlo 1990).

5 CALCULATION EXAMPLE

5.1 General assumptions

The slope body was two-layered medium described by random strength parameters (c_1, Φ_1) and (c_2, Φ_2). The unit weight was assumed as deterministic with the value $\gamma = 19.5$ kNm^{-3}. The height of the first layer was $h_1 = 2.7$ m and the height of the second one was $h_2 = 5$ m. The loading of the slope was determined by (9). The shape of the layered slope was evaluated using the method of characteristics. An example of the results is presented in Figure 2. The dashed line indicates the profile of a homogeneous slope, built of the first layer only.

The numerically obtained solution of the static problem yielded the field of characteristics and the values σ and φ in the nodal points of the characteristics lattice inside the region and along the slope profile. The step assumed in computations was x = 0.025 not exceeding the value of 0.1 m (in dimensional variables) for different values of c/γ. With the increasing density the slope profiles did not change their shapes any more.

5.2 Probabilistic assumptions

The random characteristics of soil strength parameters used in the example in the form of mean values, standard deviations and correlation coefficients are given in the Table 1. By the use of eqs. (19) and (21) – (24) we have:
$\alpha_1 = 1.311$, $\alpha_2 = 1.5344$, $\alpha_3 = 2.6171$
for the first layer and
$\alpha_1' = 1.2541$, $\alpha_2' = 1.4174$, $\alpha_3' = 2.7714$
for the second layer.
The aim of the computations was to estimate the probability distribution of the slope range R at the bottom of the slope

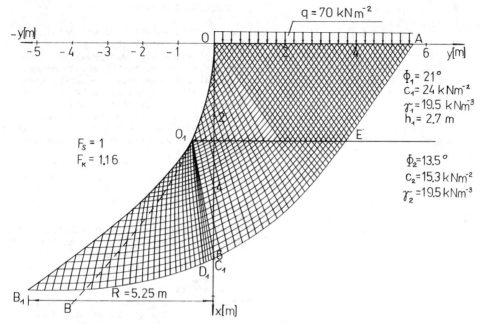

q = 70 kNm⁻²

$F_S = 1$
$F_K = 1.16$

$\Phi_1 = 21°$
$C_1 = 24\ kNm^{-2}$
$\gamma_1 = 19.5\ kNm^{-3}$
$h_1 = 2.7\ m$

$\Phi_2 = 13.5°$
$C_2 = 15.3\ kNm^{-2}$
$\gamma_2 = 19.5\ kNm^{-3}$

R = 5.25 m

Fig.2 Example of characteristics net used in computations

Table 1. Statistical characteristics of strength parameters used in the example (c = [kPa], Φ = [deg]).

Layer	$\bar{c}$ $\bar{\Phi}$	d_c d_Φ	c_{min} Φ_{min}	c_{max} Φ_{max}	$\rho_{c,\Phi}$
1st	20 20	4.2 3.15	14 15	39 32.8	−0.35
2nd	20 16	3.6 3.32	15 11	36.7 30.2	−0.32

and the probability distribution of the average inclination β of the slope, determined by the relation

$$\tan\beta = \frac{h_1 + h_2}{z} \qquad (25)$$

The simulation computations were performed for N = 170 realisations.

5.3 Results of computations

As a result of the simulation computations we are able to evaluate probability distributions in the form of histograms and cumulative distribution functions. The histogram of R is presented in Figure 3, while the cumulative distribution functions of R and β are shown in Figures 4 and 5.

The mean values and standard deviations of R and β are collected in Table 2.

Table 2. Distribution parameters of random variables R and β (R = [m], β = [deg]).

Random variable	Mean value	Standard deviation	Variation coefficient
R	3.45	0.93	0.27
β	65.89	5.81	0.09

Using the cumulative distribution function of the range R we are able to find the probability that the range will be greater than the given quantity R_o.

As an example we have:
P{ R > 5.0 } = 0.059, P{ R > 5.4 } = 0.035, P{ R > 6.0 } = 0.006 .
It is easy to see that the distribution of range R significantly differs from a normal one (cf Figure 3).

Coefficients of variations of R and β are not large in comparison with coeffi-

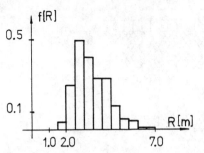

Fig.3 Histogram of the slope range R

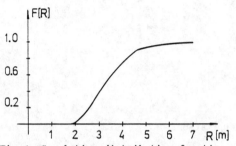

Fig.4 Cumulative distribution function of range R

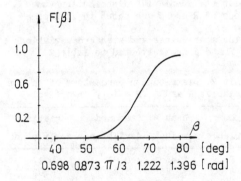

Fig5. Cumulative distribution function of average inclination β

cients of variation of c_1, Φ_1, c_2 and Φ_2. We can observe a large average inclination of the slope profile. But this large inclination is mainly caused by the small height of the slope in the example. It is well-known that slopes at the state of limiting equilibrium consisting of cohesive soils usually have large inclinations near the top of slope.

6 CONCLUDING COMMENTS

The complete solution of the problem of layered slope stability by the use of the limit states theory was presented, including static and kinematic solutions and probabilistic analysis. Comparing the slope profiles obtained in the calculation example with profiles of homogeneous slopes we can conclude that the profiles of layered slopes have greater inclination angles than homogeneous slopes composed of soil of upper layer.

Using the evaluation of the slope range R distribution we are able to find the probability that the range of the slope can be greater than the given value of R_o.

Because of significant random variation of strength parameters in natural soil, this probability can be useful as an auxiliary factor in designing embankment and slope profiles when the material consists of more than one soil layer.

The efficient probability analysis was possible owing to the use of a new pseudorandom numbers generator of Dirichlet distribution and the applcation of the simulation technique.

REFERENCES

Athanasiou-Grivas, D. & M.E. Harr 1979. A relibility approach to the design of soil slopes. Proc. Seventh European Conf. on Soil Mechanics and Foundation Engineering, Brighton: 95-99.
Derski, W & R. Izbicki & I. Kisiel & Z. Mroz 1988. Rock and soil mechanics. Amsterdam: Elsevier.
Lumb, P. 1970. Safety factors and the probability distribution of soil strength. Canadian Geotechnical Journal, vol.7 No.3: 225-242.
Pula, W. & Stilger-Szydlo 1990. Analysis of the embankment shape with random soil parameters using the theory of limit states. Studia Geotechnica et Mechanica Vol.12, No 3-4.
Sokolovsky, V.V. 1960. Statics of soil media. London: Butterworth.
Stilger-Szydlo, E. 1989. Kinematically admissible stability analysis of a stratified embankments. Arch. of Hydrotechnics Vol. XXXVI No. 1-2: 107-120.
Wilks, S.S. 1962. Mathematical statistics. New York: J.Wiley & Sons.

Landslides, Bell (ed.) © 1995 Balkema, Rotterdam, ISBN 90 5410 032 X

Applying modern technology to analyse complex landslide

T.L.Slosson, D.D.Crowther, M.B.Phipps & J.E.Slosson
Slosson and Associates, California, USA

Abstract: The multiple working hypotheses methodology proved effective in studies of the Rambla Pacifico Landslide in Malibu, California. It evolved from multiple interpretations for the origin of the landslide using seven basic methods of investigation which were analyzed as to limitations as the data were synthesized. The foremost hypothesis indicated a complex mass of moving rock material displaying both rotational and translational styles complexly interrelated within the same landslide affected by pre-existing structural conditions. Man-induced groundwater conditions have played a prominent role. The geometry, hydrogeologic conditions, and kinematic development of the complex Rambla Pacifico Landslide have been successfully modeled through the application of "multiple working hypotheses" technique in conjunction with the multiple geologic and geophysical technologies. The following were utilized to produce an extensive data base: (1) detailed geologic mapping; (2) extensive literature search; (3) subsurface exploration (bucket auger, continuous core and trenching); (4) downhole geophysics; (5) shallow seismic refraction; (6) geotechnical instrumentation; and (7) kinematic analysis displacement within the landslide complex.

1 INTRODUCTION

The active Rambla Pacifico Landslide comprises approximately 21 acres of moderate to steep terrain on the east-facing slope of Las Flores Canyon in Malibu, California. The canyon is one of a series of subparallel, south-draining canyons on the south flanks of the Santa Monica Mountains adjacent to the Pacific Ocean in the Transverse Ranges geomorphic province of California. As a result of the landslide occurrence, property damage, and ensuing litigation, the authors were retained by Los Angeles County to provide technical advice and an evaluation of conditions within and adjacent to the actively moving landslide. The study was initiated by employing the method of multiple working hypotheses (Chamberlin, 1892) to derive a viable approach for a complete geologic analysis of all aspects of the landslide.

An on-site "brainstorming" session, as well as researching several published references, produced multiple interpretations of possible conditions, causes, and effects pertinent to the origin of the active landslide. Both the U.S. Geological Survey (Yerkes, 1983) and the California Division of Mines (Weber, 1983) had mapped landslides in this area. Examples of some of the resultant ideas are:

- The active landslide is either a reactivation of an ancient landslide or it is a modern failure.
- The landslide is either a single moving mass or a complex of masses.
- The landslide possesses either a single failure surface or multiple sliding surfaces.
- Depending on the existence of a single or multiple failure surfaces, the slide is moving at the same rate in a single direction or at multiple rates in more than one direction.
- The existing bedding geometries and structural fabrics within the active landslide are a product of previous landslide activity, tectonic deformation, or both.
- Groundwater is likely to be a significant landslide-triggering mechanism

and may have multiple sources such as rainfall, irrigation, and seepage pits.

It was evident from the initial generalized concepts and possible interpretations that certain combinations of ideas tended to be more acceptable than others. Nevertheless, the groundwork for a full scale investigation was laid out, in the realization of the importance of avoiding posing more credence to one hypothesis or combination of ideas than another. Given the initial observations and formulated hypotheses, it was apparent that the use of multiple technologies were necessary to develop, test, and refine the multiple hypotheses into a working model.

2 APPLICATION OF METHODS

Aerial Photo Analysis

Aerial photos allows analysis of a geologic terrain from altitude. The use of a stereoscope to review and analyze sequential "stereo-pairs" is an unparalleled technique in identification of ancient landslide terrain. With a sufficient number of photo sets spanning many decades, consistent morphologic features -- as well as changing surficial conditions -- are observable as they evolve.

Aerial photos, both vertical and oblique and spanning six decades, were utilized during the analysis of the Rambla Pacifico Landslide Complex defining and documenting the evolution of the terrain from an obvious ancient landslide mass to a manufactured network of roads, graded pads and structures, and finally to a partial reactivation of the ancient landslide.

The most vital photo set utilized during this study was flown in 1928 (C-300 Fairchild series) and was of marginal to poor quality with regard to resolution. Nevertheless, the ancient landslide features in the photos were quite revealing and it was determined that the active Rambla Pacifico Landslide Complex is nested within a larger ancient landslide complex. These photos pointed to the morphology of a large ancient landslide with at least six shallower subordinate slides nested within its perimeters. The current active slide encompasses most of two of the largest pre-existing subordinate slides (mapped by the USGS) but remains a subordinate landslide entirely contained within the boundaries of the larger ancient landslide (mapped by CDMG). Additionally, lineations indicating faulting were noted on the aerial photographs.

Literature Review

Review of pertinent documents establishes a solid background for solving a complex problem. Existing interpretations of relevant features described in single-lot developments provided many clues to the geologic conditions as they existed before the currently active landslide failed. Many reports were also produced which documented initial structural distress occurring to the existing residences. Such documents literally became case histories, often recording data and interpretations made by engineering geologists during the early stages of landslide movement. Regional literature and geologic maps provided background data for determining the regional tectonic framework. It was noted in these documents and maps that both steeply-dipping and shallow-dipping faults existed in the area (Dibblee). These background data were considered essential for the successful unravelling of the intertwined structural fabrics present within the complex landslide.

A chronology of events leading up to the initial failure in 1978 was compiled based on all available documents. A separate focused chronology of damages from the commencement of failure through the present was also assembled from these documents and supplemented by our own data. As a result of the extensive literature review, numerous working models were investigated, providing an abundance of additional possibilities to be considered during the field investigation of the landslide complex.

Structural Analysis

The geometry and orientations of exposed earth materials in the surrounding area as well as within the Rambla Pacifico Landslide Complex were determined by field mapping and data collection. Mapping was a continually evolving process of structural data collection and refinement. The result was a series of detailed maps which graphically display the morphologic characteristics and the geologic structure of the landslide complex and vicinity.

Geologic Mapping: Geologic mapping of the landslide complex produced many generations of geologic maps utilizing four individual topographic bases. The mapping evolved from a small scale (1"=100') overview of the observable landslide-related features to detailed structural analyses in site-specific areas. Initial work consisted of careful mapping of cracks, bulges, scarps, grabens and other landslide-related features. Aerial photographs were employed to refine contact locations in covered and heavily vegetated areas. These data provided an accurate assessment of all areas which were being affected by the movement of the active landslide. The nature of the litigation required detailed large scale (1"=10') maps to be produced graphically displaying highly definitive geologic conditions in certain site-specific areas. The data collected, even from the most detailed large scale maps, were limited relative to the continuity of geologic structures (i.e., tectonic and landslide-related features). It became apparent that a large volume of data was needed to determine the structural history of the rock materials in the vicinity of the Rambla Pacifico Landslide Complex.

Detailed Structural Analysis: The complex nature of the rocks within the landslide, accompanied by a general lack of exposure, facilitated detailed analysis of the rocks that were exposed. Bedding, fracture, and shear orientations were measured in large quantities at several locations in rocks exposed by road construction. The overall abundance of fractures and secondary fracture-filling minerals (i.e., gypsum and calcite) was also documented. The data were compiled and equal area plots were constructed to determine the "statistically valid" structural fabrics contained within the rocks. The data were then compared with regional structural trends to determine the orientations of structures which could be related to fold and thrust development. Statistically valid structural fabrics, oblique to the expected tectonic orientations, were evaluated for their possible relationship to landslide activity. A number of fold sets mapped during this investigation are interpreted as landslide-related structures.

Subsurface Exploration

Perhaps the most critical data obtained were from subsurface exploration which added a third dimension of raw data. Formulating a three-dimensional model, both mentally and physically, greatly facilitated an understanding of the landslide geometry and failure mechanisms. Having the capability of obtaining an abundance of subsurface data through multiple exploration methods promoted further adoption of certain hypotheses (such as the existence of multiple slide surfaces) and consequently disproving alternate hypotheses (such as a solid intact mass failing on a single slide surface). Three methods were employed to maximize the subsurface data retrieval, including deep continuous core borings to depths of 350 feet, bucket auger borings to depths of 125 feet, and backhoe trench excavations up to 15 feet deep. Several instrumentation borings were also drilled.

Continuous Core Borings: The use of small diameter (3¼ inch) continuous core borings has several advantages over other exploration methods in landslide investigations. Namely, it permits logging below the water table and to depths three to four times deeper than most economical and safe bucket auger holes. The accurate retrieval of a nearly continuous core sample in each boring of this type permitted detailed logging and sampling of lithologic and structural features (i.e., bedding, fractures, shear surfaces, faults etc.).

Most of the ten continuous core borings were drilled utilizing a skid-mounted drill rig. Cores were recovered using a 63-inch split tube NX diameter core barrel with a diamond-studded coring bit. High fracture permeability in the upper 50 to 75 feet caused severe circulation problems necessitating the use of casing to retain circulation and preclude caving of fractured slide materials. In extreme situations, casing was extended deeper than 250 feet to permit completion of the boring. Drilling rates and percent recovery were recorded to aid in the evaluation of the materials. Upon completion of the coring operations, each boring was reamed for geophysical logging and/or instrumentation purposes. One boring was also utilized for packer testing to determine fracture permeability in the landslide materials.

Bucket Auger Borings: Six 24-inch diameter bucket auger borings were drilled at selected locations on the landslide to depths ranging from 25 to 125 feet. This drilling method facilitated hands-on observation and downhole logging of the landslide features by geologists. In downhole logging, a geologist equipped with safety harness, sampling tools, rock hammer and Brunton compass was lowered down the borehole in an aluminum cage or on a swing-like platform. A continuous air supply was provided when hydrogen sulfide gas was encountered at intolerable levels. Logging in several borings was prohibited below caving of saturated and fractured materials. Downhole logging permitted accurate measurements of bedding, fracture, and shear surface orientations as well as slickenside bearings on landslide surfaces. Bulk samples of shear surfaces were obtained in sufficient quantity to permit laboratory tests. Additionally, photographs and video tapes were taken of portions of the borings.

Trenching: A four-wheel drive backhoe with a 30-inch bucket was utilized for shallow exploration in seven locations. This method was chosen for near-surface observation of landslide boundaries, shear surfaces, displaced remnants of faults and folds, and damaged utilities such as septic tanks and seepage pits.

Instrumentation of Borings: In most of the core holes, slope inclinometer casing and staged pneumatic piezometers were grouted into place. The piezometers were located at selected depths (usually immediately above and below suspected slide surfaces) to permit periodic measurement of piezometric levels within the landslide. The slope inclinometers provided information regarding direct movements of the landslides.

Borehole Geophysical Methods

Considering the difficulties and hazards of attempting deep bucket auger borings along with the lack of oriented core samples, it became necessary to utilize additional technologies to improve the three dimensional perspective in the boring locations. Recent oil field technological advances have been utilized in the investigation of landslides to supplement continuous core information. Specifically, gamma ray logging and dipmeter logging have become extremely helpful in evaluating landslide subsurface conditions at depth and these methods were employed in two deep continuous core borings drilled in the Rambla Pacifico Landslide.

The small diameter core holes were prepared for geophysical wireline logging by reaming each boring with a 7¼ inch tri-cone rockbit inasmuch as dipmeter tools require a minimum 6-inch diameter hole to operate. While the gamma ray log could be recorded in a cased boring, the dipmeter required an open hole.

Gamma Ray Logging: The gamma ray tool provides a measurement of the natural radiation within formation materials and is especially useful in lithology identification and correlation. Radioactive elements tend to concentrate in marine shales and clays, while sands and basalts generally exhibit low radiation levels. On this basis, lithology identification was enhanced, especially in sections of the boring where little or no recovery of cores were made. These data were also supplemented by the caliper tool log (measuring borehole diameter) since zones of borehole washout in sheared clays and caving in highly fractured sandstones could readily be identified.

Dipmeter Logging: Although the dips of planar features could be easily measured in core samples, the strike or bearing of these features could not be measured without orientation of the core samples. The dipmeter log provided the missing component as well as attitudes in zones of unrecovered core. Using six equally spaced microresistivity electrodes mounted on a six-arm hydraulically actuated tool, the wireline dipmeter tool records resistivity data from the borehole materials. These data are subsequently used in a computer program which computes three-point problems from correlations in the resistivity data at selected depth intervals. Although only three points are needed, the additional three points provide a redundancy factor which permits determination of data quality for each interval in which an attitude was computed. The resultant data plot provided accurate attitudes from which inferences can be drawn regarding bedding structure, landslide surfaces or discontinuities, and faults.

Seismic Refraction

A geophysical consulting firm was subcontracted to run a series of seismic refraction lines across the landslide. The seismic array consisted of ground motion sensitive transducers placed at 25-foot intervals along 300-foot long lines. This array is considered effective for determining the depth and configuration of subsurface layers to a depth of ±100 feet. Velocity profiles were constructed and interpretations provided by the subcontracting consultant.

The profiles were utilized to derive some general conclusions relative to the nature and distribution of the subsurface materials. Seismic velocities recorded tended to decrease from the north to the south across the landslide and were interpreted as evidence for the overall increase in the number of discontinuities present in the rock materials and an overall reduction of competence. The break from relatively high to low velocities is coincidental with the change in the dominant deformational style from translational to rotational. The profiles also displayed a vertical change in seismic velocities recorded in the Las Flores Canyon stream channel. This change was interpreted as the contact between the alluvial gravels and the underlying Calabasas Formation sediments or Conejo volcanics. Valuable data regarding the character and distribution of the rock materials in and adjacent to the Rambla Pacifico Landslide Complex were supplied by the "shallow" seismic profiles.

Kinematic Analysis

The internal and surficial movements of the Rambla Pacifico Landslide were measured by use of slope inclinometers and vector analyses. Rates of movement manifested on individual slide surfaces within the landslide were measured as deflections in slope indicator casings. Internal rates of motion as high as 40" per year were recorded in the Fall of 1988. Documented deflections in most of the borings indicated sliding is manifested on multiple surfaces occurring at various rates.

Displacement vectors were constructed from observed surficial offsets and displacements of typically man-made surface features. The locations of displaced surface features such as houses, foundations, roads, and power poles were surveyed on-site using the Brunton and tape method and were plotted on a newly constructed topographic map. Numerous aerial photographs in conjunction with pre-slide topographic maps and development plans provided sufficient information to determine the original location of a majority of the mapped features. Vectors were constructed by measuring the total horizontal distance and direction of the displaced features from their original locations. These data were interpreted as absolute or total horizontal displacements of the landslide surface.

3 CONCLUSIONS

The method of multiple working hypotheses was continually utilized throughout the investigation of the Rambla Pacifico Landslide complex. The application of each method or technology was in and of itself a test of one or more hypotheses which provided a unique set of data. With the completion of each successive investigative method or technique (i.e., completion of a continuous core boring), each hypothesis was reviewed until it was proven or disproven and discarded. This process generally caused a reduction in the number of hypotheses, but at times also generated new ideas. Eventually, consistent characteristic signatures related to geologic conditions and processes emerged from the plethora of data, resulting in a highly refined, geometric and kinematic model of the landslide. The landslide model, which is the end product of a single refined hypothesis, is the most viable solution supported by the greatest percentage of the data obtained; it is that which appears closest to scientific truth.

REFERENCES

Allen, J. E. 1991. How Geologists Think – The Method of Multiple Working Hypotheses: Journal of Geological Education. Vol. 39, p. 67

Campbell, R.H., Yerkes, R.F., and Wentworth, C.M., 1966. Detachment Faults in the Central Santa Monica Mountains, California: U.S.G.S Professional Paper 550-C, pp. 1-11

Chamberlin, T.C., 1907. The Method of
the Working Hypotheses; Journal of
Geology, Vol. 5, pp. 837-848

Cronin, V.S., Slosson, J.E., and Slos-
son, T.L. 1991. Method of Multiple
Working Hypothesis - A Chimera or a
Necessity for Ethical Practice in
Engineering Geology: Proc. AEG Annual
Meeting, 1991

Dibblee, T.W., Jr., unpublished geologic
map of the Malibu Beach Quadrangle,
Dibblee Geological Foundation

Schoellhamer, J.E. and Yerkes, R.F.,
1971. Preliminary Geologic Map of the
Coastal Part of Malibu Beach Quadrangle, Los Angeles, California:
U.S.G.S. Open-File Map

Yerkes, R.F. and Campbell, R.H., 1980.
Geologic Map of East-Central Santa
Monica Mountains, Los Angeles County,
California: U.S.G.S Miscellaneous
Investigation Series Map I-1146

Weber, F.H. and Wills, C.J., 1983. Geo-
logic Map Showing Landslides of the
Central and Western Santa Monica Moun-
tains, Los Angeles and Ventura
Counties, California: California
Division of Mines and Geology, Open-
File Report #83-16LA

G3 Stabilization and remedial works
Stabilisation et mesures pour remédier aux glissements

Landslides, Bell (ed.) © 1995 Balkema, Rotterdam, ISBN 90 5410 032 X

Keynote paper: Recent advances in slope stabilization

Robert L. Schuster
US Geological Survey, Denver, Colo., USA

ABSTRACT: This paper reviews physical means of slope stabilization, particularly subsurface drainage and rock/soil control and retention systems, which have been improved and technically advanced in recent years because of innovative changes in design, analysis, and construction methods. The subjects dealt with in greatest detail are drainage, control of rockfall hazards, and the stabilization of soil slopes by means of geosynthetic retention systems. Several cases are presented in which new slope-stabilization techniques have been applied.

1 INTRODUCTION

The term "landslide" traditionally has been defined as the downward and outward movement of slope-forming materials: rock, soils, artificial fills, or combinations of these materials (Varnes 1958). In theory, landslides comprise the group of slope movements wherein shear failure occurs along a specific surface or combination of surfaces; thus, strictly speaking, it does not apply to all types of slope movements (Varnes 1978). However, because of the common usage of "landslides," particularly in civil engineering literature, the author has decided to follow common practice and to use landslides as the collective term for all types of slope movements in soil or rock.

Effective landslide hazard management has done much to reduce economic and social losses due to slope failure by avoiding the hazards or by reducing the damage potential. This has been achieved primarily by four mitigative approaches: (a) restriction of development in landslide-prone areas, (b) evolvement and application of excavating, grading, landscaping, and construction codes, (c) use of physical measures to prevent or control slope failures, and (d) landslide warning systems. The third of these methods, physical measures to prevent or control the gravitational and/or dynamic failure of slopes, is known as slope stabilization.

The use of crude slope-stabilization techniques, such as surface drainage and simple retaining walls, predates written history. However, some techniques for implementing these methods are relatively new or have undergone considerable upgrading in the past few years in terms of technology, lower costs, or increased harmony with the environment. This paper will discuss these relatively recent advances. The use of product, trade, proprietary, and company names is for clarity of expression; it does not imply endorsement or superiority of these specific procedures or of the equipment used.

2 CATEGORIES OF SLOPE STABILIZATION

The most commonly used physical approaches for control of unstable slopes are:

(a) Drainage -- Because of its high stabilization efficiency in relation to cost, drainage of surface water and groundwater is the most widely used, and generally the most successful, slope-stabilization method. Underground drainage systems and pumping wells collect and remove groundwater; surface water is diverted from unstable slopes by ditches.

(b) Slope modification -- Increased slope stability can be obtained by removing all or part of the landslide mass.

(c) Earth buttresses -- Earth buttress counterforts placed at the toes of unstable slopes often are successful in preventing failure. In many nations, this is the most common mechanical (as contrasted to hydrologic) method of landslide control.

(d) Earth retention systems -- Where

methods (a) to (c) will not ensure slope stability by themselves, structural controls, such as retaining walls, piles, caissons, fences, anchors, or internal reinforcement of the earth materials, are commonly used to prevent or control slope movements. Properly designed retention systems are useful in stabilizing most types of slope failures where these failures do not involve large volumes and where lack of space precludes slope modification.

Earth retention systems often are used in conjunction with drainage, slope modification, and/or construction of earth counterfort berms. Outstanding recent examples of large-scale use of combined remedial measures are the stabilization of slopes associated with Tablachaca Dam in Peru and the Clyde Power Project in New Zealand. In the early 1980's, Tablachaca Dam was endangered by a 3-million-m³ creeping mass of rock and colluvium on the right abutment (Fig. 1). Deere and Perez (1985) noted that approximately US$40 million was spent by the Peruvian government in landslide stabilization measures consisting of (1) a 460,000 m³ toe buttress founded on compacted river sediments, (2) 405 prestressed rock anchors, (3) 1,300 m of drainage tunnels, 190 radial drains, and 3,300 m of surface ditches, (4) 68,500 m³ of rock excavation, (5) improvement of the river-channel flow pattern, and (6) numerous inclinometers, piezometers,

Fig. 1. Creeping rock slide (outlined by arrows) endangering Tablachaca Dam and Reservoir, Mantaro River, Peru. Complex stabilization measures, consisting of a reservoir-level earth buttress, surface and subsurface drainage, and rock anchors, were used to reduce the threat of a catastrophic slope failure to Peru's largest hydropower dam. Photograph was taken in February 1982 before most mitigative measures had been installed.

extensometers, and other instrumentation (Morales Arnao et al. 1984).

The toes of several major actively creeping landslides in the Cromwell Gorge of the Clutha River on the South Island of New Zealand will be inundated by the planned 1992 filling of Lake Dunstan behind Clyde Dam (Gillon & Hancox in press). This 102-m-high concrete gravity dam is now complete, but impoundment of Lake Dunstan has been delayed pending stabilization of several large landslides that threaten the dam and reservoir. The following extensive multiple remedial measures currently are being implemented to offset the effects of reservoir filling on the landslides: (1) gravity drainage, using drainage tunnels, drill holes, and surface drainage systems, (2) free-draining rock- and gravel-fill buttresses at the landslide toes, (3) grout curtains, and (4) pumped drainage.

These types of physical control methods have been discussed at length in the landslide literature (e.g., Baker & Marshall 1958, Veder 1981, Zaruba & Mencl 1982, Hausmann 1990). All are in common use worldwide, and all are continually being improved by modern methods of analysis, design, and construction. However, in the author's opinion the greatest innovations in slope stabilization in recent years, in terms of technology, economy, and environmental improvements, have been in drainage and in rock/soil retention systems; thus, this paper will accentuate recent advances in these stabilization methods.

3 RECENT APPROACHES TO SUBSURFACE DRAINAGE AS A METHOD OF SLOPE STABILIZATION

Subsurface drainage as a method of lowering the groundwater table in an unstable slope has traditionally consisted of one or more of the following technologies: (1) drainage trenches, (2) drainage wells, (3) drainage galleries, tunnels, or adits, (4) subhorizontal (commonly called "horizontal") drains drilled either from the slope surface or from drainage wells or galleries, and (5) subvertical drains drilled upward from drainage galleries. Most often, these systems drain by means of gravity flow; however, pumps are occasionally used to lift water from low-level collector galleries or wells. This section will discuss recent advances in the above types of drainage systems; in addition, it will briefly mention less commonly used, but innovative, means of drainage, such as electroosmotic dewatering, vacuum and siphon drains, geosynthetic drains and filters, and blasting.

3.1 Drainage trenches

Trench drains, filled with free-draining materials, have been used for effective shallow subsurface drainage for several decades. If they fully penetrate firm bedrock, they are often called "counterfort" drains. Backhoe-type excavators are used to depths of 5-6 m and clamshells for greater depths. For clamshell excavation, the diaphragm method is used, undertaking excavation and concreting in alternating panels (Collotta et al. 1988). Cancelli et al. (1987) reviewed the theory behind trench drains in their excellent paper on groundwater problems in embankments, dams, and natural slopes. A systematic study of the efficiency of trench drains was carried out by Hutchinson (1977) using the finite-element method and assuming two-dimensional steady-state flow. Stanic (1984) followed, using finite-element analysis for three-dimensional flow in trench drains on sloping surfaces. Di Maio and Viggiani (1987) have considered the non-steady-state-flow case based on intermittent rainfall.

An increasingly used technique utilizes vertical drains to remove water from the bottom of drainage trenches. Lew and

Graham (1988) described a project along the Assiniboine River in Canada in which compacted sand drains were connected to the base of a drainage trench to stabilize a slope in which a high-pressure gas pipeline was buried (Fig. 2). The design used 0.75-m-diameter augered holes backfilled with sand connecting to trench drains placed on either side of the pipeline.

3.2 Drainage wells

Deep wells increasingly are being used to drain unstable slopes, particularly where the depths needed are too deep for economical construction of drainage trenches. Collotta et al. (1988), Bianco & Bruce (1991), Beer et al. (in press), Bianco and Bruce (in press), and Peila et al. (in press) have described the development and use in Italy of large-diameter (up to 2-m) vertical wells (trade name: RODREN) spaced in rows at 5-20 m, center-to-center, and reaching depths of as much as 50 m. Each shaft is connected to its neighbor by a horizontal drill hole, placed just above the base of the well. These holes are lined with 76-to-100-mm PVC pipe and serve as gravity collector drains. Two thirds of the wells are filled with 3-to-20-mm-diameter free-draining material and serve as vertical drains; every third well is kept open for (1) physical inspection and cleaning, and (2) monitoring and adjustment of flow rates. The innovation of this system is that it allows gravity discharge of water from the wells by means of the small-diameter PVC pipes that are installed using mini-drilling rigs placed on the bottoms of the wells. This method has been successfully used to stabilize a slope along the Florence-Bologna Motorway (Fig. 3) (Collotta et al. 1988) and to control the edge of a large landslide in the city of Ancona on the Adriatic Coast (Beer et al. in press).

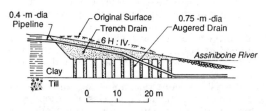

Fig. 2. Trench and augered sand-drain slope stabilization system for gas-pipeline crossing of the Assiniboine River, Canada (after Lew & Graham 1988).

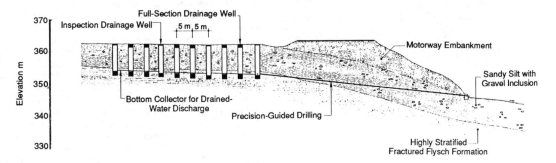

Fig. 3. Vertical drainage shafts connected to a horizontal PVC outlet drains, Florence-Bologna Motorway, Italy (after Collotta et al. 1988).

A similar drainage scheme was used for a slide area encountered in relocation of German Federal Highway B-10 where it crosses the Rhine valley near the village of Albersweiler (Wichter et al. 1988). As the main part of the stabilization scheme, 172 wells (1.5-m-diameter) and 13 maintenance shafts (1.8-m-diameter) were installed across the landslide to depths of 16-25 m. The shafts were filled with filter sand. Unlike the Italian examples noted above, these wells were spaced very closely (3 m, center-to-center); however, the outlet drain from the system required a 170-m-long subhorizontal directional borehole from the foot of the hill to one of the maintenance shafts. The short connecting holes between the wells were drilled using augers driven by compact drill rigs that fit within the casing of the vertical wells.

Gabus et al. (1988) have described an extensive stabilization effort on the Arveyes landslide in the Swiss Alps in which 16 vertical pumped drains were installed in a 1-km^2 unstable area. The deepest well reached a depth of 95 m, and the piezometric level locally was lowered more than 50 m.

In 1991, "belled" (i.e., bell-shaped) drainage caissons backfilled with fine gravel were used to stabilize a slow-moving slide in a construction cut on Colorado Highway 93 near the city of Golden (Chou, N.N.S. 1991, personal communication, Colorado Department of Transportation, Denver). As shown in Fig. 4, the caissons were closely spaced so that the bells overlapped, allowing gravity drainage along the line of caissons to the ground surface outside the perimeter of the landslide. The use of bells for horizontal drainage precluded the need for horizontal drilling and insertion of drain pipe between the bottoms of the wells.

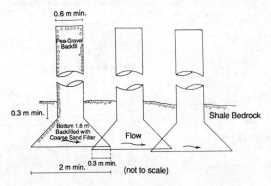

0.6 m min.

Pea-Gravel Backfill

0.3 m min.

Shale Bedrock

Bottom 1.6 m Backfilled with Coarse Sand Filter

Flow

2 m min.   0.3 m min.   (not to scale)

Fig. 4. Overlapping belled drainage caissons for deep drainage of slopes (courtesy of Colorado Department of Transportation).

In most cases, collected water flows from the bottoms of drainage wells or caissons by gravity. However, if the base of the well is low enough, pumping may be needed. Olcese et al. (1991) have reported on a slope-stabilization project near Genoa, Italy, in which submersible motor-driven pumps were used to remove water from the bottoms of small-diameter (200-mm) drainage wells. The pumps were automatically activated when the water surface in the wells reached predesignated levels.

3.3 Drainage tunnels, adits, or galleries

On large slope-stabilization projects, such as those sometimes needed for hydropower projects, deep drainage galleries are being used increasingly. Groundwater is intercepted directly by the galleries and by collection from fan drains drilled from the galleries. The galleries commonly are placed by conventional tunneling techniques. In addition to being drainage collectors, galleries and tunnels provide access for at-depth study of the landslides; they also enable at-depth installation of monitoring equipment, such as piezometers and extensometers.

As is the case for deep drainage wells, most drainage galleries are designed to drain by gravity. In some cases, however, to be effective as collector drains the wells must be placed at sufficiently low elevations that the collected water has to be pumped to the surface.

Classic recent examples of the use of drainage tunnels and galleries are the slope-stabilization projects at the Downie slide on the Revelstoke (power) Project, British Columbia, Canada, and the above-mentioned Tablachaca Dam in Peru and the Clyde Power Project in New Zealand. To prevent reactivation of the 1.5-billion-m^3 Downie rockslide by filling of the reservoir behind 160-m-high concrete-gravity Revelstoke Dam, deep drainage adits and connecting vertical drains were installed in the lower part of the slide before the reservoir was filled in 1984 (Imrie et al. in press). The drainage works were successful in lowering piezometric levels considerably below where they were before river level was raised 70 m, and thus stabilized the slide. At Tablachaca Dam, 1,300 m of gravity-drained collector tunnels were driven in the valley wall. On the Clyde Power Project, 49 tunnels (total length: 16,392 m) were driven into nine landslides around the reservoir (Gillon & Hancox, in press). Most of these tunnels were situated above reservoir level, and thus could be drained by gravity. However, to stabilize the Brewery Creek slide, the toe of which is at lower elevation than the pre-reservoir level of the Clutha River, it was necessary to install a low-level (i.e., below-river-level) collector gallery (Fig. 5) from which water must be pumped to the surface. Interesting features of the Brewery Creek landslide stabilization scheme are a zoned earthfill blanket and connected grout curtain; both will act as barriers to flow from the reservoir through the landslide to the drainage works (Gillon et al. in press). The zoned earthfill blanket is in effect a canal lining that is made up of wave-armoring, shoulder, transition, core, and filter zones. The low-level drainage tunnel and its fan drains will intercept any water that passes through the blanket and the grout curtain.

3.4 Subhorizontal drains

Conventional subhorizontal drains (henceforth referred to as "horizontal drains," a more common usage) are used worldwide. For example, in the State of California more than 300,000 m of horizontal drains were used effectively during the period 1940-1980 as an economical method of draining unstable slopes (Smith 1980).

Horizontal drains can be inserted from the ground surface or by drilling from drainage galleries, large-diameter wells, or caissons. The typical horizontal drain hole is 120-150 mm in diameter and is lined with slotted plastic casing (commonly PVC) 60-100 mm in diameter. Conventional rotary techniques are adequate for drilling most horizontal drain holes; however, it often is necessary to use precision drilling to connect a horizontal drain to a drainage shaft. In recent years, precision drilling techniques have progressed to where horizontal directional boreholes as long as 200 m have been installed (Sembenelli 1988).

In 1980, the California Department of Transportation conducted a study of the long-term effectiveness of 20 of its horizontal-drain installations (Smith 1980). The main conclusions reached were: (1) a 30-40-year life span is about the maximum that can be expected from perforated metal pipe casing, (2) slotted PVC will provide longer service life than metal pipe casing, (3) slotted PVC casing allows considerably less sediment to enter the drain than the standard U.S. 3/8-inch (9.5-mm) perforated metal pipe, (4) high-pressure water cleaning systems used for cleaning sewers and unplugging culverts can be adapted for horizontal-drain cleaning, and (5) most

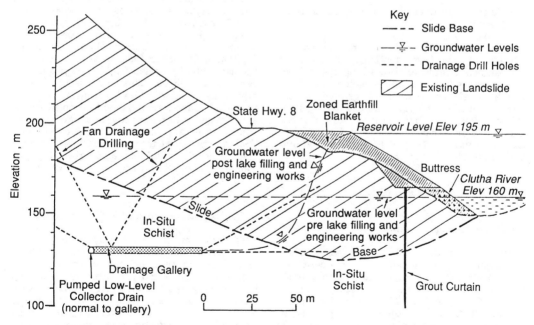

Fig. 5. The Brewery Creek landslide-stabilization scheme, Clyde Power Project, New Zealand (after Gillon et al. in press).

drains need to be cleaned once every 5-8 years (heavy root growth or exceptionally fine-grained sediments can change this requirement to every 2 years).

When using horizontal drains to dewater an unstable slope, the length, spacing, and position of the drains must be chosen. These drain-placement parameters are usually selected on the basis of engineering judgment or design charts. However, Nakamura (1988) has developed a numerical-analysis technique to evaluate time-related drop in groundwater level in a slope as a result of use of horizontal drains, and Resnick and Znidarcic (1990) have tested model slopes in a geotechnical centrifuge to determine the influence of horizontal drains on distribution of piezometric levels within the slopes.

As noted above, there is an annular space between the horizontal casing and the soil or rock of the drillhole. Filling this space with uniformly graded filter sand is difficult, and checking the actual quality of the work in the hole is impractical. Sembenelli (1988) has noted that many attempts have been made to eliminate the need for this awkward sand filter. In the past, pipes with very fine slots have been used; however, these are being supplanted by pipes with coarse slotting protected on the outside by one- or two-ply synthetic geotextile stockings.

3.5 Subvertical drains drilled upward from tunnels

Fans of subvertical drains are now commonly driven upward from drainage galleries or tunnels. Groundwater flows by gravity from these drains into the larger collectors, from which it either drains by gravity or is pumped to the surface. The use of a system of subvertical fan drains to remove groundwater from the Brewery Creek landslide in New Zealand is discussed above and illustrated in Fig. 5.

3.6 Less commonly used drainage methods

3.6.1 Electroosmotic dewatering

In the mid 1930's, Professor Leo Casagrande introduced the concept of electroosmotic drainage into geotechnical engineering practice, and many papers (e.g., Casagrande, 1948) have since been published on this subject. Despite some success in the past in slope stabilization, this process has not received wide application because of the costs of operation and remaining technical uncertainties. However,

Lo et al. (1991a, 1991b) have improved the process by using specially designed copper electrodes to prevent gas accumulation around the electrode and to allow free water to flow from the cathode without pumping. In field tests conducted on the soft, sensitive Leda clay of eastern Canada, undrained shear strength increased by about 50 percent in a period of 32 days throughout the depth of the electrodes. Because no pumping was needed, both installation and electricity costs were reduced considerably compared with earlier electroosmotic drainage installations.

3.6.2 Vacuum dewatering

Arutjunyan (1988) has reported on the use of vacuum in drill holes to dewater soils of low permeability in slopes in the Soviet Union. The technique has been applied to landslides as a quickly installed temporary measure until long-term stabilization could be effected. The vacuum treatment increases soil suction and accelerates the process of soil consolidation. It has proved to be successful to depths of 30-35 m by applying the vacuum for a period of 2-4 weeks.

3.6.3 Drainage by siphoning

Siphon drains for slope stabilization have been installed at 40 sites in France in the past 5 years (Gress in press). These drains have the advantage of being able to raise water to the surface without pumping. Siphoning of water from unstable strata is accomplished by sealed PVC pipe systems. A recent example of the successful use of siphoning to lower the groundwater table under a highway embankment on an unstable slope occurred at Venarey-Les-Laumes near Dijon, France, where five vertical siphon drains, spaced at intervals of 10 m, lowered the piezometric level from an original depth of 2 m beneath the highway to a depth of 8 m (Fig. 6).

3.6.4 Use of geosynthetics for drainage

ASTM Committee D-35 (1991) defined a "geosynthetic" as "a planar product manufactured from polymeric material used with soil, rock, earth, or other geotechnical engineering related material as an integral part of a man-made project, structure, or system." Geosynthetics include geotextiles, geogrids, geomembranes, geonets, and geocomposites (Koerner 1990). Correctly designed and installed geotextiles ("thin, flexible, permeable sheets of synthetic

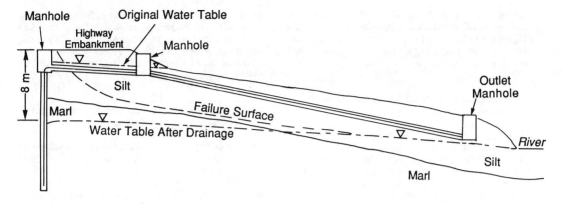

Fig. 6. Cross section of siphon-drain stabilization of highway embankment and underlying unstable slope near Dijon, France (after Gress in press).

material used to stabilise and improve the performance of soil associated with civil engineering works") have the ability to filter, drain, reinforce, and separate geologic materials (Ingold & Miller 1988)). They can be used for lateral or vertical drainage of fine-grained soil masses (silts or clays) if the required flow rates are modest. When used for drainage, flow is in the planar direction, rather than moving across the plane of the fabric as is the case when geotextiles are used as filters. Needle-punched fabrics, that often are as much as 6-mm thick and can be made much thicker, provide the most effective drainage, particularly when used in layers. Their most common usage on slopes is for drainage of embankments, within reinforced earth walls, and as pore-pressure dissipators behind retaining walls (Koerner 1990).

In-place subsurface drainage capability can also be accomplished by use of geonets, the most recently introduced members of the geosynthetics family. Koerner (1990) has defined a geonet as "a netlike polymeric material formed with intersecting ribs integrally joined at the junctions used for drainage with foundation, soil, rock, earth, or any other geotechnical-related material as an integral part of a human-made project, structure, or system." In manufacture, geonets are usually formed by a continuous extrusion process of polyethylene into a netlike configuration of intersecting ribs which is free-draining. In use for drainage, geonet mats are commonly 5-13 mm thick. Because they do not function as filters, geonets are always confined by a geotextile, geomembrane, or other material placed on their outer surfaces; such combinations of materials are known as geocomposites.

A recent example of the use of geofabric drains to help stabilize a steep slope in residual soil and weathered rock in Singapore has been presented by Broms & Wong (1986). Woven geofabric was wrapped around sand-and-gravel drains that cut across failure surfaces. The fabric functioned as both a filter around the drains and as reinforcement in the slope.

3.6.5 Blasting for drainage of rock slopes

List (1988) has described the reduction of pore pressures in unstable bedrock open-pit mine slopes by means of blasting. The procedure was used in an open-pit oil-sand mine in northeastern Alberta, Canada. During oil-sand excavation, large block slides occurred in the 60-m-high sedimentary-rock wall of the pit. Local stability was obtained by setting off explosive charges of as much as 1500 kg of Heavy AN/FO per hole in boreholes along the face. The explosive forces reduced pore pressures by forming microfractures and fractures in the rock.

4 RECENT ADVANCES IN USE OF EARTH-RETENTION SYSTEMS IN SLOPE STABILIZATION

In this paper, I will treat earth-retention systems as a broad category of remedial or control measures that include conventional retaining walls, rock fences and nets, soil and rock anchors, soil nailing, and reinforced-earth-type walls, as well as changes in physical/chemical character of the earth mass by processes other than drainage.

For clarity, the discussion will be divided into uses of retention systems on (1) rock slopes and (2) soil slopes, although there are obvious overlapping uses in these two categories of geologic materials.

4.1 Retention systems for rock faces and slopes

Although several different categories of failures occur on rock slopes, only rockfalls, rock topples, rock slumps, and rock slides can be prevented or controlled at reasonable expense. Rock avalanches, another major type of rock-slope failure, in most cases are too large to be controlled; however, rock avalanches often start as relatively small falls, topples, slides, or slumps that are controllable at the source. This section will deal mainly with stabilization of slopes that are subject to rockfall, because it is in this area of rock-slope engineering that the most recent advances have been made. Several of the new techniques considered here also are applicable to rock topples, slumps, and slides.

Increasing traffic volumes in mountainous areas have heightened public awareness of the danger of potential rockfalls, resulting in significant ongoing research and development of innovative measures for rockfall prediction and control. In much of the United States, rockfall accidents are no longer considered to be "acts of God;" the traveling public increasingly demands protection from rockfalls on mountain highways (Barrett & White 1991). An important factor in stabilization of slopes subject to rockfall has been the prioritization of slopes that are most subject to catastrophic failure; this prioritization enables unstable slopes to be stabilized before accidents occur. An example of a system of prioritization is the Rockfall Hazard Rating System (RHRS) developed by the Oregon Department of Transportation (Pierson 1991). Rockfall risks have been rated on the basis of: (1) slope height, (2) ditch effectiveness, (3) average vehicle risk (percentage of time a vehicle will be in the rockfall hazard zone), (4) sight distance for drivers, (5) roadway width, (6) geologic character of the slope, (7) block size or quantity of rockfall per event, (8) presence of water on the slope, and (9) rockfall history. The Oregon RHRS system is intended to be a proactive tool that will allow transportation agencies to rationally address rockfall hazards instead of simply reacting to rockfall accidents.

The measures most commonly used to prevent rockfall from encroaching upon a highway, railway, or other structure or development are (1) rock nets, fences, walls, attenuators, benches, and ditches, and (2) rock bolting and buttresses (Fig. 7). In addition, even though it is expensive, under extreme conditions tunneling can be used to avoid rockfall.

4.1.1 Rock nets, fences, walls, attenuators, benches, and ditches

The most important rockfall input factors in design of systems to prevent rockfall from reaching a highway, railway, or other critical structure include (1) trajectory (height of bounce), (2) velocity, (3) impact energy, and (4) total volume of accumulation. Nearly 30 years ago, Ritchie (1963) documented that by varying slope steepness and length, height of rock bounce, velocity, and distance of travel are greatly affected. In recent years, computer programs have been developed to provide statistical analysis of probable rockfall behavior for a given slope

Fig. 7. Anchored-concrete-pillar support of huge block of gneiss above high-rise buildings, Cantaglo Hill, Rio de Janeiro. The 25-m maximum-height pillars were installed by the Rio de Janeiro Geotechnical Control Office in 1967. (1990 photograph courtesy of Rio de Janeiro Geotechnical Control Office).

by modeling the slope; this is accomplished by converting slope survey data into a cartesian-coordinate system in which each change in slope angle is represented by a different line segment or cell. One of the most recent of these programs is the Colorado Rockfall Simulation Program (CRSP), which incorporates numerical input values assigned to slope and rockfall characteristics (Pfeiffer et al. 1990). Empirically derived functions relating velocity, friction, and material properties are used to model the dynamic interaction of the rock and the slope. The model utilizes equations of gravitational acceleration and conservation of energy to describe the motion of falling and bouncing rock. The statistical variation among rockfalls is modeled by randomly varying the angle at which a rock impacts the slope within limits set by rock size, slope material differences, and irregularities of slope geometry. The program provides estimates of probable bounce height and velocity at various locations on a slope (Figs. 8, 9, & 10).

Programs such as CRSP are constantly being refined as experience is gained in their application to design of rockfall retention walls, fences, benches, and ditches. The Colorado Department of Transportation is using CRSP for design of rockfall retention walls in construction of Interstate Highway 70 (I-70) through the rugged gorge of Glenwood Canyon in western Colorado (Barrett & White 1991). An example of this use was the 1990 design and construction of a rubber-tire-faced, geosynthetically reinforced soil and concrete rockfall retention wall to protect I-70 near the town of Gypsum, Colorado; the 5-m design height of the constructed wall was based on rock rebound height as estimated by CRSP (Figs. 8 & 9) (Andrew in press).

Highways and railways in the mountains of North America commonly have been protected from rockfall by traditional single-twist mesh fencing supported by fixed, rigid posts; thus, in the United States the term "rock fence" often is associated with this basic style, commonly referred to as "chain-link" fence (Barrett & White 1991). This basic fence, as a single-mesh layer, is relatively inexpensive and will effectively contain small rockfalls. However, larger and heavier duty rockfall catch fences or nets have been in use for many years in the Alpine countries of Europe to protect highways, railways, and mountain communities from rockfall events; these systems evolved from avalanche protection systems developed in the 1950's to protect Alpine villages (Yarnell 1991).

The European technology is now being used extensively in the mountainous parts of the United States in spite of its higher cost

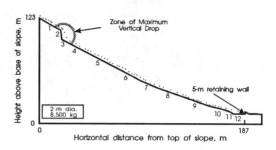

Fig. 8. Schematic rockfall trajectory produced by Colorado Rockfall Simulation Program (Pfeiffer et al. 1990) for rockfall on Interstate Highway 70 at Gypsum, Colorado. The slope was divided into 12 segments with surface characteristics determined by field study. For this analysis, individual rock spheres were chosen with a diameter of 2 m and weight of 8,500 kg each. Design of 5-m-high retaining wall shown at a horizontal distance of 187 m from the top of the slope was based on this rockfall simulation.

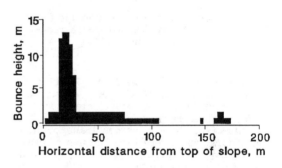

Fig. 9. Computer-produced Colorado Rockfall Simulation Program histogram of bounce heights for 25 2-m-diameter rocks on slope at Gypsum, Colorado (Fig. 8).

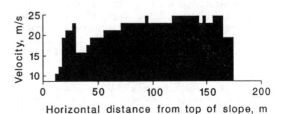

Fig. 10. Computer-produced Colorado Rockfall Simulation Program histogram for velocities of 25 2-m-diameter rocks on slope at Gypsum, Colorado (Fig. 8).

1723

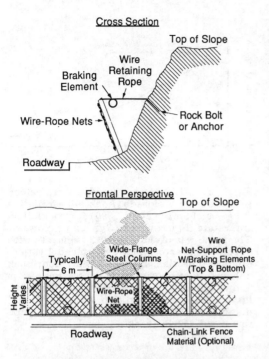

Fig. 11. Schematic cross section and frontal perspective of a typical wire-rope rock-fall barrier net (after Yarnell 1991.)

Fig. 12. Three-ton boulder impacting panel of rock net in field tests by California Department of Transportation. (Photograph by J. L. Walkinshaw, Federal Highway Administration, U.S. Department of Transportation.)

Fig. 13. Upslope side of Isomat Italia s.p.a. wire-rope rockfall-retaining-net fence. Friction brakes are enclosed in modules at lower right. (Photograph courtesy of Isomat Italia s.p.a., Milan, Italy.)

as compared to chain-link fence. Most United States applications of wire-rope safety nets are found in the eastern States (Yarnell 1991). This may be the result of limited rights-of-way along eastern transportation corridors, which preclude the use of other mitigative measures. However, rockfall nets/fences also can be used to control rock slopes of great heights, such as those found in the western United States.

Rockfall problem areas have been identified along 5,000 km of California highways (McCauley et al. 1985). The 1985 study concluded that rolling rocks up to 0.6 m in diameter can be restrained by chain-link fence; however, this type of restraining device frequently suffers severe damage when hit by rocks of this size and is inadequate to stop larger rocks. Thus, a rigorous field testing program of "European-style" rock fences (Fig. 11) was conducted by the California Department of Transportation (Caltrans) (Smith & Duffy 1990). Many of these fences are proprietary. The two types tested by Caltrans were high-impact wire-rope net systems developed by Brugg Cable Products, Inc., (Switzerland) and The Industrial Enterprise Corporation (France). Both types of fences rely on

friction braking devices. When the bouncing rocks collide with the fences, producing deformation of the nets, the nets engage the energy-absorbing friction brakes, thus extending the time of collision, and thereby significantly increasing the capacity of the nets to restrain the rocks. This approach allows the use of lighter, less costly elements, thus reducing the cost of the fences. The Caltrans field tests were conducted on an 80-m-long, 34° slope below State Highway 1 between Big Sur and San Simeon, California. Large boulders (135 to 6000 kg) were rolled down the slope into the net (Fig. 12), imparting rotational energy on individual components of the fence. Design-load rockfalls were effectively stopped by both types of rock nets. A similar fence/net (Fig. 13) with friction braking has been produced by Isomat Italia s.p.a. for rockfall control in northern Italy. Isomat fences have been used successfully to control rockfall on the relocated route of the Bormio-Sondrio highway at the toe of the 1987 35-million-m³ Val Pola landslide in the Italian Alps.

One disadvantage of systems that use friction energy-absorbing brake systems is that the friction brakes require resetting after each significant rockfall, a factor in long-term maintenance costs of the systems. For this reason, the Colorado Department of Transportation has recently developed the Colorado Flexible-Post Rockfall Fence (Barrett & White 1991) (Fig. 14). By grouting bundles of wire tendons into steel casings, posts are produced that have the ability to be flexible, yet are rigid enough to support the mesh netting. The principle for the system is: the fence catches and redirects bouncing rocks to energy-dissipating collisions with the slope; immediately after each collision, the flexible posts rebound, leaving the fence ready for the next encounter without needing maintenance (Barrett & White 1991). To insure that this innovative system is available to the public, it has been patented by the Colorado Department of Transportation; thus, its design and use are public property.

An untested variation of the Colorado flex-post fence utilizes a long tail of wire mesh that causes bouncing rocks to "mole" (i.e., go underneath) between the mesh and the slope surface until all energy has been dissipated (Barrett & White 1991) (Fig. 15). If, as a result of its rotation, a bouncing rock becomes entangled in the wire mesh, large forces may suddenly be applied to the

mesh, cable, and posts, resulting in flexing of the posts.

Another type of energy-absorbing fence has recently been used to contain rockfall on a slope above a housing development in Springdale, New Foundland, Canada (Boyd 1991). The 240-m-long fence, which was designed for a rock impact energy of 100 KNm, incorporates a series of high-strength cables that run horizontally along the slope and parallel to each other, and are supported by 3-m-high galvanized-steel posts set in concrete foundations. The fence absorbs energy from rock impacts by stretching of the cables. The cable system is covered by galvanized-steel mesh to aid in absorbing rotational energies and to intercept flying rock fragments.

Polymer grids (discussed as a means of soil reinforcement in section 4.2.1.1.3) have been used in Hong Kong as fences to protect a housing development from rockfall (Threadgold & McNichol 1984). In Norway, similar geogrid fences are being used as protection against avalanches during the spring thaw (Bush 1988).

Another approach to controlling rockfall is to partially absorb or attenuate the energy of bouncing or rolling rocks without actually stopping them. The Colorado Department of Transportation has developed an attenuation system that utilizes columns of used tires and rims mounted on vertical 75-mm-diameter steel pipes suspended from a large-diameter wire rope mounted across the rockfall chute (Barrett & White 1991; Andrew in press) (Fig. 16). Rock anchors are used to secure the ends of the wire rope to the bedrock walls of the gully. To address aesthetic concerns, a facade of wooden timbers is usually suspended from a wire rope immediately

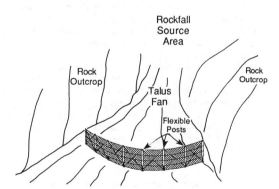

Fig. 14. Schematic diagram of Colorado Department of Transportation Flexible-Post Rockfall Fence (after Barrett & White 1991). The mesh is double-twist hexagonal wire.

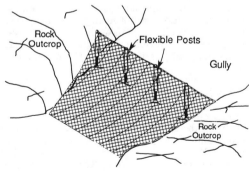

Fig. 15. Colorado "mole" fence (after Barrett & White 1991). Posts are flexible; mesh is double-twist hexagonal wire.

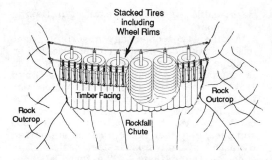

Fig. 16. Schematic diagram of Colorado Department of Transportation rockfall attenuator (after Andrew in press).

Fig. 17. Uphill view of Colorado Department of Transportation rockfall attenuator located in rockfall chute above Interstate Highway 70, Glenwood Canyon, Colorado. Note timber facing on downchute side of the attenuator.

downslope of the hanging tires (Fig. 17). The Colorado "rockfall attenuator" is designed to absorb most of the kinetic energy and to reduce rebounding heights from incoming rockfall; after a rock passes through the attenuator, the system returns to its original position without maintenance.

4.1.2 Rock bolting

Rock bolting has been used for several decades to stabilize rock slopes. Tensioned rock bolts are positioned across potential rock failure surfaces and anchored in stable rock beyond the surfaces. The application of tensile stress in the bolts results in increased normal stress in the direction of the bolts and decreased shear stress on the failure surface. Methods of securing the distal end of the bolt in the drill hole include cement

anchors). For permanent anchorage, grout, resin, and mechanical anchors (e.g., slot-and-wedge and expansion-shellgrouted bolts are preferred because they provide better bonding between the rock and the bolt, and are resistant to corrosion (Wyllie 1991). If the face plate for the bolt fails or corrodes away, the bolt force continues to be transferred to the rock by means of the grout bond. The grout can be a nonshrinking Portland-cement mix or a plastic (high-density polymer). Plastics with variable setting rates often are chosen so that the grout in the anchorage zone hardens first, allowing tensioning before the rest of the grout sets. Current world practice in design and installation of rock anchors has recently been summarized by Littlejohn (1990).

Barley (1991) has described five recent cases in the United Kingdom in which rock-bolting systems were used to stabilize rock and soil slopes. Unique methods of rotary percussive drilling, water jetting to clean the holes, and grouting were used in these installations. An interesting concept of multiple "unit anchors" within a single borehole is provided by the Single Bore Multiple Anchor (SBMA) system. Each anchor has its own tendon, and each anchor is encapsulated in a corrosion-protection system; the anchor "capsules" are located at staggered depths within the borehole, so that each anchor, loaded with its own stressing jack, transfers its load to a discreet length of the anchor bore. This system almost eliminates the effect of the progressive failure mechanism that occurs in normal anchorage systems and allows the simultaneous mobilization of almost the entire ground strength throughout the full length of the borehole (Fig. 18). The SBMA system has been used successfully to stabilize a coastal slope at Alexander Quay, Southhampton, England (Barley 1991).

Rock anchors often are combined with other methods of stabilization. One of the best-known examples was the use of 405 rock anchors to control the slide on the right abutment of Tablachaca Dam, Peru, mentioned earlier in this report. These anchors accompanied a large toe buttress, surface and subsurface drainage, rock excavation, and improvement of the river-channel flow pattern (Morales Arnao et al., 1984).

4.1.3 Tunneling as a means of avoiding rock-slope failures

Because of its expense, tunneling seldom is used to avoid rock-slope problems. However, after a section of the main line of

BC Rail in British Columbia, Canada, had been blocked for 22 days within a period of 2 years, and geologists had predicted continuing rockslide/rockfall problems for the line, the railroad company constructed a new 1.2-km-long tunnel that bypassed the unstable slopes (Leighton 1990). The tunnel was completed in 1989 at a cost of US$6.6 million. This was an expensive, but permanent, solution to a very difficult problem of rock slope instability.

4.2 Retention systems for soil slopes

Figure 19 summarizes the current methods of earth retention; these methods have been presented in two groups depending on whether they provide external or internal stabilization. Examples of externally and internally stabilized earth retention systems are presented in Figure 20. Externally stabilized systems rely on external structural walls against which stabilizing forces are mobilized. Prior to the late 1960's, external walls, mainly gravity and cantilever walls, were the predominant types of retaining structures. The principles and use of external walls are well understood, and will not be discussed here. Internally stabilized earth retention systems rely on reinforcement that is installed within the slope and extends

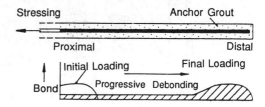

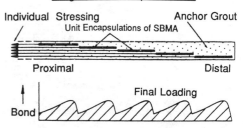

Fig. 18. Schematic illustration of load distributions for a traditional rock anchor and the Single Bore Multiple Anchor (SBMA) system (after Barley 1991).

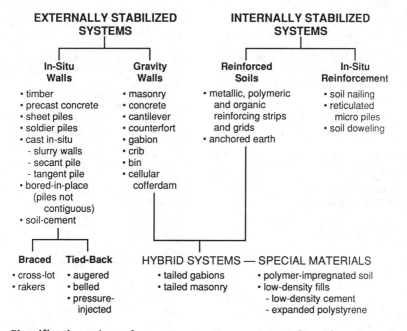

Figure 19. Classification scheme for earth retention systems (after O'Rourke & Jones 1990).

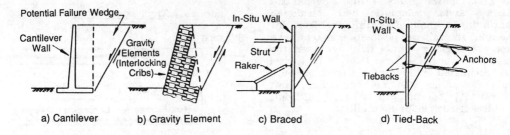

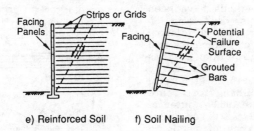

Fig. 20. Examples of externally and internally stabilized earth retention systems (after O'Rourke & Jones 1990).

beyond the potential failure surfaces into stable ground. General references on this topic are Jones (1985), Mitchell & Villet (1987), Christopher et al. (1989 & 1990), Mitchell & Christopher (1990), and O'Rourke & Jones (1990). This section will deal with advances in the use of these internally stabilized earth retention systems, which are known generically as "reinforced soil."

4.2.1 Reinforcement of soil slopes and embankments by internal stabilization

Earth reinforcement (reinforced soil), which can be defined as the inclusion of resistant elements in a soil mass to improve its overall strength, has emerged over the past 25 years as a technically attractive and cost-effective technique for extending the use of soil ⸱ a stable construction and slope-forming material. Internal reinforcement can be used to stabilize natural slopes or the slopes of embankments, or to retain excavations. Reinforced soil structures have the following advantages over traditional retaining walls: (1) they are coherent and flexible, and thus are tolerant of large deformations, (2) a wide range of backfill materials can be used, (3) they are easy to construct, (4) they are resistant to seismic loadings, (5) the variety

of available facing types makes possible aesthetically pleasing structures, and (6) they are often less costly than conventional retaining structures or piles (Mitchell & Villet 1987). Steep slopes of reinforced soil reduce the required width of new transportation rights of way and are especially suitable for the widening of existing constricted rights of way.

The modern concept of earth reinforcement was originated by Professor Arthur Casagrande, who proposed reinforcing weak soils by laying high-strength membranes between layers of soil (Westergaard 1938). Internally stabilized earth retention systems rely on transfer of shear forces to mobilize the tensile capacity of closely spaced reinforcing elements. Advances in this concept have led to increased use of internal reinforcing elements, either by incremental burial to create reinforced soils (Fig. 21) or by systematic in-situ installation of reinforcing elements, such as soil nails. The common types of inclusions are steel strips, steel or polymeric grids, geotextile sheets, and steel nails, that are capable of withstanding tensile loads, and, in some cases, shear and bending stresses as well (Mitchell & Christopher 1990). The two main mechanisms of stress transfer between the reinforcement and the soil are: (1) friction

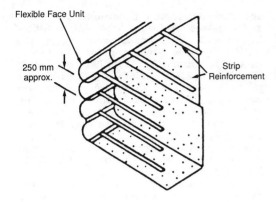

Flexible Face Unit

250 mm approx.

Strip Reinforcement

Fig. 21. Example of incremental burial of strip reinforcing elements (after O'Rourke & Jones 1990).

between the soil and the surface areas of the reinforcement, and (2) passive soil bearing resistance on reinforcement surfaces that are oriented normal to the direction of relative movement between the reinforcement and the soil.

Finite-element studies by Adib et al. (1990) and centrifuge model experiments by Jaber et al. (1990) of four types of reinforced soil walls (using steel strips, bar mats, geogrids, and non-woven geotextiles) have indicated good agreement between predicted and measured stresses in reinforcment elements.

As coherent, but flexible, gravity masses, reinforced soil structures are particularly well suited for use in seismically active areas (Mitchell & Villet 1987). They provide a high degree of structural damping that absorbs the dynamic energy associated with earthquakes. Seismic design of reinforced soil systems can be based on numerical calculations and shaking-table model tests (Bonaparte et al. 1986; Segrestin and Bastick 1988).

An important recent element in earth reinforcement is the use of geosynthetics as the reinforcing components. As noted earlier, geosynthetics include geotextiles, geogrids, geonets, geomembranes, and geocomposites. Each of these classes of geosynthetics is currently being used for soil reinforcement or slope drainage. A limitation of geosynthetics as reinforcement is that they possess low stiffness (relative to steel); consequently the amount of deformation required to attain maximum shear strength can exceed the allowable deformation of the soil structure.

All materials used as elements in soil reinforcement are subject to deterioration with time when exposed to soil, groundwater, and the elements. Elias (1990) has discussed the durability/corrosion of soil-reinforced structures in detail.

Table 1 summarizes the use of geosynthetically reinforced permanent retaining walls in North America through 1987. Most of the recent walls have used geogrids as the primary means of reinforcement (Koerner 1990).

As suggested by Mitchell & Villett (1987) and Christopher et al. (1990), this section is divided into placed soil reinforcement systems: (1) strip reinforcement, (2) sheet reinforcement (3) grid, bar, and mesh reinforcement, (4) placed soil-anchor reinforcement, and (5) fiber reinforcement, and in-situ soil reinforcement systems: (1) soil nailing, (2) inserted soil anchors, and (3) root piles.

Table 1. Summary of types of geosynthetically reinforced permanent retaining walls constructed in North America through 1987 (after Yako & Christopher 1988).

Wall height	Woven geotextile	Nonwoven geotextile	Geogrid	Other
Low < 3 m	4	3	5	1
Medium 3-7 m	5	7	14	1
High > 7 m	1	1	4	0
Total Number (Total Percent)	10 (22%)	11 (24%)	23 (50%)	2 (4%)

Fig. 22. Tiered Reinforced Earth wall on Vail Pass, Interstate Highway 70, Colorado. (Photograph courtesy of Reinforced Earth Company, MacLean, Virginia.)

4.2.1.1 Placed soil reinforcement systems

4.2.1.1.1 Strip reinforcement

In strip reinforcement systems, a coherent strengthened material is formed by placing metal or geosynthetic strips horizontally between successive backfill layers. The modern concept of soil reinforcement by means of galvanized steel strips was introduced by the French architect and engineer Henri Vidal in the early 1960's. Vidal named his development "Terre Armee," or "Reinforced Earth," terms that have become generic in many countries, being used to describe all forms of soil reinforcement. However, in some countries, including the United States and Canada, Reinforced Earth is trademarked (Jones 1985). As of 1991, 16,000 Reinforced Earth walls with a total face area of 9,600,000 m^2 have been constructed worldwide (McKittrick, D., 1991, personal communication, Reinforced Earth Company, McLean, Virginia). Schlosser (1990) has noted that 33 percent of the Reinforced Earth wall area in the world has been built in Europe and 34 percent in the United States (e.g., Fig. 22) and Canada.

The introduction of Reinforced Earth by Vidal led to rapid development of the concept of soil reinforcement. Much fundamental research was sponsored by government agencies, notably the Laboratoire des Ponts et Chaussees in France, the United States Department of Transportation, and the United Kingdom Department of Transport. Recently, Yoo and Ko (1991) have conducted centrifuge tests on Reinforced Earth models to investigate the behavior and failure mechanisms of walls subjected to self loading and to surcharge. Experiments were conducted by changing strip lengths, materials, and geometries. Test results were analyzed and compared to various current design methods to verify design feasibility for Reinforced Earth.

Vidal originally proposed the use of fiber-glass-reinforced polymers as the strips in Reinforced Earth (Schlosser 1990). However, in 1966 an experimental wall using fiber-glass-reinforced plastic strips failed after 10 months, apparently as the result of bacterial attack. This resulted in the use of stainless steel and aluminum strips for Reinforced Earth walls built in France. However, as of 1990 (10-15 years after construction), a large number of the stainless steel strips and some of the aluminum strips were corroded, indicating that these materials were not sufficiently resistant to corrosion to be used in soil. As a result, all Reinforced Earth walls currently are constructed using galvanized steel strips. However, even galvanized steel is subject to corrosion, and is thus restricted to use as reinforcement in cohesionless, granular, free-draining backfills to reduce the potential for chemical and water attack (Carroll & Richardson 1986). As one result, epoxy-coated steel reinforcements have been developed that offer potential for high resistance to corrosion.

In recent years non-metallic reinforcing materials, such as geotextiles, fiberglass, plastics, and composites, have been used extensively for soil reinforcement. These materials do not corrode, but may undergo other chemical and/or biological forms of deterioration. Many of these materials are new, and the effects of long-term burial and exposure to the elements are not well known (Elias 1990). For this reason, research is currently being undertaken on their weathering characteristics.

In 1973, the first polymeric strips were introduced in construction of a highway retaining wall in Yorkshire, United Kingdom (Holtz 1978). The reinforcing strips were made of continuous glass fibers embedded lengthwise in a protective coating of resin. An example of a recently marketed geosynthetic strip is the Paraweb strip (Fig. 23), in which the fibers are made of high-tenacity polyester or polyamarid.

4.2.1.1.2 Sheet reinforcement

Sheet reinforcement commonly consists of geotextiles placed horizontally between layers of embankment soils to form a composite reinforced soil; the mechanism of stress transfer and sheet reinforcement is predominantly friction (Mitchell & Villet 1987; Christopher et al. 1989; STS Consultants, Ltd. & Geoservices Inc, 1990). A variety of geotextiles with a wide range of mechanical properties and environmental resistance can be used, including nonwoven needle-punched or heat-bonded polyester and polypropylene, and woven polypropylene and polyester (Christopher et al. 1989). However, most of the geotextile fabrics used in earth reinforcement are made of either polyester or polypropylene fabrics. Granular soil ranging from silty sand to gravel commonly is used as

backfill. Facing elements are formed by wrapping the geotextile around the soil at the face (Fig. 24) and covering the exposed fabric with gunite, asphalt emulsion, or shotcrete, or with soil and vegetation, for long-term protection from ultraviolet light and vandalism. Typical applications of geotextile-reinforced walls include slope stabilization on mountain roads and retaining walls for temporary or permanent road widening or diversion.

The use of geotextiles in reinforced soil walls resulted from the beneficial effect of geotextile reinforcement in highway embankments over weak subgrades. The first geotextile-reinforced wall was built in France in 1971. The first full-size fabric retaining wall in the United States was a 3.3-m-high wall (Fig. 25) built by the U.S. Forest Service to reconstruct a failed road fill

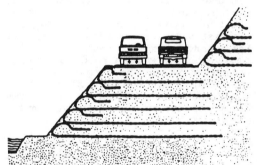

Fig. 24. Schematic cross section of a geotextile reinforced soil wall (after STS Consultants, Ltd. & Geoservices Inc 1990).

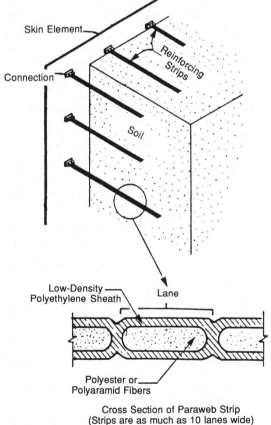

Fig. 23. Schematic diagram of a nonmetallic-strip reinforced soil wall (after Mitchell & Villet 1987).

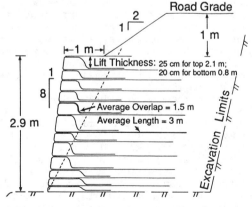

Fig. 25. Geotextile reinforced and faced wall, Siskiyou National Forest, Oregon (after Bell & Steward 1977).

in Siskiyou National Forest, Oregon; the fabric for this wall was a synthetic, nonwoven, needle-punched, spunbonded polypropylene (Bell & Steward 1977). This fabric was permeable and, when buried, was resistant to rotting. However, it was subject to deterioration when exposed to ultraviolet light; therefore, the wall was finished with a gunite facing to protect it from sunlight.

During the 1980's, studies were conducted in various countries to determine the effects of outdoor exposure on geotextiles used for slope reinforcement (Elias 1990). One of the most comprehensive of these studies was carried out by the Hong Kong Geotechnical Control Office to determine the outdoor-exposure performance of 14 geotextiles subjected to Hong Kong conditions (Brand & Pang 1991). All of the geotextiles (12 nonwoven, one woven, and one composite) lost strength and became brittle due to the combined influence of sunlight, temperature, rainfall, wind, oxygen, and atmospheric pollution. The average loss of strength in the first month was less than 16 percent. Long-term performance varied widely, with some geotextiles losing virtually all of their strength after 6 months of exposure. These data confirm that use of geotextiles for slope stabilization should be limited to short-term projects if they are not protected from the elements.

4.2.1.1.3 Grid, bar, and mesh reinforcement

Grid reinforcement systems consist of polymer or metallic elements arranged in rectangular grids, metallic bar mats, and wire mesh. The two-dimensional grid/soil interaction involves both friction along longitudinal members and passive bearing resistance against the transverse members. Because of the passive resistance developed on the cross members, grids are more resistant to pullout than strips; however, full passive resistance develops only for relatively large displacements (5-10 cm) (Schlosser 1990).

The greatest advance in development and use of rectangular grids for soil reinforcement has been in the area of polymeric "geogrids." As defined by Koerner (1990), a geogrid is "a deformed or non-deformed gridlike polymeric material formed by intersecting ribs joined at the junctions used for reinforcement with foundation, soil, rock, earth, or any other geotechnical engineering-related material as an integral part of a human-made project structure or system". Geogrids are relatively stiff, netlike materials with open spaces called "apertures," that usually

measure 1-10 cm between the ribs. The first use of polymer grids was by Japanese engineers in the 1960's to reinforce subgrade soils for railway embankments (Jones 1985). Because these original grids were made of non-oriented polymers, they were relatively fragile, lacking the necessary tensile strength to serve as slope reinforcement. In the 1970's, advances in the formulation of polymers led to significant improvements in strength and stiffness of geogrids. In the late 1970's, geogrids were developed with oriented polymers, which provided increased directional strength. These new geogrids were used in 1979 in construction of a reinforced soil wall at a railroad station in Yorkshire, United Kingdom (O'Rourke & Jones 1990). In 1981, the development of soil reinforcement advanced into a new area of application when synthetic grid materials were used to repair failures in cuts on the M1 and M4 motorways in England (Jones 1985).

In 1983, the first geogrid wall in the United States was built to stabilize a landslide on the Oregon coast (Szymoniak et al., 1984). This 9-m-high geogrid wall with a face slope of 80° was selected over other alternatives because (1) it had the lowest estimated cost, and (2) the open face of the grid wall allowed establishment of vegetation, which provided a natural appearance compatible with the surroundings of an adjacent state park. The geogrid was a high-density polethylene stabilized with carbon black to provide resistance to ultraviolet light. At about the same time, Forsyth and Bieber (1984) reported on the construction of a geogrid wall to reinforce a California slope 9.5 m high with a slope angle of 48°. This wall was built along State Highway 84 near La Honda to repair a small slide. Since the construction of these early walls, more than 300 polymeric geogrid wall and slope projects have been constructed in the United States (Mitchell & Christopher 1990).

Manufacturing processes have evolved to the point where strong and durable geosynthetic soil reinforcing elements can be mass produced. The most familiar products in earth retention systems are high density polyethylene (HDPE) and polypropylene grids (O'Rourke & Jones 1990). An example is Tensar (Fig. 26), a proprietary plastic grid reinforcement developed in the United Kingdom in the early 1980's.

Several proprietary and non-proprietary systems of bar and mesh reinforcement have been developed that rely on both frictional and passive resistance to pullout. In 1974, the first "bar-mat" system of soil reinforcement was developed by the California Department of Transportation

(Caltrans) to construct a 6-m-high wall along Interstate Highway 5 near Dunsmuir, California; these crude grids were formed by cross-linking steel reinforcing bars to form a coarse "welded-wire" bar mat (Forsyth 1978). For proprietary reasons, the Caltrans bar-mesh reinforcement technique was designated "Mechanically Stabilized Embankment" (MSE), in agreement with the Reinforced Earth Company. One of the difficulties with MSE in the field has been deterioration of the bar-mesh reinforcement. Recent centrifuge model studies by Ragheb and Elgamal (1991) have investigated the potential failure mechanisms associated with localized deterioration of MSE reinforcement strips in an effort to better understand the effects of this deterioration on long-term strength.

Evolving in the United States from the Caltrans project were other similar techniques: Hilfiker Welded Wire Wall and Hilfiker Reinforced Soil Embankment, VSL Retained Earth, and the Georgia Stabilized Embankment System.

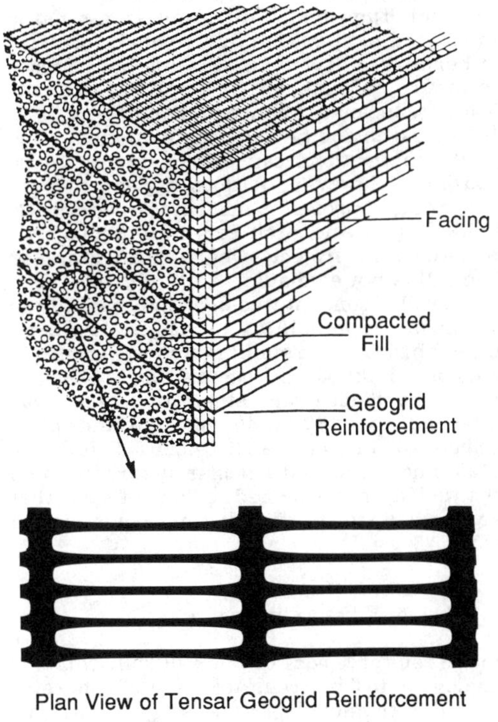

Plan View of Tensar Geogrid Reinforcement

Fig. 26. Schematic diagram of a geogrid reinforced wall, and plan view of Tensar geogrid reinforcing element as placed in the backfill.

Hilfiker Welded Wire Wall (WWW) uses welded-wire reinforcing mesh of the type that is commonly placed in concrete slabs; the facing is a continuation of the horizontal mesh reinforcement. The material is fabricated in 2.4-m-wide mats with grid spacing of 15 x 61 cm. To the casual observer, WWW may appear to be a type of gabion wall. However, gabion walls, which are gravity walls made by encasing coarse-grained fill in wire baskets, are based on the principle of confinement and gravity retainment rather than on internal tensile reinforcement (Hausmann 1990). The first commercial WWW was built for the the the Southern California Edison Power Company in 1977 for road repair along a power line in the San Gabriel Mountains of southern California. By 1980, the use of WWW expanded to larger projects, such as a 250-m-long, 5-m-high wall built by the Union Oil Company at their Parachute Creek oil-shale development in Colorado (Mitchell & Villet 1987). During the 1980's, the use of WWW for retaining structures expanded rapidly; by 1990 about 1600 WWW projects had been completed in the United States (Mitchell & Christopher 1990).

The Hilfiker Reinforced Soil Embankment system (RSE), which resembles Caltrans Mechanically Stabilized Embankment, is a continuous welded-wire reinforcement system with precast concrete facing. It was introduced commercially in 1983 on New Mexico State Highway 475 northeast of Santa Fe, where four reinforced soil stuctures were built with a total of 1600 m^2 of wall facing. By 1990, more than 50 additional RSE walls had been constructed in the United States (Mitchell & Christopher 1990).

VSL Retained Earth utilizes strips of steel-grid ("bar-mat") reinforcement that is bolted to hexagonal precast concrete panels. The first VSL Retained Earth wall in the United States was constructed in Hayward, California, in 1983. By 1990, more than 600 VSL Retained Earth walls with some 465,000 m^2 of wall facing had been built in the United States (Mitchell & Christopher 1990). The system is licensed in the United States under a Reinforced Earth patent, but it uses its own patented system for connecting the bar-mat reinforcement to the concrete facing panels.

The Georgia Stabilized Embankment, which was recently developed by the Georgia Department of Transportation, is another steel-grid, or bar-mat, reinforcing system, that has seen extensive use in the United States. It is licensed in the United States under a proprietary agreement with the Reinforced Earth Company.

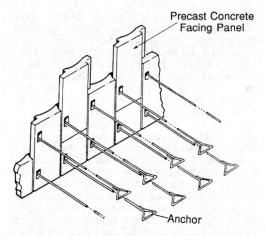

Precast Concrete Facing Panel

Anchor

Fig. 27. Schematic diagram of an Anchored Earth retaining wall (after Murray & Irwin 1981).

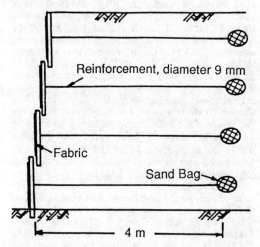

Reinforcement, diameter 9 mm

Fabric

Sand Bag

4 m

Fig. 28. Schematic cross section through proposed fabric-faced reinforced-soil wall with multiple tie bars attached to bags of sand that serve as anchors. The rods and anchors are embedded in the embankment during construction. (After Fukuoka 1986).

4.2.1.1.4 Emplaced earth-anchor reinforcement

Embankment slopes can be reinforced during construction by emplacement of slender steel rods bent at one end to form "anchors." This type of retaining wall is still in an experimental stage. Soil-to-rod stress transfer is mainly by means of passive resistance on the "anchor," which implies that the system provides stability in the same

manner as tied-back retaining structures, and thus is not truly a reinforced soil system (Mitchell & Villet 1987). However, the system is discussed here because it is analogous in placement technique to other methods of soil reinforcement in embankments.

The concept of emplaced earth-anchor reinforcement was developed and patented by the Transport and Road Research Laboratory (TRRL) of the United Kingdom as "Anchored Earth" (Murray & Irwin 1981). The reinforcement consists of 16- to 20-mm-diameter mild-steel bars. The outer end of each bar is threaded to fit into concrete facing panels; the other end is formed into an anchor in the form of a "Z" or triangle (Fig. 27). Unlike other soil reinforcement methods described here, which are based on the premise that frictional stress develops along the entire length of the reinforcement, Anchored Earth is designed to rely only on passive resistance developed against the deformed ends ("anchors") of the reinforcing bars. Centrifuge model tests of Anchored Earth were conducted recently by Craig et al. (1991) to better understand the Anchored Earth concept at prototype stress levels. In the centrifuge tests, different stress paths were used to approach failure of the models, either by gravity, by increasing body forces to induce collapse, or by applying external loading. Because Anchored Earth is still in the research and developmental stages, none of its applications can be considered to be routine. However, it does appear to be a promising approach to soil reinforcement. It is likely that Anchored Earth will prove most beneficial on projects where clean granular backfill is not available.

A similar concept was used successfully in Japan in the late 1970s to construct a 5-m-high fabric-faced retaining wall with multiple anchors (Fukuoka & Imamura 1982). Each of the 20-mm-diameter steel tie bars was attached to a 40 x 40 cm concrete plate embedded in the backfill soil. Fukuoka (1986) has proposed a similar system in which the tie bars are attached to bags of sand that are embedded in fill during construction (Fig. 28).

4.2.1.1.5 Fiber reinforcement

Fiber reinforcement of backfill soil, which is analogous to fiber reinforcement of concrete, is still in the developmental stage. Materials being investigated for possible fiber reinforcement include synthetic fibers (geotextile threads), metallic fibers (metal threads), and natural fibers (reeds and other plants) (Mitchell & Villett 1987). A recent innovation is a three-dimensional reinforcement technique that was developed in France at the Laboratoire des Ponts et Chaussees (LCPC) in 1980 (Schlosser 1990). The technique consists of polymer impregnation of granular backfill soil by mixing the soil with a small continuous polymer filament with a diameter of 0.1 mm and a tensile strength of 10 kN (Leflaive 1982). Approximately 0.1-0.2 percent of the composite material (known as "Texsol") consists of filament, resulting in a total length of reinforcement of 200 m per cubic meter of reinforced soil. The first Texsol wall was built in France in 1983 (Leflaive 1988). By the end of 1988, 85 Texsol projects had been completed in France, using 100,000 m^3 of Texsol-reinforced soil (Schlosser 1990). These walls have demonstrated high bearing capacity and resistance to erosion. However, there are difficulties associated with efficiently mixing the fibers with the backfill; the mixing process must be perfected before fiber inclusion can become an economically feasible and routinely used means of soil reinforcement.

Another material that has been suggested for use as fiber reinforcement is bamboo, which is one of the fastest growing and most replenishable biological materials. Bamboo can also be used as continuous elements in other types of reinforced soil. Fang (1991) has presented data on the strength and durability of bamboo as soil reinforcement.

4.2.1.2 In-situ soil-reinforcement systems

4.2.1.2.1 Soil nailing

Soil "nails" are steel bars, metal rods, or metal tubes that are driven into in-situ soil or soft rock or are grouted into predrilled boreholes. Together with the soil, they form coherent reinforced soil structures capable of stopping the movement of unstable slopes (Fig. 29) or of supporting temporary excavations (Fig. 30). Nailing differs from tieback support systems in that the nails are passive elements that are not post-tensioned as tiebacks are, and the nails are spaced more closely than tiebacks. Commonly one nail is used for each 1 to 6 m^2 of ground-surface area. Stability of the ground surface between the nails commonly is provided by a thin layer (10-15 cm) of shotcrete reinforced with wire mesh (Fig. 30), by intermittent rigid elements similar to large steel washers, or by prefabricated steel panels (which later may be covered by shotcrete). Soil nailing can be used to restrain two different types of unstable slopes: (1) potentially unstable slopes, where little or no movement is

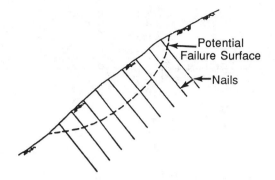

Fig. 29. Schematic cross section of soil nailing for slope stabilization.

Fig. 30. Soil nails extruding from shotcrete surface of temporary excavation for tunnel portal, Interstate Highway 70, Glenwood Canyon, Colorado.

occurring, but where safety factors are low enough to indicate a strong possibility for future movement, and (2) creeping slopes, in which movement is actually occurring.

Soil nailing has been used for slope stabilization for nearly 20 years. In North America, the system was first used in Vancouver, Canada, in the early 1970's for temporary excavation support. In Europe, the earliest reported soil-nailing projects were for retaining wall construction in Spain

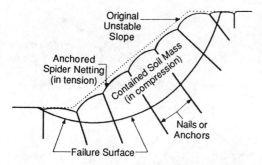

Fig. 31. Schematic cross section of anchored geosynthetic "spider netting" used with soil nails or anchors to stabilize a slope (after Koerner and & Robins 1986).

(1972), France (1973), and Germany (1976), in connection with highway or railway cut-slope construction or temporary support for building excavations (Elias & Juran 1991). Today the technique of soil nailing is widespread in France, Germany, Great Britain, Japan, and the United States.

The stability of soil-nailed reinforcement relies upon: (1) development of friction or adhesion mobilized at the soil-nail interface and (2) passive resistance developed at the face of the nail. Soil nailing is most effective in dense granular soils and low plasticity stiff silty clays. It is generally not cost-effective or practical in the following soils (Mitchell & Christopher 1990):

(1) loose granular soils with standard penetration N values lower than about 10 or relative densities of less than 30 percent,

(2) poorly graded soils with uniformity coefficients of less than 2 (nailing is not practical because of the necessity of stabilizing the cut face prior to excavation),

(3) soft cohesive soils with undrained shear strengths of less than 48 kPa, because of the inability to develop adequate pullout resistance, and

(4) highly plastic clays (PI > 20 percent), due to excessive creep deformation.

Soil nailing currently is used mainly for temporary structures because of the uncertainty of the corrosion rate of steel bars used in the process. However, new types of reinforcements and reinforcement coatings with high resistance to corrosion are being developed. For example, in 1987 fiber-glass nails were used to retain nearly vertical cuts for a freeway tunnel excavation in Reutlingen, southern Germany (Gassler in press).

Analysis of soil nailing and procedures for design have been presented by Juran et al. (1990) and Elias & Juran (1991). To increase the confidence of engineers in the potential use of this method for permanent slope stabilization, additional research is being conducted on field performance of soil-nailed structures (Plumelle et al. 1990; Stocker & Riedinger 1990). In 1986, a 4-year, US$4 million national research program, titled CLOUTERRE, was initiated by the French Minister of Transport to improve the state of knowledge and develop design and construction guidelines for soil-nailed retention systems (Schlosser & Unterreiner in press).

A new method of soil nailing uses geotextiles, geogrids, or geonets to cover the ground surface (Koerner & Robins 1986). The geosynthetic material is reinforced at distinct nodes and anchored to the slope using long rods (soil nails) at the nodes (Fig. 31). When the rods are properly fastened, they pull the surface netting into the soil, placing the net ("spider netting") in tension and the constrained soil in compression.

Soil nailing systems are both flexible and massive, and thus are resistant to seismic loading. An example of this dynamic stability was provided by the lack of damage to three California soil-nailing projects that were located within 33 km of the epicenter of the 1989 magnitude-7.1 Loma Prieta earthquake (Ferworn & Weatherby in press). However, current understanding of the dynamic behavior of soil-nailed earth structures is limited, and research is needed to develop procedures for earthquake-resistant design.

There are no proprietary restrictions on the use of soil nailing. However, some specific systems of nails and/or facing are patented. A recently patented (by Soil Nailing Limited, United Kingdom) soil-nailing technique inserts reinforcing nails into the ground by means of a compressed-air "launcher," which was originally developed in the United Kingdom for military use in shooting projectiles into the air (Bridle and Myles 1991). Under favorable conditions, the launcher can inject 38-mm-diameter nails up to 6 m long into a soil slope at a rate of one every 2-3 minutes.

4.2.1.2.2 Soil anchors

Stabilization of soil slopes by deep prestressed anchors is being used increasingly. Hutchinson (1984) has described the stabilization of a landslide in glacial deposits in southern Wales, United Kingdom (Fig. 32). Prior to treatment in 1980, movements of as much as 15 mm per year were occurring at the head of the slide and of 2-5 mm per year at the toe. Because of severe spatial constraints, anchoring into the underlying Coal Measures bedrock proved to be the most effective stabilization measure.

4.2.1.2.3 Root piles

Another recent approach to reinforcement of in-situ soils and soft rocks is the use of micropiles (commonly known as "root piles"). A root-pile system forms a monolithic block of reinforced soil that extends beneath the critical failure surface (Fig. 33). The main difference between root piles and traditional soil-nailing systems is that reinforcement provided by root piles is strongly influenced by their three-dimensional root-like geometric arrangement (Schlosser & Juran 1979).

Root piles are cast-in-place reinforced concrete piles with diameters ranging from 7.5 to 30 cm. In the smaller-diameter range, these insertions are provided with a central reinforcing rod or steel pipe, while those with larger diameters may be provided with a reinforcing bar-cage bound with spiral reinforcement (Christopher et al. 1989).

"Root Piles" and "Reticulated Root Piles" were originally developed in the 1950's by F. Lizzi and were patented by the Italian firm Fondedile of Naples, which introduced and installed the system worldwide (mainly for underpinning); the original patents have now expired (Christopher et al 1989). It has only been in the past 20 years that root piles have been used for slope stabilization, and most root-pile slope-stabilization works have been constructed within the past 10 years.

4.2.2 Stabilization by use of chemical admixtures

One means of improving a soil to meet engineering standards is to mix it with chemical admixtures, such as lime or Portland cement; this approach has been in use for many decades for highway subgrades. Lime stabilization in the United States was first used in the mid-1940's to stabilize clay-gravel base materials for highways. A major problem encountered in using lime for slope stabilization has been to obtain adequate insertion of the lime into the soil at depth. However, in recent years, insertion has been obtained by means of pressure injection of lime slurry into the soil (Fig. 34). The slurry, which follows natural fracture zones, bedding surfaces, and other surfaces of weakness, is injected through 40-mm-diameter pipes fitted with perforated nozzles (Rogers 1991). The pipes are hydraulically pushed into the ground, and the slurry is injected to refusal at depth intervals of 30-45

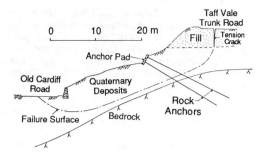

Fig. 32. Cross section of landslide in Quaternary deposits at Nangarw, south Wales, United Kingdom (after Hutchinson 1977). The slide has been stabilized by deep anchors in underlying bedrock.

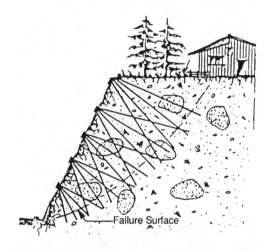

Fig. 33. Schematic cross section illustrating the use of root piles for stabilization of a slope (after Lizzi 1977).

cm. Typical injection pressures range from 350 to 1,300 kN/m^2. Depths of more than 40 m can be treated in this way. Blacklock and Wright (1986) have discussed restoration of failed soil embankments along the Interstate Highway system in Alabama, Arkansas, and Missouri using the lime- and L/FA-slurry injection method of in-situ soil stabilization. Baez et al. (1992) have evaluated the use of L/FA injection on the slope rehabilitation of a levee on the Lower Chariton River in Dalton, Missouri. Their studies showed that double injection of L/FA slurry increased the strength of the levee soil by 15-30 percent. Furthermore, since the levee was treated with L/FA injections in April 1988, there have been no slope failures in the rehabilitated stretch of the levee, whereas there have been failures in adjacent untreated areas.

A new type of retaining wall utilizing facing-panel units anchored into Portland-cement-stabilized backfill has been developed for the Texas Highway Department by the Texas Transportation Institute of Texas A&M University (Morris & Crockford 1990). Only short anchors were required, and because the strength properties of the soil were significantly improved by addition of the cement, the structure became a conventional mass-gravity structure. One of the advantages of the system is that it is non-proprietary.

5 VEGETATIVE STABILIZATION OF SLOPES

Vegetative slope stabilization can be provided: (1) directly by vegetation or (2) by biotechnical slope protection (the use of vegetation combined with structural slope-stabilization elements). The basic concepts of vegetative stabilization are not new; for example, the U.S. Forest Service made specific recommendations for the use of vegetation to stabilize slopes in the western United States as early as 1936 (Kraebel 1936). However, continuing research and new developments in design now enable more effective use of vegetation than in the past. For additional details on the effects of vegetation on slopes, refer to Gray (1970), Gray and Leiser (1982), Greenway (1987), and Wu (1991).

Vegetation contributes to stability of slopes by: (1) Restraint -- root systems physically bind or restrain soil particles, (2) Interception -- foliage and plant residues absorb rainfall energy, (3) Retardation -- above-ground residues increase surface roughness and slow the velocity of runoff, (4) Infiltration -- root and plant residues help to maintain soil porosity and permeability, and (5) Transpiration -- depletion of soil moisture by plants delays onset of saturation and runoff (Gray & Leiser 1982).

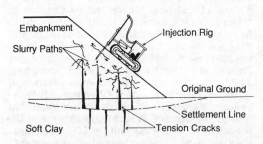

Fig. 34. Schematic cross section illustrating process of lime-slurry pressure injection for stabilization of an embankment slope (after Boynton & Blacklock 1985).

Case studies have shown that slope failures can be attributed to the loss of tree roots as slope reinforcement (e.g., Wu et al. 1979; Riestenberg & Sovonick-Dunford 1983; Riestenberg 1987). Wu (1991) has quantified this protection in terms of root reinforcement and reduction of soil moisture and pore pressures.

In recent years, trees have been planted on many slopes worldwide to increase slope stability. Examples with which the author is familiar follow:

One element of a program to correct an embankment failure on Interstate Highway I-77 near Caldwell, Ohio, was the planting of the slope with black locust seedlings at a spacing of 1.2 m. The long-term objective was to help lower the groundwater table and to develop root stabilization. As of 1987, the project appeared to be successful; however, in that short a period of time root development was not great enough to increase the factor of safety substantially (Wu 1987).

Another recent well-documented case of the planting of tree seedlings occurred as part of the stabilization program of the Cucaracha landslide in the Gaillard Cut on the Panama Canal. The historic Cucaracha slide was reactivated in 1986, almost blocking the canal (Berman 1991). As part of a comprehensive stabilization program, portions of the surfaces of the Cucaracha slide and other landslide areas in the Canal Zone were planted with 60,000 fast-growing acacia and gmelina seedlings beginning in 1987 (Rivera 1991). Although the failure surfaces in these landslides are deep-seated and generally won't be affected directly by the tree roots, the trees have significantly improved the groundwater regime, thus increasing stability.

The planting of trees to control slope failures (mostly debris torrents, flows, and avalanches) in the Vorarlberg of Austria has been documented by the Austrian Federal Service for Torrent and Avalanche Control (Olz 1990). Since 1945, this federal agency has planted 4.8 million fir and mixed-forest trees on 740 hectares of potentially unstable land. As noted by Olz, "forests grow while steel avalanche defenses grow rusty and they are and always will be an alien element in the midst of nature."

Research into the engineering role of vegetation for slope stabilization in Hong Kong may be the most comprehensive such program in the world (Barker 1991). Especially notable have been the root-reinforcement studies conducted on vegetated slopes in Hong Kong by Greenway et al. (1984), Greenway 1987), and Yin et al. (1988) under the auspices of the Geotechnical Control Office of Hong Kong. In addition, the "Geotechnical Manual for Slopes" (Geotechnical Control Office 1981) includes an excellent table noting the mechanical and hydrological effects of vegetation.

Stabilization of slopes by the combined use of vegetation and man-made structural elements working together in an integrated manner is known as "biotechnical slope stabilization." Biotechnical slope stabilization is a relatively new concept that is generally cost effective as compared to the use of structures alone; it increases environmental compatibility, and allows the use of indigenous natural materials. Vegetative treatments alone are commonly much less expensive than earth retaining structures or other constructed protection systems; however, their effectiveness in terms of arresting slope movement or preventing soil loss under extreme conditions may be much less than that of the structures (Gray & Leiser 1982).

Grasses and woody plants are used most often in biotechnical stabilization. They have a true reinforcing function and should not be considered merely a cosmetic adjunct to the structure. They may be planted on a slope above a low retaining wall, or the interstices of the structure may be planted with vegetation whose roots bind together the soil within and behind the structure. The stability of all types of retaining structures with open grid-work or tiered facings benefit from such vegetation. The following cases illustrate advances in the use of biotechnical stabilization:

Gray and Sotir (in press) have described the 1989 use of a drained rock-blanket buttress combined with an earthen brush-layer fill to stabilize a roadcut along a scenic highway in Massachusetts. The rock buttress was placed at the toe of the cut; the brush-layer fill included stems and branches of plant species, such as willow and dogwood, that rooted readily from cuttings. The branches acted as reinforcement and as horizontal drains, and rooting of the embedded stems provided secondary stabilization.

Barker (1991) has noted the recent use in the United Kingdom of a composite vegetated geotextile/geogrid reinforced structure named "Biobund." The protoype for this structure was the 1986 Schuepfen Bund along the T6 Bern-Biel Autobahn in Switzerland. This 300-m-long, 4.5-m-high visual and acoustic barrier incorporated more than 33,000 willow cuttings between pockets of geotextile-wrapped fill. Another 5,000 container-grown rooted shrubs and trees were interspersed with the willow cuttings.

Suyama (in press) has discussed the use of forests in Japan: (1) to control slope erosion, (2) to stabilize landslide scars, and (3) to absorb debris-flow and rockfall impacts. In very interesting full-scale field experiments, estimations of the resistance of trees to the impact energy of debris flows and rockfalls were determined.

6 USE OF LIGHTWEIGHT FILLS TO FACILITATE SLOPE STABILIZATION

To reduce the gravitational driving force behind slope-stabilizing retaining structures, various types of lightweight backfills have been used. Sawdust, burned coal, and fly-ash wastes have been used in areas where these waste products are readily available. In the past few years, two new types of lightweight fill have been proposed and tested for use in lightweight backfills for slope stabilization: styrofoam blocks and shredded waste car and truck tires.

The introduction of superlight expanded polystyrene (EPS; styrofoam) blocks in 1972 allowed the construction of lightweight fills for highways. In additon, it has been used successfully for road-base insulation in cold regions for several years. By 1987, the Norwegian Geotechnical Institute was involved in more than 100 projects where this superlight material had been used with great success (Flaate 1987). As superlight fill, EPS is used in the form of large blocks with a density of 0.02 t/m^3, a drastic reduction in density compared to other lightweight materials.

The most common application of EPS as superlight fill has been for highway embankments, and especially for bridge approaches. However, the Colorado Department of Transportation recently has successfully used EPS for slide correction. During the spring of 1987, an 8400-m^3 slide closed the eastbound lane of heavily traveled U.S. Highway 160 in southern Colorado. The slide area was successfully stabilized by using a counterfort berm at the toe and replacing the slide material in the highway embankment with EPS (Yeh & Gilmore in press).

Another recently applied lightweight fill for slide correction is shredded waste rubber car and truck tires. Nearly 300 million tires are discarded annually in the United States, creating a major disposal problem. Shredded tires have a compacted dry unit weight of about 0.64 t/m^3 (Humphrey & Manion in press). About 580,000 shredded rubber tires were used as lightweight fill in correction of a landslide that occurred in 1989 under a highway embankment on U.S. Highway 42 in the State of Oregon (Read et al. 1991). The force driving the slide was considerably reduced by replacing the slide material with the lightweight shredded tires.

7 FUTURE TRENDS IN SLOPE STABILIZATION

Research in analysis, design, and construction of systems for subsurface drainage, rockfall control, and soil retention will continue to provide new approaches to the development and use of these slope-stabilization systems. Particularly important is the development of new economical, strong, corrosion-resistant, and environmentally acceptable materials that can be used as elements in stabilization systems for both rock and soil slopes. For steep rock slopes, new computerized approaches will allow increased understanding of the rockfall process that will lead to better rockfall control. New experimental techniques, such as the use of the geotechnical centrifuge, will complement analytical approaches to better understand the mechanics of failure of retention systems, thus leading to improvements in design.

REFERENCES

Adib, M., J.K. Mitchell & B. Christopher. 1990. Finite element modeling of reinforced soil walls and embankments. In Design and performance of earth retaining structures, P.C. Lambe & L.A. Hansen, eds., American Society of Civil Engineers Geotechnical Special Publication 25:409-423.

Andrew, R.D. In press. Selection of rockfall mitigation techniques. Transportation Research Record, Transportation Research Board, National Research Council, Washington, D.C.

Arutjunyan, R.N. 1988. Prevention of landslide slope process by vacuuming treatment of disconsolidated soils. In Proceedings of the Fifth International Symposium on Landslides, Lausanne, 10-15 July, 2:835-837.

ASTM Committee D-35. 1991. ASTM standards on geosynthetics, second edition. American Society for Testing and Materials, Philadelphia, 104.

Baez, J.I, R.H. Borden, & J.H. Henry. 1992. Rehabilitation of Lower Chariton River Levee by lime/flyash slurry injection. Transportation Research Record, Transportation Research Board, National Research Council, Washington, D.C., 1310:117-125.

Baker, R.F., & H.E. Marshall. 1958. Control and correction. In Landslides and engineering practice, E.B. Eckel, ed., Highway Research Board Special Report, National Academy of Sciences-National Research Council, Washington, D.C., 29:150-188.

Barker, D.H. 1991. Developments in biotechnical stabilisation in Britain and the Commonwealth. In Proceedings, Workshop on Biotechnical Stabilization, University of Michigan, Ann Arbor, 21-23 August, A-83 to A-123.

Barley, A.D. 1991. Slope stabilization by new ground anchorage systems in rocks and soils. In Proceedings, International Conference on Slope Stability Engineering: Developments and Applications, Institution of Civil Engineers, Isle of Wight, U.K., 15-18 April, Thomas Telford Ltd., London, 335-340.

Barrett, R.K. & J.L. White. 1991. Rockfall prediction and control. In Proceedings, National Symposium on Highway and Railroad Slope Maintenance, Association of Engineering Geologists, Chicago, 2-3 October, 23-40.

Beer, P., U. Hegg & V. Manassero. In press. Landslide stabilization at Ancona (Italy) by deep drainage wells. In Proceedings, Sixth International Symposium on Landslides, Christchurch, New Zealand, 10-14 February 1992.

Bell, J.R. & J.E. Steward. 1977. Construction and observations of fabric retained soil walls. In Proceedings, International Conference on the Use of Fabrics in Geotechnics, Paris, 20-22 April, 1:123-128.

Berman, G. 1991. Landslides on the Panama Canal. Landslide News, Japan Landslide Society, 5:10-14.

Bianco, B. & D.A. Bruce. 1991. Large landslide stabilization by deep drainage wells. In Proceedings, International Conference on Slope Stability Engineering: Developments and Applications, Institution of Civil Engineers. Isle of Wight, U.K., 15-18 April, Thomas Telford Ltd., London, 341-348.

Blacklock, J.R., and Wright, P.J. 1986. Injection stabilization of failed highway embankments. Paper presented at 65th Annual Meeting of the Transportation Research Board, National Research Council, Washington, D.C., January.

Bonaparte, R., G.R. Schmertmann & N.D. Williams. 1986. Seismic design of slopes reinforced with geogrids and geotextiles. In Proceedings, Third International Conference on Geotextiles, Vienna, 7-11 April, 1:273-284.

Boyd, R.D. 1991. Rockfall containment measures at Springdale, Newfoundland, Canada. In Proceedings, International Conference on Slope Stability Engineering: Developments and Applications, Institution of Civil Engineers, Isle of Wight, U.K., 15-18 April, Thomas Telford Ltd., 409-414.

Boynton, R.S., & J.R. Blacklock. 1985. Lime slurry pressure injection bulletin. National Lime Association Bulletin 331:43.

Brand, E.W. and P.L.R. Pang. 1991. Durability of geotextiles to outdoor exposure in Hong Kong. Journal of Geotechnical Engineering, American Society of Civil Engineers, 117,7:979-1000.

Bridle, R. & B. Myles. 1991. A machine for soil nailing--process and design. Paper presented at the Civil Engineering European Courses Seminar, Ecole Nationale des Ponts et Chaussees, Paris, October 1991, 10.

Broms, B.B. & I.H. Wong. 1986. Stabilization of slopes in residual soils with geofabric. In Proceedings, Third International Conference on Geotextiles, Vienna, 7-11 April 1986, 2:295-300.

Bush, D.I. 1988. Recent advances in landslide stabilisation using polymer reinforcement. In Proceedings of the Fifth International Symposium on Landslides, Lausanne, 10-15 July, 2:877-880.

Cancelli, A., B. D'Elia & P. Sembenelli. 1987. General Report. In Groundwater problems in embankments, dams, and natural slopes, Session 4, Proceedings, 9th European Conference on Soil Mechanics and Foundation Engineering, Dublin, 3:1043-1072.

Carroll, R.G. & G.N. Richardson. 1986. Geosynthetic reinforced retaining walls. In Proceedings, Third International Conference on Geotextiles, Vienna, 7-11 April, 2:389-394.

Casagrande, L. 1948. Electro-osmosis. In Proceedings, 2nd International Conference on Soil Mechanics and Foundation Engineering, Rotterdam, 1:218-223.

Christopher, B.R., S.A. Gill, J.P. Giroud, I. Juran, J.K. Mitchell, F. Schlosser & J. Dunnicliff. 1989. Reinforced soil structures, Volume II. Summary of research and systems information. Federal Highway Administration, Department of Transportation, McLean, Virginia, Report No. FHWA-RD-89-043, 158.

Christopher, B.R., S.A. Gill, J.P. Giroud, I. Juran, J.K. Mitchell, F. Schlosser & J. Dunnicliff. 1990. Reinforced soil structures, Volume I. Design and construction guidelines. Federal Highway Administration, U.S. Department of Transportation, McLean, Virginia, Publication No. FHWA-RD-89-043, 285.

Collotta, T., V. Manassero & P.C. Moretti. 1988. An advanced technology in deep drainage of slopes. In Proceedings, 5th International Symposium on Landslides, Lausanne, 10-15 July, 2:887-892.

Craig, W.H., R. Sapak & K.C. Brady. 1991. Model studies of an anchored earth structure. In Centrifuge 91, Proceedings of the International Conference Centrifuge 1991, H.Y. Ko & F.G. McLean eds., Boulder, Colorado, 13-14 June, 303-310.

Deere, D.U. & J.Y. Perez. 1985. Remedial measures for large slide movements. In Proceedings of P.R.C.-[People's Republic of China] U.S.-Japan Trilateral Symposium/Workshop on Engineering for Multiple Natural Hazard Mitigation, Beijing, 7-12 January: L-7-1 to L-7-15.

Di Maio, C. & C. Viggiani. 1987. Influence of intermittent rainfall on effectiveness of trench drains. In Proceedings, 9th International Conference on Soil Mechanics and Foundation Engineering, Dublin, 1:149-152.

Elias, V. 1990. Durability/corrosion of soil reinforced structures. Federal Highway Administration Publication No. FHWA-RD-89-186, U.S. Department of Transportation, McLean, Virginia, 163.

Elias, V. & I. Juran. 1991. Soil nailing for stabilization of highway slopes and excavations. Federal Highway Administration Report No. FHWA-RD-89-198, Department of Transportation, McLean, Virginia, 210.

Fang, H.Y. 1991. The use of bamboo inclusions for earth reinforcement. In Proceedings, Workshop on Biotechnical Stabilization, University of Michigan, Ann Arbor, 21-23 August, 33-62.

Ferworn, D.E. & D.E. Weatherby. In press. A contractor's experience with soil nailing. Transportation Research Record, National Research Council, Washington, D.C.

Flaate, K. 1987. Superlight material in heavy construction. Geotechnical News, Sept., 22-23.

Forsyth, R.A. 1978. Alternative earth reinforcements. In Proceedings, American Society of Civil Engineers Symposium on Earth Reinforcement, Pittsburgh, 358-370.

Forsyth, R.A. & D.A. Bieber. 1984. La Honda slope repair with grid reinforcement. In Proceedings, International Symposium on Polymer grid Reinforcement in Civil Engineering, London.

Fukuoka, M. 1986. Fabric retaining wall with multiple anchors. Proceedings, Third International Conference on Geotextiles,

Vienna, 7-11 April, 2:435-440.

Fukuoka, M. & Y. Imamura. 1982. Fabric retaining walls. Proceedings, Second International Conference on Geotextiles, Las Vegas, Nevada, 2:575-580.

Gabus, J.H., C. Bonnard, F. Noverraz & A. Parriaux. 1988. Arveyes, un glissement, une tentative de correction. In Proceedings of the Fifth International Symposium on Landslides, Lausanne, 10-15 July, 2:911-914.

Gassler, G. In press. German practice of soil nailing. Transportation Research Record, Transportation Research Board, National Research Council, Washington, D.C.

Geotechnical Control Office. 1981, Geotechnical manual for slopes. Public Works Department, Hong Kong, 228.

Gillon, M.D., C.J. Graham & G.G. Grocott. In press. Low level drainage works at the Brewery Creek Slide. In Proceedings, Sixth International Symposium on Landslides, Christchurch, New Zealand, 10-14 February.

Gillon, M.D. & G.T. Hancox. In press. Cromwell Gorge landslides -- a general overview. In Proceedings, Sixth International Symposium on Landslides, Christchurch, New Zealand, 10-14 February.

Gray D.H. 1970. Effect of forest clear-cutting on the stability of natural slopes. Bulletin of the Association of Engineering Geologists, 7,1:45-66.

Gray, D.H. & A.T. Leiser. 1982. Biotechnical slope protection and erosion control. Van Nostrand Reinhold, New York, 271.

Gray, D.H. & R.B. Sotir. In press. Biotechnical stabilization of a highway cut slope. Submitted to Journal of the Geotechnical Engineering Division, American Society of Civil Engineers.

Greenway, D.R. 1987. Vegetation and slope stability. In Slope stability, M.G. Anderson & K.S. Richards, eds., John Wiley, New York, 187-230.

Greenway, D.R., M.G. Anderson, & K.C. Brian-Boys. 1984. Influence of vegetation on slope stability in Hong Kong. In Proceedings, 4th International Symposium on Landslides, Toronto, 1:399-404.

Gress, J.C. In press. Siphon drain: a technic for slope stabilization. In Proceedings, Sixth International Symposium on Landslides, Christchurch, New Zealand, 10-14 February 1992.

Hausmann, M.R. 1990. Engineering principles of ground modification. McGraw-Hill, New York, 632.

Holtz, R.D. 1978. Special applications. In Proceedings, Symposium on Earth Reinforcement, American Society of Civil Engineers, New York, 77-97.

Humphrey, D.N. & W.P. Manion. In press. Properties of tire chips for lightweight fill. Proceedings, Conference on Grouting, Soil Improvement, and Geosynthetics, American Society of Civil Engineers, New Orleans, 1992.

Hutchinson, J.N. 1977. Assessment of the effectiveness of corrective measures in relation to geological conditions and types of slope movement. Bulletin of the Association of Engineering Geology, 16:131-155.

Hutchinson, J.N. 1984. Landslides in Britain and their countermeasures. Journal of Japan Landslide Society, 21,1:1-24.

Imrie, A.S., D.P. Moore & E.G. Enegren. In press. Performance and maintenance of the drainage system at Downie Slide. In Proceedings, Sixth International Symposium on Landslides, Christchurch, New Zealand, 10-14 February 1992.

Ingold, T.S. and K.S. Miller. 1988. Geotextiles handbook. Thomas Telford, London, 152.

Jaber, M., J.K. Mitchell, B.R. Christopher & B.L. Kutter. 1990. Large centrifuge modeling of full scale reinforced soil walls. In Design and performance of earth retaining structures, P.C. Lambe and L.A. Hansen, eds., American Society of Civil Engineers Geotechnical Special Publication 25:379-393.

Jones, C.J.F.P. 1985. Earth reinforcement and soil structures. Butterworths, London, 183.

Juran, I., G. Baudrand, K. Farrag & V. Elias. 1990. Design of soil nailed retaining structures. In Design and performance of earth retaining structures, P.C. Lambe & L.A. Hansen, eds., American Society of Civil Engineers Geotechnical Special Publication 25:644-659.

Koerner, R.M. 1990. Designing with geosynthetics. Prentice Hall, Englewood Cliffs, New Jersey, 2nd ed., 652.

Koerner, R.M. & J.C. Robins. 1986. In-situ stabilization of soil slopes using nailed geosynthetics. In Proceedings, Third International Conference on Geotextiles, Vienna, 7-11 April, 2:395-400.

Kraebel, C.J. 1936. Erosion control on mountain roads. United States Department of Agriculture Circular No. 380:44, Washington, D.C.

Leflaive, E. 1982. The reinforcement of granular materials with continuous fibers [in French]. In Proceedings, 2nd International Conference on Geotextiles, Las Vegas, Nevada, August, 3:721-726.

Leflaive, E. 1988. Texsol: already more than 50 successful applications. In Proceedings, International Symposium on Theory and Practice of Earth Reinforcement, Japan, 541-545.

Leighton, J.C. 1990. New tunnel at Shalalth Bluff on BC Rail's Squamish Subdivision. In Tunnelling in the 90's, Proceedings of the Eighth Annual General Meeting of the Tunnelling Association of Canada, Vancouver, B.C., 31 October-2 November, Bitech Publishers Ltd., 255-266.

Lew, K.V. & J. Graham. 1988. Riverbank stabilization by drains in plastic clay. In Proceedings of the Fifth International Symposium on Landslides, Lausanne, 10-15 July, 2:939-944.

List, B.R. 1988. Stabilization of open pit slopes through blasting. In Proceedings of the Fifth International Symposium on Landslides, Lausanne, 10-15 July, 2:945-952.

Littlejohn, S. 1990. Ground anchorage practice. In Design and performance of earth retaining structures, P.C. Lambe & L.A. Hansen, eds., American Society of Civil Engineers Geotechnical Special Publication 25:692-733.

Lizzi, F. 1977. Practical engineering in structurally complex formations (the in situ reinforced earth). In Proceedings, International Symposium on the Geotechnics of Structurally Complex Formations, Capri, Italy.

Lo, K.Y., I.I. Inculet & K.S. Ho. 1991a. Electro-osmotic strengthening of soft sensitive clays. Canadian Geotechnical Journal 28:62-73.

Lo, K.Y., H.S. Ho & I.I. Inculet. 1991b. Field test of electrosoosmotic strengthening of soft sensitive clay. Canadian Geotechnical Journal 28:74-83.

McCauley, M.L., B.W. Works & S.A. Naramore. 1985. Rockfall mitigation. Office of Transportation Laboratory, California Department of Transportation, Sacramento, 147.

Mitchell, J.K. & B.R. Christopher. 1990. North American practice in reinforced soil systems. In Design and preformance of earth retaining structures, P.C. Lambe & L.A. Hansen, eds., American Society of Civil Engineers Geotechnical Special Publication 25: 322-346.

Mitchell, J.K. & W.C.B. Villet. 1987. Reinforcement of earth slopes and embankments. National Cooperative Highway Research Program Report 290, Transportation Research Board, National Research Council, Washington, D.C., 323.

Morales Arnao, B., V.K. Garga, R.S. Wright & J.Y. Perez. 1984. The Tablachaca Slide No. 5, Peru, and its stabilization. In Proceedings, IV International Symposium on Landslides, Toronto, 1:597-604.

Morris, D.V. & W.W. Crockford. 1990. Cement stabilized soil retaining walls. In Design and performance of earth retaining structures, P.C. Lambe & L.A. Hansen, eds., American Society of Civil Engineers Geotechnical Special Publication 25:307-321.

Murray, R.T. & M.J. Irwin. 1981. A preliminary study of TRRL Anchored Earth. TRRL (Transport and Road Research Laboratory) Supplementary Report 674.

Nakamura, H. 1988. Landslide control works by horizontal drainage works. In Proceedings of the Fifth International Symposium on Landslides, Lausanne, 10-15 July, 2:965:970.

Olcese, A., C. Vescovo, S. Boni & G. Giusti. 1991. Stabilisation of a landslide with submerged motor-driven pumps. Proceedings, International Conference on Slope Stability Engineering: Developments and Applications, Institution of Civil Engineers, Isle of Wight, U.K., 15-18 April, Thomas Telford Ltd., 321-326.

Olz, R. 1990. A short report about the Federal Service for Torrent and Avalanche Control in Vorarlberg. In ALPS 90, Alpine Landslide Practical Seminar, 6th International Conference and Field Workshop on Landslides, Milan, 31 August-12 September, 3 pages.

O'Rourke, T.D. & C.J.F.P. Jones. 1990. Overview of earth retention systems: 1970-1990. In Design and Performance of Earth Retaining Structures, P.C. Lambe & L.A. Hansen, eds., American Society of Civil Engineers Geotechnical Special Publication 25:22-51.

Peila, D., F. Lombardi & V. Manassero. In press. Stabilization of landslides using large diameter wells. In Proceedings, Sixth International Symposium on Landslides, Christchurch, New Zealand, 10-14 February 1992.

Pfeiffer, T.J., J.D. Higgins & A.K. Turner. 1990. Computer aided rockfall hazard analysis. In Proceedings, Sixth International Congress, International Association of Engineering Geology, Amsterdam, 6-10 August, 93-103.

Pierson, L.A. 1991. The Rockfall Hazard Rating System. In Proceedings, National Symposium on Highway and Railroad Slope Maintenance, Association of Engineering Geologists, Chicago, 2-3 October, 1-22.

Plumelle, C., F. Schlosser, P. Delage & G. Knochenmus. 1990. French national research project on soil nailing: CLOUTERRE. In Design and performance of earth retaining structures, P.C. Lambe & L.A. Hansen, eds., American Society of Civil Engineers Geotechnical Special Publication 25:660-675.

Ragheb, A. & A.W. Elgamal. 1991. Effect of

gradual reinforcement compromise on the behavior of mechanically stabilized earth walls. In Centrifuge 91, Proceedings of the International Conference Centrifuge 1991, H.Y. Ko & F.G. McLean, eds., Boulder, Colorado, 13-14 June, 333-340.

Read, J., T. Dodson & J. Thomas. 1991. Experimental project use of shredded tires for lightweight fill. Post-Construction Report, Federal Highway Administration Experimental Project No. 1, DTFH-71-90-501-OR-11, Oregon Department of Transportation, Salem, 31 plus 4 appendixes.

Resnick, G.S. & D. Znidarcic. 1990. Centrifugal modeling of drains for slope stabilization. Journal of Geotechnical Engineering, American Society of Civil Engineers, 116,11:1607-1624.

Riestenberg, M.M. 1987. Anchoring of thin colluvium on hillslopes by roots of sugar maple and white ash. Ph.D. dissertation, University of Cincinnati, Cincinnati, Ohio.

Riestenberg, M.M. & S. Sovonick-Dunford. 1983. The role of woody vegetation in stabilizing slopes in the Cincinnati area, Ohio. Geological Society of America Bulletin 94:506-518.

Ritchie, A.M. 1963. Evaluation of rockfall and its control. Highway Research Record, Highway Research Board, National Research Council, Washington, D.C., 17:13-28.

Rivera, R. 1991. Reforestation program - Gaillard Cut. Panama Canal Commission Report No. 1, January, 12 plus appendix.

Rogers, C.D.F. 1991. Slope stabilisation using lime. In Proceedings, International Conference on Slope Stability Engineering: Developments and Applications, Institution of Civil Engineers, Isle of Wight, U.K., 15-18 April, Thomas Telford Ltd., 395-402.

Schlosser, F. 1990. Mechanically stabilized earth retaining structures in Europe. In Design and performance of earth retaining structures, P.C. Lambe & L.A. Hansen, eds., American Society of Civil Engineers Geotechnical Special Publication 25: 347-378.

Schlosser, F. & I. Juran. 1979. Design parameters for artificially improved soils. In Proceedings, 7th European Conference on Soil Mechanics and Foundation Engineering, Brighton, United Kingdom, 5.

Schlosser, F. & P. Unterreiner. In press. Soil nailing in France: research and practice. Transportation Research Record, Transportation Research Board, National Research Council, Washington, D.C.

Segrestin, P. & M.J. Bastick. 1988. Seismic design of Reinforced Earth retaining walls: the contribution of the finite elemnt analysis. In Proceedings, International Symposium on Theory and Practice of Earth Reinforcement, Kyushu, Japan, 577-582.

Sembenelli, P. 1988. General report: Stabilization and drainage. In Proceedings of the Fifth International Symposium on Landslides, Lausanne, 10-15 July, 2:813-819.

Smith, D.D. 1980. The effectiveness of horizontal drains. California Department of Transportation, Sacramento, Report No. FHWA/CA/TL-80/16, 71.

Smith, D.D. & J.D. Duffy. 1990. Field tests and evaluation of rockfall restraining nets. California Department of Transportation Report No. CA/TL-90/05, Sacramento, 138.

Stanic, B. 1984. Influence of drainage trenches on slope stability. Journal of the Geotechnical Engineering Division, American Society of Civil Engineers, 110,11:1624-1636.

Stocker, M.F. & G. Riedinger. 1990. The bearing behaviour of nailed retaining structures. In Design and performance of earth retaining structures, P.C. Lambe & L.A. Hansen, eds., American Society of Civil Engineers Geotechnical Special Publication 25:612-628.

STS Consultants, Ltd. & GeoServices Inc. 1990. Geotextile design & construction guidelines. Federal Highway Administration Pub. No. FHWA-HI-90-001, Federal Highway Administration/National Highway Institute, Washington, D.C., 334.

Suyama, M. In press. Assessment of biotechnical slope stability effect for urban forest in Japan. In Proceedings, Sixth International Symposium on Landslides, Christchurch, New Zealand, 10-14 February 1992.

Szymoniak, T., J.R. Bell, G.R. Thommen, & E.L. Johnsen. 1984. A geogrid-reinforced soil wall for landslide correction on the Oregon coast. Transportation Research Record, Transportation Research Board, National Research Council, Washington, D.C., 965:47-55.

Threadgold, L. & D.P. McNichol. 1984. The design and construction of polymer grid boulder barriers to protect a large public housing site for the Hong Kong Housing Authority. In Polymer grid reinforcement, Thomas Telford, London, 212-219.

Varnes, D.J. 1958. Landslide types and processes. In Landslides and engineering practice, E.B. Eckel, ed., Highway Research Board Special Report, National Academy of Sciences, Washington, D.C., 29:20-47.

Varnes, D.J. 1978. Slope movement types and processes. In Landslides--analysis and control, R.L. Schuster & R.J. Krizek, eds., Transportation Research Board Special Report, National Academy of Sciences, Washington, D.C., 176:11-33.

Veder, C. 1981. Landslides and their stabilization. New York, Springer-Verlag: 247.

Westergaard, H.M. 1938. A problem of elasticity suggested by a problem in soil mechanics; soft material reinforced by numerous strong horizontal sheets. Harvard University report, Cambridge, Massachusetts.

Wichter, L., E. Krauter & W. Meiniger. 1988. Landslide stabilization using drainage wells, concrete dowels and anchored bore pile walls. In Proceedings of the Fifth International Symposium on Landslides, Lausanne, 10-15 July, 2:1023-1028.

Wu, T.H. 1987. Long-term strength of embankment materials: shale and colluvium. Federal Highway Administration Report FHWA-OH-87-008, Ohio State University, Columbus.

Wu, T.H. 1991. Soil stabilization using vegetation. In Proceedings, Workshop on Biotechnical Stabilization, University of Michigan, Ann Arbor, 21-23 August, A-1 to A-32.

Wu, T.H., W.P. McKinnell III & D.N. Swanston. 1979. Strength of tree roots and landslides on Prince of Wales Island, Alaska. Canadian Geotechnical Journal, 16:19-33.

Wyllie, D.C. 1991. Rock slope stabilization and protection measures. In Proceedings, National Symposium on Highway and Railroad Slope Maintenance, Association of Engineering Geologists, Chicago, 2-3 October, 41-64.

Yako, M.A. & B.R. Christopher. 1988. Polymerically reinforced retaining walls in North America. In Proceedings, Workshop on Polymeric Reinforcement in Soil Retaining Structures, P.M. Jarrett & A, McGown, eds., Kingston, Ontario, Canada, Kluwer Academic Publishers, 239-284.

Yarnell, C.N. 1991. Steel wire rope safety net systems: a brief history of application to rockfall mitigation in the U.S. In Proceedings, National Symposium on Highway and Railroad Slope Maintenance, Association of Engineering Geologists, Chicago, 2-3 October, 135-148.

Yeh, S.T. & J. Gilmore. In press. Use of Expanded Polystyrene (EPS) for slide correction. In Proceedings, Conference on Stability and Performance of Slopes and Embankments II: a 25-year Perspective, American Society of Civil Engineers, Berkeley, California, June 1992.

Yin, K.P., L.K. Heung, and D.R. Greenway. 1988. Effect of root reinforcement on the stability of three fill slopes in Hong Kong. In Proceedings, 2nd International Conference on Geomechanics in Tropical Soils, Singapore, 293-302.

Yoo, N.J. & H.Y. Ko. 1991. Centrifuge modeling of reinforced earth retaining walls. In Centrifuge 91, Proceedings of the International Conference Centrifuge 91, Boulder, Colorado, 13-14 June, 325-332.

Zaruba, Q. & V. Mencl. 1982. Landslides and their control. New York, Elsevier, 324.

Theme report

G. Ramsay
Beca Carter Hollings & Ferner Ltd, Wellington, New Zealand

1   INTRODUCTION

A total of 33 papers have been assigned to the Theme Stabilisation and Remedial Works. While some of these papers deal specifically and solely with the details of constructing or assessing the effectiveness of particular remedial works, the bulk of the papers deal with case histories. Many of these papers address the complete process of investigation, analysis, selection and implementation of remedial or stabilisation works and the instrumentation and monitoring necessary to verify the performance, effect and hence effectiveness of the remedial works. In some cases the stabilisation measures have consisted of a sequence of urgent actions taken in response to developing failures in order to maintain an operational facility or road. In other cases, stabilisation measures have been aimed at increasing the stability of slopes to prevent development of slides or reactivation of existing dormant slides by a planned activity such as road construction, powerhouse construction of filling of a reservoir. The diversity of the papers has made it difficult to produce a highly structured report.

Sections 2 and 3 of this report summarise individual papers which have been grouped as:

(i)   Papers dealing with specific techniques and remedial work design methods.

(ii)   Case histories which have been further subdivided according to whether they relate to repair of slope failures or deal with general stabilisation of a slope to prevent reactivation or to increase the margin of safety.

Because of the diverse range of situations described, it is not considered useful to attempt to categorise the papers in detail. However, in describing new techniques, these have been broadly classified into drainage, buttress and lightweight fill and reinforcement.

2   REMEDIAL WORK AND STABILISATION TECHNIQUES

2.1   Drainage Techniques

Siphon Drainage

Gress describes an innovative siphon drain arrangement that permits drainage to lower water tables using downward holes without pumping. The system includes an ingenious arrangement to ensure maintenance of the siphon operation even where flow rates are too slow to flush air bubbles from the pipe and air coming out of solution accumulates at the top of the pipe breaking the siphon. Gress reports that the technique has been used in France over the last five years with more than 40 applications. He describes two applications.

Rodren

The RODREN system described by Peila and Lombardi is a technique developed in Italy by the Rodio Co. which involves drilling a series of large (1.2 - 2.0 m) diameter wells which are interconnected by a collector drain that is drilled from the base of the hole into adjacent wells. The system is also referred to as 'modular discontinuous drainage alignments'. Individual wells can be backfilled with drainage material or cased and left open with ladders for permanent access for inspection or pumping. It is also possible to drill fans of sub-horizontal holes from within individual wells. The paper describes one case history which involved additional complications due to potential corrosion due to aggressive groundwater. This was countered by installing a cathodic protection system. The case history described by Beer et al utilised RODREN wells up to 52 m depth and 2 m diameter which are reported as being the deepest applications to date. Olcese et al present another case history involving the RODREN system.

Inclined Drainage Holes

In addition to papers specifically discussing drainage installation techniques, a number of papers describe drainage provisions installed in slides. Of note is the drilling of long inclined drainage holes up to 200 m long from tunnels to target specific zones in various Clyde slides. Details of hole lengths are reported in papers by Jennings et al and Gillon et al on the Nine Mile and Jacksons Creek slides. However, little comment is made on the techniques, progress rates and cost of drilling such long holes.

2.2   Buttress and Lightweight Fill

Pneusol - Lightweight Tyre Reinforced Fill

Bricout et al describes the use of a reinforced fill produced with used tyres and utilised as a cheap lightweight fill for reconstruction of a road embankment at the head of a landslide. The material Le Pneusol has been developed and widely used in France but this is presented as the first application as an alternative lightweight fill in slope failure repairs.

Diagrid System

Evans describes a proprietary bin crib system that permits construction of a retaining wall that can follow the slope and changes in direction of a slope. The system is thus more flexible than conventional crib wall systems (using headers and stretchers essentially at right angles) for buttressing slopes.

2.3 Soil Support and Reinforcement

Building Foundation Reinforcement

Kolev and Tzoner describe a system used to tie together the foundations of a church in Balchik which is founded on an area affected by three landslide movements. The treatment comprised casting an 8 m deep wall around the perimeter of the 15 m wide x 26 m long building and using inclined grouted anchors to tie these walls together in a triangulated fashion. The result was a reinforced block of soil isolated from slope movement and constrained by concrete walls to which a structural floor grillage was attached.

Trench Support

Lafleur et al have described a trench support system utilising geotextile incorporating horizontal reinforcing bars which span between vertical metal frames held apart by hydraulic jacks. The system has been designed to prevent the development of slope failures in the sides of deep (up to 4 m) trenches. The analysis of the support system and field validation trials are reported.

2.4 Loads on Piles

Anagnostopoulos et al describe laboratory model testing and numerical analyses undertaken to determine the effect of a row of piles in cohesionless soils to resist down slope loads from a slope failure. On the basis of the work the authors have proposed an approach for determining the load transferred to piles in a cohesionless soil and the pressure distribution on the pile.

Maugeri and Motta reported on the capacity of piles in cohesive materials to resist loads from landslides as passive piles. Numerical analyses were undertaken and the results compared with available field data. The paper concludes that bearing capacity factors Kc of between 3 and 3.5 were appropriate for a cohesive soil 'flowing' past the pile. For the soil beneath the failure surface where the pile is actively being pushed through the soil, Kc factors of between 5.6 and 8 gave best fit with published results.

2.5 Reinforcement by Tree Roots

Suyama commented on testing carried out in Japan to determine the strength of tree roots for various species and the effect of roots in increasing soil strength by an increased apparent cohesion. The paper refers to the general detrimental effects of deforestation of urban forest in Japan. He notes the progressive loss of strength in the soil as tree roots decay after filling of forest. Evans and McKelvey in a paper on the failure of a mudstone slope after an intensive storm, noted that revegetation was recommended 'as root strength adds significantly to slope strength' and commented that one of the destabilising factors was the harvesting of trees planted on the slope 'without early replacement'.

3 CASE HISTORIES

3.1 General Practice and Regional Experience

Popescu has presented a comprehensive paper on coping with landslide problems in Romania. He describes the type of landslide problems encountered in Romania and the approach to investigation analysis and stabilisation. In particular, he has noted the successful use of pile restraining structures, a technique apparently more greatly favoured in Eastern Europe. In Western Europe high capacity anchors tend to be used in such cases and in US/Canada 'containment' is far less common. He presents a number of case histories involving piles, drainage and localised excavations and fills on the slope to redistribute weight about the "neutral" line. The paper restates a number of axioms of slope stability assessments including: most landslides arise from a number of contributory factors; the efficacy of simple monitoring devices in situations where sophisticated and expensive high precision devices are not required; while too little site investigation has been a cause of failure of some slope stabilisation, the converse of excessive investigations and grossly uneconomic designs is also unprofessional. He notes that some stabilisation may be overdesigned as a result of inevitable mismatch between the state of theoretical knowledge and practical applications.

Baldassarre and Radina has reported on the general instability of a 360 km^2 drainage basin in Italy and describes the type of stabilising works associated with dam, road and aquaduct projects in the area.

Chummar reports the use of grouting to stabilise slopes in the Andaman Islands where deforestation has resulted in some slopes at relatively low angles failing during monsoons. Investigations suggest that failures occur on dip slopes on 'layering' joints containing 2 - 3 mm of fine silt between silty clay and sandy clay beds. Stabilisation works have comprised grouting of these silt layers with cement slurry from vertical holes at 10 m spacing. The effectiveness of this work is not reported. Evans and McKelvey noted in addressing a failed mudstone slope that they had considered stabilisation with lime slurry injection. This solution was rejected primarily as it was 'unproven technology'. They noted that subsequent research shows lime injection might be an option in the area considered.

3.2 Repair of Slope Failures

Alonso et al report on the analysis and stabilisation of a slope failure resulting from quarry operations above a reservoir. Excavations during remedial works uncovered a 4 mm thin clay layer which is considered the failure surface. Stabilisation was by recontouring the slope. Of interest is the use of finite element analyses to determine the local factors of safety along the failure surface and the change in margin of safety (difference between available shear strength and acting shear stress).

Beer et al describe extensive drainage works utilising the RODREN system aimed at stabilising a part of a large landslide which occurred in 1982 following very heavy rain. The slide covers 220 hectares, has a head to toe elevation difference of 170 m, width of 1 800 m and a depth below ground level of up to 100 m. The section of the landslide treated had a failure surface up to 50 m below ground level and comprised about 7 hectares. The paper describes the drainage works, the construction techniques and subsequent monitoring. Even with the three years of monitoring information available, the authors could not 'conclude on any tendency for the slope to stabilise'. The response of piezometers has been slow but significant and the general trend of movement indications has been repeatedly obscured by seasonal fluctuation. This paper illustrates the difficulty often experienced in gauging the effectiveness of stabilisation works for large slides. The amount of water extracted has typically amounted to 10 - 20 litres per minute per 350 m long row of 20 wells.

Bryant et al describe slope stability considerations for a new road across a number of old landslides in the Cromwell Gorge. The paper refers in particular to stabilisation of a slide that occurred on one old landslide when the road bench was constructed and a failure occurred on major shear zone 15 m thick, activating 0.5 million cubic metres of material. Head unloading of 0.6 million cubic metres was undertaken after analyses assuming a dry slope. Six months later movement was reactivated. Subsequently boreholes indicated a relatively thin (7 m) water table perched on the active failure surface. Installation of 55 m long inclined drains was undertaken to intercept this water table and resulted in a 1.5 m lowering of the water table and an associated reduction in movements.

Diyaljee describes a succession of repair measures undertaken over a six year period to maintain a road and railway across the head of an active slope failure moving on a formation interface dipping towards the adjacent river. The paper demonstrates a reactive approach to a developing failure where the aim is to maintain thoroughfare. Methods adopted included surface drainage control measures, horizontal drains, a pile wall, deep pump wells, tiebacks for the pile wall and lightweight fill of sawdust and trees.

Evans and McKelvey describe the use of a 3½ m deep cut off drain for stabilisation of a 3 - 8 m deep surface colluvium slope in weathered mudstone. The slope had been affected by failures (following extreme rainfall) which blocked a roadway at the toe and damaged a sheet pile retaining wall between the road and the river. The authors took into account in their analyses the effect of tree roots and the new slope profile on stability. The cut off drains were designed to tap existing springs and limit rising of groundwater levels during heavy run-off. Construction constraints of limited stand up time of the deep trenches affected the selection and detailing of the drain. The authors noted that the installed drains had performed well but that ultimate success "would require the test of another extreme storm event".

Gordon and Phillips describe the failure of the face of a cutting in lateritic weathered materials containing clays, core stones and relict joint planes. The cutting contained a conveyor from a bauxite mine to an alumina refinery. Failure of the face in the form of rilling, slabbing and erosion posed a risk to the conveyor which was close to the toe of the slope. Initial attempts to halt the slope surface problems by covering with PVC sheeting and mesh were of limited effectiveness. The paper describes the mechanisms of the surface failures and various treatment options considered. The adopted solution was to recut the slope leaving a 5 m wide roadway between the toe and the conveyor, accepting continued failure of the slope but removing the risk to the conveyor and allowing easy access.

Kolev and Tzonev described the use of a 1 275 m long drainage gallery with drain holes to stabilise a large slide moving at 30 - 700 mm per year. The groundwater recharge was from karstic limestone.

Mostyn and Adler have described the design and implementation of remedial works for reinstatement of a railway embankment south of Sydney, Australia. The stabilisation took the form of rockfill buttress constructed against the toe of the slope and keyed down to interrupt the failure surface up to 10 m below the existing ground surface. A major issue addressed by the paper was the procedure for excavating the key without causing failure of the slope above. The excavation for the key was held open by anchored sheet pile walls and constructed in hit and miss panels. The overall analysis of the existing slope and proposed remedial measures was carried out using conventional limit equilibrium methods and then the entire construction sequence was analysed using the programme FLAC to verify the adequacy of the sheet pile sections and predict deflections and bending moments in the piles. The paper demonstrates the input of construction personnel in developing an practical economical design and the use of modern analysis techniques to predict the performance of staged construction.

Nguyen Van Tho described the use of revetments to prevent failures in the banks of the Mekong River which have resulted from undermining of overlying cohesive soils as the river erodes a sand layer that is exposed beneath the water.

Tran Vo Nheim et al describe stabilisation of a slide in volcanic tuff overlying andesite on the Island of Martinique. The slide which had an elevation difference of 150 m from toe to head scarp involved about 1 million cubic metres and destroyed a road. The slope was reconstructed by "judicious remodelling" of the slope into terraced platforms with drainage works in the form of sub-horizontal drains and surface drains. Slight residual displacements have continued to be observed during rainy periods and the authors note that the stability is sensitive to groundwater level.

Wei et al describe reconstruction and stabilisation of a 4 m high slope failure using a combination of toe loading, buttress drains and embankment piles. The reconstruction was constrained by space and the requirement to retain the original slope angle. In this case the piles are provided to support the weight of the slope and reduce the overturning movement rather than to provide resistance to the displacement of the soil over the failure surface. For this function closely spaced timber piles designed as friction piles within the failure mass have been used.

3.3 Repair of Rock Slope Failures

Hall and Hanbury describe the treatment of a 30 m high rock slope failure in South Africa. The slope comprised residual diabase soil overlain by very hard diabase rock. The work included removal of a remaining precarious overhang by blasting, anchorage of part of the slope by double corrosion protection ground anchors soil nailings and shotcrete as well as horizontal drains. The paper contains a comprehensive discussion of construction consideration and experience with blasting of the face and the drilling and installation of the anchors. The methods of grouting the fixed anchor lengths were devised to prevent excessive takes in the fractured rock and avoid affecting the drainage of the slope. The paper also discussed contractual and tendering aspects of this type of slope stabilisation work.

Baumer and Schindler discuss removal of an unstable rock mass in Switzerland resulting from movement of a 43° slope on stress relief joints infilled with sand and silt. After removal of the moving mass the exposed rock face was restrained where necessary using prestressed anchors holding back buttresses and concrete ribs. The remedial works involved the removal of 97 000 m³ of material over an elevation difference of 60 m by step by step blasting and removal of rock. Consideration was given to a proposal to remove the unstable mass with one large blast. The paper sets out the reasons why this was discarded.

3.4 Stabilisation for Reservoir Filling

A number of papers deal with the stabilisation of large rock slides on the Revelstoke Project in British Columbia, Canada and the Clyde project in NZ. In both cases, analyses indicated that lake filling would inundate the toe of slides with a resultant reduction in stability and the possibility of reactivating the slides with potentially catastrophic results.

The papers by Gillon et al, Jennings et al and Newton and Smith describe the stabilisation of very large landslides in schist terrain along the margin of the future Lake Dunstan reservoir in the Cromwell Gorge. The papers describe stabilisation works on three slides using combinations of toe buttresses and drainage using both surface holes and long holes drilled from tunnels within and beneath the slides. The slides are notable for their scale, the complexity of groundwater due to compartmentalisation of the slides by both high and low angle faults and shears and the marginal stability of at least some of the slides. For the three slides reported a total of 10.5 km of drainage drives were driven from which nearly 50 km of drainage drill holes were installed in vertical up, down and sub-horizontal directions. In addition, some 11 km of surface sub-horizontal holes were drilled and nearly 4 million cubic metres of buttress material placed.

The three slides described (Jacksons, 9 Mile Creek and Brewery) while all in the same geologic setting had particular features that affected the selection of remedial measures. The scale of the landslides was such that investigation drives provided the most positive means of confirming geologic structure and were incorporated into the stabilisation drainage works. Drainage holes up to 200 m long are reported as being drilled from underground chambers to target specific areas of perched or trapped water. Instrumentation and monitoring indicated removal of substantial volumes of water, lowering of piezometers and settlements of the ground surface.

At the Brewery Slide existing groundwater tables are low and an innovative approach has been used with shafts and a low level drainage drive (below existing river level and 70 m below final reservoir level) to reduced groundwater levels in the slope to below reservoir level. A zoned earthfill blanket was placed on the exposed face to be flooded by the reservoir with a grout curtain extending down below existing riverbed level. The aim was to limit flow from the existing riverbed and the sides and base of the reservoir back into the slope. This approach results in the hydrostatic pressure of the reservoir water effectively buttressing the toe of the slope.

Jacksons Creek Slide is a 5 million cubic metre scarp slope slide that was stabilised by buttressing and drainage drilling from tunnels. Of particular note was the immediate reactivation of the slide by the stripping of some 4 000 m³ of topsoil and loess from the toe of the slide in preparation for the placing of a 0.7 million m³ toe buttress. The stripping was assessed as reducing the factor

of safety by 2% and demonstrated the marginal stability of some of the slides in the Cornwall Gorge.

The Nine Mile Creek slide is noticeable for its size (300 million cubic metres) dimensions and the complexity of its groundwater systems as a result of barriers provided by faults, shears and failure surfaces. Stabilisation works included more than 8 km of tunnels, 35 km of drainage holes from tunnels and 10 km of surface drilled drainage holes. Over 1 million cubic metres of water has been drained from within and below the landslide.

Imrie et al have described experience with the performance of the large (1 500 million cubic metre) Downie Slide the toe of which was inundated by the filling of the reservoir behind the Revelstoke Dam in 1983. The slide was stabilised by drainage using holes drilled from drainage adits with the objective of at least reducing groundwater pressures sufficiently to offset the calculated reduction in slope stability (7%) due to reservoir fillings. This objective had been set by the Licensing Authority as a condition of the granting of approval to proceed with the project.

The experience at Downie is similar in some respects to that reported at Clyde but different in others. Difficulties with driving of the drainage drives in areas of "soft .... schist [with] .... large volumes of water under heads of up to 210 m" are reported and yet it is noted that in hydrogeological terms the slide was more homogeneous than other slides in schist in British Columbia.

A major aspect of the paper is the reporting of the instrumentation systems used for monitoring of the slide after lake filling and the response of the slide to the reservoir filling. The authors noted that the measure of stabilisation has been the maintenance of the lowered groundwater pressures after reservoir filling. They comment that while measurement of movements is important it may not be possible to stop movements completely and the goals should be described in terms of improvements to factor of safety.

The drainage works described and the associated instrumentation and ADAS (Automatic Data Acquisition System) have been in service for 10 - 15 years. The authors noted that the ADAS and some of the instrumentation is currently being replaced by advanced telemetry, hardware and software systems. They also noted the blockage of some drainage holes with time and the necessity for secure permanent portal structures and support/lining in the drives.

3.5 General Slope Stabilisation

Heiner and Nilsson have described the analysis and development of slope profiles and drainage works for a 90 m high open cut required to accommodate the power station on the Mrica project in Indonesia. The paper describes the original investigations and analysis and the re-evaluation and modification of slope profile as a result of observations during construction. The slope is cut into a sequence comprising two 10 m thick layers of halloysitic clays (formed by weathering of tuffs) separated by sedimentary deposits and overlying bedrock of breccia, tuffs and lavas. The paper reports on the performance of the slope and the value of instrumentation during construction which provided data from which the need to modify the design was established. The authors referred to this as an "active design" process and commented on its success in this project.

Martin and Warren have reported on stabilisation works

undertaken to ensure the safety of a new trunk roading South Wales. The Taren slide involves an estimated 7 to 8 million cubic metres and is believed to be a complex translational rock slide on weak mudstone layers with a coal measures formation. The slide has complex groundwater conditions and is affected by faults which result in the subdivision of the slide into three zones. After investigation and analysis, stabilisation works including a toe embankment and two separate drainage systems were designed. In the central section of the slide a system of gravity wells discharging into an adit was detailed while in the toe area, where artesian pressures had been detected, pressure relief wells were constructed to lower piezometric head to ground level.

The paper describes in detail comprehensive monitoring and testing that was carried out to ascertain the effectiveness of the proposed drainage systems, estimate insitu hydraulic transmissivity and determine general geohydrology. The information gained was used to assess the appropriateness of the proposed measures and determine drain spacing. In particular pumping tests were carried out on drainage wells in the central zone prior to construction of the adit. In the lower zone tests included Brine Tracer Tests to assess the movement of groundwater from the underlying bedrock into the artesian zone and an Airlift Pumping Test.

Details are given of the construction of the various drainage systems and the authors reported on monitoring of the drawdown in piezometric levels and the confirmation by analysis that these represented acceptable improvements in stability factors of safety.

This paper demonstrates the value of systematic insitu pump testing with appropriate instrumentation to determine the geological and hydrogeologic information necessary for design of drainage for slope stabilisation.

Olcese et al describe the stabilisation of a large quiescent landslide in Northern Italy using 20 m deep RODREN system drains. The slide is crossed by a gas pipeline laid in a trench and it had been necessary to periodically uncover the pipe and replace it in the trench to relieve stress in the pipeline due to the slope movement. The paper describes the drainage system installed and its effectiveness as measured by installed instrumentation including piezometers, extensometers, inclinometers and drainage discharge measurement. A finite element analysis based on a viscoplastic algorithm was used to assess the viscoplastic shear strains with the original and lowered water tables.

4   TRENDS AND DISCUSSION TOPICS

In reviewing the papers in this Theme, a number of trends were noted, as were areas where a number of authors identified the need for further development and/or the adoption of particular approaches in dealing with particular aspects.

4.1   Trends

4.1.1   General Stabilisation Technique Developments

In terms of stabilisation or remedial work methods the most significant trends appear to be in development of techniques for drilling both deep wells from the surface and long inclined holes from tunnels. Papers in the symposium report vertical wells by the RODREN system up to 52 m drop and 200 m long drainage holes drilled from tunnels.

4.1.2   Analysis Technique Developments

Two papers were presented describing the analysis of piles restraining slides in cohesive and cohesionless materials respectively. There is also a noticeable trend to employ Finite Element techniques to provide more information on local stability along a failure surface then is available with traditional limit equilibrium analysis methods which produce a single factor of safety. Several papers referred to use of sophisticated analyses to ascertain the stability of a slope during progressive construction activities.

While reference was made in a number of papers to analysis of groundwater flow there still appears to be a limited ability to model geohydrology and analyse the effect of various drainage works. This area is one worthy of discussion.

4.1.3   Philosophy

A number of authors emphasised the need to review the appropriateness, adequacy and details of remedial works throughout construction and service life and the consequent need for appropriate instrumentation. This philosophy is a consequence of the realisation of the frequent difficulties in determining and adequately modelling complex geological and geohydrological conditions. Comment was made in a number of papers to the assessed improvement in factor of safety (based on measured piezometric level changes and geometric changes) being a more appropriate measure of stabilisation than the change in movement patterns.

4.2   Discussion Topics

4.2.1   Stabilisation by Drainage

For large slides and many small slides drainage may be the only practical method of stabilisation.

Drainage differs from most other forms of stabilisation in that the quantification of the effect of particular remedial work component may be difficult. Whereas the effect of placing a buttress (of readily measurable location and assessable mass) or installing a pile or anchor of known structural capacity can be directly input in to a stability analysis, the effect of installing drainage system has to be expressed in terms of the change in groundwater pressure achieved. The current approach to drainage works is to select a general drainage system and system on the basis of judgement and sometimes trials or pump tests. The selected system is then incrementally constructed and its effect on groundwater pressures and movements observed.

While a number of papers report sophisticated and comprehensive stability analyses only a few (e.g. Gillon et al and Imrie et al) refer to groundwater analyses or report (e.g. Martin et al) substantial investigation stage testing to assess effectiveness of drains. In most cases the drainage system is not analysed but simply reviewed after installation on the basis of measured discharges and groundwater changes. To further complicate matters, several authors (Evans et al and Vo Nhiem et al) reported cases where drainage, installed to prevent reoccurrence of a slip in extreme rainfall event, cannot be tested until a reoccurrence of a similar event. Where the slide materials are of low permeability response of piezometers and water tables may be slow with some time elapsing before effectiveness can be ascertained. The case presented by Beer et al falls into that category.

While the difficulties in measuring geohydrological parameters and creating realistic models may favour the observational approach, there may be grounds for pursuing further methods of analysing the effectiveness of drainage systems prior to installation.

Discussion is invited on means of analysing groundwater models and the effectiveness of drainage system proposals.

4.2.2 Effect of Vegetation

The removal of vegetation and forest clearance are recognised as contributing factors in many slope failures both deep-seated and surficial. Similarly planting of suitable species of trees is often included as a component of a stabilisation programme. However, there appear to be limited guidelines for quantitatively including the effects of tree roots in stability analysis. Suyama presented information on strength of roots of specific species.

Discussion is invited on means of assessing the effect of vegetation on slope stability.

4.2.3 Stabilisation by Grouting

Chummar describes the grouting of a thin silt layer with cement slurry to 'make [it] impermeable and increase its shear strength'. Evans and McKelvey also referred to their consideration and rejection of lime injection as a means of stabilising a mudstone colluvium.

Stabilisation by grouting does not appear to be a very common technique and appears to have potential difficulties. The influence of grouting on drainage patterns is difficult to predict and the degree of penetration in fine grained soils must be questionable. In NZ in the 1950s a number of unstable mudstone slopes and unconsolidated sidling fills were treated by cement grouting. Initially the movement of the slopes was halted but subsequently accelerated movement occurred and continued until drainage drilling was installed to relieve high groundwater pressure ponded behind the grouted sections. Comment is invited on the use of grouting of various forms as a slope stabilisation technique.

4.2.4 Stabilisation by Pile Structures

Popescu commented on the different philosophies and practice within Eastern and Western Europe and the US/Canada towards the use of piles and other restraining structural elements in stabilisation of soil slopes.

Discussion is invited on the use and relative cost-effectiveness of restraining systems for slope stabilisation.

4.2.5 Repair of Rock Faces

Two papers addressed the repair of rock slopes where blasting was required to remove hard rock. Both contained discussions on the selection of a sequential blasting process and the rejection of removal of rock by a single blast.

Discussion is invited on the relative merits of various blasting techniques for removal of rock during stabilisation.

4.2.6 Monitoring of Remedial Works

Papers by Gillon et al, Imrie et al and Olcese et al refer to the use of remote recording and telemetry systems for collection of critical monitoring information. It is apparent that while systems in use are sophisticated, they can be expected to be obsolete in 10 - 15 years as a result of changing technology.

Popescu refers to the efficacy of simple monitoring devised where sophisticated and high precision devices are not required.

Discussion is invited on the determination of the necessary level of monitoring and selection of appropriate level of technology in instrumentation for monitoring the effectiveness of slope stabilisation works.

5 CONCLUSIONS

The papers presented in this Theme cover a diverse range of geological situations and slides of varying scale and type. A common factor in the majority of papers is the frequent complexity and non-homogeneity of ground conditions and inherent problems in predicting and modelling groundwater conditions. While many of the papers described comprehensive investigations and analyses prior to designed stabilisation works, the authors recognised the inherent limitations of investigation and analyses and emphasised and demonstrated the need for observations and monitoring both during construction to detect changes from assumed conditions that might indicate a need to modify the remedial or stabilisation works; and after construction to check the effectiveness of the works.

Landslides, Bell (ed.) © 1995 Balkema, Rotterdam, ISBN 90 5410 032 X

Field study of the behaviour of prestressed grouted rock anchors subjected to repeated loadings

Brahim Benmokrane, André Beaudet, Gérard Ballivy & Mohamed Chekired
Laboratory of Rock Mechanics, Department of Civil Engineering, Faculty of Applied Sciences, Université de Sherbrooke, Que., Canada

SYNOPSIS

Prestressed grouted rock anchors are generally designed based on anchor behaviour studies under static loading, the most frequent and widespread type, extending to most applications, such as slopes, underground excavations, pylons, and retaining walls. Sometimes, however, natural conditions render the situation more complex. For example, cable swing engenders oscillations in cable-car and high-tension pylons, hydraulic loading variations affect slope stress states and thermal/seismic loads can also generate repeated loading. This paper presents the results of repeated loading tests performed on prestressed grouted rock anchors installed in the field. External repeated loads lesser and greater than the initial prestress load values were applied on the anchors. It is shown that the higher the prestress value, the greater the anchor life, provided that the applied repeated loads are less than the initial prestress load values.

INTRODUCTION

Grouted prestressed anchors are extensively used in rock or soil slope stabilization projects. The main advantages of a prestressed anchor over a dead anchor are that the movements of the anchor and surrounding ground are greatly reduced, the anchor exerts a permanent stabilizing force on the surrounding ground, and the anchor is also load tested during the prestressing and lift-off operation.

Grouted anchors are designed based on anchor behaviour studies under static loadings the most frequent and wide-spread type extending to most applications, such as slopes, pylons, under-ground excavations and retaining walls.

Sometimes, however, natural conditions render the situation more complex. For example, cable swing engenders oscillations in cable-car and high-tension pylons, hydraulic loading variations affect slope stress states, or thermal/seismic loads can also generate repeated loadings.

Determining the effect of repeated loading on a grouted anchor under load (prestressed anchor) is of significant interest. To this end, an experimental program involving field prestressed anchors which constitutes a continuity to that performed earlier on dead anchors (Ballivy et al., 1988a, b), was developed. The anchors used were 36-mm-diameter Dywidag bars cement grouted into 72-mm holes drilled into rock.

This experimental study, the results of which are reported in this paper, had two main objectives: (i) to investigate the effect of load level and amplitude on the behaviour of a prestressed anchor subjected to repeated loads; and (ii) to examine the effect of prestress load level on the life span of an anchor.

EXPERIMENTAL PROGRAM ; CHARACTERISTICS OF THE ANCHORS AND TESTING PROCEDURE

The three anchors considered in this study were cement grouted to lengths varying between 500 and 2000 mm, Table 1, in a local quarry in a sound rock

TABLE 1

Geometric characteristics of the anchors

Anchor No	Anchored length AL (mm)	Free length FL (mm)	PL* (mm)	NL* (mm)
A	2000	2465	355	670
B	1000	2440	355	670
C	500	2405	355	670

* See Fig. 1

TABLE 2

Summary of mechanical properties of rock and cement (Ballivy et al., 1986)

	Compressive strength C_o (MPa)	Tensile strength T_o (MPa)	Modulus of elasticity E (GPa)	Poisson ratio υ
Rock	212 ± 15 (7)	14 ± 2.3 (5)	59.9 ± 3 (7)	0.25 (7)
Cement grout	49.5 ± 3.2 (10)	3.2 ± 0.2 (6)	11.1 ± 2.2 (10)	0.18 (3)

( ) : number of tested specimens

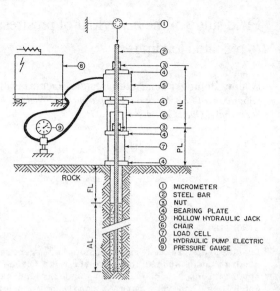

Fig. 1   Field experimentation: Test setup used for the anchors' stressing and load application.

mass belonging to the Appalachian geological formation; this site has been described elsewhere (Ballivy et al., 1986; Benmokrane and Ballivy, 1991).

The steel bars (type Dywidag) employed (Grade 1035 MPa (150 ksi)) have a nominal diameter of 36 mm and elastic limit and ultimate loads of 860 and 1050 kN, respectively. Holes of diameter of 76.2 mm were drilled vertically by percussive and rotatory drilling processes.   The bars were inserted into the holes, then anchored with a cement-based grout prepared with a water/cement ratio of 0.40.

A Type 10 Portland cement was used together with an expansive agent (aluminum powder) at a ratio of about 0.005% of cement weight.   The installation of the anchors and the preparation of the cement grout have been described elsewhere (Ballivy et al., 1986). Table 2 presents the values of the mechanical properties of the rock and the grout.

The anchors were loaded by using a hollow hydraulic jack, having a capacity of 1100 kN, operated with the aid of an electrical pump allowing static and repeated loadings.   Figure 1 illustrates the test setup used for the anchors' stressing and load application.   This setup allows the anchor to be prestressed and an external repeated loading to be applied also.

Details of the repeated loading tests performed on the three anchors (A, B and C) are given in Tables 3 to 5, respectively.   The lock-off of the desired initial prestress load, P_i, was accomplished by tightening the nut through the hole made in the chair with the help of a wrench (Fig. 1).   A load cell, having a capacity of 1150 kN, was used to monitor any change in the initial prestressed load during the application of the repeated loadings on the anchor. Elsewhere, during the application of the repeated loadings, the displacement at the head of the anchor was noted with the aid of a comparator (Fig. 1). This allows to investigate the displacement - number of load cycle relationships. The applied loading frequency was four load cycles per minute.

RESULTS AND DISCUSSION

Displacement at the head of the anchors stressed under monotonic loading

Before the initiation of the repeated loading tests, the three anchors were submitted to uplift static tests, (monotonic loading) according to the test method recommended by the International Society for Rock Mechanics (1985). These tests have been conducted on the anchor under a prestress load (prestressed anchor) and under no prestress load (dead anchor), in order to compare their performances in terms of external uplift load versus displacement at the head of the anchor. The value of the applied tensile load, increased at the rate of about 45 kN. min^{-1}, and the displacement at the head of the bar were noted at each increase of 45 kN in load. The obtained results showing the variations of the tensile load against the displacement obtained at the head of the anchor, under a prestress and under no prestress load, are plotted in Figure 2.

The results show clearly that the displacements observed at the head of the anchors under an initial prestress load (prestressed anchors) are greatly reduced in comparison with those noted on anchors under no initial prestress load (dead anchors). This reduction of displacements has resulted in an increase of the axial stiffness of prestressed anchors. This aspect demonstrates the benefit of the application of a prestress load on grouted anchors, in order to enhance the long-term performance of the anchor-retained structure. It can be noted that the axial stiffness is defined by the slope of the more-or-less straight line portion of the applied tensile load-anchor displacement curve.

For the experimental conditions described herein, the axial stiffnesses of the prestressed anchors are more than five times greater than those shown by the dead anchors. One can observe, however, from Fig. 2, that when the external applied tensile load becomes superior to the initial prestress load, the two types of anchors (prestressed or dead) show a similar behaviour resulting in analogous axial stiffnesses. In this situation, the anchor which was initially prestressed behaves as a dead one.

The difference noted here between the axial stiffnesses of prestressed and dead anchors, in the range of external applied tensile loads less than the initial prestress loads, is due to the elongation of their free lengths which are different, depending on whether the

anchor is prestressed or dead. This elongation is more important for the dead anchors because it originates from a great free length (law of Hooke).

TABLE 3

Repeated loading tests performed on anchor A

Test No.	Initial Prestress Load, P_i (kN)	(% P_u)*	Load levels (% P_i)	Number of cycles
1	225	22.5	70-40	1500
2	225	22.5	100-40	1500
3	220	22.0	170-40	1500
4	500	50.0	60-10	1500
5	500	50.0	80-10	1500
6	490	49.0	120-10	1500
7	470	47.0	150-10	1000
8	735	73.5	60-10	1000
9	735	73.5	80-10	1000
10	735	73.5	120-10	1000

* P_u = ultimate load of the anchor (1000 kN)

TABLE 4

Repeated loading tests performed on anchor B

Test No.	Initial Prestress Load, P_i (kN)	(% P_u)*	Load levels (% P_i)	Number of cycles
1	225	22.5	40-10	1000
2	225	22.5	80-10	1000
3	220	22.0	120-40	3000
4	215	21.0	150-10	2000
5	620	62.0	100-10	1500
6	520	52.0	120-10	1500
7	410	41.0	150-10	1500

* P_u = ultimate load of the anchor (1000 kN)

TABLE 5

Repeated loading tests performed on anchor C

Test No.	Initial Prestress Load, P_i (kN)	(% P_u)*	Load levels (% P_i)	Number of cycles
1	210	28.0	40-10	1000
2	220	29.0	80-10	1000
3	210	28.0	120-40	1000
4	210	28.0	150-10	1000
5	525	70.0	40-10	1000
6	525	70.0	80-10	1000
7	510	68.0	120-10	1000

* P_u = ultimate load of the anchor (750 kN)

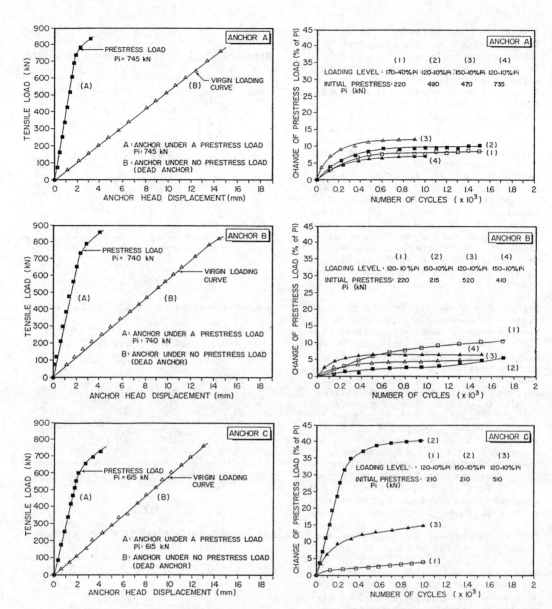

Fig. 2 Anchor head displacement versus tensile load (monotonic loading).

Fig. 3 Loss in prestress as function of load cycles for loading levels more than that of the initial prestress values.

Indeed, from the test setup illustrated in Fig. 2 and the corresponding data in Table 1, the free length which deforms under stressing is around 3500 mm for the dead anchors (summation of FL, PL and NL) comparatively to 670 mm for the prestressed anchors (only NL), which represents a ratio of 5,2 (3500/670).

Change of prestress load for anchors under repeated loading

Figure 3 illustrates the loss in prestress as function of load cycles for loading levels more than that of the initial prestress values. It is important to note that when the applied repeated loads were less than the initial prestress loads, there was no loss in prestress, as it can be seen from Figure 4. Moreover, for loading levels less than the initial prestress values, the displacement at the head of the anchors

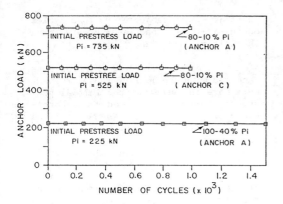

Fig. 4 Anchor load versus number of cycles for loading levels less than the initial prestress values.

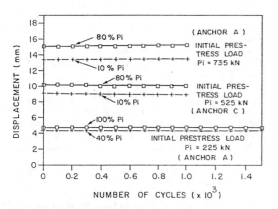

Fig. 5 Anchor head displacement versus number of cycles for loading levels less than the initial prestress values.

has remained unchanged, as function of the number of cycles, as illustrated in Figure 5 for the maximum and the minimum load levels.

The loss in prestress, recorded when the applied repeated loads were more than that of the initial prestress values, appears to be related to a continuous shortening of the anchors as the number of load cycles increases.

This shortening of the anchors is presented in Figure 6 which illustrates the displacement at the head of the anchors, noted at unloading (minimum load level), as function of the number of cycles.

The examination of the results, shown in Figure 3, reveals that as repeated load-

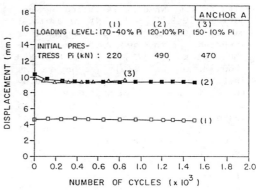

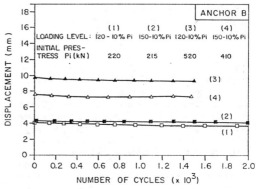

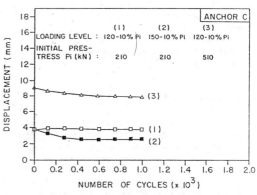

Fig. 6 Anchor head displacement as function of load cycles for loading levels more than that of the initial prestress values.

ings continue, there is a gradual loss of the initial prestress load. In general, this loss of load has remained less than 10% of the initial applied prestress load, after about 1500 loading cycles, with the exception of anchor C, which undergoes losses of prestress as high as 40%, when it is cycled between 150% and 10%P_i (P_i = 215 kN). As a matter of fact, the anchor C has the shorter anchored length, 500 mm, compared to those of anchors A and B (2000 mm and 1000 mm, respectively), and before the initiation of the repeated loading tests, it has been submitted to an uplift static test in which the applied maximum load was near the failure load of the anchor. This high loading level should generate debonding of the grout over a big part of the anchored length (Benmokrane and Ballivy, 1991b). In view of this finding, it is reasonable to suggest that an important part of the losses in prestress noted for anchor C is due to the debonding of the grout generated before the repeated loading tests.

Also, it is very clear from Figure 3 that the behaviour of prestressed grouted rock anchors subjected to external repeated loadings, above that of the initial prestress values, manifests two distinct phases: Phase I occurring within the first four hundred load cycles, where rapid losses of prestress are recorded, Phase II where rate of loss of prestress becomes small and uniform. This behaviour is similar to that described earlier (Benmokrane and Ballivy, 1991) dealing with the long-term behaviour of prestressed grouted rock anchors under static loads (losses of load due to the relaxation of the steel tendon and creep of the grout and the rock).

CONCLUSIONS AND RECOMMENDATIONS

A field experimentation including repeated loading tests on prestressed cement grouted rock anchors was undertaken. External repeated loads lesser and greater than the initial prestress load values were applied on the anchors. The study resulted in identifying certain factors pertaining to the behaviour of prestress grouted rock anchors subjected to repeated loadings. In particular, it was observed that:

• when the applied repeated loads were less than the initial prestress values, there was no loss in prestress and, in such cases, the displacement noted at the head of the anchors has remained independent of the number of cycles. As a matter of fact, there was a loss in prestress only when the prestress anchors have been subjected to load cycles having a maximum load level greater than the initial prestress values.

• As expected, the loss in prestress recorded when the applied repeated loads were more than that of the initial prestress values, is related to a continuous shortening of the anchors as the number of cycles increases. If Δ_S is the shortening of the anchor, the loss in prestress due to this shortening would equal (Δ_S/FL) x E x A where E and A are respectively the modulus of elasticity of the steel tendon and its area. Therefore the loss in prestress is inversely proportional to the free length of the anchor (FL).

• As repeated loading continued, there was a gradual loss of the initial prestress load in the anchor. This gradual loss of prestress seems to be directly related to the following parameters: 1) the maximum and minimum levels of loading; 2) the initial prestress load and 3) the amplitude.

• Consequently, this study has shown that the loss in prestress is important and it must be taken into account in the design of prestressed grouted rock anchors where repeated loadings can occur.

ACKNOWLEDGEMENTS

This study was made possible by a financial support from the Natural Science and Engineering Research Council of Canada (NSERC) and from the ministère de l'Education du Québec (Fonds pour la formation de chercheurs et l'aide à la recherche, FCAR; programmes: Action structurante et nouveaux chercheurs).

REFERENCES

Ballivy, G., Benmokrane, B. et Saioudi, A. (1988a). Endommagement des ancrages injectés dans le roc soumis à des chargements cycliques, 15tg Canadian Rock Mech. Symp. Toronto (Canada) 49-60.
Ballivy, G., Benmokrane, B., Sage, A.-P. et Saioudi, A. (1988b). Stabilisation des talus rocheux à l'aide d'ancrages injectés: influence des chargements cycliques, Proc. of the Fifth Int. Symp. on Landslides, Lausanne (Switzerland), A.A. Balkema, Vol. 2, p. 839-846.
Ballivy, G., Benmokrane, B. and Aïtcin, P.-C. (1986). Rôle du scellement dans

les ancrages actifs scellés dans le rocher, Revue canadienne de géotechnique, 23(4): 481-489.

Benmokrane, B. and Ballivy, G. (1991a). Five year monitoring of load losses on prestressed cement grouted rock anchors, Canadian Geotechnical J., 28(5): 8 p.

Benmokrane, B. and Ballivy, G (1991b). Distribution expérimentale des contraintes le long long d'ancrages scellés dans un massif rocheux sous charges de tensionnement, Bull. CIM, 84(951), 45-52.

International Society for Rock Mechanics (1985). Suggested method for rock anchorage testing. "Commission on Testing Methods", Int. J. of Rock Mech., Min. Sci., Vol. 22, No 2, p. 71-83.

Landslides, Bell (ed.) © 1995 Balkema, Rotterdam, ISBN 90 5410 032 X

Ochre clogging in subsoil drains

Kevin Forrester
Formerly: Roads and Traffic Authority, New South Wales, Australia

ABSTRACT: Ferro-bacterial clogging, commonly referred to as ochre clogging, is a frequent cause of subsoil drain failure. The conditions necessary for ochre to form are described here, together with how the intensity of its formation may be predicted. Ways of dealing with the problem, applicable to landslide control projects, are also described. When ochre formation is expected, the only clogging prevention method known to have a chance of success is to design the drainage system so that it is permanently sealed against the atmosphere, except at its outlet. Alternatively, there is an advantage in having large diameter inlet holes in the pipes. Where these two methods are not practicable, the only effective and feasible treatment known at present is regular sluicing.

1 INTRODUCTION

With adverse groundwater conditions a major cause of slope instability, it follows that subsoil drainage in some form is included as a vital component in many control schemes. Further, the drainage system is required to function effectively for a considerable time, if long term stabilisation is to be achieved. The factors involved in granular clogging in subsoil drains have been known for about fifty years, and ways of dealing with the problem have been well documented; see for example Hausmann (1990). What is not so well known among civil engineers is the problem of chemical and bacterial clogging. Difficulties may arise in particular areas with, for example, manganese, but the most widespread material that causes clogging, possibly resulting in the total failure of a drain, is ferric oxide hydrate, associated with certain bacteria. Ferric oxide is the main constituent of the material referred to in the literature as ochre. Ochre clogging has been the subject of considerable investigation in agricultural engineering. Much of the information set out in this paper has been gathered from publications in that discipline, and a particularly comprehensive statement has been given by Kuntze (1982).

2 NATURE OF OCHRE CLOGGING

The presence of two components is necessary for ochre to develop: filamentous (hair-like) aerobic bacteria, and ferrous iron (Fe^{++}) in solution in the groundwater in sufficient concentration. Ferrous iron alone may cause problems, but far more serious results arise when it is associated with the bacteria.

There are several types of bacteria in existence that fit the above description. Their filamentous shape enables them to join together to form a colloidal slime. Since they are aerobic, they may appear on any air/water interface. Some types can be seen as an oil-like multi-coloured film of linked organisms floating on water puddles, but may be distinguished from oil by not coalescing after being broken up.

In contact with air, any ferrous iron dissolved in the groundwater is oxidised to ferric (Fe^{+++}) oxide, which is almost insoluble. The bacteria also act as an oxidising agent. Ferric oxide is thus precipitated on an air/water interface, and together with the bacteria, forms there a rust-coloured expanding slimy mass. In a drainage system, the slots in the drain pipes provide the air/water interfaces that are most vulnerable to obstruction, and it is there that the most serious clogging occurs; in severe cases, much of the pipe's cross sectional area may eventually be choked. Obstruction is increased by the tendency of the slime to entrap granular material that would otherwise pass freely through the drain. Growth of the slime may be quite rapid, depending mainly on ferrous iron concentration, but also on temperature and water pH. The bacteria exist in the pH range of 5 to 8, and oxidise most actively at pH 6.5; ref Kuntze (1982).

For as long as there is a flow of water in the drain and the slime remains moist, it is in many situations possible to clear the obstructions by sluicing, although the slime adheres tightly to any solid surfaces. This may be compared to ferric oxide alone, which does not adhere well. However, if the slime is allowed to become dry, it crystallises to a hard insoluble rust-coloured mass that is even more difficult to dislodge. This is not so much due to adhesion, which actually decreases with time, but to the solidifying of the material to the shape of the pipe, particularly within the slots. With sufficient pipe cross sectional area lost through solid clogging, it may in extreme cases become impossible to pass a sluicing nozzle up the pipe, and the drain then becomes virtually useless. It is unfortunate that the most effective drains will produce substantial water discharge in wet weather, but will cease flowing in dry weather, and are thus most likely to be clogged by solidified ochre, provided there is sufficient ferrous iron present.

The quantity of ferrous iron in the groundwater may be limited, due to ground conditions in the immediate vicinity of the drainage site. Provided drainage is effective, this ferrous iron will enter the drain and oxidise, and cease appearing after a limited time, which may be of the order of three to eight years. Ochre formation is then only a temporary problem. On the other hand, the ferrous iron may originate some distance away, from a widespread source. The supply of this ferrous iron may then be regarded as continuous and permanent; for most landslide projects, this is the situation that must be anticipated. These two cases are referred to in the literature as "autochthonous" and "allochthonous" respectively.

3   DETERMINATION OF FERROUS IRON CONCENTRATION

If the concentration of ferrous iron in the groundwater is determined as part of the investigation of a site, the degree of risk of clogging can be predicted, and a decision may then be made on how any anticipated problem is to be dealt with.

On an elementary level, an indication of possible ochre clogging can be obtained by examining the site for rust-coloured stains on rocks and below seepage outlets. Ochre deposits and bacteria films may also be seen in puddles. Kuntze (1982) put this on a systematic basis with a table linking various surface clues to the degree of risk of ochre formation. However, the method was devised for agricultural applications, and is therefore only concerned with surface conditions, whereas the groundwater at significant depths is of interest in landslide work. Further, it was admitted to be variable with the season. Some way of testing aquifer water samples taken from boreholes or the sides of exploratory trenches is therefore needed.

ASTM D 1068-84 (Method A) and BS 6068:Section 2.2 are laboratory tests that involve, essentially, adding orthophenanthroline to a water sample. Any ferrous iron in the sample will produce an orange-red colour, the intensity of which is a measure of Fe^{++}

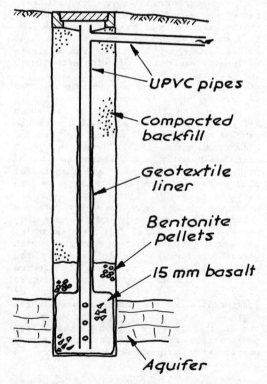

Fig 1.   Relief well

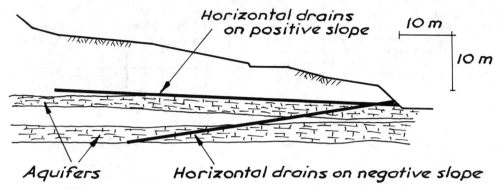

Fig 2. Horizontal drains on positive and negative slopes

concentration. Intensity is measured with
a photometer.

A version of these tests, suitable for
field use, was described by Kuntze (1982).
Since most people involved in landslide
control work do not have easy access to
this reference, the procedure is set out
in the Appendix, together with Kuntze's
table for interpreting the results.

Ford (1982) described a similar field
test for the determination of ferrous iron
concentration. The method is claimed to
give results that are independent of
differences in pH.

4 METHODS OF DEALING WITH OCHRE

An ochre problem may be dealt with either
at the design/construction stage, or after
the drainage system has been put into
service. With present knowledge, unless
the problem has been treated
satisfactorily as part of construction,
the only option then is to manage it by
regular long term maintenance. Most of
the control methods have been tried with
agricultural applications in mind; here,
only those appropriate to landslides will
be treated in any detail.

4.1 Design/construction stage methods

4.1.1 Chemical methods

Various control methods that involve
building the drain to include a chemical
material of some kind have been tried.
The aims have been either to oxidise and
precipitate the ferrous iron before it can
enter the drain, or raise the pH of the
surrounding soil, or inhibit or kill the
bacteria. At best, any success has been
temporary, lasting only until the material
has been finally leached out of the drain;

for temporary clogging conditions, this is
all that is required. At worst, some
chemicals are environmentally undesirable,
or they may form clogging of a different
kind, or they may reduce the permeability
of the ground adjoining the drain. For
these reasons, a major difficulty with any
chemical method is determining the
appropriate "dose". A comprehensive
account of the various chemical methods
has been given by Kuntze (1982).

4.1.2 Drain pipe inlet holes and
geotextile socks

If a hole in a drain pipe is large enough,
it is difficult for bacteria to form a
slime bridge across it. Taylor (1977)
therefore suggested using holes between 12
and 19 mm dia; these should be along the
top of the pipe. This would only be
acceptable in rock, or aggregate bigger
than the hole diameter, to avoid the
intrusion of fine material. Further, Ford
and Altermatt (1986) found that if the
holes are cleanly drilled with no burrs,
the slime cannot easily form around a
hole. Geotextile envelopes around the
pipe have also been tried, and the same
authors found that knitted polyester socks
were less subject to ochre formation than
those made of needle punched fabrics. The
risk with methods such as large holes and
geotextile socks is that they may move the
air/water interface, and hence the point
at which ochre forms, further back into
the drain system, where it is less
accessible for maintenance sluicing.

4.1.3 Reduced slime adhesion

If the ochre slime's ability to adhere to
surfaces is reduced, there is a chance
that groundwater flow will carry it

through the drain, or at least sluicing
will more easily dislodge it. Kuntze
(1982) described tests made with various
additives included in the pipe polymer; at
the time, these had not been very
successful, but the concept has merit.

4.1.4 Air-sealed system

This method is referred to in agricultural
literature as the submerged outlet method.
If a drainage system can be kept
continuously and completely closed off
against the atmosphere except at its
outlet, then oxidation within the system
cannot occur, either by air or by aerobic
bacteria. Ochre can only form at the
outlet, where it may easily be removed.
It can form within the drain only if air
enters, and success in preventing clogging
depends on this not occurring. There are
three dewatering methods used in landslide
work for which this concept can be
applied:

  a. Wellpoints. Wellpoints and similar
pumping systems are mostly applied to
temporary construction dewatering
projects, but they may also be used for
permanent landslide control. An example
may be seen on a railway cutting
stabilisation project near Wollongong, New
South Wales. A wellpoint screen will be
substantially free of ochre clogging
provided its drillhole collar is
effectively sealed against the atmosphere;
ref Elliott (1991).

  b. Relief wells. A relief well is used
to drain a confined aquifer under artesian
pressure, and is sealed against the
atmosphere by having its intake
permanently below water level; see Fig 1,
which is the type of well used on a
landslide near Wollongong, New South
Wales, ref Forrester and Nyland (1980).

  c. Horizontal drains on negative slope.
According to rules laid down many years
ago by California Division of Highways,
horizontal drains have always been placed
on positive slopes of between 3 and 20%;
ref Smith and Stafford (1957). This is to
ensure that they are self-cleansing of
granular material. At a landslide near
Newcastle, New South Wales, one aquifer
could only be drained by placing
horizontal drains through it on negative
slopes, in effect forming relief wells;
ref Forrester (1987) and Fig 2. From
completion of the job in January 1986
until recently (1991), these drains were
constantly full of flowing water, and were
free of ochre except at their outlets. In
contrast, nearby drains placed on positive
slopes continually showed signs of

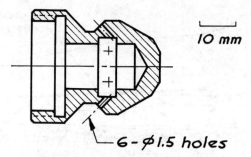

Fig 3. Sluicing nozzle, for
        normal applications

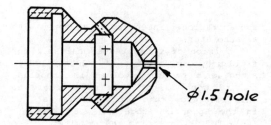

Fig 4. Sluicing nozzle, with
        forward-facing jet

internal ochre formation. During a recent
drought, however, all the drains ceased to
flow, and whether this will lead to ochre
clogging in the drains on negative slopes
has yet to be determined.

4.2 Post-construction methods

4.2.1 Chemical methods

Ochre deposits have been removed
successfully by filling drains with
sulphur dioxide solution for three or four
days; ref Grass and MacKenzie (1972).
There are probably other chemicals that
would also be effective. Care has to be
taken against causing environmental
problems when using this method, and there
are also risks for the people involved in
the work.

4.2.2 Sluicing

Wet ochre slime, and some solidified
ochre, may be removed by sluicing. Roads
and Traffic Authority use a sluicing
nozzle with six water jets pointing
radially and backwards; ref Forrester
(1980) and Fig 3. These draw the nozzle
into the drain, and at the same time spray

Table 1. Clogging risk

Ferrous iron content - ppm		Clogging risk
pH < 7	pH > 7	
< 0.5	< 1.0	None
0.5 - 1.0	1.0 - 3.0	Slight
1.0 - 3.0	3.0 - 6.0	Moderate
3.0 - 6.0	6.0 - 9.0	High
> 6.0	> 9.0	Very high

water into the drain slots. When sluicing relief wells and horizontal drains on negative slopes, a similar nozzle is used, with an additional single jet hole drilled through the front; see Fig 4. This mobilises granular material lying at the lower end of the pipe, thus allowing the radial jets to flush the material out of the drain. Similar nozzles are used by University of Florida; ref Ford (1974).

5 RECOMMENDATIONS AND CONCLUSION

When landslide control work involves drainage and ochre clogging occurs, the project is at risk of having its effectiveness substantially reduced, often within a relatively short time. An assessment of the ochre risk is therefore strongly advisable at the investigation stage. This requires measurement of the ferrous iron content of the groundwater, and its pH.

The most effective way of dealing with an ochre problem is to use one of the air-sealed methods. Where none of these is appropriate, clogging may be reduced by allowing water into the drain pipe only by means of cleanly drilled holes, 12 to 19 mm dia, along the top of the pipe. Otherwise, a sluicing program, consistently applied and using an appropriate nozzle, is the only procedure that is both easy to apply, and has a chance of keeping drains serviceable.

Little significant work appears to have been published on the effects of ochre on geotextile drain filters. This could be at least as critical as granular clogging, which has been the subject of very extensive examination.

6 REFERENCES

Elliott, J. 1991. Personal communication. Cumec Pty Ltd, Sydney.

Ford, H. 1974. Low pressure jet cleaning of plastic drains in sandy soil. Trans Amer Soc of Agricultural Engrs: 895-897.

Ford, H. 1982. Estimating the potential for ochre clogging before installing drains. Trans Amer Soc of Agricultural Engrs: 1597-1600.

Ford, H. & Altermatt W.E. 1986. Hancor Bio-Flow tubing, an aid against biological clogging of underdrains. Hancor Inc, Findlay, Ohio.

Forrester, K. 1980. Discussion, Proc Int Symp on Landslides, Vol. 3, New Delhi: 256-257.

Forrester, K. 1987. The Carisbrooke Avenue landslide. In: Walker, B.F. & Fell, R. (eds) 1987. Soil Slope Instability and Stabilisation. Sydney: 337-346. Rotterdam: Balkema.

Forrester, K. & Nyland, G. 1980. Two landslides on New South Wales highways. Proc Int Symp on Landslides, Vol. 1, New Delhi: 181-184.

Grass, L.B. & MacKenzie, A.J. 1972. Restoring subsurface drain performance. J Irrig and Drnge Div, ASCE, Vol. 98, No. IR1, March 1972: 97-106.

Hausmann, M.R. 1990. Engineering principles of ground modification. New York: McGraw-Hill.

Kuntze, H. 1982. Iron clogging in soils and pipes, analysis and treatment. Deutscher Verband fur Wasserwirtschaft und Kulturbau, Bull.10. English translation from Hamburg: Paul Parey and London: Pitman.

Smith, T.W. & Stafford, G.V. 1957. Horizontal drains on California highways. J Soil Mech and Foundns Div, ASCE, Vol. 83, No. SM3, July 1957: Proc Paper 1300.

Taylor, G.S. 1977. Ochre is the ogre. Drainage Contractor 3: 1: 79-80.

APPENDIX

The field method given by Kuntze (1982) for assessment of the risk of clogging is as follows:

a. Take a fresh sample of the water and filter it.

b. Flush out a small reagent bottle several times with the sample.

c. Place 5 ml of the sample in the bottle.

d. Add 5 drops each of:

(i) Buffer solution - glacial acetic acid (50 ml to 100 ml water) and 40 g ammonium acetate.

(ii) 0.5% aqueous solution ortho-phenanthroline hydrochloride.

e. Shake the mixture well after each addition.

f. After at least 10 minutes, or up to 30 minutes in air temperatures below $8^{\circ}C$, determine the ferrous iron concentration in parts per million by comparing the colour of the sample with a standard colour scale.

g. Complete the test as quickly as possible, so as to minimise oxidation of the ferrous iron through contact with the air.

h. To determine the concentration of total iron, both ferrous and ferric, add 20% aqueous hydroxylamine hydrochloride solution as a reducing agent to a new sample before adding reagents (i) and (ii), then proceed as before.

i. Determine the pH of a new sample with a portable pH meter.

j. Estimate the clogging risk from Table 1.

Landslides, Bell (ed.) © 1995 Balkema, Rotterdam, ISBN 90 5410 032 X

Stabilization of a slide movement near Urbino, Italy*

U.Gori & F.Veneri
Institute of Applied Geology, Urbino University, Italy

ABSTRACT: The present paper deals with a slide case provoked by an anthropic intervention for a slope digging. The immediate execution of a diaphragm wall of reinforced concrete with a large diameter made in order to avoid damages to buildings, has not produced any slide containment because it has partially turned over. One can relate this phenomenon to the inadequate deepth of the diaphragm in the stable earth, due to the objective difficulty to drill the compact rock of the substratum. A little further up a new diaphragm wall of reinforced concrete piles has been made with a small diameter, fixed in the rock trough rotary-percussive drilling. The work has been completed by the execution of anchors on the horizontal steel beam connecting the piles and a draining intervention.

1 INTRODUCTION

In the last twenty years, in the area studied in this paper many residential buildings have been built, with loft buildings, roads, squares, etc.

During the digging of a declivity a slide has taken place with a fast and dangerous evolution. The translated rockfall formed an eluvial- colluvial detrital accumulation almost 7 m thick. The underlying rock formation, located with dipslope strata attitude and layers immersion like the angle of natural declivium, seems to be constituted by very fractured layers of marly limestone, with sandy silts and small levels of smectic clays of millimetric thickness (Bisciaro formation). In order to cope with the progress of the accident, which, in just a few days had provoked deep cleats along the slope as to threaten some buildings, a diaphragm wall was built at the bottom of the slide with a 120 cm diameter and connected on top by a horizontal steel beam.

Because of the objective difficulties to perforate a large diameter of the calcareous-marly bedrock and having the necessity to intervene urgently, the perforation was interrupted at a depth of about 8.5 m, that is at 1-2 m within the stable rock.

The diaphragm wall so made and not fixed in enough at the bottom, after one month from the completion, began to be deformed and to translate some metres for artificial digging, as to overturn almost completely in correspondence with the main lateral split of the slide. Very quickly, a new diaphragm wall near the damaged one, 9.5 m deep and with a 28 cm diameter, was executed by

(*) National Research Council G.N.D.C.I. Publication n 390 – Responsible U.O. 2/47 U. Gori

air-compressed rotary-percussive drilling. Soon, the horizontal steel beam of the diaphragm wall was anchored and the water flows were stabilized by a 4 m-deep drainage. After ten years, the new retaining work has proved to be very efficient and statically perfect.

2 GEOLOGICAL AND MORPHOLOGICAL FEATURES

In the slide area previously described and in the surrounding ones the following outcropping formations have been recognized, which are part of the umbro-marchean stratigraphic succession: Bisciaro, Schlier, Marnoso-arenacea (Fig. 1).

The following, main lithologic characteristics will be described, starting with the most ancient terms of the series. Furthermore, the Scaglia cinerea will be also analysed because, though it does not outcrop in the studied area, it is the stratigraphic base of the Bisciaro fmt. and, as will be seen afterwards, has strongly influenced its stability.

- Scaglia cinerea Formation (Priabonian-Early Aquitanian).

This formation is constituted by a rotation of calcareous marls, marls, clayey marls and, in a minor degree, marly limestones. The marly-calcareous litofacies, like the marly limestones, can be prevalently found in the lower portion of the unit, whereas the marly-clayey deposits prevail in the upper part. The thickness of the Scaglia cinerea, which, as already said, never outcrops, according to published data, is about 150-200 m.

- Bisciaro Formation (Aquitanian p.p. - Langhian p.p.).

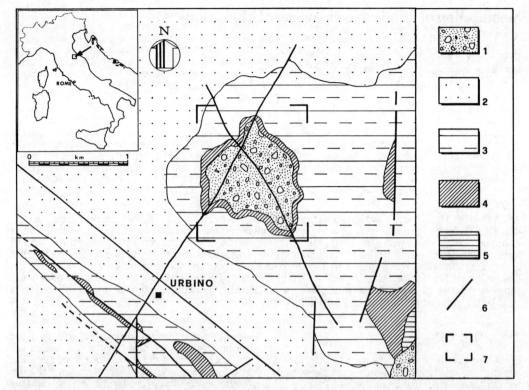

Figure 1. Geological setting of the Sasso area. 1) Alluvial, colluvial and detrital covering; 2) Marnoso-arenacea formation (Tortonian p.p.- Early Messinian); 3) Schlier formation (Langhian p.p.- Tortonian p.p.); 4) Bisciaro formation (Aquitanian p.p.- Langhian p.p.); 5) Scaglia Cinerea formation (Priabonian-Early Aquitanian); 6) Faults; 7) Studied area.

This formation outcrops, with continuity, all along the slope edging the studied area.

It is constituted by medium layers of limestone and greyish marly limestones, marls, calcareous marls, and alternating grey clayey marls.

In the calcareous strata at the base there can be frequently found black or grey chert layers and/or nodules. Furthermore, the whole formation shows numerous volcaniclastic levels having a thickness from 10 to 50 cm. These are mostly constituted by fragments of volcanic glass with rhyolitic rhyodacitic composition, often altered into clayey minerals (smectites) or replaced by calcium carbonate (Guerrera, et al., 1986).

The thicknesses and the lithologic associations of the Bisciaro fmt. change locally. As regards the thicknesses, complete data cannot be furnished, because the base of the formation is never outcropping in the studied area. Nevertheless thickness measures in some areas not far from the one analysed show values varying between 40 and 60 m. As to the associations (Guerrera, 1977), three distinct members can be recognized on the basis of lithological associations, which are characterized as follows:

- lower member: alternating chert-rich marly limestones ($CaCO_3 = 80\%$), and greyish marls, with a limestones/marls ratio of about 1; the thickness of each calcareous bed varies from about 10 cm to 1 m; moreover, there can be found numerous volcaniclastic levels.

- intermediate member: it is constituted by greyish marls ($CaCO_3 = 60\%$) containing sporadic calcareous levels (limestones/ marls ratio = 0.2).

- upper member: lithologically very similar to the lower member, it differs for the total lack of chert lists, the smaller thickness of the calcareous layers (from 10 to 30 cm) and the numerous levels with a "tripolaceous" aspect.

The stratigraphic succession with formations less resistent to erosion, that is Scaglia cinerea and Schlier, respectively at the base and at the top of the Bisciaro fmt., has caused a rigid behaviour of this latter whose most evident effect, due to tectonic stresses, is given by a diffuse and deep fracturing in the whole formation.

- Schlier Formation (Langhian p.p.-Tortonian p.p.)

It is constituted by alternating marls, calcareous marls, clayey grey marls, and, subordinately, whitish marly limestones, at times finely detrital, but always having a small thickness. In the upper part of the formation, the calcareous-marly levels tend to progressively diminish so as to disappear completely near the boundary with the overlying Marnoso-Arenacea fmt.

The stratification is not very evident because of the scarce differentiation of lithologic types and the deep fracturing and diffuse cleavage of the formation.

The thickness of the Schlier fmt. seems to be smaller than the average values found in other areas and can be estimated around 80 m, and however it is always less thick than 100 m.

Detritic accumulations are frequent and abundant, because of the particular marls' alterability provoked by weathering.

The Schlier fmt. outcrops close by the upper edge of the slope delimiting the Sasso area.

- Marnoso-Arenacea Formation (Tortonian p.p.- Early Messinian)

This formation consists of various lithotypes which, associating in a different way from one area to the other, give to this unit dissimilar aspects and physico-mechanic behaviours.

They are coarse to fine grained turbiditic sandstones, with thin to very thick and/or amalgamated layers (2 to 5 m) and tabular geometry, with interbedded grey-blue pelites or hazel marly siltstones, partly with a torbiditic origin, with thicknesses varying from a few centimetres to some metres. The sand/mud ratio varies, but it is always in favour of the former ones. The total thickness of the Marnoso-arenacea fmt. is not measurable since, in the studied sector, the top of the formation is never visible. However, one can locally calculate the various partial thicknesses, by considering the attitude and the structural situation. Thus, as to the Marnoso-arenacea fmt. of the studied area, it is possible to calculate thicknesses not lower than 350 m.

The formation outcrops extensively close by the examined area.

From a structural point of view, the studied area can be located on the north-western end of the Cesane anticline. This structure is a large and short anticline, with walls with small declivity and a weak south-westward vergence. The anticline begins a little NW of Urbino within the Marnoso-arenacea fmt. and has its summit on the Cesana Mountains, continuing south-eastwards and disappearing near the Cesano River.

Two faults, almost perpendicular to each other, can be also found in the Sasso area. The first one, with apenninic direction, (NW-SE), seems to cause, as a most evident phenomenon, a relative movement raising the south-western part of the fault and lowering the north-eastern one, as one can clearly see both from a limited outcrop affected by the fault itself, and, more generally, from the matching, within the Bisciaro fmt., of lithologies usually overlying each other.

The second fault develops SW-NEwards, comprehending also the historical centre of Urbino and provokes the lowering of its north-western part, as one can easily understand by observing the

shift of geological limits across of the fault itself. This movement occurs over the whole northern Marchean sector, where there can be noted the drowning and immersion of the two main carbonatic ridges, to which the Cesane anticline belongs itself, costituting the framework of the Umbro-Marchean Apennine.

In particular, Fig. 2 shows the geomorphological setting of the area, delimited by morphologies differentiated and separated by the fault with appenninic pattern. The south-western sector shows a more uneven morphology where there can be recognized big formational detrital blocks keeping, in limited extensions, the stratimetric characters of the Bisciaro fmt. in that area. These disarticulated rock masses are the result of several overturning and translation phenomena, probably connected with two concomitant factors:

a) presence of intense and deep fracturing in the Bisciaro fmt. with fluid percolations;

b) presence of a formation (Scaglia cinerea) at the base of Bisciaro fmt. with strain-plastic behaviour.

Both factors have determined a slope evolution similar to that of the lateral expansions (m 2 - m 3, Varnes, 1978).

The presence of diffused reverse slope testify to the gravitational action of the past, whereas the present processes are well visible along the southern edge of the area and along the Apsa torrent course. In this, traces of paleobed are evident, which was probably abandoned after the described processes of lateral expansion. It should be also noted the presence of a consistent detrital stream of a complex slide-type (s 1, Varnes, 1978), in the most southern portion of the area, by now stabilized without any particular intervention. At last the two stressed slopes delimiting this sector show local rock downfalls amassing at the bottom; the presence of a small talus cone is not connected with any natural phenomenon, but with an anthropic action (burrow of heterogeneous soils).

The north-eastern sector of the area shows morphological characteristics better differentiated from each other than in the previous sector. In detail, one can note a greater morphologic uniformity, characterized by a very regular declivity, which is delimited by two pseudoparallel streams joining into the Apsa torrent (northern edge of the area). The southern portion of the slope does not show any clear morphologic delimitations. The most plastic character of the present eluvial-colluvial covering (thickness up to 7-10 m) is probably connected with a different lithology of the Bisciaro fmt. (marly facies referrable to the intermediate member), together with a layer component in dipslope strata.

The deformative processes are remarkable, with solifluction phenomena and translative creeps (q 2, j 3, Varnes, 1978). The anthropic action for the urbanization has caused considerable slopes and detrital accumulations, sometimes about 10 m thick. The NE slope separates the above mentioned area from the rock formation through a

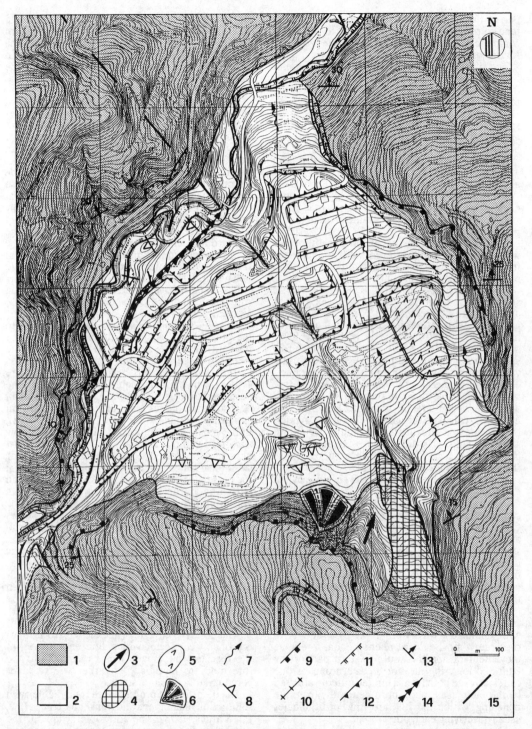

Figure 2. Geomorphological setting of the Sasso Area. 1) Rock Formations; 2) Alluvial, colluvial and detrital covering; 3) Lateral spreads; 4) Complex slide; 5) Translational slide described in this paper; 6) Talus cone; 7) Solifluction; 8) Reverse slope; 9) Rock fall slope; 10) Tension fracture; 11) Edge of torrential erosion escarpment; 12) Anthropic slope; 13) Dip and strike of strata; 14) Trace of paleobed; 15) Fault.

Figure 3. The tensional fracture of the slide (left flank)

very marked ditch. In the central part of this sector there is a slide movement with rapid evolution, which will be dealt with in detail afterwards. In the underground of that area there is a groundwater with a depth between 7 and 10 m from the top surface. This sensitively feels the effects of the meteorologic and seasonal vicissitudes, due to its strong connection with meteoric flows (an average of 800 mm of annual rainfalls), which feed it directly because of both the high permeability for fracturing, and the direct absorption of the detrital and eluvial-colluvial covering. The nappe exploitation occurs discontinually and through a few private wells, used for agricultural usage. The scarce diffusion of these water wells has not allowed, in this phase of the study, to draw up an even, schematical map of the aquifer geometry.

3 KINEMATICS OF THE SLIDE MOVEMENT

The studied area shows coplex evolutive characters which are sharply distinguished from each other in the south-western sector and in the north-eastern one. While the slide phenomena which can be noticed at present are a result of past events and one can reconstruct their story by observationa and geomorphologic interpretation of the place, in the

autumn of 1979 a rapid crumbling event occurred with immediate effects on the industrial and residential buildings in that area. This movement began after a digging made in order to enlarge a building area and at the beginning it occurred on about 5,000 square metres with a depth from 5 to 7 m. The slip surface has resulted to be pseudo-plane and almost correspondent to the bedding plane one of the rock formation.

At the beginning, after about 30 days from the anthropic intervention, there could be noted a swelling of the slip about 2 m deep. Soon after, a main NW-SEwards cleavage formed, going from the bottom of the bank 200 m upwards and having a maximum width of 1.5 m (Fig. 3). Afterwards a ground downstream translation was noted, with a displacement of a few metres and the involving of a mobilized volume calculated to be about 140,000 mc.

A new balance phase reached by the slope for the ground translation, allowed us the reconstruction of a part of the stripped base of slope, thus determining a quiescent phase of the movement. During this standstill period it was possible to intervene with a first urgent consolidation work, in order to avoid subsequent evolutions of the phenomenon which would have endangered some private buildings and firms.

In the wide survey of the various intervention types practicable in unsettled areas, the movement components and the geometrical data of the phenomenon can be easily recognized. The following aspects have been adopted as leading criteria:
- necessity of a rapid and accurate intervention only in the area of the slide threatening directly the buildings;
- an immediate checking of the effect of the intervention on the stability of that place;
- a reduced tampering of the slide front with sample interventions in order to avoid further gravitational evolutions;
- restrained diggings in depth in order not to cause further spatial discontinuities in the underground.

This series of criteria conditioning the success of the reinforcement intervention, led us to adopt a support consisting of a large diameter reinforced concrete diaphragm wall, anchored at the bottom and free at the top.

The slide phenomenon has resulted to be clearly connected with the digging of the slope and has developed along the point of contact between the formation at the base (Bisciaro) and the detrital-colluvial surface covering. The digging occurred on a slope with a 10 degrees acclivity, and probably affected, in the slip zone, some small levels of smectic clays made plastic by the presence of water.

In fact, there exists a kind of free will in ascribing a physical, sedimentological and mineralogical significance to both a complex and heterogeneous formation such as the Bisciaro one. Some published granulometric data (Ardanese & Grandi, 1978) mention the following percentages for the calcareous marly fraction and for the volcaniclastic levels, both dissolved in diluted HCl:

Lithology	sand %	silt %	clay %	CaCO₃ %
Calcareous marls	11.1	26.2	62.7	62-79
Volcaniclastites	15.5	19.0	65.5	2-42

As to the mineralogical composition referred to marls, the prevalence can be noted of montmorillonite in the clayey fraction, and subordinately, the illitic component partially in mixed layers with montmorillonite. In the same way in the clayey fraction of volcaniclastites there can be pointed out the constant presence of montmorillonite (50 to 70%), mixed layers of illite-montmorillonite (30%) and 10-15% illite (Guerrera et al., 1986).

Some analyses of geotechnic characterization have been carried out on some samples collected at various depths. The presence of numerous lithologic heterogenities (muds, clays, sands and calcareous-marly detrituses) found in the tests has allowed us to obtain some data on samples rearranged and reconstituted in the laboratory. The following list summarizes the average values obtained from the analysis of 18 samples.

N. of samples	W %	γ KN/m³	γd KN/m³	LL %	IP %	A %
18	32	18	13.6	34.7	16.6	1.1

Tests of direct shearing C.U. of the remoulded samples in the tensional field up to 200 KPa and deformation speed of 0.005 mm/min have showed values of frictional angle extremely variable between 4 and 31 degrees, with cohesive values from 2 to 8 KPa. For the estimation of the residual frictional angle to be used in the calculation of earth pressure, a value of 14 degrees has been employed, corresponding to the inclination of the slip surface of the substratum. The same value seems to be in agreement with what is deducible from the correlations with the index properties (Kanji, 1974; Cancelli, 1979).

4 METHOD OF STABILIZATION

In the following the two different types of works realized, since the first intervention was unsuccessful, will be separately described.

4.1 Diaphragm wall with large diameter

The intervention consisted in the realization of a discontinuous reinforced concrete pile structure with a 120 cm diameter, spaced 150 cm and connected on top by a horizontal steel beam having

Figure 4. View of the diaphragm wall partially traslated

a rectangular section of 120 x 80 cm (Fig. 5). The expected depth for the work, in a projection phase, was 11 m; of which 4 m was used for the anchoring into the stable bedrock. For economical reasons, the extent of the intervention has been restrained into 25 m in order to avoid the spreading of the phenomenon to the area of the artisans' buildings. For topographic reasons, the support has been located by the upper edge of the digging slope. As a drilling technique for the execution of the diaphragm wall, many different excavation tools were used, according to the different lithologies found, always by rotating and moving away the ground. The excavation was stopped at a depth of about 8.5 m, for the objective difficulty to go on to a major depth for the presence of very compact calcareous and calcareous-marly levels. We also tried to demolish manually, by using a pneumatic drill in the hole, the calcareous lithoid and calcareous-marly component of the layers, without any appreciable result, but with some risk for the safety of people working these.

Thus having to intervene rapidly and for the approaching of the winter season, we decided to finish the work, without an adequate anchor in the bedrock. Soon after, we planned to make some anchors of the horizontal steel beam of the diaphragm wall, so as to make an even partially static scheme of anchoring at the base and of support at the top. In order to estimate the underground tensions, a 14 degrees residual frictional angle and no cohesion was adopted for the unstable 7 m thick cover. Because of the high fracturing and permeability of the material present in the slide area and because of the spatial discontinuity of the diaphragm wall (piles spaced 30 cm apart), the contribution of the water pressure probably present underground was not considered in our calculation hypothesis.

About 30 days after the completion of works, some splits appeared on the horizontal steel beam of the piles, which at first were considered as a settling phenomenon. After a few days we witnessed a progressive rotation of 2/3 of the diaphragm wall, till its partial topple towards the artisans' building (Fig. 4).

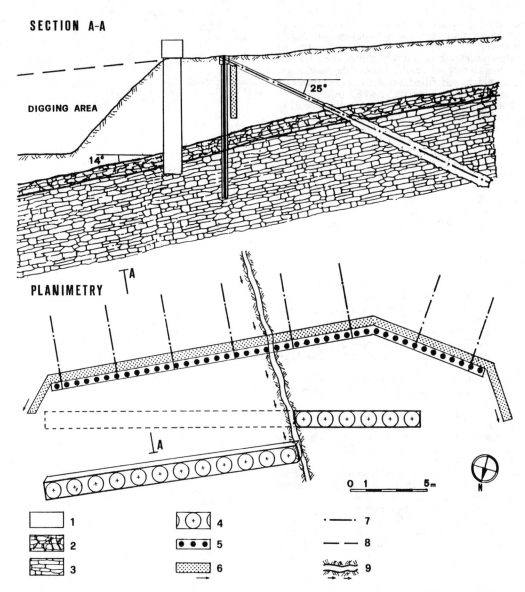

SECTION A-A

DIGGING AREA

25°

14°

PLANIMETRY

A

A

0 1   5 m

N

	1			4			7
	2			5			8
	3			6			9

Figure 5. Intervenctions of stabilization of the slide movement. 1) Detrital covering; 2) Fractured Bisciaro formation; 3) Bisciaro formation; 4) Large diameter diaphragm wall; 5) Small diameter diaphragm wall; 6) Drainage trench; 7) Anchors; 8) Original top surface; 9) Principal crack.

The lack of anchorage at the base of the structure and the quick and unexpected evolution of the slide, which did not allow the realization of the bearing anchors of the horizontal steel beam, still seem the only causes of the failure.

4.2 Small diameter anchored diaphragm wall

A little higer than the preceding work (Fig. 5),

using the same earth pressure calculation scheme, a new diaphragm wall 30 m length with piles of 28 cm of diameter, spaced 55 cm apart and 9.5 m depth has been realized. The reinforced concrete reinforcement of the diaphragm wall consisted in vertical bored piles reinforced with steel H beams (HE 140 B), connected on top by a very rigid horizontal steel beam. This time, the excavation technique for this intervention has used a rotary-percussive drill with air-compressed, so to be able to easily pierce the calcareous and calcareous-marly layers present beyond 7 m depth. In order to control the possible presence of underground waters and to stabilize their level, a draining cutting was realized above the diaphragm wall, which was filled up with coarse grained material, down to 4 m depth and 0.5 thickness.

Having finished the diaphragm wall, 8 bearing anchors of the horizontal steel beam were made spaced 4 m apart. The Tirsol I.R.P. anchors were 20 m long and divided into 10 m of fixed anchor length and 2 m inflatable packer bag. The anchors inclination on the horizontal direction was 25 degrees, with nominal tension of 60 tons each.

Ten years ago, the described work proved to be extremely efficient and able to oppose the slide movement, and no other interventions have been made, even of water surface regimation.

REFERENCES

Ardanese R.L. & Grandi L. (1978) - *Caratteri sedimentologici, mineralogici e chimici di una serie del Bisciaro (Miocene inferiore e medio - Appennino centrale)*. Boll. Serv. Geol. d'It., 98: 3-60.

Cancelli A. (1979) - *Sulla misura della resistenza residua dei terreni mediante prove di taglio lungo un contatto suolo-roccia*. Geol. Appl. e Idrog., 14 (3): 351-365.

Guerrera F. (1977) - *Geologia del Bisciaro dei Monti della Cesana (Urbino)*. Giorn. Geol., 42 (1): 109-132.

Guerrera F., Tonelli G. & Veneri F. (1986) - *Caratteri litosedimentologici e mineralogico-petrografici di vulcanoclastiti mioceniche presenti nella successione umbro-marchigiana*. Boll. Soc. Geol. It., 105 (3-4): 307-325.

Kanji M.A. (1974) - *The relationship between drained friction angles and Atterberg limits of natural soils*. Geotechnique, 24: 671-674.

Varnes D.J. (1978) - *Slope movements types and processes*. In "Landslides: analysisand control". Transportation Research Board, National Academy of Science, Special Report 176, chapter 2.

Landslides, Bell (ed.) © 1995 Balkema, Rotterdam, ISBN 90 5410 032 X

The development and implementation of a landslide management strategy for the Ventnor Undercliff, on the southern coast of the Isle of Wight, England

Robin G. McInnes & Graham McIntyre
Engineering Services, South Wight Borough Council, Isle of Wight, UK

ABSTRACT: The Isle of Wight, situated off the south coast of England, is an area of outstanding interest to the geologist and geotechnical engineer. Rocks vary in age from Cretaceous to Recent and have been subjected to uplift, folding and compression. Subaerial and in particular coast erosion has encouraged instability and landslippage in a number of locations.

South Wight Borough Council is the local authority responsible for some 88 km of cliffs of general weak and often poorly lithified rocks. As coastal protection authority, it is involved in a major ongoing programme of civil engineering works. An area of particular interest is the Undercliff along the south coast between Blackgang in the west and Bonchurch in the east which includes the town of Ventnor. This area of coastal instability has recently been the subject of detailed study by the Department of the Environment as part of their planning research programme developing improved methods of landslide hazard and risk assessment.

The Council has been closely involved with this project and has produced a landslide management strategy for the Ventnor Undercliff. This paper describes the development and implementation of this strategy and some of the questions it has posed for the Council and local residents.

INTRODUCTION

The Isle of Wight is situated midway along the south coast of England and is separated from mainland England by the narrow strait of water known as the Solent. With an area of approximately 385 km^2 the Isle of Wight is renowned for its extraordinarily varied geology and coastal scenery. With a coastline of 105km, the Island is 37km from east to west and 21 km from north to south.

Administration at a local level is by the two Borough Councils. Medina Borough Council occupies the north-eastern quadrant and includes the county town of Newport and the towns of Cowes and Ryde. South Wight Borough Council administers the larger but slightly less populated remaining area which includes the popular holiday resorts of Sandown, Shanklin and Ventnor. With responsibility for a coast of 88km, South Wight has one of the longest of any Borough in the United Kingdom.

The folded, uplifted and compressed rocks range from Cretaceous to Recent and have been subjected to agressive erosion along the southern coasts. This has resulted in spectacular land forms such as the Needles rocks and has also promoted coastal instability along the Ventnor Undercliff.

Although South Wight Borough Council is responsible for much of the northern coast of the Island the contents of this paper is restricted to that part of the southern coast known as the Ventnor Undercliff which lies within the Undercliff belt between Blackgang in the west and Bonchurch in the east. There are numerous examples of coastal instability within the Isle of Wight, these have discussed elsewhere (McIntyre and McInnes) and this paper concentrates on experience gained in recent months on landslide management in the Ventnor area.

Responsibility for management of the coastline and dealing with problems of coastal instability rests with the Department of Engineering Services of South Wight Borough Council based in Ventnor. This location is particularly

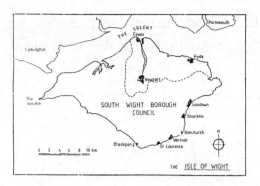

FIG.1

FIG.2 Ventnor looking inland over the town

appropriate as the Council has played a major role in developing an understanding of the coastal landsliding and in producing a landslide management strategy.

1 GEOLOGICAL BACKGROUND

1.1

The rocks to be found in the Ventnor Undercliff lie within the Upper Cretaceous system, the prominent sequences being the Upper Greensand and the Gault Clay. The scenery of the Undercliff is dominated by the dramatic rear scarp face in which the near horizontal stratification of the chert beds of the Upper Greensand are particularly well displayed. From this scarp a series of blocks or terraces have slipped forward on lower sequences aided by the processes of erosion. In other areas mudslides have occured whilst reconsolidated landslide debris forming much of the cliffline along this frontage.

1.2

The geology and geomorphology of this fascinating area has been described by a number of authors (including Hutchinson, Chandler, Bromhead, Brunsden, Lee and Moore). The recently published report by the Department of the Environment entitled 'Coastal Landslip Potential Assessment - The Ventnor Undercliff' has linked this research into a comprehensive working document.

1.3

The scenery that results from the remarkable geological history of the Undercliff was a major factor in the development of Ventnor. The series of

landslide benches were considered ideal sites at that time for terraces of properties, each allowing magnificent sea views. By the 1880s the town had developed into a popular health and seaside resort but this rapid expansion brought with it haphazard planning and little attention to the need for proper public services. This resulted in considerable problems with human activity aggravating existing landslide conditions and leaving a situation which is only now being tackled for the first time.

2 THE LOCAL AUTHORITY ROLE

2.1

South Wight Borough Council is the authority responsible for a variety of statutory duties and other functions which relate to landslide management including:
   a) the role as local planning authority
   b) responsibility for building control and protecting the public from dangerous structures.
   c) coast protection authority
   d) Agents for Southern Water Services Ltd with respect to sewerage

   e) owners of considerable areas of public land and property within the Undercliff
   f) being the point of contact for the local community.

2.2

As a result of the involvement of the Director of Environmental and Engineering Services in these aspects he was requested by the Department of the Environment (DOE) to aid with the selection of suitable consultants for their two year study into coastal landslip potential in the Ventnor Undercliff.

2.3

The purpose of the study was not necessarily to produce a site specific appraisal of land movement in the Ventnor area but to develop a national strategy for 'Development on Unstable Land' (now published as Planning Policy Guidance Note 14). Ventnor was an ideal choice for this research, being a moderately sized town subject to coastal landsliding. However, in view of the amount of detailed factual information gathered and the positive response from the Council, their final report fulfilled both objectives. This research has now enabled the Council to develop a landslide management strategy for the Ventnor area.

3   A LANDSLIDE MANAGEMENT STRATEGY FOR VENTNOR

3.1

In the broadest terms the findings of the DOE study revealed that whilst Ventnor had a reputation for land movement, large areas of the town had remained virtually unaffected. In many areas buildings of considerable age had survived with negligible or only slight damage. Many of the Victorian properties were poorly built and were constructed in ways that took no account of ground conditions. The result of this was that some instability problems appeared to be worse than was actually the case. There are of course some unstable areas in the town where building should be avoided but developers should have confidence in the town hand in hand with a sensible landslide management strategy.

3.2

The steering committee for the Ventnor study realised that it was of prime importance to ensure that their findings were published in a carefully coordinated way; particularly in view of the sensitivity of the subject for some property owners. The first of seven main objectives for the strategy was therefore education and information.

  a) Education and Information

The Ventnor report was published in three formats of varying complexity to suit respective audiences. For general information in the broadest form there was an attractive, eye catching, four page colour leaflet which was available free of charge. For the educated layman, a 65 page book entitled 'Ground Movement in Ventnor' was availalbe at a price of £6. This edition included photographs, sections and maps.

The full technical report ran to over 60,000 words and was published with four sets of maps on a scale of 1:2500 covering land use, geomorphology, ground behaviour and planning guidance.

Education involved explaining the findings of the report to a variety of audiences including Councillors, the general public, insurance companies, estate agents, the press, solicitors, architects, builders and planners. Each of these groups would be interested in differing aspects of the report to a lesser or greater degree and appropriate emphasis had to be given at these presentations.

In the past, the problems of land movement and resulting damage to property and services had been suppressed to some degree. Sensational reporting of land movement events in the past and their linkage to other far more active sites at Blackgang and Luccombe at the extremities of the Undercliff had been a problem. Markedly differing processes were taking place there but the result was that many insurance companies blighted the whole area and ill-informed speculation by some local residents increased concern.

Unfortunately a lack of full understanding of the problems meant that such comments were not countered by the local Council over the years.

Subsequently any structural problem within the Undercliff was automatically assumed to be as a direct result of land movement regardless of the evidence. It can be seen therefore that the steering committee and the Council viewed the correct presentation of the report as a matter of prime importance.

The support of the insurance companies was regarded as an essential first step in restoring confidence in the town of Ventnor. As a result the first of a series of presentations over one week was made to the leading insurance companies in London. The format for each presentation was as follows:
  1) Introduction by DOE Steering Committee Chairman
  2) Presentation by consultants who undertook the study
  3) The local Council view and role presented by the Director of Environmental

FIG.3 Ventnor from the Western Cliffs.
This section of cliffline will
receive rock protection shortly

FIG.4 A £1.3 million rock breakwater and
beach nourishment scheme will
commence shortly along this frontage
at Monks Bay, Bonchurch. This
section of coast is affected by
severe landslippage

& Engineering Services
  4) Summing up by the DOE
  5) Questions
  Following a successsful presentation to
insurance companies a series of meetings
were arranged for the same week on the
Isle of Wight. These were made to local
councillors from South Wight Borough, the
Isle of Wight County Council and Ventnor
Town Council, the local press and radio,
and finally building societies, estate
agents, solicitors and consulting
engineers; the illustrated presentations
being well received locally. A further
series of more specialised workshop
sessions have since been held with
builders, architects and staff from the
planning and building control departments.
  Perhaps the most interesting and
innovative idea for developing public

awareness of the findings of the Ventnor
report was the suggestion made to the
Council by the Director of Engineering
Services that a Ventnor Landslip
information centre should be opened in the
town. This would give local people the
opportunity to view a display and discuss
in confidence with consultants the
findings of the report and any concerns
they might have about their own
properties.
  The Council was somewhat concerned about
the public reaction to the opening of the
information centre in Ventnor High Street.
There was initially some resistance from
an extremely small number of residents but
the vast majority welcomed the initiative
and regarded it as long overdue.
The publication of the planning guidance
and other maps on a scale as large as
1:2500 where individual properties could
be clearly identified and were allocated
to zones was obviously a sensitive
subject. However, for those property
owners in the most vulnerable areas the
problem was often already known to them.
By the end of the first week some 200
residents had visited the information
centre and a number of insurers, loss
adjusters, consulting engineers and estate
agents also returned for more detailed
discussions. After 8 weeks the volume of
visitors had reduced significantly and
therefore opening hours were reduced from
a 5 day week to 2½ days. However it is
believed that most households in Ventnor
will have visited the centre during this
period.

b) Coastal Protection

Protection of the toe of the Ventnor
Landslip from marine erosion is of course
of prime importance and to its credit the
Council has recognised the need for this
and made a substantial budget provision.
Removal of rocks from the foreshore to
construct the ill-fated Ventnor Harbour in
1863 had a devastating effect on Ventnor
Bay through accelerated erosion at that
time. This was a reminder of the need for
proper foreshore management as well as
coastal protection.
  Severe landslippage within cliffs of the
Carstone series at Monks Bay, Boncurch as
a result of winter/spring storms in 1990
presented a hazard to adjacent properties
as well as significant land loss.
A £1.3 Million scheme for this 0.4 km
frontage has now been prepared with work
to be commenced before winter 1991. This
will involve the construction of a
substantial rock breakwater parallel with

the coast at low water mark protecting a tombola-shaped artificially nourished beach. Existing groynes will be given added protection with rock haunching. The Gault Clay, which overlies the Carstone and forms the coastal slope, will require cliff reprofiling and counterfort drainage.

The other remaining unprotected frontage along the Ventnor Undercliff is the Western Cliffs immediately to the west of Ventnor Esplanade. Here also, rapid undercutting of the 25-30 m high cliffs of reconsolidated Chalk and Greensand landslide debris has taken place. Uncontrolled erosion could result in the loss of amenity land, the coastal footpath, and a car park, but more important have a destabilizing effect on the land behind which is heavily developed with residential properties. A £1.2 million rock protection scheme is scheduled for completion during the coming winter.

c) Control of water in the ground

Reduction of water entering the ground is a particularly important aspect of the landslide management strategy. The prevention of leakage from water pipes and sewers will help to lower ground water levels and the consequent risk in times of heavy rain. The link between antecedent rainfall and landsliding events has been established and documented in the DOE report. Therefore elimination as far as possible of uncontrolled water entering the ground can only improve the situation.

The Isle of Wght had been selected as a trial area in the UK for water metering. This involved renewal of a number of service pipes and also gave residents the incentive to be careful with their use of water and to be alert for leakage. It was originally intended that as part of the trial, half the water meters should be sited inside properties and half outside, usually on property boundaries. The Council suggested to Southern Water that in view of ground conditions in Ventnor, all meters should be sited outside the properties. The onus would then rest on householders for ensuring that the supply pipes within their gardens are not leaking.

The reporting and rapid repair of pipes was much more likely if the property owner was responsible. Leakage from water pipes is generally in excess of 25% and could be considerably greater in a landslip area such as the Undercliff. Southern Water has been requested to look carefully at the questions of leakage and urgently consider a programme of replacement for old cast iron mains by new mains of flexible construction.

The rapid development of Ventnor between 1830-1870 following the breakup of the Ventnor estate meant that little attention was given to town planning. Sewers were laid in a haphazard manner and many have suffered as a result of ground movement.

The Council as sewerage agents for Southern Water Services Ltd has promoted and commenced a major programme of sewer renewals on Ventnor Esplanade and adjoining streets. Other sewers are likely to be renewed as part of a rationalisation of the system required for construction of a new sea outfall to comply with EC regulations.

Closed circuit television film of most main sewers in the town is now available for detailed examination enabling a programme of priorities for renewal to be drawn up. Residents will be encouraged to renew sewer services pipes and ensure the adequate disposal of roof water at the same time. The County Surveyor has agreed to examine all road drainage in the town to ensure that the system is sealed, replacing brick pits with preformed flexible jointed gully pits where necessary.

Other areas of concern include the need to extend the system of main sewers to include the whole of Bonchurch and St Lawrence providing first time sewerage for an additional 350 properties. Finally a regular system of inspection is required for swimming pools to ensure that pools are properly sealed and perhaps more important, a suitable point of discharge for emptying.

d) Code of practice for repairs of property and new development

The Council is concerned that the blight on land and property in the Ventnor area which has existed as a result of a lack of knowledge and understanding of the problem should now be lifted. New development can take place in most areas of the town if sensible note is taken of the contents of the report with building developments taking due account of ground conditions. The planning guidance maps offer advice to developers on the likely requirement for geotechnical investigation prior to development in different parts of the town.

As there are few 'green field' sites remaining, much of the building work will be refurbishment or re-development. In

FIG.5 Sites of former properties in the
vicinity of the Lowtherville graben,
Upper Ventnor have been acquired by
the Council and landscaped

the past, Ventnor like many seaside
resorts of the period (1840-80), suffered
from rapid (sometimes poor quality)
building whilst other properties were
constructed to resist rather than
accommodate ground movements. As a result
some damaged buildings appear to have
suffered far worse structural damage than
is really the case.

To assist builders and developers, the
Council will be publishing shortly a code
of practice which it is hoped will be
helpful to those involved with works in
the Undercliff. If the recommendations
are followed, a better quality of
development should result with the
capability of accommodating minor ground
movements.

A number of developments including
swimming pools and certain small buildings
and extensions are regarded as Permitted
Development under the Town and Country
Planning Act. However, such works carried
out without formal approval and inspection
procedures can sometimes have a serious
local effect on ground conditions. Ill
-advised excavations into rear scarp
slopes in order to improve plot sizes can
de-stabilize the ground above.
Consideration is being given to seeking a
direction under the General Development
Order 1988 to require planning approval
for such works.

Tree growth can also have a stabilizing
effect on slopes and can reduce soil
erosion. A Tree Preservation Order for
wooded areas of the Undercliff would
ensure that this natural stabilizing agent
is maintained.

e) Avoidance of highly susceptible areas

It is accepted that in those limited areas
that are highly susceptible to ground
movement neither new or re-development
should be permitted.
At locations such as the Lowtherville
graben, the Council has been actively
acquiring sites particularly where
property has become a danger to the
public. Such plots are then landscaped to
form attractive amenity areas.

f) Monitoring

A programme of monitoring of rainfall,
ground water levels and structural damage
is planned. More direct recording of land
movement by on site measurement or
photogrammetric techniques are being
costed and considered. The appointment of
a member of staff with specific
responsibility for all such aspects within
the Ventnor Undercliff is also a
recommendation in the Undercliff
Management strategy

g) Further investigation and research

Ongoing research is recommended on a range
of topics to improve understanding of the
Ventor Undercliff. As part of the
sewerage scheme and ground water study a
borehole was taken to 25m on Ventnor
Esplanade in the hope of finding a basal
shear surface; unfortunately this must be
at a greater depth. A programme of deep
boreholes in section through the town
would be a costly exercise but would
compliment the DOE findings which are
based at present on geomorphological study
and some limited borehole information.

An extension of the DOE study area is
proposed to include the remainder of the
village of Bonchurch to the east of
Ventnor, and St Lawrence to the west. The
same level of understanding can then be
gained for these residential areas and
development guidelines can be produced as
for Ventnor. Finally a beach and coastal
management plan is required to ensure
ongoing maintenance takes place and beach
levels are monitored. Rising sea levels
could cause increased erosion of the
foreshore and toe of the Ventnor landslip
(which extends some way out to sea) and
may have a de-stabilizing influence.

CONCLUSION

Considerable progress has been made by the
Council already in developing its
landslide management strategy for Ventnor
although there are many areas of research
still to follow. The Council has taken
the advice of its officers and directed

funds towards coast protection and
research in Ventnor at a time when there
are considerable financial pressures to
spend money in other directions. It is
hoped that the public works being
undertaken on sewers, coast protection and
surface water drainage will re-establish
confidence in the town and further improve
its economic wellbeing.

Finally, the authors hope that this
paper may be of practical use to those
seeking to develop a landslide management
strategy in other areas.

REFERENCES

Geomorphological Services Ltd. Report for
    the Department of the Environment:
    'Coastal Landslip Potential Assessment,
    Isle of Wight Undercliff, Ventnor.
    March 1991.
McIntyre G & McInnes R G. 'A review of
    instability on the southern coasts of
    the Isle of Wight and the role of the
    local authority'. International
    conference on Slope Stability
    Engineering, Shanklin, Isle of Wight,
    Uk. April 1991.

Landslides, Bell (ed.) © 1995 Balkema, Rotterdam, ISBN 90 5410 032 X

Stabilization of Dutchman's Ridge

Dennis P. Moore & Alan S. Imrie
BC Hydro, Vancouver, B.C., Canada

ABSTRACT: Dutchman's Ridge is a 115 million cubic metre potential rockslide along the reservoir of Mica Dam on the Columbia River in British Columbia, which was stabilized by drainage between 1986 and 1988. The observational method was used to identify the natural barriers which isolate groundwater compartments and the natural plumbing needed to drain these compartments efficiently and thoroughly.

1 INTRODUCTION

The lower portion of Dutchman's Ridge is one of the few large, potential rockslides stabilized through drainage. It was investigated between 1961 and 1970 by the designers of Mica Dam, a 245 m high embankment dam being built only 1.5 km downstream [Mylrea et al 1978]. A tectonic fault, dipping towards the reservoir, discovered beneath the lower part of the ridge forms the base of a potential slide. The rock overlying the fault is weathered and loose, indicating a substantial amount of ancient, downslope creep had occurred. At that time, rather than stabilize the potential slide, the dam crest was widened to increase its erosion resistance to overtopping and monitoring of the ridge was instituted. If movements were detected, the stability of the potential slide was to be re-examined.

Shortly after reservoir filling, slow movements were detected along the underlying fault (Fig. 1). Additional investigations and upgrading of the monitoring system confirmed that about 115 million cubic metres of rock were creeping downslope along the fault at a rate of about 10 mm per year and that reservoir filling had triggered this resumption of the ancient movements.

Were these movements acceptable? Many mountainslopes creep and can be considered to be at the limit of their equilibrium: the real question is "Can the slope be triggered by a small disturbance and release its potential energy quickly, i.e. is it metastable, or is it not?". After

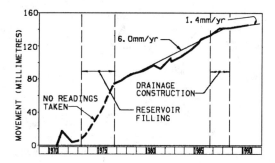

Fig. 1 - Rate of Movement on the Basal Fault

considerable deliberation, with the guidance of a review board consisting of Dr. C. Allen, Dr. E. Hoek and Dr. R. Peck, it was concluded in late 1985 that it would be prudent to stabilize the rock slope [Moore 1990].

Extensive investigation and monitoring enabled much to be learned about drainage of this type of rock mass. This paper describes the geological setting , the drainage system design and the drainage characteristics of this rock mass.

2 GEOLOGICAL SETTING

Dutchman's Ridge rises 1450 m from the floor of the Columbia Valley at an overall slope angle of about 30 degrees (Fig 2). The lower slope is crossed by several ridges and troughs which trace subparallel

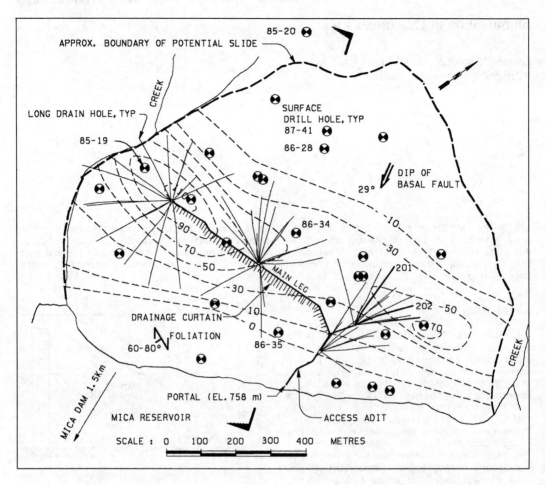

Fig. 2 - Drainage system layout with contours showing the reduction in water levels in metres

to the topographic contours. Similar features are common in B.C. and elsewhere and are often, but not always, associated with downslope movement. At Dutchman's Ridge, these features typically occur along tectonic fault or fracture zones which dip steeply into the slope. In a small area near the upstream end of the potential slide, open cracks and large, angular rock blocks indicate considerable local sliding has occurred. Other than the ridges and troughs and the rock block area, the potential slide shows little evidence of instability.

Foliation in the hornblende and quartzite gneisses and interbedded mica schists forming the ridge strikes into the slope and dips 60 to 80 degrees southwest (Fig. 2). The underlying tectonic fault, the Basal Fault, dips about 29 degrees

towards the reservoir and is composed of a 1.5 to 40 m thick layer of fractured and decomposed rock encasing an essentially continuous, clayey, gouge layer (Fig. 3). The gouge is typically less than 2 m thick and has a residual strength between 26 and 30 degrees. Several steep, gouge-filled fault zones cross the potential slide area trending parallel to the surface contours and intersecting the Basal Fault almost at right angles.

Above the Basal Fault, the rock is extremely fractured and weathered and contains numerous decomposed and sheared zones. Below the Basal Fault, the rock is generally much less fractured, but contains some zones of closely spaced fractures and steep faults. Near the western end of the adit many of the joints are open, some as much as 10 mm.

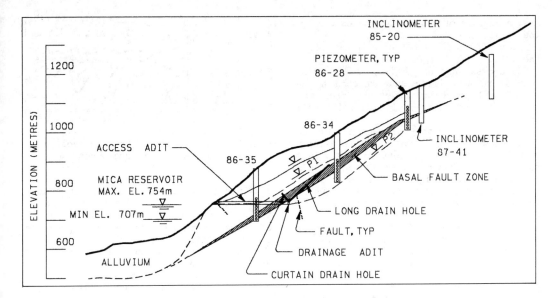

Fig. 3 - Cross section. Water levels before drainage and after drainage below (P2) and above (P1) the Basal Fault

3 DRAINAGE SYSTEM DESIGN

3.1 Drainage objectives

The objective of the drainage was to restore the stability of the slope to that which existed prior to reservoir filling. Once this stability was restored, the slope was expected to behave as it had in the past. In particular, the slope was expected to withstand any future earthquakes in the same manner it had withstood past earthquakes. Thus, the possibility of a catastrophic slide would be reduced to acceptable levels.

The effect of the drainage on the movement rates was to be monitored closely, but completely stopping the movements was not a design criterion. Reduced movement rates were considered likely but continued slow creep in the soft materials in the Basal Fault were expected to continue as the shear stresses on the layer would remain high. No sudden deterioration in strength was expected as a result of ongoing slow movements as they would be very small in comparison to the metres of ancient movement.

Water pressure at the Basal Fault where it was above reservoir elevation ranged from 0 to 126 m and averaged about 36 m. The objective of the drainage was to reduce this average to as little as 5 m to 10 m which is a very stringent requirement. This was expected to at least offset the destabilizing effect of reservoir impoundment.

3.2 Drainage system design

The design of the drainage system was strongly influenced by the experience gained in stabilizing Downie Slide [Imrie et al 1992]. Nevertheless, every large rockslide is unique so it was considered necessary to use the "observational method" to allow the results of early portions of the drainage work to be used to tailor subsequent work. Prior to construction, 267 piezometers, mainly of the Westbay type, were installed to enable the observations to be made.

The Basal Fault is a barrier to flow and has water pressures on both sides; therefore, drainholes or tunnels were necessary both below and above the fault. A drainage system which relied on gravity rather than pumps was considered desirable so the tunnel and drainhole collars were located above reservoir level.

The drainage system as designed consisted of five components:

1. an access adit through the potential slide material;

2. the main leg of the adit located just beneath the Basal Fault;

3. a second crossing of the fault;

4. a curtain of short drain holes through the Fault, and;

5. long holes drilled from chambers along the adit.

The access adit was located in gneissic rock where tunnelling was expected to be easier and where the distance from the surface to the Basal Fault was relatively short.

The main leg of the adit was positioned just below the Basal Fault at an elevation of a few metres above reservoir level. This position was chosen because it was along the axis of the highest drainable water pressures and it was below the fault where tunnelling conditions were expected to be better.

The upstream end of the main leg was chosen to be in reach of the most upstream area having significant water pressures and the downstream end was chosen to be in reach of the downstream boundary of the potential slide.

A downstream crossing of the fault was designed at the western end of the adit to provide a drilling chamber on the top side of the Basal Fault to facilitate drilling long holes to drain this side, should it be too difficult to drain from below the fault.

The curtain holes were to be drilled upward from the main leg to cross the Basal Fault on about 10 m centres (Fig. 2). The long holes generally were to be drilled in fans angled upward from drilling chambers to pierce the potential slide base on about 50 m centres. Most long holes were designed at their maximum practical length of 250 m to effect widespread drainage. Throughout construction, the observational method was to be used and modifications and additional holes were anticipated.

3.3 Modifications during Construction

During construction, the access adit portal was relocated as a result of surficial slides at the original portal, two double bends were added to the adit alignment to minimize tunnelling in adverse geological conditions which trended subparallel to the adit. An 82 m leg was driven upstream to enable drainage of some high pressure water not known to exist prior to the start of construction (Fig. 2). In addition, the second crossing of the Basal Fault was eliminated in favour of drilling long drainholes through the Fault in this area.

Also during construction, the locations, orientations and numbers of drainholes were adjusted to suit the drainage characteristics of the rock mass. Although

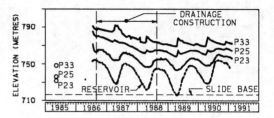

Fig. 4 - The effects of the reservoir on piezometric pressures

distinct phases of drilling were initially anticipated, these were unnecessary due to the rapid response of the pressures to drainage, combined with the slow progress of the drilling.

A total of 872 m of adit and 17,000 m of drainholes were ultimately constructed. The construction techniques are described elsewhere (Lewis and Moore 1989).

4 DRAINAGE CHARACTERISTICS

4.1 Effect of the weather and the reservoir on groundwater pressures

The weather at Dutchman's Ridge is characterized by fall rains turning to snow by about mid-November followed by spring rains and snow melt beginning in about March. At the higher elevations, the snowmelt can extend well into the summer. This results in a characteristic sawtooth shape to individual water pressure graphs (Fig. 4). There appears to be little influence at depth from individual rainstorms or short dry periods. The peak pressures within the potential slide area can occur weeks apart, presumably because of the influence of the elevation change on snowmelt and the range of permeabilities.

Although the annual fluctuation of the reservoir is usually more than 30 m it has surprisingly little effect on water pressures in the slope except below the current reservoir level or within about 20 m of it (Fig. 4). Both before and after drainage, the piezometers typically have shown that piezometric levels more than about 20 m above the current reservoir level follow the characteristic annual sawtooth cycle apparently unaffected by reservoir level changes. This is observed even in piezometers with their tips below the current reservoir level. As the reservoir level rises toward the unaffected piezometric level, the sawtooth cycle gradually becomes dominated by the reservoir level cycle. Although these

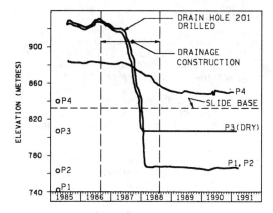

Fig. 5 - Contrasting water level reductions
below and above the slide base due
to drainage

observations were made later, they imply
that the effect of the initial reservoir
filling on pressures in the slope also was
limited.

The difference in timing of the reservoir
induced peak water pressures and the
weather induced peaks has to be considered
in the stability assessment as does the
effect of reservoir drawdown. There
appears to be no significant lag between
the groundwater pressures and the reservoir
level at the drawdown rates of less than
1 m per day that have been experienced.

4.2 Groundwater barriers and the "natural plumbing"

The rock mass at Dutchman's is
characterized by clayey, gouge layers in
the faults which act as barriers to
waterflow and divide the rock mass into
isolated compartments and by zones of open,
interconnected fractures which act as
carriers of water i.e. the "natural
plumbing". During investigation it was
necessary to install very closely spaced
piezometers to detect these compartments
and monitor pressures in them. During
construction it was necessary to penetrate
each compartment with drainholes or the
adit to achieve thorough drainage and to
tap into the "natural plumbing" within
these compartments to achieve efficient
drainage.

Early evidence of barriers was from
measurements in drillholes which showed
abrupt changes in piezometric level across
short sections of hole. These changes were
commonplace at Dutchman's Ridge and in one
extreme case a 50 m head drop was measured

across a 150 mm layer of clay gouge.

The Basal Fault was the most important of
these barriers as it divided the pressures
affecting the potential slide into two
separate compartments. Drainage below the
Fault was only partially effective in
lowering pressures above the fault
(Fig. 5).

A second extensive barrier is formed by a
steep fault which appears to extend right
across the potential slide below the Basal
Fault dividing this area into two
compartments for a total of three main
compartments. These three main compartments
are subdivided into numerous other
compartments of lesser extent. In marked
contrast to the lack of drainage across
barriers are the widespread effects when
the natural plumbing in the rock is
intersected. For example, an intersection
of water carrying fractures in drillholes
201 (Fig. 2) caused the initial rapid drops
in pressure below the Basal Fault in drill
hole 85-19 (Fig. 5) almost 700 m away.

Thus, drainage of rock masses such as
that at Dutchman's Ridge must take into
account the vast contrasts in permeability
which exist between the water carriers and
barriers and the numerous, isolated
compartments that can be formed. It is
unlikely that all of the important
pressures and compartments could have been
detected if closely spaced piezometers and
short completion zones were not used.

4.3 Groundwater flows and pressures

Peak flows from some of the long drainholes
were in the 1000 to 1800 litre per minute
range but the average peak flow for the 52
long holes was about 300 litres per minute.
Flows typically declined to less than a
half of the initial flow within a few weeks
of drilling. A similar pattern was observed
in the tunnel where flows up to about 7000
litres per minute dropped to half of their
initial flow within two weeks as tunnelling
progressed. The total flow from the adit
since completion has varied from 1500 to
3000 litres per minute seasonally.

Peak pressures at the Basal Fault were
reduced from an average of about 36 m to
about 10 m and pressure reductions were
widespread (Fig. 2).

5 EFFECT OF WATER PRESSURES ON STABILITY AND MOVEMENT

The calculated change in factor of safety
due to reservoir filling is very sensitive
to assumptions of failure surface geometry

and pre-reservoir water pressures at the toe of the slope. As there is ..ttle data on these pressures, a wide range of changes to the safety factor could be calculated within the range of reasonable assumptions.

The pre-drainage and post-drainage water pressures are well known compared to the pre-reservoir pressures. The increase in factor of safety due to drainage averages 6% which is considered greater than the decrease due to reservoir filling.

The seasonal fluctuation in water pressures has had only a very subtle influence on the rate of movement of the potential slide. Some movement zones show a period of greater rates in the spring and early summer.

The drainage has resulted in a reduction of movement rates throughout the potential slide (Fig.1).

6 CONCLUSIONS

A 115 million cubic metre potential rockslide in fractured and faulted gneissic and schistose rock with its toe inundated up to 180 m by a reservoir has been successfully stabilized through construction of a drainage system.

The groundwater pressure increase due to the initial reservoir filling was enough to trigger renewed downslope movement on an underlying fault but the drainage has offset this effect.

Annual fluctuations in reservoir level before and after the drainage have had little or no effect on piezometric levels where they are more than about 20 m above the current reservoir level. By inference, increased piezometric levels resulting from the initial reservoir filling were also limited.

The clayey gouge in faults commonly forms barriers to flow and divides the mass into isolated compartments. In contrast, if the "natural plumbing" is intersected, widespread and thorough drainage within a compartment can occur within a few days or weeks of intersection.

During construction the "observational method" was used successfully in conjunction with the extensive piezometric and geological data to identify the compartments and the natural plumbing useful in draining them. This allowed efficient and effective drainage to be implemented.

Only 872 m of adit and 17,000 m of drainholes were required to lower the average piezometric head at the potential slide base to 10 m, reducing the rate of movement and stabilizing 115 million cubic metres of rock.

REFERENCES

Imrie, A.S., D.P.Moore & E.G. Enegren. 1992. Stabilization and performance of Downie Slide. *Proceedings of ISL Symposium, Christchurch, NZ.*

Lewis,M.R. & D.P. Moore. 1989. Construction of the Downie Slide and Dutchman's Ridge drainage adits. *Canadian Tunnelling.*

Moore, D.P. 1990. Stabilization of Dutchman's Ridge. *Geotechnical News, Vol. 8, No. 4, December 1990.*

Mylrea, F.H., R.C. Dick & D. Hay. 1978. Symposium on the Mica Project, Vol. 40, *Paper No. 5, Proc. American Power Conference.*

Landslides, Bell (ed.) © 1995 Balkema, Rotterdam, ISBN 90 5410 032 X

Recent changes in the approach to landslip preventive works in Hong Kong

G. E. Powell
Geotechnical Engineering Office, Hong Kong Government, Hong Kong

ABSTRACT : The Hong Kong Government has an extensive ongoing public safety programme to bring existing steep slopes which pose a high risk to life to current standards. With increased confidence in the understanding of the behaviour of slopes in deeply weathered tropical soils a change in the required standards has led to design solutions being adopted which improve the soil insitu. Soil nailing is now the most commonly adopted solutions. These solutions are cheaper and safer to construct than those used earlier, which concentrated on reducing the stresses in the slopes through cutting back or transferred the loads out the slopes by structural means.

1 INTRODUCTION

In Hong Kong's crowded conditions land is at a premium, particularly in residential areas. Prior to the formation of the Geotechnical Control Office in 1978 slopes for development platforms were cut into deeply weathered hillsides to form slopes at angles of 60° or steeper and up to 30 m in height. These slopes are liable to failure in periods of intense rainfall, and since 1977 the Hong Kong Government has had a long term public safety programme to carry out landslide preventive & remedial works (LPM programme) to bring these slopes to current standards (Brand, 1988). Each year preliminary studies are carried out on about 1500 slopes and retaining walls, detailed investigations are undertaken on 70 to 80 slopes, and preventive works completed on 30 to 40 slopes, with the annual capital expenditure on the programme running at approximately $9 million U.S. Preventive works have now been completed on more than 600 slopes and retaining walls under this programme.

The majority of unsatisfactory existing slopes on which work is carried out pose a direct and high risk to life. Slopes and retaining walls are often directly behind or below occupied buildings. Typical height and steepness at the critical section of slopes on which preventive works have been completed are shown on Figure 1 and their distance from the nearest occupied structures is shown in Figure 2.

Working space is severely limited and access is often difficult, through narrow lanes or over steep slopes which places considerable constraints on the size and type of plant that can be used and contractors' working methods.

2 SLOPE FORMING MATERIALS

The vast majority of preventive works are carried out to slopes formed of deeply weathered granitic or volcanic saprolites. The granites range from fine to coarse grained with the slightly older volcanic suite containing a wide range of rock types including fine to coarse grained pyroclastic tuffs and less dominant lavas (McFeat-Smith et. al., 1989).

There is generally a relatively thin or non-existent residual soil layer overlying the saprolite, which retains much of its original structure and rock fabric. Relict joints are extensively present and participate or control failure mechanisms over the full range of weathering (Irfan & Woods, 1988). Corestones of less weathered rock can be present at all depths and are particularly prevalent in the granitic saprolytes.

Material properties vary widely over short distances (Irfan, 1988) and characterisation of mass strength from small samples obtained for identification and testing purposes is difficult and unreliable (Massey et. al., 1989).

Table 1 : Factors of safety for analysis and design of remedial and preventive works to existing slopes

Risk to Life	High	Low	Negligible	Remarks
Minimum F of S against loss of life (1 in 10 year return period rainfall)	1.4	1.3	1.2	GCO, 1979
	1.2	1.1	≥ 1.0	GCO, 1984*

* Applies only where detailed geotechnical study has been conducted, the slope has been standing a considerable time and where the loading condition, groundwater regime and basic form of the modified slope remain substantially the same as those of the existing slope.

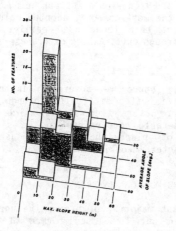
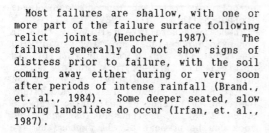

Figure 1 : Maximum slope height vs. average slope angle for completed LPM soil cut slopes

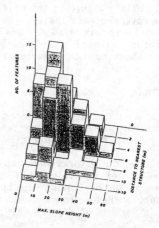

Figure 2 : Maximum slope height vs. distance to nearest structure for completed LPM soil cut slopes

Most failures are shallow, with one or more part of the failure surface following relict joints (Hencher, 1987). The failures generally do not show signs of distress prior to failure, with the soil coming away either during or very soon after periods of intense rainfall (Brand., et. al., 1984). Some deeper seated, slow moving landslides do occur (Irfan, et. al., 1987).

3 DESIGN STANDARDS

Following disastrous landslides in 1972 and 1976 the Geotechnical Control Office was established. One of the earliest priorities was to establish and promulgate standards for the design of all slopes in

Hong Kong. These were set out in the Geotechnical Manual for Slopes (GCO, 1979). The same standards were adopted for the design of both new and existing slopes. Detailed guidance was also provided in this publication on investigation, testing and analytical methods which were appropriate for Hong Kong's conditions.

By 1984, it was appreciated that considerable confidence could be obtained from the successful performance of existing slopes through known intense rainstorm events. If a slope did not show signs of distress the margins of safety allowing for uncertainties associated with the development of a geological model, prediction of ground water response and estimation of mass shear strength could be significantly reduced. The changes in minimum required factors of safety adopted in the revised Geotechnical Manual for Slopes in 1984 as a result of this appreciation are given in Table 1 (GCO, 1984).

A critical factor in these changes of standards was that the existing condition of a slope which had not failed under a 1 in 10 year storm rainfall event could be conservatively taken as equivalent to a factor of safety of one. Hence the requirement to achieve a factor of safety of 1.2 for a high risk existing slope is equivalent to achieving a 20% improvement in the factor of safety over the condition existing under the design rainstorm event.

4 INVESTIGATION APPROACH

The normal approach to investigation of existing slopes in the early to mid-80's was to undertake an intensive geotechnical investigation, typically consisting of a comprehensive desk study of available data, of which a considerable amount is readily available in Hong Kong through the Geotechnical Information Unit of the Geotechnical Engineering Office, followed by a ground investigation consisting of a number of boreholes with alternate standard penetration testing (SPT) and triple core barrel (Mazier) sampling (GCO, 1987). Water pressures would be monitored for a minimum of one wet season in open Casagrande standpipe piezometers at 1 to 2 week intervals. Peak water levels were obtained using "Halcrow" buckets (Brand & Phillipson, 1984). Little reliance was placed on back analysis of the existing condition of the slope, with existing factors of safety being calculated using strengths obtained from back saturated samples tested in consolidated undrained triaxial tests. It was not uncommon for this approach to lead to calculated factors of safety of standing slopes of less than unity.

The adoption of the new standards in 1984 allowed designers to place much more reliance on back analysis of the existing slope condition. Detailed investigations are still required to establish a reliable geological model and to estimate likely ground water levels under the 1 in 10 year return period design rainstorm event. However, once these are obtained back analysis can provide a much more realistic and reliable estimate of the operating mass strength than that obtained from small material samples tested in the laboratory.

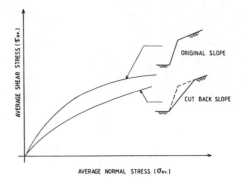

Figure 3 : Resistance envelopes for slope cutback

5 DESIGN SOLUTIONS

The preferred design solution has been, and continues to be, to cut back the slope to a flatter angle. Provided the geological conditions are similar for the cut back and original slopes an improvement through stress reduction of 20% can be readily obtained by the resistance envelope approach (Janbu, 1954) (Figure 3). Cutting back provides a positive improvement to the slope stability and allows the slope to be readily grassed, in accordance with government policy of reducing the visual impact of such work.

However for many of the slopes in Hong Kong cutting back is not possible as the slopes are located between buildings and no land is available.

Early solutions for providing support to these severely constrained sites relied on transferring the loads out of the slope to flatter ground or to stronger strata. This was achieved by the use of conventional retaining walls, or inclined retaining walls where space was available, and cantilevered caisson walls into either soil or rock where space was severely limited (Koirala & Tang, 1988). Typical examples are given in Figure 4.

The cost of such structures was high, particularly cantilevered walls which had to be carried to considerable depths to achieve the required factors of safety on the passive pressures in front of the walls. Hand dug caissons of 1.2 to 1.5 m diameter, either continuous or at spacings up to two diameters were commonly used.

The caissons were initially designed to standards for structural concrete, i.e. serviceability, whereas the slope was designed for the ultimate condition. this resulted in excessively heavy reinforcement being required for bending stresses. Later

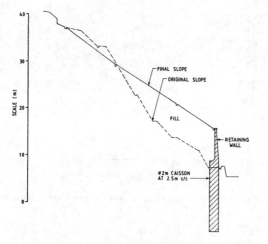

(a) Retaining wall supported on caisson

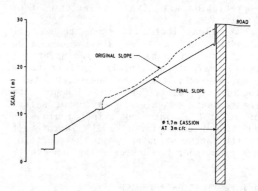

(c) Soil socketed caisson wall

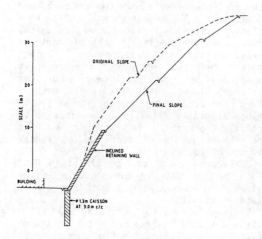

(b) Inclined retaining wall

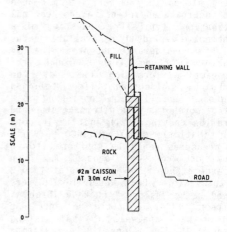

(d) Rock socketed caisson wall

Figure 4 : Design solutions which transfer loads out of the slope

designs introduced refinements including the use of tie backs, dowelling into rock, passive tie bars etc. which resulted in more efficient designs (Yim et. al., 1988).

The change in philosophy towards an improvement approach to factors of safety embodied in the 1984 changes of standards gradually led to a change where solutions which improved the soil in place were favoured (Figure 5).

The initial designs with this in mind involved micropiling and soil dowelling (Powell et. al., 1990). These were rapidly replaced by soil nailing which now dominates the solutions adopted for

landslip preventive works (Figure 6) (Powell & Watkins, 1991).

Soil nails act with the insitu soil to form a composite material. Relict joints and other weaknesses inherent in the slope are effectively bound to the stronger material so that the full mass-strength is available, as well as the strength added by the soil nails.

The designs which strengthen the soil insitu soil are much more robust than the previous types of solutions, in that they are not dramatically affected by changes on site from the soil conditions assumed during design.

6 CONSTRUCTION

Both cut back and structural solutions, such as retaining walls and cantilever caisson walls, involve opening the slope during construction and often require steep temporary cuts. Both these operations increase the risk of failure during construction from oversteeping or increased infiltration through exposed soil or from ponding during periods of intense rainfall. Any failure during construction, no matter how minor, is a matter of considerable concern as the slopes are very close to occupied buildings etc. Very close supervision and stringent contractual controls are used to minimise the risks associated with the temporary works condition. Where the risks to areas outside the works site are high restrictions on working methodology and sequencing are incorporated in the contracts. Such control measures do not come cheaply and this is reflected in the rates for this type of work which are in the range of 1.5 to 2.5 times the rate for similar work in normal civil engineering contracts. In addition, working in deep caissons involves the workers in considerable health and safety risk from silicosis, noise and falling objects.

In contrast, the use of soil nailing to strengthen the slopes avoids the need to open or alter the slope from its current condition. Bamboo scaffold can be readily erected to provide access and drilling platforms for the soil nails installation, and the work is generally clean, safe and fast (Figure 7). From comparisons made of different options for particular sites the cost of soil nailing is considerably cheaper than structural solutions, typically one half to one fifth of the cost.

With the trend to maintaining steep profiles attention has recently been paid to vegetating slopes which were previously covered with a hard cover, such as soil/cement plaster (chunam). With the use of proprietary erosion protection mats it has been possible to successfully grass slopes at angles up to 55°. Tree seeds are included in the hydroseeding mix to allow for the establishment of a long term tree cover to the slopes (Forth & Leung, 1989).

7 CONTRACTUAL FORMS

With cut-back and structural forms of remedial works it has been necessary to provide strict control as outlined above. For these types of works slopes are

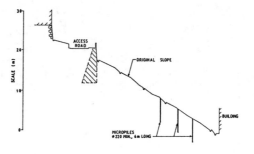

(a) Micropiling

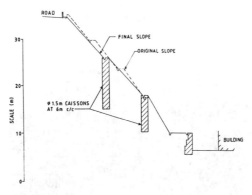

(b) Soil dowels

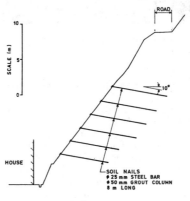

(c) Soil nailing

Figure 5 : Design solutions which improve the soil in place

packaged together into groups of fifteen to twenty and full working drawings produced for each slope. A programme is specified which sets the start and end date for each slope and all necessary constraints are stated in the contract.

Designs based on insitu soil strengthening provide solutions which are

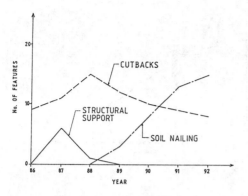

Figure 6 : Changes in design solutions with time

Figure 7 : Soil nail installation from bamboo scaffold

robust and which can be made suitably conservative whilst remaining cost competitive (Watkins & Powell, 1992). The increased confidence that results, together with the reduced risks associated with construction of works which do not interfere significantly with the stable existing slope profile has allowed contracts to be run with considerably less supervision, and design based on the observational approach. Preliminary designs are prepared using minimal ground investigation and a limited topographical survey. The extent and type of works likely is normally apparent to an experienced engineer, although design proposals are checked independently before being passed to the contractor as site instructions. The site is then observed by the designer as work proceeds and changes to the design scope or methodology are made if conditions on site are found to be significantly different to those previously assumed. The cost of a full site investigation for minor slopes can be of a similar order to the cost of the permanent remedial works. Reducing initial investigation and being more conservative in design and construction can result in considerable overall savings. Currently this approach is limited to soil slopes less than about 10 m in height and masonry retaining walls, which together constitute more than half of the features currently being worked on under the LPM programme. This approach has worked well when used with contractors experienced in slope stabilization works.

8 CONCLUSIONS

The improvement of the stability of steep slopes in saprolitic tropical soils on very constrained sites in urban Hong Kong continues to provide challenges.

The introduction of design solutions which improve the insitu soil in place has led to more robust designs which are easier to construct and at lower cost. Soil nailing is now the most commonly used method for such improvement works.

9 ACKNOWLEDGEMENTS

This paper is published with the permission of the Director of Civil Engineering, Hong Kong Government.

REFERENCES

Brand, E.W. (1988). Landslide risk assessment in Hong Kong. (Special Lecture). Proceedings of the Fifth International Symposium on Landslides, Lausanne, vol. 2, pp 1059-1074.

Brand, E.W. & Phillipson, H.B. (1984). Site investigation and geotechnical engineering practice in Hong Kong. Geotechnical Engineering, vol. 15, pp 97-153.

Forth, R.A. & Leung, K.W. (1989). Use of geotextiles to prevent erosion of steep slopes in Hong Kong. Proceedings of the Symposium on Application of Geosynthetics and Geofibre in South East Asia, Petaling Jaya, Malaysia, pp 2.1-2.5.

Geotechnical Control Office (1979). Geotechnical Manual for Slopes. (First edition). Geotechnical Control Office, Hong Kong, 242 p.

Geotechnical Control Office (1984). Geotechnical Manual for Slopes. (Second edition). Geotechnical Control Office, Hong Kong, 295 p.

Geotechnical Control Office (1987). Guide to Site Investigation (Geoguide 2). Geotechnical Control Office, Hong Kong, 353 p.

Hencher, S.R. (1987). The implications of joints and structures for slope stability. Slope Stability: Geotechnical Engineering and Geomorphology, edited by M.G. Anderson & K.S. Richards, pp 145-186. John Wiley & Sons, Chichester, UK.

Irfan, T.Y. (1988). Fabric variability and index testing of a granitic saprolite. Proceedings of the Second International Conference on Geomechanics in Tropical Soils, Singapore, vol. 1, pp 25-35.

Irfan, T.Y. Koirala, N.P. & Tang, K.Y. (1987). A complex slope failure in a highly weathered rock mass. Proceedings of the Sixth International Congress on Rock Mechanics, Montreal, vol. 1, pp 397-402.

Irfan, T.Y. & Woods, N.W. (1988). The influence of relict discontinuities on slope stability in saprolitic soils. Proceedings of the Second International Conference on Geomechanics in Tropical Soils, Singapore, vol. 1, pp 267-276.

Janbu, N. (1954) Stability analysis of slopes with dimensionless parameters. Harvard Soil Mechanics Series No. 46, 81 p.

Koirala, N.P. & Tang, K.Y. (1988). Design of landslip preventive works for cut slopes in Hong Kong. Proceedings of the Fifth International Symposium on Landslides, Lausanne, Switzerland, vol. 2, pp 933-938.

Massey, J.B., Irfan, T.Y. & Cipullo, A. (1989). The characterization of granitic saprolitic soils. Proceedings of the 12th International Conference on Soil Mechanics and Foundation Engineering, Rio de Janeiro, vol. 1, pp 533-532.

McFeat-Smith, I., Workman, D.R., Burnett, A.D. & Chau, E.P.Y. (1989). The geology of Hong Kong. Bulletin of the Association of Engineering Geologists (USA), vol. 26, pp 17-107.

Powell, G.E., Tang, K.W. & Au-Yeung, Y.S. (1990). The use of large diameter piles in landslip prevention in Hong Kong. Proceedings of the Tenth Southeast Asian Geotechnical Conference, Taipei, Taiwan, vol. 1, pp 197-202.

Powell, G.E. & Watkins, A.T. (1990). Improvement of marginally stable existing cut slopes by soil nailing in Hong Kong. Proceedings of the International Reinforced Soil Conference, Glasgow, pp 241-247.

Watkins, A.W. & Powell, G.E. (1992). Soil nailing to existing slopes as landslip preventive works. Hong Kong Engineer, in press.

Yim, K.P., Watkins, A.T. & Powell, G.E. (1988). Insitu ground reinforcement for slope improvement in Hong Kong. Proceedings of the International Geotechnical Symposium on theory and Practice of Earth Reinforcement, Fukuoka, Japan, pp 363-368.

Landslides, Bell (ed.) © 1995 Balkema, Rotterdam, ISBN 90 5410 032 X

The role of geogrid reinforced embankments in landslide stabilization: Theory and practice in Italy

P. Rimoldi
Tenax SpA, Italy

A. Ricciuti
Tenax Geosynthetics Testing Laboratory, Italy

ABSTRACT: The paper deals with the application of geosynthetics, particularly reinforcing geogrids, in landslide stabilization. The Authors describe the various categories of geosynthetics and summarize their applications in landslide stabilization, pointing out the functions required and the proper product for each problem. A method for designing geogrid-reinforced slopes for landslide stabilization is described. Tables have been prepared showing the variation of the factor of safety versus geogrids layout. Few selected italian case histories furtherly illustrate the concepts.

1. INTRODUCTION

Landslides are sub-aerial mass movements on natural or man made slopes that threath human lives or goods.

Landslide can be an ambiguous term: Hutchinson (1988) lists eigth major types of slope movements and some thirty subclasses, calling "landslides" only relatively rapid downslope movements of soil and rock which take place on one or more discrete slip surfaces which define the sliding mass. However, different kinds of slope movement refer to different failure mechanisms and triggering events, consequently, different slope movements need different counter measures, taking into account the acting natural phenomena. Three main critical features can be connected to slope instability (Bromhead, 1986): variation of the geometry of the slope (e.g. due to natural erosion or cut and fill operations); decay of strength parameters (e.g. due to soil softening); build up of pore water pressure. The adequate countermeasures corresponding to to these features are: re-shaping of the slope; soil stregthening; drainage.

Since a long time civil engineers had only traditional materials, like soil, rock, concrete and iron to design and carry out these countermeasures. Starting from the late Seventies, due to the development of the technology, new synthetic products appeared to help design and construction of engineering works. Among these new products, an important role was gained by a class known as geosynthetics.

Geosynthetics gave new solutions to a number of old problems in civil engineering: separation, drainage, filtration, waterproofing, stabilization, erosion control and reinforcement.

Among geosynthetics, a group named geogrids is taking an important place in geotechnical engineering. Recently, geogrids have been used in landslide stabilization in a number of projects in Italy and their use in this field is growing very quickly.

2. GEOSYNTHETICS IN LANDSLIDES: A SHORT CLASSIFICATION

According to the classification represented in Fig.1, the following geosynthetics can be described.

Geotextiles can be defined as permeable textile-like materials, usually synthetic; many fibres and different fabric types have been developed for various applications. There are four major uses of geotextiles in civil and geotechnical engineering: separation of different materials; filtration across the geotextile layer; drainage along the geotextile layer; reinforcement of soft soils (for woven geotextiles). Geotextiles application in landslide stabization covers almost all the above mentioned items.

Geogrids can be defined as two-dimensional, geometrically defined, high tensile modulus structures, manufactured with extruded and stretched High Density Polyethylene or Polypropylene sheets, having elliptical or rectangular holes regularly distributed; they are used for soil reinforcement.

Geonets can be defined as three-dimensional, geometrically defined, bulky structures, composed of thermally or mechanically bonded synthetic filaments, used mainly for drainage and transport of liquid or gases.

Geocomposites can be defined as three dimensional combinations of two or more basic geosynthetics, like a geonet, and one or two geotextiles, or a geomembrane and a

Tab. 1 - Application of geosynthetics in landslide stabilization.

APPLICATION	REQUIRED FUNCTIONS	SPECIAL REQUIREMENTS	SUITABLE GEOSYNTH.
-Horiz. drains to prevent pore pressure buildup	Separation, filter, drainage	High compression resistance, high flow rate	Geocomposites
-Vert. drains to accelerate soil consolidation	Filtration, drainage	High flow rate, resistance to installation	Vert. strip drains
-Drain behind a diaphragm to collect water leaking through the diaphragm	Separation, filter, drainage	High compression resistance, high flow rate	Geocomposites
-Separation of different soils	Separation, filter	Resist. to damages, good permettivity	Nonwovens or wovens geotex., geonets
-Slope reinforcement, to steepen the slopes and/or increase the Safety Factor	Soil reinforcement	High tensile strength and modulus, good friction properties	Mono-oriented geogrids, woven geotextiles
-Base reinforcement, to minimize settlements	Soil stabilization	High tensile strength and modulus, good friction properties and/or confinenent	Geogrids, woven geotextiles, geocells
-Base reinforc. in case of local collapsible soils	Membrane effect	Same as previous one	Geogrid, woven geotextiles
-Stabilization of gunite on slopes	Reinforcement	High tensile strength and modulus, open structure	Bioriented geogrids
-Erosion control on slopes	Erosion control	Soil confinement, micro reinforcement, gripping, surface protection.	Geocell, geomats, biomats

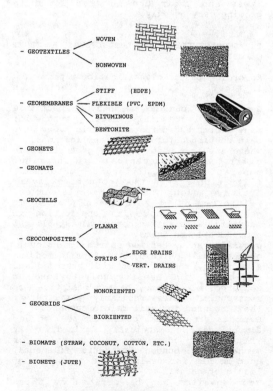

Fig.1 - Classification of geosynthetics based on their physical aspect.

geotextile, or a geomembrane and a geonet, and other possible combinations. Two layers of geotextiles comprising Bentonite granules can also be considered a geocomposite. Vertical strip drains are considered geocomposite too.

Geocells are honeycomb tridimensional structures for soil confinement and erosion control. They are manufactured or through a single extrusion process from HDPE granules or glueing together strips of Polyesther nonwoven geotextile or HDPE membrane.

Geomats are tridimensional structure used mainly for surface protection and erosion control on slopes. They are usually constituted of a mat of randomly distributed filaments or with a 10-20 mm thick core contained between two lightweight grids.

Biomats are manufactured using straw or coconut fibres held between two layers of fotodegradable, lightweight Polypropylene grids. It is possible to include in this class also the jute nets; both products are used for erosion control and slope vegetation.

Given the characteristics shortly described above, Tab.1 shows the main applications of geosynthetics in landslide stabilization, enhancing the required functions and the special requirements, thus selecting the most suitable products for each case.

3. DESIGN OF GEOGRID REINFORCED SOIL FOR LANDSLIDE REPAIR WORKS

The most important application of geosynthetics in landslide management is the use of geogrids as soil reinforcement in landslide repair works. In this case geogrids layers are laid horizontally into the fill soil to provide interlocking and tensile resistance as shown in Fig.2.

The design of geogrid reinforced slopes can be carried out using a trial and error procedure: an initial layout of the reinforcement is set on the bases of experience; then a stability calculation using a Bishop modified method is run; if the Factor of Safety results too low, the layout is changed and the process continues until a suitable FS is reached.

The layout of the geogrids reinforcement consists of setting the length and the vertical spacing of the reinforcing layers, and the allowable resistance of the geogrids. A geogrid lengtt equal to half the slope height is usually reputed to be the minimum, but sometimes a geogrid length/slope height ratio in excess of 1.0 is needed. The vertical spacing of the geogrids can be uniform for all the slope height, which greatly facilitates the installation works, or the geogrids can be more closely spaced at the bottom of the slope, in order to increase the tensile strength available in the most critic zone. The vertical spacing is usually in the range of 0.30-1.50 m. The geogrids on the market nowadays present a maximum tensile resistance in the range of 40-110 kN/m, but taking into account the tensile creep suffered by the grid under the constant load produced by soil thrust, the allowable tensile strength of High Density Polyethylene geogrids (the most commonly used) is usually reduced to 35-40% of the peak value. Geogrids producers run specific laboratory tests to determine the long term allowable tensile strength Fa of their products.

Together with the initial layout, the designer needs to know the geotechnical characteristics of the fill soil which will be used for the remedial works : namely the angle of internal friction ϕ, the cohesion c and the unit weight γ (usually the value corresponding to 95% Standard Proctor density is specified).

With reference to Fig.2, the Factor of Safety of the geogrid reinforced slope can be computed as :

$$FS = \frac{MS + Mg}{Mo} = FSo + \frac{Mg}{Mo} \qquad (1)$$

where :
Ms = stabilizing moment due to soil shear resistance;
Mg = stabilizing moment due to geogrids tensile forces;
Mo = destabilizing moment,
FSo = Factor of Safety without geogrids.
Ms, Mo and FSo, can be computed using a

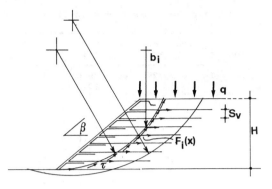

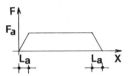

Fig. 2 - Scheme for the stability analysis

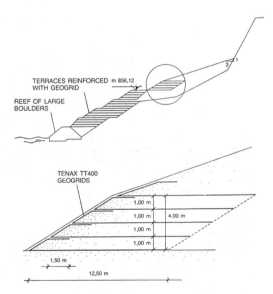

Fig. 3 - Remedial works at La Boscaccia landslide.

Bishop of Fellenius or Janbu method.
The stabilizing moment due to geogrids tensile forces can be computed as :

$$Mg = \sum_i Fa_i(X) \cdot b_i \qquad (2)$$

where :
$Fa_i(X)$ = tensile force in i th geogrid layer, at the point where the failure surface cut the geogrid;
b_i = arm of the geogrid tensile force.

Tab. 2a - FS versus geogrid layout

Φ Deg	c kPa	Ru -	Fa kN/m	Sv m	L m	FS -
0	0		20	0.6	5.0	0.612
					10.0	0.896
				1.5	5.0	0.521
					10.0	0.570
			40	0.6	5.0	0.672
					10.0	1.029
				1.5	5.0	0.617
					10.0	0.737
		0.25	20	0.6	5.0	0.423
					10.0	0.674
				1.5	5.0	0.341
					10.0	0.400
			40	0.6	5.0	0.472
					10.0	0.784
				1.5	5.0	0.434
					10.0	0.605
18	25	0	20	0.6	5.0	1.308
					10.0	1.623
				1.5	5.0	1.258
					10.0	1.325
			40	0.6	5.0	1.369
					10.0	1.776
				1.5	5.0	1.262
					10.0	1.439
		0.25	20	0.6	5.0	1.108
					10.0	1.399
				1.5	5.0	1.047
					10.0	1.125
			40	0.6	5.0	1.162
					10.0	1.529
				1.5	5.0	1.060
					10.0	1.340

Tab. 2b - FS versus geogrids layout

Φ Deg	c kPa	Ru -	Fa kN/m	Sv m	L m	FS -
0	0		20	0.6	5.0	0.836
					10.0	0.981
				1.5	5.0	0.705
					10.0	0.705
			40	0.6	5.0	0.932
					10.0	1.370
				1.5	5.0	0.771
					10.0	0.814
		0.25	20	0.6	5.0	0.578
					10.0	0.898
				1.5	5.0	0.454
					10.0	0.454
			40	0.6	5.0	0.650
					10.0	1.040
				1.5	5.0	0.512
					10.0	0.612
25	25	0	20	0.6	5.0	1.526
					10.0	1.742
				1.5	5.0	1.464
					10.0	1.532
			40	0.6	5.0	1.621
					10.0	2.104
				1.5	5.0	1.500
					10.0	1.656
		0.25	20	0.6	5.0	1.246
					10.0	1.462
				1.5	5.0	1.198
					10.0	1.263
			40	0.6	5.0	1.333
					10.0	1.777
				1.5	5.0	1.214
					10.0	1.377

Still with reference to Fig.2, the envelope of available tensile force along each geogrid layer can be assumed to be trapezoidal, with a central uniform value corresponding to Fa and allowing for anchorage zones at the face and the end of each layer. The anchorage length can be assumed to be in the range of 0.5-1.0 m for the actual geogrids.

In order to facilitate the initial layout, calculations have been made for a rather common problem, obtaining tables which give the Factor of Safety for different length, vertical spacing and allowable tensile strength of the geogrids layers.

Precisely, the calculations reported in Tab.2 are referred to a 10 m high slope having an inclination of 60°. The following hypothesis have been made: the soil is considered omogeneous, with a density of 18 kN/m3; there is no surcharge on top of the slope; the failure surface pass through the toe of the slope; the geogrids have a uniform vertical spacing.

Tab.2 reports the Factor of Safety obtained using a Bishop modified method to calculate Ms and Mo, for different values of the friction angle Φ and the cohesion c of the fill soil, of the pore pressure coefficient Ru, of the allowable geogrid tensile strength Fa, of the vertical spacing Sv and of the length of the geogrids layers.

Tab.2 can be easily used for setting the inital layout of the geogrid layers: as an example, given a fill soil having a friction angle of 18° and a cohesion of 25 kPa, provided that an adequate drainage is provided to guarantee that Ru=0, it is possible to see from the Table that a Factor of Safety of about 1.30 (as usually required for this kind of projects) can be achieved using geogrids with 20 kN/m of allowable tensile strength, vertically spaced of 0.60m and with a length equal to 5.0m, or alternatively using a geogrid with 40 kN/m allowable tensile strength, spaced of 1.50m and with a length equal to 10.0m.

4. THE ROLE OF GEOGRIDS IN LANDSLIDE STABILIZATION : SEVERAL CASE HISTORIES

The following case histories describe several typical applications of geogrids in landslide stabilization projects in Italy.

La Boscaccia landslide (Valtellina, Italy). In July 1987, following heavy rainfalls and exceptionally large quantities of melting snow, the Adda river overflowed, bursting the river banks. This severe flooding triggered several landslides through Valtellina. One of such landslides occurred

Tab. 2c - FS versus geogrid layout

ō Deg	c kPa	Ru -	Fa kN/m	Sv m	L m	FS -
0	0		20	0.6	5.0	1.002
					10.0	1.113
				1.5	5.0	0.812
					10.0	0.812
			40	0.6	5.0	1.142
					10.0	1.629
				1.5	5.0	0.937
					10.0	1.067
		0.25	20	0.6	5.0	0.682
					10.0	0.803
				1.5	5.0	0.504
					10.0	0.504
			40	0.6	5.0	0.777
					10.0	1.226
				1.5	5.0	0.617
					10.0	0.769
30	25	0	20	0.6	5.0	1.732
					10.0	1.916
				1.5	5.0	1.628
					10.0	1.661
			40	0.6	5.0	1.797
					10.0	2.304
				1.5	5.0	1.667
					10.0	1.819
		0.25	20	0.6	5.0	1.352
					10.0	1.569
				1.5	5.0	1.299
					10.0	1.337
			40	0.6	5.0	1.449
					10.0	1.954
				1.5	5.0	1.327
					10.0	1.483

at La Boscaccia site, in the Municipality of Sondalo. The landslide involved gravelly silty soils which formed part of an alluvial cone: the total volume of the landslip was approximately 1.1 million m3 and the height of the main scarp was around 50 m. The problem involved both the stabilization of the slope and the protection against toe erosion, taking into account the great scenic and touristic value of the area. The erosion control was carried out by means of a rockfill reef 160 meters long and a groin. The slope was stabilized constructing 5 geogrid-reinforced embankments 4 m high and 130 m long (see fig.3). The construction of the embankments reinforced with HDPE monoriented geogrids permitted: the construction of a 25 m high reinforced structure to stabilize the toe of the slope, reaching a safety factor of 1.9; the re-use and strengthening of the slipped material; the creation of a 1:1 gradient terraced slope, completely stable and grassed.

Fiuggi landslide (Lazio Region, Italy).
Fiuggi is a small town in Central Italy which has for centuries been famous for its thermal springs. It is situated in a seismic area on sloping terrain, with abundant groundwater. These factors cause several problems to the stability of the slopes in the surrounding areas. Along the Prenestina Nuova state highway a landslide occurred on a slope with a gradient of some 20° in soft volcanic clays. The failure surface was nearly plane and at 10-15 m of depth. The landslip was stabilized

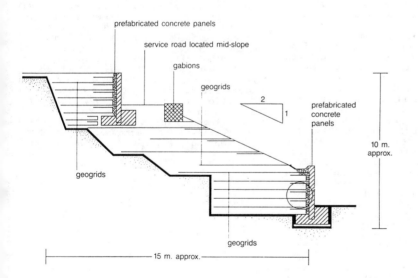
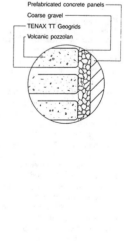

Fig. 4 - Remedial works at Fiuggi landslide

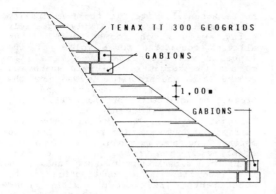

Fig. 5 - Remedial works for the Orvieto landslide.

using the reinforced soil technique: the failed soil was strengthened by means of HDPE monoriented geogrids. The installation of the geogrids was performed using the wrap around method: the geogrids were laid down on the soil and fixed using pegs; then a layer of soil was spread and compacted; the geogrids were then turned over and re-fixed. The job was extremely simple and carried out by unskilled labour using traditional earth moving equipment. The job was completed with the addition of non structural precast concrete panels (used for their attractive appearance). The drainage was made by putting coarse gravel between the reinforced soil wall and the concrete panels (see fig.4).

Orvieto Cliffs (Umbria Region, Italy).
Orvieto is one of the most beautiful towns in Italy: it stands on a tufa cliff, emerging from the surrounding clay hills. The rain and sewage water is drained from the town and collected into radial ditches starting at the base of the cliff. One of these ditches, the Civetta ditch, eroded the toe of both the side clay slopes, just at the base of the cliff, producing a large landslide 100m long and 15m high. Since all the Orvieto area is preserved for its historic importance and scenic beauty, the engineers had to face the hard task to reconstruct the slope to its original geometry. The need to stabilize the slope and to allow for fast revegetation called for soil reinforcement using mono-oriented HDPE geogrids. The use of such reinforcement allowed to use the same failed clay soil to reconstruct the slope, resulting in very large savings compared to other possible solutions involving the substitution of the clay with another soil having better characteristics. The cross section of the final reinforcement design if shown in Fig.5. The toe of the reinforced slope was protected with gabions and the face was hydroseeded so that the grass grew fast through the geogrids apertures. Few weeks after the end of the reconstruction, the slope was grassed and it was back to its original appearance.

5. CONCLUSIONS

As it can be seen from the above data, geogrids progressed from the initial experimental stage, gaining recognition as construction materials in all respects. As time goes by, more extensive and reliable data concerning performances and durability (i.e. safety) are being obtained from different products: a number of records now go back 15 years: operating lives of 30 years now can be forecast with reasonable confidence. Small slips, first-time landslides and earthflows, are the typical cases for more designs, applications and experiences. A number of such solutions are being built today in a variety of environments and involving a large spectrum of design solutions. The great advantage of geogrids in landslide stabilization, today, is that they offer both performance and uniformity of quality at very low price levels compared to the very high costs of the construction industry.

6. REFERENCES

Bromhead, E.N. (1986) "The Stability of Slopes", Surrey University Press, N.Y.

Giroud, J.P. (1990) "Functions and Applications of Geosynthetics in Dams", International Water Power and Dam Construction, Vol.42, No.6, pp.16-25.

Hutchinson, J.N. (1988) "Morphological and Geotechnical Parameters of Landslides in Relation to Geology and Hydrology", V Int. Symp. on Landslides, Lausanne.

Jewell, R.A., Paine, N., Woods, R.I. (1985) "Design methods for steep reinforced embankments", Polimer grid reinforcement, Thomas Telford, London, 1st Edition.

Koerner, M.R. (1988) "Designing with Geosynthetics", Prentice Hall, Englewood Cliff, New Jersey, U.S.A., 2nd Edition.

Rimoldi, P. (1988) "Le applicazioni dei geosintetici nell'ingegneria civile", SEP Pollution Symposium, Padova, Italy (in Italian language).

Veldhuijzen van Zanten, R. (1986) "Geotextiles and Geomembranes in Civil Engineering", Balkema, Rotterdam/Boston.

G4 Landslide hazard assessment
    Évaluation des risques de glissements de terrain

Landslides, Bell (ed.) © 1995 Balkema, Rotterdam, ISBN 90 5410 032 X

Keynote paper: Landslide hazard assessment

J. N. Hutchinson
Imperial College of Science, Technology and Medicine, London, UK

ABSTRACT: After defining landslide hazard and indicating its global scale, the main physical factors bearing on landsliding are outlined. Attention throughout is concentrated on sub–aerial landslides, though the great and growing importance of sub–aqueous landsliding is recognised. The existing approaches to landslide hazard assessment, principally on a regional scale, are then reviewed and discussed. While in the future, remote sensing and indirect approaches, using GIS, are likely to continue to play a valuable role in such assessments, the main thrust of the paper is to urge that maximum use also be made of geological and geomorphological insights, particularly with regard to temporal and spatial patterns in the incidence of landslides, and that greater attention be paid to geotechnical factors.

1 FOREWORD

"In her thirteenth Year a prodigious Earthquake hapned [*sic*] in the East parts of Herefordshire, near the little Town call'd Kynaston. On the 17th of February, at six o–clock in the Evening, the Earth began to open, and a Hill with a Rock under it (making at first a great bellowing noise, which was heard a great way off) lifted itself up a great height, and began to travel, bearing along with it the Trees that grew upon it, the Sheep folds and Flocks of Sheep abiding there at the same time. In the place from where it was first mov'd it left a gaping distance forty foot broad, and fourscore ells long; the whole Field was about 20 acres[*]: Passing along, it overthrew a Chapel standing in the way, remov'd a Yew–tree planted in the Church–yard from the West into the East; with a like force it thrust before it High–ways, Sheep–folds, Hedges, and Trees, making tilled Ground Pasture, and again turning Pasture into Tillage : Having walk'd in this sort from Saturday in the Evening till Monday Noon; it then stood still "(Baker 1674).

This account from 1571, the thirteenth year of Queen Elizabeth's reign, is one of the earliest reliable English landslide records. In retrospect we may note that it was probably a first–time slide in clayey and rocky Palaeozoic strata, it occurred in about the wettest part of a winter, within the Little Ice Age, it had a considerable, though fairly slow, run–out and consequently, although an appreciable hazard, it constituted a risk to property rather than to life.

* 1 foot = 0.305m; 1 ell = $1\frac{1}{4}$ yards = 1.14 m; 1 acre = 4840 yd^2 = 4047 m^2; fourscore = 80.

Much older evidence of human awareness of landslide hazard is provided by the burial by a debris flow of the Roman mining town of Risa, in Carinthia, in 328 A.D. (Eisbacher & Clague 1984), and the earthquake–induced rock and debris avalanche of 1789 B.C. at Wudu, Gansu Province, China (Li 1990). Geologically, of course, landslides have formed an integral part of the cycle of erosion and degradation for hundreds of millions of years. A fossil landslide in the Silurian of Wales, for example, is described by Jones (1937).

2 FRAMEWORK

The paper is arranged as follows. After an indication of the global scale of landslide hazards, the physical factors bearing on sub–aerial landslides are briefly summarised. Following an indication of the differing constraints involved in local and regional landslide hazard assessment, the methodologies currently used in regional landslide hazard assessment are described fairly briefly, as there are several good recent reviews of this area. In the following main part of the paper, the aim is to identify areas where future improvements in our landslide hazard assessment capability can be made. These consist essentially of improving our understanding of the dynamic and systematic aspects of landscape development, particularly in the Quaternary, of recognising spatial and temporal patterns of landslide occurrence and of increasing geotechnical imputs. Methodologically, the wider application of simple geotechnical techniques and the use of a terrain evaluation approach supported by detailed geotechnical/engineering geological case records of characteristic landscape units are recommended. More work is needed to find the best way of fully incorporating such geological/ geomorphological/geotechnical insights and inputs with the results of computerised factor mapping.

3 DEFINITION AND SCALE OF LANDSLIDE HAZARD

Landslide hazard refers to the probability of a landslide of given magnitude occurring within a specified time period and within a given area. The associated risk is the consequent damage to or loss of lives, property and services (Varnes et al., 1984). Corresponding hazard and risk maps can be made. Landslide hazard clearly has to be assessed before landslide risk can be estimated.
Studies of landslide hazard have naturally been pursued chiefly in areas of appreciable associated risk, such as the Andes, the Himalaya, the Alps, Hong Kong, Japan, Italy and the United States and parts of the former USSR. Landslide hazards may arise naturally or through the actions of man. They generally damage structures or property resting on the moving mass, will exert pressures on obstacles in their path and may undercut or damage land and property nearby (Hansen 1984).

3.1 Sub-aerial landslides

Landslide hazard assessments are normally directed at sub-aerial landslides. In these, the most severe losses of life have generally been caused by earthquake-induced landslides, for example that of 1920 in Khansu Province, China, which killed approximately 200,000 people, that of 1949 in the Tien Shan Mountains of Tadzhikistan which killed some 20,000 people and that of 1970 in the Peruvian Andes near Huascaran which caused over 18,000 deaths. Non-earthquake induced landslides killed nearly 1,200 people in the city of Kure, Japan, in 1945 and approximately 450 people in the landslide at Mayunmarca, Peru (Sheko et al. 1988).
Some estimates of the direct and indirect costs of sub-aerial landslides are given by Arnould & Frey (1978), Schuster (1978), Varnes et al. (1984) and Sheko et al. (1988). For example, these amounted to over U.S. $1,100 million in Italy in 1977, when the figure in the United States was around $1,000 million. By 1983 the U.S. figure had probably risen to nearly $1,500 million. A similar figure is estimated for landslide and mudflow losses in Japan in 1982. Annual landslide costs of $100 million are quoted for the 80,000 km of roads alone, in the mountainous parts of northern India. Annual losses from natural disasters generally, to which landslides contribute significantly, are estimated by the UN Disaster Relief Co-ordinator to amount to 1 or 2% of the gross natural products in many developing countries, including most of those in South-East Asia. In some cases, this is enough to make the difference between economic progress and stagnation (Fournier d'Albe 1976).

It is clear that in many regions "slope failures constitute a continuing and serious impact on the social and economic structure, of which the true measure is not in monetary units but rather in the disruption and attendant misery of human lives" (Varnes et al. 1984). A recent world review of the problem is provided by Brabb & Harrod (1989).
More recently, chiefly in connection with the United Nations International Decade for Natural Disaster Reduction, efforts to come to grips with this problem have been intensified. As part of this work, an International Geotechnical Societies' UNESCO Working Party on World Landslide Inventory has been established. A progress report is given in the present Symposium by Cruden & Brown (1992).

3.2 Sub-aqueous landslides

In both areal and volumetric terms, sub-aqueous (mainly submarine) landslides greatly exceed those which occur sub-aerially. Individual sub-aqueous slides can have enormous volumes; for example the Agulhas slide on a sheared continental margin off the coast of southeast Africa has an estimated volume of about 20,300 km^3 (Dingle 1977). About 40% of the submarine slopes of the Mississippi delta are reported as being occupied by landslides (Prior & Coleman 1984). Sub-aqueous slides damage submarine cables, destroy offshore structures founded in the sea-bed and cause damage to marine life. They may also retrogress to damage shore-line structures and can cause destructive waves. Useful reviews are given by Moore (1978), Saxov & Niewenhuis (1980), Prior & Coleman (1984) and Lee (1989).

3.3 Secondary hazards

Secondary hazards, that arise subsequent to the slide movement and its run-out, may pose a greater hazard than the failure itself. For sub-aerial slides, this is illustrated by the failure at Vaiont, N. Italy, in 1963, when a very rapid landslide from the flancs of Mt Toc entered the reservoir at its foot and displaced much of the reservoir water over the arch dam, in a wave about 100m high. This entered the town of Longarone and killed more than 2,000 people (Kiersch 1964, Hendron & Patton 1985). Some of the dangerous waves which have been generated by rockfalls into Norwegian fjords are described by Jørstad (1968). Other sub-aerial secondary hazards are brought about by landslide dams and their breaching, as discussed later. Secondary hazards from sub-aqueous landslides are illustrated by the non-seismic submarine flow slides of 1888 in Trondheim harbour, which produced a damaging wave up to 7m high in the fjord (Andresen & Bjerrum 1967). Earthquake generated submarine landslides may contribute to the production of tsunamis, which can have very destructive effects onshore.

Table 1.   Main physical factors bearing on subaerial landslides.

1. Bedrock geology
   a. Lithostratigraphy & sedimentology
   b. Structure: folds, flexural shear, faults & joints
   c. Fabric and layering
   d. In situ stresses

2. Quaternary geology
   a. Glacial and proglacial
   b. Glacitectonics
   c. Pluvial
   d. Periglacial
   e. Glacio-eustasy &-isostasy

3. Geomorphology
   a. Slope morphology & gradient
   b. Slope aspect
   c. Former landslides & other mass movements
   d. Energy & state of development of landscape

4. Weathering
   a. Physical, chemical & biological
   b. Endogene & exogene, past & present
   c. Regolith thickness, rate of formation

5. Erosion & deposition
   a. General erosion, knick points, fronts of aggression
   b. Erosion of toe & face of slopes
   c. Deposition at head of slopes
   d. Surface erosion, gullying
   e. Seepage (internal) erosion

6. Climate
   a. Precipitation
   b. Evapo-transpiration
   c. Freeze-thaw
   d. Heat expansion & cracking

7. Vegetation & pedology
   a. Vegetation types, root strength, etc.
   b. Palaeosols, datable indicators of past stability & instability

8. Hydrogeology
   a. Run-off & infiltration
   b. Snow drifting, snow melt
   c. Groundwater pressures in fissures, soil pipes, burrows
   d. Artesian & perched groundwater
   e. Positive & negative pore-water pressures
   f. Groundwater pressure variations with depth
   g. Groundwater chemistry

9. Geotechnics
   a. Index properties, mineralogy, clay content & cementation
   b. Geochemistry
   c. Shear strength: peak, fully softened & residual, anisotropy, stress history, progressive failure
   d. Presence or absence of pre-existing shears
   e. Brittleness, rate effects on shear strength
   f. Metastable structure, porosity
   g. Swelling & shrinkage
   h. Permeability: profile and anisotropy
   i. Unit weight, variation with rain infiltration

10. Volcanic activity
    a. Lava flows, diversion of drainage
    b. Ash accumulation
    c. Steam emissions, forming intense rainfall
    d. Hydro-thermal alteration

11. Neotectonics & seismicity
    a. Tilting, uplift & enhanced erosion
    b. Earthquake shaking

12. Natural dams
    a. Landslides
    b. Lava flows
    c. Glaciers & ice sheets
    d. River ice

13. Human activity & land-use
    a. Cuts & fills
    b. Mining
    c. Forestation & deforestation
    d. Irrigation and leakage
    e. Other hydrogeological modifications
    f. Impounding, draw-down & critical pool effects

4   PHYSICAL FACTORS BEARING ON SUBAERIAL LANDSLIDES

Landslides are generally a natural accompaniment of the geological cycles of uplift, weathering and erosion. These underlying, long-term preparatory factors for landslides and the more local, much shorter-term effects which trigger a particular failure have been reviewed by many authors, for example; Terzaghi 1950, Cooke & Doornkamp 1974, Pasek et al. 1977, Carrara & Merenda 1976, Cotecchia 1978, Kienholz 1978, Varnes et al. 1984, Brabb, 1984, Hansen 1984, Brunsden 1985,

Sidle et al. 1985, Twidale 1985, Crozier 1986 and Mulder 1991.

The main factors involved, summarised briefly in Table 1, underlie the discussion in the rest of the paper.

5 LOCAL AND REGIONAL SUBAERIAL LANDSLIDE HAZARD ASSESSMENT

In principle, all the physical factors bearing on landsliding (Section 4) should be considered in any landslide assessment. In practice, local assessment of individual landslides or restricted areas tend to concentrate on the site–specific, particularly sub–surface features, and may fail to take sufficient account of the more general, spatial and temporal aspects of the environment. Regional landslide assessment, on the other hand, has generally to rely heavily on features that can readily be observed and measured at the ground surface and is thus commonly deficient with regard to information on sub–surface features which, if available, is often based on extrapolation from a few scattered sites.

5.1 Local Assessment

Most geotechnical landslide investigations are essentially local and site specific in character, being concerned with establishing the nature and degree of stability of a particular slope or slope failure or an associated group of these. This is normally achieved through well–established, preferably step–wise procedures. These should comprise desk studies, geomorphological mapping in the field and from stereo aerial photographs or even satellite imagery for surveys of large areas, surface movement monitoring, shallow subsurface investigations (e.g. trial trenches), deeper subsurface investigation (e.g. boreholes, shafts, adits), location of slip surfaces, monitoring including especially the installation of piezometers to measure the ground–water pressures acting on the slip surface and their seasonal variations, sampling, index and special testing (particularly of shear strength), back–analysis and, if appropriate, design of stabilisation works and installation of long–term monitoring. Each step in this process should be designed on the basis of the information yielded by the previous steps. Such spatial determination of the degree of stability of a slope may be supplemented by a temporal study to determine when it might fail. The latter technique, using pre–failure movement measurements, is treated by Saito (1969), Varnes (1983), Fuzukono (1985), Vibert et al. (1988), Salt (1988) and Voight (1989).

5.2 Regional assessment

The current cost, in the UK, of detailed local geotechnical site investigations range between from below 0.5 to over £5.0/m^2 (Table 2). Hence it is generally not feasible to apply such methods to the assessment of landslide hazard over large regions, possibly covering some hundreds of square kilometres and even a whole country.

However, more use should be made generally of the simpler geotechnical approaches, while in smaller, more urban areas, some detailed geotechnical input will often be appropriate, as noted below.

One aim of the present paper is to review briefly the principles and methodologies (in which the scope for sub–surface investigation is generally

restricted) which can be employed in such regional landslide hazard assessments. In this I have been much helped by previous reviews, and particularly those of Ollier (1977), Cotecchia (1978), Varnes et al. (1984), Brabb (1984), Hansen (1984), Griffiths (1986), Crozier (1986), Hartlén & Viberg (1988), Mulder (1991), Siddle et al. (1991), Hansen & Franks (1991) and Youd (1991). Reference to these and the original papers should generally be made for more details and for illustrations of the cartography and symbols used.

It is clearly preferable to make regional landslide assessment before development is carried out rather than after. However, in very many cases this is, of course, no longer possible and such assessments have to be made in both non–urban and urban environments. There is much common ground as far as the methodologies are concerned. The most important difference is that the elements at risk are normally a much more considerable factor in the urban environment. Accordingly there is even more pressure to produce high quality and reliable landslide hazard assessment maps there. This is aided by the generally greater availability of pre–existing surface and sub–surface data : in addition the financial environment makes the carrying out of supporting geotechnical work both more necessary and more feasible.

6 APPROACHES TO REGIONAL SUBAERIAL LANDSLIDE HAZARD ASSESSMENT

Outside China, where these matters were being considered in antiquity, it is in the populated and landslide–prone areas of the Old World where the earliest efforts to record and assess landslides appear to have taken place. Cotecchia (1978) singles out Sarconi's map of 1783, of landslides and landslide–dammed lakes produced in Calabria by the earthquake of that year, as the earliest example of a regional landslide map in Italy (Sarconi 1784). He also notes that the first landslide map of the whole country, at a scale of 1:500,000, was published by Almagia (1910).

In the high Alps, notable contributions to the understanding of landslides were made by Stini (e.g. 1910, 1938), Mougin (1914) and Heim (1932), not least with regard to documentation and classification. The considerable losses of life through landsliding in the Alps affected Heim deeply, and led him to make very sensible proposals to reduce this. These were, initially, for a landslide–experienced geologist to walk around the entire area giving reason for concern about instability, to prepare a large scale map and install and monitor a system of appropriately placed survey stations. In addition, he urged the creation of a surveillance service with, at the same time, the passing of information to the affected public about the danger zone, the best direction for flight and, eventually, the timing of its evacuation. Post–failure, Heim saw the need to

Table 2.   Approximate costs of investigations in landslipped ground (D.R. Norbury, pers. comm.).

Level of investigation	Engineering	Range of cost £/m²
Walkover/Desk Study	Establish presence, extent and nature of landslip terrain by remote sensing, desk study and walkover	<0.01
Reconnaissance	Confirm extent and nature of landslip terrain with mapping, simple field investigations, and initial assessment of stability	0.01-0.10
Design	Establish stability state of landslip and obtain parameters to allow design of stabilisation works and/or allow engineering works to proceed without reactivating movements	0.10-1.00 (simple) 0.50-5.00 (complex)
Post-failure	Engineering works or other man made (re) activation of landslip having occurred, detailed investigation for re-establishment and other remedial works	1.00-10.00 (simple) 10.00-100.00 (complex)

Notes:
1)   'Simple' decides straightforward ground profile, low drilling costs, average sophistication and extent of sampling, testing and instrumentation.

2)   'Complex' denotes complex and/or variable ground profile, high drilling costs due to nature of ground, extensive/high sophistication of sampling, testing and instrumentation.

3)   Larger sites generally have economies of scale and so verge toward lower end of cost ranges.

check for dangerous remnants of the slide, to specify decide on any remedial measures and whether and when it would be safe for the evacuees to return. Heim acknowledges Tscharner, the mayor of Mayenfeld, as his forerunner, in 1807, in promulgating such advice.

6.1 Principles and postulates

The most fundamental of the principles on which landslide hazard assessment is based is:
   a) the past and present are keys to the future. Thus natural slope failures in the future are most likely to occur in geological, geomorphological and hydrological conditions similar to those that have led to past and present failures (Varnes et al. 1984).   This is a useful axiom in geology, being an inversion of Hutton's concept that the past history of our globe must be explained by what can be seen to be happening now.   As pointed out by Varnes et al. (1984), it does not therefore follow that the absence of past or present failures means that failures will not occur in the future. This is particularly true in relatively young Post–glacial deposits like the glacio–isostatically heaved, former marine clays of Scandinavia. Similarly, it is evident that application of this principle could be disturbed or even falsified by significant changes in the conditions of the past and present, for instance by a significant shift of climate or by major intervention in the landscape by man.

The second basic element identified by Varnes is:
   b) the main conditions that can cause landsliding can be identified.
   This is a general principle, distinct, for example, from that which categorises landslides as Acts of God, but it also has elements of a postulate.   It might be better to write:
   b) the main conditions that cause landsliding are controlled by physical factors and are therefore, in principle, identifiable.   With our present technology and resources, some of these can be recognized and evaluated.

The third basic proposition of Varnes et al. (1984) is:
   c) degrees of hazard can be estimated.

1809

Similar comments apply to this as to b). It is again true in principle, but the extent to which the degree of hazard can be estimated depends on how much success is achieved in identifying and evaluating the conditions causing landsliding, in b).

Although doubtless inherent in the above three statements, it seems appropriate to distinguish a fourth:

d) the various types of landsliding can generally be recognised and classified, both morphologically, geologically and geotechnically.

It is clearly valuable, for example, when extrapolating data to an as yet stable area of similar physical setting, to be able to indicate also the type of landslide most likely to occur, with implications for size, speed, run-out, etc. Recent reviews of landslide classifications are given by Crozier (1986) and Flageollet (1988). Those of Varnes (1978) and Hutchinson (1988) are perhaps most commonly used.

In addition to the above, it is important to classify slides geotechnically, into whether they are first-time slides or slides on pre-existing slip surfaces and whether they occur under undrained, intermediate, drained or drained-undrained ground-water conditions (Skempton & Hutchinson 1969; Hutchinson 1988). In the present context the presence or absence of pre-existing shears and the degree of brittleness is particularly important. The latter largely controls run-out. Brittleness and run-out are generally greatest in first-time failures, expecially if they also involve drained-undrained behaviour. Brittleness on pre-existing shears is generally low or zero, so renewal of movement on such shears are usually slow, with little run-out. Some important exceptions to this, including negative rate effects on shear strength, are indicated by Hutchinson (1987).

6.2 Data collection

In all methods, the first and generally most expensive step is that of data collection. In this, standardisation of format is necessary to allow storage in some form of databank and to facililate subsequent computerised data handling and analysis (Hansen 1984). Examples of registration forms, embodying check lists, are provided by Cooke & Doornkamp (1974), Carrara & Merenda (1976), and Pašek et al. (1977).

6.3 Available methodologies

The discussion, in the following sections, of available methodologies for regional landslide hazard assessment refers primarily to the natural environment. The important group of landslides brought about by man's activities is dealt with separately in Section 6.9, partly as these tend to fall outside the first principle of Varnes et al. (1984), given earlier.

Although there is some overlap between the various methods used and they are sometimes combined, they fall essentially into three groups: the geotechnical approach; direct methods, based principally on geomorphological mapping; and various indirect methods, usually based on factor mapping. A more detailed break-down is provided by Hansen (1984). The aim is to produce a landslide hazard map, which shows and categorises all existing landslides and zones both these and the currently stable areas into degrees of susceptibility to landslide hazards. Not all the maps discussed fully reach this objective, but even simple landslide maps represent an enormous improvement on having no assessment of landslide hazard at all.

Over the past decade or so there has been an impressive degree of activity in the field of landslide hazard assessment in some countries. It has been feasible to review only part of this work here. Work on the related problem of snow avalanche hazard assessment is generally more advanced, as exemplified by the work of Hutter et al. (1987), Abe et al. (1987) and by other papers in the Davos Symposium.

6.4 Geotechnical approach

This approach, discussed above under local assessment of landslide hazard, is highly developed but, as noted, is for the most part too expensive to use for regional assessments. However, the simpler geotechnical techniques of ground assessment, for example involving surface or near-surface logging, sampling and index testing, lend themselves well to regional landslide hazard assessment, while in more urban areas of moderate size some borings and subsurface instrumentation may be required. An example of an application of this type to clay slopes in Tasmania, using index properties and outline sub-surface data, is given by Stevenson (1977). As noted earlier, in cases where the elements at risk are such that an improved accuracy and reliability of hazard assessment is required, further use of geotechnical methods will tend be both appropriate and financially justifiable.

6.5 Landslide inventories and landslide maps

All landscapes may, in principle though not always in practice, be divided into areas where landslides have already occurred and areas where landslides have not yet occurred. Under favourable conditions, the existing landslides can be mapped directly, chiefly by some form of remote sensing, with such field checks as are feasible. Satellite and radar imagery are particularly valuable for small-scale regional studies of large-scale instability. Vertical stereoscopic aerial photographs generally provide the most useful basis for direct landslide mapping at both the local and regional scales (Hansen 1984). An early example of a small–scale, regional map is that for the whole of Czechoslovakia, produced at a scale of 1:1,000,000 (Rybář et al. 1965, Nemčok & Rybář 1968). In this, natural landslides are distinguished from those induced by mining and quarrying. An excellent landslide inventory map is that produced by the Geographical Survey Institute, Japan (1983). More detailed maps of this type (e.g. Wieczorek 1982, Oyagi et al. 1982) may also show zones and areas of erosion and deposition, both with respect to landslides and generally, the type and estimated depth of sliding and some data on the nature of the material involved. It is important to subdivide existing landslides into active and inactive or preferably by their sensitivity to the expected meterological, erosional and seismic triggers. This is usually done on the basis of historical data, evidence of recent movement and freshness of form (Del Prete et al. 1992).

In practice, even in thickly vegetated areas, fresh landslides can readily be mapped. With the passage of time, however, the landslide scar may cease to be recognisable, either through degradation or erosion of its features or through the growback of vegetation (Brabb 1984). For contemporary slipping, this problem can be overcome, as in Hong Kong, by taking aerial photographs annually (A. Hansen, pers. comm.). Depending on circumstances, the longevity of scars may range between a few years and many thousand years. Further problems of recognition arise if ancient Pleistocene landslides are present. Some of these, like certain solifluction sheets, have little surface expression while others may be completely hidden by a subsequent slope mantle (Fig. 1).

The great potential usefulness of even the most basic landslide map is illustrated, unfortunately in a negative sense, by events at Sevenoaks, Kent, S.England in the early 1960s, where a new road was planned on sidelong ground, below a Lower Cretaceous scarp. A high quality, photogram–metric contour map was made, but used solely for cut and fill balances. Construction was started but immediately ran into great difficulties, mainly through the reactivation of 'fossil' periglacial landslides and solifluction lobes, formed in the Late–glacial (Skempton & Weeks 1976). This forced a major re–alignment of the route at a (then) cost of around £2,000,000. This loss could have been saved by the most basic recognition of the landslide origin of these features, which were generally evident on the ground, on air photographs and on the photogrammetric map.

Landslide inventory maps are not landslide hazard maps, but even if only this first stage is carried out, the social, planning and economic benefits can be very considerable.

6.6 Hazard zonation within a landslide area

In some cases, virtually the whole of the region of concern is occupied by landslides. The

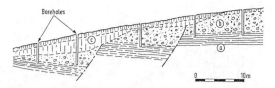

Figure 1. Landslides in terrace gravels (b) overlying Cretaceous marls (a), mantled and concealed by an intact Pleistocene loess cover (c) (after Záruba & Mencl 1982, Fig. 6-7, p. 153).

landslide map, with appropriate subdivision and zoning of activity, then becomes the landslide hazard map. An example of such a situation is provided by the Undercliff of the Isle of Wight, England. This is a coastal slide area around 12km long and half a kilometre wide, with a population of more than 6,000 people. An initial slide hazard assessment of the most built–up area, using a combination of geology, geomorphology, historical data, information on changes in level of Ordnance Survey benchmarks during the preceding 90 years and observations of distress to structures, was made by Chandler & Hutchinson (1984). This work has been greatly expanded by Geomorphological Services Ltd/Rendel Geotechnics under contract to the Department of the Enviroment (Lee et al. 1991). A preliminary landslide zonation of the whole Undercliff has been made by Hutchinson & Chandler (1991).

6.7 Landslide susceptibility and hazard maps

The making of landslide susceptibility, hazard or potential hazard maps involves the very difficult task of assessing the stability of the areas where landslides have not yet taken place, with generally very restricted subsurface information. Some methodologies for attempting this are reviewed below.

Primarily cartographic methods. Perhaps the simplest way of extending a landslide map to a landslide susceptibility map is to assess the percentage of the area of each litho–stratigraphical unit that is occupied by landslides and to rank the units generally on this basis (Radbruch–Hall et al. 1976, and later updates). An improvement is to make landslide orientated geomorphological maps, showing all existing landslides and land forms of an area, plus any geological and hydrogeological information. They thus constitute more than just landslide maps, as some inferences can be made about the status of the non–slid areas. Three examples of maps of this type, which have proved invaluable in both the general and detailed planning of roads and other public works are: a map of 65km of proposed road alignment in Eastern Nepal; and a map of a 10km length of the Taff valley, S.Wales, for a road improvement (Brunsden et al. 1975); also a map of landsliding and erosion in the Lattarico area of Italy (Carrara & Merenda 1976). In addition Varnes et al.

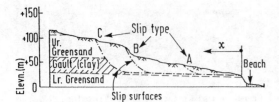

Figure 2. Preliminary section of coastal landslips in Cretaceous strata at Bonchurch, Isle of Wight (after Hutchinson et al. 1981).

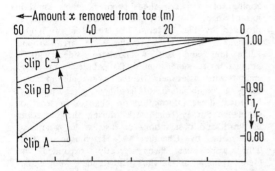

Figure 3. Reductions in stability caused by removing x metres of ground from the toe of slips A, B, & C (Fig.2). F_0= original factor of safety; F_1= that after x metres of toe removal (after Hutchinson et al. 1981).

(1984) describe a very detailed geomorphology and geology based slope map, at scale 1: 50,000, of unstable landforms in Kyushu, Japan, which also makes some indications with regard to landslide zonation (Hatano et al. 1974).

Notable work in this area of geomorphology and process mapping has been carried out over the past decade and a half in France. This has led to the production of the important landslide hazard maps termed ZERMOS (Zones exposées aux risques liés aux mouvements du sol et du sous–sol), generally at a scale of 1: 25,000 (Humbert 1977: Antoine 1977: see also Varnes et al. 1984). Such mapping has been developed further in the more slide–prone areas of France into POS (Plans d'occupation des sols), which have legal force in relation to the use and occupation of land (Porcher & Guillope 1979), and into PER (Plans d'exposition aux risques) in response to a law of 1982 concerning the indemnification of victims of natural catastrophes. The history of the technical and legal developments which have led from ZERMOS to PER is reviewed by Champetier de Ribes (1987). This general approach has been carried further by Kienholz (1978) in 1:10,000 scale maps of the Grindelwald area, Switzerland, and by Malgot & Mahr (1979) in maps of similar scale in the Handlova and Kordiky areas of Czechoslovakia.

Combined approaches. A main theme of this paper is that landslide hazard assessments should be based on as fully integrated multi–disciplinary approach as is feasible. Some work of this general type has already been done: three examples are given here.

For clay slopes in Tasmania, Stevenson (1977) developed a method of assessment which depends partly on geomorphological and other surface factors, namely slope angle, slope complexity and land use, and partly on geotechnical factors, namely clay factor (PI in lower third, mid–third or upper third of range) and water factor (highest seasonal position of piezometric surface relative to typical failure surface either below the latter, between this failure surface and mid–depth of slide mass or above mid–depth of the slide). These factors are given scores, which are then weighted and combined to give a numerical rating for each site, found very useful in planning decisions. The method is feasible because the geological situation is relatively simple, being dominated by slab sliding in clay slopes, and the environment is sufficiently urban for some sub–surface data to be available.

The second example concerns a coastal landslide in near–horizontal Cretaceous strata, again in the Undercliff of the Isle of Wight, England. There, although the lithologies involved are varied, the geological structure is simple. Thus, with a combination of geomorphological mapping and the surveying of exposed geology to determine the structure, it was possible to make preliminary estimates of the sub–surface conditions (Fig. 2). A landslide hazard assessment within this extensive landslide area was then carried out, partly on the basis of the geology and geomorphology and partly through deterministic stability analyses of typical failure modes, A, B, & C. In addition the sensitivity to further coastal erosion of the various landslides considered (Fig. 3) was calculated (Hutchinson et al. 1981).

A final example is provided by the multi–disciplinary research programme "Detection and Use of Landslide–prone Areas", at Lausanne. This involves a truly combined approach, using geology, geomorphology, historical data, accurate measurement of surface movements and sub–surface investigation (Bonnard & Noverraz 1984, Ecole Polytechnique Fédérale de Lausanne 1985).

Terrain evaluation. A joint geomorphological/ engineering geological approach is that of terrain evaluation (also termed analysis or classification), reviewed by Way (1973), Ollier (1977) and in the Geological Society Working Party Report in Land Surface Evaluation for Engineering Practice (1982). In this technique, the total area to be mapped is divided into a number of basic terrain units (components, facets), of similar morphology, geology and landslide (or other) characteristics, enabling data for a particular unit to be utilised more generally. Recurrent groups of such units

Table 3. Some physical factors currently considered in indirect landslide hazard assessment (based largely on Carrara & Merenda 1976 and Carrara et al. 1990).

Geology
  Lithological characterisation
      (I_p sometimes used)
  Lithostratigraphical conditions,
      monolithological, alternating, etc.
  Slope/structure relationships
  Degree of weathering
  Nature & thickness of regolith

Erosion
  Monocyclic or polycyclic
  Stream erosion, lateral, downward
  Sheet, rill erosion
  Gullying
  Bad-lands
  Seepage (internal) erosion

Hydrogeology
  Climate
  Permeable beds capping impermeable
  Impermeable beds throughout slope
  Inferences from vegetation
  Rough estimate of piezometric level
      relative to slip surface

Geomorphology
  Presence or absence of former slides
  Landslide type & morphometry
  Degrees of activity of slides
  Slope inclination
  Slope roughness
  Slope aspect
  Mean elevation of slope unit
  Slope form and size in plan
  Slope form in downslope profile
  Slope form in cross-slope profile
  Size of sub-drainage basin

Seismicity
  Seismic zoning & micro-zoning
  Liquefaction potential

Land-use, human activities
  Forestation, deforestation
  Rooting depth, root strength
  Cultivation
  Irrigation
  Earth-moving operations
  Nature of any corrective measures

(systems) can sometime also be usefully distinguished. The approach appears to have been developed independently in Australia (Christian & Stewart 1953: Grant 1965) and the U.S.S.R. (Solentsev 1962). It has been widely used in Australia where the PUCE (Pattern – Unit – Component – Evaluation) system is now well developed (Finlayson 1984). In Hong Kong, terrain evaluation forms an important part of the GASP (Geotechnical Area Studies Programme), in which GLUM (Geotechnical Land Use Map) classes define areas of varying geotechnical limitations to development (Brand et al. 1982, Brand 1988a, Styles & Hansen 1989). It has been applied in Sweden by Viberg & Adestam (1980).

A possible criticism of the terrain evaluation approach is that it is often based on the present situation in a landscape and may thus lead to neglect of its dynamic and evolutionary aspects (J.S. Griffiths, pers. comm.). This need not be the case however, as illustrated in the subsequent discussion of abandoned cliffs.

With careful choice of which geological/ geomorphological entities qualify as terrain units, supported by representative detailed geotechnical/ engineering–geological case records, and accompanied by an appreciation of the past and likely future development of the landscape, the terrain evaluation approach seems the best way of extrapolating available sub–surface data and ensuring that physical insights are fully considered in arriving at a landslide hazard assessment.

Indirect factor mapping and numerical methods. In this widely used, and flexible approach,

landslide susceptibility is estimated from an often weighted combination of factor maps, covering the statistically more significant of the measurable physical factors summarised in Section 4. Current practice with regard to these is summarised in Table 3. By reason of being limited virtually to surface observations, the list of factors in Table 3, while rather comprehensive, is inevitably incomplete in relation to the physical factors summarised in Table 1.

In the early maps of this type, produced in California in the 70s, the three factors of the landslide distribution, slope declivity and nature of the bedrock were shown on line maps and used to define the various land units by manual compilation, in pioneering work by Brabb et al. (1972). It was later found advantageous to generalise and quantify the landslide deposits in contour form. For this purpose, isopleth maps of these and other deposits, showing lines of their equal percentage coverage, were prepared at scales of 1:24,000 (Campbell 1973) and 1:125,000 (Wright & Nilsen 1974). The procedure is described by Wright et al. (1974). Fulton (1986) points out that this technique can become misleading at large scales.

With increasingly complexity (Neuland (1976), for example, used 31 initial parameters), the technology of the time required that, for clarity and convenience, multivariate landslide susceptibility maps be based on grid cells and be handled by computer. Cell sizes are usually square between about 50 and several hundred metres, depending upon the mapping scale. The end result of the above operations is usually a numerical rating for each cell, which forms the basis of the landslide hazard assessment. However, grid cells of fixed size have the disadvantage of often relating poorly to the geomorphologically meaningful slope units distinguishable in the landscape. Accordingly, later work (e.g. Carrara et al. 1990), facilitated by the use of GIS, associates land characteristics with these units rather than with grid cells. Discriminant analysis is sometimes employed, using empirical combinations of measured attributes, to help in distinguishing stable from unstable slopes (Payne 1985, Carrara et al. 1990). This is found unsatisfactory by Mulder (1991). Assessments of the above nature have been made, for example, by Neuland (1976), Carrara et al. (1978), Newman et al. (1978) and Siddle et al. (1991). Further examples of indirect factor mapping methods are given by Hansen (1984) and by Jennings et al. (1991) and an assessment of factor representation at different scales by Hansen & Franks (1991).

6.8 Reliability of landslide hazard assessment maps

An unsatisfactory feature of the present situation is that landslide hazard assessment maps are being produced in large numbers, but with a generally uncertain degree of reliability. Many of the authors of the maps are acutely aware of this and have made efforts to check them. For example Brabb (1984) reports that in California in 1982 thousands of debris flows occured in extreme rainfall in areas mapped earlier, by Brabb et al. 1972, as not susceptible to landslides. Further study showed that many of these areas had been occupied by earlier old slides, but that these were missed because of their shallowness, subsequent degradation and vegetation cover. Similar difficulties attended the mapping of earthquake-induced landslides as most of those from the last major shock, in 1906, were not visible on the 1968 air photography. Brabb (1984) also draws attention to failures, elsewhere in the U.S., to estimate the volume and run-out of debris flows.

In the case of the stable areas, their susceptibility to future landsliding, as shown on landslide hazard assessment maps, does not so readily lend itself to being checked. If the existing landslides have not moved or broken up too much, it is sometime possible to reconstruct their pre-slide attributes, particularly their profile, and to incorporate these attributes in an indirect factor analysis. The degree to which this predicts the actual existing landslides then provides some degree of check (Griffiths 1986). Another measure which can provide a degree of internal check, but not an absolute one, is to assess the landslide hazard of an area by several different methods, comparing the results. This has been done in South Wales by Jennings et al. (1991), who found an empirical landslide hazard alogorithm marginally simpler and more effective than a logistic or a discriminant statistical

approach. Success rates in excess of 70% are claimed for all the methods. Data from other checks on statistical multivariate models are quoted from the literature. For six of these, rates of prediction success are claimed which vary from 67 to 94% and average more than 70%. An internal check by Carrara et al. (1990) on their assessment of the Tescio basin, Italy, indicated that 83% of grouped cases were correctly classified. A related study has been carried out in the area of Wellington, New Zealand, by Gee (1991). As the nine different methods of assessment compared there give maps of similar quality, it is recommended that the simplest and cheapest methods be used.

Further such comparisons, but preferably employing two or more independent teams, should be made. However, the true reliability of landslide hazard assessment maps can clearly only be properly checked through the passage of time. Such checks should be made and published as soon as meaningful, so that our methodologies can be improved. Preliminary checks of this nature are reported by Siddle et al. (1991). Of 8 events in 4 years, 5 occurred in areas mapped earlier as old slides and 3 took place in areas mapped as having some landslide potential (H.J. Siddle, pers. comm.).

6.9 Effects of human activity

As noted earlier, the earth-moving and constructional power now available to humans is such that very large scale interferences with slopes are possible. These usually take the form of cuts, fills or changes in the erosion, external water or groundwater regime. The larger projects are normally, though not always, carefully investigated. This is less commonly the case with small projects, which thus generally give rise to the most trouble (D. Brook, pers. comm.). It seems best to make landslide hazard maps for natural conditions and triggers and to cover possible human interference, which may take many forms and be of various scales, by warning notes calling for detailed geotechnical/ engineering geological investigations before any significant works are carried out. In some cases, it may be possible also to be more specific about what is meant by "significant". For example, for some regolith mantles within a certain range of slope angles, it may be justified to state that they can tolerate cuts or fills not greater than a given height, possibly a metre or so. On artificial reservoir perimeters it is generally advisable to specify a maximum permissible draw-down rate, while on coastlines cutting across sensitive deep-seated landslides, attention should be drawn to the danger of reactivating these if the littoral drift is interrupted.

Many hazardous landslides, usually of flow slide type, have occurred from industrial waste tips (Bishop et al. 1969, Bishop 1973) and from tailings dams (Seed 1968). With the continuing increase in the numbers and scale of such structures, particular care must be exercised in their siting, design, construction, maintenance and monitoring.

Man's activities with regard to the planting and clearance of vegetation have important effects on stability. There seems to be general agreement that a cover of vegetation improves protection against shallow slides and erosion, chiefly by reducing infiltration, reducing pore-water pressures, by increasing evapo-transpiration and by the reinforcing action of roots (Cotecchia 1978, Varnes et al. 1984, Greenway 1987). There is similar agreement that destruction of the cover of vegetation by man is extremely damaging with regard to both these factors. In the case of forest clearance, time needed for the roots to rot may delay the main onset of sliding by a few years (Sidle et al. 1985). The effect of tree cover on deep-seated landslides is more problematical. Some workers suggest that the stability of these can be reduced by the weight of the trees and associated wind forces.

7 REGIONAL SUBAQUEOUS LANDSLIDE HAZARD ASSESSMENT

The enormous scale of subaqueous landslides has already been referred to. With the great increase in offshore structures, particularly oil and gas platforms, on continental shelves and even the continental slopes, there is a clear and increasing need for subaqueous landslide hazard assessments.

There are advantages and disadvantages in dealing with subaqueous, in comparison with subaerial, landslides. The general exploration method, of seismic reflection profiling, gives a fairly good indication of the internal structures and depths of the landslides. On the other hand, sampling, dating, morphological survey and movement measurements are more difficult than on land. The activity or inactivity of subaqueous landslides usually has to be judged from the freshness of their features or changes in topography as revealed, for example, by side-scan sonar or by repeated seafloor surveys. These techniques are reviewed by Prior & Coleman (1984). Some fundamental aspects concerning the incidence of submarine landslides: their concentration on continental margins and their relationship to plate tectonics, sedimentary processes and Quaternary sea level changes, are discussed by Moore (1978). Submarine landslide hazard assessment is implicit in much recent work but, as noted by Brabb (1984) and Cruden et al. (1989), this generally comprises small scale, landslide inventory maps of specific sites. The continuing improvement of survey techniques, particularly side-scan sonar of broad coverage and high resolution, is making submarine hazard assessments increasingly possible.

8 FURTHER DEVELOPMENTS

8.1 General

The earth's surface is part of a dynamic system, evolving both progressively and episodically through the actions of geological, geomorphological and meteorological processes. Landslides form one aspect of this evolution, controlled by various physical laws operating amidst the great variety of geology, landforms and climate. Weathering tends to weaken the rocks and prepare them for failure : the actual triggers of failure are predominantly erosive, climatic or seismic. Despite the undoubted complexities, significant spatial and temporal patterns of landslide incidence and behaviour do exist (e.g. Hutchinson 1973, Cotecchia & Melidoro 1974, Palmquist & Bible 1980, Nemčok 1982, Brunsden 1973, 1987). To appreciate fully the geological, erosive and climatic settings, to discover and define the patterns of landsliding and to build these physical insights into our methodology are considered to be crucial to the further development of landslide hazard assessment.

In view of the physical factors bearing on sub-aerial landslides, outlined in Section 4, the task of assessing the propensity to these, particularly on a regional scale, is formidable, or even daunting when Terzaghi's (1929) dictum concerning the importance of "minor geologic details" is borne in mind. Nevertheless, given the very considerable suffering and economic losses caused globally by landslides, we have a clear responsibility to do what we can to anticipate and mitigate these hazards.

This undertaking will be best served by bringing to bear all the relevant disciplines in a fully integrated manner. Approaches to this have been made by some groups, but all too often local assessments are made chiefly by geotechnical engineers, with the geological and geomorphological factors not infrequently neglected, and regional assessments are made principally by geomorphologists and geologists, with rarely much geotechnical input. It behoves us, especially in this International Decade of National Hazard Reduction, to examine the processes of landslide hazard assessment in our own countries and to deploy our particular skills where we can help to reduce present inbalances of approach.

8.2 Remote sensing and data treatment

The most rapidly developing parts of landslide hazard assessment at present are those concerning remote sensing and the treatment of the resulting data, as outlined above. Remote sensing systems range from hand-held cameras to satellites (Hansen 1984). The resolution of the latter is improving and systems like SPOT are now useful in the sensing of large areas. Thick vegetation cover, such as equational forest, remains a problem and there appear to be few, if any techniques as yet, which will yield reliable quantitative information on the ground surface morphology under such circumstances. Air-borne radar holds some promise and is currently being explored (M.G. Culshaw, pers. comm.). It has the advantage of being unconstrained by light and weather conditions. A side-looking radar scanner was used with some success in forested areas of Brazil. While not yielding quantitative results, it did give a useful qualitative impression in particular the roughness of the terrain. Some maps of this type, "Radam Brazil", are commercially available (J.C. Doornkamp, pers. comm.).

More quantitative results in forested areas appear to be attainable through the use of air-borne vertical laser sensing. A system developed in Canada, known as ALTP (air-borne laser topography profiler) is operated from a helicopter travelling at 160 km/hour. With a lazer pulse frequency of 2000/second, this gives a reading about every 2 cm or so along the flight path. The aim is to penetrate the leaf cover with a proportion of the readings, the ones giving the lowest elevation being used (J.S. Griffiths, pers. comm.).

The application of aerial photography is discussed by Crozier (1986) and details of earth observing satellites are given by the National Remote Sensing Centre (1990). Digitisation of the resulting landform data enables them to be handled rapidly and conveniently in Geographical Information Systems, through which the observable, multivariate landslide controlling factors can be readily explored, analysed, processed and presented (Wadge 1988, Carrara et al. 1990, Van Driel 1991). These approaches usually lead to estimation of a probability of failure (Brand 1988b). The relationship of factor of safety to this is not uniform but depends on the degree of conservatism inherent in the various assumptions of the stability analysis. In parallel, expert systems and artificial intelligence are also being deployed (Wislocki & Bentley 1991). Such advances are welcome. In combination with the disciplines of probability and statistics, they enable the most to be made of the available data (Carrara et al. 1977, Chowdhury 1984).

It is clear, however, from the foregoing review that whatever the power and complexity of such remote sensing-based techniques, there is likely always to be a number of relevant sub-surface factors remaining unknown, for example, depths of slip surfaces and the groundwater pressures acting on them. To a greater or less degree, depending upon the particular circumstances, this will limit the reliability of the landslide hazard assessment that can be achieved. The danger that this self-evident fact will be overlooked as the sophistication of these techniques increases can be avoided by operating in balanced, multi-disciplinary teams giving proper attention to all relevant factors.

The shallower mass movements, involving principally superficial deposits and related to ground slope rather than the underlying geology, are likely to be handled more satisfactorily by remote sensing-based techniques, particularly when the movements are of short return period. Debris flows epitomise the above category and, of these, lahars are in some ways the simplest as the material involved, often mainly tuff, is finer-grained and more uniform than in debris flows generally. On the other hand, particularly in a volcano discharging large quantities of tephra, the surface form of the cone may change too rapidly for this important parameter to be established. In cases where the terrain could be adequately defined, lahar modelling would also embrace the intensity and distribution of the rainfall, the incorporation of this with the surficial volcanic debris, the physical properties controlling the flow of the resultant mixture and hence its track, speed, discharge and run-out. The recently active and much studied lahars of the Ruiz Volcano, Colombia, may be a possible place to try out this approach. Work of this type has already been done, for example by Wadge & Isaacs (1988) who used a digital terrain model, with image processing, to predict future hazards from pyroclastic flows on Soufriere Hills Volcano, Montserrat, and by Griffiths & Richards (1989), who used GIS in conjunction with terrain evaluation techniques and the Universal Soil Loss Equation in soil erosion and soil sonservation studies in Ethiopia. It may prove possible to extend the approach to debris flows generally, and possibly to other relatively shallow mass movements.

Geophysical methods are little used in regional landslide hazard assessment. Some examples of the application of ground techniques, at scales ranging from the local to lengths of proposed motorway, are given by Culshaw et al. (1987). The use of air-borne geophysics, as distinct from normal remote sensing, does not appear to have been explored.

8.3 Geotechnical inputs

It is accepted that it is normally financially and logistically impossible to apply the usual geotechnical, site specific investigation techniques to landslide hazard assessment on a regional scale. However, some of the techniques are simple and inexpensive and should be more widely used. Index tests are a prime example. These have been developed precisely for such a reconnaissance role, to give simple measures of soil and rock properties at small expense. There are often good opportunities to take weathered samples from surface exposures. These should be supplemented, where feasible, by the logging and sampling of machine or hand-dug trial trenches (Bell & Pettinga 1986). Furthermore, many useful correlations exist between index properties, such as water content, liquid and plastic limits, clay fraction, granulometry and porosity or dry unit weight, and other physical properties such as shear strength, compressibility and permeability. Such correlations are particularly good between residual shear strength, clay fraction and I_p (e.g. Vaughan et al. 1978, Chandler 1984 and Skempton 1985). Where the terrain is simple enough, approximate deterministic and probabilistic estimates of stability can then be obtained using, for instance, infinite or finite slope analyses or the Bishop & Morgenstern (1960) dimensionless charts in conjunction with an anticipated range of shear strength values and pore−water pressures.

It follows from the geotechnical classification of landslides, outlined above, that an important question in any landscape is to what extent pre−existing shear surfaces or zones are present and may affect potential modes of slope failure. These are particularly significant where the Brittleness Index, I_B, is high. Bishop et al. (1971) quote I_B values in excess of 0.75 for some highly plastic clays. Pre−existing shears will, of course, generally be present in the existing landslides, but have they been formed, by causes other than landsliding, in the intervening, un−slid areas? If such shears are deep−seated, they are likely to be difficult to locate without subsurface exploration, unless they are exposed in outcrop. Fortunately, in some cases, non−landslide induced pre−existing shears are quite shallow. This applies particularly to those produced by the periglacial solifluction of clayey soils (Fig. 4), forming widespread and dangerous fossil features in much of NW Europe and elsewhere, which extend

generally to depths of about 1 to 3 metres (Hutchinson 1991) and are easily reached by trial pitting. More generally, landslides affecting regoliths are perhaps the most common failure type. As such regoliths are frequently shallow (Iida & Okunishi 1983), they also lend themselves sometimes to exploration by trial pitting, to establish particularly their thickness and its variability and whether or not pre−existing shears are present.

In other cases, however, non−landslide induced shears can be very deep. In the western prairies of Canada, for example, glacitectonics or "ice shove" has displaced the Tertiary and Cretaceous sediments there to depths of up to 180m, generating significant pre−existing shears (Cruden & Tsui 1991). The zone affected in this way can be defined broadly from its relationship to the former edge of the Wisconsin ice sheet, and from the geology.

A further, probably still more important, generator of pre−existing shears is flexural slip along bedding as a result of tectonic folding. The power and potential influence of this process tend to be underestimated (Hutchinson 1988). Such shears are found in clayey strata with dips down to only a few degrees (Fell et al. 1988). A superb example of a continuous smooth shear in a mudstone around 2.5 m thick and dipping at only 4° was exposed and logged for a length of 58 m in a nullah at Kalabagh (Fig. 5), in the Siwaliks of Pakistan (Professor A.W. Skempton, pers. comm., 1988, by courtesy of Kalabagh Consultants). As noted in the Figure, it is possible to infer from the exposure that the

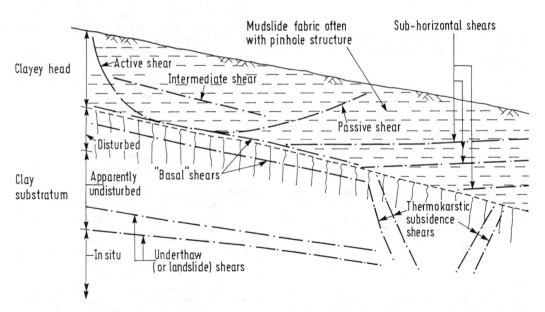

Figure 4. Main types of down-slope shear surfaces in clayey periglacial solifluction sheets over a clayey sub-base. Minor, Riedel and other shears and cross-slope shears are not shown (after Hutchinson 1991, Fig. 11, p. 292).

1817

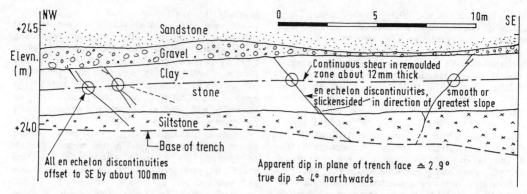

Figure 5. Log of middle part of northern face of Trench No. 52 at Kalabagh, Pakistan, showing continuous flexural shear surface at about mid-thickness of the claystone. Note that this shear consistently displaces by about 0.1 m to the SE several en echelon discontinuities, inclined at about 50°, thus giving an indication of the displacement on the shear (Prof. A.W. Skempton, pers. comm. 1988, by courtesy of Kalabagh Consultants).

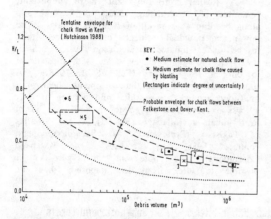

Figure 6. Plot of H/ L (inverse mobility) against log. of debris volume for chalk flows in Kent (acknowledgements to Trans-Manche-Link).

displacement on this flexural slip surface is about 0.1 m to the SE.

The other important consideration which arises from the geotechnical classification of landslides refers to the ground water pressure conditions, which are generally undrained, intermediate, drained or drained–undrained. A majority of slopes are in the drained condition (in the sense of having no excess ground water pressures) but note should be taken of any which involve undrained loading (Hutchinson & Bhandari 1971) or undrained unloading (Vaughan & Walbancke 1973), both of which can trigger failures, though with some delay in the latter case. Of the greatest importance however, are the highly brittle drained – undrained failures causing flow slides. These most commonly result from a collapse of loose, saturated metastable structure, with consequent sudden generation of high excess pore–water pressures, loss of strength, rapid

movement and long run–out. Once the debris of a flow slide has come to rest at a low–angle and all its excess pore–water pressures have dissipated, it will tend to have a very high factor of safety. The range of materials prone to flow sliding is indicated in Table 4.

To predict the run–out of flow slides, or of other very mobile mass movements such as rock avalanches (sturzstroms), is difficult and this matter is often neglected in hazard evaluations. Various theoretical models have been developed which can be fitted retrospectively to the phenomena (e.g. Edgers & Karlsrud 1982, Hutchinson 1986, Evans et al. 1989, Sousa & Voight 1991). It is difficult to use these for prediction, however, as the results are highly sensitive to the failure volume and flow thickness assumed. At present, the most reliable approach is to base run–out estimates on previous observed behaviour in cognate circumstances. Fig. 6 shows the results of such an approach to estimating the run–out of chalk flows from the coastal cliffs between Folkestone and Dover and Fig. 7 illustrates such a flow.

The run–out distance of rock falls is treated by Toppe (1987). He states that 50% of all rock–falls will stop inside a point on the sectional profile making an angle of 45° with the highest point from which the rock–fall came, while 95% will stop inside a point defined by a related line inclined at 32° to the horizontal. This may apply to low porosity rocks. However, as indicated by Hutchinson (1988) and outlined above, much greater run–outs occur in high porosity rocks in which excess pore–water pressures can be generated through impact–collapse. In these, the above angle (essentially the Fahrböschung) can be as low as 12 to 13° (tan⁻¹ H/L in Fig. 6).

A further important type of brittleness may be induced on pre–existing slip surfaces by rapid shearing. Lemos et al. (1985) have shown in rapid ring shear tests that, after an initial peak,

Table 4. Flow slide family

All involve generation of high $+ \Delta u$ (liquefaction) produced by collapse of metastable, saturated or near-saturated structure in all or part of the mass (largely after Hutchinson 1988).

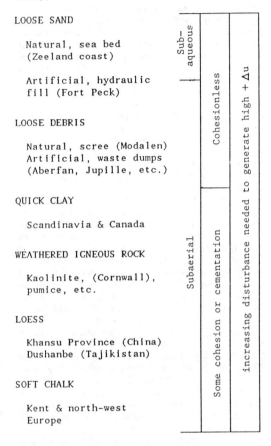

LOOSE SAND		
Natural, sea bed (Zeeland coast)	Subaqueous	
Artificial, hydraulic fill (Fort Peck)		
LOOSE DEBRIS	Cohesionless	
Natural, scree (Modalen) Artificial, waste dumps (Aberfan, Jupille, etc.)		
QUICK CLAY		increasing disturbance needed to generate high $+ \Delta u$
Scandinavia & Canada		
WEATHERED IGNEOUS ROCK	Subaerial	Some cohesion or cementation
Kaolinite, (Cornwall), pumice, etc.		
LOESS		
Khansu Province (China) Dushanbe (Tajikistan)		
SOFT CHALK		
Kent & north-west Europe		

Figure 7. Photograph of the "Great Fall" of 1915 at Folkestone Warren, England. This chalk flow had a debris volume of over one million m^3 and ran seaward for about 4.5 times the cliff height (photo by courtesy of the Geolological Survey, London).

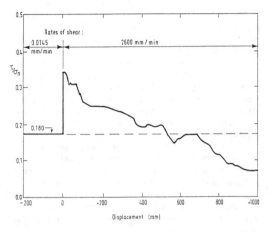

Figure 8. Slow and fast ring shear tests at $\sigma_n = 980$ kPa on Sample 3 of gouge from the Vaiont slip surface. The fast stage was carried out at a displacement rate of 2600 mm/ min after the slow residual strength of the sample had been established (after Tika-Vassilikos & Hutchinson, in prepn.).

most clays exhibit a positive rate effect with increase of displacement rate while others show a negative rate effect. The latter are of particular concern. Recent rapid ring shear tests on gouge from the slip surface of the Vaiont slide (Tika-Vassilikos & Hutchinson, in prepn) reveals that this has a high negative rate effect, the residual shear strength at displacement rates faster than 100 mm/min being only about 50% of that in normal drained slow tests. The results for Sample 3, for which $w_p = 30\%$, $w_L = 49\%$ and the clay fraction, $< 2\mu = 27\%$, give a residual strength at a 2600 mm/min displacement rate about 60% below the slow residual value after a shear displacement of less than a metre (Fig. 8). This effect, if representative, is sufficient to explain the high speed of the Vaiont slide without the need to involve slip surface heating.

It is clearly important to be aware of the presence of materials with this behaviour in the area of any landslide hazard assessment. So far they appear to have been found at a handful of sites only, with Vaiont being the sole example, to date, with an associated landslide. Much more work needs to be done to define the nature and range of such materials. Materials of strong negative rate effects in shear are unlikely to be as rare as this very preliminary information might suggest. Hutchinson (1988) points out that, if it had not been stopped by the opposite valley wall, Vaiont would have became a sturzstrom. It is suggested, therefore, that the head areas of sturzstroms that commenced with a slide rather than with a rockfall would make promising sites at which to seek more such materials.

The general lack of data on the location and

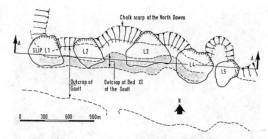

Figure 9. Plan of the five old landslides on the scarp of the North Downs, near Folkestone, England, showing the relationship between the slip toes and the outcrop of Bed XI of the Gault clay. (Slip L1= Danton Pinch; L2= West Cheriton; L3= East Cheriton; L4= Cherry Garden Hill; L5= Castle Hill) (acknowledgements to TransManche-Link).

shape of pre-existing slip surfaces is likely to render the influence line approach of limited applicability in regional landslide hazard assessment. It does emphasize, however, the danger of loading the head of a slide and the even greater one of unloading its toe. In the latter connection, it should be borne in mind that flattening the toe of a slide can lead to its destabilisation and that this is particularly dangerous in cases where the slip surface in the toe area is inclined into the valley slope (Hutchinson 1984).

Overall, one of the most useful geotechnical contributions to landslide hazard assessment is the provision of a number of well explored case records for typical slopes and landslides in the region. If these can be linked with characteristic geomorphological units in a terrain evaluation approach, as discussed subsequently, so much the better.

8.4 Recognition of spatial patterns of landsliding

It is clearly helpful, in regional landslide hazard assessment, to recognise any patterns of landslide behaviour, both spatial and temporal, which may exist in the region. Stini (1938) was one of many workers to have sought these patterns. They are termed regularities of landslide behaviour in much Russian work. Zolotarev & Novoselsky (1977), for example, find that major landsliding in the Carpathians occurs where high slopes run parallel or at a small angle to the trend of marked tectonic features, particularly steep folds and faults, and at the intersections of such disturbances.

Globally, the strongest concentration of landslides is in the vicinity of certain plate boundaries, particularly around the Pacific rim, through Indonesia and along the southern margin of Eurasia. At these, orogenic activity leading to high relief and active fluvial erosion and accompanied by frequent earthquake shaking are highly conducive to landsliding. Seismic zoning is fairly well developed and is discussed subsequently.

Volcanic activity, producing its own suite of characteristic mass movements, is frequently associated with these phenomena. Another very important trigger of landslides is rainfall: failures caused in this way can be predicted in space (and time) to the extent that storm tracks and intensities can be forecast.

Another strong spatial association is that between landslides and clayey strata, for example, in the ridges and cuestas formed in marls and clays of Jurassic, Cretaceous and Eocene age of the Paris basin (Flageollet 1989). Similar associations are found in the soft rock areas of New Zealand (Crozier et al. 1982) where, in addition, many landslides occur on thin, montmorillonite-rich bedding planes in conjunction with joints and faults which reflect the regional tectonic patterns (Bell & Pettinga 1988). In some cases, particular slide-prone horizons can be recognised, within clay layers or other strata. An example from the Gault clay near Folkestone, approximately 45 m thick, where at least five old landslides are seated in a particular zone of the Gault only about 3 m thick, within Bed XI, is illustrated in Figs. 9 and 10. Further west, the Gault changes in facies and, in the slips of the Undercliff of the Isle of Wight, exhibits another clear slide-prone horizon, 15 to 18 m above its base (Bromhead et al. 1991). Other slide-prone horizons exist, for example, in the seat-earths and underlying mudstones associated with the Cefn Glas, Brithdir, No. 2 Rhondda and Tormynydd seams of the South Wales coalfield (Conway et al. 1980) as well as with the Brithdir Rider seam. Slide-prone horizons may exist too between foraminiferal zones in Post-glacial marine sediments in Norway (Hutchinson 1960). A further well-established, preferred location for landsliding is in clays just beneath spring lines which tend to irrigate them continuously and maintain high ground-water pressures. An example from the Claygate Beds/London Clay contact in Essex is described by Denness & Riddolls (1976).

The association between clays and landslides is indeed strong, but it should be borne in mind that weathered regoliths, particularly in residual soils, contain little or no clay and are nevertheless affected by great numbers of landslides, as in Hong Kong, for example.

The incidence of landsliding is also strongly controlled by the pattern of fluvial and coastal erosion (Fig. 11). Shifts of base level, whether through eustatic, isostatic, tectonic or other causes, are thus of fundamental importance in landslide hazard assessment, as are other changes in the erosive power of rivers, for example through palaeoclimatic and palaeohydrological influences and through Quaternary river diversions.

In formerly glaciated and periglaciated countries like Britain, a majority of the present rivers are misfits, their valleys having been formed chiefly in earlier periods of much stronger glacial and/or fluvial erosion. It follows that, in such regions,

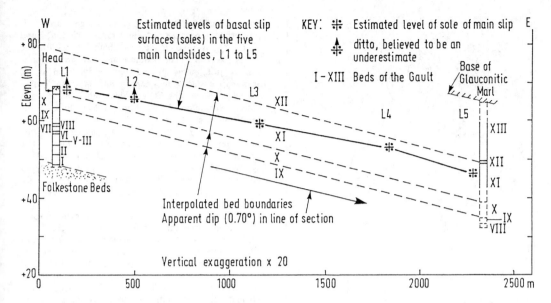

Figure 10. Sectional elevation of chalk scarp near Folkestone, England, showing the relationship between the soles of the main landslides and the stratigraphy and structure of the Gault clay (acknowledgements to TransManche-Link).

the majority of the existing landslides will be old ones, inherited from earlier phases of the Quaternary. Such old landslides and solifluction features may occupy the slopes backing older, higher river terraces (Chandler 1976; Palmquist & Bible 1980) and those behind raised beaches (Hutchinson 1991). Similarly, old rock slides in the Scottish highlands and in the Alps tend to be concentrated in the valleys occupied by the latest Late–glacial glacier advance (Holmes 1984, Ballantyne 1984, Abele 1974, Eisbacher & Clague 1984). In these circumstances, direct landslide mapping is likely to provide a close approach to landslide hazard mapping. The main exceptions to this, in Britain for example, are the areas of present–day strong erosion on the coast and along rivers, such as the Severn, which were diverted and rejuvenated during the late Quaternary.

Particularly in areas of the above general type, abandoned cliffs are a common feature of the landscape. These exhibit spatial and temporal patterns, useful in the present connection, which are discussed separately below.

Various thresholds exist in landscapes (e.g. Carson & Petley 1970, Schumm 1973, Hutchinson 1973) and many are relevant in the present context. The simplest example is the angle of ultimate stability against landsliding, β_{ult} (Skempton & De Lory 1957). Because soils are frictional, for given groundwater and seismic conditions, there is a minimum slope gradient upon which sliding can occur. In temperate, non–seismic regions, this is frequently assumed to be controlled by residual shear strength, and by groundwater located at and flowing parallel to ground level. As, for such conditions, failure tends to be shallow and

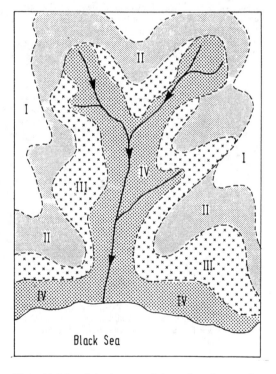

Figure 11. Map of development of slopes through successive generations of erosion and landsliding on the Caucasus coast of the Black Sea (after Kyunttsel' 1988, Fig. 2.11, p. 51).

1821

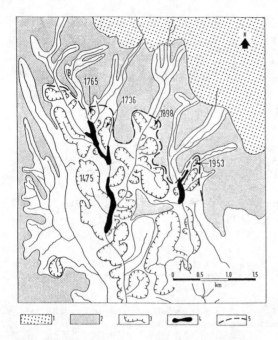

Figure 12. Map of area NE of Oslo, Norway, showing stream erosion and quick clay slides in the raised marine clays of the former sea bed (after Bjerrum et al. 1969, Fig. 2, p. 532). 1= bedrock; 2= uneroded former sea bed; 3= scars of quick clay slides; 4= redeposited quick clay masses; 5= "front of aggression".

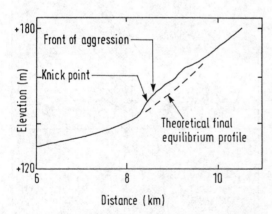

Figure 13. Longitudinal profile of the Hynna stream, branch B (Fig. 12). Horizontal axis gives distance from its confluence with the River Römua (after Bjerrum et al. 1969, Fig. 7b, p. 537).

translational, the infinite slope analysis (Skempton & Hutchinson 1969) often applies with reasonable accuracy.

Thus, for the London Clay under present climatic conditions, taking $\gamma_{sat} = 19 kN/m^3$, $c'_r = 0$ and $\varphi'_r = 14.5^o$ (difficult to measure at the very low normal effective stress range that is appropriate, because of apparatus friction) and $F=1.0$, β_{ult} is predicted to be about 7°. With the slightly more realistic assumption that groundwater level is just below ground level (say 0.2m deep, for a slip surface 2m deep, giving m = 0.9), β_{ult} = 7.9^o. Field checks on this value have been made. Skempton & De Lory (1957), looking for the steepest stable slopes of London Clay, found β_{ult} = 10°. Hutchinson (1967), looking for the lowest unstable slopes, found β_{ult} = 8°. The difference these two values is believed to reflect slightly different groundwater conditions. Upper and lower values of critical slope angle are reported in granular Alpine regoliths by Moser and Hohensinn (1983).

Where the geological and hydrogeological circumstances are sufficiently uniform for the value of β_{ult} to be established reasonably closely, it can with advantage be taken account of in landslide hazard assessment; with respect to both fully and partially developed abandoned slopes and to assess whether the accumulation zone is normal, eroding, deficient or extended, as discussed subsequently. Where periglacial solifluction is, or has been active, the situation is more complicated. For instance, for the London Clay, the high excess pore-water pressures generated in multiple cycles of freezing and thawing in the Late Pleistocene have enable slide movements to take place at slopes well below the "temperate ultimate" value of 8°. For Mesozoic clays generally in southern England, pre-existing basal shears beneath fossil periglacial solifluction sheets, noted earlier, are found down to slope angles of 4 to 5° and discontinous shears down to 2 to 3° (Hutchinson 1991). Thus, in such areas, it is important to distinguish between the Late Pleistocene slopes which have been affected by periglaciation and the Late-glacial slopes, which have not. In harder rock terrains, more complex systems of thresholds are found, such as the three-fold one reported from Carboniferous areas by Carson & Petley (1970).

Scandinavia, with its strong glacio-isostatic crustal heaving, well in excess of the Late- and Post-glacial eustatic recovery of world sea levels, provides striking examples of the effects of shifts in base level. In the Oslo region, Bjerrum et al. (1969) describe the fluvial erosion of a former sea bed consisting of marine clays as a result of being heaved above sea level. A "front of aggression" (Fig. 12) is identified, with which the main quick clay landslides are associated. This front is located in the vicinity of the knick points of the Post-glacial streams. Together with the associated landslides, it tends to migrate landwards with time. The potential for future erosion and landsliding can be judged from the difference between the present stream profile above the knick points and its long-term equilibrium profile (Fig. 13). These concepts have also been used by Brunsden & Jones (1976) and others. The spatial distribution of landslides for other incision models are illustrated by Palmquist & Bible (1980).

The areas occupied by marine clay in the heaved areas of Scandinavia and Canada are well defined. The distribution of quick clays and potential quick clay slides within these areas has received much attention, summarised by Viberg (1984). In addition to the possible presence of slide–prone horizons and the effects of erosion patterns, discussed above, an association is found between quick clay and shallow bedrock giving rise to concentrated groundwater seepage (Hutchinson 1961). In locating leached areas of higher electrical resistivity, there may be scope for the application of direct conductivity sounding or indirect geophysical electrical resistivity surveying. Although probably valid in the main quick clay areas, this approach would have to be used with caution as leaching is not the only cause of quick clay formation : it is also formed in association with peat and other organic materials (Söderblom 1974).

The well–known tendency for erosion and landsliding to be concentrated on the outside of meanders and other river bends, leaving more stable slip–off slopes on their insides, should not be overlooked. Other mechanisms for lateral fluvial migration include homoclinal shifting, the tendency to move down–dip in a situation where a river flows along the strike of inclined bedded strata (Thornbury 1966). A possible example is provided by the valleys of the Rivers Stour, Orwell and Deben in East Anglia, England (Fig. 14). These flow mainly in the London Clay, broadly parallel to the minor, Ipswich – Harwich anticlinal axis, which appears to have been active from the late Mesozoic to the Quaternary (Boswell 1915). As shown by Fig. 15, there is a clear pattern in the valley cross sections : that of the central river, the Orwell, is symmetrical, while those of the two outer rivers are asymmetrical in opposite senses, with the Stour valley steepest on its S side and the Deben on its N. It is inferred that, although the antiform is very subdued, these forms result from the homoclinal shifting of the latter two rivers away from the anticlinal axis. This process appears to be aided by the numerous thin sheets of claystone interbedded with the London Clay there. The distribution of major landslides on these three rivers (Fig. 14) does not fit a simple outside–of– bend explanation but is consistent with the mechanism outlined above (Hutchinson 1965).

Where rejuvenation of rivers and streams is particularly strong, as in the Andes and other plate–margin orogenic belts, the rate of downward erosion dominates over that of sub–aerial processes and steep–sided gorges are formed in the bottom of former V or U – shaped valleys. These materially increase the associated incidence of landslides (Fig. 16). Related gorges are common in the Alps in the lower reaches of lateral hanging valleys, as at Vaiont, where the lateral stream is rejuvenated by the greater erosion in the main valley.

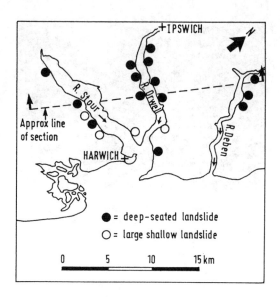

Figure 14. Map of three rivers in East Anglia, England, showing the distribution of the main shallow and deep-seated landslides (Hutchinson 1965).

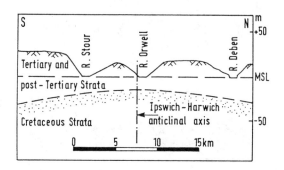

Figure 15. Section across the three rivers in Fig. 14, showing the relation between their valley profiles and the underlying geological structure (Hutchinson 1965).

Although there has not been opportunity to check these in any detail, there are indications that, in mountainous terrain, many of the largest and most dangerous landslides orginate in lateral, rather than in main valleys (Fig. 17). This would appear to arise chiefly through the higher gradient and erosive power of the lateral streams and because their valleys penetrate to the highest parts of the massif, so that the highest and steepest slopes, and potentially greatest failure volumes are often found on the walls of the inner parts of the lateral valleys, in the area A of Fig. 17a, as in the Huascaran (Plafker et al. 1971) and Mayunmarca (Kojan & Hutchinson 1978) slides in Peru. In contrast, the main valley bluffs, B, provide considerably less landslide potential. In cases where strong erosion in the outer part of

Figure 16. Photograph of intense fluvial erosion, with gorge, in the valley of the Rio Huaura, Peru (acknowledgements to Dr. C.M. Clapperton, pers. comm.).

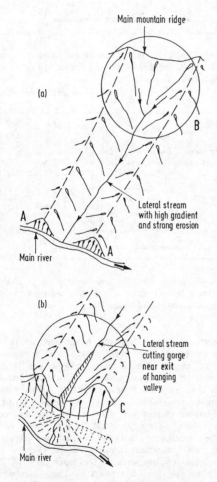

Figure 17. Diagrams of mountainous areas showing a tendency for the larger landslides to occur (a) at the heads of lateral valleys, or (b) in association with a gorge cut near the exit of the lateral valley, especially where this is hanging in relation to the main valley, usually through differential glacial erosion.

the lateral valley has created a gorge, as at C (Fig. 17b), the focus of major landsliding may move to this area, again as in the prehistoric failure at Vaiont. It should be noted that, with the topography illustrated in Fig. 17, only slides from the head of the lateral valleys have space in which to achieve a long run-out.

In the case of many failures of flow type, the debris tends to follow a pre-existing track, determined by the topography: mudslides and debris flows are obvious examples. Hence, to a considerable degree, the location and path of the next failure can be anticipated by mapping the deposits of earlier ones. This was done, unfortunately retrospectively, for the catastrophic failure of 1970 from Huascaran, Peru (Plafker et al. 1971). Mapping of old deposits showed that the valley leading to Ranrahirca was very prone to debris flows. It also revealed the existence of a larger pre-Columbian precedent for the flow which destroyed Yungay. Similar mapping of the deposits of Post-glacial lahars from Mount Rainier, U.S., was carried out by Crandell (1971).

Natural dams, which may be formed by landslides (Schuster 1986), glacier ice (Kendall 1902), river ice (Hutchinson 1961) or lava flows (Novosad 1990) are important in the present connection. Landslide dams are fairly common, especially in high mountain areas affected by neotectonic movements. Such dams also occur in more gentle Scandinavia and Canada, where the presence of quick clays can give landslides the requisite energetic run-out.

Some of the destabilising, secondary effects of landslide dams are illustrated by that which formed in 1974 on the Rio Mantaro, in the Peruvian Andes (Kojan and Hutchinson 1978). In that case, a lake of eventually 600×10^6 m^3 of water was impounded. During impounding, great numbers of slides were caused around the perimeter of the growing lake, particularly in partially saturated granular slopes. After 44 days, the slide dam was over-topped and breached. Being composed chiefly of rather fine granular materials, it eroded very rapidly, producing a draw-down rate in the lake of over 40m in 24 hours. This, in turn, led to a large number of rapid draw-down failures around the lake perimeter. The breaching of the slide dam also led to the formation of a flood wave, initially more than 35m high and travelling at more than 15 to 30km/hour. This caused great destruction, including many landslides in the river banks, for tens of kilometres downstream.

Natural dams can also have significant, longer-term sedimentological effects, particularly when they do not breach fully. In the pool of stiller water which then remains, finer-grained sediments, including clays, can accumulate and form potential future seats of landsliding.

Useful discussions of the distribution of landslide-susceptible terrain, from the regional to the local scale, are given by Rib & Liang (1978), Nemčok (1982), Crozier (1984). Fookes et al.

(1985), Sidle et al. (1985) and Fookes (1987). The early signs of landslide movement, visible in the field and often also in aerial photographs, are also described, particularly by Rib & Liang (1978), Crozier (1984) and Aulitzky (1990). It is emphasised that such signs are the best evidence for the location of a forthcoming landslide and may also give a general indication of its possible time of occurrence. Fig. 18 shows the scarps at the head of the Quebrada Ccochacay, Rio Mantaro, Peru less than a year before the catastrophic Mayunmarca slide there (Kojan & Hutchinson 1978).

8.5 Recognition of temporal patterns of landsliding

Temporal patterns of landsliding generally derive from regularities in behaviour of the three main triggers which, excluding man's activities, are erosion, rainfall and earthquakes. Fluctuation in external water levels, even with no erosion, are also significant. Examples of some of these effects are given below.

In many parts of the world, erosion on near-present coastlines was halted by the glacio-eustatic depressions of world sea levels during the Pleistocene and recommenced on the recovery of these. Radiocarbon dates on some English coastal landslides indicate that they were last reactivated in this way by the rising Flandrian sea level, some 4000 to 8000 years B.P. With a fairly constant sea level, erosion continues steadily, often tending to cause coastal landslides. A cyclic situation then develops, in which a fresh landslide is slowly eroded away, in situ cliff is over-steepened and a further landslide is brought about. A characteristic of such cylic behaviour is that extended periods of slow movement and adjustment are followed by sudden, potentially catastrophic, first-time failures. There is thus a need to appreciate this situation when making landslide hazard assessments, particularly when the landslide cycle is long.

On the approximately 40m high London Clay coastal cliffs of Warden Point, Sheppey, England, the landslips have a cycle length of about 40 years (Hutchinson 1973). In the more complex and harder Cretaceous rocks of the Undercliff of the Isle of Wight, England the coastal landslide cycle occupies around 6000 years (Hutchinson et al. 1991). In the related, but softer Cretaceous rocks of the Folkestone Warren landslides, Kent, England, major renewals of movement took place in 1877, 1896 and 1915, in an apparently 19-year cycle, followed by a smaller failure in 1937. Terzaghi (1950) drew attention to these events in his discussion of periodicity of landslides and ascribed them to rainfall-induced variations in piezometric level in the Lower Greensand aquifer immediately beneath the slides. Subsequent work (Hutchinson 1969; Hutchinson et al. 1980) has shown that while the periodicity of the Folkestone Warren landslides was caused partially by piezometric fluctuations (in the slipped masses

Figure 18. Photograph in 1973 of scarps at the head of the Quebrada Ccochacay, Peru, which eventually became the rear scarp of the Mayunmarca slide of April 1974 (acknowledgements to Ing. J.G. Bustamente).

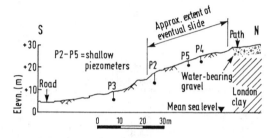

Figure 19. Section of defended cliffs of London Clay, with capping of gravel, at Southend-on-Sea, Essex, England, showing locations of piezometers and subsequent landslide (at Grid Ref. TQ878851).

rather than in the Lower Greensand), discussed below. Enhanced marine erosion resulting from blockage of the eastward littoral drift by successive extensions to the West Pier of Folkestone Harbour, just west of the Warren, during the 19th & early 20th centuries, also played a major role.

Not surprisingly, individual coastal landslides in the softer rocks have often occurred at or about the same time as major surges and storms. Examples in the London Clay are the failures of December 1897 at Walton Cliffs, Essex, and of 31st January/1st February, 1953, near Herne Bay, Kent (Anon 1898; Robinson 1953).

Rainfall-induced increases in ground-water levels, through either short-lived intense precipitation or prolonged rainfall of lower intensity, are a major cause of landslides. The relevant ground water body is frequently perched on the failure surface of the landslide. Precise monitoring of these effects is, not surprisingly, rare. Piezometers have, however, sometimes been carried away by landslides, as on the Kimola Canal, Finland (Kenney & Uddin 1974) and in the Southend cliffs, England. The piezometric readings obtained in the latter case are shown in Figs 19 & 20. A more general correlation between high pore–water pressures in the slipped masses at Folkestone Warren and the incidence of major landslides there is given in Fig. 21. A general discussion of mechanisms leading to groundwater fluctuctions is given by Freeze & Cherry (1979). A recent study of the relationships between rainfall and movement for a big landslide in the English Pennines is provided by Skempton et al. (1989).

There is fairly good evidence that the incidence of landslides in the northern hemisphere was markedly increased during the Little Ice Age, though historical evidence for the early part of this period is lacking. A similar landslide

response is also likely to have taken place at other times of cool climate, with low evapotranspiration, and increased rainfall. Palaeoclimatic studies are in progress which, when related to dated landslides, should lead to an improved understanding of climatic triggering, the role of antecedent weather and soil conditions and of associated thresholds.

In many of the drier areas of the world, the upper layers of soil (often residual) are partially saturated, with predominantly negative pore–water pressures. Landslides can be triggered in such slopes by rainfall if some threshold intensity is exceeded, so that pore–water pressures are increased to the necessary degree. This process is also affected by such factors as slope angle, the distribution of shear strength and permeability within the regolith, the effects of soil pipes, variations in regolith thickness and antecedent weather and pore–water pressure conditions (e.g. Fukuoka 1980, Crozier & Eyles 1980, Brand et al. 1984, Anderson & Pope 1984, Vaughan 1985, Crozier 1986, Brand 1988b, Capecchi & Focardi 1988, Shimizu 1988, Gostelow 1991). The specific problem of the climatic triggering of debris flows is dealt with, for example, by Keefer et al. (1987), Church & Miles (1987) and Sheko (1988 b & c).

Related behaviour is also found in the drier parts of south–eastern England. Soil Moisture Deficit is a convenient measure of the degree of dryness of a soil. As indicated by Fig. 22, the SMD over most of the northern and western parts of Britain falls to zero early in the winter. However, in the south–east of the country, the SMD values are considerably greater and these deficits may even be maintained right through the winter. Southend, a coastal town on the London Clay in south–east Essex (Fig. 22), has about 5 kilometres of sea cliffs up to 30m in height, which have been abandoned or defended for 60 or 80 years. Search of the local newspapers has yielded records of over 70 landslips on these cliffs during a period of several decades. A characteristic of the slips is their highly episodic nature: none will occur for several years and then a considerable number will fail more–or–less

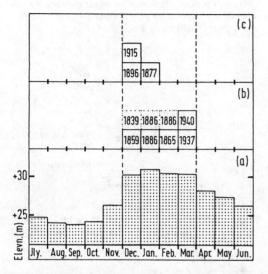

Figure 21. Correlation between landslides and high groundwater pressures at Folkestone Warren, Kent, England (Hutchinson 1969, Fig. 15, p. 17).
(a) Max. monthly piezometric levels in slipped masses between 1954 & 1963,
(b) Incidence of dated type R landslips (at seaward edge of old slips),
(c) Incidence of dated type M landslips (involving the whole mass of old slips).

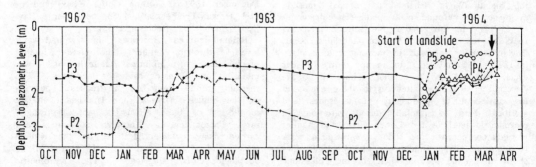

Figure 20. Piezometer readings in cliffs at Southend-on-Sea (Fig. 19) for the 16 months leading up to the failure of March 22nd to 30th, 1964 (ED18d in Hutchinson 1965).

together. A selection of these data is shown in Fig. 23 for the period mid-1967 to mid-1976 in comparison with the SMD values and dates when the soil was at Field Capacity. The SMD first fell to zero in the winter of 1968/69, when 8 landslides occurred. From then for the next 6 years, the SMD remained above zero and no landslips occurred apart from 3 in the winter of 1969/70, when the SMD fell to about 7mm, its second lowest winter value during the period of observations. A further 4 landslides occurred in the winter of 1974/75, when the SMD again fell to zero. Corresponding piezometric readings, not available at this site, are reported for the period 1973 to 1975 at nearby Hadleigh Castle, in relation to SMD, by Hutchinson & Gostelow (1976).

The general spatial predictability of debris flows has already been referred to. Recent work in the former USSR has attempted to classify debris flows and landslides and to predict the time of their occurrence in the Caucasus, the Tien Shan range and other mountain regions (Sheko 1988a). For long-term predictions, the effects on rainfall of both sunspot activity as expressed in Wolf numbers (Herman & Goldberg 1978) and, less importantly, circulation mechanisms, using the typology of Vangengeim – Girs (Mal'neva & Kononova 1988;

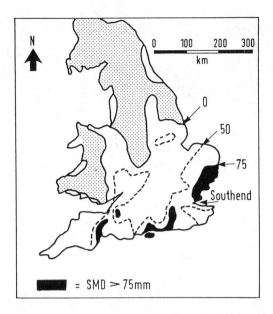

Figure 22. Distribution of estimated Soil Moisture Deficit over England and Wales on 12th October, 1966 (based on Grindley 1967, Fig. 4, p. 106).

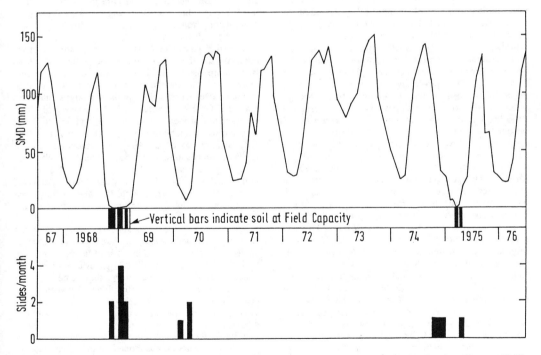

Figure 23. Plot of landslide occurrence in relation to variation of Soil Moisture Deficit and periods when the soil was at Field Capacity for defended London Clay cliffs at Southend-on-Sea, England, 1967-1976.

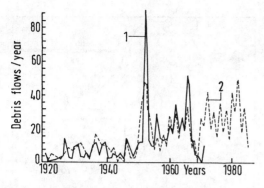

Figure 24. Comparison between (1) the actual numbers of debris flows occurring in the Caucasus, Tien Shan, etc., in the period 1921-1971 and (2) backward- and forward-predictions of their numbers (after Sheko & Sergeeva 1988, Fig. 15.6, p. 188).

Barry & Perry 1973), have been made use of and some fairly encouraging results (Fig. 24) obtained (Sheko 1988b; Sheko & Sergeeva 1988). Both regional and local, short-term forecasts are made, taking into account, inter alia, seismic activity, the volumes, distribution and condition of loose fragmental material, the amount, character and distribution of antecedent and predicted precipitation and the character and distribution of snow cover. In addition, consideration is given to possible spates orginating from slide dams and artificial dams, from the thawing of glaciers & frozen moraines, from the sudden release of water bodies trapped by moraines and glaciers and, for lahars, to the possibility of volcanic activity generating debris flows, generally either through intense rain on the slopes of loose eruptive deposits forming the volcano or as a result of magma spill onto a glacier (Sheko 1988c).

Changes in external water level at the toe of a slope or slip, even if they are slow enough to maintain drained conditions, affect its stability through changes in the external support and pore-water pressures. If the changes in external water level are more rapid than this, futher transient pore-water effects are produced, which can be very destabilising on draw-down. Through excellent monitoring by the late Mr A.H.T. Todd, reported by Hutchinson (1988), it is evident that the old landslide at Sandgate, on the coast of Kent, England, is reactivated at each low tide and virtually halted at each high tide. The tidal range is approximately 5.5m. There are many related examples of the response of reservoir perimeter slopes to man-made changes in reservoir level.

Landslides triggered by undrained loading tend to occur immediately the requisite fill or other load is placed, as seen in the Rissa slide, Sweden (Gregersen 1981). With undrained unloading, negative pore-pressures tend to be set up initially. These improve the stability in the short term, but a slide often occurs as these decay. Such failures,

controlled in part by the value of c_s, the coefficient of swelling, are termed delayed failures and are a feature of railway and road cuttings in the London Clay, where they occur typically between about 20 and 100 years after the formation of the cutting (Vaughan & Walbancke 1973, Skempton 1977).

8.6 Terrain evaluation/case record approach.

As noted earlier, the concept of terrain evaluation has been found useful in landslide hazard assessment and adopted fairly widely. This approach is most valuable when used on carefully defined geological/geomorphological entities, for which at least one case record of a thorough geotechnical investigation is available. Different such entities will be appropriate in the various widely differing landscapes for which landslide hazard assessments are to be made. One example, which has been found useful in both England and Papua New Guinea, is described below.

Abandoned cliffs result from a period of erosion which then ceases, so that no material is then removed from the slope toe. Such cliffs, formed in stiff fissured clays, have been much studied in southern England. A simple case, with no capping stratum, is exemplified by the abandoned cliff of London Clay at Hadleigh, Essex. The main features are shown in Fig. 25. An irregular degradation zone, of average inclination 12.6°, can be distinguished, formed of successive rotational landslides which are still active, especially at the top of the slope. Beneath these is a more gently inclined, smoother and slightly more stable accumulation zone, with an average inclination of nearly 8°. Back-analysis indicates that the factors of safety of the degradation and accumulation zones are about 1.0 and up to 1.05, respectively (Hutchinson & Gostelow 1976). The accumulation zone is built up predominantly of repeated mudslides, of translational habit and exhibiting mudslide fabric. The degree of break-down of the London Clay fabric decreases towards the slope crest, with the age of the mass movements. The latest failure is that nearest the crest.

With time, the degradation zone degrades further and will eventually reach the ultimate angle of stability against landsliding for the London Clay of about 8°, provided that the climate remains approximately the same. Thus the difference in inclination between the degradation and accumulation zones is a measure of the age of the abandoned cliff. When this difference is zero, the feature is fully developed. The inclination, α, of the accumulation zone in a partially developed abandoned cliff (Figs 25 & 26a) anticipates closely the eventual inclination of the fully developed cliff. In the latter the successive rotational slopes become quiescent, and form cross-slope undulations (Hutchinson 1973).

Subsurface, the orginal slope toe and an

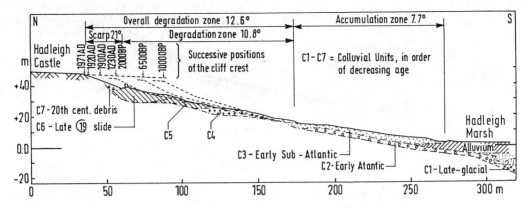

Figure 25. Detailed section of the partly developed abandoned cliff of London Clay at Hadleigh, England, showing its degradation and accumulation zones and their development over the past 10,000 years (Hutchinson & Gostelow 1976, Fig. 2.8, p. 593).

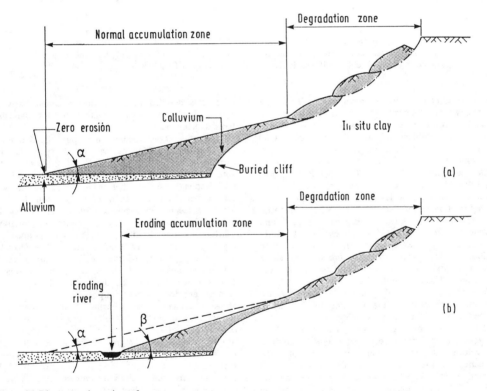

Figure 26. Diagrammatic sections of:
a) partly developed, abandoned cliff of stiff plastic clay i.e. of Hadleigh type, with zero erosion and a normal accumulation zone, of inclination α.
b) a partly developed, degrading cliff of stiff plastic clay with an eroding accumulation zone, of inclination β > α.

associated buried cliff can usually be recognised. The colluvium masks this feature and projects well forward over the littoral or alluvial deposits left by the last phase of erosion. The colluvium, softened and broken as noted above, is interlaced by numerous slip surfaces at residual strength, and is readily reactivated by cuts or fills or by adverse changes in groundwater conditions.

In the London Clay terrain of Essex and N. Kent such abandoned cliffs, in various stages of

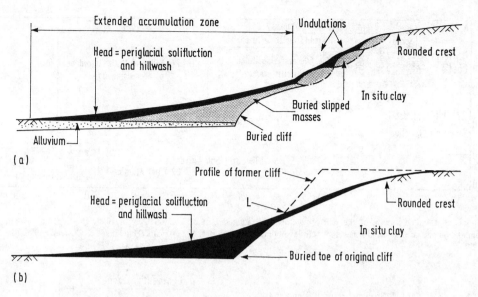

Figure 27. Diagrammatic sections of:
a) partly developed, abandoned cliff of Hadleigh type, subsequently modified by periglacial solifluction.
b) a previously stable cliff of stiff plastic clay, modified by periglacial solifluction.

development, can readily be recognised. They are all Post–glacial features and it is important to distinguish from these the earlier slopes which have been subjected to severe periglacial conditions. The latter have lost their sharp crest and tend to have a smooth, convexo–concave profile (Hutchinson 1991). Two main categories of soliflucted London Clay slopes are found: those previously exhibiting a Hadleigh–type profile, when the solifluction often partially or wholly mantles the former successive rotational slips and accumulation features but leaves these intact within the slope (Fig. 27a); and those which were previously stable, where the only pre–existing shear surfaces found are those resulting from the periglacial solifluction itself (Fig. 27b).

Note that the area of greatest landslide hazard is at the slope crest in the situation of Fig. 26a), but in or near the steepest mid–slope facet, L., in the situation of Fig. 27b). The latter slides are generally of translational slab type, doubtless representing a reactivation of part of the solifluction mantle. The above comments for the situation in Fig. 26a, apply only for monocyclic erosion. A fresh phase of basal erosion, arising from neotectonics, for example, would introduce a polycyclic situation, leading to a shift of the greatest landslide hazard to the slope foot, in the area of latest fluvial or marine erosion.

The Post–glacially eroded slopes are naturally fairly closely associated with the present rivers and streams. The soliflucted slopes are found slightly further inland, where Post–glacial fluvial erosion has scarcely penetrated. The above concepts, backed by the detailed Hadleigh case record and

other less detailed ones, provide a sound basis, in conjunction with terrain evaluation, for landslide hazard assessment of the area.

A further, very common type of abandoned cliff, of Post–glacial or soliflucted type, is that formed chiefly of a thick bed of stiff clay, but with a rigid capping stratum. These follow closely the Hadleigh model, but have more pronounced rear scarps and successive rotational slip features. The Jurassic escarpment of the Cotswolds exemplifies a rather more complex situation, soliflucted and with several alternations of argillaceous and more rigid strata. The detailed engineering geology of this scarp is only partially determined as yet. The most complete sections to date are provided by Chandler et al. (1976).

Another obvious and important variant to the above models is partial or interrupted abandonment, when some erosion, probably considerably less than that which formed the original cliff, is in progress. Then, as illustrated by Fig. 26b), the depleted accumulation zone is maintained at a steeper angle, β, than would be the case for the fully abandoned cliff, and is corresponding even less stable, with a factor of safety at or close to 1.0. The lower slopes of the left bank of the Ok Ma River, Papua New Guinea, at the site of a proposed tailings dam provide a good example. They, not surprisingly, failed comprehensively when the excavation for the dam footprint was made in their toe in 1984 (Newman 1985).

Steeper than normal accumulation zones can also exist, more subtly, where there is no present–day erosion. The term "deficient" is proposed for accumulation zones of this type. The classic case is at Lympne, Kent, England, for the accumulation zone of an abandoned cliff of Weald Clay, capped by a thin bed of arenaceous Hythe Beds. The slope was investigated by Hutchinson et. al. (1985). The latest phase of strong marine erosion took place around 5000 years ago as the sea level recovered from its last glacio-eustatic depression. This formed an old sea cliff (Fig. 28a). With the cessation of erosion, the cliff was too strong to degrade appreciably itself and the earlier slip masses from upslope slipped down over it to rebuild the accumulation zone. The volume of these masses was insufficient to form a fully developed accumulation zone, at an inclination of about $\alpha = 9^0$, so a deficient zone resulted, with an inclination of around $\theta = 10^0$ (Fig. 28b)). Eventually, probably during a time of wet weather, a major slide of the accumulation zone took place, moving the toe about 40m forward and reducing the overall slope of the accumulation zone to about 9^0, its normal value. This slide destroyed Portus Lemanis, the Roman fort of the Saxon Shore, which was built on the deficient accumulation zone.

Other geological/geomorphological entities, useful in landslide hazard assessment, exist in each type of terrain. They are best identified by those most familiar with the locality.

8.7 Hazard assessment for earthquake-triggered landslides

The strong association between neotectonics, earthquakes and landslides is brought out by Cotecchia (1978). The important phenomenon of landslides involving the seismically induced liquefaction of sands is dealt with by Seed (1968), Ishihara (1985) and Youd (1991). It is helpful to distinguish between landslides that occur synchronously with an earthquake (direct failures) and those that occur some time, even days, after the seismic shock (indirect failures). There is some evidence that the former type (for which a dynamic analysis is appropriate) tend to be limited in size to around a valleyward dimension of half the wavelength of the seismic stress waves (Ambraseys 1977: Hutchinson 1987). The latter type (for which a static analysis including the seismically induced groundwater pressures is generally appropriate), seems to be characteristic of renewals of movement on pre-existing slip surfaces in clays. Work by Ishihara (1985) and by Lemos et al. (1985), with rapid triaxial and ring-shear tests respectively, indicates that in some clays, a reserve of strength over and above the

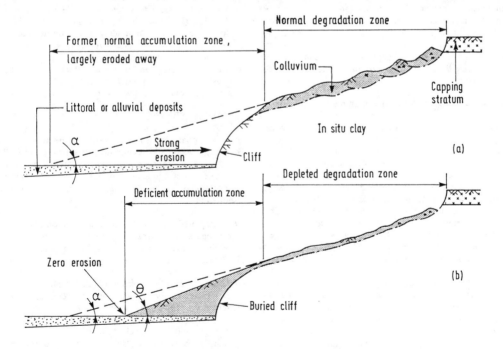

Figure 28. Diagrammatic sections of:
a) previously fully developed cliff of Hadleigh type, plus a relatively thin capping stratum, after a strong second phase of erosion.
b) situation resulting after the second abandonment, with a deficient accumulation zone, of inclination θ, steeper than α for a corresponding normal accumulation zone, and thus even less stable.
Note that very similar models would apply if the capping stratum were absent.

slow value is mobilised during rapid loading, which will tend to prevent sliding in many earthquakes. As noted above, after this peak the strength falls again, to values that may be above (positive rate effect) or below (negative rate effect) the slow value of residual strength. The important implications of these phenomena in landslide hazard assessment are evident

Seismic zonation is fairly well established in many countries (e.g. Kárnik & Algermissen 1978, Aki & Irikura 1991). The seismic triggers needed to cause or reactivate landslides naturally vary with the type of failure produced (Keefer et al. 1978, Cotecchia 1987). Recently, earthquake-induced landslide suseptibility maps have begun to be produced, notably in California by Wieczorek et al. (1985). In the latter, slopes steeper than 35^0 are regarded as susceptible to falls, areas susceptible to liquefaction (based on Youd 1991) are regarded as leading to lateral spreads and wet-sand flows, and various types of translational and rotational landslides are assessed by means of the Newmark dynamic stability analysis to find their expected displacement for an estimated design earthquake. A general review of such mapping, in a sub-aerial context, is given by Hansen & Franks (1991). Shear rate effects do not appear to be allowed for.

Occasional, surprisingly large intra-plate earthquakes occur in connection with glacio-isostatic land heaving as the associated stresses are relieved. The Post-glacial fault break at Pärve in N. Sweden, which has a throw of up to about 10 metres and a length of about 150km (Lundqvist & Lägerback 1976), provides striking evidence of this phenomenon.

8.8 Socio-political, legal and planning matters

These matters are ultimately of the greatest importance, as unless they are dealt with effectively even the best landslide hazard assessment cannot reach its aim of reducing human suffering and loss. The translation of the finding of such assessments into public policy is not easy, involving as it does the very fabric of society, and even danger to life itself. Proper communication and cooperation with the public, the decision-makers and others is thus essential. The wise remarks of Varnes et al. (1984) and Doornkamp & Griffiths (in prepn) on this score should be heeded.

The degree to which legal provisions have been made in connection with the incorporation of landslide hazard assessments into decision making varies greatly from country to country. Austria, France, New Zealand, Japan and the United States are among those to have gone furthest in this direction, in some cases also with regard to insurance arrangements. The generally satisfactory results in France are described by Champetier de Ribes (1987) and by Flageollet (1989).

In Britain, the need to regard matters of geology and instability as material considerations in the planning process is not specifically identified in the legislation. However, guidance provided by the Department of the Environment (1990) clarifies this situation; "The stability of the ground in so far as it affects lands use is a material consideration which should be taken into account when deciding a planning application". In the area of land and property purchase, the principle of caveat emptor means that purchasers may not be informed of a past land movement unless they specifically raise the question. Only recently, the writer stayed at a Victorian hotel in Sandgate, Kent, which had moved 1 to 2 metres seaward in the landslide of 1893 and found that the new owner was quite unaware of this: hardly a satisfactory state of affairs.

As noted above, the problems of evacuation in the face of landslide hazard, were tackled by Heim (1932). Rather successful operations of this nature were carried out for the Mount St Helens event, USA (Decker 1981) and for several large landslides in China (Wang Sijing, pers. comm.). In the case of Mayunmarca slide, Peru, the failure was predicted six months before it occurred by a mining gelogist, J. G. Bustamente, largely on the basis of the considerable ground cracking around the village and the large scarps at the head of the lateral valley (Fig. 18). Unfortunately his report was not acted upon. On the other hand, once the slide became known to the relevant Peruvian authorities (Defensa Civil), they acted very firmly with regard to the impending over-topping of the huge landslide dam. This breached 44 days after the slide, releasing hundreds of millions of cubic metres of water in a great flood wave but, because of a prudent and comprehensive evacuation policy, no lives were then lost.

9 CONCLUSIONS

9.1 Regional landslide hazard assessment is of human, social and economic importance globally. It is most appropriate that we redouble our efforts in this area in the present United Nations International Decade for Natural Disaster Reduction.

9.2 Regional landslide hazard assessment is a most challenging task, for which all our relevant disciplines need to be brought to bear in an integrated fashion.

9.3 The nature and scale of the problem is such that we shall have to continue to rely, to a great extent, on remotely sensed surface observations. This limitation means that knowledge of ground and groundwater sub-surface conditions will generally be sparse or lacking. As a consequence, a degree of uncertainty will be inherent in the assessments produced. This should be explicitly stated.

9.4 This uncertainty will tend to be least for the shallower mass movements of short return period and greatest for the more deep-seated, less frequent failures.

9.5 Such uncertainties may be reduced in several ways. The most important is to ensure that the best geological and geomorphological data and insights are incorporated into the assessment. A holistic approach to the landscape, with an awareness of its dynamic evlution, will be necessary. Among matters to be considered are lithostratigraphy and structure, Quaternary events, changes of base level, neotectonics and earthquakes and the existence of spatial and temporal patterns of erosion, landform and landsliding.

9.6 The uncertainties can also be reduced by making greater use of geotechnical methods. Although detailed, site specific site investigation techniques will generally be too expensive for regional landslide hazard assessment, it is important to obtain and build in to such assessments as much geotechnical information as feasible. Index tests, leading to correlations with other physical properties, lend themselves ideally to such use. The necessary sampling can be carried out in surface exposures and in machine- or hand-dug trial trenches. The presence or absence of pre-existing shears, produced other than by landsliding, of metastable sediments and of materials with negative rate effects are among other matters that should be addressed.

9.7 In urban environments, the socio-political, environmental and economic impact of a landslide hazard assessment will tend to be greater than in non-urban, undeveloped areas. Greater demands are thus likely to be put upon the reliability of the assessment in the former case. It is then both appropriate, and justifiable to carry out a more detailed assessment, with a greater degree of geotechnical input, at necessarily higher cost.

9.8 The best technique for incorporating geological, geomorphological and geotechnical inputs into a landslide hazard assessment would appear to be provided by use of a well - focussed, evolutionary terrain evaluation approach, in combination with detailed geotechnical/ engineering geological case records of each of the terrain elements chosen.

9.9 The power and sophistication of the remote sensing techniques and the means of handling the resultant data are increasing rapidly, particularly through the development of Geographical Information Systems and methods of modelling the response of surficial soils to various triggers. These developments are most welcome. A possible danger is that the specialists in this area may become increasingly detached from the overall field situation and forgetful of point 9.3, above. For this, and many other reasons, it will be advisable to carry out landslide hazard assessment in multi-disciplinary teams, always including some members with a detailed knowledge of the particular terrain. This practice is already followed by a number of groups.

9.10 Efforts are being made to improve our ability to remotely sense the ground surface and its morphology through thick vegetation. Useful qualitative results have been obtained by vertical and side-scanning air-borne radar. The most promising new quantitative technique appears to be vertical laser profiling.

9.11 With a few partial exceptions, current landslide hazard mapping is unchecked and, particularly in view of 9.3 above, its reliability is uncertain. Some degree of internal check can and, should be applied, preferably using independent teams. However, a true check on the reliability of such mapping can only be effected after the passage of some considerable time.

9.12 The degree to which legal provision has been made to support the introduction of landslide hazard assessment into the planning process varies considerably from country to country. This process tends to have an effect on property values and may thus meet resistance. What is needed, particularly in an already built-up area, is a balanced approach which deals satisfactorily with the level of hazard and takes due account of financial interests of both property owners and future purchasers.

9.13 Sub-aqueous regional landslide hazard assessment seems generally to be at an early stage, involving chiefly direct mapping. Current developments in side-scan sonar and other techniques are such that more comprehensive assessments are now essentially within reach.

9.14 While we must continue to strive to improve our methodologies, it should be borne in mind that the most telling step globally is probably to make even a simple landslide or landslide hazard map where none existed before.

9.15 The essential aim of landslide hazard assessment is to reduce human suffering and loss from landslides. Proper communication and cooperation with the public and with the decision-makers is thus of crucial importance.

10 ACKNOWLEDGEMENTS

I am grateful to Professor D. Brunsden, Dr J.S. Griffiths, Mr A. Hansen, Professor M. Del Prete and Professor E.N. Bromhead for their very helpful comments on early draft of the paper and to Dr E.E. Brabb, Dr R.L. Schuster. Mr D.J. Varnes, Dr H.G. Owen, Mr H.I. Siddle, Dr D. Brook, Mr D. R. Norbury, Dr J.C. Doornkamp, Mr D.A.B. Baker, Mr. M.G. Culshaw, Mr L. Viberg, Mr P. Fagan, Mr W.H. Theakstone and numerous other colleagues for their assistance. I wish to thank also Dr. T. Tika-Vassilikos for her agreement to the inclusion, in Fig. 8, of results from our forthcoming paper, Dr C.M. Clapperton and Ing. J.G. Bustamente for the photographs reproduced in Figs 16 & 18, respectively, TransManche-Link for permission to publish Figs. 6,9 & 10, and The Royal Society, The Institution of Civil Engineers, The Norwegian Geotechnical Institute, Elsevier and Posford Duvivier for permission to reproduce Figs. 25, 21, 12 & 13, 1, 2 & 3, respectively. Finally, I am most grateful to Miss Diane Ryan, Miss Selina Bourne and Miss Fionnuala Donovan for their word-processing, to Mrs Alison Davis for the graphics, to Mr Richard Packer for the photography, and to Mr C. G. Wedgwood and Mr S. Vinyard for their general support.

11 REFERENCES

Abe, O., T. Nakamura, T.E. Lang & T. Ohnuma. 1987. Comparison of simulated runout distances of snow avalanches with those of actually observed events in Japan. In : B. Salm & H. Gubler (eds), *Avalanche Formation, Movement and Effects*, IAHS Pub. No. 162 (Proc. Davos Symp.), 463–473.

Abele, G. 1974. Bergstürze in den Alpen. *Wissenschaftlige Alpenvereinshefte*, Heft, 25. München.

Aki, K. & K. Irikura 1991. Characterization and mapping of earthquake shaking for seismic zonation. *Proc. 4th Int. Conf. on Seismic Zonation*, Stanford, 1 : 61–110.

Almagia, R. 1910. Studi geografici sulle frane d'Italia. *Mem. R. Soc. Geogr.*, 2, Roma.

Ambraseys, N.N. 1977. On the response of structures to travelling waves. *Proc. CENTO Seminar on Earthquake Hazard Minimization*, Tehran: 410–414.

Anderson, M.G. & R.G. Pope. 1984. The incorporation of soil water physics models into geotechnical studies of landslide behaviour. *Proc. 4th Int. Symp. Landslides*, Toronto, 1 : 349–353.

Andresen, A. & L. Bjerrum. 1967. Slides in subaqueous slopes in loose sand and silt. In : A.F. Richards (ed.), *Marine Geotechnique*, 221–239. Urbana, Chicago, London: University of Illinois Press.

Anon. 1898. Serious subsidence of the cliff at Walton-on-the-Naze. *Essex Nat.*, 10: 236–237.

Antoine, P. 1977. Réflexions sur la cartographie ZERMOS et bilan des expériences en cours. *Bull. Bur. Rech. geol. min.*, Sec. III, No. 1–2: 9–20.

Arnould, M. & P. Frey. 1978. Analyse des réponses a une enquête internationale de l'Unesco sur les glissements de terrain. *Bull. Int. Ass. Engng. Geol.*, No. 17: 114–118.

Aulitzky, H. 1990. About methods of judgement of slope stability and torrents in Austria. In: A

Cancelli (ed.), *Alps 90. 6th Int. Conf. & Field Workshop on Landslides*, Milano : 9–15.

Baker, Sir Richard. 1674. A chronicle of the Kings of England, etc. (6th impression). London, Ludgate-hill : George Sawbridge, Hosier Lane : Thomas Williams.

Ballantyne, C.K. 1984. The Late Devensian periglaciation of upland Scotland. *Quaternary Sci. Reviews*, 3: 311–343.

Barry, R.G. & A.H. Perry. 1973. *Synoptic Climatology – Methods and Applications*. London: Methuen & Co. Ltd.

Bell, D.H. & J.P. Pettinga. 1986. Engineering geology and land – use planning in New Zealand. *Proc. 5th Int. Congr Int. Ass. Engng Geol.*, Buenos Aires, 4 : 2329–2338.

Bell, D.H. & J.R. Pettinga. 1988. Bedding – controlled landslides in New Zealand soft rock terrain. *Proc. 5th Int. Symp. on Landslides*, Lausanne, 1: 77–83.

Bishop, A.W. 1973. The stability of tips and spoil heaps. *Q. J. Engng Geol.* 6 : 335–376.

Bishop, A.W., G.E. Green, V.K. Garga, A. Andresen & J.D. Brown. 1971. A new ring shear apparatus and its application to the measurement of residual strength. *Géotechnique*, 21: 273–328.

Bishop, A.W., J.N. Hutchinson, A.D.M. Penman & H.E. Evans. 1969. Aberfan Inquiry : geotechnical investigation into the causes and circumstances of the disaster of 21st October, 1966. A selection of Technical Reports submitted to the Aberfan Tribunal, 1–80. London: Welsh Office.

Bishop, A.W. & N.R. Morgenstern. 1960. Stability coefficients for earth slopes. *Geotechnique* 10 : 129–150.

Bjerrum, L., T. Løken, S. Heiberg & R. Foster. 1969. A field study of factors responsible for quick clay slides. *Proc. 7th Int. Conf. Soil Mechs & Foundn Engrg*, Mexico, 2: 531–540.

Bonnard, C.H. & F. Noverraz. 1984. Instability risk maps: from the detection to the administration of landslide – prone areas. *4th Int. Symp. on Landslides*, Toronto, 2: 511–516.

Boswell, P.G.H. 1915. The stratigraphy and petrology of the Lower Eocene deposits of the north-eastern part of the London Basin. *Q. J. Geol. Soc. Lond.*, 71: 536–588.

Brabb, E.E. 1984. Innovative approaches to landslide hazard mapping. *4th Int. Symp. on Landslides*, Toronto, 1: 307–323.

Brabb, E.E. & B.L. Harrod (eds). 1989. *Landslides : Extent and Economic Significance*. Rotterdam/Brookfield: A.A. Balkema.

Brabb, E.E., E.H. Pampeyan & M.G. Bonilla. 1972. Landslide susceptibility in San Mateo County, California. *U.S. Geol. Surv. Misc. Field Studies Map*, MF–360 (Scale 1: 62,500).

Brand, E.W. 1988a. Special Lecture: Landslide risk assessment in Hong Kong. *Proc. 5th Int. Symp. on Landslides*, Lausanne, 2: 1059–1074.

Brand, E.W. 1988b. Predicting the performance of

residual soil slopes. *Proc. 11th Int. Conf. Soil Mechs & Foundn Engrg*, San Francisco, 5 : 2541–2578.

Brand, E.W., A.D. Burnett & K.A. Styles. 1982. The geotechnical area studies programme in Hong Kong. *Proc. 7th S.E. Asian Geotech. Conf.*, Hong Kong, 1: 107–123.

Brand, E.W., J. Premchitt & H.B. Phillipson. 1984. Relationship between rainfall and landslides in Hong Kong. *Proc. 4th Int. Symp. Landslides*. Toronto. 1 : 377–384

Bromhead, E.N., M.P. Chandler & J.N. Hutchinson. 1991. The recent history and geotechnics of landslides at Gore Cliff, Isle of Wight. In : R.J. Chandler (ed.), *Slope Stability Engineering*: 189–196. London: Thomas Telford.

Brunsden, D. 1973. The application of systems theory to the study of mass movement. *Geologia Applicata e Idrogeologia*, 8: 185–207.

Brunsden, D. 1985. Landslide types, mechanisms, recognition, identification. In : C.S. Morgan (ed.), *Landslides in the South Wales coalfield*, Proc. Symp. Poly. of Wales: 19–28.

Brunsden, D. 1987. Principles of hazard assessment in neotectonic terrains. *Mem. Geol. Soc. China*, No. 9 : 305–334.

Brunsden, D., J.C. Doornkamp, P.G. Fookes, D.K.C. Jones & J.M.H. Kelly. 1975. Large scale geomorphology mapping and highway engineering design. *Q. J. Engng Geol.*, 8: 227–253.

Brunsden, D. & D.K.C. Jones. 1976. The evolution of landslide slopes in Dorset. *Phil. Trans. R. Soc. Lond.* A283 : 605–631.

Campbell, R. 1973. Isopleth map of landslide deposits, Point Dume Quadrangle, Los Angeles County, California; an experiment in generalising and quantifying areal distribution of landslides. *U.S. Geol. Surv. Misc. Field Studies Map*, MF–535 (Scale 1: 24,000).

Capecchi, F. & F. Focardi. 1988. Rainfall and landslides: research into a critical precipitation coefficient in an area of Italy. *5th Int. Conf. on Landslides*, Lausanne, 2: 1131–1136.

Carrara, A., M. Cardinali, R. Detti, F. Guzzetti, V. Pasqui & P. Reichenbach. 1990. Geographical Information Systems and multivariate models in landslide hazard evaluation. In : A. Cancelli (ed.), *Alps 90. 6th Int. Conf. & Field Workshop on Landslides*, Milano: 17–28.

Carrara, A., E. Catalano, M. Sorriso Valvo, C. Realli & I. Ossi. 1978. Digital terrain analysis for land evaluation. *Geologia Applicata e Idrogeologia*, 13: 69–127.

Carrara, A. & L. Merenda. 1976. Landslide inventory in northern Calabria, southern Italy. *Geol. Soc. Amer. Bull.*, 87: 1153–1162.

Carrara, A., E. Pugliese Carratelli & L. Merenda. 1977. Computer–based data bank and statistical analysis of slope instability phenomena. *Zeitschr. für Geomorph.* N.F., 21 : 187–222.

Carson, M.A. & D.J. Petley 1970. The existence of threshold hillslopes in the denudation of the landscape. *Trans. Inst. Brit. Geogrs*, 49 : 71–95.

Champetier de Ribes, G. 1987. La cartographie des mouvements de terrain Des ZERMOS aux PER. *Bull. liaison Labo. P. et Ch.*, 150/151, Réf. 3226: 9–19.

Chandler, M.P. & J.N. Hutchinson. 1984. Assessment of relative slide hazard within a large, pre–existing coastal landslide at Ventnor, Isle of Wight. *4th Int. Symp. on Landslides*, Toronto, 2: 517–522.

Chandler, R.J. 1976. The history and stability of two Lias clay slopes in the upper Gwash valley, Rutland. *Phil. Trans. R. Soc. Lond.* A. 283: 463–491.

Chandler, R.J. 1984. Recent European experience of landslides in over–consolidated clays and soft rocks. *4th Int. Symp. on Landslides, Toronto*, 1 : 61–81.

Chandler, R.J., G.A. Kellaway, A.W. Skempton & R.J. Wyatt. 1976. Valley slope sections in Jurassic strata near Bath, Somerset. *Phil. Trans. R. Soc. Lond.* A. 283: 527–556.

Chowdhury, R.N. 1984. Recent developments in landslide studies : probabilistic methods. *4th Int. Symp. on Landslides*, Toronto, 1 : 209–220.

Christian, C.S. & G.A. Stewart. 1953. General Report on Survey of Katherine–Darwin Region, 1946. C.S.I.R.O., *Aust. Land Res. Ser.*, No. 1.

Church, M. & M.J. Miles. 1987. Meteorological antecedents to debris flow in British Columbia: some case histories. In: J.E. Costa & G.F. Weiczorek (eds), *Debris flows/avalanches : process, recognition and mitigation. Geol. Soc. Amer. Reviews in Engrg Geol.*, 7: 63–79.

Conway, B.W., A. Forster, K.J. Northmore & W.J. Barclay. 1980. South Wales Coalfield Landslip Survey. *Inst. Geol. Sci. Engng Geol. Unit*, Report EG 80/4.

Cooke, R.U. & J.C. Doornkamp. 1974. *Geomorphology in Environmental Management*. Oxford: Oxford University Press.

Cotecchia, V. 1978. Systematic reconnaissance mapping and registration of slope movements. *Bull. Int. Ass. Engng Geol.*, No. 17: 5–37.

Cotecchia, V. & G. Melidoro. 1974. Some principal geological aspects of the landslides of southern Italy. *Engng Geol.* 9 : 23–32.

Crandell, D.R. 1971. Postglacial lahars from Mount Rainier volcano, Washington. *U.S. Geol. Surv. Prof. Paper* 677 : 1–75.

Crozier, M.J. 1984. Field assessment of slope instability. In : D. Brunsden and D.B. Prior (eds), *Slope Instability*: 103–142. Chichester : John Wiley & Sons.

Crozier, M.J. 1986. *Landslides : causes, consequences & environment*. London : Croom Helm.

Crozier, M.J. & R.J. Eyles. 1980. Assessing the

probability of rapid mass movement. *3rd Australian – New Zealand Conf. Geomechs*, Wellington, 2 : 2.47–2.53.

Crozier, M.J., M. Gage, J.B. Pettinga, M.J. Selby & R.J. Wasson. 1982. The stability of hill slopes. In : J.M. Soons & M.J. Selby (eds). *Landforms of New Zealand* : 45–66. Auckland: Longman Paul Ltd.

Cruden, D.M. & W.M. Brown III. 1992. Progress towards the World Landslide Inventory. In : D.H. Bell (ed.), *Proc. 6th Int. Symp. on Landslides*, Christchurch, 1 : 59–64.

Cruden, D.M. & S. Thomson, B.D. Bornhold, J-Y. Chagnon & J. Locat, S.G. Evans & J.A. Heginbottom, K. Moran & D.J.W. Piper, R. Powell, D. Prior & R.M. Quigley. 1989. Landslides: extent and economic significance in Canada. In: E.E. Brabb & B.L. Harrod (eds). *Landslides: Extent and Economic Significance*: 1–23. Rotterdam/Brookfield: A.A. Balkema.

Cruden, D.M. & P.C. Tsui. 1991. Some influences of ice thrusting in geotechnical engineering. In : A. Forster, M.G. Culshaw, J.C. Cripps, J.A. Little & C.F. Moon (eds). *Quaternary Engineering Geology, Geol. Soc. Engng Geol. Spec. Pub.* No. 7: 127–134.

Culshaw, M.G., P.D. Jackson & D.M. McCann. 1987. Geophysical mapping techniques in environmental planning. In : M.G. Culshaw, F.G. Bell, J.C. Cripps & M. O'Hara (eds). *Planning and Engineering Geology, Geol. Soc. Engng Geol. Spec. Pub.* No. 4 : 171–177.

Decker, R.W. 1981. The 1980 activity – a case study in forecasting volcanic eruptions. In : P.W. Lipman & D.R. Mullineaux (eds). *The 1980 eruptions of Mount St. Helens, Washington. U.S. Geol. Surv. Prof. Paper* 1250 : 815–820.

Del Prete, M., T.P. Gostelow & J. Piniska. 1992. The importance of historical observations in the study of climatically controlled mass movement on natural slopes, with examples from Italy, Poland and UK. *Proc. 6th Int. Symp. on Landslides*, Christchurch, 3: (in press).

Denness, B. & B.W. Riddolls. 1976. The influence of geological factors on slope stability in the London Clay of South Essex, England. *Bull. Int. Ass. Engng. Geol.*, No 14: 37–40.

Department of the Environment. 1990. Planning policy guidance : development on unstable ground, PPG 14. London: HMSO.

Dingle, R.V. 1977. The anatomy of a large submarine slump on sheared continental margin (southeast Africa). *J. Geol. Soc. Lond.*, 134: 293–310.

Doornkamp, J.C. & J.S. Griffiths. Slope Stability: an outline of the legislative framework, planning and insurance issues in England and Wales. (in prepn).

Ecole Polytechnique Fédérale de Lausanne. 1985. Detection et utilisation des terrains instabiles. Rapport Final, Rapport General. Lausanne: Ecole Polytechnique Federale.

Edgers, L. & K. Karlsrud. 1982. Soil flows generated by submarine slides – case studies and consequences. *Proc. 3rd Int. Conf. Behaviour of Off-shore Structures*, 2: 425–437.

Eisbacher, G.H. & J.J. Clague. 1984. *Destructive Mass Movements in High Mountains: Hazard and Management.* Paper 84–16, Ottawa: Geological Survey of Canada.

Evans, S.G., J.J. Clague, G.J. Woodsworth & O. Hungr. 1989. The Pandemonium Creek rock avalanche, British Columbia. *Can. Geotech. J,* 26 : 427–446.

Fell, R., T.D. Sullivan & J.P. MacGregor. 1988. The influence of bedding plane shears on slope instability in sedimentary rocks. *Proc. 5th Int. Symp. on Landslides*, Lausanne, 1: 129–134.

Finlayson, A.A. 1984. Land surface evaluation for engineering practice : applications of the Australian PUCE System for terrain evaluation. *Q.J. Engng Geol.*, 17: 149–158.

Flageollet, J.C. 1988. *Les mouvements de terrain et leur prévention.* Paris: Masson.

Flageollet, J.C. 1989. Landslides in France : A risk reduced by recent legal provisions. In: E.E. Brabb & B.L. Harrod (eds). *Landslides: Extent and Economic Significance*, 157–167. Rotterdam/Brookfield: A.A. Balkema.

Fookes, P.G. 1987. Land evaluation and site assessment (hazard and risk). In : M.G. Culshaw, F.G. Bell, J.C. Cripps & M. O'Hara (eds). *Planning and Engineering Geology, Geol. Soc. Engng Geol. Spec. Pub.* No. 4 : 273–282.

Fookes, P.G., M. Sweeney, C.N.D. Manby & R.P. Martin. 1985. Geological and geotechnical engineering aspects of low-cost roads in mountainous terrain. *Engng Geol.* 21 : 1–152.

Fournier d'Albe, E.M. 1976. Natural disasters. *Bull. Int. Ass. Engng Geol.*, No. 14: 187.

Freeze, R.A. & Cherry, J.A. 1979. *Groundwater.* New Jersey: Prentice-Hall Inc.

Fukuoka, M. 1980. Landslides associated with rainfall. *Geotech. Engrg*, 11: 1–29.

Fulton, A.R.G. 1986. The supply and use of engineering geology information relating to slope instability hazards for planning purposes. In : M.G. Culshaw, F.G. Bell, J.C. Cripps & M.O'Hara (eds). *Planning and Engineering Geology, Geol. Soc. Engng Geol. Spec. Pub.* No. 4: 295–302.

Fuzukono, T. 1985. A new method for predicting the failure time of a slope. *Proc. 4th Int. Conf. & Field Workshop*, Tokyo : 145–150.

Gee, M.D. 1991. Classification of landslide hazard zonation methods and a test of predictive capability. *Proc. 6th Int. Symp. on Landslides*, Christchurch, 2: 947–952.

Geographical Survey Institute (Japan). 1983. 1978 Near Izu-Oshima earthquake, relationship between damage and topography (Scale 1:50,000, in Japanese).

Geological Society Working Party. 1982. Report

on land surface evaluation for engineering practice. *Q. J. Engrg Geol.*, 15: 265–316.

Gostelow, T.P. 1991. Rainfall and landslides. In: M.E. Almeida–Tixeria, R. Fantechi, R. Oliveira & A. Gomes Coelho (eds). *Prevention and control of landslides and other mass movements*: 139–164. Luxembourg: Commission of European Communities (EUR 12918).

Grant, K. 1965. Terrain features of the Mt. Isa–Dajarra Region and an assessment of their significance in relation to potential engineering land use. *Tech. Paper 1, CSIRO Aust., Div. Soil Mechs.*

Greenway, D.R. 1987. Vegetation and slope stability. In: M.G. Anderson & K.S. Richards (eds). *Slope Stability*: 187–230. Chichester: John Wiley & Sons.

Gregersen, O. 1981. The quick clay landslide in Rissa, Norway; the sliding process and discussion of failure modes. *Proc. 10th Int. Conf. Soil Mechs & Foundn Engrg*, Stockholm, 3: 421–426.

Griffiths, J.S. 1986. Landslide Hazard Analysis. Geomorphological Services Ltd., Review of Research into Landsliding, Series D, Vol. 1. Department of the Environment, U.K.

Griffiths, J.S. & K.S. Richards. 1989. Application of a low–cost database to soil erosion and soil conservation studies in the Awash Basin, Ethiopia. *Land Degradation & Rehabilitation*, 1 : 241–262.

Grindley, J. 1967. The estimation of Soil Moisture Deficits. *Meteorological Mag.*, 96, No. 1137: 97–108.

Hansen, A. 1984. Landslide Hazard Analysis. In: D. Brunsden & D.B. Prior (eds). *Slope Instability*: 523–602. Chichester : John Wiley & Sons.

Hansen, A. & C.A.M. Franks. 1991. Characterisation and mapping of earthquake triggered landslides for seismic zonation. *Proc. 4th Int. Conf. on Seismic Zonation*, Stanford, 1 : 149–195.

Hartlén, J. & Viberg, L. 1988. Evaluation of landslide hazard. *Proc. 5th Int. Symp. on Landslides*, Lausanne, 2: 1037–1057.

Hatano, S., F. Okabe, Y. Watanabe & T. Furukawa. 1974. Morphometrical map of large–scale landslide landforms in Hokusho District, north–western Kyushu, Japan. Japan Nat. Res. Center for Disaster Prevention, Report of Cooperative Res. for Disaster Prevention, No. 32.

Heim, A. 1932. *Bergsturz und Menschenleben.* Beiblatt zur Vierteljahrsschrift der Naturforschenden Gesellschaft in Zürich, 77: 1–217. (*Landslides & Human Lives*, Vancouver, B.C. : Bi Tech Publishers Ltd. Translator N. Skermer).

Hendron, A.J. Jr & F.D. Patton. 1985. The Vaiont Slide, a geotechnical analysis based on new geologic observations of the failure surfaces. *Tech. Rept. U.S. Army Waterways Expt. Stn*, GL–85–5.

Herman, J.R. & R.A. Goldberg. 1978. *Sun, Weather, and Climate*. Washington, D.C.: National Aeronautics and Space Administration.

Holmes, G. 1984. Rock slope failure in parts of the Scottish Highlands. Ph.D. Thesis (unpub.), University of Edinburgh.

Humbert, M. 1977. La cartographie ZERMOS. Modalités d'établissement des cartes des zones exposées à des risques liés aux mouvements du sol el du sous–sol. *Bull. Bur. Rech. geol. min.*, Sec. III, No. 1–2: 5–8.

Hutchinson, J.N. 1960. Guiding principles for future ground investigations in Namdalen. Norwegian Geotechnical Inst. Rept. F.169–5.

Hutchinson, J.N. 1961. A landslide on a thin layer of quick clay at Furre, Central Norway. *Géotechnique*, 11: 69–94.

Hutchinson, J.N. 1965. The stability of cliffs composed of soft rocks, with particular reference to the coasts of south–east England. Ph.D. Dissertation (unpub.), University of Cambridge.

Hutchinson, J.N. 1967. The free degradation of London Clay cliffs. *Proc. Geotech. Conf. Oslo*, 1: 113–118.

Hutchinson, J.N. 1969. A reconsideration of the coastal landslides at Folkestone Warren, Kent. *Géotechnique*, 19: 6–38.

Hutchinson, J.N. 1973. The response of London Clay cliffs to differing rates of toe erosion. *Geologia Applicata e Idrogelogia*, 8: 221–239.

Hutchinson, J.N. 1984. An influence line approach to the stabilisation of slopes by cuts and fills. *Can. Geotech. J.*, 21: 363–370.

Hutchinson, J.N. 1986. A sliding–consolidation model for flow slides. *Can. Geotech. J.*, 23: 115–126.

Hutchinson, J.N. 1987. Mechanisms producing large displacements in landslides on pre–existing shears. *Memoir Geol. Soc. of China*, No. 9: 175–200.

Hutchinson, J.N. 1988. Morphological and geotechnical parameters of landslides in relation to geology and hydrogeology. *Proc. 5th Int. Symp. on Landslides*, Lausanne, 1: 3–35.

Hutchinson, J.N. 1991. Periglacial and slope processes. In : A Forster, M.G. Culshaw, J.C. Cripps, J.A. Little & C.F. Moon (eds). *Quaternary Engineering Geology*, Geol. Soc. Engrg Geol. Spec. Pub. No. 7: 283–331.

Hutchinson, J.N. & R.K. Bhandari. 1971. Undrained loading, a fundamental mechanism of mudflows and other mass movements. *Géotechnique*, 21: 353–358.

Hutchinson, J.N., E.N. Bromhead & M.P. Chandler. 1981. Report on the coastal landslides at Bonchurch, Isle of Wight. Report to Lewis & Duvivier (unpub.), p.22.

Hutchinson, J.N., E.N. Bromhead & M.P. Chandler. 1991. Investigations of landslides at St. Catherine's Point, Isle of Wight. *Slope Stability Engineering*: 169–179. London: Thomas Telford.

Hutchinson, J.N., E.N. Bromhead & J.F. Lupini. 1980. Additional observations on the landslides at Folkestone Warren. *Q.J. Engng Geol.*, 13: 1–31.

Hutchinson, J.N. & M.P. Chandler. 1991. A preliminary landslide hazard zonation of the Undercliff of the Isle of Wight. In: R.J. Chandler (ed.), *Slope Stability Engineering*: 197–205. London : Thomas Telford.

Hutchinson, J.N. & T.P. Gostelow. 1976. The development of an abandoned cliff in London Clay at Hadleigh, Essex. *Phil. Trans. R. Soc. Lond.*, A283: 557–604.

Hutchinson, J.N., C. Poole, N. Lambert & E.N. Bromhead. 1985. Combined archaeological and geotechnical investigations of the Roman Fort at Lympne, Kent. *Britannia*, 16: 209–236.

Hutter, K., F. Szidarovszky & S. Yakowitz. 1987. Granular shear flows as models for flow avalanches. In : B. Salm & H. Gubler (eds), *Avalanche Formation, Movement and Effects*, IAHS Pub. No. 162 (Proc. Davos Symp.), 381–394.

Iida T. & K. Okunishi. 1983. Development of hillslopes due to landslides. *Zeitschr. für Geomorph.* 46: 67–77.

Ishihara, K. 1985. Stability of natural deposits during earthquakes. *Proc. 11th Int. Conf. on Soil Mechs. & Foundn Engrg*, San Francisco, 1: 321–376.

Jennings, P.J., H.J. Siddle & S.P. Bentley. 1991. A comparitive study of indirect methods of landslip potential assessment. In: R.J. Chandler (ed.), *Slope Stability Engineering*: 143–148. London: Thomas Telford.

Jones, O.T. 1937. On the sliding or slumping of submarine sediments in Denbighshire, North Wales, during the Ludlow Period. *Quart. J. Geol. Soc. Lond.*, 93: 241–283.

Jórstad, F. 1968. Waves generated by landslides in Norwegian fjords and lakes. *Norwegian Geotech. Inst. Pub.* No. 79: 13–32.

Kárník, V. & S.T. Algermissen. 1978. Seismic Zoning. *The assessment and mitigation of earthquake risk*: 11–47. Paris: UNESCO.

Keefer, D.K., G.F. Wieczorek, E.L. Harp & D.H. Tuel. 1978. Preliminary assessment of seismically induced landslide susceptibility. *Proc. 2nd Int. Conf. on Microzonation*, San Francisco, 1: 279–290.

Keefer, D.K., R.C. Wilson, R.K. Mark, E.E. Brabb, W.M. Brown III, S.D. Ellen, E.L. Harp, G.F. Wieczorek, C.S. Alger & R.S. Zatkin. 1987. Real-time landslide warning during heavy rainfall. *Science*, 238 : 921–925.

Kendall, P.F. 1902. A system of glacier-lakes in the Cleveland Hills. *Q.J. Geol. Soc. Lond.*, 58 : 471–571.

Kenney, T.C. & S. Uddin. 1974. Critical period for stability of an excavated slope in clay soil. *Can. Geotech. J.*, 11 : 620–623.

Kienholz, H. 1978. Maps of geomorphology and natural hazards of Grindelwald, Switzerland, Scale 1: 10,000. *Arctic and Alpine Res.*, 10: 169–184.

Kiersch, G.A. 1964. Vaiont reservoir disaster. Civ. Engng (New York), 34: 32–39.

Kojan, E. & J.N. Hutchinson. 1978. Mayunmarca rockslide and debris flow, Peru. Chap. 9 in: B.Voight (ed.), *Rockslides and Avalanches* (Developments in Geotechnical Engineering 14A): 315–361. Amsterdam, Oxford, New York: Elsevier.

Kyunttsel', V.V. 1988. Landslides. In: E.A. Kozlovskii (ed.). *Landslides and mudflows*: 35–54. Moscow: UNESCO/UNEP.

Lee, E.M., R. Moore, D. Brunsden & H.J. Siddle. 1991. The assessment of ground behaviour at Ventnor, Isle of Wight. In : R.J. Chandler (ed.), *Slope Stability Engineering*: 207–212. London : Thomas Telford.

Lee, H.J. 1989. Undersea landslides: Extent and significance in the Pacific Ocean. In: E.E. Brabb & B.L. Harrod (eds). *Landslides: Extent and Economic Significance*: 367–379. Rotterdam/Brookfield: A.A. Balkema.

Lemos, L. A.W. Skempton & P.R. Vaughan. 1985. Earthquake loading of shear surfaces in slopes. *Proc. 11th Conf. Soil Mechs & Foundn Engrg*, San Francisco, 4: 1955–1958.

Li, T. 1990. Landslide management in mountain areas of China. *Int. Centre for Integrated Mountain Development*, Kathmandu, Nepal. *Occasional Paper* 15.

Lundqvist, J. & R. Lägerback. 1976. The Pärve fault; a late-glacial fault in the Precambrian of Swedish Lapland. *Geol. fören. förhand. Stockholm*, 98: 45–51.

Malgot, J. & T. Mahr. 1979. Engineering geological mapping of the West Carpathian landslide areas. *Bull. Int. Ass. Engng Geol.*, 19: 116–121.

Mal'neva, I.V. & N.K. Kononova. 1988. Climatic conditions. In: E.A. Kozlovskii (ed.). *Landslides and mudflows*: 20–22. Moscow: UNESCO/UNEP.

Moore, D.G. 1978. Submarine slides. In: B. Voight (ed.). *Rockslides and Avalanches*. Developments in Geotechnical engineering, 14A: 563–604. Amsterdam, Oxford, New York: Elsevier.

Moser, M. & F. Hohensinn. 1983. Geotechnical aspects of soil slips in Alpine regions. *Engng Geol.*, 19: 185–211.

Mougin, P. 1914. *Les torrents de la Savoie*. Grenoble.

Mulder, H.F.H.M. 1991. Assessment of landslide hazard. Proefschrift ter Verkrijging van Graad van Doctor aan de Rijksuniversiteit te Utrecht. Utrecht : Faculty of Geographical Sciences.

National Remote Sensing Centre. 1990. *A Guide to Earth Observing Satellites*. Farnborough, Hampshire : National Remote Sensing Centre Limited.

Nemčok, A. & J. Rybář. 1968. Synoptic map of Czechoslovak landslide areas. *Geol. Ustar Ceskoslov, Akad.* Ved : Praha.

Nemčok, A. 1982. *Zosuvy v Slovenských*

Karpatoch. Bratislava : Veda.

Neuland, H. 1976. A prediction model of landslips. *Catena*, 3: 215-230.

Newman, E.B., A.R. Paradis & E.E. Brabb. 1978 Feasibility and cost of using a computer to prepare landslide susceptibility maps of the San Francisco Bay Region, California. *U.S. Geol. Surv. Bull.* No. 1443.

Newman, I.C. 1985. The Ok Tedi Project. *Bull. Australian Institution of Mining & Metallurgy*, 290, August 1985: 67-72.

Novosad, S. 1990. Evaluation and mitigation of geologic hazard related to the construction of Slezska Harta dam on the Moravice river in Czechoslovakia. *Proc. 6th Int. Congr. Int. Ass. Engng Geol.*, Amsterdam: 1941-1948.

Ollier, C.D. 1977. Terrain classification : methods, applications and principles. In : J.R. Hails (ed.), *Applied Geomorphology*: 277-316. Amsterdam, Oxford, New York: Elsevier Scientific Pub. Co.

Oyagi, N., F. Shimizu & T. Inokuchi. 1982. Landslide map of Kiyokawa. Natl Res. Center for Disaster Prevention, Japan, Map NJ-54-20-16 (Scale 1: 50,000).

Palmquist, R.C. & G. Bible. 1980. Conceptual modelling of landslide distribution in time and space. *Bull. Int. Ass. Engrg Geol.* No. 21 : 178-186.

Pašek, J. J. Rybář & M. Špůrek. 1977. Systematic registration of slope deformations in Czechoslovakia. *Bull. Int. Ass. Engrg. Geol.*, 16: 48-51.

Payne, H.R. 1985. Hazard assessment and rating methods. In : C.S. Morgan, *Landslides in the South Wales coalfield*. Proc. Symp. Poly. of Wales: 59-71.

Plafker, G., G.E. Eriksen & J.M. Concha. 1971. Geological aspects of the May 31, 1970, Peru earthquake. *Bull. Seism. Soc. Amer.*, 61: 543-578.

Porcher, M. & P. Guillope. 1979. Cartographie des risques ZERMOS appliquée à des plans d'occupation des sols en Normandie. *Bull. liaison Labo. P. et Ch.*, 99, Réf. 2277: 43-54.

Prior, D.B. & J.M. Coleman. 1984. Submarine slope instability. In: D. Brunsden & D.B. Prior (eds). *Slope instability*: 419-455. Chichester: John Wiley & Sons.

Radbruch-Hall, D.H., R.B. Colton, W.E. Davies, B.A. Skipp, I. Lucchitta & D.J. Varnes. 1976. Landslide overview map of the conterminous United States. *U.S. Geol. Surv. Misc. Inv. Map*, I-982 (Scale 1: 7,500,000).

Rib, H.T. & T. Liang. 1978. Recognition and Identification. Chap 3 in : R.L. Schuster & R.J. Krizek (eds), *Landslides Analysis and Control*, *Special Report* 176 : 34-80. Transportation Research Board. Washington, D.C. : National Academy of Sciences.

Robinson, A.H.W. 1953. The storm floods of 1st February 1953. V. The sea floods around the Thames Estuary. *Geography*, 38: 170-176.

Rybář, J., J. Pašek & L. Repka. 1965. Dokumentation der systematischen Untersuchung der Rutschungsgebiete in der Tschechoslowakei. *Engrg Geol.* (Amsterdam) 1: 21-29.

Saito, M. 1969. Forecasting time of slope failure by tertiary creep. *Proc. 7th Int. Conf. Soil Mechs & Foundn Engrg*, Mexico, 2: 677-683.

Salt, G.A. 1988. Landslide mobility and remedial measures. *Proc. 5th Int. Symp. on Landslides*, Lausanne, 1 : 757-762.

Sarconi, M. 1784. Osservazioni fatte nelle Calabrie e nella frontiera di Valdemone sui Fenomeni del tramoto del 1783 e sulla Geografia fisica di quelle regioni. *R. Acc. Sc. e Belle Lett.*, Napoli.

Saxov, S. & J.K Nieuwenhuir (eds). 1980. *Marine slides and other mass movements*. New York : Plenum Press.

Schumm, S.A. 1973. Geomorphic thresholds and complex response of drainage systems. Chap. 13 In : M. Morisawa (ed.), *Fluvial Geomorphology*: 299-310. Binghampton : Publications in Geomorphology.

Schuster, R.L. 1978. Introduction. In : R.L. Schuster & R.J. Krizek (eds). *Landslides – Analysis and Control*: 1-10. Washington, D.C. : National Academy of Science (*Transportn Res. Board Spec. Rept* 176).

Schuster, R.L. (ed.). 1986. Landslide dams : processes, risk and mitigation, *Geotech. Spec. Pub.* No. 3. New York: Am. Soc. Civ. Engrs.

Seed, H.B. 1968. Landslides during earthquakes due to soil liquefaction. *J. Soil Mechs. & Foundn Div.*, Proc. A.S.C.E., 94 (SM5) :193-260.

Sheko, A.J. 1988a. Mudflows. In: E.A. Kozlovskii (ed.). *Landslides and mudflows*, 1: 54-74. Moscow: UNESCO/UNEP.

Sheko. A.I.; 1988b. Time prediction of landslides and mudflows. 1. Cyclicity in nature. In: E.A. Kozlovskii (ed.). *Landslides and Mudflows*, 1: 181-184. Moscow: UNESCO/UNEP.

Sheko. A.I. 1988c. Time prediction of landslides and mudflows. 3. Short-term forecasts. In: E.A. Kozlovskii (ed). *Landslides and Mudflows*, 1: 195-199. Moscow: UNESCO/UNEP.

Sheko, A.I., R.L. Schuster & R.W. Fleming. 1988. Introduction. 1. Socio-economic significance of landslides and mudflows. In E.A. Kozlovskii, (ed): *Landslides and mudflows*, 1: 5-10. Moscow : UNEP/UNESCO.

Sheko, A.I. & N.S. Sergeeva. 1988. Time prediction of landslides and mudflows. 2. Long-term forecasts. In: E.A. Kozlovskii (ed.). *Landslides and Mudflows*, 1: 184-195. Moscow: UNESCO/UNEP.

Shimizu, M. 1988. Prediction of slope failures due to heavy rains using the tank model. *Proc. 5th Int. Symp. Landslides*, Lausanne, 1: 771-777.

Siddle, H.J., D.B. Jones & H.R. Payne. 1991. Development of a methodology for landslip potential mapping in the Rhondda Valley. In: R.J. Chandler (ed.), *Slope Stability Engineering*: 137–142. London: Thomas Telford.

Sidle, R.C., A.J. Pearce & C.L., O'Loughlin. 1985. *Hillslope stability and land use. Water Resources Monograph* 11. Washington, D.C.: Amer. Geophys. Union.

Skempton, A.W. 1977. Slope stability of cuttings in brown London Clay. *Proc. 9th Int. Conf. Soil Mechs & Foundn Engrg*, Tokyo, 3 : 261–270.

Skempton, A.W. 1985. Residual strength of clays in landslides, folded strata and the laboratory. *Geotechnique*, 35 : 3–18.

Skempton, A.W. & F.A. DeLory. 1957. Stability of natural slopes in London Clay. *Proc. 4th Int. Conf. Soil Mechs & Foundn Engrg*, London, 2: 378–381.

Skempton, A.W. & J.N. Hutchinson. 1969. Natural slopes and embankment foundations. *Proc. 7th Int. Conf. Soil Mechs & Foundn Engrg*, Mexico, State-of-the-Art Vol: 291–340.

Skempton, A.W., A.D. Leadbetter & R.J. Chandler. 1989. The Mam Tor Landslide, N. Derbyshire. *Phil. Trans. R. Soc. Lond.*, A 329: 503–547.

Skempton, A.W. & A.G. Weeks. 1976. The Quaternary history of the Lower Greensand escarpment and Weald Clay vale near Sevenoaks, Kent. *Phil. Trans. R. Soc. Lond.* A283 : 493–526.

Söderblom, R. 1974. Organic matter in Swedish clays and its importance for quick clay formation. Swedish Geotech. Inst. Proc. No. 26: 1–89.

Solentsev, N.A. 1962. Basic problems in Soviet landscape science. *Sov. Geogr.*, 3: 3–15.

Sousa, J. & B. Voight, 1991. Continuum simulation of flow failures. *Géotechnique*, 41 : 515–538.

Stevenson, P.C. 1977. An empirical method for the evaluation of relative landslide risk. *Bull. Int. Ass. Engng Geol.*, No. 16 : 69–72.

Stini, J. 1910. *Die Muren.* Innsbruck: Wagner.

Stini, J. 1938. Über die Regelmässigkeit der Wiederkehr von Rutschungen, Bergstürzen, und Hochwasserschäden in Österreich. *Geologie und Bauwesen*, 10 : 9–31 & 33–48.

Styles, K.A. & A. Hansen. 1989. *Geotechnical Area Studies Programme : Territory of Hong Kong*, GASP Report XII. Hong Kong: Geotech. Control Office.

Terzaghi, K. 1929. Effect of minor geologic details on the safety of dams. *Amer. Inst. of Mining & Metallurgical Engrs. Tech. Pub*, 215: 31–44.

Terzaghi, K. 1950. Mechanism of Landslides. *Engineering geology*, Geol. Soc. Amer., (Berkey) Vol: 83–123.

Thornbury, W.D. 1966. *Principles of Geomorphology*. New York : Wiley.

Tika–Vassilikos, T. & J.N. Hutchinson. Ring shear tests on soil from the Vaiont landslide slip surface (in prepn).

Toppe, R. 1987. Terrain models – a tool for natural hazard mapping. In : B. Salm & H. Gubler (eds), *Avalanche Formation, Movement and Effects*, IAHS Pub. No. 162, (Proc. Davos Symp.): 629–638.

Tscharner, J.B. von. 1807. Etwas über Bergstürze, Bergfälle, Erdstürze, Schlipfe und Erdsinken. *Der Sammler, ein gemeinnützziges Archiv für Bünden*: 3–30. Chur.

Twidale, C.R. 1985. Ancient landscapes their nature and significance for the question of inheritance. In : R.S. Hayden (ed.), *Global Mega-geomorphology*: 29–40. NASA Conf. Pub. 2312.

Van Driel, N. 1991. Geographic Information Systems for earth science applications. *Proc. 4th Int. Conf. on Seismic Zonation*, Stanford, 1 : 469–485.

Varnes, D.J. 1978 . Slope Movement Types and Processes. In : R.L. Schuster & R.J. Krizek (eds), *Landslides-Analysis and Control*: 11–33. Washington, D.C. : Nat. Acad. Sciences. (*Transportation Res. Board Spec. Rept* 176).

Varnes, D.J. 1983. Time-deformation relations in creep to failure of earth materials. *Proc. 7th Southeast Asian Geotech. Conf.*, Hong Kong, 2 : 107–130.

Varnes, D.J. & the International Association of Engineering Geology Commission on Landslides and Other Mass Movements on Slopes. 1984. *Landslide hazard zonation : a review of principles and practice.* Paris : UNESCO.

Vaughan, P.R. 1985. Pore pressures due to infiltration into partly saturated slopes. *Proc. 1st Int. Conf. Geomechanics in Tropical Lateritic and Saprolitic Soils*, Brasilia, 2 : 61–71.

Vaughan, P.R., D.W. Hight, V.G. Sodha & H.J. Walbancke. 1978. Factors controlling the stability of clay fills in Britain. *Clay Fills*: 203–217. London : Institution of Civil Engineers.

Vaughan, P.R. & H.J. Walbancke. 1973. Pore pressure changes and the delayed failure of cutting slopes in overconsolidated clay. *Geotechnique,* 23: 531–539.

Viberg, L. 1984. Landslide risk mapping in soft clays in Scandinavia and Canada. *Proc. 4th Int. Symp. Landslides*, Toronto, 1 : 325–348.

Viberg, L. & L. Ademstam. 1980. Geotechnical terrain classification for physical planning – a Swedish research project. *Bull. Int. Ass. Engrg Geol.*, No. 21: 174–178.

Vibert, C., M. Arnould, R. Cojean & J.M. Le Cleac'h. 1988. Essai de prévision de rupture d'un versant montagneux a Saint-Etienne-de-Tinée, France. *Proc. 5th Int. Symp. on Landslides*, Lausanne, 1 : 789–792.

Voight, B. 1989. A relation to describe rate–dependent material failure. *Science*, 243 : 200–203.

Wadge, G. 1988. The potential of GIS modelling of gravity flows and slope instabilities. *Int. J. Geographical Infn Systems*, 2: 143–152.

Wadge, G. & M.C. Isaacs. 1988. Mapping the volcanic hazards from Soufriere Hills Volcano, Montserrat, West Indies, using an image processor. *J. Geol. Soc. Lond.*, 145: 541–551.

Way, D.S. 1973. *Terrain Analysis*. Stroudsburg, Pa: Dowden, Hutchinson & Ross.

Wieczorek, G.F. 1982. Map showing recently active and dormant landslides near La Honda, central Santa Cruz Mountains, California. *U.S. Geol. Surv. Misc. Field Studies Map*, MF–1422 (Scale 1: 4,800).

Wieczorek, G.F., R.C. Wilson & E.L. Harp. 1985. Map showing slope stability during earthquakes in San Mateo County, California. *U.S. Geol. Surv. Map* I–1257–E.

Wislocki, A.P. & S.P. Bentley. 1991. An expert system for landslide hazard and risk assessment. *Computers & Structures*, 40: 169–172.

Wright, R.H., R.H. Campbell & T.H. Nilsen. 1974. Preparation and use of isopleth maps of landslide deposits. *Geology* (Geol. Soc. Amer.), 2: 483–485.

Wright, R.H. & T.H. Nilsen. 1974. Isopleth map of landslide deposits, southern San Francisco Bay Region, California. *U.S. Geol. Surv. Misc. Field Studies Map*, MF–550 (Scale 1: 125,000).

Youd, T.L. 1991. Mapping of earthquake–induced liquefaction for seismic zonation. *Proc. 4th Int. Conf. on Seismic Zonation*, Stanford, 1 : 111–138.

Záruba, Q. & V. Mencl. 1982. *Landslides and their control*. Amsterdam, Oxford, New York: Elsevier.

Zolotarev, G.S. & F.A. Novoselsky. 1977. Ancient and recent landslides and rockfalls in the skiby zone of the south–east Carpathians and geological regularities of their formation. *Bull. Int. Ass. Engng Geol.*, No. 16 : 74–76.

Landslides, Bell (ed.) © 1995 Balkema, Rotterdam, ISBN 90 5410 032 X

Theme report

M.J.Crozier
Victoria University, Wellington, New Zealand

ABSTRACT: A review of 37 papers on the landslide hazard assessment theme shows a predominance of papers concerned with methodologies for ranking different terrain units in terms of their susceptibility to the landslide hazard. Most of these require little in the way of geotechnical information and employ instead other readily obtained stability factors selected deductively or inductively. A smaller number of papers have attempted to predict frequency of occurrence in time. Even fewer have provided schemes for predicting the frequency of a given magnitude of landslide activity in different places. Commendably some papers have set out to test the predictive capacity of their methodology. Some novel approaches to problem solving are evident, particularly in the use of probability distributions for handling variability of landslide related factors.

1 OBJECTIVES

There are seven broad objectives that can be identified in Landslide 'Hazard' Assessment (Table 1). They are all legitimate and worthwhile pursuits in their own right: each one providing a certain level of useful information. Objective (C) 'Characteristics' is essentially explanatory description and inventory while the other objectives listed in Table 1 are predictive. Only objective (CST) addresses the 'complete landslide hazard' and hence should be the ultimate goal of comprehensive landslide hazard research. However this goal is dependent on achievement of most of the other listed objectives and requires the greatest range of skills. Nevertheless, land managers and planners require a measure of the 'complete landslide hazard' so that it can be combined with VULNERABILITY to assess fully the RISK; i.e. the expected costs per unit of time arising from living in a place subject to landslides.

If the nature of the papers offered to this section is anything to go by, landslide hazard assessment still lags behind the ability of our colleagues to predict, for example, flood hazard.

There is yet a long way to go before landslide hazard assessment in terms of frequency and magnitude becomes common practice. Just as the magnitude (volume, depth, velocity, and duration) and its recurrence interval is a goal of flood prediction, so it should also be a goal of landslide hazard assessment. Thus, the conventional hazard equation (UNDRO)

HAZARD (H) x VULNERABILITY (V) = RISK (R)

can be factorised for different areas (S) as

[HAZARD FREQUENCY (T)/MAGNITUDE (C)] x V = R

where magnitude for landslides is measured in terms of its impact characteristics (C) e.g. volume, depth, velocity, area, runout distance.

As reference to Table 1 indicates, only two of the 37 papers contributed to this theme have assessed risk (R) in these terms. A further seven have addressed the time element (T) specifically or for different areas (ST). By far the largest number of papers (13 out of 37) are concerned mainly with occurrence in space (S). Thus it would appear, invoking hydrological analogy again, that much of our effort is still largely

Table 1. Objectives of landslide 'hazard' assessment

	Objective	Paper reference (see Table 2)
C	Characteristics of Occurrence	
	C1 characterisation of land units by their existing landslide activity	1,8,9,33[+],35
	C2 description and explanation of existing landslides	5,13
	C3 relationship between existing landslide characteristics and specific stability factors	12,15,28
S	Occurrence in Space	
	S1 prediction of susceptibility for different land units, or:	2,3,11,16,17,18,21,22,24,25, 26,30,33[+]
	S2 relative susceptibility of a site to landslides, specific site study	
T	Occurrence in Time	
	T1 prediction of landslide occurrence in time	6,19[+],23,36
	T2 prediction of rate of movement	19[+]
CS	Character of Occurrence in Space	CSR4,
	prediction of the different characteristics of landslides for different land units	10,32,34,37
CT	Character of Occurrence in Time	
	CT1 prediction of the evolution and behaviour of a landslide	14,29
	CT2 frequency of occurrence for different forms or magnitude of landslide	31
ST	Occurrence in Time and Space	
	prediction of the frequency of any landslide event for different land units	
CST	Character in Time and Space	CSTR20,
	determining the probability of landslide occurrence and characteristics in a unit of time, for different land units	27

NOTE 1. If the 'vulnerability' or consequences of predicted landslide activity to social, economic, or personal well-being are assessed along with the probability of landslide occurrence then 'R' is added to the objective classification, denoting that 'risk' has been addressed.

NOTE 2. The superscript '+' indicates the paper qualifies for more than one class.

devoted simply to locating the floodplain.

It is evident then that landslide hazard assessment, in terms of current practice, more often than not involves the mapping and ranking of different terrain units in terms of their susceptibility to landslides. Thus it is useful to reflect for a moment on the different methodologies employed (Table 3).

2 METHODOLOGIES

The determination of spatial differences in landslide 'hazard' can be approached in three ways: inductively, deductively and geotechnically by application of some form of stress analysis of limiting equilibrium (factor-of-safety) analysis. Because of the cost of

Table 2. Numerical reference for papers presented to 'Landslide Hazard Assessment' (Theme G4) by first author.

1	Alloul, B	14	Favre, J L	26	Mickelson, D M
2	Anbalagan, R	15	Focardi, P	27	Moon, A T
3	Barros, W T	16	Gee M D	28	Moore, R
4	Berggren, B	17	Ghosh, A	29	Nicoletti, P G
5	Bertocci, R	18	Hammond, C J	30	Olds, R J
6	Bhandari, R K	19	Hayashi, S	31	Omura, H
7	Brown, W M	20	Hearn, G J	32	Rao, P J
8	Carmassi, F	21	Huang, R	33	Sinclair, T J E
9	Cascini, L	22	Kasa, H	34	Tsiambaos, G
10	Chang, S-C	23	Kim, S K	35	Yoon, G G
11	Choubey, V D	24	Kingsbury, P A	36	Zhang, N
12	Corominas J	25	Mehrotra, G S	37	Zhang, Y-F
13	Culshaw, M G				

geotechnical investigation, regional landslide hazard assessment generally (but not always) employs mappable stability factors derived from existing sources or field observations and measurements that can be obtained with little subsurface investigation. Indeed only four of the 24 theme papers dealing with regional assessments have employed geotechnically determined stress analysis.

The inductive and deductive methodologies for regional landslide hazard assessment are more or less equally represented by contributions to this theme (seven inductive and nine deductive). The inductive approach involves the direct investigation of the relationships between active or previously active landslides and stability factors. The significance of stability factors and their recognition as causative factors is determined empirically, post event.

Arguably this is a more scientific (at least less subjective) method of selecting and weighting factors for extrapolation throughout a region. Three additional papers offered, while not producing a scheme for regional hazard zonation, systematically investigate the relationship between the occurrence of landslides or landslide characteristics (e.g. runout behaviour) and one or more causative factors, thus contributing a useful basis for future zonation schemes.

The deductive (indirect) approach to regional hazard assessment involves the selection of causative factors to be employed in zonation procedures without direct reference to existing landslides. The selection of factors and their relative weighting is determined by judgement (theory, experience, and intuition), pre-event. In practice, there is often not a clear distinction between the deductive and inductive approaches, neither should there be. Indeed two of the most comprehensive schemes presented (papers 33 and 20, see Table 2), while predominantly deductive in concept, employ the presence and characteristics of existing landslides as an influential weighting factor in the overall determination.

3 HANDLING UNCERTAINTY

By nature landslide hazard assessment is a complex, multivariate problem involving extrapolation of point data to wider dimensions of space and time. Inherently the practice involves a high level of uncertainty. Because of the important consequences of drawing conclusions on stability, zonation schemes must address and express this level of uncertainty.

There are three ways of expressing uncertainty. The first is based on the principle: that the less that is known, the greater the level of care which should be exercised. Thus if the parameters used for zonation purposes include a measure of input data quality and quantity, then the poorer the input the greater the hazard. This approach has been nicely illustrated by paper 4. A second approach expresses uncertainty in terms of

Table 3. Focus of papers presented to the landslide hazard assessment theme

Theme	Paper reference (see Table 2)
LANDSLIDE HAZARD ZONATION	
- Prediction by inductive (direct) method:	4^+,21,22,24,25,27,28
- Prediction by deductive (indirect) method:	2,3,10,11,20^+,30^+,32,33^+,34^+
- Geotechnical stress analysis:	
i Deterministic	4^+,17^+
ii Probabilistic	18,26
- Evaluation, testing and criticism	6^+,16,34^+
- Policy of implementation	30^+
CONDITION or OCCURRENCE/INVENTORY:	8,9^+,20^+,33^+,35
INVESTIGATION OF RELATIONSHIPS BETWEEN STABILITY FACTORS AND LANDSLIDE OCCURRENCE AND CHARACTER	
- Inherent factors	1,12^+,15
- Triggering factors	6^+,17^+,23,31,36
- Landslide behaviour	12^+,14,19,29
DESCRIPTION AND EXPLANATION	5,9^+,13
INFORMATION BASE	7
THEORY	37

zonation resolution. Here the principle is simply: the less you know the less you say. A consequence of this approach may be a two or three class classification - 'landslides possible', 'landslides unlikely', 'further investigation required'.

Increase in the quality and quantity of input data does not necessarily remove uncertainty but it does allow its qualification. Thus the third way of handling uncertainty is through statistical and verbal qualification, commonly through use of probability concepts. In this approach uncertainty may be handled by including the variability characteristics of the input data and then deriving a definitive classification (e.g. paper 18) or by defining the ultimate classification in terms of probability.

The degree of uncertainty generally prevents answering the critical questions of: can landslides occur here? if so what type, what magnitude, how often, or when? Instead of answering these questions many so-called prediction schemes divide up a region into some form of terrain classes, and then rank them in order of susceptibility. This is a low order of prediction, which avoids answering any of the critical questions outlined above but nevertheless offers some form of rational guidance to planning or further investigation.

4 SUBJECTIVITY/OBJECTIVITY

'Subjectivity' I define as the application of undisclosed judgement criteria and 'objectivity' as the application of rules dictated by established relationships. While clearly there is a place for both approaches, depending on the constraints of the exercise, for the achievement of communication of knowledge, verification and reproducibility, subjectivity should be kept to a minimum.

Subjectivity generally most commonly occurs at the beginning and end of the landslide hazard assessment process. In the deductive approach (9 papers) subjectivity is used to select the initial classification parameters and it may be continued with the selection of weighting factors (e.g. paper

Table 4. The main contribution of each paper within the landslide hazard assessment theme: or, 'why I would read them again!'

Contribution	Paper reference (see Table 2)
Clearly explained, highly appropriate hazard/risk zonation scheme	20
Clearly explained, highly appropriate condition/hazard zonation scheme	33
As above, with novel approach to addressing uncertainties in data base	4
Clearly explained debris flow hazard assessment	27
Methodology for evaluating success of zonation prediction	16
Useful statistics for controls of landslide size	12
Clear account of probability measures for variability in input factors	18
Experimentation to verify rainfall effect	23
Novel approach determining effect of storm rainfall on landslide distribution	31
Probability of triggering rainfall	36
Statistical weighting of factors	21
Runout prediction	29
Shoreline stability	26
Using GIS for zonation	24
Rockfall activity and history	13
Experience in implementing a zonation scheme	30
Rock mass strength for stability	34
Predicting size submarine landslides	14
Time of failure for creeping slide	19
Predicting types and occurrence	10
Local urban stability problems	5
weathering grades and stability	9
Weighting factors by landslide occurrence	25
Another zonation scheme	3,2,11,32
Geothermal site investigation	8
Application of remote sensing	22
Proposals for global database	7
Questioning rainfall relationships	6
Process and slope evolution	15
Visit to Aurora Province	28
Sketch of landslides, Korea	35
Rock structure/stress theory	37
Rocks and landslides, Algeria	1

32). Alternatively a degree of objectivity may be employed to weight selected parameters (e.g. paper 21).

The last stage of hazard assessment, i.e. the conversion of results into a hazard classification, virtually always involves some level of subjectivity. For example, an objective scheme based on factor-of-safety analysis (paper 26, Table 2) in the end resorts to a subjective interpretation of stability. One point that should be made clear to end users is that quantification does not always equate with objectivity.

Even if judgement is employed to select and weight stability factors, it is important that a rational explanation is provided on how that judgement has been exercised. Given a rational explanation, future changes in an area can then be introduced to reassess stability rankings. It follows that while the final classification of a region may be expressed in simple terms, there needs to be access and reference to the way in which procedural decisions have been made.

5 TESTING

It is good scientific practice to test any predictive scheme before application, whenever possible. A trial application of a scheme tests its implementation feasibility not its predictive capabilities.   Whereas feasibility trials were carried out on a number of schemes presented, it is commendable that five of the 21 papers (3, 16, 24, 34, 22) involved some form of testing.

Within the time frame available it is often unlikely that landslide occurrence will permit testing in the area for which the zonation scheme was developed. However, tests can sometimes be run on analogous areas affected by landslides (24) or by developing a scheme using pre-event data sources and then applying it to event conditions (16).

6 HIGHLIGHTS

The main contribution of each paper is listed in Table 4. This is a highly subjective selection listed more or less in the order of how they appealed to me. The listing relates partly to my own research interests, but primarily to the substance of the contribution. I was impressed by papers that attempted to assess all aspects of hazard (character, spatial distribution, and frequency), those that presented clear, unambiguous methodology, and those that involved testing or some innovative approach to problem solving.

Among the papers offered there are a number of excellent accounts of hazard zonation schemes and of these, papers by Hearn (20) and Sinclair (33) are particularly clear examples of comprehensive methodology tailored to specific objectives. The paper by Berggren et al. (4) nicely combines, uncertainty, hazard and risk assessment in a region where there is an unusually high level of information on failure behaviour of clay slopes. Moon et al. (27) in their contribution show an innovative use of geomorphic factors in an area without much geotechnical information.   Commendable

attempts to assess the variability of input parameter in factor-of-safety analysis are illustrated by Hammond et al. (18) and another novel approach to probability distribution of rainfall-triggered landslides is offered by Omura and Hicks (31). Corominas et al. (12) provide a useful example of the statistical selection of factors that may be of value to hazard assessment.

Indeed it may be argued that a practitioner is now confronted with a baffling array of zonation methodologies from which to chose. Thus it is important that the employment of any one particular scheme can be rationally justified. In this regard it is commendable that methodologies are being developed for evaluating predictive capacities (e.g. Gee, 16).

Landslides, Bell (ed.) © 1995 Balkema, Rotterdam, ISBN 90 5410 032 X

Landslide hazards and their mitigation in the Himalayan region

Vishnu D.Choubey
Engineering Geology Division, Indian School of Mines, Dhanbad, India

ABSTRACT : In recent years there have been several major landslides in Himalayas and following these disasters, a comprehensive research programme of slope stability investigation was initiated. The programme has involved (i) detailed inspection and documentation of landslides and related deformations (ii) monitoring of slides and related deformation to structures (iii) a few cored small diameter diamond drill holes with piezometers installed for obtaining pore water pressures (iv) laboratory studies mainly the shear strength characteristics of slide materials and (v) critical examination of documented case studies by other organisations. Key factors for the landslide hazard are discussed with the principal aim of identifying potential risk areas. The results make it possible to assess the stability and to understand the mechanism of slide developments and for making recommendations of remedial measures based on analytical calculations of stress deformation conditions and of rock slope stability computations.

Regarding the mechanism of the landslides, it is concluded that the slope stability problems are caused mainly for three reasons : (i) a tropical climate is conducive to deep weathering causing strength losses and permeability changes that lead to a reduction in stability of natural slopes; (ii) the intense rainfall results in rapid erosion, raising the groundwater level and saturating the overburden materials; (iii) the seismotectonic activity and the presence of several major faults and lineaments in Himalayas are responsible for placing these areas at high risk level for slope failures.

Among the recommendations made, the principal one is that a major programme of shear strength testing and back analysis of failed slopes be implemented in order to provide reliable values of shear strength characteristics for designing the remedial measures. A brief review is also included for the Terrain Classification and Landslide Hazard Zonation approach. This paper summarizes the engineering geological and geotechnical investigations, assesses the slope stability and make recommendations on a combination of remedial measures; specially the drainage control for improving the slope stability in complex seismotectonic environment. These investigations assume a special significance in context with the currently on-going United Nations Decade of Disaster Reduction Programme.

1 INTRODUCTION

The widespread occurrence of landslides together with the potential for catastrophic earthquakes and regional economic impact emphasise the need for the development and implementation of loss reduction and natural hazard mitigation strategies.Landslide losses continue to grow despite the availability of successful techniques for landslide management and control and the overwhelming evidence that landslide hazard mitigation programmes will enormously benefit the nation. Landslides occur frequently and their impact on several regions in north and north-east India is evident. These landslides are triggered by seismological events. Landslides cause direct losses of about $ 2 billion per year.

The landslides and other mass movements are the serious geoenvironmental hazards in the Himalayas. There was a 10 km long landslide into the Bhagirathi in 1978. There have also been massive landslides in the Alaknanda valley in Garhwal Himalayas. In Teesta river valley in Sikkim, there were 20,000 landslides killing 33,000 people in 1968 (Chopra 1977, CRRI 1984; and Choubey 1989) and in Mizoram, Nagaland, Meghalaya and Kumaon-Garhwal Himalayas, there has been considerable increase in the events of the slope failures (Bhandari 1987) and because of exceptionally high rainfall and high relief. Catastrophic damages took place during 1968 and 1973 in Sikkim Himalayas,(CRRI 1984; Natrajan et.al.1980) involving one million cubic metre materials, and more than 18 km of road was blocked, about 100 km road suffered regional subsidence, 10 briedges were washed away and 3 bridges were completely buried under debris. Massive landslides have occurred in Kumaon Himalayas recently. Human suffering and economic losses have been rising during the last two decades (Cruden & Lugt. 1990).

In Nepal Himalayas, landslides are also the principal hazard (Wagner et.al. 1988). In developing countries like India, with limited resources and marginal economy, it is specially important that such hazards be adequately evaluated and to prevent or mitigate the landslides natural hazard controlling measures should be developed.

The study of stability of slopes is basically a problem of engineering geology (Choubey 1976 and Patton 1970) and assumes that the rock is (i) anisotropic (ii) heterogeneous and (iii) discontinuous in nature and that the failure tends to be confined to structural discontinuities (Richards, 1975). Analytical analysis of slope in earth material (Janbu 1980) requires that certain geological parameters (Hutchinson 1977) and engineering properties of the slide material (Sassa 1988; Brabb 1991)should be studied and this approach has been followed in the present studies. The key factors are :

* Structural discontinuities are detectable and their physical characteristics can be quantified.

* Within the whole mass it is possible to define smaller masses of similar geotechnical characteristics.

* The surface of failure will be a plane or combination of planes.

* A reliable model representing jointing of the rock mass can be constructed (Barton 1973).

The pre-requisite for such an analysis is the qualitative and quantitative data (Piteau 1970 and Goodman 1976) of the area and in particular of the (i) attitude (ii) geometry and (iii) spatial distribution of the rock discontinuities (Choubey 1976). The basic principles on which these studies depend are (i) the system of jointing (ii) their relationship to the possible failure surface and (iii) the strength parameters of the rock discontinuities (Hoek and Bray 1977). In the present studies, as will be evident from the case studies discussed these parameters have been studied in detail and analysed.

1.1 Seismotectonic environment

Considering the distribution of recent major earthquakes in India (West 1937 and Gupta et.al. 1986) it is clear that they are confined almost entirely to the southern boundary of the belt of Tertiary folding passing through northern India and Burma and is in close association with the great belt of Tertiary folding, a convulsion that has not yet died out, especially at its eastern end. The formation of the Himalayan mountain ranges is inevitably accompanied by fracturing of the rocks along weak planes and over thrusting and that earthquakes are obviously correlated with movements along faults explains the association of earthquakes with Himalayas (Figure 1). The north-east India region (Figure 2) is seismically one of the most active region in the world (Gupta et. al, 1988). The major earthquakes in Himalayas i.e. Kangra 1905, Bihar-Nepal 1934, Quetta 1935, Assam Meghalaya 1897, Assam 1950, and the bitter memory of recent earthquake on 20th Oct. 1991 in Garhwal Himalaya, fresh in our mind are considered as part of ongoing geodynamic process that has build the Himalayas over the past 40 million years and will continue to do so for sometime (Windley 1988 and Gansser,1964).

The landslides are the consequence of diverse and complex effect of earthquakes and some of the massive landslides have been triggered by the earthquakes and the earthquake generated landslides are quite common in Himalayas. Some excellent detailed geological accounts are readily available (Wadia 1953, West 1937, Valdiya 1988 and Nandy 1980) and this introductory explanation is added here (Figure 3) for completing the picture of the landslide hazard and their relation with seismotectonics.

1.2 Shear strength of rock discontinuities

The overall controlling factor in the stability of rock slopes is the presence of structural features, mainly the joint planes and faults in the rock mass. The most important factors relating to rock slope stability are (i) orientation and shear strength characteristics (Barton and

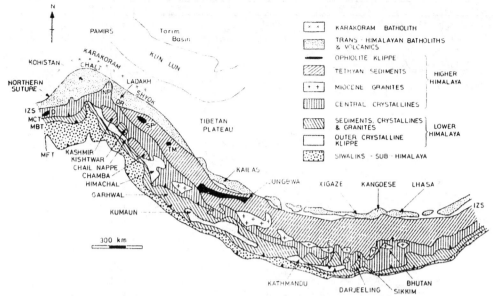

FIGURE 1 Map of the Himalaya showing the main tectonic zones and key localities. D = Dargai ophiolite, Dr = Dras, E = Everest NP = Nanga Parbat, SP = Spongtang ophiolite, TM = Tso Morari, IZS = Indus-Yarlung Zangbo Suture, MCT = Main Central Thrust, MBT = Main Boundary Thrust, MFT = Main Frontal Thrust. After Windley (1988).

Choubey 1977) of the planes along which the failure is possible. In this regard, the structural resistance caused by two rock surfaces sliding relative to each other and the resistance caused by the geometry of the irrigularities on the surfaces of sliding (Richards, 1975). The basic angle of friction of the surface depends on the frictional resistance between the two surfaces. The irregularities on the two surfaces give rise to an apparent cohesion on the rock discontinuity.

As the relation between shear strength and normal stress is generally non-linear (Barton and Chobey 1977), it is convenient in rock slope stability calculations to simplify the shear strength envelope into linear failure envelope by means of tangent to the curve at the specific normal stress level. But in the present studies, where the weathered rock discontinuities are mostly involved, the shear strength is affected (i) a decrease in peak strength due to a decrease in the compressive strength and (ii) a change in the residual strength due to a change in the surface mineralogy (Choubey 1976). Further the basic friction angle of a plane rock discontinuity surface may be considered as independent of the size of the surface for stability calculation purposes (Richards, 1975). Water pressures reduce the shear resistance considerably

by decreasing the normal stress acting across the rock discontinuities (Choubey, 1976).

While performing the shear tests (Barton and Choubey 1977) it has been observed that for a constant normal stress, the shear increases rapidly with displacement for a small amount (Bandis 1980) and finally the surface reaches a state where shear strength becomes independent of any further displacement on that rock discontinuity. The value of shear strength at this point is the residual shear strength and this criterion has been followed in the present rock testing programme.

The stability of the Himalayan slopes is markedly reduced as the piezometric levels within the slope rise following the rainfall. It is possible that the apparent decrease in overall stability and the corresponding increase in slope movement rates are related to the transition in the strength of the failure surfaces from peak to residual strength. In the slide zones, there are sections that have undergone sufficient movement to be very close to residual strength. The slide materials around the edges and crown sections have undergone relatively smaller movements and are still above the residual strength and during this stage, slope movement will increase as the piezometric surface rises above a certain level and once that

1851

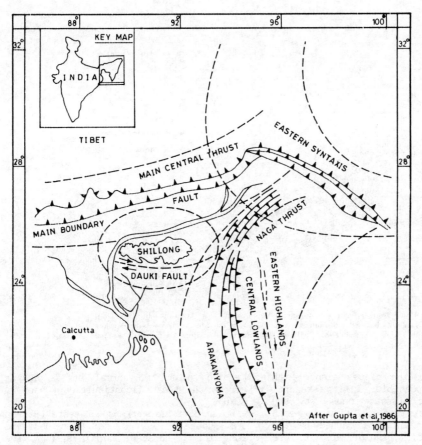

FIG. 2 TECTONIC MAP OF NORTH-EAST HIMALAYAS

critical level is exceeded, faster movements result.

The importance of in-situ effective stress is directly considered in the effective stress analysis. In-situ pore pressure assumes special significance when the total unit weight of the slide material is low. The in-situ pore pressure varies considerably with time due to seasonal variations and therefore the possibility of higher pore-pressure should be considered.

The landslides are mostly governed by :

1. Location of possible planes of failure and its mode of failure, and,

2. Present in-situ pore pressure, stresses and seismotectonic history of the area.

Although considerable progress has been achieved during the last two decades, and a fair assessment of the landslide risk can be made (Brabb 1991), the following key factors are yet to be developed.

i. Further development of procedures for analysing the complete state of stress in natural slopes, and also in-situ field measurements.

ii. The inter-relationship and interdepency between shear strength characteristics and Himalayan landslide mechanism including development of theoretical failure models.

iii. The effect of time-dependent properties of weathered rocks and how the stresses applied to remedial support measure varies with time.

1.3 Factor of safety for Himalayan slopes

A simple method of assessing the degree of stability of a slope is to consider the ratio between the forces tending to cause instability and the forces acting against these disturbing forces. The main forces tending to cause instability in these natural slopes are the weight of the rock above the potential slide surface and

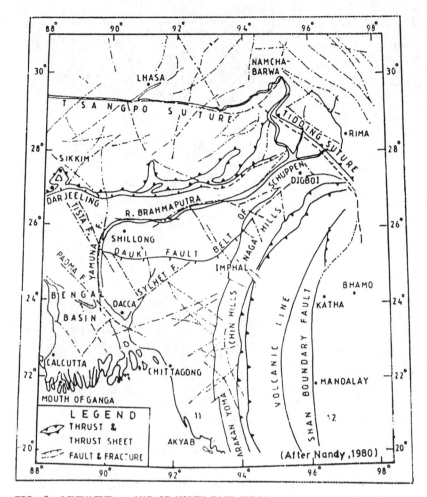

FIG. 3 LINEAMENT MAP OF NORTH-EAST INDIA

those forces that tend to promote stability is the shear resistance that may be mobilised on the failure surface.

It has been considered necessary to give the above brief discussion to highlight the role of shear strength parameter for designing and suggesting the remedial measures. The shear strength behaviour of rock joints has been excellently discussed by Barton (1973), Richards (1975) Barton & Choubey (1977) and Bandis (1980) and for a long time stability, and remedial design should be based on residual strength. Regarding the shear strength at soil rock interface, the studies suggest that the shear strength along soil rock interface is considerably lower than that of the soil mass alone and the reduction in strength is greater at low normal stress levels than for higher stresses.

For the design and analysis of the Himalayan slopes, the factor of safety seems to be a function of the uncertainty attached to the slope geometry, water pressure distribution and rock joint orientations and joint shear strength. Whether peak or residual shear strength should be used depends on the particular individual situation. In the case of designing rock bolting and retaining wall supports, peak strength values are recommended. However, peak strength combines unknown amounts of dilational and intact shear resistance with residual friction angle and consequently a higher factor of safety is preferred for designing the remedial measures in

Himalayan regions. In particular,

1. The factor of safety has been computed for a number of potential failure surfaces with different orientations and extent.

2. If minimum safety factor is found to be less than F = 1, the slope has been considered unstable but this also takes into account the factor of land use.

Some other factors considered are :

* In-situ stresses and pore-pressures: This factor is directly taken account in the effective stress analysis. This becomes more significant when the unit weight of the slide material is low, and varies considerably with time due to seasonal temperature variations.

*The exact locations of clay filling and fault gauge material zones etc., in relation to overall surface topography, have a decesive impact on landslide risk and on landslide mechanism.

2 INVESTIGATIONS AND CASE STUDIES

2.1 Kaliasaur Slide (Garhwal Himalaya)

The slide is located at 18 km east of Srinagar (Garhwal) on Haridwar-Badrinath road in a sharp bend of river Alakananda (Figure 4). This zone has been active since 1920. The total area of the slide is 86000 m^2 (above and below the road levels). Every year, about 5000 tonnes of debris is added to the river Alaknanda. The area contains white and light green quartzites interbedded with maroon shales belonging to Garhwal Group of rocks (Figure 5).

The site poses both hydrological and instability problems. River Alaknanda takes almost 90^0 turn just below the slide zone. As a result when the river water carries huge quantities of discharge (especially in the rainy season), the river water straight way hits the toe of the slide and causes extensive toe erosion and leads to landslide activity. At the same time, the rain water seeps into slopes to deeper levels because of presence of highly jointed, fractured and sheared rocks in the slide zone and reduces the shearing resistance of the discontinuity planes and as a result the landslide takes place. The attitudes of three sets of joints are:

1. N 120-300^0/dipping 25-35^0 NNE(J_1)

2. N 106-286^0/dipping 55-65^0 SSW (J_2)

3. N 62-242^0/dipping 45-55^0 SE(J_3)

The stereogram (Figure 6) shows that the J_1 is almost parallel to the slope

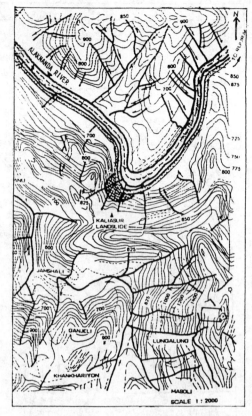

FIG. 4 CONTOUR MAP SHOWING THE LOCATION OF KALIASAUR LANDSLIDE (After Bhandari,R.K.,1987).

indicating plane of failure. At the same time the J_1 and J_3 are intersecting each other and the dip of line of intersection of J_1 and J_3 is less than the dip of slope face. Therefore it is this line (line of intersection of J_1 and J_3 having trend N 80 - 260^0/21^{0}E) which is making a wedge and also contributing to sliding activity but the dip of line of intersection of J_1 and J_3 is low (i.e. 21^0) therefore it may not be playing a major role in sliding.

The study of surface geology of the slide area has been carried out by C.B.R.I. Roorkee, (Bhandari 1987) and the observations are being taken regularly during pre-monsoon and post-monsoon period. The results clearly indicate two directional lateral movements in NE and NW direction. The extent of lateral and vertical movements range from 1 - 764 mm and 0 - 204 mm respectively. The pedestals located on the eastern side of the slide area have shown the maximum

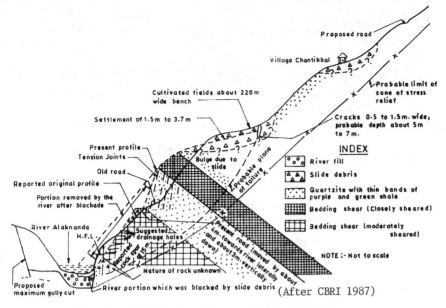

(After CBRI 1987)

FIG. 5 SKETCH GEOLOGICAL SECTION OF KALIASAUR LANDSLIDE

lateral and vertical movements indicating more instability in this horizon of the slide area. Therefore, eastern side of the slide area is a zone of intensive landslide movement involving both crustal and surfacial movements. In the central and western side of the slide area, the movement is confined predominantly to surface only.

2.1.1 Laboratory Testing

The shear strength of all the samples of Kaliasaur slide Sample numbers K_1 to K_7) are shown in Figure 7. The maximum shear strength values of these rock samples ranged between 60 to 130 kg/cm²
and angle of internal friction (ϕ_r) was found to be 69°. Also, a very large number of point load tests on quartizite specimens were performed which reveal the uniaxial compressive strength of the order of the 1800 kg/cm² (Figure 8). The samples of maroon shales were found to be so soft that even undisturbed sample was difficult to obtain. The laboratory tests clearly indicated wide variations in the shear strength and uniaxial compressive strength values confirming the observations made during the geological investigations of the slide area.

2.1.2 Discussion

The Kaliasaur slide area may be considered as a multilinear retrogressive

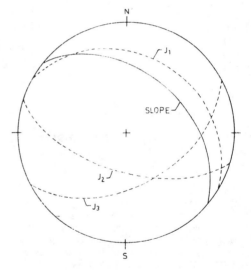

FIG. 6 STEREOGRAPHIC REPRESENTATION OF MAJOR DISCONTINUITY PLANES IN KALIASAUR SLIDE AND THEIR RELATION WITH SLOPE

landslide in complex rock formations and a combination of various factors is responsible for the landslide activity in the area. The main factor at Kaliasaur site is extensive toe erosion by the river at the point of bend in its course and the flow of underground and rain water through concealed channels which is causing vertical subsidence and further

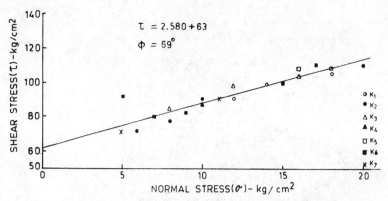

FIG. 7   SHEAR STRENGTH OF KALIASAUR ROCK SAMPLES

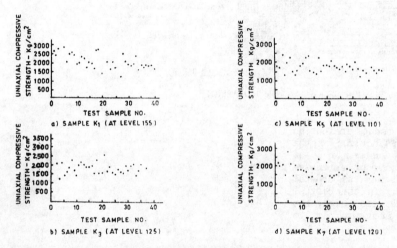

FIG. 8   UNIAXIAL COMPRESSIVE STRENGTH OF KALIASAUR
ROCK SAMPLES

this is giving rise to planar type of sliding of the rock material. The role of the erosion is corroborated by Krishnaswamy (1980) who studied the topomaps prepared in 1920, wherein a 30 m high scarp above the river level was shown which has now been reduced nearly to half. The sliding usually takes place along the repose slope and Alkananda flows below the toe of these repose slopes.

The sliding appears to be a translational (planar)-cum-rotational type. The inherent stability of the closely jointed and highly sheared rock mass is another factor, as the apertures in joints, parallel to the valley are as much as 2 cm – 10 cm wide. The rain water which percolates down through highly jointed and sheared rocks, saturates the joint planes and reduces their shearing resistance. Also, this is aided by the presence of crushed slates which forms mud on wetting. It is likely that the present slide zone may extend further.

2.1.3 Control Measures

Based on the results obtained through site and laboratory investigations the following remedial measures are suggested:-

1. Modification of surface drainage system and river training
2. Grading of slope
3. Sealing of cracks and joints.
4. Anchored drum diaphragm wall (Bhandari 1987)
5. Gabion walls using geogrid.
6. Anchored masonry walls.
7. Stitching of debris and
8. Vegetative turfing on slope

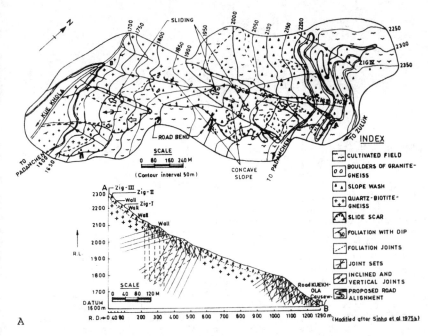

FIG. 9 GEOLOGICAL MAP AND CROSS SECTION OF PADAMCHEN SLIDE WITH PROPOSED
ROAD ALIGNMENT

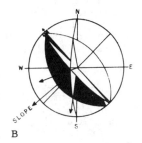

FIG. 9   FAILURE DIAGRAM

2.2   Padamchen Slide (Sikkim Himalaya)

A huge landslide extended from 13 km to 16 km located north-east of Rangli town on Rishi-Kupkup road (Figure 8). The entire affectd stretch of the road is devided into two portions, lower one called as Keukhola slide, while upper one is Padamchen slide. The Padamchen portion is reported to be active since 1968, when heavy slide occurred which jeoperdised ropeway station and also endangered bridge across the Keukhola (CRRI, 1984).

The rock types observed in the area are weathered to fresh garnetiferous biotite-gneiss and augen gneiss. The general trend of the foliation is NW-SE with 40° dip towards NE. Four prominent sets of joints (Figure 9) have rendered the rock mass blocky. These joints are generally open near surface. The surface water passes through these joints, particularly in the gneisses. The entire slope at Padamchen village is covered up with a thick mantle of the material containing boulders of granite embedded in clay to fine sandy matrix (Sinha et.al.1975).

2.2.1 Geomorphological study

The slide has developed from the left bank of Keukhola and spreaded laterally towards Beusa terrace in the south. The entire slope at this location is covered up with a thick mantle of the material containing boulders of granite embedded in clay to fine sandy matrix. The material is derived from glacial action (Sinha et.al. 1975). The slide is approximately 90 metre deep and equally broad at the widest portion. The gradient of the sliding zone varies between $40^{\circ} - 60^{\circ}$. Due to continuous sliding, the area has been devoid of vegetation, the main drainage of the area is Rangli chu and its tributaries, which are perrinial in nature, the tributaries feeding Rangli chu from north with parallel drainage pattern along the southern slope

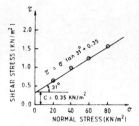

FIG. 10 SHEAR STRENGTH DIAGRAM

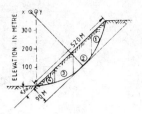

FIG. 11 Circular Failure Diagram

of the Padamchen ridge, and guided mostly by the structural pattern of the country rock. Rangli chu and Keukhola are both deeply incised valleys and flanked on either side by steep walls. The area receives heavy intermittent showers, of a shorter duration, specially during the months of May to August. The average rainfall in the area is around 800 mm and its major part is drained out as a surface runoff. Due to high altitude of the slide location from MSL about 3000 m, the area comes in temperate climatic zone with temperature ranging between $22^{\circ}C$ - $68^{\circ}C$.

2.2.2 Seismicity

The sliding area has suffered severe earthquake shocks in the past and entire zone lay within the iso-seismals of VIII and VI on the modified Mercali Scale. Besides, the area is also prone to minor tremors, but these could not be properly recorded (Sinha et.al. 1975).

2.2.3 Laboratory Investigations

Series of testing were performed to determine the physico-mechanical parameters of the slide material. The results of direct shear tests indicate that the material is cohesionless (C=0.35 KN/m^2) and having internal friction angle (ϕ_r) 31°, indicating lower shearing resistance of the material (Figure 10). The consistency limit tests also indicated, the narrow plasticity limit range of the materials. Further the field moisture was also determined, 17.39%.

2.2.4 Stability Analysis

The stability analysis for the sliding zone was also performed using Janbu's non-circular failure approach. The value of calculated factor of safety (i.e. 0.15) is found much below the unity, indicating that the slide is in critical stage of equilibrium (Figure 11).

2.2.5 Mechanism of Failure

The sliding at this location is caused due to interaction of several factors, which have contributed to the decrease the shearing strength of the slope forming materials. The chief contributing factors for slope failure are:
1. Adverse climatic conditions and intense precipitation, (75 mm/hour), which have caused deep weathering of rocks and then over saturation within short span of time, leading to loss of shearing resistance due to excessive pore pressure within the slide material and along the failure plane.
2. High percentage of fines in the slope material, causing reduction of free drainage capacity of the material.
3. Slope disturbances caused by human activities, and natural hazards like earthquakes.
4. High seismicity of the region.
5. Typical nature of the material (i.e. the glacial fill material) which is heterogenous in nature with loose soil structure with poor compaction. These conditions help in initiating the slide movements with the slightest provocation.

2.2.6 Remedial Measures

The following remedial measures are recommended on the basis of conducted field investigations, laboratory tests and stability calculations. The emphasis was placed broadly on better drainage facilities, flattening of the slope and restraining of the toe.
1. A concrete catch water drain is proposed at the crown of the slide to check the water entering into the slide materials.
2. Two series of partly lined trench-cum-surface drains are proposed between the zigs affected by the sliding. The walls of the drains on hillward side should be provided with limited number of weep holes. All the drainage should be connected with the main water course.

3. The sides and beds of the nallah courses should be sealed with boulder pitching. Further to reduce the velocity of flowing water through these courses, regrading the bed of water course by providing benches and berms; is also suggested.

4. Tension and shear cracks located in and around sliding zones are proposed to seal with suitable impervious materials and to provide bituminous coat on the top surface after proper compaction of fill material (CRRI 1984).

5. The joints observed on the exposed outcrops in the slide affected zone should be sealed with suitable grout material.

6. To drain out subsurface water, two series of subhorizontal drains are proposed to install between zigs above the proposed trench-cum-surface drains. Thus, water discharged from the horizontal drains can be collected with surface drains. The suggested horizontal drains should be concentrated within the central part of the slide zone parallel to the existing road.

7. A reinforced concrete curtain wall having two sections each of about 2 m. in height are proposed above and below the road stretches passing through the sliding zone. The bottom of the each wall should be provided with a concrete box drain, connected with the main water course.

8. Soil nailing may be effective on the uphill slope.

9. Concrete piles are suggested to provide on the downhill slope.

10. Extensive afforestation with fabric geogids is suggested over the denuded zones. The plants selected for afforestation should be fast growing, deep rooted and water absorber.

2.3 Vaivakawan landslide (Mizoram State)

Out of several mass movements which have taken place in recent years, the slope in Vaivakawan area has affected the road transportation, thereby causing great inconvenience to the community in the densely inhabited area. The boundary of the Vaivakawan slide zone (Figure 12) runs just by the side of a newly constructed RCC house founded on the bed rock. In one corner of the house a small wall constructed on the overburden material has been affected by the slide with development of cracks. This clearly proves that the slide is restricted to the overburden and there is no disturbance in the bed rock.

The toe of slide in Vaivakawan lies on the left bank (southern bank) of Vivakawan

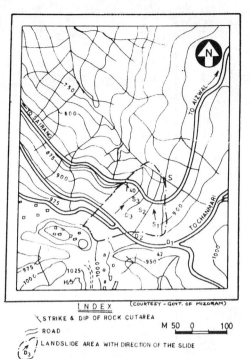

I N D E X   (COURTESY - GOVT. OF MIZORAM)

↖ STRIKE & DIP OF ROCK CUTAREA

∼ ROAD   M 50 0   100

LANDSLIDE AREA WITH DIRECTION OF THE SLIDE

FIG. 12 PLAN SHOWING LANDSLIDE AREA OF VAIVAKAWN IN AIJAWAL, MIZORAM.

Lui, a master nalla (drainage) of the area. The Vaivakawan Lui below the Aizawl – Sairang road has very steep bank with a number of local gullies. This stream is actively eroding its left bank near the toe of slide zone. North of the Vaivakawan Lui in its lower reaches of the slope, is however, gentle and covered with natural vegetation. As such the northern slope does not require any additional protection work at present.

The slide affected area rests on Vivakawan Line bounded by the gullies and nallas from the upper reaches of slides. This slide has been originated from the culverts on road of Vaivakawn-Luangmual. The water from market square Vaivakawn also drains to Vaivakawn Line through built up areas. The water can be diverted below the Vaivakawn market on southern slopes. The toe erosion needs to be prevented immediately to check the further landslides (Choubey & Lallenmawia 1987).

2.3.1 Laboratory Investigations

A number of samples collected from different horizons of Vaivakawn slide, were subjected to laboratory testing to assess the natural moisture content, field

Table 1A: Geotechnical laboratory testing of landslide material – Vaiva Kawn.

Sample No.	Natural Moisture Content (%)	Dry Density (Yd) (gm/cc³)	Saturated Density (Y_{sat}) (gm/cc)	Moisture content at Saturation.(%)	Void ratio	Liquid limit (L.L.) (%)	Plastic limit (P.L.) (%)	Plasticity Index (P.I.)	Liquidity Index (L.I.)	Consistency index (C.I.)	Permeability k(x10⁻⁵)	C' (kg/cm²)	Ø' (deg.)	M.C. Av.(%)
V_1	14.5	1.58	1.92	21.5	0.56	24.3	10.3	14.0	0.3	0.7	2.2	0.45	34	22.7
V_2	14.8	1.57	1.98	26.1	0.68	27.8	11.6	16.2	0.19	0.81	2.7	0.38	32	27.3
V_3	16.4	1.59	1.90	19.5	0.50	25.2	12.3	12.9	0.31	0.69	2.1	0.47	37	21.8
V_4	15.5	1.63	2.10	28.8	0.75	30.3	10.9	19.4	0.24	0.176	3.3	0.35	28	29.6

Table 1B: Geotechnical laboratory testing of landslide material – Vaiva Kawn.

Sample	TRIAXIAL TEST									UNCONFINED COMPRESSION TEST (U)		
	Unconsolidated Undrained (UU)			Consolidated								
				Undrained (CU)			Drained (CD)					
	C	Ø	M.C.	C	Ø	M.C.	C	Ø	M.C.	C	Ø	M.C.
V_1	0.65	0.00	24.6	0.43	17.0	20.9	0.38	37.4	20.1	0.61	0.00	23.8
V_2	0.56	0.00	29.6	0.37	19.7	25.4	0.33	35.2	24.3	0.52	0.00	28.4
V_3	0.68	0.00	23.6	0.45	22.6	20.1	0.41	40.7	19.7	0.65	0.00	23.7
V_4	0.51	0.00	28.9	0.34	17.3	27.3	0.31	30.8	26.8	0.48	0.00	28.1

NOTE : C = Cohesion (Kg/cm²); Ø = Angle of internal friction (deg.); M.C. = Moisture Content (%).

dry density, saturated density, liquid limit, plasticity index, permeability and soil parameters, effective cohesion, effective angle of internal friction for different combinations of drainage and consolidation conditions (Figure 13 & Table 1A, 1B).

2.3.2 Discussion

Slope failure in Aizawl is the combined effect of several factors. In overconsolidated clays, the failure has taken place due to the fact that the shear strength has been reduced to residual values along critical failure surfaces. The frequent and intense rainfall has led to deep weathering, seepage forces, undermining of underlying materials and erosion of toe zone. Due to various weathering processes and consolidation settlement in past have led to (i) high horizontal stresses (ii) recoverable strain in soil has been generated to accommodate expansion to post peak strength. Lack of adequate drainage measure is responsible for the accumulation of water influencing the effective cohesion and effective angle of internal friction. At many places the permeability has been influenced and the pore-pressure has become the main factor controlling the slope stability.

2.3.3 Remedial Measures

Satisfactory results in terms of stability can be produced by regulating drainage system and by profile modifications of these areas. It is recommended to make cuts in the slope upto maximum safety angle and to lower the ground water level by making rock filled trenches. Escarpments can be secured by flexible support at the toe by constructing drained buttress.

3 TERRAIN CLASSIFICATION & LAND SLIDE HAZARD ZONING (North Western Himalaya)

Several landslides and other mass movements have been occurring in the Himalayas. These on-going hazardous processes pose serious geo-environmental problems to inhabitants, traffic and river valley projects. To reduce the damage caused by such hazards, their prediction is necessary, which can be done only by detailed mapping, risk evaluation and zonation of such hazards in order of their severity. With this view, an attempt has been made to prepare a terrain classification map (TCM) of the Garhwal area, assuming slope as a basic factor. Further a detailed physiographic

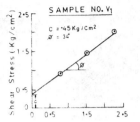

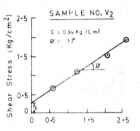

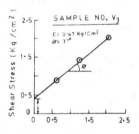

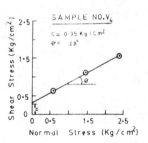

FIG. 13 SHEAR STRENGTH DIAGRAMS (Vaivakawn area, Aijawal)

and land-use map (PLM) has also been prepared using medium scale aerial photographs which is supplemented by demarcation of categories of landslides.

Terrain analysis requires the compilation of various terrain characteristics and qualities to evaluate and classify pieces of land. The TCM was supplemented by the information obtained from aerial photographs while preparing

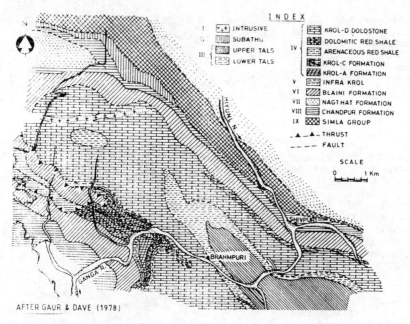

FIG. 14 GEOLOGICAL MAP OF THE RISHIKESH - NARENDRA NAGAR AREA

FIG. 15 TERRAIN CLASSIFICATION MAP OF RISHIKESH-SHIVPURI AREA

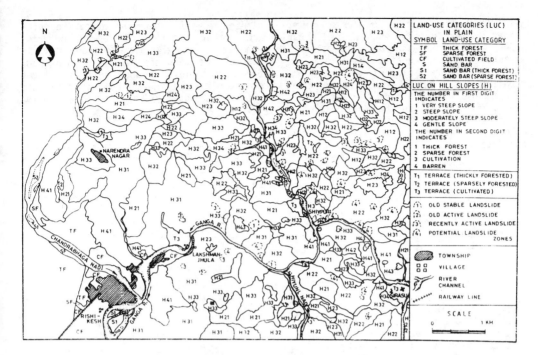

FIG.16 PHYSIOGRAPHIC AND LAND-USE MAP OF RISHIKESH-SHIVPURI AREA

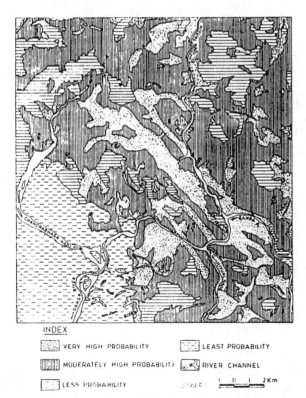

INDEX

VERY HIGH PROBABILITY   LEAST PROBABILITY

MODERATELY HIGH PROBABILITY   RIVER CHANNEL

LESS PROBABILITY   SCALE: 1   0   1   2 Km.

FIG. 17 LANDSLIDE HAZARD ZONATION MAP OF RISHIKESH-SHIVPURI AREA

1863

the PLM. Further both the TCM (showing geology, slope categories and old land slides) and PLM (showing forest and vegetative cover, active land slides, cliffs and barren slopes) are used in preparing the hazard map considering seismic data.

In order to evaluate the proneness of a particular facet to landslide hazards, the following factors are taken into consideration:

- Facet category
- Distance of each facet from nearest ridge.
- Distance of each facet from nearest thrust/fault.
- Closeness of the facet to river/stream.
- Lithology
- Forest/vegetative cover

Case History :

This work plan has been followed and illustrated here.

Rishikesh-Shivpuri area of Garhwal Himalayas is having an altitudnal variations ranging from 300 to 2350 m above mean sea level constituting the part of Garhwal syncline of Krol belt. Several landslides and other mass movements have been occurring along the roads and high hills constantly posing a serious geo-environmental problem.

The area forms a part of the Siwaliks and Lesser Himalayan belt, demarkating the contact near Lakshman Jhula. The various litho-stratigraphic and structural boundaries given in the geological map (Figure 14) are used to demarcate pattern boundaries on TCM. TCM of the area (Figure 15) has been prepared using topographic map (scale 1:50,000) assuming the slope as 'basic unit'.

Medium scale parchromatic black and white aerial photographs are used for the physiographic and land-use map of the area which is shown in Figure 16.

Detailed analysis of slopes using TCM and PLM, has been obtained about the hazard situation where aerial photographs have been extensively used. It is noticed that most of the landslides and cliff escarpments are situated either on the river banks or in the vicinity of rivers, therefore the river valleys with steep slopes having sparse/no vegetation are found to be more susceptible to future landslides. All the facet categories (Table-2) marked as 2 & 3 (slope range 45° 60° and above) are assessed as highly prone zones for landslides. The facet categories 4 & 5 (slope range 31 - 45°) are assessed having moderately high probability to

sliding when it contains vegetative cover and other factors are unfavourable, otherwise, in the reverse case, it is assessed as less probable to sliding. The facet categories 6 and below are least probable to sliding if the other factors are favourable (Choubey and Litoria, 1990).

Table 2. Facet categories in the Rishikesh-Shivpuri area

Facet	Facet category	Slope in degree
1	Ridge top	-
2.	Cliff/Escarpment	> 60
3.	Very steep slope	46 - 60
4.	Steep slope	36 - 45
5.	Moderately steep slope	31 - 35
6.	Less steep slope	26 - 30
7.	Moderately gentle slope	21 - 25
8.	Gentle slope	16 - 20
9.	Very gentle slope	11 - 15
10.	Extremely gentle slope	6 - 10
11.	Flat or plain	0 - 5
12.	River channels and banks	-

Figure 17 shows four landslide hazard prone zones on the Landslide Hazard Zonation Map (LHZM) of the area.

Zone Number	Probability of Landslide hazard
I	High
II	Moderately high
III	Less
IV	Least

The LHZM is, in fact, a snapshot of the hazard situation in the Rishikesh-Shivpuri area, portraying a regional picture of landslide hazards, especially while aligning new highway in the Garhwal Himalaya.

4. REMEDIAL MEASURES

Designing the suitable stabilization remedy requires determination of an appropriate geometry for the newly developed rock slope. This in turn requires the provision of an absolute safety factor which reflects the uncertainty of strength parameters and future pore-pressure parameters. In the sections where landslides have already taken place, it is recommended that some of remedial measures should be taken to ensure the immediate stability of the slide mass for ensuring a form of equilibrium with their environment. The various remedial measures which may improve the stability of the Himalayan slopes are :

1. Drainage control
   (a) Horizontal drains
   (b) Intercept drains
   (c) Surface drainage
2. Load re-distribution
3. Restraint on civil constructions and human activities.

Here special emphasis is placed on two measures :
* the drainage control with particular attention to horizontal drains
* modification of the slope profiles.

Drainage to lead away surface water immediately (Zaruba and Mencl 1969) for stopping the build-up water pressure in tension cracks has the potential for becoming most succcssful measure for improving the stability. The stability of the Himalayan slopes would be improved by taking the following measures:

(a) Horizontal drains : Horizontal drains are most suitable on a rising gradient so as to discharge by gravity and is considered most suitable in the case of Himalayan slopes where groundwater is deep and such conditions are common in relatively steep slopes and deep seated slip surfaces.

Subsurface horizontal or gently inclined drains have been successfully used to stabilise many landslides (Kenny et.al. 1976). These drains with perforated or porous liners should be installed in such a way that the water flow from each drain can be monitored following installation and the out flowing water is discharged off the slide area. Additional drains higher in the slide may also be effective for lowering the groundwater levels in the slide area. The main purpose of the horizontal drain is to lower the pore pressures within the slope, and this parameter is very important in stabilizing the slopes. Monitoring of the horizontal drains is important for ensuring that they do not become blocked. If the movement of the slide continues following drain installation, the drains would be damaged

and deformed and will need to be reinstated.

(b) Deep intercept drain across the landslide, also has considerable effect on the stability of the slide. If groundwater level in the toe remained high then more horizontal drains would be required in the toe area. Some time is required for fully assessing the impact of an intercept drain on the piezometric levels in the toe zones. The vertical drains are recommended for discharging the water by gravity through horizontal surface drainage.

Careful control of the surface water is required to prevent water entering the slide mass. It is desirable to collect all surface water and to discharge this away from the slide area. Based on past experience, surface drainage control is considered the most practical method for improving the stability in the Himalayas. However further modifications may achieved by:

(a) Additional piezometers should be installed to confirm the groundwater model in the central and toe sections of the landslides. Some horizontal drains should also be installed into the retaining walls in the toe areas for improving the stability of the toe area and for lowering groundwater pressures behind the retaining walls.

(b) If continued slide movement is noticed, more horizontal drains should be installed in the toe of the slide.

(c) As a long term comprehensive measure, a programme of regular monitoring should be initiated to check that all drains are operating effectively and if any slope movements are noticed, the reasons for the movements should be investigated for taking suitable remedial measures.

(d) Retaining structures like retaining walls, anchored to sound rocks by pre-stressed rock anchors may also be very effective on translational rock slides.

(e) Another promising measure seems to be erosion control (Bjerrum et.al. 1969 and Huchinson 1973). Modification of slope profile by cuts and fills may be very effective in deep-seated landslides.

We therefore need analytical tools and instruments (Bhandari 1988 and Kovari 1988) with which the landslide risk can be evaluated and quantified (Chowdhary, 1988). A suitable interpretation of deformaton monitoring can lead to forewarning of landslides in Himalayas. It is strongly recommended that a major multi-disciplinary research programme involving the measurements of movements, stresses, pore-water pressure and drain discharge etc. should be initiated for managing and predicting the landslides risk and for taking appropriate remedial measures. The success of the remedial measures will largely depend on the development of our understanding of complex geological set up, seismotectonics and slope failure mechanism involved (Varnes 1978). More attention should be paid towards the landslide susceptible zones because the landslides are generally more easily predictable and managable compared to other natural hazards like earthquakes.

5 RECOMMENDATIONS

1. Drainage control is considered as being the most practical method of controlling the landslides and improving the slope stability in Himalayas.
2. Landslide risk in Himalayas can reasonably be assessed with comprehensive site investigations.
3. A comprehensive shear strength testing programme including the scale effects on landslide materials and back analysis of failed slopes should be initiated.

Geologists, seimologists and civil engineers have important role in evaluating and devising and implementing remedial measures that will reduce the landslide risk in Himalayas. With regard to landslide research at present, such work is being done independently by several organisations, particularly Geological Survey of India, Border Roads Organisation,CSIR Laboratories and Universities and these should be co-ordinated and expanded by developing a special Landslide Hazard Reduction Programme Institute with its headquarter preferably at Shillong in North-East India.

ACKNOWLEDGEMENTS

The author thanks Dr Nick Barton of Norwegian Geotechnical Institute, Oslo for his valuable suggestions and stimulating discussions. He is also thankful to Border Roads Organisation, Govt. of India (Swastik Project, Gangtok and CRRI, New Delhi) for providing facilities for field work. Financial assistance provided by Department of Science and Technology, Govt. of India and University Grants Commission, New Delhi is thankfully acknowledged. The assistance from ISM Authorities and of R K Rawat, P K Litoria, S Choudhari, S M Pandey, T Mazumder and other research students who made multidisciplinary project possible is thankfully acknowledged.

REFERENCES

Bandis, S.C.1980. Experimental studies of scale effects on shear strength and deformation of rock joints. Ph.D. Thesis (Unpublished), Univ. of Leeds : 385.

Barton, N. 1973. A model study of behaviour of steep excavated rock slopes. Ph.D. Thesis (Unpublished), Univ. of London : 376.

Barton, N. & Choubey, V. 1977. The shear Strength of rock joints in Theory and Practice. Rock Mechanics, 10 : 1-54.

Bhandari, R.K. 1974. Landslides on the North Sikkim Highway Unpublished Report. Border Roads Organisation, Swastik Project, Sikkim.

Bhandari, R. K. 1987. Slope stability in the fragile Himalaya and strategy for development. Special lecture. Indian Geotechnical Journal: 1-81.

Bhandari, R. K. 1988. Practical lessons in investigation, instrumentation and monitoring landslides 5th. Inter. Symp. landslide. Laussane 3.

Bishop, A. W. 1954. The use of the slip circle in the stability analysis of slopes, Proc. European Conf. on stability of Earth slopes. Geotechnique. 5 : 7-17

Bjerrum, L. & Jorstad, F. 1968. Stability rock slopes in Norway. Norwegian Geotech. Inst. 70: 1-11.

Brabb, E. E. & Harrod, B. L. 1989. Landslides extent and economic significance, (eds.), A.A. Balkema, Netherlands.

Brabb, E.E. 1991. The world landslide problem. Episodes. 14(1): 52-61.

Chaudhary, R. N. 1988. Analysis methods for assessing landslide risk - Recent developments. Special lecture 5th Inter. Symp. landslides, Laussan. 1: 515-524.

Chopra, B.R. 1977. Landslides and other Mass Movements along Roads in Sikkim and North Bengal. Bull. IAEG. 16: 162-186.

Choubey, V.D. 1976. Fracture Pattern and their relevance to Geotechnical Engineering. Norwegian Geotechnical Institute, Oslo. Spl. Report 5410-7: 1-121

Choubey, V. D. & Lallenmawia, H. 1987. Landslides and other Mass Movements in Aizawal, Mizoram State, NorthEast India. Proc. 5th Int. Conf. and Field Workshop on Landslides: Chirstchurch, New Zealand : 113 - 120.

Choubey, V.D. 1989. Stability Assessment and Remedial Measures of landslides in Himalayas, Northeast India. 28th Int. Geol. Cong., Washington DC., 1: 288.

Choubey, V.D. & Litoria, P.K. 1990. Landslide Hazard zonation in the Garhwal Himalaya : A Terrain Evaluation Approach. Proc. 6th. Int. Cong. IAEG, Amsterdam (The Netherlands). 1: 65-72.

Choubey, V. D. & Litoria P.K. 1990. Terrain classification and land hazard mapping in Kalsi-Chakrata area (Garhwal Himalays) India. I.T.C. Journal.-1: 58-66

Choubey, V.D. & Rawat, R.K. 1990. Engineering appraisal of the major landslides and their stabilization in north Sikkim, Region, India. 6th. Int. Cong. IAEG, Amsterdam (The Netherlands) 3: 1547-1554.

Choubey, V.D. & Rawat, R. K. 1990. Geotechnical Investigation of landslides in Sikkim region, India. Proc. 6th. Int. Conf. & Field Workshop on Landslides. Italy, Milan, Austria. (ALPS-90): 135-144.

Costa, N.A.J. 1969. Landslides in soil of decomposed rock due to intense rainstorms. Proc. 7th. Int. Conf. Soil Mech. & Found Engg., Mexico 2: 547-554.

C.R.R.I. 1984. A comprehensive report on major landslides in Sikkim Region and their stabilization. A report furnished to Border Roads Organization, Swastik Project, Sikkim (Unpublished).

Cruden, D.M. & Lugt, J. de 1990. The world inventory of historic landslides. 6th. IAEG. Amsterdam/Netherland. 3: 1573-1578.

Gansser, A. 1964. Geology of the Himalaya, London Interscience Publishers: 289.

Gaur, G.C.S. & Dave, V.K.S. 1978. Geology and structure of a part of Garhwal Syncline, Rishikesh, Garhwal Himalaya. Him. Geol., 8(1): 524-549.

Goodman, R.E. 1976. Methods of geological engineering in discontinuous rock, West N.York: 472.

Gupta, H.K., Kusala, R. & Singh, H.N. 1986. Seismicity of the North-East region, Jour.Geol.Soc.India, 28: 345-365.

Hoek, E. & Bray, J.W. 1977. Rock Slope Engineering (2nd.Revised edn.). The Institution of Mining and Metallurgy, London.

Hutchinson, J.N. 1977. Assessment of the effectiveness of corrective measures in relation to geological condition and types of slope movement. Bull. IAEG., No. 16: 131-155.

Janbu, N. 1980. Critical evaluation of the approaches to stability analysis of landslides and other mass movements. Proc. Int. Symp. Landslides, New Delhi, 2: 109-128.

Johnson, J.D. 1987. Stability assessment of the Albany str.slide. DSIR Report: 25.

Kenny, T.C., Pazin, M., & Chai, W.S. 1976. Horizontal drains in homogenous slopes, 29th Canadian Geot. Conf. 1: 1-15.

Kovari, K. 1988. Methods of monitoring landslides. 5th Inter. Symp. Landslides Laussane.

Krishnaswamy, V.S. 1980. Geological aspects of landslides with particular reference to Himalayan region. Proc. Int. Symp. on Landslides, New Delhi 2: 61-71.

Nandy, D.R. 1980. Tectonic patterns in northeastern India. Indian Jour.Earth Sci., 7: 103-107.

Natarajan, T.K., Bhandari, R.K., Rao, E.S., & Singh, A. 1980. A major landslide in Sikkim - Analysis, correction and efficasy of corrective measures. Proc. Int. Sym. landslides, New Delhi 1: 397-402.

Natarajan, T.K., Bhandari, R.K., and Tolia, D.S. 1980. Some landslides on North Sikkim highway - their analysis and correction. Proc. Int. Symp. on Landslides, New Delhi 1: 455-460.

Piteau, D.R. 1970. Geological factors significant to stability of slopes cut in rock. Proc. planning open pit mines, Johannes. burg:33-51.

Richards, L.R. 1975. The shear strength of rockjoints in weathered rocks. Ph.D. thesis (Unpublished), Univ. of London.

Sassa, K. 1988. Geotechnical model for the motion of landslides. Spl. lecture, 5th Inter. Sym. Landslides, Laksanne 1: 37-55.

Schuster, R.L. 1978. Introduction. In Landslides: Analysis and control. National Academy of Science, Washington, D.C., Transportation Research Board special report. No. 176: 1-10.

Schuster, R.L., & Krizek, R.J. 1978. Landslides: analysis and control. National Academy of Sciences, Washington, D.C., Transportation Research Board Spl. Report. 176: 234.

Sinha, B.N., Pradhan, S.R. & Sinha, P. 1975. Geological analysis of Padamchen slide. Seminar on landslides and toe erosion problems with special reference to Himalayan region. ISEG, Gangtok, India: 227-234.

Valdiya, K.S. 1988. Tectonics and evolution of the central sector of Himalaya. Tectonic evolution of the Himalaya and Tibet. Phil. Trans. R. Soc. London. 326: 151-175.

Vernes, D.J. 1978. Slope Movement: Types and processes. In Landslides: Analysis and control Schuster, R.L. and Krizek, R.J. (Eds). National Academy of Science, Washington, D.C., Transportation Research Board Spl. Report. No. 176: 11-33.

Varnes, D.J. 1984. Landslide hazard zonation: A review of principles and practice. UNESCO Pub., Paris: 63.

Wagner, A., Lieite, E., & Olivier, R. 1988. Rock and debris slides risk mapping in Nepal. 5th. IAEG. A.A.Balkema: 1251-1958.

West, W.D. 1937. Earthquakes in India. Presidential address, 24th. Indian Sci. Congress: 39p.

Windley, B. F. 1988. Tectonic framework of the Himalaya, Karakoram & Tibet and problems of their evolution. Phil. Trans.R. Soc., Lond. 326: 3-16.

Zaruba, Q. & Mencl, V. 1969. Landslides and their control. Elsevier Academia, Prague: 205.

Landslides, Bell (ed.) © 1995 Balkema, Rotterdam, ISBN 90 5410 032 X

Suggestion on the systematical classification for slope deformation and failure

Wang Lansheng, Zhang Zhuoyuan, Cheng Mindong, Xu Jin, Li Tianbin & Dong Xiaobi
Chengdu College of Geology, People's Republic of China

ABSTRACT: A systimatical classification of the rok masses deformation and failure on slope is suggested in this paper. The classification is based upon the characteristics of three different evolution stages of slope, i.e., slope deformation, slope faiture and post-failure continuing movement. Summarizing the evolution process and mechanic mechanism of slope systematically, this classification reveals some important relationship between the geomechanical models of slope deformation, slope failure types and the types of slope structure. Study hows that this classification system is useful to applicate in research on slope stability.

1. INTRODUCTION

In the fifth International Symposium on Landslide, 1989, the IAEG Commission on Landslides (Cruden, 1989) suggested working out an international standard classification of slope deformation and failure, and recommended Varnes' s classification (Varnes, 1978) which was based on the suggestions by other scholars (Zischinsky, 1966; Zaruba and Mencl, 1969; Skempton and Hutchinson, 1969; Nemčok, 1972; de Freitas and Watters, 1973). This suggestion by COL of IAEG has played an important role in promoting the study and international exchange on hazards from slope failure, being of important and stratigic significance.

Two principles, i. e., the motion mode and the material composition are considered in Varnes' s classification. Classified by motion mode, there are five basic types of slope failure—falls, topples, spreads and flows, as well as some combinations such as rockfall-debris flow, slide-mud flow etc..

Some complementary suggestions are presented in this paper based on the following understanding:

1. The slope must undergo a process of deformation prior to final failure and this process is characterized by evident stages. The formation and evolution modes of the internal deforming-cracking structures and the corresponding external morphological features of slope in this process thus can be taken as the basis of classification, because they are the important evidences for analysing the development stage of slope rock mass and predicting its developing tendency, as well as the basis for designing the artificial slope. Some motion modes defined by Varnes, such as sliding, toppling and flowing (for describing slope creep) actually belong to the deformation modes of slope before failure.

2. During the evolution of slope, once the failure surface is penetrated with outside and the separated body breaks away from the bed rock at a certain acceleration, the slope develops into failure stage; this process is called the slope failure. Therefore, in the classification, it is better to make a distinction between the starting mode of the separated body or the motion mode at the time of its breaking away from the bed rock and its mode of continuing motion, and the conversion of motion mode as well. The combination of modes in Varnes' s analysis embodies this kind of distinction and conversion, which is important in the assessment and prediction of the geological hazards caused by slope failure.

3. The classification based on the characteristics of three different evolution stages of slope—slope deformation, slope failure and post-failure continuing movement, should reflect not only the relationship among them, but also its certain correlation with the calssification of slope, so that the evolution mechanism in the whole process of slope development can be analysed systematically. This is of importance in the assessment of slope stability.

Based on the geomechanical models of slope deformation proposed by author (Wang & Zhang, 1979 and 1984), some suggestions and discussion about the classification are given as follows.

2 CHARACTERISTICS AND CLASSIFICATION OF SLOPE DEFORMATION

2.1 Types and characteristics of deformation fractures in slope rock mass

According to its mechanical behavior, the slope deformation is divided into unloading rebound and creep.

1. Unloading rebound and the resulted deforming-cracking in slope rock mass

In the process of unloading the re-distribution of stress in rock mass near the free face of slope would bring about local stress concentration effect and the differential rebound would bring about a restrained residual stress system in rock mass. It is by these two kinds of change of stress state that the deformation fractures are produced in slope rock mass in the process of unloading. Therefore, according to the mechanical mechanism of their formation, the deformation fractures in deformed rock mass (epigenetic discontinuities) which are related to the effect of unloading is divided into the main types shown in figure 1.

The stress differentiation induced cracks can be divided into three types as shown in figure1 according to their formation mechanism. Their occurring position are matched with the distribution state of stress in slope. For instance, the tensile cracks usually occur in the tensile stress concentrated marginal zone of slope (Fig. 1 ①). Near the free surface of slope where because the major principal stress is approximately parallel to slope surface, and the closer to the slope surface, the closer to zero the minor principal stress is, the slope is actually in a tensile stress state. In this case the compression cracks approximately parallel to slope surface can be produced (Fig. 1 ②、③、④). This kind of cracks are often seen in high steep slopes. If there exist originally the discontinuities oblique to the direction of major principal stress at acute angles in slope, the compression cracks stemmed from these discontinuities and parallel to the direction of major principal stress, which are similar to the Griffith's tensile cracking, are possibly produced (Fig. 1 ②、④). In addition, in the area of fairly high earth stress, the shear fractures similar to the upthrow faults are produced at the bottom of valley due to the stress concentration (Fig. 1 ⑤).

The cracks induced by differential unloading rebound can also be classified as tensile cracks and shear fractures (Fig. 1 ⑥-⑨). The differences of the constitutional elements of rock mass either in mechanical property, stress history (loading history) or in structure can bring about the differential rebound and consequently the residual tensile stress (Lajeta, 1977). The cracks appoximately parallel to the plane of denudation can be produced by this kind of stress (Fig. 1 ⑥).

Type		Graphic representation
fractures produced by stress redistribution	tensile crack	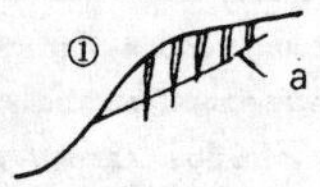
	compression crack	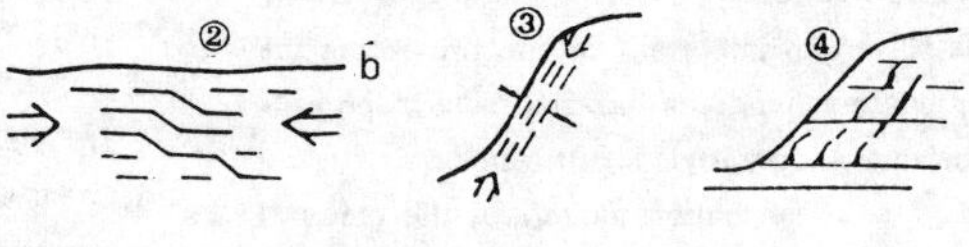
	shear fracture	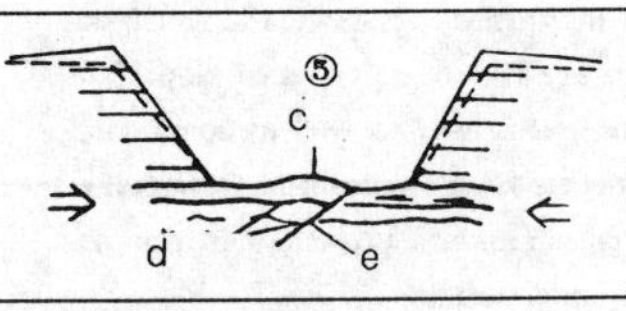
fractures produced by differential rebound	tensile crack	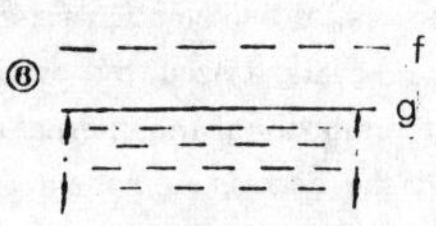
	shear fracture	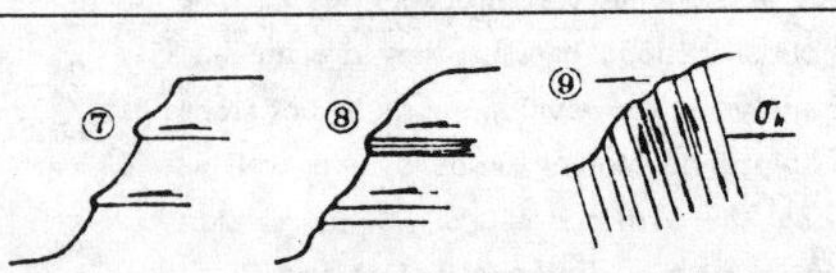

Fig. 1 Main types of superficial fracture related to the unloading rebound in rock mass

a—tensile-stress zone; b—denudation plan; c—upheaval; d—hole; e—upthrow fault; f—initial ground; g—denudation plan

The unloading rebound can also bring about the residual shear stress in the slope rock mass and induce the shear fractures thereby. In the region of high earth stress, during the process of downcutting of river valley or man-made excavation, the gently inclined cracks can be formed at the foot of slope due to the differential rebound of rock masses above and below the restricted planes there, particularly when the major principal stress is approximately perpendicular to slope surface (Fig. 1 ⑦). This phenomenon takes place more easily if there exist gently-inclined weak planes in rock mass (Fig. 1 ⑧). If there are steeply-inclined weak planes striking approximately parallel to slope surface, the rebound unloading towards free surface can also bring about residual shear stress along weak planes and the reverse slip thereby (Fig. 1 ⑨).

2. Slope creep and deformation units

Although the effect of unloading can bring about the above deformation fractures in the process of formation of natural or artificial slopes, not all of them can develope into slope creep stage. Only in those slopes that possess the specific internal and environmental conditions, can the slope creep go on.

Along with the progress of creep deformation, the original slope structure and the superficial structures described above will be transformed further and tend to being more complicated. They are the products as well as the signs and evidences of slope deforming, and they play controlling roles in the evolution of slope to different degree. The influence of the environment on slope evolution is exerted by means of changing the property and occurrence characteristics of these superficial structures to a great extent. Therefore such kinds of superficial structures can be regarded as the constitutive units of slope deformation, i. e. , the deformation units. According to their mechanical effects and their self evolution modes in the process of slope evolution, the superficial structures are generalized into four types of deformation units [1].

(1) Cracking   It includes the cracking induced by tensile stress and the cracking induced by compressive stress through dilating of slope in free direction (compression cracking).

(2) Sliding   It is the shear deformation along a certain plane or a certain zone. The shear deformation along a certain plane includes the "island" sliding along shear plane or fracture and the one-by-one shearing-off of the locked segments or discontinuous segments of weak plane. The shear creep along a certain zone is called creep sliding. In the siltized intercalation, superficial intercalated mud, weathering film and calc-sinter deposits along slide plane, are often seen the striae of which the slip direction are evidently determined by the slope structure features and free condition, in this way the striae are different from the structural ones.

(3) Bending   It is the "folding" deformation of rock mass under the effect of self weight, including transverse bending, longitudinal bending and cantilever bending etc. . It can be distinquished from a structural one by the direction of "interlayer-slip", the inclining of axial plane of bending and the cracking characteristics of bending beds etc. .

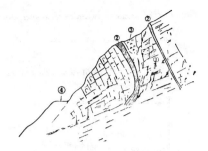

Fig. 2 Deforming granite mass at the bank of Yalong River
① compression cracks;   ② diabase dike;
③ tensile crack with debris deposite;   ④ high way.

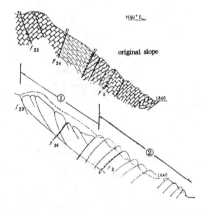

Fig. 3 Deforming rock mass on the open pit in Jinchuan (Sun Yuke 1976)
① sliding-cracking area;
② bending-cracking area.

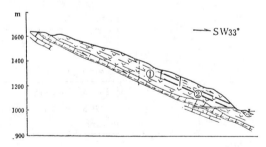

Fig. 4 Deforming rock mass at the bank of Jinlong mountain, Yalong river
① sliding-bending area;   ② sliding-cracking area

(4) Plastic flowing   It is the compressive deformation of soft beds and plastic flowing (squeezed out) of soft rock or crashed material in the free or decompressed direction.

Of the above four basic deformation units, the cracking belong to brittle fracture; while the latter three are elasto-plastic, plastic or visco-elastic deformation, of which the

Table 1. Comparision between the characteristics of slope deformation and failure and the types of slope structure

Type of slope structure		Geomechanical model of deformation and typical diagram	Possible failure mode
I. homogeneous or semi- homogeneous slope	approximately regarded as continuum	(CSC)	rotary slide with curved rupture surface
II. stratified rock mass slope, controlled by one set of discontinuities or prominent weak plane	II$_1$ subhorizontal $\alpha = 0 \sim \pm \varphi_r$ $\alpha < \beta$	(SCC)	horizontally-pushing slide triggered by heavy rainfall, rotary slide
	II$_2$ outwards gently-dipped $\alpha = \varphi_r \sim \varphi$ $\alpha \leqslant \beta$	(SC)	block-slide or "mazy" slide
	II$_3$ outwards moderately-dipped $\alpha = \varphi \sim 40°$, $\alpha \geqslant \beta$	(SB)	consequent slide, rotary slide or rockfall
	II$_4$ outwards steeply-dipped $\alpha = 40° \sim 60°$ $\alpha \geqslant \beta$	(BC)	topples and rockfalls, rotary slide
	II$_5$ vertical $\sim$ inwards steeply-dipped $\alpha \approx 90°$, $\alpha > \beta$	(BC or CSC)	topples and rockfalls, slide, rotary slide (in depth)
	II$_6$ inwards moderately-dipped $\alpha = -30° \pm$	(CSC or BC)	surface slide, rotary slide (in depth)
	II$_7$ outwards dipped with varying dip angle (chair curve) $\alpha \approx \beta$	(SB)	consequent and rotary slide
III. blocky mass slope	two or more sets of prominent discontinuities cutting the rock mass into blooks	(SC)	wedge slide or flowslide
IV. broken mass slope	numerous sets of joints, no prominent weak plane, similar to homogeneous mass	(SC or CSC)	rotary slide, flow slide

			block slide, horizontally pushing slide triggered by heavy rainfall
V. slope with soft foundation	V₁ with sub-horizontal soft foundation	(PFC)	
	V₂ with inwards dipped soft foundation		rock fall, rotary slide

Note: α—dip angle of prominent weak planes; β—slope angle; φ and φ_r—basic and residual friction angle of the weak planes

time-dependant feature can be described by the models of e-lasto-plastic medium or visco-elastic medium. The time-dependant features of rock mass deformation in slope are generally determined by the latter three units.

2.2 Geomechanical models of slope deformation

1. Basis for classifying geomechanical models of slope deformation

Although a deformed rock mass can contain several deformation units, it is common that a couple of them which are causes and restraints for each other and play a dominant role in the process of deformation can be selected to reflect the inherent mechanical mechanism of slope evolution. Therefore the deformation can be summarized as five basic geomechanical models (see Table 1):

(1) sliding (or creep sliding)—cracking (SC or CSC);
(2) sliding—compression cracking (SCC);
(3) sliding—bending (SB);
(4) bending—cracking (or toppling) (BC);
(5) plastic flowing—cracking (PFC).

2. Compound and conversion of geomechanical models of slope deformation

Under some circumstances, two or more deformation models can exist simultaneously on a slope and they compound together in a certain manner, this is called compound of geomechanical models of deformation. The rock mass controlled by a compound model is called compound deformed rock mass. The compound includes associating, joining and superposing etc. of deformation models. As in figure 2, at the front part of slope is bending-cracking, while at the rear part is sliding-compression cracking, both promoting each other. As in figure 3, at the front part is bending-cracking, while at the rear part is sliding-cracking; the observation has shown that the latter promotes the formation and development of the deformation of front part. As in figure 4, at the lower part of slope is sliding-bending, while at the upper part is sliding-cracking; the intensification of bending will accelerate the deformation of upper part, while the loosening or falling of the upper deformed rock mass will favour the development of bending at lower part.

During its evolution, the deformation of some slopes may convert from one model into another, because the original structure of rock mass is transformed and keeps with changing and some new deforming fractures are formed. For instance, in Huangya deforming rock mass, Wujiang river, as in figure 5, the differential compression deformation of the underlying soft layers brings about the cracking of overlying rock mass, that is, the deformation model is plastic flowing—cracking. Due to the developing of cracks in depth, the stress concentration takes place at the foot of steep crack. Based on the data of exploratory tunnel, which revealed that the shear cracks have been formed discontinuously at the foot of rock pillar, it can be predicted that the further developed deformation may convert into creep sliding—cracking along the potential slip surface formed by deep-seated shear cracking zone. Once this conversion takes place, the upper part of crack may turn closed, which will indicate that the stability of rock mass will have been worsened further. The model conversion itself represents a leap in the development of deformation, often predicting that the deformation develops into the stage of progressive fracture.

Each geomechanical model of deformation has its own laws of formation and evolution [1].

3 STRUCTURE TYPE OF SLOPE ROCK MASS AND ITS RELATION WITH THE TYPES OF SLOPE DEFORMATION AND FAILURE

The research results indicate that on the slope composed of different types of rock, the deformation mode and formation of deforming rock mass are largely determined by the appearance feature and the occurrence characteristics of the original weak plane (or zone) playing controlling role in slope, and the latter factor is predominant. The investigation proves that, if containing relatively weak planes or zones, the soil slope also shows the deforming process similar to the above examples. Therefore the slopes can be classified by their governing structure characteristics and then compared with their possible deforming modes and failure types, so that we can make prediction and assessment of the possible deforming and evolution modes and final failure modes for the slopes with differ-

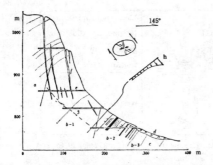

Fig. 5 Deforming rock mass on Huangya Slope, Wujiang river
a — b₁ limestone; b₂ — b₃ shale and coal; c limestone; d rock avalanch; e, f tensile cracks; g
shear fractures; h settlement curve with depth.

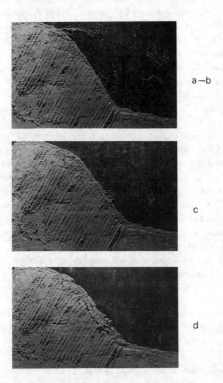

a—b

c

d

Fig. 6 Base friction modeling for the slope type II_5

ent structure types (see table 1)

It's worth to explain following concepts.

1. The boundary angle between moderately (II_3) and
steeply (II_4) dipped one

The boundary angle of 40° is roughly corresponding to the
nature repose angle (35° ~ 42°) of fragmental deposites or
weathered cataclastic rockmasses, it means that as a dip slope
($α≈β$) the slope surface can't keep them.

2. Relation between α and β

Investigation shows that in nature slope, mostly the type of
II_1 and II_2 are consequently stepped slopes (over dip
slopes $α<β$), type II_3 and II_4 are consequent slopes (dip
slopes $α≈β$) or reversedly stepped slopes (under dip slopes $α$
$>β$)

3. Consequently, reversedly and tangentaily stratified
slope

As shown in table 1, type $II_1 ~ II_4$, II_7 belong to the
consequent ones and type II_5, II_6 belong to the reversedly
one. The tangentaily one defined as that the angle between
the dip direction of the slope and strata in consequent slope is
over 30°, in fact which belongs to type III because of the
slope failure there always being wedge slide.

4. Multiple "nature"

The slope structure may show dual or multiple "nature",
which results from its complexity, the difference between
whole mass and local part, or the two-way or muti-way al-
ternation of the free surface direction. Nevertheless it is still
possible to select a principal structure type which plays a controlling role in the stability of slopes.

4 APPLICATION OF SYSTEMATIC CLASSIFICATION IN ANALYSIS OF SLOPE STABILITY

The above discussion indicate that a certain mode of failure
can result from a variety of deformation models, so the deformation models, failure modes and the post-failure motion
modes may be connected togather to describe a series of deformation—failure types of slope, such as sliding—bending
slide—mudflow, bending—cracking (toppling) rockfall—debris flow, plastic flowing— cracking spread etc.. This is
called the systematic classification of slope deformation and
failure. Summarizing the evolution process and mechanical
mechanism of slope systematically and revealing the relation
between geomechanical models of deformation and the types
of slope structure, this classification keeps some relationship
with the slope types and thereby can be applied at least in the
following aspects of slope stability research.

4.1 Geological criterion of evolution stage of slope deforma-tion

It has been proved by investigation that each deformation
model has its own particular evolution process and it can be
divided into several stage by the slope structure and appearance feature. The evolution process can not only be observed
in in-situ investigation, but also be reproduced by modelling.
Taken as an example, the result of modelling the evolution
process of bending-cracking for a slope type II_5 composed
of thinly stratified rock mass is shown in figure 6, which includes 4 stages:

a) surface bending; b) bending zone developing in depth

and cracking at rear edge being intensified; c) bending intensified, consequent snaping and shear cracking zone formed at the deep, cracks at rear widened, surface slide taking place possibly; d) getting through of deep snaping and shear cracking zone, larger slide taking place. These characteristics described above can be taken as the geological criteria for judging the state of slope deformation.

4. 2 Determination of starting criterion of slope failure

On the basis of the analyses of mechanical mechanism of deformation evolution, the quantitative criteria for the starting of slope failure can be made. Here the sliding—bending deformation is taken as an example. Proved by both in-situ observation and modelling study (Fig. 7), the bending of multi-layer plates is a progressive process of deforming-failure, developing from surface layer to the deep gradually. In this process the plates may be parted and voided between them. Based on this feature, the longitudinal critical stress σ_{cr} of plates bending is (by Cheng Mingdong, 1990).

$$\sigma_{cr} = \frac{E\pi^2}{12y^2} + \frac{\gamma h y^2}{\pi^2}\cos\alpha \cdot y^2 \\ + \frac{n+1}{2}\gamma h \cdot tg\varphi \cdot \cos\alpha \cdot y + Cy \quad (1)$$

where E—elastic modulus of plates;

    y—the ratio of thickness of single layer to the length of bending segment;

        γ—density of plates;

        h—mean thickness of single layer of plate;

        n—number of layers;

        φ—friction angle between plates;

        α—dip angle of layer;

        c—interlayer cohesive force.

The longitudinal stress σ_s exerted by slope on the segment where bending may possibly take place is

$$\sigma_e = (L - L_0)[\gamma(\sin\alpha - \cos\alpha \cdot tg\varphi) - C/nh] \quad (2)$$

where L is total length of slope plate,

    L_0 is the length of bending segment.

Then the ratio between two stress

$$k_a = \frac{\sigma_{cr}}{\sigma_e} \quad (3)$$

is defined as stability factor against deformation.

The back calculation for the landslides resulted from sliding-bending, which are located on bank slopes between Fongjie and Yunyang in Three Gorges of Yangtze river, shows that their k_a values are mostly less than 3. So $k_a=3$ can be taken as the starting criterion of this kind of deformation, although the deformation may take a very long time. The less k_a is, the less time it takes. The existence of original slight bending at the lower part of plates and the action of

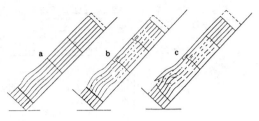

a. Monolithic sliding and gentle bending

b. Dislocated sliding and strong bending

c. Shear fracturing and slope failure

Fig. 7 Geomechanical modeling for the type II₄ slope (the test spended about 50 mouths)

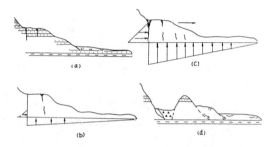

Fig. 8 Evolution mode of horizontally-pushing landslide a—plastic flow-crack deformation b,c—pore water pressure distribution in common and havy rainfall d—landslide

surface water or underground water favour the forming and developing of this kind of deformation. Once bending starts, the force state will change, as a general tendency, to be more favorable to promoting the evolution of slope.

4. 3 Quantitative evaluation on the promoting effect of internal and external dynamic agents on slope evolution

The superficial fractures produced in the process of slope deformation are the places where the action of various kinds of dynamic agents are relatively concentrated, and also the key places for slope evolution. For example, the heavy rainfall triggered horizontally-pushing landslides on the slope (II₁) composed of subhorizontal stratified rock mass, in Sichuan Basin, 1981, were all produced on the slopes which had shown the signs of plastic flowing—cracking or sliding—compression cracking deformation, under the joint action of pore uplift pressure along slip surfaces and the anomalous hydrostatic pressure due to water filling in the vertical tensile cracks at rear edge in the conditions of extraordinary heavy rain (figure 8). The critical level h_{cr} of water filled in the tensile crack at rear edge can be taken as the starting criterion of slide:

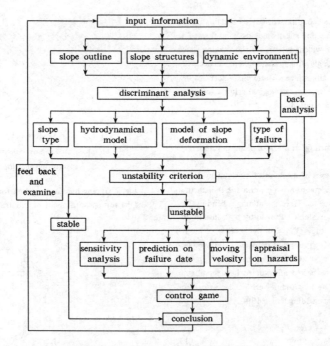

Fig. 9 Block diagram of slope stability research

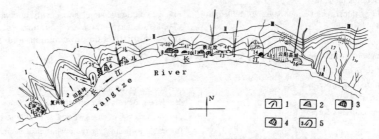

Fig. 10 Results of stability evaluation for the Three Gorges bank near Yunyang County

1. stable landslides; 2. under-water landslides at 150m water-level; 3. unstable landslides at 150m water-level; 4. unstable landslides at 180m water-level; 5. number of landslides   I —stable slope; II —basic stable slope; III —potential unstable slope ① ② ③—potential unstable slope with some deformation phenomena

$$h_{or} = \frac{\frac{1}{2}[l^2 \mathrm{tg}^2\varphi - 8W(\cos\alpha\mathrm{tg}\varphi - \sin\alpha)(\cos\alpha - \sin\alpha\mathrm{tg}\varphi)]^{1/2}}{\gamma_w(\cos\alpha + \sin\alpha \cdot \mathrm{tg}\varphi)}$$

$$- \frac{\frac{1}{2}tg\varphi}{\gamma_w(\cos\alpha + \sin\alpha \cdot \mathrm{tg}\varphi)} \quad (4)$$

if $\alpha = 0$, $\gamma_w = 1$, then

$$h_{or} = \frac{1}{2}[l^2 \cdot \mathrm{tg}^2\varphi + 8w \cdot \mathrm{tg}\varphi]^{1/2} - \frac{1}{2}\mathrm{tg}\varphi \quad (5)$$

where w— weight (unit width) of slide mass; γ_w— unit weight of water; α—slope angle of slip surface; l—length of slip surface; φ—friction angle of slip surface.

Once the slide mass starts to move, the level of water filled in rear tensile crack drops down rapidly and at the same time the uplift pressure along slip surface also decreases rapidly; so the slide mass brakes by itself due to losing the effect of pore pressure.

The study in this aspect thus can provide the basis for working out the prevention and treatment measures.

4. 4 Prediction of motion mode after slope failure

The deformed rock mass with different deformation models

also differ in the modes of energy reserve as well as energy release at the time of failing. The motion mode of the failed rock mass is determined by this fact to a great degree. For example, the sliding—cracking type slide generally behaves as slow-slide or block-slide and are sensitive to the rainfall and underground water. The sliding-bending type slide, being able to release a large number of strain energy, mostly behave as rapid-slide or high-speed slide, and some even can convert into "flowing"; the slide mass can come to a super-stable state. The bending-cracking type landslide or rockfall are mostly fully disintegrated to become detrital deposits or debris flow. In addition the study proves that the revival mechanism of slide mass is also related to its formation and evolution history.

4. 5 Working out of systematic engineering analytical programm and diagnosis system for the landslide stability assessment and prediction

By means of the systematic analysis on the complete courses of deformation based on the relationship among slope, deformation and failure (see Fig. 9), on the basis of the determination of startion criteria for various types of deformation-failure and on the basis of quantitative evaluation of the promoting effects of environmental dynamic agents, the computer programm is available to make diagnoses automatically for the stability of slope or landslides and to make prediction and forecast as well. This method was used tentatively in the research of bank slope stability in the projected reservoir of Three Gorges of Yangtze River and the results were fairly satisfactory (Fig. 10).

CONCLUDING REMARKS

To work out the systematic classification of slope deformation and failure and to make it perfect gradually will make a significant impact on the development of both theory and practice of the slope problem.

REFERENCES

Wang Lansheng and Zhang Zuoyuan 1984. Fundamental geomechanical model of rock masses deformation on slope in China. Proc. of 4th Inter. Symposium on landslide, Toronto, Canada.

Varnses, D. J. 1978. Slope movements and types and process. Landslide analysis and control, Transportatian Res. Board Nat. Ac. Sci. Washington Spec. Rep.

Nemčok, A. 1972. Classification of landslides and other mass movement. Rock mechanics.

Voight B. (Editor) 1978. Rockslides and avalanches 1. Elsevier Scientific Pub. Co.

Zaruba, Q and Mensl, V. 1976. Engineering Geology. Elesevior Sciemtific Pub. Co.

Hoek, E. 1977. Rock slope engineering. Insti. Min. Metal. London.

Müller L. and Hofmann, H. 1970. Selection, compilation and assesment of geological data for the slope problem. "Planning Open Pit Mines".

Zhang, Z. Y., Huang, R. Q and Wang, S. T. 1990. Epigenetic recreation of rock mass structures and time dependent deformation. Pro. VIth Inter. Congress of IAEG, Amsterdam.

Zhang, Z. Y. etal. The formation and Kinematic mechanism of landslides in the Pleistocene lacustrine clays beds near Lonyang Gorge damsite on the Yellow River, Inter. Geomorphology.

Wang Lansheng, Zhang Zuoyuan etal. 1986. Rainstorm-rockslides and their instability problems in the construction at the Sichuan Basin. Proc. of Vth Inter. Cong. of Engineering Geology, Buenos Aires.

G5 Instrumentation and monitoring
Auscultation et instrumentation

Landslides, Bell (ed.) © 1995 Balkema, Rotterdam, ISBN 90 5410 032 X

Keynote paper: Monitoring and instrumentation of landslides

John Dunnicliff
Bedford, Mass., USA

ABSTRACT: The goal of this paper is to describe the role of geotechnical instrumentation in addressing geotechnical questions that may arise during the diagnosis of a landslide, during construction of remedial treatment, and during subsequent behavior. Brief sections describe the factors that influence the stability of landslides, the steps in planning and executing a monitoring program, and the principal geotechnical questions that are likely to arise when dealing with monitoring and instrumentation of landslides. The bulk of the paper describes methods for monitoring deformation, groundwater pressure, stress in slope reinforcement, drainage flow and ambient conditions. The paper concludes with some additional suggestions for good practice.

1 INTRODUCTION

The birth of geotechnical instrumentation as a tool to assist with field observations occurred more than half a century ago. Since then, instruments have been used to observe conditions and monitor behavior during design, construction, and operation of civil engineering works that require engineering consideration of soil or rock.

At an active or potential landslide instrumentation can be used to assist with diagnosing the characteristics of the landslide, monitoring the effectiveness of remedial treatment, and monitoring subsequent behavior.

This paper is not intended as an exhaustive summary, a state-of-the-art paper, or a "cook-book." It is intended merely to open the minds of readers to the possible role of geotechnical instrumentation and monitoring at landslides, and to guide them towards implementation.

2 FACTORS THAT INFLUENCE STABILITY

Analysis of slope stability is the principal geotechnical task when monitoring landslides. Factors influencing stability include stratigraphy, groundwater levels, seepage gradients, strength of the soil or rock mass, geometry and driving moments.

Stability of slopes in soil is controlled by the ratio between the available shearing resistance along a potential failure surface and the shear stress on the surface. Circular or wedge-shaped surfaces are often used in analyses that attempt to model actual conditions. Available strength includes cohesion and frictional components. For long-term considerations, the contribution of cohesion is often reduced significantly.

Stability of slopes in rock is usually controlled by the presence of discontinuities in the rock mass and the presence of water under pressure in these discontinuities. Failures most frequently occur as a result of sliding or separation along discontinuities.

3 STEPS IN ACCOMPLISHING A SUCCESSFUL MONITORING PROGRAM

Planning and execution of a monitoring program using geotechnical instrumentation is similar to other engineering design and construction efforts. A typical engineering design effort begins with a definition of an objective and proceeds through a series of logical steps to preparation of plans and specifications. Similarly, the task of planning a

Table 1. 25 steps to follow in planning and executing a successful monitoring program

---

Planning Phase
1. Predict mechanisms that control behavior
2. Define the purpose of instrumentation
3. Define the geotechnical questions
4. Select parameters to monitor
5. Predict magnitudes of change
6. Devise remedial actions
7. Assign tasks for monitoring program
8. Select instruments
9. Select instrument locations
10. Plan recording of factors which influence measurements
11. Establish procedures of ensuring data correctness
12. Prepare budget
13. Write instrument procurement specifications
14. Plan installation
15. Plan regular calibration and maintenance
16. Plan data collection and management
17. Write contractual arrangements for field services

Execution Phase
18. Procure instruments
19. Install instruments
20. Calibrate and maintain instruments
21. Collect data
22. Process and present data
23. Interpret data
24. Report conclusions
25. Implement data

---

monitoring program should be a logical and comprehensive engineering process that begins with defining the objective and ends with planning how the measurement data will be implemented.

Unfortunately, there is a tendency among some engineers and geologists to proceed in an illogical manner, often first selecting an instrument, making measurements, and then wondering what to do with the measurement data.

The analogoy of a chain with many links is apt when planning and executing a monitoring program. The chain -- or the monitoring program in this case -- is only as strong as its weakest link, and breaks down with greater facility and frequency than most other tasks in geotechnical engineering.

Systematic planning requires special effort and dedication on the part of responsible personnel. The planning effort should be undertaken by personnel with specialist expertise in applications of geotechnical instrumentation. Recognizing that instrumentation is merely a tool, rather than an end in itself, these personnel should be capable of working in a team-player capacity with the project design team.

Table 1 summarizes the steps that should be followed when planning and executing a monitoring program. Details of each step are given by Dunnicliff (1988). It is imperative that, prior to considering instrumentation for a landslide, an engineer first develop one or more working hypotheses for a potential behavior mechanism. The hypotheses must be based on a comprehensive knowledge of the locations and properties of discontinuities.

4 PRINCIPAL GEOTECHNICAL QUESTIONS

Every instrument on a project should be selected and placed to assist with answering a specific question: if there is no question, there should be no instrumentation. Before addressing measurement methods themselves, a listing should be made of geotechnical questions that are likely to arise. The following are the two primary goetechnical questions when dealing with landslides.

4.1 What are the existing conditions?

When evaluating existing conditions, the primary need is to know whether the landslide is active and, if yes, to know what are the locations and rates of movement. However, this knowledge is not sufficient, because it tells nothing about the causes of the movement. If remedial treatment is required, the engineering effort to select appropriate treatment will involve a comprehensive understanding of the relationships between causes and effects, hence causal factors must also be examined. Many of the potential causal factors are examined by use of conventional site investigation procedures, sometimes supplemented by in situ testing.

Geotechnical instrumentation plays a role in defining groundwater pressures, which can be a significant causal factor when slope stability is in question. Groundwater pressure measurements will usually include determination of any perched or artesian water pressures.

4.2 Has remedial treatment arrested the landslide?

When the evaluation of existing conditions shows that the landslide is active, or potentially unstable, a choice must be made among:

(a) doing nothing, accepting the consequences of slope failure,

(b) monitoring to provide a forewarning of instability, so that remedial treatment can be implemented before critical situations arise, and

(c) stabilizing the slope, perhaps including a monitoring program to verify that stability has been achieved.

The choice will be based on many factors, including the consequences of failure and the economics of stabilizing.

5 MONITORING METHODS

When using geotechnical instrumentation to help with providing answers to the above geotechnical questions, the two primary parameters of interest are deformation and groundwater pressure. In addition, if slope reinforcement has been used for remedial treatment, there may be a need to monitor stress in the reinforcement. If remedial treatment has included drainage provisions, measurements of drainage flow may be appropriate. Finally, because rainfall and temperature change can be a direct cause of deformation, and because some instruments are sensitive to changes in temperature and barometric pressure, these parameters will usually be monitored. The various available monitoring methods are described below, together with some suggestions for good practice. Table 2 summarizes the monitoring methods. More details of monitoring methods are given by Dunnicliff (1988).

6 MONITORING SURFACE DEFORMATION

Methods for monitoring surface deformation include surveying methods to determine horizontal and vertical deformation over the area of concern, crack gages across scarps, tiltmeters, and multi-point liquid level gages.

6.1 Surveying methods

Measurements of surface deformation by surveying methods should extend beyond the uppermost limit of any possible movement zone to an area which is known to be

Table 2. Summary of monitoring methods

Parameter	Instruments
Surface deformation	Surveying methods Crack gages Tiltmeters Multi-point liquid level gages

Subsurface deformation	Inclinometers
	Fixed borehole
	  extensometers
	Slope extensometers
	Shear plane
	  indicators
	In-place
	  inclinometers
	Multiple
	  deflectometers
	Acoustic emission
	  monitoring

Groundwater pressure	Open standpipe
	  piezometers
	Vibrating wire
	  piezometers
	Pneumatic
	  piezometers
	Multi-point
	  piezometers

Stress in slope reinforcement	Load cells Strain gages

Drainage flow	Bucket and
	  stopwatch
	Water level behind
	  weir
	Pipe flowmeters

Ambient conditions	Rainfall gages Temperature gages Barometric pressure gages

stable, so that possible surface strain in advance of cracking can be monitored. Any toe heave should also be monitored.

Electronic distance measurements (EDM) are the primary method, using stable control points, trilateration and trigonometric leveling. Typical accuracy for both horizontal and vertical deformation measurements is +10-20 mm. Greater accuracy can be obtained by observing directly along a single line from a control point to a monument, typically giving an accuracy of + 3 mm for deformation perpendicular to the line.

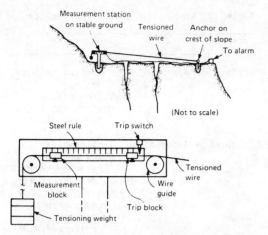

Figure 1 Tensioned wire crack gage

Automated slope monitoring procedures, using total station surveying instruments, are now available. These instruments can be left unattended, and have a computer controlled measuring system which can be programmed to measure distances and horizontal and vertical angles at predetermined intervals to a number of prism targets mounted on the slope. Tran-Duc et al (1992) describe an EDM-based automated landslide monitoring system.

The satellite-based Global Positioning System (GPS) is also capable of accuracies in the ± 10-20 mm range. Plans for this system called for 18 satellites to be operational in 1989, but these plans have been delayed by the interruption to the U.S. space program. Until the system is complete, availability is limited, and cost is high.

6.2 Crack gages

Tension cracks at the crest of the slope may be the first sign of instability. If cracks appear at the crest of the slope or elsewhere, their widths and vertical offsets should be monitored. Crack measurements give clues to the behavior of the entire slope, and often the direction of movement may be inferred from the pattern of cracking, particularly by the matching of the irregular edges of the cracks.

Numerous types of crack gages are available. A stake can be driven into the ground on either side of a crack or scarp, and relative movement measured with a survey tape or a tape extensometer. Alternatively a wire can be attached to

one stake and passed over a pulley attached to the other, with a weight on its end, as in Figure 1. Relative movement is measured with a mechnical or electrical gage, and the system can be fitted with a trip switch to sound an alarm if a predetermined movement is reached. Clearly such a system must be protected from adverse weather conditions and animals.

If protection from weather conditions and animals is impracticable, a "surface extensometer" can be installed on the ground surface or in a shallow trench. This device will typically be a sleeved rod spanning between an anchor point on each side of the crack or scarp, with a mechanical or electrical transducer. The electrical transducer can be connected to a datalogger, and data can be telemetered to a central location. Various practical considerations need to be addressed when designing details. First, an extensometer is designed to monitor extensions and compressions along its own length, and the rod will usually bind in its sleeve if significant deformation occurs perpendicular to its length. The alignment of the extensometer should therefore be as parallel as possible to the expected direction of movement, significant clearance should be provided between rod and sleeve, and it may need to be isolated from shear movement by cushioning above and below with a compressible material. Arrangements should be made so that the rod can be disconnected at any time from its anchor point, by using a bayonet or threaded disconnect, so that it can be checked for free-sliding. Second, temperature sensitivity should be evaluated. A carbon fiber/vinyl ester composite material is available for the rod, with a very low thermal coefficient. Third, the anchor points must be designed so that they are not affected by frost heave and seasonal moisture changes. Fourth, if an electrical transducer is used, the need for lightning protection should be evaluated: this is an economic decision, based on the likelihood of damage, the cost of protection, and the cost of transducer replacement. Incorporation of a datalogger will tend to favor adoption of lightning protection. Fifth, if an electrical transducer is used, the extensometer head should include a method for making a mechnical reading, for checking purposes.

In cases where shear deformation across cracks is important, various mechanical and electrical gages are available. If an unobtrusive remote-reading gage is

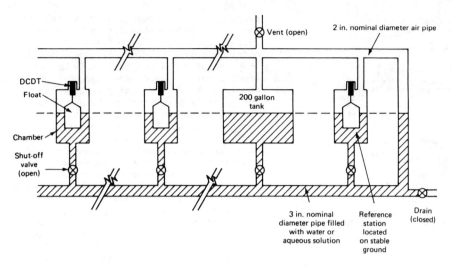

Figure 2 Schematic of multi-point liquid level gage

required, an adaptation can be made to the above "surface extensometer." The rod/ sleeve system can be very stiff, and hinged at its attachments to the anchor points. An electrical tiltmeter, mounted on the sleeve, can provide data for determination of shear deformation.

6.3 Tiltmeters

Single-point tiltmeters can be used to monitor surface deformation, provided the surface of the landslide has a rotational component. Transducer types are normally electrolytic levels or force-balance accelerometers. This approach has been useful to track the rate of tilt, either using a portable tiltmeter periodically mated with a series of reference plates, or by leaving a few tiltmeters in place, connected to a datalogger. However, because a landslide can move by trans- lation, with no rotational component, tiltmeter measurements can be misleading unless accompanied by subsurface deformation measurements. A second caution: on several landslides involving litigation, tiltmeter measurements have been made over a very short time period, and extrapolated to a "tilt per year". This approach is valid only if tiltmeter fixity is adequately precise, and if data are corrected for all extraneous causal factors such as temperature changes, and ground deformation caused by seasonal moisture changes.

6.4 Multi-point liquid level gages

Multi-point liquid level gages, consisting of a series of interconnected liquid- filled pots, mounted on a bench, have been used for monitoring stability of land- slides. As shown in Figure 2, each pot contains a float, with a deformation transducer linking the float and the head of the pot. If a pot moves vertically, the liquid level in the pot moves with respect to the head, causing a change in transducer reading. The transducers can be arranged to sound an alarm in the event that a predetermined settlement is exceeded at any pot.

7 MONITORING SUBSURFACE DEFORMATION

Subsurface deformation measurements will be required if sliding occurs, and if the depth of sliding is not readily apparent from surface measurements and visual observations. Measurements of subsurface horizontal deformation are more important than measurements of subsurface vertical deformation.

Methods for monitoring subsurface deformation include various borehole instruments that provide deformation data parallel or perpendicular to the alignment of the borehole. For slopes in soil, inclinometers are the instruments of choice, although shear plane indicators can be used for crude measurements, and slope extensometers may be preferred if deformation is predicted to occur at well- defined zones. Critical movements of

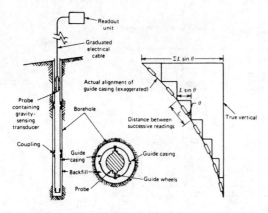

Figure 3 Principle of inclinometer operation

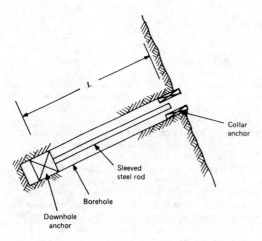

Figure 4 Operating principle of single-point fixed borehole extensometer

slopes in rock are often smaller than critical movements of slopes in soil, and therefore the required accuracy of deformation measurements is generally greater. High accuracy inclinometer measurements often provide the primary data, but fixed borehole extensometers may also play an important role. Multiple deflectometers and in-place inclinometers can provide real-time monitoring of sub-surface deformation, and these instruments can be connected to alarms if required. In addition, acoustic emission monitoring can provide an indication of subsurface deformation.

7.1 Inclinometers

Inclinometers (Figure 3) have been the primary tool for monitoring subsurface horizontal deformation for the past 35 years. Significant improvements have occurred during the last decade, since the adoption of space-age force balance accelerometer transducers, and the development of recording readout systems. Remote automatic winching systems have recently been developed in Italy and Canada (Lollino, 1992; Stevens et al, 1989), allowing the inclinometer probe to be lowered and raised at a predetermined frequency and the data transferred either to truck-mounted recording equipment or via a telephone modem to a remote location.

For monitoring at landslides, large diameter inclinometer casing should be used wherever possible, so that longevity will be maximized when shear deformation occurs and eventually prevents passage of the inclinometer probe. 85 mm diameter ABS plastic casing is recommended. When monitoring for small rates of movement, data should be processed to screen out errors (Green and Mikkelsen, 1988). For installations deeper than about 50 m, a survey should be made to determine groove spiral.

When using both surveying methods and inclinometers to monitor a landslide, there will typically be many more survey monuments than inclinometer casings. It is worthwhile to install a survey monument alongside key inclinometer casings, so that comparisons between inclinometer and survey data can be used to add to the evaluation of survey data accuracy.

7.2 Fixed borehole extensometers

Fixed borehole extensometers, often termed multi-point borehole extensometers (MPBXs), can be used to monitor deformations parallel to the axes of the boreholes in which the instruments are installed. Figure 4 shows the operating principal of a single-point version. They are subject to the same limitation described above for surface extensometers: the rods will usually bind in their sleeves if significant deformation occurs perpendicular to their length. Their use is therefore limited to cases where the borehole is nearly parallel to the expected direction of movement, hence it is rarely practicable to install them from the ground surface. However if tunnels have been driven within the landslide for investigation or drainage purposes, this alignment limitation can often be over-come, and MPBXs can be expedient and accurate tools for monitoring subsurface deformation.

As for surface extensometers, it is important to provide significant clearance between rods and sleeves, to arrange for a disconnect between rods and anchors to check for free-sliding, and to provide for backup mechanical readings.

7.3 Slope extensometers

The slope extensometer is a multi-point borehole extensometer with tensioned wires instead of rods, arranged for monitoring deformation perpendicular to the axis of the borehole (Figure 5). At the instrument head each wire passes over a pulley, and is attached to a weight. If shear deformation occurs between two anchors, no vertical movement of the weights occurs for all anchors that are above the shear. However, weights attached to anchors below the shear move downward as the wires are dragged across the shear. Hence by measuring movement of the weights relative to the head, shear deformation can be monitored, and the shear is known to be within a certain depth band. Up to about 10 anchors and wires can be installed in a borehole.

Initial shear deformation will not cause an equivalent reading change, owing to lateral movement of the wires within the borehole, but after the borehole has been separated completely, the reading change will equal the shear deformation. This "dead" spot can be reduced by installing each wire within a plastic tube, and grouting between the tubes.

When compared with conventional inclinometer measurements, advantages of the slope extensometer include a simple and rapid reading procedure, the option to provide an alarm by inclusion of limit switches, and the ability to monitor much larger shear deformations. However, the device is suitable only for monitoring distinct shear planes or thin shear zones. Precision is obviously much less than for inclinometers, but in addition to possible use as described above, they may be useful when installed in inclinometer casings or piezometer standpipes that have sheared to a point where they can no longer be read.

7.4 Shear plane indicators

Examples of shear plane indicators are rupture stakes, shear probes, and shear strips.

Rupture stakes can be used in soft clays to examine the areal extent and depth of a landslide that is continuing to move.

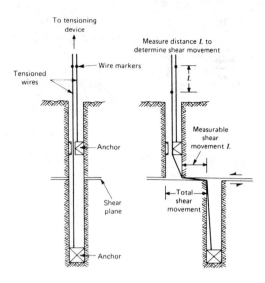

Figure 5   Schematic of slope extensometer

Wooden stakes are installed to a depth beyond the anticipated shear plane. Shearing will break the stakes, and the depth to the shear plane can be determined by pulling out the upper part of each stake. This is an economical procedure if installations can be made by hand, but if a drill rig is required, the shear probe described below may be the preferred approach for crude and inexpensive measurements. Because there is a significant risk of breakage when removing stakes, a large number should be used so that false data can be discarded.

The shear probe, also referred to as a poor man's inclinometer, slip indicator, and poor boy, consists of plastic tubing or thin-wall plastic pipe, installed in a nominally vertical borehole. The depth to the top of the shear zone is determined by lowering a rigid rod within the tubing or pipe, and measuring the depth at which the rod stops at a bend. The depth to the bottom of the shear zone can be measured by leaving a rod with an attached graduated nylon line at the bottom of the tubing or pipe and pulling on the line until the rod stops. Approximate curvature can be determined by inserting a series of rigid rods of different lengths and noting the depth at which each rod will not pass further down the tubing or pipe. As for the slope extensometer, the shear probe may also be useful in inclinometer casings or piezometer standpipes.

The shear strip (Figure 6) consists of a parallel electrical circuit made up of

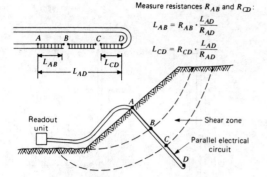

Measure resistances R_{AB} and R_{CD}:

$$L_{AB} = R_{AB} \cdot \frac{L_{AD}}{R_{AD}}$$

$$L_{CD} = R_{CD} \cdot \frac{L_{AD}}{R_{AD}}$$

Figure 6 Schematic of shear strip

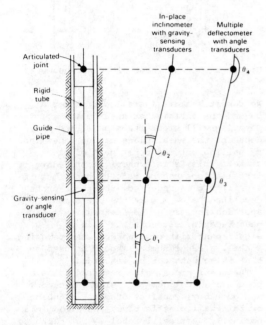

Figure 7 Schematic of in-place inclinometer and multiple deflectometer

resistors that are mounted on a brittle backing strip and waterproofed. The strip is installed in a borehole, and the locations of up to two breaks in the strip are determined by measuring resistances at the head of the instrument. The device can be connected to an automatic recording system and also arranged to sound an alarm if the strip breaks. Its primary limitation is that it is only useful as it breaks: the maximum information provided is the location of the top and bottom of the shear zone, one time only.

7.5   In-place inclinometers

An in-place inclinometer is generally designed to operate in a near-vertical borehole and provides essentially the same data as a conventional inclinometer. As shown in Figure 7, the device consists of a series of gravity- sensing transducers (tiltmeters) joined by articulated rods. Downhole deformation data are calculated from the known distances between the transducers and the measured inclination changes. The transducers are positioned at intervals along the borehole axis and can be concentrated in zones of expected shear deformation.

The device generally uses standard inclinometer casing as guide pipe and can be removed for repairs. However, data continuity will be interrupted when the device is removed and replaced. When compared with conventional inclinometers, advantages include more rapid reading, an option for continuous automatic reading, and an option for connection to a console for transmission of data to remote locations or for triggering an alarm if deformation exceeds a predetermined amount. Disadvantages include greater complexity and expense of the hardware. Lightning protection is required in regions of thunderstorm activity. When conventional inclinometers are read, any long-term drift of the gravity-sensing transducers is removed from calculations by taking a second set of readings with the inclinometer rotated 180 degrees (the "check-sum" procedure), but this is not possible with the in-place version. Although the transducers are generally stable, there is always the possibility of a "rogue transducer," and this possibility should be recognized if one is planning to use an in-place inclinometer for long-term applications where high precision is required.

In-place inclinometers can be used effectively in combination with a conventional inclinometer. An in-place version can first be installed to define the location of any transverse deformation, with minimal labor costs for reading. If deformation occurs, the in-place system can be removed and the moving zone monitored with a conventional inclinometer. Alternatively, a conventional inclinometer can be used first to indicate any deformation and an in-place version later installed across a critical deforming zone to minimize subsequent effort and perhaps to provide an alarm trigger. As an additional option, some engineers prefer to install two casings near each other, one for an in-place inclinometer, the other for occasional readings with a conventional inclinometer, for checking purposes.

Lollino (1992) describes a "fixed sensor" arrangement, whereby a series of tiltmeters are fixed within inclinometer casing, without the articulated rods that form part of the in-place inclinometer system. As discussed by Lollino, this arrangement is likely to give misleading data, because data reduction assumes that each measured tilt can be applied to the entire interval between adjacent tiltmeters: this assumption is usually unwarranted.

7.6 Multiple deflectometers

Multiple deflectometers operate on a similar principle to in-place inclinometers, but rotation is measured by angle transducers instead of tilt transducers (Figure 7). Downhole deformation data are calculated in the same way as for an open survey traverse.

Until recently this instrument was commercially available in two versions in USA and Germany: articulated rods with full bridge bonded resistance strain gage transducers attached to cantilevers, and a tensioned wire passing over knife edges, with induction transducers. However, both versions have been discontinued, but a new version is now available in USA, with vibrating wire strain gages attached to cantilevers.

Multiple deflectometers are usually installed within inclinometer casing. As for the in-place inclinometer, the system can be removed from the borehole at any time for maintenance and calibration, but data continuity will be interrupted when the device is removed and replaced.

Although advantages and limitations are generally the same as for in-place inclinometers, there are three important differences between the two systems. The first favors multiple deflectometers, while the second and third favor in-place inclinometers. First, multiple deflectometers are not limited by borehole inclination, because their transducers are not referenced to gravity. They can therefore be used to sense shear deformation across a borehole at any inclination. Second, because deflectometer data are calculated by determining the position of one arm of the instrument relative to another, and not with respect to gravity, the device has no means of sensing rotation of the system as a whole.

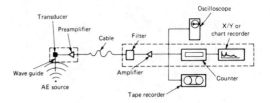

Figure 8 Schematic of basic single-channel acoustic emission monitoring system

Third, deflectometer errors accumulate exponentially, whereas inclinometer errors accumulate arithmetically.

Portable borehole deflectometers can also be used, but commercial availability is again limited. Kumbhojkar (1991) provides useful guidance for minimizing measurement errors.

7.7 Acoustic emission monitoring

Acoustic emissions are sounds generated within a soil or rock material that has been stressed and subsequently deforms. A piezoelectric transducer can be used as a "pickup" to detect the acoustic emissions and produces an electrical signal proportional to the amplitude of sound being detected. A basic system is shown in Figure 8. Usually, if no acoustic emissions are present, the material is in equilibrium and therefore stable. If emissions are observed, the material is not in equilibrium and may be in a condition that eventually leads to failure.

Acoustic emission monitoring can sometimes be used by experienced personnel over a wide area in shallow drillholes to determine landslide deformation trends and locations. It is most effective when the amplitude of the signals is high and thus is more effective for rock and cohesionless soil than for cohesive soil.

7.8 Future Innovations

Lord et al (1991) described three innovations that are being researched for measurement of subsurface horizontal deformation of highwalls at an open pit oil sand mine in northern Alberta, Canada. At present the primary stability monitoring method is with inclinometers, but for this application inclinometers have some safety and operational concerns. The first innovation is being developed, but the second and third, although showing promise, are currently on hold. The three

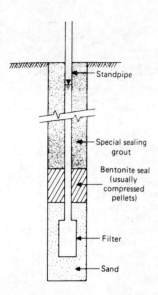

Figure 9 Schematic of open standpipe piezometer

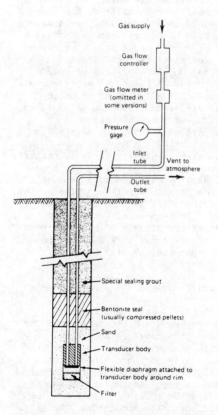

Figure 10 Schematic of pneumatic piezometer

are described briefly below.

First, "electrolytic bubble sensor strings" are being developed. Conceptually each sensor unit will consist of an electrolytic level which will be attached to a rod, and the inclination of the level will be integrated over the rod length. A string of sensors will be grouted into a borehole.

The second innovation uses time domain reflectometry (TDR), consisting of coaxial electrical cable grouted into a borehole, and a standard TDR cable tester. The cable tester is used to transmit a voltage pulse along the cable and to monitor the return signal. Electromagnetic wave theory indicates that a voltage pulse travelling down a coaxial cable will be reflected to some extent when it encounters a change in impedance in the cable. A series of ultra-fast voltage pulses are used to determine both the location and magnitude of the changes of impedance caused by a change in the cross-sectional geometry of the cable.

The third innovation uses fiber-optic cables. Fiber-optic sensors depend on the ability of the fibers to carry light from a source to a photosensitive detector. They can be used to sense the relative position between an object and the end of a fiber or the distance between two points along a fiber. They can also indicate bending, hence have a good potential for monitoring deformation both along and transverse to the fiber.

8 MONITORING GROUNDWATER PRESSURE

Groundwater pressure is typically monitored with open standpipe, vibrating wire or pneumatic piezometers (Figures 9-11). Advantages and limitations of the options are given by Dunnicliff (1988). Observation wells, in which the annulus between a perforated pipe and the ground is filled with a permeable material, are rarely appropriate, because they create a hydraulic short circuit between different groundwater pressure regimes.

8.1 Slopes in soil

Open standpipe piezometers are normally selected for slopes in soil, but vibrating wire or pneumatic piezometers are appropriate if more rapid response is required. Also, longevity of vibrating wire and pneumatic piezometers is likely to be longer than open standpipe

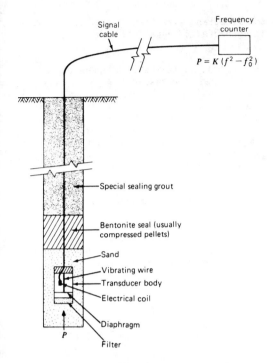

$$P = K(f^2 - f_0^2)$$

Signal cable

Frequency counter

Special sealing grout

Bentonite seal (usually compressed pellets)

Sand

Vibrating wire

Transducer body

Electrical coil

Diaphragm

P

Filter

Figure 11 Schematic of vibrating wire piezometer

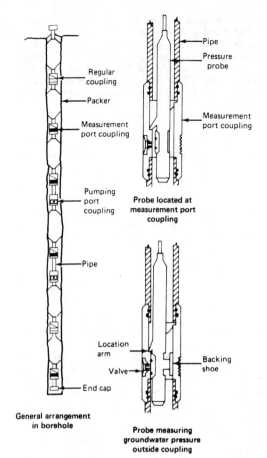

Regular coupling

Packer

Measurement port coupling

Pumping port coupling

Pipe

End cap

General arrangement in borehole

Pipe

Pressure probe

Measurement port coupling

Probe located at measurement port coupling

Location arm

Valve

Backing shoe

Probe measuring groundwater pressure outside coupling

Figure 12 Westbay multiple piezometer system

piezometers if the piezometers are located below the sliding surface, because of shear deformation across the borehole.

Open standpipe piezometers can be converted to remotely-read instruments by inserting a pressure transducer within the standpipe, and this adaptation also allows a datalogger to be used. 11.1 mm diameter vibrating wire piezometers have been proven for this application.

If any sliding is occurring, pore water pressures at or near the sliding surface must be measured to enable an effective stress analysis to be performed.

8.2 Slopes in rock

Open standpipe, vibrating wire or pneumatic piezometers can also be used for slopes in rock. However, the hetero-geneous nature of most rock masses results in a need for comprehensive monitoring of joint water pressure along, above and below each possible failure plane. Such intensive monitoring is often not possible with single-point piezometers, and multi-point piezometers will often be the instruments of choice. Figures 12 and 13 show two commercially available versions.

The Westbay multiple piezometer system consists of pipe, couplings and packers permanently installed in a borehole, a portable pressure measurement probe, and installation tools. Measurement port couplings are installed in the pipe wherever groundwater pressure measurements are desired. Each of these couplings has a hole through its wall, with a filter on the outside and a spring-loaded check valve on the inside. The remainder of the pipe is connected with sealed couplings, and a packer is installed around the pipe above and below each measurement port coupling. The assembly is lowered into the borehole, and the packers inflated one at a time with water, using a probe temporarily inserted within the pipe. To take readings, a pressure measurement probe is lowered within the pipe to locate a measurement port, jacked against the opposite wall of the pipe to open a check valve, and a measurement made. The procedure is repeated at each measurement port. Alternatively, pressure monitoring modules can be left and placed at selected ports. Separate probes are also available for taking pressurized or unpressurized samples of groundwater through the measurement ports. Additional valved couplings, referred to as pumping port couplings, can be included in the pipe for permeability testing or decontamination pumping.

The Waterloo System uses a chemical sealant, which swells when in contact with water that is poured down the central PVC pipe. The figure shows standpipe piezometers, but alternatively pneumatic or vibrating wire piezometers can be used. The instrument can also be adapted for groundwater sampling. The system is supplied in modular form, with threaded connections in the PVC pipe between modules.

9 MONITORING STRESS IN SLOPE REINFORCEMENT

If rockbolts have been installed for slope reinforcement, the opportunity arises to measure load in the rockbolts, and these measurements may be useful for verifying assumed design loads. However, recognizing that the purpose of rockbolts is to restrain deformation, it is usually more effective to monitor deformation rather than stress in rockbolts, using fixed borehole extensometers. A typical approach is use of single-point fixed borehole extensometers, anchored deeper than the rockbolts. If measurements of load in the rockbolts are also required for more comprehensive monitoring, load cells can be used on end-anchored rockbolts but strain gages must be used on fully grouted rockbolts.

10 MONITORING DRAINAGE FLOW

Methods for monitoring drainage flow include use of a bucket and stopwatch, pipe flowmeters, and various techniques for measuring the level of water behind a weir. Two well-proven techniques for weir monitoring are first the Geonor remote-reading weir (Figure 14), consisting of partially submerged cylinders suspended from vibrating wire force transducers and second, ultra-sonic systems. Ultra-sonic pipe flow meters have been used successfully for monitoring flow from individual drainage holes. Ultra-sonic

and vibrating wire transducers can readily be connected to dataloggers.

11 MONITORING RAINFALL, TEMPERATURE AND BAROMETRIC PRESSURE

Numerous instruments for monitoring rainfall, temperature and barometric pressure are available from manufacturers of weather stations and from manufacturers of geotechnical instrumentation, with a wide range of datalogging options.

12 SOME ADDITIONAL SUGGESTIONS FOR GOOD PRACTICE

Some suggestions for good practice have been included in earlier sections of this paper. Additional suggestions are included here.

12.1 Reliability of data

Very often the cost of drilling instru-

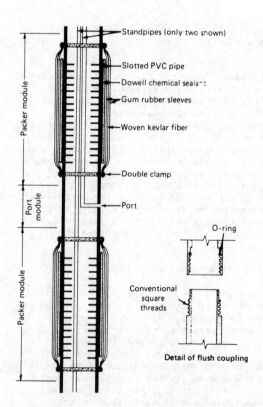

Figure 13 Waterloo multilevel groundwater monitoring system

1892

mentation boreholes is significantly larger than the cost of the instruments themselves. There is therefore a temptation to install as many instruments as possible in a single borehole but, as a general rule, this practice leads to data of questionable reliability. One cannot see and inspect the condition of an instrument after installation and back-filling in a borehole.

Two examples will be given. First, installation of several piezometers in a single borehole can lead to uncertainties about seal adequacy. When annular space is limited, it is often difficult to verify that grout has not run into the ground, that granular material intended to surround each piezometer is in the correct place, and that other sealants such as bentonite are in place at the planned locations. Hydraulic short circuits are a common problem when several piezometers are installed in a small diameter borehole, particularly when the distance between piezometers is small. Second, installation of one or more piezometers in the same borehole as an inclinometer casing can lead to hydraulic short circuits between piezometers, incomplete base fixity for the inclinometer casing, and a zone of poor fixity at each piezometer sand zone.

While a certain amount of questionable data may be acceptable during the first phase of answering the question, "what are the existing conditions?", this is unacceptable while answering the question "has remedial treatment arrested the landslide?" In this second case data must be highly reliable, because data provide input to serious engineering decisions concerning safety and economics. Hence the emphasis must be on verifiable data of high reliability, despite the cost impact of such an emphasis. Usually this will mean one instrument per borehole (an exception can be applied to single-point piezometers, for which two can usually be installed in a single borehole if its diameter allows for adequate annular space, and if the depth interval between the two piezometers is more than about 10 m), and a carefully planned and quality assured installation procedure to verify that all is as it should be. "Quality assured" means that no installation step is taken without a means of verifying that the step has achieved its purpose: no "closing the eyes, crossing the fingers and hoping" is allowed!

12.2 Grout-tightness of boreholes

Certain instruments, such as inclinometer casings, piezometers, and fixed borehole extensometers with groutable anchors, will not provide reliable data unless grout is at the planned locations. To ensure this, there must be certainty that grout will not run into the ground. The key to the approach is "prove it", before installing the instrument. This will usually mean pressure testing with water or grout, followed by grouting and redrilling if the borehole fails the test, and re-testing.

12.3 Grouting in near-horizontal boreholes

When grouting instruments such as fixed borehole extensometers in near-horizontal boreholes, a downhole elevation survey should be made before grouting, so that grout inlets can be positioned at low points, and air vents at high points. The borehole must of course have adequate diameter to contain any required grout and vent tubes. It is usually worthwhile to locate the borehole so that drilling starts about 5° below horizontal, to minimize this problem.

12.4 Combined inclinometer casing and open standpipe piezometer

Gillon et al (1992) describe use of a combined inclinometer casing and open standpipe piezometer at the Cromwell Gorge landslides. The system consists of

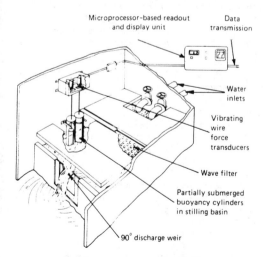

Microprocessor-based readout and display unit

Data transmission

Water inlets

Vibrating wire force transducers

Wave filter

Partially submerged buoyancy cylinders in stilling basin

90° discharge weir

Figure 14   Geonor remote reading weir

Table 3. Some components of a quality assurance program

Factory calibrations
Personnel training for all tasks listed in Table 1
Preparation and review of written procedures for tasks 19 thru 25 listed in Table 1
Pre-installation acceptance tests
Acceptance tests at "hold points" during installation
Post-installation acceptance tests
Comprehensive installation records
Regular maintenance and calibrations of readout units
Duplicate or triplicate initial readings
Rapid identification of whether a significant change has been measured
Ongoing re-evaluation of data collection frequency
Ongoing verification of the correctness of data processing, presentation and interpretation tasks
Ongoing correlations among measured parameters, to develop relationships between causes and effects

---

a Westbay packer, casing components and pumping port at the bottom of inclinometer casing, so that the inclinometer casing also functions as a piezometer standpipe. The packer provides good base fixity for the inclinometer casing, and the hole is deliberately drilled deeper to ensure that the piezometer filter and packer are well below any zone of deformation. Special care is taken to ensure that couplings in the inclinometer casing are watertight against unbalanced pressures in both directions, and ABS plastic casing is used. The authors comment that these installations are a compromise for use in very limited circumstances and are not intended to be used as instruments that provide primary warning of adverse events.

12.5 Quality assurance

Gillon et al (1992) comment that good installations do not happen by chance. This view can be broadened to: reliable data are not obtained by chance. An important component of reliable data is rigorous quality assurance throughout all phases of the monitoring program. Table 3 summarizes some components of a quality assurance program. More details are given by Dunnicliff (1988).

12.6 Automatic data acquisition and processing

Great efficiency can be achieved by judicious use of electronic field books, dataloggers and computerized data management procedures. However, "judicious" should recognize that no automatic system can replace engineering judgment. Direct personal observations and direct personal evaluations remain an essential component of a successful monitoring program.

13 RECENT AND FUTURE PUBLICATIONS DESCRIBING MONITORING AND INSTRUMENTATION OF LANDSLIDES

Two publications that contain detailed and valuable information relating to monitoring and instrumentation of landslides are Cavers et al (1991) and Transportation Research Board (1978). The latter publication is currently being revised, and an updated version is expected during 1993 or 1994.

14 ACKNOWLEDGEMENTS

My recent experience while assisting with monitoring aspects of the Cromwell Gorge landslides in New Zealand has given me the opportunity to focus my thoughts on the subject of this paper. I acknowledge the large technical and personal benefits that I have received during interactions with my New Zealand colleagues.

Finally, what New Englander could express other than delight at being invited to present this paper here and now, in the summer, when there's probably a blizzard blowing at home? Thank you for inviting me to be with you.

REFERENCES

Cavers, D.S., L.B.B. Peer & C.E. Rea 1991. Methods of Monitoring Waste Dumps Located in Mountainous Terrain. Report to CANMET by Hardy BBT Ltd, 2227 Douglas Road, Burnaby, BC, Canada, V5C 5A9.
Dunnicliff, J. 1988. Geotechnical Instrumentation for Monitoring Field Performance. John Wiley & Sons Inc., New York, 577 pp.
Gillon, M.D., P.F. Foster, G.T. Proffitt & A.P. Smits 1992. Monitoring of the Cromwell Gorge Landslides. Sixth International Symposium on Landslides, Christchurch, New Zealand.
Green, G.E. & P.E. Mikkelsen 1988.

Deformation Measurements with
Inclinometers. Transportation Research
Record 1169, Transportation Research
Board, National Research Council,
Washington, DC.

Kumbhojkar, A. 1991. Elimination of
Initial Rotation Error in Portable
Borehole Deflectometers. J. Geotech.
Eng. ASCE, Vol. 117, No. 12, Dec.,
pp. 1949-1955.

Lollino, G. 1992. Automated Inclinometric
System. Sixth International Symposium on
Landslides, Christchurch, New Zealand.

Lord, E., D. Peterson, G. Thompson & T.
Stevens 1991. New Technologies for
Monitoring Highwall Movement at Syncrude
Canada Ltd. The Petroleum Society of CIM
and AOSTRA, Technical Conf. Our Energy
Future, Banff, Alberta, April, pp. 97-1
thru 97-8.

Stevens, T.G., P. Brown & E.R. Lord 1989.
Development of a Prototype Automated
Inclinometer Winching System for Slope
Stabililty Monitoring. Colloquium V,
Petroleum Surveying in the 90's
Environment, Canadian Petroleum
Association, Calgary, Oct.

Tran-Duc, P.O., M. Ohno & Y. Mawatari 1992.
An Automated Landslide Monitoring System.
Sixth International Symposium on
Landslides, Christchurch, New Zealand.

Transportation Research Board 1978.
Landslides, Analysis and Control.
Special Report 176, R.L. Schuster and
R.J. Krizek (Eds.) National Academy of
Sciences, Washington, DC, 234 pp.

Landslides, Bell (ed.) © 1995 Balkema, Rotterdam, ISBN 90 5410 032 X

Theme report

P. Erik Mikkelsen
Landslide Technology, Bellevue, Wash., USA

ABSTRACT: Twelve papers were submitted for this theme of the Symposium. Authors from seven countries submitted: four from Italy, two each from Japan and New Zealand, and one each from Canada, Sweden, Sri Lanka and France. Nine papers relate to case histories, most of which include detailed background geological and geotechnical information in addition to the instrumentation approach. Several projects are still in progress. The most universally adopted monitoring approaches for landslides were surface surveying, borehole inclinometers and piezometers. Other monitoring utilized, in the order of frequency mentioned, were extensometers, rainfall measurements and other purpose designed instrumentation. Automation for case histories were discussed briefly by three papers, and in more detail by the two papers from New Zealand that also cover data management and computer applications. Most papers discussed current practice, with four papers covering design philosophy. Four papers brought up new instrument designs, and another three introduced interesting analytical methods.

1 INTRODUCTION

This report summarizes the contents of 12 papers submitted to the G5 Theme from authors representing seven countries on the subject of monitoring and instrumentation. Nine case histories with detailed background information were presented. Five papers dealt with monitoring for research type projects or new instrument designs. A numerical list of the papers are presented in alphabetical order by author at the end of this report. In the text, the papers will be referred to by first author and/or the number of the paper in parentheses. Table 1 classifies the subjects covered, and Table 2 summarizes the countries represented.

Included in this summary are common trends in the approach to landslide monitoring and notable instrumentation advances which are discussed in the papers. Since the emphasis in this Theme G5 is on monitoring and instrumentation, the subjects under environmental/ geological/geotechnical background, associated laboratory testing and routine analytical studies are considered secondary in this context.

2 SURFACE SURVEYING

Surface surveying (or geodetic control) is mentioned by ten of the papers, implying that it is the most fundamental of all landslide monitoring techniques. Settlement of benchmarks and downslope monitoring of surface monuments are the first and most extensive methods used to define landslide activity. Angeli (1) used 78 benchmarks for high accuracy leveling (+/- 0.2 mm) in an Italian village at the top of a high, steep limestone rock cliff on the Adriatic Sea. Antoine (2) put in 20 concrete monuments at the Mas d'Avignonet landslide near Grenoble in France and observed about 70 mm of movement per year since 1985. Triangulation and trilateration were accomplished over distances of 2.5 to 6.5 km. In New Zealand, Gillon (7) reports that 250 monuments plus 280 leveling

Table 1. Summary of subject categories

Subjects	Technical Paper												Σ
---	1	2	3	4	5	6	7	8	9	10	11	12	
Case History	●	●	●	●	●	●	●	●			●		9
Research/Testing			●		●				●	●		●	5
Current Practice	●	●	●	●		●	●	●		●			8
New Technology			●	●					●			●	4
Rock Landslide	●		●	●		●	●	●			●		7
Soil Landslide		●	●		●					●			4
Stability Analysis	●	●		●						●			4
Earthquake Influence	●	●		●									3
Laboratory Tests		●		●						●			3
Analytical Work		●			●					●			3
Monitoring Philosophy			o			o	o		o				4
Survey/Geodetic	o	o	o	o		o	o	o		o	o	o	10
GPS							o	o					2
Inclinometers	o	o	o	o		o	o	o	o	o	o		10
Piezometers	o	o	o	o	o	o	o			o	o		9
Extensometers	o		o				o	o			o		5
Rainfall Measurement	o		o	o			o	o			o		6
Seepage Measurement							o				o		2
Sediment Accumulation								o					1
Special Instr./Other			o				o	o	o	o	o	o	7
Automation	o		o	o			o		o		o	o	7
Early Warning			o				o				o	o	4
Data Management							o				o		2
Long Mon. Record					●	●		●					3
Project in Progress	●	●					●		●		●		5
Remedial Work	●	●			●		●						4

Table 2. List of papers by country

Country	Number	Paper No.
Canada	1	6
France	1	2
Italy	4	1, 4, 5, 9
Japan	2	8, 12
New Zealand	2	7, 11
Sri Lanka	1	3
Sweden	1	10

points (benchmarks) are used in monitoring 18 large landslide areas in the Cromwell Gorge for the Clyde Hydroelectric Project. GPS is starting to be utilized, but mostly on an experimental basis since it is fairly new and the accuracy is more in question than current surface methods. Tran-Duc (12) from Japan presents another recent development, combining EBM and CCD camera with image processing capability and robotics to give the coordinates of the targets in a landslide area directly at any given time.

Enegren (6) in Canada reports employing ten primary monuments and two control monuments for the Branham Ridge Slide totaling 2.5 million cubic meters of rock where 4 to 35 mm per year of displacement has occurred since 1972. This 20-year long history with a moving slide affirmed that comprehensive investigation and monitoring were key in mitigating this geologic hazard. In this case the design and implementation of the survey network seems to have been the most important of all the monitoring.

3 INCLINOMETERS

Inclinometers are used extensively and are perhaps as commonplace as test borings. An effective extension of the exploration program is to install inclinometers and piezometers in the borings. Some (1, 7) report both types of instruments installed in the same boring, or perforating the inclinometer casing for observa-

tion of groundwater. Detailed inclinometer results are discussed by five papers (1, 2, 4, 7, 11). Data presentations vary from cumulative plots (1, 4) and incremental or unit strain plots (2), to surface vectors (4) and time displacement of the zone of shear (11). Proffitt (11) also mentions the importance of efficient software and the ability of the program to have systematic error diagnostics and correction capabilities.

Of special note, possibly the most massive inclinometer monitoring program in the world today is discussed in two papers on the current Cromwell Gorge/Clyde Dam Project in New Zealand by Gillon (7) and Proffitt (11). An excess of 12,000 m/month in 180 casings (averaging 90 m deep) are monitored on some 20 large landslides. Movement rates are reported to be on the order of 2 to 30 mm/year. Some of the monitoring in known shear zones is done with in-place, or fixed, continuously recording inclinometers.

Lollini (9) discusses the limitations of in-place (fixed) inclinometers where the zones of movement are not known. As a more suitable alternative, the author suggests that a traversing borehole inclinometer could be made to run as an unattended machine, the AIS. This machine has apparently not been designed or built, nor is it the first time the idea has been suggested.

4 PIEZOMETERS

Observations of piezometric levels are fundamental to landslide investigations and remediation, and are mentioned by nine papers. The most notable contribution is from Cascini (5) who discusses a mathematical model which links up rainfall and piezometric levels in a case study. Gillon (7) mentions the problems with combining standpipe piezometers with inclinometer casings, and suggests using dedicated borings for the installation of multi-point piezometers at critical locations.

5 EXTENSOMETERS

Extensometers are generally being used for landslides occurring in rock, not soil. This is also reflected by the papers for this Theme.

Notable is a wire extensometer installed as a retrofit to an inclinometer casing that had experienced too much lateral deformation to accept the probe. No details are described, but apparently the wires are drawn into the casing as the shear at depth occurs, making it possible to measure the movement of the wires. Similar monitoring has been reported for basal movement of glaciers and high displacement landslides.

Hiura (8) introduces a surface extensometer array that reportedly can measure movements in three orthogonal directions. These three-dimensional, shear displacement meters have a 17-year monitoring record for the Zentoku Landslide in Japan. In addition, the paper discussed underground erosion mechanisms (piping?) and associated sediment discharge measurements--somewhat uncommon monitoring parameters.

Bhandari (3) also discusses the use of special wire surface extensometer arrays with automatic monitoring and warning.

6 RAINFALL AND SEEPAGE MEASUREMENTS

Rainfall monitoring is reported by half the authors, but seepage measurements are only mentioned by two. Perhaps this is an indication that rainfall is an easier parameter to measure. As indicated in the piezometer section, Cascini (5) covers the subject of rainfall in the most depth. Rainfall and its intensity are noted as the most notorious trigger mechanism for landslides. Better access to continuous, detailed rainfall and seepage data (where possible) will allow for better quantification of these transient events, particularly for rapid flow slides.

7 AUTOMATION, EARLY WARNING AND DATA MANAGEMENT

Automation is becoming more commonplace and is mentioned by seven of the papers. Most systems are small with a data logger at a central monitoring point, frequently connected by telephone or radio link to an office. The papers on the Clyde Project (7, 11) have a good discussion of all aspects of their massive data collection program. More than 2,000 data points from piezometers, inclinometers, survey points and flow (seepage) measurements are being monitored. Hardware and software considerations are also discussed.

The subject of early warning is touched on by four of the papers. Very interesting observations from research on automation and early warning is made by Bhandari (3) relative to public perception and the value of information from such instrumentation.

8 INSTRUMENTATION AND MONITORING PHILOSOPHY

Four papers discuss in depth the ideas behind monitoring, with two of these papers (3, 7) having the most comprehensive discussion on the subject. Bhandari (3) points out the need and the difficulties in working out meaningful, long-range monitoring and warning methods for massive landslide areas in Sri Lanka. Gillon (7) lays out how the monitoring philosophy of the giant Clyde landslides project changed through the various phases of the project from investigations and construction to the first filling of the lake behind Clyde Dam.

9 CONCLUSIONS

The 12 papers submitted present good case histories representative of the current state-of-practice in monitoring and instrumentation. The papers reflect that current practice has advanced further because of easier access to hardware and software. More instrumentation and more timely acquisition of data are possible, making instrumentation programs more meaningful when presented quickly and efficiently with good software. Electronic data acquisition can help mold the monitoring into a highly disciplined exercise, reducing to a minimum human error and missed opportunities to monitor significant events. Rapid as well as slow creeping landslides can be monitored in time domains from one second to months. With the miniaturization of electronics with steadily lower power requirements, portability of computers and proliferation of excellent tele-communications, the monitoring of landslides has become highly improved. It is up to us as professionals to use the best of this technology for the benefit of those who are exposed to the dangers of landsliding.

10 LISTING OF PAPERS

Angeli, M.G., Barbarella, M. & Pontoni, F. Instability of a sea cliff: Sirolo Landslide (Italy).

Antoine, P., Monnet, J., Rai, N.E., Moulin, C. & Meriaux, P. Resultats de cinq annees d'auscultation sur un glissement dans les argiles glacio-lacustres du Trieves (Sud-Est de la France).

Bhandari, R.K., Thayalan, N. & Fernando, A.T. Relevance of simple innovative and economical instrumentation in landslide monitoring.

Cancelli, A., Crosta, G., Nardi, R. & Pochini, A. An example of combined geological and geotechnical investigations for a landslide in seismic area.

Cascini, E., Cascini, L. & Gulla, G. A back-analysis based on piezometer's response.

Enegren, E.G. & Moore, D.P. Branham Ridge Slide - twenty-two years of slope monitoring.

Gillon, M.D., Foster, P.F., Proffitt, G.T. & Smits, A.P. Monitoring of the Cromwell Gorge landslides.

Hiura, H., Sassa, K. & Fukuoka, H. Monitoring system of a crystalline schist landslide.

Lollino, G. Automated inclinometric system.

Moller, B. & Ahnberg, H. Stability analysis and observations of a failure test slope.

Proffitt, G.T., Fairless, G.J., Grocott, G.G. & Manning, P.A. Computer applications for landslide studies.

Tran-Duc, P.O., Ohno, M. & Mawatari, Y. An automated landslide monitoring system.

Landslides, Bell (ed.) © 1995 Balkema, Rotterdam, ISBN 90 5410 032 X

Mécanisme des glissements de terrains argileux – Bilan de surveillance sur plusieurs années

C. Azimi, J. Biarez & P. Desvarreux
Association pour le Développement des Recherches sur les Glissements de Terrains, Gières, France

Y. Giuliani & C. Ricard
Compagnie Nationale du Rhône, Lyon, France

ABSTRACT : The authors describe the monitoring of the Leaz landslide which develops in silty clays in the french Alps. The continuous displacement and piezometric measurements since 1977 allowed the establishment of relations between rainfall and water level, and between water level and ground movements. The mechanism is that of the Bingham's solid with a critical piezometric level.
Such relations are extended to other landslides in silty clays and seem to be representative for a certain type of landslide. Conclusions are drawn about the way to realize a good monitoring of such landslides.

1. INTRODUCTION

La plupart des glissements de terrains naturels (particulièrement ceux affectant les matériaux argileux) sont des phénomènes géomécaniques qui évoluent dans le temps avec des phases d'arrêt, des phases à vitesse moyenne faible, des phases d'accélérations dont certaines peuvent mener à des catastrophes.

La surveillance de tels glissements lents a pour objectif de répondre aux questions suivantes :
- le glissement peut-il accélérer et passer à une phase catastrophique, et ceci dans quelles conditions
- ces conditions peuvent elles se produire ?
- connaissant les mécanismes du glissement, comment en déduire les moyens de stabilisation les plus efficaces.

L'objet de l'article est de réaliser le bilan de plusieurs surveillances de glissements en matériaux argileux, parmi lesquelles figure le site de Leaz et de montrer comment utiliser les mesures continues de pluviométrie, de piézométrie et de déplacements pour déterminer les mécanismes et élaborer un critère de danger.

2. CAS DU GLISSEMENT DE LEAZ

2.1. Description du glissement

Ce glissement de terrain naturel domine la retenue de Genissiat sur le Rhône. Il était connu dès 1934 avant établissement de la retenue, mais n'avait pas fait l'objet d'études particulières. A partir de 1964, suite à une réactivation des mouvements, une surveillance a été progressivement mise en place, parallèlement avec des reconnaissances géologiques classiques. Le but de ces études était de préciser le volume en mouvement, le rôle éventuel de la

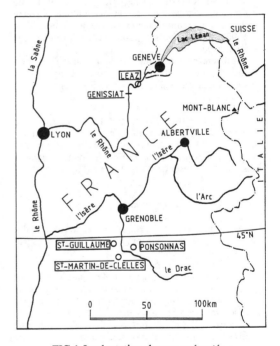

FIG.1 Implantation des cas présentés.

retenue dans ces mouvements, de prévoir dans quelles conditions des masses importantes de matériaux pourraient arriver dans la retenue et à quelles vitesses (ceci dans le but d'apprécier les conséquences du phénomène et en particulier les caractéristiques de l'onde engendrée).

La synthèse des reconnaissances géologiques est

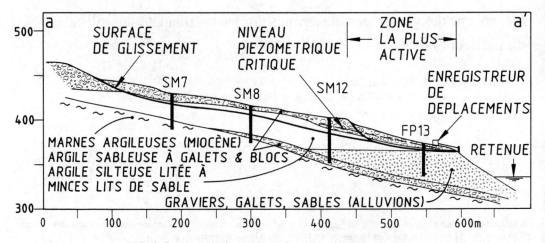

FIG. 2 Coupe géologique schématique du glissement de Léaz.

présentée en Fig.2 et on peut retenir les points suivants :
- le substratum de marnes situé à 30-40 m de profondeur est stable, de même qu'une terrasse d'alluvions anciennes (aucun mouvement entre 1964 et 1990). La retenue n'a donc aucune influence sur le glissement.
- le glissement s'effectue au sein d'une formation d'argiles litées d'origine glacio-lacustre, très répandue dans la région et caractérisée par :
Wl = 35 - 50%
Ip = 20 - 30%
$\emptyset$' = 19,5° c' = 0 (cisaillement à 1 μ/mm)
- le volume total des matériaux en mouvement est de $1,4.10^6$ m³, caractérisés en 1969 par des vitesses de 5 - 10 cm/an et de 10-30 cm/an dans la zone la plus active représentant 100 000 m³ (fig. 3).

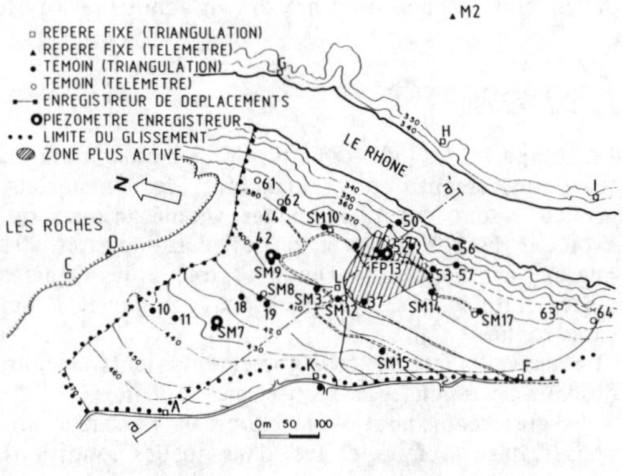

2.2. Principe de la surveillance

Celle ci comporte des mesures de déplacements, des mesures de niveaux piézométriques et des mesures de pluviométrie locale.
Les mesures de déplacements ont commencé en 1964 avec la mise en place de 44 témoins mesurés en triangulation et par alignements. Les mesures, de fréquence mensuelle à trimestrielle, ont permis de préciser les vitesses moyennes et l'existence de la zone plus active. Elles ont permis de montrer que l'activité du glissement était saisonnière mais pas de préciser le mécanisme exact.
C'est pourquoi depuis 1974 on a adopté le système de surveillance suivant (fig. 3).
- mesures en triangulation sur 25 témoins à partir des bornes fixes A, C... M2. La fréquence est annuelle et permet de vérifier l'activité moyenne des diverses zones en mouvement,
- mesures au distancemètre entre le point fixe M2 et 11 témoins du glissement, avec une fréquence mensuelle qui peut être resserrée en cas d'accélération des mouvements. Le but est de

FIG. 3 Plan du réseau de surveillance.

vérifier qu'aucune zone du glissement ne dépasse en vitesse la zone la plus active,
- enregistrement en continu des déplacements d'un point de la zone la plus active avec, depuis 1978, transmission automatique des mesures à la centrale de Génissiat par ligne téléphonique. Le système enregistreur est constitué d'un fil invar de 30 m de long tendu entre une pilier fixe en béton et l'enregistreur proprement dit. Le mouvement de translation de l'enregistreur est transformé en mouvement de rotation, lui même transmis à un codeur numérique angulaire. Le pas de mesure est de 1 mm. Le fil invar est protégé par un tube métallique de 800 mm de diamètre posé sur le terrain.
Les mesures piézométriques qui sont conservées actuellement sont effectuées mensuellement dans 8 piézomètres et 4 cellules de pression interstitielle. De plus dans 4 piézomètres on dispose depuis 1987 d'enregistrements en continu du niveau d'eau.
Comme on avait, entre 1977 et 1987, réalisé des mesures 3 fois par semaine dans ces 4 piézomètres,

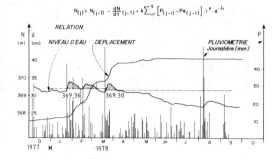

$$N_{(j)} = N_{(j-1)} - \frac{dN}{dt}(j-1) + k \sum_{i=1}^{5} \left[P_{(j-i)} - Pe_{(j-i)} \right] \cdot i^P \cdot e^{-\lambda i}$$

FIG. 4 Corrélation pluviométrie journalière P (mm), niveau piézométrique N (m) au FP 13 et déplacements à l'enregistreur.

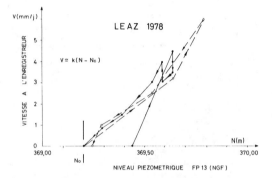

FIG. 5 Corrélation / vitesse instantanée niveau piézométrique au FP 13.

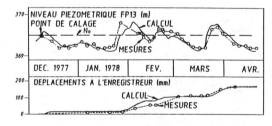

FIG. 6 Comparaison des niveaux piézométriques et des déplacements calculés avec les mesures en 1977-78.

on peut considérer qu'on dispose de mesures piézométriques continues depuis 1987.

La pluviométrie a été enregistrée sur le site de 1970 à 1978, puis à partir de 1988.

2.3. Principaux résultats - Modèles de comportement.

Sur la fig. 4 on a représenté un exemple de 10 mois d'enregistrements. De ces données on a pu tirer une modèle incluant

- d'une part la réponse hydraulique du sol à une précipitation donnée,
- d'autre part la réponse cinématique du glissement à un niveau piézométrique donné.

Pour déterminer le modèle hydraulique, on a cherché à déterminer quelle était la part d'alimentation en eau par la surface du glissement lui-même, par la surface du bassin versant hydrographique et éventuellement par d'autres bassins versants plus lointains. Dans le cas présent, la réponse piézométrique est très rapide et on a admis que l'alimentation locale était prépondérante. Le modèle hydraulique formalisé sur la fig.4 prend donc en compte une influence de la pluviométrie limitée à quelques jours, un tarissement naturel avec un niveau de base constant et l'évapotranspiration. Le tarissement naturel a été approximé par une fonction du type

$$\frac{dN}{dt} = a_0 + a_1 (N-N_1) + a_2 (N-N_1)^2 + a_3 (N-N_1)^3,$$

où N1 est le niveau de base. Le calage est effectué sur les couches piézométriques réelles.

Le modèle mécanique prend en compte les observations fondamentales suivantes qu'on peut faire sur les fig 4 et 5.
- les périodes d'activé sont réduites dans le temps alors qu'on a plusieurs mois par an d'arrêt des mouvements
- les mouvements ont lieu lorsque le niveau piézométrique dépasse un seuil critique No au sondage FP13. On a vérifié par la suite que ce niveau pouvait varier légèrement autour de 369.20-369.40 NGF,
- au cours d'une phase d'activité la vitesse instantanée des mouvements est :

$V = k (N-No)$ si $N > No$
$V = 0$ si $N < No$

Ceci montre que, dans le domaine des niveaux piézométriques mesurés, le matériau argileux se comporte comme le corps viscoplastique de Bingham avec seuil.

2.4. Conséquences pour l'exploitation de la retenue.

Les 2 modèles ci dessus ont été déterminés en 1978. On a d'abord constaté qu'à partir de la pluviométrie journalière locale on pouvait reconstituer avec une bonne approximation le niveau piézométrique au FP 13, puis, par intégration, les déplacements à l'enregistreur. Ceci est illustré sur la fig. 6 où le calage est simplement effectué au départ sur le niveau piézométrique au FP 13 en date du 9.12.1977.

On a ensuite effectué des calculs de simulation en utilisant des séquences pluviométriques enregistrées à Génissiat (8 km de distance) entre 1965 et1979. On a constaté que, si le comportement du glissement restait identique, les vitesses passaient par un maximum de 10 mm/jour puis diminuaient. Il n'y

1905

aurait donc pas de risque d'atteindre de très grandes vitesses tant que la vitesse resterait liée au niveau piézométrique par les lois établies (comportement considéré "normal").

C'est pourquoi la consigne de sécurité prend comme référence la vitesse de 10 mm/j à partir de laquelle le niveau de la retenue sera abaissé préventivement. Cette consigne a été appliquée une seule fois en mai 1983 lors d'une période active où les vitesses ont atteint 10 mm/j durant 2 jours. Lorsque la vitesse a diminuée la consigne a été levée.

On indique en fig. 7 les déplacements en 1982-83 et on note deux particularités :

- au cours des phases d'activité successives et rapprochées, le niveau critique No à tendance à diminuer,
- entre le 15 mars et le 17 juin 1983, le niveau piézométrique est resté relativement constant à 10 cm plus phaut que le niveau critique. Durant cette période, la vitesse a augmenté, ce qui constitue déjà un comportement "anormal".

Ces deux remarques suggerent que la résistance du matériau peut diminer lorsque les mouvements ont une durée importante.

Le détail sur les études hydrauliques et la détermination des effets de l'onde sont indiqués dans l'article de Selmi et Fruchart (1990).

2.5. Critique de ces modèles

Le modèle hydraulique prend en compte l'évapotranspiration de manière simplifiée. D'autre part il ne prend pas en compte le stockage des précipitations sous forme de neige. Néanmoins, dans le cas de Léaz, il nous a paru acceptable.

Le modèle mécanique est simplifié car il prend en compte le niveau d'un seul piézomètre situé en partie basse de la zone active (fig. 2). Pour affiner le modèle on a pris en compte un niveau piézométrique déterminé à partir des enregistrements dans les 2 piézomètres SM 12 et FP 13.

On a calculé, pour la zone la plus active, le coefficient de sécurité F par la méthode des perturbations et on a examiné comment il était corrélé avec les vitesses. Sur la figure 8 les résultats sont donnés pour 3 périodes : 1977-78, 1982-83 1990. On peut faire plusieurs remarques :

- pour $F > 1,015$ les vitesses sont pratiquement nulles,
- pour des diminutions du coefficient de sécurité de 2 à 3 % au dessous d'un seuil correspondant à $F \sim 1$ les vitesses augmentent de 1 à 20 et passent de 0,5 à 10 mm/j,
- on est actuellement dans l'ignorance, faute d'essais de fluage spéciaux, sur le comportement du matériau à des coefficients de sécurité $F < 0,97$,
- pour un même coefficient de sécurité, la réponse du glissement n'est pas univoque. Si on prend en compte les déplacements (non enregistrés) de la partie haute de la zone active, on constate qu'en 1982-83 ils ont été plus importants que ceux de

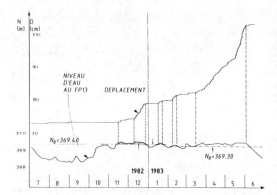

FIG. 7. Enregistrements du niveau piézométrique au FP 13 et des déplacements en 1982-83.

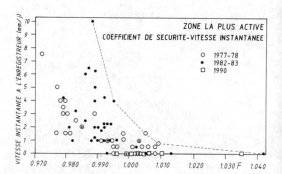

FIG. 8. Corrélations vitesse instantanée / coefficient de sécurité F de la zone la plus active.

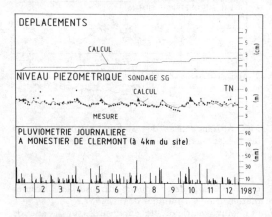

FIG. 9. Glissement de St Guillaume. Application de modèles hydraulique et mécanique de type Leaz.

l'enregistreur. Au contraire en 1977-78 ils ont été équivalents à ceux de l'enregistreur.

Ceci montre que la réponse cinématique au niveau de l'enregistreur est non seulement fonction du

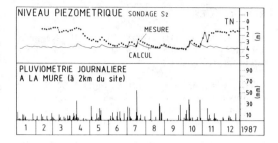

FIG. 10. Glissement de Ponsonnas. Application du modèle hydraulique de type Léaz.

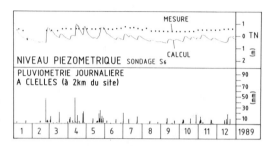

FIG. 11. Glissement de St Martin de Clelles. Application du modèle hydraulique de type Léaz.

Tableau 1. Principales caractéristiques des glissements étudiés.

		ST GUILLAUME	PONSONNAS	ST MARTIN DE CLELLES	LEAZ
ARGILES	W_L%	35-45	32-38	37-45	35-50
	I_p%	19-27	13-21	19-26	20-30
	$\phi'_r °$		18-21		19.5
PENTE (°)		8-12	8-18	8-16	12
PROFONDEUR DU GLISSEMENT		20m	17-45m	20-40m	12m
VOLUMES (m³)		5×10⁶	10×10⁶	5×10⁶	1.4×10⁶
VITESSES MOYENNES (cm/an)		2-3	1-2	1-2	1970: 5-20 1990: 1

niveau piézométrique mais aussi des poussées exercées par la partie amont, car le glissement ne se déplace pas en bloc. Le modèle mécanique devrait donc pouvoir intégrer cette redistribution des masses au cours du glissement.

3. AUTRES CAS DE GLISSEMENTS SURVEILLES.

3.1. Description

Ces 3 cas sont situés dans le Trièves, au Sud de Grenoble et intéressent une formation d'argiles litées, déposées dans un lac à une époque interglaciaire. Les caractéristiques de ces glissements sont indiquées au tableau 1 en comparaison avec celles du glissement de Léaz.

On peut ajouter que sur ces glissements ou à proximité immédiate se trouvent des villages habités. Ces 3 glissements ont fait l'objet entre 1985 et 1990 d'une surveillance. Cette dernière à comporté, après une reconnaissance géologique, des mesures piézométriques hebdomadaires, des mesures inclinométriques trimestrielles (à Ponsonnas), des mesures topographiques semestrielles sur 5 à 10 repères.

3.2. Résultats

Etant donné qu'il n'y a pas d'enregistrements en continu des déplacements, on ne peut établir de corrélation vitesse-niveau piézométrique. Au contraire, comme on dispose de la pluviométrie journalière à proximité on peut comparer les niveaux piézométriques calculés à l'aide d'un modèle hydraulique analogue à celui de Leaz avec les niveaux mesurés. Les résultats sont donnés aux fig. 9, 10 et 11.

Dans le cas de St Guillaume (fig.9) le modèle donne de bons résultats. Il permet même, moyennant une estimation du niveau critique (1.5 m de profondeur), de reconstituer les ordres de grandeur des déplacements annuels.

Dans le cas de Ponsonnas (fig. 10) on ne peut reconstituer les variations réelles de niveau piézométrique par ce type de modèle. Ceci montre qu'il existe, en plus de l'alimentation directe par la surface, une alimentation latérale provoquant une variation du niveau de base, de période annuelle et d'amplitude 2 m dans le cas présent.

Dans le cas de St Martin de Clelles (fig. 9) on ne peut pas reconstituer le niveau piézométrique à partir de la pluviométrie journalière. Ceci confirme l'existence d'une alimentation profonde par le substratum rocheux sous jacent.

Il est évident que dans les 2 derniers cas l'établissement du modèle hydraulique est plus délicat et demande un temps d'observation assez long pour intégrer les composantes de variation de niveau d'eau de longue période.

4. CONCLUSIONS

La mise en place d'une surveillance de mouvement

de terrain lent en matériaux argileux doit être associée à une étude géotechnique destinée à préciser la nature des matériaux, la position du substratum stable, celle de la surface de glissement et les caractéristiques générales du mouvement. On pourra ainsi fixer la position des points d'enregistrement pour qu'ils soient significatifs. En effet les mesures discontinues présentent un intérêt limité et toute surveillance devrait comporter au moins 1 point d'enregistrement quasi continu de la pluviométrie, du niveau piézométrique et des déplacements.

On cherchera ensuite à établir un modèle hydraulique prenant en compte les caractéristiques locales, et l'origine des alimentations en eau. Pour cela il faut disposer de mesures sur une période comportant au moins 1 étiage et 1 niveau de hautes eaux. L'établissement du modèle mécanique nécessitera des enregistrements de déplacements en surface ou en profondeur. On cherchera s'il existe un seuil piézométrique critique et comment se font les déplacements au dessus de ce seuil.

L'ensemble des modèles hydraulique et mécanique permettra de réaliser des simulations de comportement au cours d'épisodes de précipitations importantes. Ils permettront également de préciser les types de travaux de stabilisation appropriés. Dans des cas comme Leaz ou St Guillaume, un réseau de drainage superficiel peut-être très efficace car il intercepte les eaux avant qu'elles ne s'infiltrent plus en profondeur. Le réseau de drainage superficiel de Leaz, réalisé en 1969 et complété en 1987, a permis de ramener l'ensemble des mouvements à 1 cm/an depuis 1988.

Enfin l'ensemble de ces 2 modèles permettra de définir un comportement "normal" du glissement. Le critère de danger pourra être défini comme la réalisation d'une des conditions suivantes :
- des vitesses supérieures à ce qui a pu être déterminé par des simulations reposant sur des données pluviométriques réelles antérieures,
- une modification défavorable dans la loi de comportement sans explication valable (par exemple une accélération des vitesses pour un niveau piézométrique constant).

Nous remercions :

- la Compagnie Nationale du Rhône pour la réalisation de l'ensemble des mesures sur le glissement de LEAZ depuis 1964,

- le Service de Restauration des Terrains en Montagne de l'Isère pour les mesures sur les sites du Trièves,

- le Pôle Grenoblois d'Etudes et de Recherches pour la Prévention des Risques Naturels pour la subvention à la recherche sur les lois de comportement des glissements en terrains argileux.

REFERENCES.

Azimi C. et P. Desvarreux 1972-1991. Rapports annuels de surveillance du glissement de Leaz (archives A.D.R.G.T.).

Comité Français des Grands Barrages 1982. Etudes et travaux réalisés en France en raison de l'instabilité du versants de retenue - 14e Congrès des Grands Barrages, Rio de Janeiro, 563-589.

Selmi J. et F. Fruchart 1990 - Etude sur modèle réduit du glissement de Leaz dans la retenue de Génissiat. La Houille Blanche 1-1990 : 61-71.

Landslides, Bell (ed.) © 1995 Balkema, Rotterdam, ISBN 90 5410 032 X

Some aspects of geodetic monitoring and deformation analysis of landslides Demonstrated by the example of the landslide 'Wißberg'

O. Heunecke
Geodetic Institute, University of Hanover, Germany

ABSTRACT: The geodetic monitoring and deformation analysis of landslides is a center of research in Hanover since years. The used geodetic methods of monitoring, quasi–static and kinematic deformation analysis are shown. A brief overview of the dynamic modelling is given. In all this three kinds of analysis the technique of KALMAN–filtering can be applied. Results of the quasi–static analysis occured at the landslide "Wißberg" near Mainz in Germany are presented.

RÉSUMÉ: La surveillance et l'analyse de déformation des glissements de terrain sont depuis des années un des points de recherche en Hannovre. Nous présentons les mèthodes utilisées en surveillance et en analyse de déformation quasi-statique et cinématique. Une première ébauche d'analyse dynamique est donnée. La technique des filtres KALMAN peut être appliquée à ces trois possibilités d'analyse. Nous présentons les résultats de l'analyse quasi–statique sur le le glissement de terrain "Wißberg" près des Mayence en Allemagne.

1 Introduction

The worldwide damages caused by mass movements may be placed at third position after other natural hazards such as inundations and earthquakes (KRAUTER 1986). Minimizing the damages to human lives and properties, the investigations of the possible causes and the potential areas of landslides with the help of engineering geodesy and engineering geology is an important interdisciplinary task.

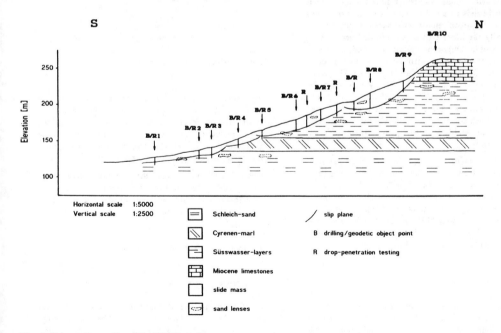

Fig. 1: Schematic profile of the "Wißberg"

Such an interdisciplinary approach exists between the Geologic Institute, University of Mainz (Prof. Dr. KRAUTER) and the Geodetic Institute, University of Hanover (Prof. Dr. Dr. PELZER).

Landslides occurred at the common research area "Wißberg", located in the Tertiary basin of Mainz ca. 8% which has been affected by mass movements, where studied in detail applying geodetic methods from Hanover. These mass movements have taken place on relative low angle slopes where the mean annual rainfall with about 600 mm/a is quite low.

One main interest of research in Hanover is the development of kinematic and dynamic analysis models. Quasi–static comparisions of epochs are already in practical operation since years. In all three kinds of analysis the technique of KALMAN–filtering can be applied and provides several advantages compared with other approaches of analysis.

The description of the geological situation and applied engineering geological methods from Mainz are treated very briefly only. Mainly some engineering geodetic ideas and points of view are discussed.

2 Geology

The development of the Tertiary Mainz basin is closely correlated to the formation of the Upper-Rhine Graben. Both were formed when the European crust collided with the Austro-alpine-adriatic crust. The earliest detectable sedimentation of the Mainz basin took place at the beginning of the Tertiary.

Different transgressions from the north and south deposited marine, limnic and brackish sediments, which today still exists as an interbedding of limstones and clayish to calcareous marls with sandy zones. Stratigraphically the lower part of the studied profile consists of Middle Oligocene "Schleich-"sand, a succession of clay and marl with lenses of fine grained sands. This horizon is overlain by Upper Oligocene "Cyrenen"-marl and "Süßwasser"-layers (marls deposited in fresh water). The top of the profile is underlain by Miocene Limestones, which are the most competent lithological unit of this studied area. These limestones are highly fractured providing paths for infiltrating water.

The inclination of the slope in the Oligicene sediments varies from $10 - 15^0$, whereas that in the limestone units goes up to 40^0. The landslides appear to have occurred on either material of earlier mass movements or through a reactivation of fossile slip planes or by a combination of both. These earlier mass movements, which are responsible for the present morphology, probably took place in Pleistocene.

3 Engineering geological examinations

Engineering geological examinations started with an airphoto-analysis in order to localize the extensions of the different mass movements. In addition to this drillings, soundings and seismic-refraction surveys

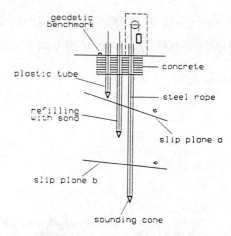

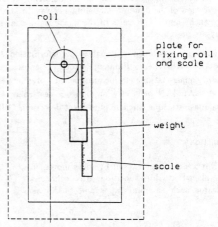

⇨: direction of movement

Fig. 2: Princip of extensometer points

were carried out to determine depth, inclination and nature of the slip planes. The main features of an extensometer developed in Mainz are shown in the following figure. The evaluation of the extensometer measurements allows the determination of displacements along one or more slip planes below the point.

Meterological data, especially rainfall, precipitation and potential evaporation were recorded to investigate the relationship between the studied landslides and these external factors.

4 Engineering geological implications

Geology of the area plays a considerable role in the occurrence of mass movements. The highly fractured Miocene limestones provide an exellent fractured-aquifer which is widely used for the domestic water supply. The rain water which is infiltrating through these fractures reaches the contact between the limstone and the nearly impervious Oligocene sediments.

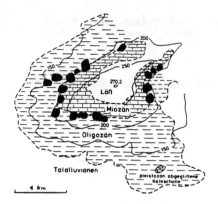

Fig. 3: Slide areas at the "Wißberg"

		$\overline{x}$	s
liquid limit	[%]	42.8	12.2
plastic limit	[%]	19.0	2.8
plasticity	Ip	23.6	9.5
consistency	Ic	0.7	0.25
density	[g/cm³]	2.0	0.1
dry density	[g/cm³]	1.6	0.1
angle of internal friction	[°]	19.2	3.0
cohesion	[KN/m²]	7.4	1.7

Fig. 4: Geotechnical parameters of Oligocene sediments

A considerable part of water which reaches the contact seeps along it and springs out where most of the landslides have taken place.

It could be argued that a part of the infiltrated water might have migrated through the water routings of the Tertiary sediments into the fine-grained sand lenses, which are more porous and permeable. During the drilling for sampling it was observed that the water within these sand lenses has risen to a considerable height revealing that water has developed a high pore water pressure. Some of the important geotechnical parameters of the examined sediments are listed in fig. 4, where $\bar{x}$ is the mean value and s the standard deviation.

It appears that the above described percolation processes of rain water were responsible for reducing the already lowered shear strenght of the material. The stability analysis of slopes, underlain by Oligocene sediments, shows low values of angle of internal friction ($\varphi_r = 8^0$) and low cohesion ($c_r = 0 \, KN/m^2$), revealing the instability of the area.

5 Monitoring network

The monitoring network "Wißberg" consists of three reference points outside of the unstable slide area and 13 object points placed in a profile down the slope. The stable points give the frame for the comparison of different epochs, which allows the computation of absolute moving rates refering to this frame. 10 of the object points were already installed as pillars in 1981 at the beginning of the geodetic monitoring (no. 1 - 10 in fig. 1). Three more were installed in spring 1991 in order to monitor the behaviour of critical and most interesting zones of the slide. Two of the new points are extensometers (see fig. 2) and one is a inclinometer. One of the extensometers is located between point no. 9 and no. 10, the other and the inclinometer between no. 7 and no. 8 (see fig. 1).

The length of the profile is nearly 1 km, the average intervall between the older points is about 100 m with the mentioned densification by the new points in the critical upper part of the landslide. After 5 geodetic campaigns in 1981 and 1982 the monitoring was interrupted until September 1988. With the start of a new research project financed by the German Research Community (DFG) in 1989, the network is observed 4 times a year since then. At the moment there are 15 epochs available.

Mainly the electronic tacheometer WILD T3000/ DI2000 is used in connection with the note book GRE4. Some of the longer distances are also measured by the range-finder WILD DI3000, some of the shorter lines also by 2m subtense bar measurements. Some points are connect by precise levelling. Typical accuracy characteristics of the geodetic measurements are:

$$\sigma_h \sim \quad 0.2 \text{ mm}$$
$$\sigma_r \sim \quad 1 \text{ mm} + 1 \text{ ppm for DI2000}$$
$$\sigma_r \sim \quad 1 \text{ mm} + 2 \text{ ppm for DI3000}$$
$$\sigma_r \sim 2 \text{ mm for subtense bar measurements}$$
$$\sigma_d \sim \quad 0.3 - 0.5 \text{ mgon weather depending}$$
$$\sigma_v \sim \quad 1.0 - 2.0 \text{ mgon weather depending}$$

The index h (height) stands for precise levelling, r (range) for the distance measurements, d (direction) for horizontal direction and v (vertical direction) for the zenith angle. The accuracy in vertical direction is distinctly lower than in horizontal direction because of the refraction. A calibration of the whole equipment, especially the range finders, the subtense bar and the meteorological instruments for athmospheric correction of the distances is self-evident before starting a campaign. For following computations the observations and their accuracies are collected in a vector l with a corresponding covariance matrix Σ_{ll} .

$$l \quad : \quad \text{vector of observations}$$
$$\Sigma_{ll} \quad : \quad \text{corresponding covariance matrix}$$

All observations can be combined to local three-dimensional coordinates by least square adjustment.

The well kown algorithmn is programed in the program system HANNA (HANover Network Adjustment) of the Geodetic Institute in Hanover. An automatic dataflow from the regristration in the field to this system is realized, only the results from precise levelling and subtense bar measurements must be supplemented manual. The resulting standard deviations

1911

(68% probability level) from the free net adjustment for the coordinates are:

standard deviation in horizontal position
$$\sigma_x = \sigma_y \sim 0,5 - 1,5 \, mm$$

standard deviation in vertical position
$$\sigma_z \sim 1 - 4 \, mm$$

The adjusted coordinates are gathered in a vector $\hat{x}$ and its covariance matrix $\Sigma_{\hat{x}\hat{x}}$.

$\hat{x}$: vector of coordinates

$\Sigma_{\hat{x}\hat{x}}$: corresponding covariance matrix

The accuracy in elevation is lower because of σ_v and the fact that only a few points are connected by levelling. Levelling is a very time consuming procedure, especially in landslide areas, so that a complete levelling loop normally is not possible. One campaign with 8 observed stations takes about three working days.

6 Quasi–static deformation analysis

Quasi–static deformation analysis in geodetic sense is the detection of point movements from two corresponding epochs. The main problem is to seperate the "real" displacements (signals) from inaccuracies of geodetic observations (measurement noise) and other residual inaccuracies (system noise), like unavoidable errors in centering or deformations caused by thermal effects only (KRAUTER 1988). Normally the results are given on 95% probability level. For this purpose a lot of different approaches are published in geodetic literature. The variety by KALMAN–filtering developed in Hannover (PELZER 1986, 1988) shall be explained. The formular set is realized in the system HANNA. Beside other aspects the main advantage of this algorithmn is its sequential basic structure.

Let assume that there is an old set of coordinates with covariance information for the epoch t_k:

x_k : given coordinates from last epoch

$\Sigma_{xx,k}$: given covariance matrix from last epoch

$s_{0,k}^2$: variance factor of the old state of network

In a new campaign at time t_{k+1} one gets:

l_{k+1} : new geodetic observations

$\Sigma_{ll,k+1}$: corresponding covariance matrix

n_{k+1} : number of new observations

The problem now is, how do these new observations fit to the old state of network, represented by x_k and $\Sigma_{xx,k}$. Basic idea of the modified KALMAN–filter algorithmen by PELZER is to take pseudo observations for the point movements

$$\Delta x = 0 \quad : \text{functional part}$$

with a corresponding covariance matrix

$$\Sigma_{ss,k} = \begin{vmatrix} \Sigma_{\xi\xi,s} & \\ & \Sigma_{\xi\xi,o} \end{vmatrix} \quad : \text{stochastic part}$$

The matrix of system noise $\Sigma_{ss,k}$ has only elements on the main diagonal. The values of $\sigma_{\xi,s}$ and $\sigma_{\xi,o}$ shall

represent the expected movements of the points. Index s is for stable points and o for object points. The old coordinates x_k are extrapoled identical to $\bar{x}_{k+1}$.

$$\bar{x}_{k+1} = x_k + \Delta x = x_k$$

The law of error propagation give the uncertainty of $\bar{x}_{k+1}$.

$$\Sigma_{\bar{x}\bar{x},k+1} = \Sigma_{xx,k} + \Sigma_{ss,k}$$

This thoughts lead to a linearized GAUSS-MARKOV model:

functional model

$$\begin{vmatrix} \bar{x}_{k+1} \\ l_{k+1} \end{vmatrix} = \begin{vmatrix} I \\ A_{\bar{x}} \end{vmatrix} \hat{x}_{k+1} + \begin{vmatrix} v_{\bar{x},k+1} \\ v_{l,k+1} \end{vmatrix}$$

stochastic model

$$\Sigma_{k+1} = \begin{vmatrix} \Sigma_{\bar{x}\bar{x},k+1} & \\ & \Sigma_{ll,k+1} \end{vmatrix}$$

$v_{\bar{x},k+1}$ and $v_{l,k+1}$ are the observation residuals because of the over determination, which means more observations than unkown parameters. With the formula

$$\bar{l}_{k+1} = \varphi(\bar{x}_{k+1})$$

one can computate now predicted observations $\bar{l}_{k+1}$, which can be compared with the real observations l_{k+1}. The difference is called innovation d_{k+1}.

$$d_{k+1} = l_{k+1} - \bar{l}_{k+1}$$

Because of law of error propagation again the belonging covariance matrix of d_{k+1} is

$$D_{k+1} = \Sigma_{ll,k+1} + A_{\bar{x}}\Sigma_{\bar{x}\bar{x},k+1}A_{\bar{x}}^T,$$

where

$$A_{\bar{x}} = \left(\frac{\partial\varphi(\bar{x})}{\partial\bar{x}} \right)_{k+1}$$

is the JACOBI matrix. Without derivate the details, the network can be updated with:

$$\hat{x}_{k+1} = \bar{x}_{k+1} + K_{k+1}d_{k+1}$$

$$\Sigma_{\hat{x}\hat{x},k+1} = \Sigma_{\bar{x}\bar{x},k+1} - K_{k+1}D_{k+1}K_{k+1}^T,$$

where K_{k+1} is the important gain matrix. The variance factor of this approach is

$$s_{0,k+1}^2 = \frac{d_{k+1}^T D_{k+1}^{-1} d_{k+1}}{n_{k+1}}.$$

This updating is only allowed if there is no significant difference in d_{k+1}, which lead to the hypothesis

$$E\{d_{k+1}\} = 0.$$

($E :=$ expected value in stochastic sense) This hypothesis had to be tested, for instance by FISHER test. For further details PELZER is recommended. The network "Wißberg" is evaluated by this algorithmn, some results shall be presented.

In Figure 5 the cumulated displacements for 6 points over the last 10 years are shown. The points no. 1 - 3 and no. 10 (see fig. 1) are nearly stable, the three new points are not regarded here. The x–direction is the direction of slope gradient, z–direction is elevation. As

point no.	Δx [m]	Δy [m]	Δz [m]
9	1.3171	0.0022	-0.4394
8	3.0820	-0.1796	-0.4927
7	1.7015	-0.2320	-0.1756
6	1.8530	-0.2724	-0.2275
5	0.9747	-0.2354	-0.1517
4	0.2282	0.0551	-0.0180

Fig. 5: Cumulated displacements since 1981

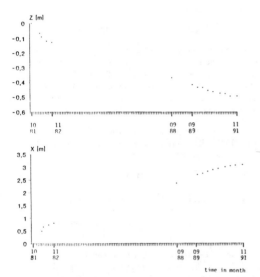

time in month

Fig. 6: Movements of object point no. 8

a representative example for the motion versus time, the movements of point no. 8 are shown in a graphical way in figure 6.

The point moved more than 3 m down the slope and subsided about 0,5 m. One can see that the rate of motion is decreasing since 1981, but movements still going on during the summertimes. The angle

$$arctan\left(\frac{\Delta z}{\Delta x}\right) \cong 16\%$$

is a reference to the inclination of the slip plane(s) on which the point is sliding.

7 Kinematic deformation analysis

The kinematic approach in geodetic sense means to regarde movements as a function of time t. In this case the state of a network can be described for instance by coordinates, velocities and accelerations.

$$x_{k+1} = x_k + \Delta x = x_k + \psi(x, \dot{x}, \ddot{x}, t)$$

Kinematic models can be used if the motion of the points is relative continual, e. g. creep or sliding. The unknown parameters of this approach are

$$y = \begin{vmatrix} x \\ \dot{x} \\ \ddot{x} \end{vmatrix} \begin{matrix} : & \text{coordinates} \\ : & \text{velocities} \\ : & \text{accelerations} \end{matrix}$$

y is called the state vector, its corresponding covariance matrix is Σ_{yy} .

$$\Sigma_{yy} = \begin{vmatrix} \Sigma_{xx} & \Sigma_{x\dot{x}} & \Sigma_{x\ddot{x}} \\ \Sigma_{\dot{x}x} & \Sigma_{\dot{x}\dot{x}} & \Sigma_{\dot{x}\ddot{x}} \\ \Sigma_{\ddot{x}x} & \Sigma_{\ddot{x}\dot{x}} & \Sigma_{\ddot{x}\ddot{x}} \end{vmatrix}$$

This matrix must be filled up because of the correlations of x, $\dot{x}$ and $\ddot{x}$. If y_k is valid for time t_k, a prediction to t_{k+1} is possible with the transition matrix T.

$$\bar{y}_{k+1} = T \quad y_k$$

$$\begin{vmatrix} \bar{x}_{k+1} \\ \bar{\dot{x}}_{k+1} \\ \bar{\ddot{x}}_{k+1} \end{vmatrix} = \begin{vmatrix} I & I\Delta t & \frac{1}{2}I\Delta t^2 \\ & I & I\Delta t \\ & & I \end{vmatrix} \begin{vmatrix} x_k \\ \dot{x}_k \\ \ddot{x}_k \end{vmatrix}$$

The predicted state vector may be influenced by some disturbances, which can be modeled by an extension of the transition equation:

$$\bar{y}_{k+1} = Ty_k + Sa$$

The expected value for a is

$$E\{a\} = 0,$$

but its covariance matrix Σ_{ss}, whose diagonal elements have to be estimated, lead to the important extension of the stochastic prediction

$$\Sigma_{\bar{y}\bar{y},k+1} = T\Sigma_{yy,k}T^T + S\Sigma_{ss}S^T .$$

Let consider now, that there are n_{k+1} new geodetic observations available at time t_{k+1} . The combination of prediction and new information give formal a GAUSS–MARKOV approach again like already used in section 6:

functional model

$$\begin{vmatrix} \bar{y}_{k+1} \\ l_{k+1} \end{vmatrix} = \begin{vmatrix} I \\ A_{\bar{y}} \end{vmatrix} \hat{y}_{k+1} - \begin{vmatrix} v_{y,k+1} \\ v_{l,k+1} \end{vmatrix}$$

stochastic model

$$\Sigma_{k+1} = \begin{vmatrix} \Sigma_{\bar{y}\bar{y},k+1} & \\ & \Sigma_{ll,k+1} \end{vmatrix}$$

With

$$A_{\bar{y}} = |A_{\bar{x}} \quad 0 \quad 0| \;, \; A_{\bar{x}} = \left(\frac{\partial\varphi(\bar{x})}{\partial\bar{x}}\right)_{k+1}$$

The KALMAN–filter formulars applied one gets:

- innovation -

$$d_{k+1} = l_{k+1} - \bar{l}_{k+1}$$

- covariance matrix of d_{k+1} -

$$D_{k+1} = \Sigma_{ll,k+1} + A_{\bar{y}}\Sigma_{\bar{y}\bar{y},k+1}A_{\bar{y}}^T$$

- gain matrix -

$$K_{k+1} = \Sigma_{\bar{y}\bar{y},k+1} A_{\bar{y}}^T D_{k+1}^{-1}$$

- updated state vector -

$$\hat{y}_{k+1} = \bar{y}_{k+1} + K_{k+1} d_{k+1}$$

- covariance matrix of $\hat{y}_{k+1}$ -

$$\Sigma_{\hat{y}\hat{y},k+1} = \Sigma_{\bar{y}\bar{y},k+1} - K_{k+1} D_{k+1} K_{k+1}^T$$

- variance factor of approach -

$$s_{0,k+1}^2 = \frac{d_{k+1}^T D_{k+1}^{-1} d_{k+1}}{n_{k+1}}$$

As already mentioned in section 6 the difference between predicted and real state can be regarded and tested. Strategies for this are developed in Hanover. The sequential base structure allows a quick updating with the new observations and irregularities in the kinematic state are detected immediately. Kinematic evaluation of the "Wißberg" is made by SCHMIDT (1992). With the actual parameters for velocities and accelerations, which have to be tested for significance, prognoses of the geometrical behaviour in the near future are possible. Of course this forecasting loses its validation if there is a drastic change in the causative factors of the slide. These limits can be bypassed by a dynamic model.

8 Dynamic modelling

Quasi–static and kinematic models are both only geometrical approaches in deformation analysis. This means, that a relation between measured point movements and causing forces is not included in the mathematical formulas. In both models a qualitative correlation between acting forces, others causative parameters and detected movements is possible in a second step of analysis, the physical interpretation. Obviously the physical interpretation of landslides needs a very close cooperation between engineering geology and engineering geodesy.

The best case for the investigation of landslides and especially for the forecasting of behaviour or stabilization purposes would be a dynamic model, establishing a functional connection between forces and motions. Due to missing information or not possible quantitative modelling of this connection, dynamic modelling is rare up to now. But the progress for a better unterstanding of landslides would be enormous. The ideas on this purpose developed in Hanover are as follows. If there is a mechanic model for the explanation of the behaviour of a specific landslide available, then the results from this model must lead to the same point movements as known from the geodetic observations if both models are correct. This mechanic model could be every model which gives a quantitative relationship between causing forces and expected movements. The most suitable way of mathematical formulation seems to be the use of the method of finite elements (FEM).

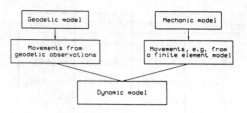

Fig. 7: Idea of dynamic modelling

There are two main difficulties for applicate such a model:

- formulation of material behaviour
- modelling of variations of causative forces

In a combination with the geodetic observations this mechanic model, even if it is just an approximation, can be controlled and calibrated. The final result is a better model with identified parameters. With this corrected values for instance a more reliable forecasting is possible.

The idea described in Fig. 7 is originally from BOLJEN (1983). Let consider now that such a mechanic model ψ is available.

$$\Delta x \quad = \quad \psi(x, x_\alpha, f)$$

Δx : displacements from mechanic model ψ
x : geometry of the slide body
x_α : parameters of material behaviour
f : acting forces

Mechanic parameters are for instance angle of internal friction and cohesion and acting forces e. g. gravity or pore water pressure. With such a model the coordinates can be predicted from time t_k to t_{k+1}.

$$x_{k+1} = x_k + \Delta x = x_k + \psi(x, x_\alpha, f)$$

Some of the coordinates must be identical with geodetic object points. In this case, the predicted coordinates must fit to the new observations carried out at time t_{k+1}. The dynamic state of such a monitored object can be described with

$$y = \begin{vmatrix} x \\ x_\alpha \end{vmatrix} \quad \begin{matrix} : & \text{coordinates} \\ : & \text{mechanic parameters} \end{matrix}$$

If the innovation is significant and there are no gross errors in the model, the mechanic part of this combined model must be changed until there is no difference anymore. The final result will be not only adjusted coordinated but especially adjusted mechanic parameters.

9 Conclusions

The major goal of this paper was to show some possibilities and ideas for monitoring and analysis of landslides from geodetic point of view. Futhermore it should be pointed out that a closer collaboration of engineering geology and engineering geodesy will help to come to a better understanding of the mechanism of slides. This cooperation begins already with the planning of the geodetic net and the choice of the suitable monitoring methods, required time span between epochs etc. .

Until today the geodetic standard analysis is the quasi–static comparison of two following epochs, which is always possible. In a second step of analysis the detected movements must be seen in context with causative forces (physical interpretation). Here again cooperation is requested.

Kinematic and dynamic approaches are higher sophisticated analysis models. Their development is in progress up to now, but there are already first results available for the kinematic model. The investigations of dynamic modelling are just at the beginning. In all approaches the KALMAN–filter technique seems to be a very powerful tool.

References

BOLJEN, J. (1983). *Ein dynamisches Modell zur Analyse und Interpretation von Deformationen*, Wiss. Arb. Uni. Hannover Nr.122, Hanover

HEUNECKE, O. (1989) *Untersuchungen zur Anwendung der KALMAN-Filtertechnik in der kinematischen Deformationsanalyse*, Diploma Thesis No.1179 (unpublished), Geodetic Institute, Hanover

HEUNECKE, O. / MATTHESIUS, H.-J. (1990) *Neue Aspekte zur Modellierung und Analyse von Hangrutschungen*, FIG–Symposium "Digital Technologies in Surveying and Mapping", Sofia

KRAUTER,E. / STEINGÖTTER,K. (1980) *Kriech- und Gleitvorgänge natürlicher und künstlicher Böschungen im Tertiär des Mainzer Beckens*, 6. Danube European Conference on Soil Mechanics and Foundation, Varna

KRAUTER, E. / STEINGÖTTER, K. (1983) *Die Hangstabilitätskarte des linksrheinischen Mainzer Beckens*, in: Geol. Jahrbuch 1983, C34, S. 3–31, Hanover

KRAUTER, E. (1986) *Phänomenologie natürlicher Böschungen (Hänge) und ihrer Massenbewegungen*, Grundbau Taschenbuch Teil 3, Berlin

KRAUTER, E. (1988) *Applicability and usefulness of field measurements on unstable slopes*, in: BONNARD, C. (Ed.) : Landslides, Vol. 1, p. 367–373, Rotterdam

NIEMEIER, W. (1988) *Deformationsanalyse – Aktueller Stand in Theorie und Praxis*, in: SCHNÄDELBACH/EBNER (Eds.): Ingenieur-vermessung 88, Vol. 2, Bonn, chap. E9

PELZER, H. (1986) *Deformationsanalyse – Zwei- und Mehrepochenvergleiche mit dem Programmsystem HANNA*,(unpublished),Hanover

PELZER, H. (1987) *Deformationsuntersuchungen auf der Basis kinematischer Bewegungsmodelle*, in: Allgemeine Vermessungsnachrichten, No. 2, S. 49–62

PELZER, H. (1988) *Anwendung der KALMAN-Filtertechnik auf die Deformationsanalyse*, in: SCHNÄDELBACH/EBNER (Eds.): Ingenieur-vermessung 88, Vol. 1, Bonn, chap. B4

SCHMIDT, A. (1992) *Anwendung der kinematischen Deformationsanalyse mittels KALMAN-Filterung auf den Rutschhang "Wißberg"*, Diploma Thesis No. 1272, (unpublished) Geodetic Institute, Hanover

S1 Seismicity and landslides

Seismicité et glissements de terrain

Keynote paper: Access to the dynamics of landslides during earthquakes by a new cyclic loading high-speed ring shear apparatus

Kyoji Sassa
Disaster Prevention Research Institute, Kyoto University, Japan

ABSTRACT : A new cyclic loading high-speed ring shear apparatus has been developed to simulate soil behavior in slopes under earthquake loading. Tests on the pumice which formed the sliding zone in the Ontake landslide triggered by the 1984 Naganoken-Seibu earthquake in Japan were carried out. The tests reproduced the behaviour of landslides such as acceleration and fast motion after failure in the peak strength state of soils. They suggested that the pumice should have failed by an acceleration of 210–260 gal less than the estimated acceleration of 300–360 gal produced by the earthquake in the landslide area. Undrained cyclic triaxial compression tests were also performed on the same sample for comparison. They could not reproduce any accelerated fast motion, but slow repeated upward-downward motions were observed.

1. INTRODUCTION

Landslides triggered by earthquakes often move rapidly in volcanic sediments, loess and well weathered rocks, while in residual state landslides (reactivating landslides) the motion is slow and stops in a short distance. Rapid landslides in the peak strength state (first-time landslides) cause much damages during earthquakes. Some of the recent earthquake-induced rapid landslides are reported in the international newsletter "Landslide News, No.1–5". They are the Ontake landslide on the Ontake volcano, Japan, 1984 (Landslide News No.1), the Bairaman river landslide in highly weathered limestone in Papua New Guinea, 1985 (Landslide News No.2), many landslides on the Reventador volcano, Ecuador, 1987 (Landslide News No.2), the Gissar liquefaction-induced landslide in loess, Tajikistan, 1988 (Landslide News No.3), many slope failures in the loess area, China, 1989 (Landslide News No.4), the Fatalok landslides in weathered limestone, Iran, 1990 (Landslide News No.5), many slope failures in highly weathered volcanic sedimentary rocks, Philippines, 1990 (Landslide News No.5). These suggest that rapid landslides triggered by earthquakes often take place in volcanic areas, loess and highly weathered limestone. In Japan, the best documented large landslide due to earthquake is the Ontake landslide ($37 \times 10^6 m^3$), (Sassa, 1988) where the sliding surface was formed in weathered pumice. Fig. 1 shows the landslide source area soon after its occurrence. The sample used in this research was taken from the pumice at the place indicated by a circle in the photo. Historically the most disastrous landslide triggered by earthquakes in Japan is the 1792 Mayuyama land-slide at the Unzen volcano. The landslide mass moved rapidly and 80 percent of the total volume entered the sea, causing a large tsunami, which devastated coastal parts of Kumamoto prefecture and the Amakusa islands across the sea. 17,891 people were killed. The landslide debris is found in the sea forming many small islands, the so-called 99 islands (Takahashi, 1987). Two big groups of residual state landslides are known in Japan; one in the Tertiary weathered mudstone areas, and the other in weathered crystalline schist along tectonic lines. These residual state landslides are comparatively resistant against earthquakes. Small motion or cracking has been reported, but rapid motion rarely results.

Geotechnical approach to landslide behavior before and after failure

The triaxial test and the ring shear test are often used for landslide research. Both tests are performed in this research. They have different characteristics as follows;

1) The triaxial test is designed to study the stress-strain relation before failure and at failure, while the ring shear test is designed to study the shear strength at failure and after failure.

2) The triaxial test deforms a specimen in a relatively homogeneous mode, while the ring shear test deforms a specimen along a surface (especially after failure). When angular grains such as volcanic sediments, loess are first subjected to shear deformation, grain crushing should occur. The effect of grain crushing must be greater in deformation along a surface than that in homogeneous deformation.

3) The shape of specimen does not change as a whole in the ring shear test, while the shape of specimen changes from a cylindrical column to an oval column in the triaxial test. Therefore, energy consumption takes place inside the whole specimen in the triaxial test, while energy consumption concentrates along a sliding surface in the ring shear test. Figs. 2, 3 show specimens of the Ontake pumice after the ring shear test and the triaxial test. We observed a crushed silty zone along a shear surface in the specimen of the ring shear test, but not observe such a zone inside the specimen of the triaxial test, although two faint shear surfaces are visible from the outside. Grain crushing along a sliding surface in a constant normal stress ring shear test, during a long shear displacement,

was studied by Fukuoka (1991). Fig. 4a shows the mobilized friction angle and the variation of sample height during a 100 m shear displacement of Shirakawa river sand. Fig. 4b shows grain distributions of the sample before the test and those taken from the shear zone after the tests. Grains were much crushed and fine grains were formed. Fig. 5 is the result of constant-volume constant-speed ring shear test on the Denjo river sands over which the Ontake landslide traveled. Due to grain crushing along the sliding surface and its resulting negative dilatancy, the shear strength much decreases along the failure envelope. The peak shear stress appeared at 1.88 cm shear displacement, and it was almost the same with the shear displacement at the peak of positive dilatancy in

Fig. 1 The Ontake landslide (3.6×10^7 m^3) induced by the Naganoken-Seibu earthquake in Japan, 1984.
◯: Sampling point.

Fig. 2 Failure mode of the pumice specimen after the cyclic loading ring shear test.

Fig. 3 Failure-mode of the pumice specimen after the cyclic loading triaxial test.

Fig. 4a. Figs. 2–5 suggest that grain crushing has an important role in the mechanical behavior of sandy soils subjected to shear deformation along a sliding surface.

2. STRUCTURE OF A NEW CYCLIC LOADING HIGH-SPEED RING SHEAR APPARATUS

A high-speed ring shear apparatus was developed for debris flow study by Sassa (1984) and Sassa et al. (1984) (the maximum normal stress is 0.4 kgf/cm², and the maximum shear speed is 90 cm/sec), at almost the same time as the development of a similar high-velocity ring shear apparatus by Hungr and Morgenstern (1984). A high-speed high-stress ring shear apparatus was developed by Sassa et al. (1989a,b) (the maximum normal stress is 3.8 kgf/cm², the maximum shear speed is 100 cm/sec). These series of ring shear apparates used constant-speed shearing similar to conventional ring shear apparates (Bishop et al., 1971). However, to simulate the behavior of landslides in earthquakes, cyclic shear stress load-a)

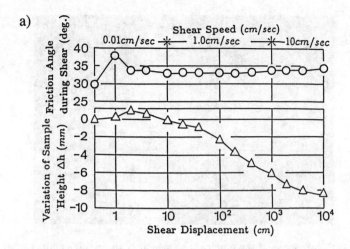

b)

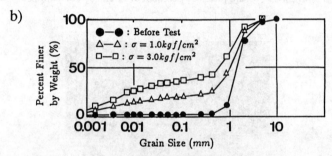

Fig. 4   Grain crushing along a slip plane during constant normal stress ring shear test on the Shirakawa river sands (Fukuoka, 1991).

a)   Variation of sample height and friction angle for the test of σ =3.0 kgf/cm^2 (Void ratio e=0.56–0.76).

b)   Grain size distributions of the sample before tests and those taken from the shear zones after tests for the tests of σ =1.0 kgf/cm^2 and 3.0kgf/cm^2.

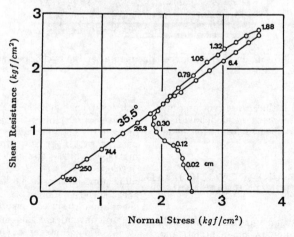

Fig. 5   Constant-volume constant-speed ring shear test on the Denjo river sands.   Void ratio e=0.61 (Sassa & Fukuoka).

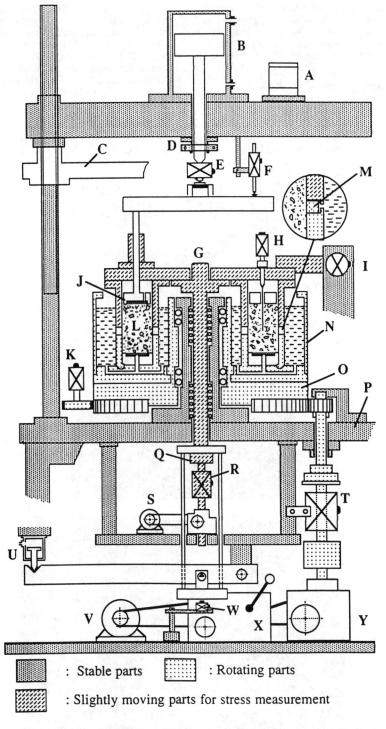

Fig. 6 Structure of a new cyclic loading high-speed ring shear apparatus. Parts identified by letters are referred to in text.

ing and cyclic normal stress loading are necessary. Sassa developed a cyclic loading high-speed ring shear apparatus in 1991. This paper presents the results of the first series of tests using this new apparatus. Fig.6 shows the structure of the apparatus.

General structure

With reference to Fig. 6, the sample (L) is set in the ring shear box, its lower half is rotated by a servo-motor (marked as a rotating part in the figure), while its upper half (marked as a slightly moving part in the figure) is restrained by a load cell (I) to measure shear load. The front view of the apparatus and the shear box are presented in Fig. 7. The lower half of the shear box is shown in Fig. 8a. The inside diameter of the shear box is 21 cm, the outside diameter is 31 cm, the shear area is 408.4 cm^2, the depth of shear box beneath the shear plane is 2.9 cm, the height of shear box above the shear plane is 3–6 cm depending on the consolidation.

Loading system

Cyclic normal stress is given by a air piston (B) in Fig. 6 controlled by EP (electric air-pressure

a)

b)

Fig. 7 Front view of the apparatus (a) and the shear box (b).

converter) to which an electrical control signal is input from a personal computer. The EP gives a certain air pressure corresponding to the control signal. (Feed back from the load cell for normal stress is not given to the computer.) Cyclic shear stress loading was applied by use of a torque control servo-motor (V). An electrical control signal from the computer is input to the servo motor, and the motor gives a certain torque corresponding to the signal. (Feed back from the torque meter of the apparatus is not given to the computer.) The servo-motor gives a constant speed of shearing by a control signal for the constant speed test. In the constant volume test, the stopper (D) fixes the axial rod to the main stable frame. It is the simplest way, though a small change of sample height occurs due to the elastic deformation (0.02 mm for the difference of 1,000 kgf) of the load cell (E).

Monitoring system

Vertical load is measured by the load cell (E), and the side friction between the sample and the upper shear box is measured by the load cell(R). To keep the gear of the gap control system below the load cell (R) and the rotary joint (Q) always in compression, a dead weight is loaded by an air regulator (U). The difference of values of the load cell (E) and the load cell(R) gives the correct vertical load on the sample. Shear load is measured by the load cell (I) and the shear force is transmitted through the shear plane from the rotating lower half of the shear box. The shear torque, given by a servo-motor, is monitored by a torque converter (T). The horizontal displacement is monitored by a rotary transducer (K).

Pore water is drained through six porous metals built in the loading plate (J) and the base plate

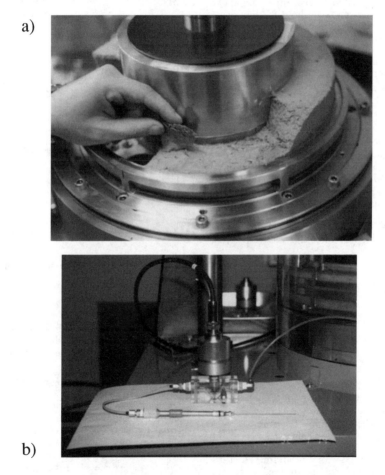

a)

b)

Fig. 8 Lower half of the shear box and a sheared specimen after test (a), and needle for pore pressure measurement (b).

of shear box, respectively. However, during cyclic loading and fast shearing, pore pressure can remain in the central part of the sample. Therefore, three needles connected to pressure cells are inserted into and around the center of sample as shown at H. The needle for pore pressure measurement is shown in Fig. 8b. The sample height is monitored through the loading plate by a linear transducer (F).

Gap control

The load cell(R) deforms 0.003 mm during cyclic loading due to the change of friction of the sample on the upper shear box. The gap between the upper ring and lower ring is automatically maintained constant by a servo-motor (S) through a feed back signal from the high precision gap sensor (W) which detects a change of 0.001 mm. A rubber edge (50 in the rubber hardness index) of 9 mm thickness is pasted to the upper ring. The rubber edge is always pressed about 0.2 mm using the total weight of 25 kgf during cyclic loading and high speed shearing to prevent the leakage of sample. The surface of the rubber edge (M in Fig. 6) is coated with teflon spray to decrease the friction. The shear friction caused by the contact of rubber edge is 0.045 kgf/cm^2 in term of shear stress.

Shear speed

The shear speed of the apparatus can be changed from 0.0002 cm/min to 150 cm/sec using three steps of gears and one set of pulleys. Even in the torque control state, the maximum shear speed is limited by the capacity of the motor. The maximum shear speed is 150 cm/sec at 1:1 pulleys and 37 cm/sec at 1:4 pulleys. In this test series, 0.1 Hz cyclic loading was adopted. It is one or two order less than that in actual earthquakes. Therefore, a maximum speed of 37cm/sec was selected, which is one or two order less than the speed of actual fast landslides.

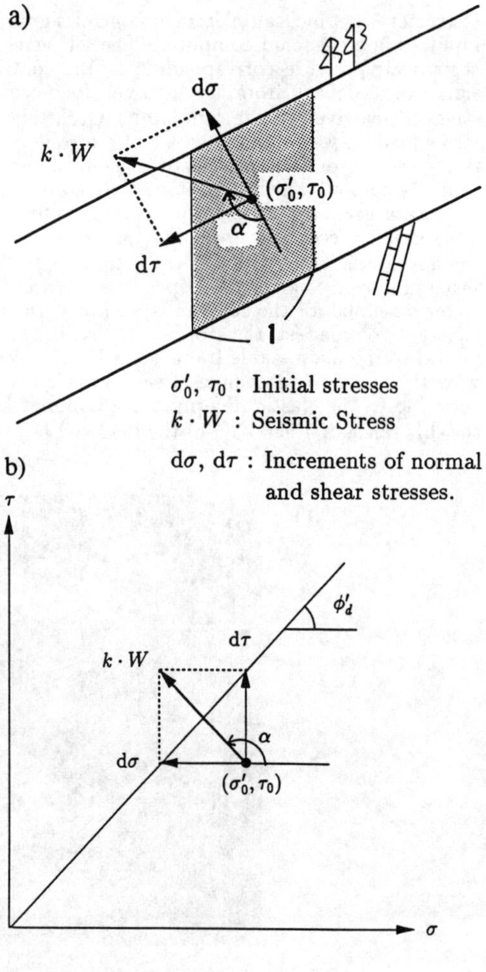

3. TEST CONDITIONS AND PROCEDURES

Stress in slopes under earthquakes

It is desirable to reproduce the stress in slopes under earthquake loading in the ring shear apparatus. Fig. 9 illustrates the seismic stress (kW, k:seismic coefficient, W: weight over unit area on the sliding surface) in slopes and in a $\sigma-\tau$ diagram. The direction of seismic stress is expressed by the angle (α). When the slope layer is in the undrained state, the effective normal stress is not changed. When the direction of seismic stress is parallel to the slope, the seismic stress gives the maximum shear stress increment. This case (A) is the most dangerous case for the undrained state. When the slope layer is in the drained state, the most dangerous direction of seismic stress will be normal to the failure envelope, because the stress can reach the envelope in the shortest stress path. This is case (B).

σ'_0, τ_0 : Initial stresses

$k \cdot W$: Seismic Stress

$d\sigma, d\tau$: Increments of normal and shear stresses.

Fig. 9 Directions of seismic stress in a slope (a) and $\sigma-\tau$ diagram (b).

Test conditions

Case (A) in the ring shear test corresponds to the condition in which cyclic shear stress is added to the initial stress under the constant-volume state.

$$d\tau = kW \sin \omega t, \quad \omega : \text{angular velocity.}$$

Case (B) in the ring shear test corresponds to the condition in which cyclic shear stress and cyclic normal stress are simultaneously added to the initial stress. But the phase is opposite.

$$d\tau = kW \cos \phi'_d \sin \omega t$$

$$d\sigma = -kW \sin \phi'_d \sin \omega t$$

here, ϕ'_d : dynamic effective friction angle.

In the triaxial test, the effective radial stress is not changed due to pore pressure generation in the fully saturated undrained condition, even if cyclic radial stress is loaded. Therefore, only cyclic axial stress is added to the initial stress. In this case, cyclic shear stress and cyclic normal stress on the failure plane are expressed by the cyclic axial stress and the friction angle.

$$d\tau = kW \cos(45° + \phi'_d/2)\sin\omega t$$

$$d\sigma = kW \sin(45° + \phi'_d/2)\sin\omega t$$

here, $kW = d\sigma_a \cos(45° + \phi'_d/2)$

And the direction of seismic stress in the normal stress-shear stress diagram is;

$$\tan^{-1}\frac{d\tau}{d\sigma} = 45° + \phi'_d/2$$

Test procedures of ring shear tests

Samples tested were alluvial soils from the Kyoto University campus and pumice from the Ontake landslide. Samples were well mixed in water and left for one to three days for saturation, and then put in the shear box filled with water. To push entrapped air bubbles out of the sample, they are prodded by a rod. Three needles for pore pressure measurement are inserted around the shear surface before loading. Then, the initial normal stress and shear stress are applied so that the stress ratio is slightly below the failure envelope. After the pore pressure dissipation, cyclic loading is applied under constant-volume conditions or both cyclic shear stress and cyclic normal stress conditions. Frequency of cyclic loading is about 0.1 Hz. When failure did not take place in 20 cycles of loading, another 20 cycles of a greater stress was loaded. In addition to these cyclic loading tests, the constant-speed / constant-volume tests and also the constant-speed / monotonic loading tests were performed.

Test procedures of triaxial tests

Dry samples were placed in the triaxial cell, CO_2 gas is entered into the sample and de-aired water is supplied slowly for saturation. Cell pressure with back pressure of 2.0 kgf/cm^2 is applied. B value was greater than 0.95. An initial axial stress smaller than its static strength is applied in the drained condition. The cyclic axial stress is then applied in the undrained condition. Frequency of cyclic loading is 0.1 Hz the same as with the ring shear test. When failure did not take place in 20 cycles, another 20 cycles of a greater stress was applied. Pore pressure is monitored at the center of the sample through a needle horizontally inserted through a rubber membrane as well as at both ends through porous stones. Pore pressure measured at the center of the sample was used

in calculating effective stress during cyclic loading because it is more representative of the pore pressure in the shear zone.

4. TEST RESULTS

Test number, samples, test conditions and some results are listed in Table 1.

Constant-volume cyclic loading test

Fig. 10a shows the stress path of the first cycle and the final (4th) cycle continuing to failure in the constant-volume cyclic loading test on alluvial soils from Kyoto University campus. Fig. 10b presents the total normal stress, shear stress, pore pressure, and shear velocity for shear displacement. The shear displacement (L) reached 20 m. The result of constant-speed constant-volume test on the same sample shows a friction angle in the static loading state (ϕ'_s) of 35.3 degrees. Cohesion seems to be almost zero, which is reasonable since the sample is a disturbed, normally consolidated soil. The constant-volume cyclic loading test gave a dynamic friction angle of 36.4 degrees, regarding the sample as a cohesionless material. Both friction angles are similar. Failure took place at the fourth cycle. After failure, shear velocity reached the maximum speed of 37 cm/sec, normal stress and shear stress decreased, and the stress finally reached the failure envelope at a low stress level. It may have resulted from the effects of grain crushing along the sliding surface as previously suggested in Fig. 5.

Because this test was under constant volume condition, decrease of effective normal stress was detected as the variation of the difference of values between two load cells (E and R in Fig. 6). However, grain crushing in the shear zone might cause a negative dilatancy and generate pore pressure even if the specimen volume was kept constant as the whole. It may be the reason why the stress path of the acceleration process after failure was much below the failure envelope. Pore pressure was monitored in this test by setting the top of the pore pressure needle at 14.8mm above the shear surface (the specimen height was 52.8mm above the shear surface). The monitored pore pressure slightly increased reflecting this pore pressure generation during the acceleration process in a delayed and decreased way.

It is noted that, when normal stress was rapidly increased from 0.5 kgf/cm^2 to 3.0 kgf/cm^2 in this test, the monitored pore pressure increment was 0.9 kgf/cm^2 equivalent to B_D (B value under one dimensional consolidation) of 0.36.

Fig. 11 is the result of the constant-volume cyclic loading test on the Ontake pumice. The first cycle and the final (10th) cycle is shown in Fig. 11a, and monitored parameters are presented in Fig. 11b. The general trend is very similar to Fig. 10. After failure shear speed was smoothly accelerated to the maximum speed of the servo-

Table 1 Test Conditions and Results.

Test No.	Test Type	Sample γ_d (gf/cm³)	Test Conditions	Initial Stress $\tan^{-1}(\tau_0/\sigma'_0)$	r_u in 27° slope	Cyclic Stress $\tan^{-1}(d\tau/d\sigma')$	Seis. Coeff. k_{cv}	k_{bc}	Friction Angle $\tan^{-1}(\tau/\sigma')_{max}$
1	Ring	Alluvial 1.58	C-Vol. C-Speed (0.01cm/sec)	0°	—	—	—	—	35.3°
2	Ring	Alluvial 1.61	C-Vol. Cyclic L. (0.096Hz)	33.2°	0.175	61.4°	0.148	0.166	36.4°
3-1	Ring	Pumice 0.90	C-Vol. Cyclic L. (0.096Hz)	32.5°	0.159	58.7°	0.217	0.254	38.6°
3-2	Ring	Pumice 0.90	Monot. L. C-Speed (0.16cm/sec)	0°	—	—	—	—	29.9°
3-3	Ring	Pumice 0.94	Both Cyclic L.	28.0°	0.033	126.3°	—	0.205	36.0°~37.5°
4	Ring	Pumice 0.96	Both Cyclic L.	34.8°	0.212	162.8°	—	0.261	49.5°
5	Triax.	Pumice 0.73	undrained C-Speed (0.15cm/min)	0°	—	—	—	—	45.6°
6	Triax.	Pumice 0.72	undrained Cyclic L. (0.096Hz)	41.6°**	0.338**	68°**	0.403**	—	48.5°
7	Triax.	Pumice 0.63~0.73	undrained Repeated loading	0°	—	68°**	—	—	46°

τ_0, σ'_0 : Initial shear stress, effective normal stress before test, $d\tau, d\sigma'$: Stress increment by cyclic loading, k_{cv} : Seismic coefficient under constant volume state, k_{bc} : Seismic coefficient under both cyclic loading, γ_d : Dry unit weight just before cyclic loading, constant speed shearing, z : Vertical depth of soil layers, r_u : Pore pressure ratio, * : Values on the failure plane in which the friction angle is assumed to be 46°, Specific gravity G_s =2.55 for alluvial soils, G_s =2.66 for pumice, coefficient of permibeality $k = 2.6 \times 10^{-5}$cm/sec for alluvial soils, $k = 1.83 \times 10^{-4}$cm/sec for pumice. Test No. 7 are four tests on the undisturbed pumice (taken from the same area) done by Kawakami et al. (Shinshu University, 1985).

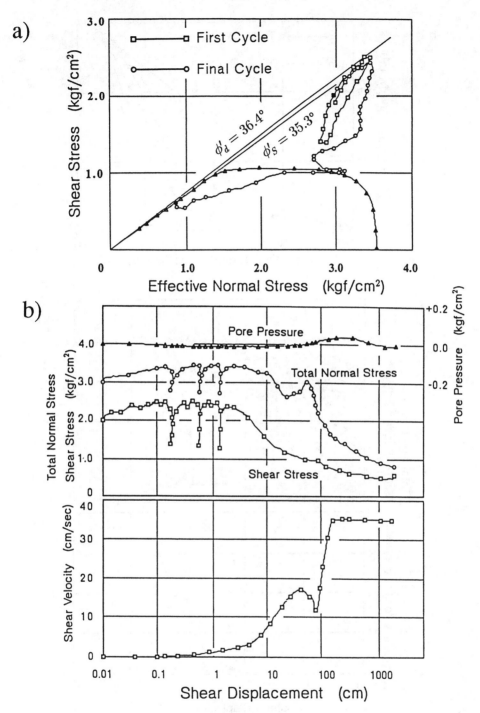

Fig. 10 Constant-volume constant-speed test (No.1) and constant-volume cyclic loading test (No.2) on the alluvial soils from Kyoto University campus.

a) Effective stress path (first and final cycles of stress are shown).
   ▲ : Stress path of the constant-volume constant-speed test.
   □ o : First and final cycles of stress of the constant-volume cyclic loading test.

b) Total normal stress, shear stress, pore pressure and shear velocity for shear displacement.

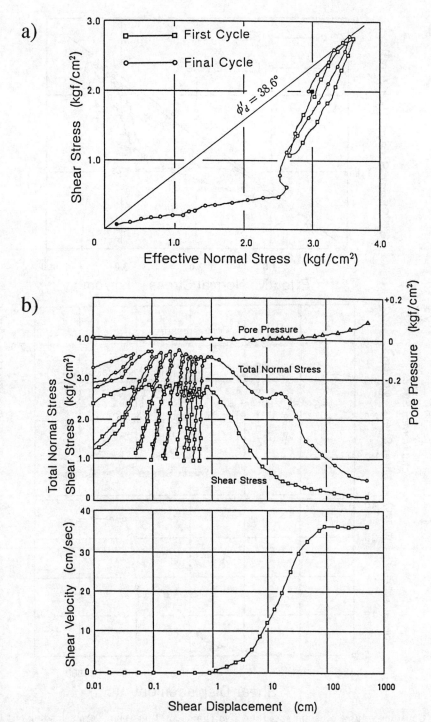

Fig. 11 Constant-volume cyclic loading test (No.3–1) on the Ontake pumice.
a) Effective stress paths (first and final cycles of stress are shown).
b) Total normal stress, shear stress, pore pressure and shear velocity for shear displacement.

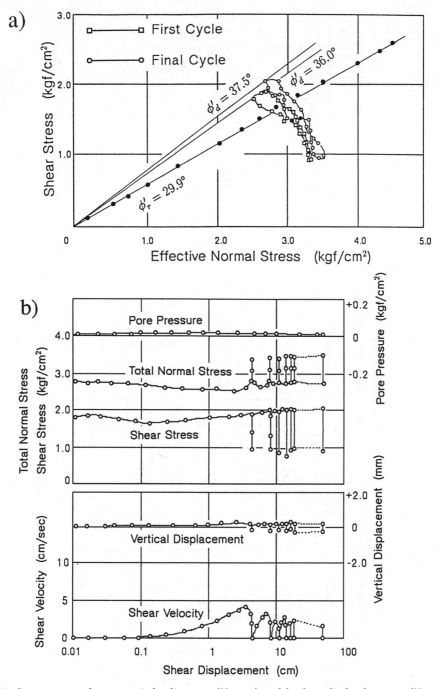

Fig. 12 Constant-speed monotonic loading test (No.3–2) and both cyclic loading test (No.3–3) on the Ontake pumice in the residual state.

a) Effective stress paths.
    ● : Stress path of the constant-speed monotonic loading test.
    □ ○ : First and final cycles of stress of the both cyclic loading test.

b) Total normal stress, shear stress, pore pressure, shear velocity and vertical displacement for shear displacement.

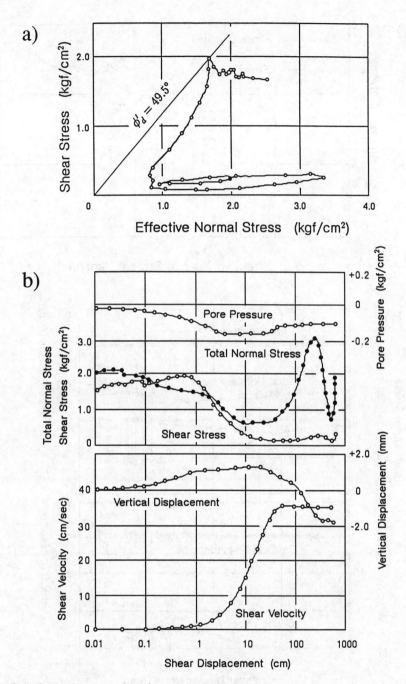

Fig. 13 Both cyclic loading tests (No.4) on the Ontake pumice.

a) Effective stress path.

b) Total normal stress, shear stress, pore pressure, shear velocity and vertical displacement for shear displacement.

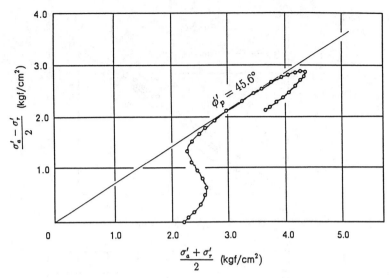

Fig. 14 Undrained constant-speed triaxial compression test (No.5) on the Ontake pumice.

motor. The shear stress became almost zero. The sample was apparently not liquefied, but the effective normal stress became very small due to grain crushing along the sliding surface. The friction angle in the dynamic state is 38.6 degrees. As in Fig. 10, pore pressure might be generated in the shear zone during the acceleration process though it was under constant volume condition. The stress path was deviated from the failure envelope. While the stress was near or on the failure envelope at a low shear velocity and a high but constant shear velocity. Monitored pore pressure around the shear surface (about 1.0mm under the shear surface) increased reflecting pore pressure generation during acceleration also in a delayed and decreased way.

Cyclic loading test in the residual state sample

To study shear behavior under cyclic loading in the residual state sample, tests were performed without replacing the specimen after the tests summarized in Fig. 11. Firstly, the constant-speed monotonic loading test was carried out, the test results are plotted by filled circles (•) in Fig. 12a. The test gave a residual friction angle of 29.9 degrees.

Dilatancy during shear is not expected in the residual state sample. Therefore, we tried to apply cyclic loading of both of normal stress and shear stress so that the direction of stress increment became almost normal to the failure envelope. It corresponds to case B) previously described in Section 3. Motion took place from the first cycle as seen in Fig. 12b. The dynamic friction angle for the first cycle is 36.0 degrees and it increased slightly to 37.5 degrees at the 20th cycle. Both of the dynamic friction angles are greater than the residual friction angle. The difference between the static friction angle and the dynamic friction angle in

the case of Fig. 12 is much greater than the difference between static and dynamic friction angles in Fig. 10. This reason is not identified at the present though there are some possible causes.

The motion of this test is quite different from the previous two tests. It was very slow (2–4 cm/sec) and repeatedly started and stopped. The motion was limited to the state where stress moved across the static failure envelope. The total shear displacement during 20 cycles of loading was less than 50 cm. This test result has experimentally illustrated presented the reason that residual state landslides (reactivating landslides) are resistant against earthquakes and their motion is usually very limited despite the fact that they repeatedly move without earthquakes.

Both cyclic loading tests

According to preliminary tests of both cyclic loading conditions, it was known that, if failure does not take place in the first cycle of both cyclic loading, it would not occur in the following cycles. It is in contrast to the constant-volume test in which effective normal stress gradually decreases during cyclic loading with the progress of grain crushing and negative dilatancy. Hence, in the test shown in Fig. 13, cyclic loading was applied so that failure took place in the first cycle of loading. The stress path is shown in Fig. 13a. The dynamic friction angle at failure was 49.5 degrees, which is much greater than the dynamic friction angle in the constant-volume cyclic loading test (38.6 degrees). This value is almost same as the dynamic friction angle (48.5°) obtained in the undrained cyclic loading triaxial compression test described below.

In this both cyclic loading test, the volume of specimen changed as shown in Fig. 13b. Initially the specimen dilated due to a decrease of nor-

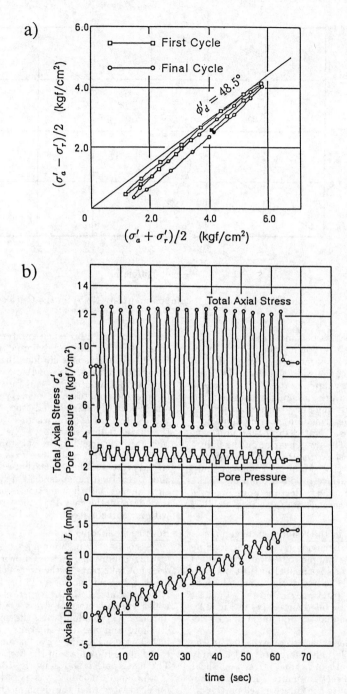

Fig. 15 Undrained cyclic loading triaxial test (No.6) on the Ontake pumice.

a) Effective stress path (first and final cycles of stress are shown).

b) Total axial stress, pore pressure, axial displacement for time.

mal stress and shear deformation, but later showed negative dilatancy due to grain crushing along the shear zone at a shear displacement of over 30cm. The monitored pore pressure showed a negative value corresponding to the dilatancy at the shear displacement of less than 2–3cm. However, in large displacements, pore pressure needles were deformed and they did not work well. The shear stress remained at a small value at a shear displacement of over 10cm though normal stress was greatly changed. The stress path was much lower than the failure envelope, as a result of pore pressure generation along the shear zone.

When the pore pressure measurement was not correct at the peak shear stress, the dynamic friction angle was not correct, though the peak shear stress itself was correct.

Cyclic loading triaxial tests

Before the triaxial cyclic loading tests, an undrained constant-speed triaxial compression test was performed on the Ontake pumice (Fig. 14). The friction angle at the peak stress was 45.6 degrees. Cohesion was regarded as zero as in the ring shear test because the samples were disturbed sandy materials. Fig. 15a presents the first and final (20th) stress paths in the cyclic loading test of 0.1 Hz. Fig. 15b shows the total axial stress, pore pressure, and the axial displacement (variation of sample height). The trend of motion is quite different from that of the ring shear test. The velocity is very small in the whole range, the average speed is 0.023 cm/sec. It repeats upward and downward movement. As found in Fig. 11b, the ring shear test also showed the recovery of shear displacement in the range of small shear displacement of 0-0.2 cm before failure. However, shear displacement progressed monotonically after failure in the ring shear test. Since the repeat of upward and downward motion cannot be imagined in actual landslides, the triaxial test is not good in representing shear behavior after failure under cyclic loading conditions. Fig. 16 shows the test results of repeated cyclic triaxial compression and extension tests and monotonic compression tests on the undisturbed pumice taken from the same area by Kawakami et al. (1985). These were static tests using conventional triaxial apparatus. The stress paths of monotonic compression and repeated compression and extension tests reached almost the same static failure line. The value is around 46 degrees. It is nearly the same value obtained in the undrained static compression test (Fig. 14).

5. DISCUSSION

Test conditions and results are listed in Table 1. The slope angle was about 27 degrees in the Ontake landslide. We can calculate the pore pressure ratio r_u on the slope of 27 degrees.

$$r_u = \cos^2 \theta - \frac{\sigma_0'}{\tau_0} \cdot \sin \theta \cos \theta$$

here, $\theta = 27$ degrees.

We can calculate the seismic coefficient of each test from $d\tau$, $d\sigma$, τ_0.

$$k = \frac{\sqrt{(d\sigma)^2 + (d\tau)^2}}{\tau_0} \cdot \sin \theta$$

here, $\theta = 27$ degrees.

In Test No.3–1 (Table 1), two seismic coefficients are obtained. One is obtained when only $d\tau$ is applied and the observed $d\sigma'$ is the result due to the constant-volume condition (pore pressure generation). It is the seismic coefficient written as k_{cv}. The second is obtained when $d\tau$ and $d\sigma$ are applied and the constant-volume state was kept as its result. The seismic coefficient is written as k_{bc}. The results of Test No.3–1 and No.4 indicate that the slope with a pore pressure ratio of 0.16–0.21 will fail under the earthquake loading represented by values of the seismic coefficient of 0.22–0.26. Pore pressure ratio of around 0.16–0.21 are probable. The seismic acceleration in the Ontake landslide area during the Naganoken-Seibu earthquake was estimated to be 300–360 gal by Ishihara et al. (1986). Values of the seismic coefficient of 0.22–0.26 correspond to 216-255 gal. Thus, it appears that the earthquake was greater than the seismic stresses which were necessary to cause failure.

In the case of the cyclic triaxial test of test (Test No.6), the seismic coefficient at failure was 0.403, equivalent to 395 gal. It gave a greater value than those obtained in the cyclic loading ring shear tests, and the acceleration to cause failure exceeded the estimated acceleration at the Ontake landslide site.

Ishihara et al. (1986) estimated that the landslide was caused by earthquake accelerations of less than 300 gal, but tests on the undisturbed Ontake pumice were carried out by the undrained static triaxial compression test without pore pressure measurement, namely in the total stress concept. So, comparison between the tests is not easy.

This is the first series of tests by a newly developed cyclic loading ring shear apparatus.

6. CONCLUSION

1) A new cyclic loading high-speed ring shear apparatus has been developed in order to study the motion of landslides under cyclic loading conditions.
2) The cyclic loading ring shear test can reproduce an acceleration of shear displacement and a rapid displacement in the peak strength state of soils, and also a slow and short repeated start–stop motion in the residual state of soils.
3) The test results on the Ontake pumice suggest that the pumice should have failed by an acceleration less than the estimated acceleration at the Naganoken-Seibu earthquake in the Ontake landslide area.
4) The dynamic friction angle, $\tan^{-1}(\tau/\sigma')_{max}$ in

1935

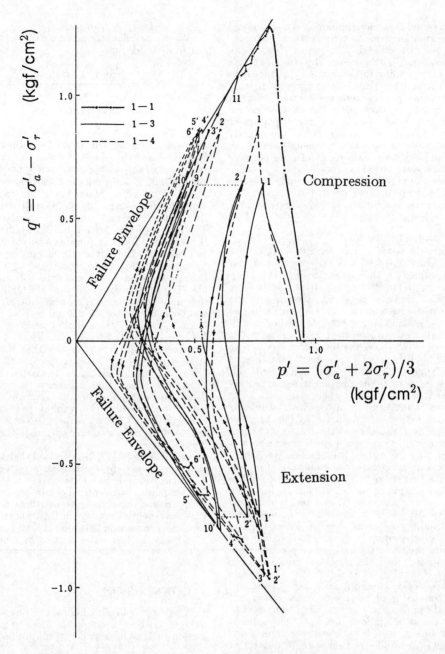

Fig. 16 Undrained repeated loading and monotonic loading triaxial tests (No.7) on the Ontake pumice (Kawakami *et al.*, 1985).

the cyclic loading of normal stress and shear stress, whose direction is almost normal to the failure envelope was greater by about 10 degrees than that in constant-volume (shear stress) cyclic loading, and its value was near the value obtained from the undrained cyclic (axial stress) loading triaxial test.

The reason has not been identified at the present stage of investigations.
5) The motion observed in the undrained cyclic triaxial compression test was very slow and short. It repeated downward and upward movement, namely a positive and negative velocity in the peak strength

state of soils. It was quite different from the motion of real landslides in the qualitative sense as well as the quantitative sense.

ACKNOWLEDGEMENT

The author acknowledges Prof. M. Shima of the Disaster Prevention Research Institute, Kyoto University for his cooperation, and thanks to Mr. H. Fukuoka and Mr. Z. Shoaei (Iran) of post graduate students and Mr. D. Zhang (China) of visiting joint researchers for their cooperation in improvement of the apparatus and the cyclic ring shear tests and the triaxial tests. Prof. K. Ishihara of Tokyo University and Prof. H. Kawakami of Shinsyu University are appreciated for their kind offer of information for this research.

REFERENCES

Arboleda, R.A. , R.S. Punongbayan (1991). Landslides induced by the 16 July 1990 Luzon, Philippines, Earthquake, International Newsletter " Landslide News," No.5, 5–7.

Bishop, A.W., G.E. Green, V.K. Garga, A. Andresen and J.D. Brown (1971). A new ring shear apparatus and its application to the measurement of residual strength, Géotechnique 21, No.4, 273–328.

Fukuoka, H. (1991). Variation of the friction angle of granular materials in the high-speed high-stress ring shear apparatus, – Influence of reorientation, alignment and crushing of grains during shear –, Bulletin of the Disaster Prevention Research Institute, Kyoto University, Vol. 41, Part 4, 243–279.

Hungr, O., N.R. Morgenstern (1984). High velocity ring shear tests on sand, Géotechnique, Vol. 34, 415–421.

Ishihara, K. , H.L. Hsu, I. Nakazumi, K. Sato (1986). Analysis of landslides during the 1984 Naganoken-Seibu earthquake, Proc. Int. Symp. on Engineering Geology Problems in Seismic Areas, Vol. 2, 175–215, Bari.

Ishihara, K. (1989). Liquefaction-induced landslide and debris flow in Tajikistan, USSR, International Newsletter "Landslide News," No.3, 6–7.

Ishihara, K. (1991). Manjil earthquake of 21 June 1990, Iran, International Newsletter " Landslide News," No.5, 2–4.

Kawakami, H. , J. Konishi, Y. Saito (1985). Mechanism of slope failures by the Naganoken-Seibu earthquake 1984 and the characteristics of pumice, Tsuchi-to-Kiso (Journal of the Japanese Society of Soil Mechanics and Foundation Engineering), Vol. 33, No. 11, 53–58 (in Japanese).

King, J.P., I. Loveday, R.L. Schuster (1988). Bairaman river landslide and natural dam, Papua New Guinea, International Newsletter "Landslide News," No.2, 10–12.

Natural Disaster Research Group of Shinshu University (1985). Disasters due to the Naganoken-Seibu earthquake, 148 pages (in Japanese).

Nieto, A.S. , R.L. Schuster (1988). Mass wasting and flooding induced by the 5 March 1987 Ecuador earthquake, International Newsletter " Landslide News," No.2, 5–8.

Sassa, K., M. Shima, H. Hiura, A. Nakagawa, A. Suemine (1984). Development of ring shear type debris flow apparatus, Report of Grant-in-Aid for Scientific Research by Japanese Min. of Education, Science and Culture (No.57860028).

Sassa, K. (1984). The mechanism starting liquefied landslides and debris flows, Proc. 4th ISL, Vol. 2, 349–354, Toronto.

Sassa, K. (1987). The Ontake debris avalanche and its interpretation, International Newsletter " Landslide News," No.1, 6–8.

Sassa, K. (1988). Special lecture: Geotechnical model for the motion of landslides, Proc. 5th ISL, "Landslides," A.A. Balkema / Rotterdam, 37–55.

Sassa, K. , H. Fukuoka, C. Vibert, M. Shima (1989a). Development of a high-speed high-stress ring shear apparatus and shear strength of reduction at rapid loading in landslides, Annuals, Disaster Prev. Res. Inst., Kyoto University, Vol. 32, No. B–1, 165–182 (in Japanese).

Sassa, K. , H. Fukuoka, C. Vibert (1989b). A new high-speed high-stress ring shear apparatus and the undrained shear strength during motion, Proc. The Japan-China Symposium on Landslides and Debris Flows, 93–97, Niigata and Tokyo.

Takahashi, R. (1987). Collapse of Mayuyama, Shin-Sabo (Journal of the Erosion-Control Engineering Society, Japan), Vol. 40, No. 2, 34–36 (in Japanese).

Zhu, H.Z. (1990). Disasters caused by the 1989 Datong-Yanggao Earthquakes in Shanxi Province, China, International Newsletter "Landslide News," No.4, 15–16.

Landslides, Bell (ed.) © 1995 Balkema, Rotterdam, ISBN 90 5410 032 X

Theme report

M.J.Pender
Civil Engineering Department, University of Auckland, New Zealand

1 INTRODUCTION

There are eight papers assigned to this session. The topics covered can be classified as follows: Paleoseismicity and landslides, 2 papers; recent earthquake events and landslide triggering, 2 papers; analysis of the effect of earthquake accelerations on the stability of slopes, 2 papers; model studies of the effect of accelerations on slope stability, 2 papers. As requested by the conference organisers this report summarises the main points raised in the papers, sets out common themes, and raises some points for discussion. The various paper titles and authors are set out in Table 1.

2 PALEOSEISMICITY

Two papers cover aspects of landslides and paleoseismicity and attempt to seek explanations for landforms indicative of major instability in the past. Paleoseismicity is a difficult subject. Even when we are witnesses to a landslide event, or get to the site shortly afterwards, there are usually many more questions than answers about the mechanism and cause of the event. Finding ourselves distanced by periods of hundreds or thousands of years multiplies the uncertainties immensely.

Crozier (paper 1) considers ancient landslides in Taranaki, New Zealand, and proposes that these are a consequence past earthquakes in the region. The paper opens with a useful discussion of the methodologies that are available for relating ancient landslides and seismicity. This includes some appropriate cautions and emphasises the difficulties inherent in the work. In particular one comment by the author in reviewing previous New Zealand work of this type seems to be very appropriate. It had been suggested that the MM X isoseismal could be regarded as a threshold for seismic triggering of landslides. As Crozier points out this cannot be universal. The geotechnical explanation is quite simple: the ground acceleration needed to take a slope to a failure state depends on the prior static degree of stability of the slope. Sarma and Bhave (1974) have shown that there exists an approximately linear relation between static factor of safety and the acceleration required to reduce the factor of safety to unity, greater accelerations (therefore larger MM intensity at a given site) are required for slopes with larger static factors of safety. An example of this effect is shown in Fig. 5 of the paper by Zhuping and Nianxue discussed below as paper 5.

The study area has been mapped from aerial photographs and field survey and over 100 landslides identified. It is stated that there appears to be no relationship between the landslides and relative relief or hillslope. This suggests that the mechanisms controlling the instability are not visible at the ground surface or in aerial photographs. In this regard the seismic triggering mechanism is appealing, although it is incomplete as we are not able to explain why some slopes have failed and others not. Consequently I find Crozier's hypothesis that seismic events are responsible for the landslides plausible but not completely convincing. Is it not possible that other mechanisms might be at work, either in conjunction with seismic activity or quite independent of it? The study area is not far from the famous Utiku slip which occurs in similar materials on the Rangitikei river valley. It is well documented that the explanation for this event is a thin clay seam with an unfavourable orientation. It would be of interest to know if there is any

Table 1. Details of the papers assigned to theme S1: Seismicity and Landslides

Topic	No.	Short title	Authors
Paleo-seismicity	1	Paleoseismicity in Taranaki	Crozier
	2	Rock avalanches in British Col.	Naumann & Savigny
Observations of recent EQ landslides	3	Recent EQ landslide in Iran	Anvar, Behpoor & Ghahramani
	4	Puriscal landslide in Costa Rica	Mora
Analysis of EQ effects on slope stability	5	Stability of old landslides	Zhuping & Nianxue
	6	EQ activation of landslides	Kostantinov, Angelov, Lakov, Stojnev & Kostantinov
Model studies of dynamic effects on slopes	7	EQ displacement of slopes	Nomachi, Sawada, Chen & Kida
	8	Dynamics of rock slopes	Shimizu, Aydan & Kawamoto

evidence for the presence of clay seams in the sequence of Pleistocene and Pliocene sediments in the Wanganui Basin.

Naumann and Savigny (paper 2) discuss large rock avalanches in southwestern British Columbia and review the evidence for seismic triggering. Five major rock avalanche events in the vicinity of Hope in southwestern British Columbia have been dated and the dates compared with the paleo-seismic record for the area. If the two sets of dates correlated well then there would be a case for associating the rock avalanches with the seismic activity. The authors found that the correlation was not good and decided to examine one slide, that at Cheam which occurred about 5000 years ago, in depth. It appears that the mechanism of this was easily identified as a wedge bounded on one surface with a clay rich gouge. The Newmark method was used to assess displacements of the slide mass. A parametric study is reported in the paper in which the effect of pore pressure ratio, earthquake return period (horizontal acceleration of 0.15g for a 475 year return period and 0.3g for a 1800 year return period), on the behaviour of the slide was calculated. It is concluded that although seismic activity is a plausible trigger for landsliding in the southwestern British Columbia, which has a wet climate, the slides are more likely to be a consequence of pore pressure effects.

3 RECENT EVENTS

The paper by Anvar et al (paper 3) details a landslide at Roodbar in Iran which occurred after the Manjil-Roodbar M7.3 earthquake in 1990. The slide started one day after the earthquake and continued for several weeks. The length of the slide mass is about 4km, with width varying between 300 and 800 metres and an average thickness of 10 metres. The region is used for the cultivation of olive trees the irrigation of which is done from surface ponds. The earthquake caused sufficient displacement to break water and petroleum pipelines which fed water and oil into the slope. The landslide proper then occurred one day after the earthquake. In this case then the earthquake could be regarded as the trigger for the landslide but the physical understanding of the slide rests in the effect of the water (and perhaps petroleum) discharged into the slope material from the ruptured pipelines.

The authors use a simple infinite slope analysis to illuminate the effect of saturation and to estimate the horizontal acceleration required to initiate failure. Values in excess of 0.5g were obtained.

Mora (paper 4) discusses an active landslide at the site of the town of Puriscal in Costa Rica. As well as seismic activity the region experiences high rainfall. The geology of the slide area is complex and the local soils are derived from weathering and hydrothermal alteration of volcanic deposits.

The landslide has been moving at several centimetres a year and is known to have been active since at least 1918. Water levels in the slide material are high and there are several springs within the slide area. Since March 1990 the region has been subject to seismic activity and seismic swarms culminating in the M7.5 Telire earthquake of 1991. This caused additional movement of the landslide but at the time of writing the paper the author did not have any data on the magnitude of these movements.

The impression gained by the Reporter from these four papers is that seismic activity might well be the cause of landslide activity but it is more likely that other factors, which act over much longer time frames than an earthquake, will also be involved. The earthquake could then act as a trigger for a landslide event at a time when other factors happen to be poised to have destabilising effects. If the earthquake was to be the sole cause of a landslide then sufficient energy would have to be supplied to initiate movements which would cause a loss in the strength of the materials at the site. With a rock mass this could be caused by loosening with a consequent deterioration in shear strength. (This is possibly the explanation for the only recent example of a major earthquake associated landslide in New Zealand: the rock avalanche which temporarily blocked the Buller River after the M7.0 Inangahua earthquake in 1968, Lenson and Suggate (1969)). With soils sufficient displacement would be needed to take the soil into a post-peak reduced shear strength condition. In both of these cases further displacement would follow if the reduced strength of the material was no longer adequate to maintain the static stability of the slope. Thus in both of these situations the earthquake event acts a trigger leading to a reduction in the shear strength of the soil or rock mass with subsequent failure under gravity. The geometry of the lower boundary of the sliding mass could also contribute to instability once an initial disturbance has occurred. Usually the lower surface of a sliding mass is concave or approximately concave. Convexity on the other hand would be expected to be decrease the stability of the slope and the tolerance of the slope to movement.

4 ANALYSIS OF THE DESTABILISING EFFECT SEISMIC EVENTS

Zhuping and Nianxue (paper 4) consider three ancient landslides along the Yangtze gorge in China. It is proposed to construct various water control structures which will raise substantially the river level. The authors investigate the stability of these landslides under seismic actions and also evaluate the effect of rising river level. The geometry of the slide masses are reasonably complex and involves differences in vertical elevations between the top and bottom of 200 to 400 metres. The authors use the Sarma method of analysis to assess the stability. As the rock in the region is closely jointed, the shear strength of the joints is used as the rock mass strength; laboratory shear tests on the soils in both saturated and unsaturated conditions provided the strength parameters for the analysis. A sensitivity analysis was done on the effect of water level, earthquake acceleration, cohesion and friction angle on the factor of safety calculated with the Sarma algorithm. The effects of seismic acceleration, cohesion and friction angle are as expected. The effect of raising the water level 80 m is quite different; for one of the slopes rising river level reduces the factor of safety, for another there is insignificant change in the calculated factor of safety, for the third slope the calculated factor of safety increases as the water level increases. These are interesting and most unexpected results, yet the authors offer no explanation.

Kostantinov et al (paper 5) consider the stability of landslides in Bulgaria which may be re-activated by seismic excitation. They consider the landslide at the town of Kotel which moved during the Vrantcha earthquake in 1977. Insufficient detail is given of the movements of the slope; it is not clear if the movement occurred during the earthquake only, if it was initiated during the earthquake and continued after, or if it was initiated after the event like that described by Anvar et al for the Roodbar slide in Iran. Their approach is unusual in that they mix site response calculations with stability analysis. The stability of the slope was analysed by the Janbu method. The normal stresses on the slice boundaries as well as the pore water pressures were estimated by considering the propagation of P waves through the materials in the slope. The shear stresses were obtained from considering the propagation of S waves. The layered soil profile at the site was converted to an equivalent single layer by conventional site response methods. The authors justify their approach by calculating a static factor of safety of 1.17 for the slope and 0.79 during the earthquake. Perhaps more justification is required as the stresses calculated with the wave propagation method assume elastic behaviour of the materials whereas the stability analysis is based on a failure mechanism and a pseudo static analysis. Also the wave propagation calculations assume undrained behaviour of the materials whereas the stability analysis seems to be based on drained strength parameters for the clay in the failure zone.

5 MODEL STUDIES OF THE EFFECT OF CYCLIC ACCELERATION ON SLOPE STABILITY

Nomachi et al (paper 7) describe a theoretical and experimental investigation of the displacements in a soil slope during earthquake excitation. They consider the stability of a slope in which a failure surface in the form of a logarithmic spiral is analysed using the upper bound theorem of limit analysis. The logarithmic spiral surface is chosen on theoretical grounds as this is the critical failure surface shape for a homogeneous frictional material. It is of interest to note that the papers reviewed herein which give landslide morphology (papers 3, 5 & 6) all indicate failure surfaces very different in shape from the logarithmic spiral. The rigour of the upper bound theorem of limit analysis is reason for the choice of this technique rather than conventional limiting equilibrium calculation. Unfortunately the assumption of an associated flow rule for the Mohr-Coulomb failure envelope means that at the end of the process these calculations, just as for the conventional limit equilibrium method, are not without criticism. In addition Chen (1974) shows that the stability numbers calculated with limit analysis and limiting equilibrium methods are for all practical purposes identical. This suggests that refinement in analytical techniques for landslide stability analysis are unlikely to lead to improvement in judgements of stability in practice. The real area of difficulty in assessing the stability of natural slopes is in understanding failure mechanisms, gaining information about groundwater conditions, and obtaining representative values for strength parameters.

The paper outlines how rotations of the soil above the logarithmic spiral can be calculated for a given earthquake input motion. The displacements predicted by the method were compared with measured displacements of a model slope mounted on a shaking table and subject to sinusoidal excitation. The authors conclude that the agreement is good.

Shimizu et al (paper 8) address the very difficult question of the dynamic response of a closely jointed rock mass. They measure the response of a slope made by staking small wooden blocks to form a regular joint pattern. The experiments provide insight into mechanisms of deformation in the model slope. The difficulty in applying this insight in anything more than a qualitative manner relates to inevitable limitations of 1g models of jointed rock masses. In the opinion of the reviewer there are two problems in the work reported in the paper. Firstly, the properties of jointed rock masses are known to be size dependent. As the size of the rock mass increases the strength decreases. This makes application of the model results to the prototype scale difficult. Secondly, Fig. 3a of the paper gives shear stress - displacement curves for the joint surfaces between the wooden blocks. These results show a peak strength - residual strength effect that becomes more apparent with increasing normal stress. The reality of real joint surfaces in rock is for this effect to be suppressed with increasing normal stress. This difficulty could be circumvented by using a modelling material for a jointed rock mass like that developed by Barton (1972).

6 REFERENCES

Barton, N. 1972. A model study of rock joint deformation, *Int. J. Rock Mech. Min. Sci.*, Vol. 9, pp. 579-602.

Chen, W. F. 1975. *Limit analysis and soil plasticity*, Elsevier, Table 9.3, p. 413.

Lenson, G. J. and Suggate, R. P. 1969. A preliminary report on the Inangahua earthquake New Zealand: Geology, May 24, 1986. *Bulletin of the New Zealand National Society for Earthquake Engineering*, Vol. 2 No. 1, pp.19-23.

Sarma, S. K. and Bhave, M. N. (1974) Critical acceleration versus static factor of safety in stability analysis of earth dams and embankments, *Geotechnique*, Vol. 24 No. 4, pp. 661-664.

Landslides, Bell (ed.) © 1995 Balkema, Rotterdam, ISBN 90 5410 032 X

Earthquake damage to a low cost road in Nepal

J. R. Boyce
Loughborough University of Technology, UK

ABSTRACT: The Dharan-Dhankuta Road was seriously damaged by the Nepal Earthquake of 21 August 1988 and subsequently by a locally intense monsoon storm on 12 September.

Earthquake effects on slopes varied significantly according to rock type. The worst affected were the steep rock slopes and cliffs of brittle fractured quartzite along the Sangure ridge crest. The weathered Siwalik sandstones/siltstones and the meta-sandstones/phyllites did not experience the same intensity of slope failure, however, detailed field inspections revealed extensive cracking of natural slopes above and below the roadline. Along the southern half of the road, many major road retaining structures suffered severe earthquake damage including 30 masonry and gabion retaining walls which showed significant signs of distress.

The monsoon storm affected primarily the Leoti Valley area causing extensive rock, soil and mud slides with 14 complete road blockages. Breaches of river walls and over-topping of tributary bridges caused further damage. There is evidence to suggest a temporary damming of the Leoti Khola by reactivation of an existing major slip, subsequently breached by the rising water caused a flood surge. Given the history of monsoon related damage in 1984 and 1987 and the detrimental effect of the 1988 earthquake, it is anticipated that this section of the road is vulnerable to further damage.

The paper presents details of the slope movements and retaining wall distress on the road. It is concluded that the earthquake has created instabilities in the already fragile environment of the Dharan-Dhankuta road which may well cause increased maintenance problems from future monsoon storms.

1 INTRODUCTION

The Dharan-Dhankuta Road is a low cost mountain road constructed in the period 1976 to 1982 from Dharan, a market town on the Nepalese plain, to Dhankuta, a regional capital in the Himalayan foothills to the North. The road crosses extremely rugged terrain with extensive slope retaining structures as shown in Figure 1. The design of this road has been the subject of several published papers including Brunsden et al (1981) and Fookes et al (1985).

The Road was seriously damaged by the Nepal earthquake of 21 August 1988 and subsequently by a locally intense monsoon storm on 12 September 1988. The aims of this paper are to describe and assess the most significant aspects of this damage.

The earthquake occurred at 0454 hours on the 21st August. According to HMG's Department of Mines and Geology, the epicentre was located at latitude 26.78°N, longitude 86.61°, in the region of Udayapur, some 65 km to the west of Dharan. The focal depth was relatively shallow, of the order of 60km to 10km. The surface wave magnitude has been assessed at Richter M 6.6. It is of interest to note that the epicentre lies close to one of the two concentrations of seismic energy release along the southern boundary of the Himalaya (Figure 2). Its position at Udayapur is within the low Siwalik foothills, about 5km NE of the edge of the Terai and some 12km south of the main Himalayan Boundary Fault. From eye-witness accounts by staff at Base Camp in Dharan, the earth tremors occurred in a series of pulses over a period of approximately 40 seconds. Over 200 people were killed in Dharan and many old brick and wooden buildings suffered major structural damage. The evidence suggests intensities of classes VIII to IX on the Modified Mercalli scale. The last Nepalese earthquake of comparable magnitude was in 1934.

On the night of 12th September, an electrical storm occurred in the Dharan/Leoti Valley area, producing heavy rain during the late evening and

Figure 1. Part of Khamingtar Hairpins

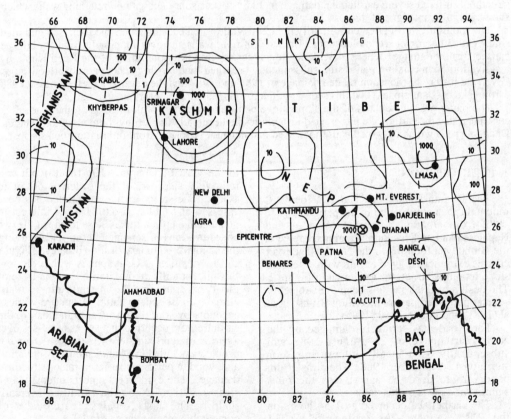

Figure 2. Seismicity of the Himalayan region (after Lomnitz, 1974) showing epicentre of the M6.6 earthquake on 21:8:88 (Seismicity contours are numbered in ergs per sq. km. per year)

continuing with lighter, steady rain until the early morning hours. Rainfall measurements taken the following morning indicated up to 150mm in the Dharan area but none at the Tamur river crossing, which is not exceptional. However it's believed that the storm was a highly localised event, not uncommon in this region, with rainfall in the Lesti Valley considerably in excess of available surrounding measurement and with a damaging effect considerably enhanced by the preceding earthquake. This is evident by observations of the degree of damage within the Leoti Valley and surrounding hillslopes, the observed flood levels and extent of water-borne deposits in the Leoti, and the absence of rainfall at the Tamur Bridge site. Further features of the storm were the inundation of the drainage system with water-borne material, resulting in numerous complete blockages of major inlet structures, culverts, side-drains and outlet channels, the extensive loss of previously stable vegetated side slopes and the backing-up of the Tamur River at Tamur Bridge, the confluence with the Leoti Khola, for some 15 minutes before midnight on 12th/13th September.

2 SLOPE DAMAGE

Serious earthquake damage to the road was confined to the first 26km from Dharan, i.e. on either side of the Sangure Ridge (Figure 3). The effects on slopes varied significantly according to rock type. The worst affected were the steep rock slopes and cliffs of brittle fractured quartzite along the Sangure Ridge crest. Instability was widespread on these slopes, ranging from small rockfalls of a few cu.m to large rockslides of several thousand cu.m. The largest slope failures directly affecting the road occurred in the major quartzite rock cuts. Elsewhere, the weathered Siwalik sandstones/siltstones and the meta-sandstones/phyllites of the Sangure Series did not experience the same intensity of slope failure. However, detailed field inspections revealed that extensive cracking of natural slopes has occurred in some areas above and below the roadline. In most cases the cracking was only observed in the upper 2-3m of colluvial soil overlaying the fractured rock, but in some locations the ground disturbance clearly extends deeper into the near-surface rock layers.

There was also some tentative evidence that much deeper-seated disturbance had occurred. For example, the Seoti Khola water supply pipeline was displaced at frequent intervals by up to 200mm over a 0.5km length running around a spur just to the east of the Base Camp. This suggests that large-scale spreading and relaxation of the whole spur had taken place.

The implications of general ground disturbance in terms of the scale of further major slope failures were very difficult to assess at the time of inspection (October - November 1988). The worst distress appeared to be concentrated just upslope of steep ridge crests, sharp convex breaks of slope (e.g. above the crest of cut slopes), and on the outside of road hairpins where the ground falls away steeply below.

It was not possible in the time available to undertake a comprehensive assessment of all aspects of damage along the roadline. Of necessity, field investigations were selective and only the main areas of concern were assessed in detail. It is likely that there is significant potential for future damage arising from the effects of the earthquake, particularly during subsequent monsoon seasons.

2.1 Khamlingtar Hairpins

No part of this section of the roadline failed completely during the earthquake or monsoon storm but some large cut slope failures caused complete blockage. There was also extensive pavement cracking along many of the retaining walls and at several hairpin sites. Off the road line, severe ground disturbance occurred in several areas and serious erosion damage took place below some of the drainage outlets. In view of the high vulnerability of the hairpin stack, a detailed field inspection was carried out in this area.

Three large cut slope failures stripped off a 2-3m thick layer of thin colluvium and highly fractured, weathered sandstone/siltstone, leaving bare scars running for up to 120m upslope. Part of a 6m high gabion revetment failed at the first of these sites. The debris from the other two slopes blocked the two main culverts in this section.

Above these slips, ground distress extended all the way to the ridge crest running ESE. The degree of cracking in this area was more severe than at any other location inspected during the investigations. The most disturbing feature was a 100m long, 3-5m high, bare scare running approximately parallel to the road some 150m upslope. There was extensive cracking in the slope below the scar and above the head of the cut slope failures. Other large cracks extended up to the ridge crest.

The worst signs of distress within the hairpins were pavement cracks behind many of the larger retaining walls and at the hairpin sites.

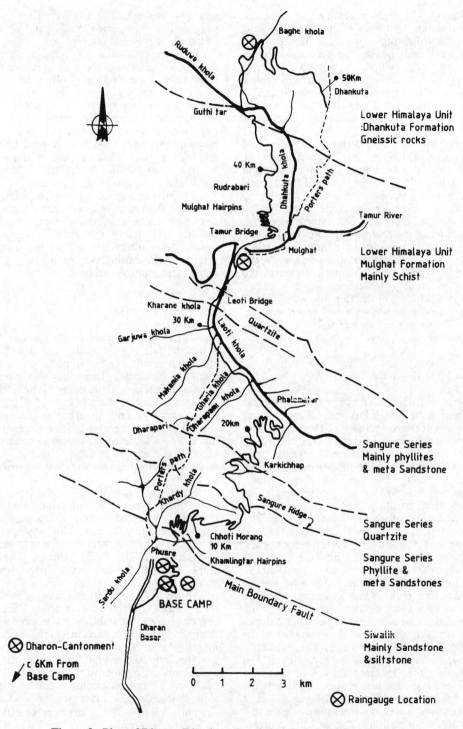

Figure 3. Plan of Dharan-Dhankuta Road showing simplified geology.

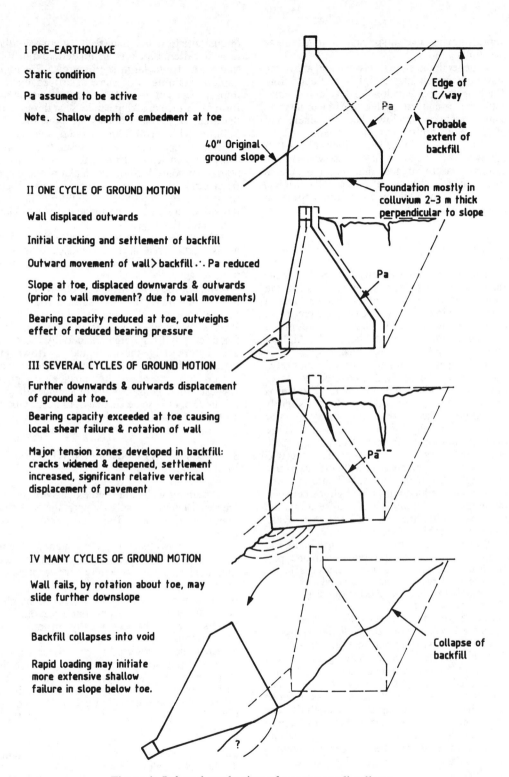

I PRE-EARTHQUAKE

Static condition

Pa assumed to be active

Note. Shallow depth of embedment at toe

40" Original ground slope

Pa

Edge of C/way

Probable extent of backfill

Foundation mostly in colluvium 2–3 m thick perpendicular to slope

II ONE CYCLE OF GROUND MOTION

Wall displaced outwards

Initial cracking and settlement of backfill

Outward movement of wall > backfill ∴ Pa reduced

Slope at toe, displaced downwards & outwards (prior to wall movement? due to wall movements)

Bearing capacity reduced at toe, outweighs effect of reduced bearing pressure

Pa

III SEVERAL CYCLES OF GROUND MOTION

Further downwards & outwards displacement of ground at toe.

Bearing capacity exceeded at toe causing local shear failure & rotation of wall

Major tension zones developed in backfill: cracks widened & deepened, settlement increased, significant relative vertical displacement of pavement

Pa

IV MANY CYCLES OF GROUND MOTION

Wall fails, by rotation about toe, may slide further downslope

Backfill collapses into void

Rapid loading may initiate more extensive shallow failure in slope below toe.

Collapse of backfill

?

Figure 4. Inferred mechanism of masonry wall collapse.

1947

Some damage was also caused to drainage structures.

In general, the slopes between and above the hairpins were in surprisingly good condition. The probability of a major slip developing which could undermine the whole stack of hairpins was considered to be very low. The most likely type of future instability would be shallow (2-3m deep) movements in the steep slopes (35º - 40º) but even here the signs of distress were not so extensive as to suggest that major problems would occur in the next monsoon. The original choice of route in this area therefore appears to be sound.

2.2 Karkichhap

This area suffered serious damage during the earthquake and to a lesser extent during the monsoon storm. In view of the complete loss of the roadline in part of this area and the precarious condition of many of the other retaining structures, this area was given the highest priority during the investigations.

The slopes are formed mostly in intensely-fractured, weathered phyllite covered by a layer of colluvial soil from 1-3m thick. The soil is generally a loose to medium dense, slightly silty/clayey gravel containing many angular fragments of phyllite and occasional large quartzite boulders. A thick band of quartzite occurs part way down the slope, forming two large pinnacles with near-vertical cliff faces up to 60m high. A few thin quartzite bands can also be seen interbedded with the phyllites between the pinnacles. An amphibolite dyke forms the prominent ridge line and has been extensively quarried for gabion stone.

Slope damage caused by the earthquake and monsoon storm can be split into four main types:
a) shallow cut and natural slope failures in soil;
b) deeper failures in soil and weathered rock;
c) rockfalls from the quartzite pinnacles; and
d) surface cracking.

There were two deeper failures in soil and weathered rock. At one the slide debris moved over 100m downslope into the breach between the two quartzite pinnacles, leaving a large bare rock scar. Below the breach, the debris formed a jumbled mass of soil, rock fragments and broken trees, mixed with large blocks of fallen quartzite. The large slip took place during the monsoon storm, causing complete blockage of two roadlines. The scar extended for about 120m upslope, almost to the ridge crest marking the boundary of the Dharapani Khola catchment (see Figure 3).

The earthquake dislodged many large blocks of quartzite from the pinnacles. The mass of the largest fallen block is estimated to be 300 tonnes, and two large sections of 5-7m high gabion retaining wall were destroyed by the impact. Signs of general slope disturbance in the form of surface cracking can be seen throughout the area. Individual cracks are typically 50-300mm wide, 100-300mm deep, and show relative vertical displacement of up to 500mm. They vary in length from a few metres up to 50m (commonly 5-20m) and in plan shape range from straight to sharply concave downslope. Most are aligned parallel or slightly oblique to the slope contours. The intensity of cracking varies considerably.

About 70% of the total length of retaining walls in this area showed significant signs of distress and substantial damage was also caused to revetments.

Most of the instability occurred in the top 2-3m of colluvial soil. Also, surface cracking in general appeared to be confined to this soil layer. Where exposed, the underlying intensely-fractured phyllites were considerably loosened and disturbed by the earthquake. However, there are relatively few signs of incipient deeper-seated movement. The surface evidence suggested that the probability of a major hillslope failure (i.e. on a scale sufficient to destroy both sections of roadline) was low.

3 ASSESSMENT OF RETAINING WALL STABILITY

Retaining walls are used extensively on the Dharan-Dhankuta Road to support the pavement over sloping ground. There are essentially two types:
a) Masonry walls (up to 5m retained height) constructed of local stone bonded with cement mortar. These walls generally have construction joints every 6m.
b) Gabion walls (up to 14m retained height) constructed of local stone hand-packed into interlocking wire boxes generally 1m square and 2 or 3m long.

Revetments are used to provide support to steep soil and weak rock cut slopes above the road line. They are designed to stabilise the surface only and do not act as full retaining walls. Most are formed in gabion 2-3m deep, typically up to 8m high and stepped back 0.5m on each course.

Retaining wall stability was reduced throughout the area by two main factors; ground displacement below the wall toes and the effects seismic loading on walls and backfills. The influence of additional loading by slips above the road is thought to be negligible. In most cases the slip debris did not run out across the full width of the backfill. Even where this occurred it seems unlikely that upslope ground movements would have preceded movement of the walls.

Apart from the two failed sections many masonry walls were displaced outwards by 0.5-3.0m. In general, the worst affected walls were not the highest. Walls with retained height as low as 2m have failed while others as high as 6m have moved but remained serviceable. From inspection of as-built elevation drawings, the failed and badly-distressed walls are generally those with a shallow depth of foundation (0.5-1.5m at the toe). Those with deeper embedment (2.0-2.5m) generally remained in a serviceable condition.

The inferred mechanism of masonry wall collapse is shown in Figure 4. Whilst the relative timing of displacement of the wall, the backfill and the ground below the toe is uncertain, it seems clear that the problem is essentially one of inadequate toe embedment and high toe bearing pressures. Both seismic loading on the wall and ground displacement below the toe would cause outward displacement of the toe into a region of reduced bearing capacity. The effect of seismic loading on the backfill is more difficult to assess. Densification and settlement as a result of vibration appears to be a common feature of all the backfills because of poor compacting during construction. In a few cases, cracking is concentrated at the rear of the backfill, but more commonly the largest cracks have occurred in the outer half of the pavement close to the wall crest. Deep tension cracks have developed within the backfill in these areas, implying that the backfill displacement as a whole has been less than that of the wall itself. In such cases, earth pressures on the wall are probably reduced compared with the static condition. In critical cases, this reduction in earth pressure has not been sufficient to outweigh the effects of reduced bearing capacity at the toe.

Many of the gabion retaining walls, especially those over 6m high, have suffered from gradual settlement over a number of years since the road was completed. This has caused settlement of the crest together with some outward movement, and also formation of a longitudinal crack and rut in the pavement some 1-2m from the outer edge. In higher walls the lower courses of gabion boxes are noticeably compressed (by up to 20%). Gabion boxes filled with platy fragments of schist or mudstone have generally suffered worse than those filled with quartzite, although the variable quality of

stone packing is also a significant factor.

In many cases, the settlement was increased by the earthquake. This is to be expected in an uncemented structure with predominantly granular fill behind, especially since backfill construction was carried out without the use of heavy compaction plant. None of these walls appear to be at risk from immediate collapse and it is likely that settlement in the future will be less than has occurred in the past.

At a number of wall sites the pavement surface has suffered cracks up to 100mm wide which cannot be readily accounted for by settlement of the wall or ground movement below the foundation. Most of these walls are on the upper part of the Sangure Ridge and it appears that the seismic shock has caused some outward movement of the crest of the walls and consequent cracking of the pavement.

An analysis of maximum crest movement of distressed walls compared to their height showed that settlement of gabion walls generally increased with height, whereas other types of movement showed no correlation. However, it should be appreciated that overall wall movement is likely to be due to a combination of several causes in many cases. In general gabion walls have been found to be preferable to masonry walls being cheaper and easier to construct in the difficult terrain and also suffering less damage from the earthquake. Several gabion walls remain serviceable after considerable movement (up to 1m).

4 LEOTI KHOLA

7km of the road running alongside the Leoti Khola was not affected by the earthquake on 21st August. However, the monsoon storm on 12/13th September caused exceptionally high flood levels in the Leoti which damaged river protection works and road retaining walls. Also, three bridges over tributaries of the Leoti were blocked by large fans of alluvial sediment which accumulated during this one storm.

Previously, damaging floods in the Leoti have occurred in 1984 and 1987. The 1987 flood destroyed 42 metres of gabion retaining wall and removed half of the road width at one location and some 25 metres of retaining wall at another. Both walls were reconstructed in masonry during the 1987/88 dry season and protected by gabion groynes to deflect flood flow from these vulnerable locations.

The 1988 flood appears to have reached a similar or higher level than in 1984, but the damage caused was less severe. A 150m length

Figure 5. Rockslide debris causing blockage of Leoti Khola (note, human figure on left hand side giving scale).

of road has been washed away and several large retaining walls have been undermined. No bridges have been damaged, however. Local reports speak of water levels above road level but the observed flood marks do not support this.

The exceptional flood level in 1988 cannot be accounted for by rainfall records. There would appear to be two possible causes, which may be linked, viz:
a) a localised cloudburst of extremely high intensity in the headwaters of the catchment;
b) temporary blockage of the Leoti Khola upstream from the road by a landslide dam, followed by a flood surge when the dam was overtopped.

The second possibility is suggested by the existence of a large fan upstream of the road which has clearly impeded flow in the Leoti (see Figure 5). The rockslide and debris flow on the northeast side of the Leoti feeding this fan has been active for a number of years. It was observed to have increased in extent substantially after the 1988 monsoon.

Having now experienced three exceptional floods in recent years, it would appear that normal analysis to determine flood return periods are not applicable in this type of terrain especially where seismic disturbance is a possibility.

CONCLUSIONS

The effects of this earthquake in 1988 on the Dharan-Dhankuta road in Eastern Nepal provide further evidence of the extent to which roadworks constructed across steep slopes are vulnerable to seismic induced damage. Full investigation was not possible but useful lessons can be drawn for the design of future roads in this type of terrain.
1. Gabion retaining walls are easier to build than masonry walls and are less susceptible to earthquake damage.
2. Soil and rock loosened by an earthquake will have a serious effect on drainage ranging from blockage of culverts to complete damming of rivers.
3. The need for long term maintenance on such roads cannot be avoided and must be considered at the planning stages.

ACKNOWLEDGEMENT

The author is grateful to Roughton and Partners International and to Kingdom of Nepal, Department of Roads for the opportunity to assess the earthquake damage to the Dharan-Dhankuta Road

REFERENCES

Brunsden, D., Jones, D.K.C., Martin, R.P. and Doorkamp, J.C. "The Geomorphological character of part of the Low Himalaya of Eastern Nepal." Z. Geomorph. N.F., Suppl. BD 37, 25-72 Berlin Stuttgart, Feb. 1981.
Fookes, P.G., Sweeney, M., Manby, C.N.D. and Martin, R.P. "Geological & Geotechnical engineering aspects of low cost roads in mountainous terrain." Eng Geol., 21: pp 1-152. (Elsevier Science Publishers), 1985.
Lomnitz, C. Global Tectonics and Earthquake Risk. Elsevier, Amsterdam, 320p, 1974.

Landslides, Bell (ed.) © 1995 Balkema, Rotterdam, ISBN 90 5410 032 X

Earthquake-induced landslides in the island of Ischia (southern Italy)

F.M.Guadagno & R.Mele
Dipartimento di Geofisica e Vulcanologia, Università di Napoli 'Federico II', Italy

ABSTRACT: Most of the landslides occurred at Ischia in the last 2000 years, were connected with seismic events. These slope failures, falls, slides and debris flows, have influenced the physiographic aspect of the island, and can constitute a secondary hazard of considerable importance in the case of seismic events.

1 INTRODUCTION

Seismic activity is a very important factor in triggering landslides in rocks and soils. These earthquake-induced slope failures can involve land masses that were never influenced by landslides or reactivate old landslides (Keefer, 1984).

In volcanic areas, characterized by seismic activity, seismogenetic landsliding plays an important role in the morphological evolution of the volcanic structures.

On the volcanic island of Ischia the geomorphological investigations as well as historical and archaeological evidence, suggests that the presence of landslides is generally connected to the volcano-tectonic evolution and therefore also to the earthquakes.

The aim of this paper is to present the preliminary results of a research carried out to identify the areas affected by landslides triggered by earthquakes, the main mechanisms involved, and the geotechnical properties of the involved terrains. This basic aspect is vital in defining the risk in an area with intense urbanization.

2 GEOLOGICAL AND STRUCTURAL SETTING

Ischia is a volcanic island at the western edge of the gulf of Naples (Fig.1). Its history is very complex, even if it developed in a quite short time (Rittman and Gottini, 1980; Vezzoli, 1988). The beginning of the volcanic activity is not precisely known, but the oldest dated rock erupted 250,000 B.P. The volcanic activity has been subdivided into five periods and the

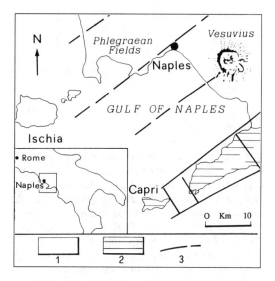

Fig.1. Geological sketch map of the Naples area. Legend: 1) Plio-Pleistocene and volcanic deposits; 2) Mesozoic carbonate unit; 3) Principal faults.

last volcanic event was The Arso eruption dated 1301 A.D. Fumaroles, the well known thermal springs, and recent shallow earthquakes, the last one in 1883, testify a still persisting state of activity of the magmatic system.

The most important structure of the island is Mt. Epomeo, a volcano-tectonic horst (787m), defined by a quadrangular shaped fault system (Fig.2). Its resurgence is connected to an increase in pressure in the magma chamber. The uplift began less than 33,000 years before present but movements along the

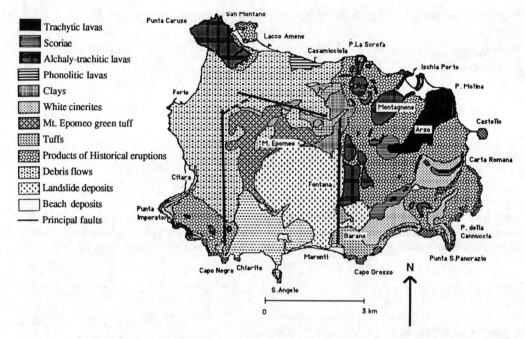

Fig.2 Semplified geological map of the Island of Ischia.

Legend:
- Trachytic lavas
- Scoriae
- Alchaly-trachitic lavas
- Phonolitic lavas
- Clays
- White cinerites
- Mt. Epomeo green tuff
- Tuffs
- Products of Historical eruptions
- Debris flows
- Landslide deposits
- Beach deposits
- — Principal faults

perimetical faults, causing earthquakes, have occurred in historical times.

A differential uplift tilted the Mt. Epomeo block towards the S-SE. The homoclinal ridge, primarily constituted by Mt. Epomeo Green Tuff, is limited by marginal and subvertical faults along the northern and western slopes. Otherwise along the eastern side there are numerous volcanic structures coeval to the resurgence of the horst. During the uplift, debris flows occurred on the dip slope of the homocline structure and their well-consolidated deposits, which reach 200 m in thickness, hide the green tuff.

The North and East structural escarpments show steep inclinations and it is along these slopes that landslides occurred during the uplift of the Epomeo horst. This paper refers, in particular to those landslides that have occurred in the previous twenty centuries.

At the base of these slopes there are landslide deposits which can reach a considerable thickness (>50m). From the morphological point of view, in these areas there are numerous accumulation lobes which indicate prevalent mechanisms of debris slides and debris flows.

3 GEOTECHNICAL PROPERTIES OF THE TERRAIN INTERESTED BY LANDSLIDES.

The Mt. Epomeo Green Tuff is mainly composed of a typical welded ignimbritic deposit, in which pumices and biotite and alkalifeldspar cristals are identified. The dry unit weights vary from 11 and 14 kN/m^3. The water content is generally 8-10%. Unconfined compressive tests show strength values in the field of 25-55 MPa. The tuff can be classified as a weak rock with a low unit weight and limited strength.

The tuff masses of the Northern outcrops show subvertical fractures which follow the main tectonic directions (N-S, E-W). Their spacing varying between 50 cm and 4-5 m. These discontinuities have subdivided the mass into polyhedric blocks whose dimensions vary between a few cubic centimetres to about ten cubic meters. Along the fractures, which are generally open, there are often traces of fumarolic activity.

The tuff escarpments of Mt. Epomeo are covered, by a significant mantle of deposits, with variable grain sizes, deriving from the alteration of the green tuff. The weathering is caused from the normal process of rock degradation, such as exfoliation, and also from the effects of the fumaroles. The slope deposits together with the tuff blocks are those

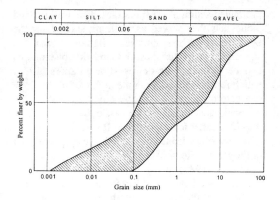

Fig.3 Grain-size distribution of the matrix of the slope debris
mantle and landslide deposits.

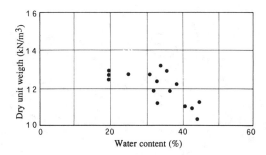

Fig.4 . Unit dry weight vs. water content for the landslide
deposits.

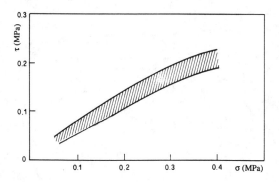

Fig.5. Peak shear strength of undisturbed sample.

usually affected by landslide events which induce the
retreat of the escarpments.

Classification tests were carried out on samples of
the matrix of the debris mantle and the landslide
deposits. Shear strength parameters were
determined for the landslide deposits from laboratory
tests on undisturbed samples.

Grain size analyses show that debris mantle and
landslide deposits rarely contained a significant
proportion of argillaceous material (Fig.3), and,
therefore, the sandy fraction generally prevails. The
sandy elements are generally composed of tuff
fragments with different degrees of weathering.

The Atterberg limit tests classified the clayey
materials as low plasticity soils with average liquid
and plastic limits of 35.4 and 16.7 % respectively.

The laboratory measurements show that the dry unit
weight varies between 10 and 13.5 kN/m^3 and
natural water content has a variation field of 20-45%
(Fig.4).

Notwithstanding the difficulty in obtaining
representative block samples, consolidated drain
shear box tests were performed on undisturbed
specimens and the results are shown in Figure 5. In
connection with the grain-size characteristics, there
is a wide variation of the shear strength parameters.
On average from these data a peak shear strength of
c' = 0.015 MPa and φ'= 32 degrees was calculated.

4 LANDSLIDE PHENOMENA

Historical analysis, supported by field surveys and
archaeological data, allowed a reconstruction of the
major landslide phenomena triggered by
earthquakes.

We shall now deal with the events, whose location
has been determined with a good degree of
approximation along the North and West slopes of
Mt. Epomeo. Going backwards in time, landslides
induced by earthquakes occurred as follows (Table
1).

Table 1. The earthquahes of the Island of Ischia. The
asterisks indicate the landslide events.

Year	Sites	Max Intensity Degree MCS	
1228	Casamicciola	IX-X	*
1302	Eastern Side	VIII	
1557	Campopagano	VII-VIII	
1762	Casamicciola	VII	
1767	Northen side	VII-VIII	
1796	Casamicciola	VII	°
1828	Casamicciola	VIII-IX	*
1841	Casamicciola	VIII	
1863	Casamicciola	VII	*
1867	Casamicciola	VI-VII	
1881	Casamicciola	IX	*
1883	Casamicciola	X	*

i) Three mayor landslides occurred on the North side of Mt. Epomeo as a result of the disastrous earthquake of Casamicciola on the 28th July 1883. Two slides occurred near the town and one close to Mt. Rotaro (Fig. 6). As pointed out by Mercalli (1884) and Johnston-Lavis (1885), there were a few falls along the marine cliffs, so that an underwater eruption was hypothesized due also to the great number of pumices floating on the water.

The two major landslides involved about 150,000 m^3 of material consisting mainly of highly alterated green tuff blocks of varying dimensions and of the above-mentioned matrix. The landslides continued to expand in the following days, due to the heavy rainfalls. Of great interest is Mercalli's (1884) observation that the failure area of the landslide was partly reddish and white in colour due to the

protracted action of the fumaroles. This well defined plane has thus served as a sliding surface for the landsliding debris mantle. Because of these significant characteristics, the landslide phenomena can be described as debris slide (Varnes, 1978) which found their origin in the high part of the slope characterized by high slope angle ($\alpha=38°$). The materials accumulated at the base of the slopes where they have latter fed slight debris flows.

The area of the landslides borders the epicentral area of the earthquake of 28th July.

Cubellis (1985), using macroseismic techniques, has estimated the focal parameters. The seismic source was placed at a depth of 1 km and was characterized by a horizontal dimensions of 2 km. The average magnitude (M) value, computed with different parameters and methods, was of 4.6 (4.2-

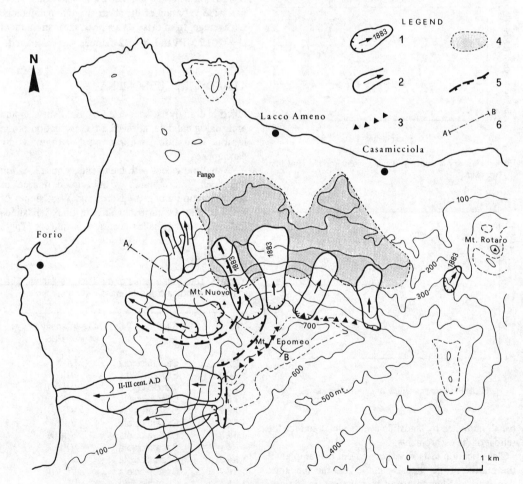

Fig. 6. The landslides of the North-West sector of the Island of Ischia. Legend: 1 Debris slide and year; 2) Debris flow; 3) Fall scarps (1881, 1863, 1828, 1797); 4) Area affected by surficial landsliding in the earthquakes of 1881 and 1828. 5) Supposed rear scarp for the 4th century collapse; 6) Section (Fig. 8)

5.5). Moreover, a high intensity in the epicentre area (Io = X degree MCS) was estimated and justified by the shallow depth of the source. The area of landslide failures can be considered at a minimun distance of 500 meters from the epicentre, on the basis of the reconstruction of the earthquake parameters.

Figure 7 shows the relation between magnitude and distance of landslides from the epicentre. It should be noted that the Casamicciola landslides caused by the earthquake of the 28 July, agree favourably with the Keefer historical world-wide data.

ii) In the years 1863 and 1881 earthquakes in the same epicentral area as in 1883, had already caused falls along the steeper slopes of Mt . Epomeo (Fig.6)

In particular, the two earthquakes of estimated intensity between the VIII and IX degree MCS, caused the rupture of large green tuff blocks which rolled down the Northern slopes of Mt. Epomeo (Johnston-Lavis, 1885). There is no trace today of such materials, due probably to the intense anthropic activity. Referring to the 1881 earthquake only, Mercalli (1884) says that in the Casamicciola area "wherever the soil was in steep inclination there where landslides". These widespread events are due to debris slides involving surface soil.

iii) In 1828 following the earthquake of the 2nd February (VIII-IX degree MCS) along the fault slopes of Mt Epomeo between Casamicciola and Fango, there were falls of large tuff blocks (Morgera, 1890, and Mercalli, 1884). Also, the chronicles of the time report several fractures in the soil along the slopes between Casamicciola and Lacco Ameno. These phenomena can be related to the partial mobilization of landslide debris on which there are the built-up urban areas of Casamicciola and Lacco Ameno (Figs. 1 and 6).

iv) On the 18 March 1796, again in the Casamicciola area, there was an earthquake of the VII degree MCS. No destructive landslide events were reported on this occasion. But the following years, on the 14th of December 1797 there were falls of large tuff blocks and debris slides along the steeper slopes of Mt. Epomeo. De Siano's report (1798) shows surprise that the landslides were not the result of rainfall or an earthquake. These phenomena, described as rather widespread, can be connected to the seismic event of the previous year, and therefore considered as "retarded"phenomena.

v) Between 1796 and 1303 earthquakes of varying intensity between the VII and VIII degree MCS were reported, and they did not cause important

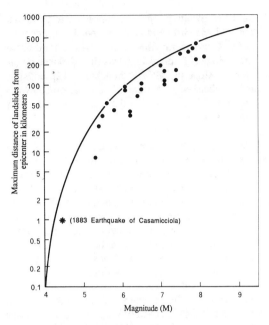

Fig.7. Keefer (1984) correlation between earthquake magnitude and distance from epicenters to landslides for slides and falls.

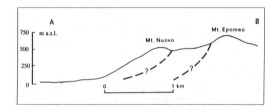

Fig.8. Supposed slip surfaces during the 4th century AD earthquakes.

landslides. In fact, the few existing chronicles do not report destructive events.

vi) In the year 1228 an earthquake of X degree MCS caused a serious landslide briefly described by Ryccardi de Sancto Germano (Muratori, 1937): " In the month of July, the mount of Ischia slipped and buried in their homes about seven hundred people, men and women". This event, in the North-western sector of Mt. Epomeo, can be attributed to a slide phenomenology of large tuff blocks and probably to debris flows. In this sector of Mt. Epomeo it is easy to identify the morphologies of source and accumulation areas.

vii) Historical chronicles also reported considerable

landslide events in previous times and in particular the 2nd and 3rd century A.D. and the 4th century B.C. Concerning the first two phenomena, archeological findings in debris flow deposits on the western slope of Mt. Epomeo, show that landslide events occurred at that time. In this case such events can probably be linked to the coincidence of heavy rainfalls and earthquakes.

With regard to the events of 4th century B.C. it is interesting to note that both Strabo and Plinius (Buchner, 1943 and Buchner,1986) report extensive landslides and earthquakes. In this case, considering the present morphological conditions of the area (freshness of the rear scarps), and the great relevance of the phenomena at that time, it is possible that the reactivation of the perimetrical faults caused a linear eruption and consequent landslides (Fig.6). In particular, in the Monte Nuovo area tuff masses have probably slided on deep concave surfaces (Fig.8) and relevant debris flows have flooded the Forio area. In fact, in this area, archaeological findings area always subsequent to 3rd century A.D (Monti, 1980).

CONCLUSIONS

Geomorphological surveys as well as historical and archaeological data show some fundamental aspects useful in defining the risks in the Island of Ischia.

i) The major landslide events of historical times occurred on the Northern and Western slopes of Mt. Epomeo triggered always by earthquakes. Infact, in historical times, analogous phenomena do not appear to have occurred in the area during the period of seismic quiescence.

The perimetrical faults of the horst, considered active, caused, in historical times earthquakes of intensity between VI and X degree MCS. The magnitude of Casamicciola earthquake in 1883 was estimated as 4.6. The shallow depth of ipocente (<1km) caused considerable damage. It is probable that this magnitude can be considered the maximum value for this volcanic system.

ii) The historical and geomorphological analysis show the recurrent fall of blocks from the tuff cliff of Mt. Epomeo. These phenomena are favoured by the type of fracture present in the tuff rocks, as well as by the degradation phenomena which tend to increase the fracture opening. Recurrent landslide mechanisms are also those related to the sliding of debris masses along the slopes. These deposits, due to the alteration process of the green tuff appear to be in a metastable condition (friction angle next to slope angle) as a result of which a seismic shaking can induce landslide phenomena.

iii) The analysis of the data at present available shows that on the Island of Ischia, events between the VII and VIII degree of Modified Mercalli scale induce falls along the cliffs of Mt Epomeo. Earthquakes of higher energy (M>4) can cause debris slides and deep translational slides. These considerations are also confirmed in the Keefer (1984) correlations in which deep events need greater energy.

REFERENCES

Buchner, P. 1943. Formazione e sviluppo dell'isola d'Ischia. - Rivista di Scienze Naturali "Natura" , 34: 40-62 .

Buchner, G. 1986. Eruzioni vulcaniche e fenomeni vulcano-tettonici di età preistorica e storica nell'Isola d'Ischia. Publ. Centre Jean Berard Naples, 7: 145-188.

Cubellis, E. 1985. Il terremoto di Casamicciola del 28 luglio 1883: analisi degli effetti, modellizzazione della sorgente ed implicazioni sulla dinamica in atto. Boll. Soc. Natur. in Napoli, 94 :157-186.

De Siano, F. 1798. Brevi e succinte notizie di storia naturale e civile dell'Isola d'Ischia. Napoli

Johnston-Lavis, H.J. 1885. Monograph on the earthquakes of Ischia: 1881/1883. London and Naples, 1885.

Keefer, D.K., 1984. Landslides caused by earthquakes. Geol. Soc. Am. Bull., 95:406-421.

Mercalli, G. 1884. L'isola d'Ischia ed il terremoto del 28 luglio 1883. Mem. R. Istituto Lombardo Sci. Mat. Nat., 15:99-154

Monti, P., 1980. Ischia: archeologia e storia. Napoli: Porzio.

Morgera, V., 1890. Le terme dell'Isola d'Ischia prima e dopo gli ultimi terremoti distruttivi (4 marzo 1881 e 28 luglio 1883). Napoli: Lanciano & D'Ordia.

Muratori, L.A., 1937. Rerum Italicarum Scriptores. 7, II, Bologna:Garufi.

Rittmann, A. & V. Gottini 1980. L'isola d'Ischia : Geologia. Boll. Serv. Geol. Ital. Vol. 101:131-274.

Varnes, D.J. 1978. Slope movement types and processes in Landslides, Analysis and Control (R.L.Schuster and R.J. Krizeh ed.) Special Report 176, Transportation research Board, Nas, Washington.

Vezzoli, L (Editor) 1988. Island of Ischia. CNR, Quaderni de "La ricerca Scientifica" , 114, 10, Roma.

Landslides, Bell (ed.) © 1995 Balkema, Rotterdam, ISBN 90 5410 032 X

Design aspects of pier foundations in landslip prone seismic areas

G. F. Rocchi & T. Collotta
GEI Geotechnical Engineering Italia srl, Italy

A. Masia & M. Trentin
Del Favero SpA, Italy

C. Fittavolini
Italimpianti SpA, Italy

ABSTRACT: The paper deals with the geotechnical aspects related to design and construction of pier foundations for motorway viaducts in unstable slopes located in the Irpinia region (Italy). The unstable slopes are composed of large ancient slide bodies overlying structurally complex clayey formations. Because of the seismicity of the area and of the large extension and thickness of the unstable materials, solutions meant to stabilize the slopes were abandoned in favour of deep pier foundations. Design criteria to evaluate appropiate diameters and lengths of the piers were controlled by allowable displacements considerations. In this respect the importance of an appropiate choice of the "equivalent" elastic mudulus is emphasized. The construction was carried out according to well controlled procedures in order to minimize swelling phenomena during excavations. Horizontal pressures against the steel supporting sets were measured by means of strain gauges; instrumentation results and their interpretations are presented and discussed.

1 INTRODUCTION

The authors have been in charge of design and construction of two viaducts of the new motorway connecting S.S. 400 and S.S. 164 roads in the Irpinia region (Italy).

The viaducts have a length of about 360 m (viaduct 4) and 460 m (viaduct 6) and cross slopes characterized by the presence of large ancient sliding bodies (up to 25 m thick) that can partly be reactivated by water table fluctuations and/or by seismic actions (the area belongs to a highly seismic district, only few km away from the epicenter of the 1980 destructive earthquake, see fig.1).

The "bedrock" is constituted by highly deformed, structurally complex formations in which the original sedimentary sequence has been deeply modified by tectonics and gravitational phoenomena and where the predominant, cahotic clayey matrix contains lapideous relicts.

The extension and thickness of the actually unstable materials, instead of preventive stabilization (drainage) of the entire area, pushed toward solutions consisting of deep pier foundations, capable to sustain both the permanent and live loads, coming from the superstructure, and the thrust of the unstable materials, free to flow around the embedded shafts (see fig.2).

Main scope of the paper is to illustrate the design criteria used in order to cope with safety considerations, feasibility, rapidity and, in the last istance, economy of piers construction.

2 GEOLOGICAL SETTING

The Irpinia region is part of the Campania-Lucania Apennine, which represents the area of tectonic contact between two paleogeographic domains where terrains of different age and facies outcrops. The most supported theory (see Ogniben 1969, 1986) recognizes the presence in this area of one carbonatic auto-

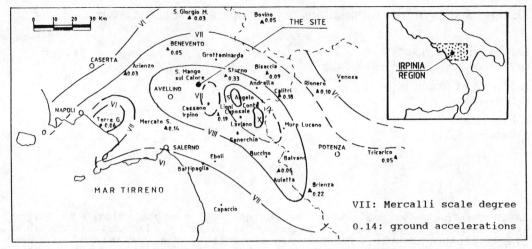

Fig. 1 — ISOSEISMIC LINES OF THE 1980 EARTHQUAKE (from Chiocchini — Cipriani, 1986)

chtonous platform covered by several different nappes of western provenance and by Miocene turbidites, which sedimented while Apennine Chain was still forming.

Deformations occurred during the Miocene orogenesis caused morphological evolutions and changes in the sedimented materials, which may vary from clayey to clastic turbidites.

Tectonic dislocations of allochtonous materials and several gravitational phenomena (landslides) disarranged the original sedimentary sequence of the turbidites.

As result the area is now interested by the presence of structurally complex formations, where the cahotic clayey matrix may contain lapideous relicts of various dimensions.

3 GEOMORPHOLOGICAL ASPECTS

In the area of the two viaducts the slopes are characterized by moderate inclinations (8-10°).

Main geomorphological aspects were detected by means of aerophotogrammetric analyses followed by in situ surveys. Results of geomorphological studies permitted to draw a detailed map of all potential and actual instability phenomena, making light to the possible causes such as erosion, shallow stream, water stagnation, etc. (see for exemple fig.3).

The studies clearly indicated the potential for landslide hazard and the necessity to carry out, besides conventional geotechnical in situ and laboratory tests, specific site investigations (inclinometric tubes and piezometers) finalized to monitor the slopes.

4 GEOTECHNICAL INVESTIGATIONS

Geotechnical investigations consisted of geophysical surveys and of several deep boreholes.

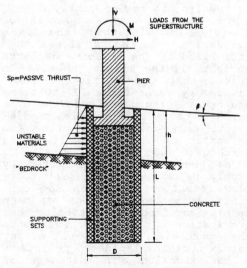

Fig. 2 — PIER FOUNDATIONS SCHEME IN UNSTABLE SLOPES

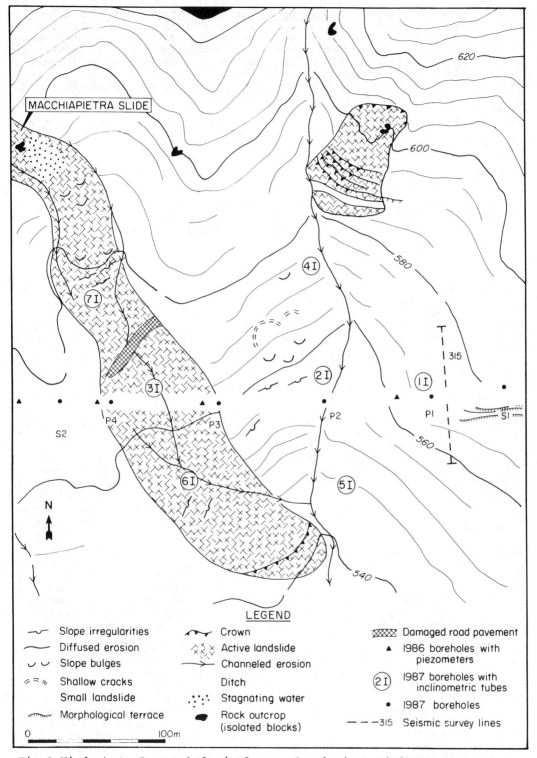

Fig.3 Viaduct 4- Geomorphological map of main instability phenomena

LEGEND

Slope irregularities
Diffused erosion
Slope bulges
Shallow cracks
Small landslide
Morphological terrace
0   100m

Crown
Active landslide
Channeled erosion
Ditch
Stagnating water
Rock outcrop
(isolated blocks)

Damaged road pavement
1986 boreholes with
   piezometers
1987 boreholes with
   inclinometric tubes
1987 boreholes
—— 315 Seismic survey lines

MACCHIAPIETRA SLIDE

N

The boreholes were carried out with continuous recovery of the materials encountered; where of significance inside them "undisturbed" samples were collected and SPTs carried out. Because of the geomorphological evidences, some of the boreholes were equipped with piezometers (Norton type) and inclinometric tubes.

Due to the particular nature of the sampled materials (see detailed descriptions in chapter 5) it was considered not useful to characterize the relevant "units" basing only on the results of conventional laboratory tests; the technical literature related to the behaviour of stiff, fissured clays or to structurally complex formations makes in fact aware on that (see for exemple Marsland 1971, Simpson et al. 1979, Pellegrino et al. 1985, Burland 1989).

Laboratory tests were essentially finalized to classify the materials in terms of their grain size and plasticity characteristics. Shear strengths and deformabilities were instead evaluated, depending on their specific use in engineering computations, basing on indications from in situ tests (geophisical survey and SPTs), empirical correlations with physical properties and analogies with the behaviour of similar instrumented underground structures recently published in the technical literature (see the discussions made in chapter 7).

5 SOIL PROFILES

As expected, boreholes logs revealed a great variety of lithological facies, in heterogeneous, chaotic association.

In the stratigraphical schematization of the slopes, finalized to the interpretation of the instability phenomena and to the definition of the "large scale" mechanical properties of the unstable materials and of the "bedrock", it was not considered important to localize the exact spatial position of each lithotype present in the boreholes; making reference on average physical, mechanical and structural properties, it was considered more appropiate to assemble the encountered materials into relevant "units" that certainly have relevant influence on the large scale behaviour of the slopes and of the foundations to be founded on them.

Simplifying the following "units" were defined:

-A.1: shallow covering consisting of grey silty clays with scattered cobbles (diameter of the cobbles less than 10 cm);

-A.2I: weathered bedrock consisting of hard marly clays with lapideous relicts (marls, marly limestones) of various dimensions;

-A.2II: chaotic complex consisting of clays, silts, sands and lapideous relicts (ancient slides);

-B: bedrock consisting of hard to very hard marly clays and lapideous relicts (marls, marly limestones).

In the clayey materials of units A.2I and B two different level of structural features were recognized; the first is represented by several small shear planes (scaly clays); the second superimposed to the first, is represented by larger shear planes, which may cross the entire formation.

In correspondence of viaduct 4 only units A.1 and B are present; the thickness of unit A.1, clearly put in evidence by means of both the boreholes stratigraphies and the geophisical survey, is in the range of 5-12 m; the velocity of the elastic longitudinal waves (v_p), were in fact very different in the two units [0.35-0.7 km/sec in units A.1 and 1.8 (top of unit)-2.6 (40 m below g.l.) km/sec in unit B].

Unit A.2II is present between abutment S1 and pier P5 of viaduct 6 (see fig.4); the thickness is variable from 20 to 28 m; measured v_p were in the range of 0.35 (4-5 m below g.l.) and 1.9 (20 m below g.l.) km/sec; according to boreholes stratigraphies, SPTs and geophisical results the materials of unit A.2II result in dense states below elevations 12-15 m.

Unit A.2I is located between pier P6 and abutment S2 of viaduct 6 (see fig.4) at elevations of 5-11 m; it was distinguished from unit B because of its less ordered structure; measured v_p are equal to 1.5

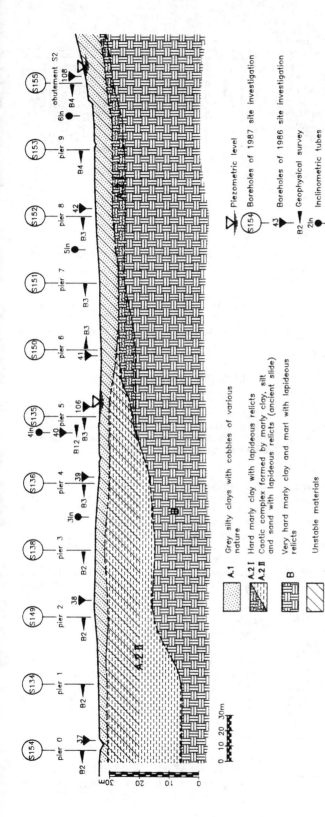

Fig. 4 – SCHEMATIC LONGITUDINAL SOIL PROFILE

A.1 Grey silty clays with cobbles of various nature

A.2 I Hard marly clay with lapideous relicts

A.2 II Caotic complex formed by marly clay, silt and sand with lapideous relicts (ancient slide)

B Very hard marly clay and marl with lapideous relicts

Unstable materials

▽ Piezometric level

(S154) Boreholes of 1987 site investigation

43 Boreholes of 1986 site investigation

▼ B2 Geophysical survey

● 2ln Inclinometric tubes

km/sec. The covering unit A.1 has maximum thicknesses of about 5 m.

Fourteen inclinometers were installed and read for one year. An exemple of the obtained high quality results is shown in fig.5.

The displacements detected allowed to identify maximum and minimum thicknesses of the unstable materials in proximity of the viaducts; they resulted in the range of 5-12 m (viaducts 6) and 8 m (viaducts 4).

Always the instability phenomena were limited to the materials of units A.1 and A.2.II. The average displacements velocity were in the range of 2.5-4 mm/month; such velocities are typical of "creep" movements induced in slopes characterized by safety factors near to unity.

6 HYDROGEOLOGY

The piezometers were installed to define the piezometric levels and their variations with time inside unit A.1 (viaducts 4 and 6) and unit A.2II (viaduct 6). The permeability of the materials of units A.1 and A.2II is generally low; in spite of that the overall structure of the slope is highly permeable, allowing the water to pass through fractures and shear planes and to flow downward. This was clearly visible during the excavation of the pier foundations. As a consequence, depending on seasonal rainfall conditions, temporary aquifer are generated inside the covering and the chaotic units.

The piezometric survey confirmed seasonal occurrences of piezometric levels close to the ground surface.

No measurements were made in units A.2I and B; in spite of the opening of the many fissures present in the undisturbed samples taken from the boreholes, the saturation degree measured in laboratory was very near to the 100%. Measurements of the piezometric levels carried out by other Authors in similar formations revealed that high water pressures may be present also inside the deep impervious materials and that such pressures may respect the hydrostatic distribution (see for exemple Cotecchia F.1989). Basing on

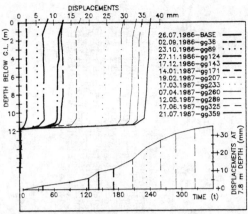

Fig. 5 — VIADUCT 6 — READINGS OF THE INCLINOMETRIC TUBE N. 3

these considerations, for design pourposes the hypothesis was made of only one aquifer, with piezometric levels close to the ground surfaces.

7 GEOTECHNICAL PROPERTIES

The discussion related to the geotechnical properties is necessarely limited to the ones considered of interest for the engineering design of pier foundations.

7.1 Unit B

Considering the high compactness of the formation (values of N_{SPT} higher than 100 blows/30 cm, elastic longitudinal wave velocities equal to 1.8-2.6 km/sec) and the embedment of the pier foundations (see chapter 8), assuming conventional undrained and drained strength parameters equal to:
- s_u > 0.4 MPa (see Stroud and Butler 1975, Stroud 1988);
- c' > 15 kPa
ϕ' = 26-30° (see Skempton 1977, Padfield and Mair 1984, fig.6 and fig.7)
bearing capacity was never the dimensioning factor. The diameter and the embedded length of the pier foundations were instead conditioned by horizontal displacements considerations.

An appropiate selection of the deformabilities characteristics of

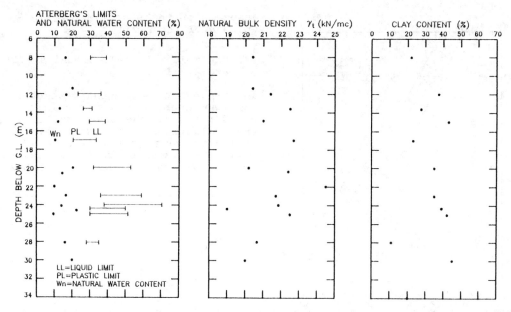

Fig. 6 — PHYSICAL PROPERTIES OF UNIT B

the "bedrock" materials became the more important aspect for the design.

In line with the more recent findings related to the behaviour of highly non linear materials such as stiff clays (see Simpson et al.1979, Jardine et al.1986, Burland 1989) the choice of the "equivalent" Young's moduli, used for computational pourposes, was made considering that:

-under working conditions (safety factors against failure higher than 2-3 and allowable displacements less than few centimeters) the large mass of the ground beneath and around large diameter piers foundation experiences strains less than 0.1%;

-in correspondence of such limited values of strains the secant Young's moduli are considerably higher than the ones obtainable from conventional laboratory tests; furthermore at these strain levels the difference between the undrained and drained Young's moduli of hard, highly overconsolidated materials can be considered not higher than 10-20%; in fact, the treshold shear strain for excess pore pressures build-up is less than 0.1% (see for exemple Chung et al.1984, Matsuda and

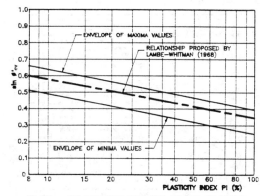

Fig. 7 — RELATIONSHIP BETWEEN ANGLES OF SHEAR RESISTANCE AT CONSTANT VOLUME AND PLASTICITY INDEXES (adapted from Lambe and Whitman, 1968)

Ohara 1989) and the drained and undrained shear moduli at such level of strains may be considered practically the same; the difference between drained and undrained Young's moduli is only dependent on the different values of the Poisson's ratio (0.3-0.4 vs 0.5);

-direct applicability of the results coming from several back analyses of the behaviour of similar underground well controlled instrumented structures (see for exemple fig.8 and fig.9).

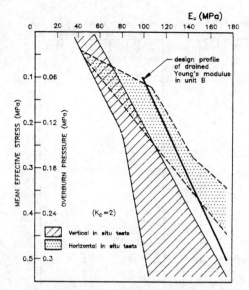

Fig. 8 – VARIATION OF UNDRAINED YOUNG'S MODULUS WITH MEAN NORMAL EFFECTIVE STRESS DERIVED FROM VARIOUS SITES ON LONDON CLAY (from St. John, 1975)

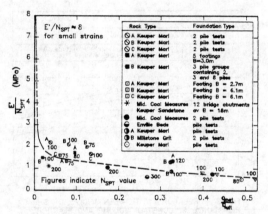

Fig. 9 – VARIATION OF E'/N_{SPT} WITH DEGREE OF LOADING FOR INSENSITIVE WEAK ROCKS (from Stroud, 1988)

Basing also on in house cross holes tests results carried out in similar formations, adequately corrected by means of the Seed and Idriss 1970 relationship (see fig.10) in order to take into account the difference in the level of induced strains (see fig.11), it was decided to assume for the design the drained Young's moduli profile shown in fig.8.

Smaller values, in direct proportion to average measured elastic longitudinal waves and to local effective overburden pressures, were assumed for the stable materials of units A.2I and A.2II.

7.2 Unstable materials of units A.1 and A.2II

The relevant strength parameters for the problem under examination are limited to the ones useful for the evaluations of the soil thrusts acting on the pier foundations.

As the failure surfaces developing in the flowing unstable

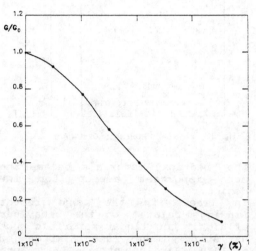

Fig. 10 – DEGRADATION OF THE SHEAR MODULUS WITH STRAIN-CLAYEY MATERIALS (adapted from Seed and Idriss, 1970)

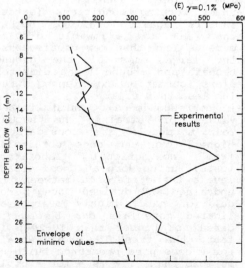

Fig. 11 – EQUIVALENT YOUNG MODULUS OF THE MONGHIDORO FORMATION (CENTRAL APENNINES)

1964

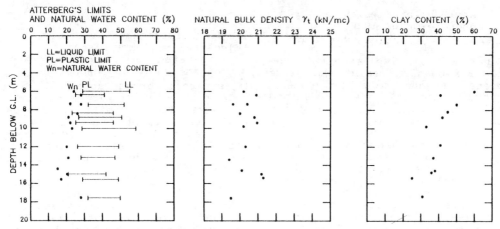

Fig. 12 – PHYSICAL PROPERTIES OF UNITS A.1 AND A.2 II

material immediately behind the pier foundations are certainly different from the deeper ones localized by means of the inclinometric tubes, it was considered not correct to make reference to the residual strength values achievable from back analyses of the observed instability phenomena.

Because of their high weathering and fracturing degree, the unstable materials were better characterized making reference to the angle of shear resistance at constant volume (ϕ'_{cv}) (see Atkinson and Bransby 1978, Simpson et al.1979, Bolton 1986). The latter was determined by means of the correlation shown in fig.7, taking into account the physical properties of the encountered materials (see fig.12).

8 PIER FOUNDATION DESIGN

Because of the compactness of unit B and of the difficulties associated with its excavation by means of traditional mechanical tools (digger, pneumatic hammers, etc.) main design aspects were related to evaluate the feasibility of solutions consisting of piers with diameter 10 m and relatively limited embedments into the "bedrock", capable to support the thrust of the unstable materials without the help of additional, long active anchors.

Such an evaluation was based on computations of:
-the forces and moments acting on the piers at the interface between stable and unstable materials;
-the interaction between the piers and the surrounding stable soil.

In absence of specific dynamic analyses, seismic loading conditions were investigated by means of static equivalent approaches, according to the prescriptions given by the Italian laws. Specifically the thrust of the unstable materials and the pier-soil interaction were evaluated as described below.

8.1 Thrust of unstable materials

The thrust (S_p) of the unstable materials was extimated basing on the following equation:

$$S_p = D \cdot h \cdot k_{p3D} \cdot k1 \cdot k2 \cdot \sigma'_{vm} \qquad (1)$$

being:
D   =diameter of the pier
h   =thickness of the unstable soil
k_{p3D}=three-dimensional static
      coefficient (see for exemple
      Reese et al.1974, Brinch
      Hansen 1961)
k1   =correction factor (>1) for
      ground surface inclination,
      calculated making reference
      to available two-dimensional
      solutions (see for exemple
      Caquot and Kerisel 1948)
k2   =correction factor (>1) for
      inertia forces, calculated
      making reference to available
      two-dimensional solutions in
      static and seismic conditions

(see Mueller-Breslau 1924 and
Mononobe-Okabe 1926,1929)
σ'_{vm}=average effective overburden

Equation (1) implies that the
failure mechanism behind the piers
happens contemporarely to the seismic excitation and in fully drained
conditions. Both the assumptions
should be considered on the safe
side.

In fact the experience gained
with the 1980 destructive earthquake is of instability phenomena occurring only some days after the
seismic events, when the inertia
forces are negligible (see D'Elia
et al.1985); hence the correction
factor k_2 is unlikely higher than
1.

Possible explanations of this
fact can be related to non homogeneous excess pore pressures
build-up inside the shaking mass,
followed by its re-distribution
once the main shock is finished.
The effective stress field inside
the sliding soil mass after the
seismic event, and consequently the
total thrust behind the piers,
should be hence lower than the one
assessable applying equation (1).

Maximum calculated horizontal
forces and moments, applied at the
interface between stable and unstable materials, including the thrust
of the unstable materials, were respectively equal to 82 MN, 340 MNm
(viaduct 6) and 43 MN, 139 MNm
(viaduct 4).

8.2 Pier-soil interaction

Solutions based on Winkler approach were used. The pier is assumed as a rigid prismatic body,
subdivided into slices, while the
soil around the shaft and at the
base is schematized by independent
springs. Each spring is subdivided
into two components, one normal to
the pier surface and one tangential; it is represented by elasto-
plastic relationships, the so called p-y and t-z curves. Solutions
are obtained by solving the linear
system:

$$\{U\} \cdot [K] = \{F\} \qquad (2)$$

being:
$\{U\}$=pier displacement vector
$[K]$=global rigidity matrix,
    evaluated for exemple by means
    of energetic approaches
$\{F\}$=external forces vector

Once the pier displacement vector
is calculated, normal and tangential displacements and forces, at
each slice location, are automatically determined by means of
geometric considerations and by entering p-y and t-z curves.

In the case under examination
bilateral curves were assumed.

According to Elson 1984 and Poulos and Davis 1974, basing also on
the geometry of the problem, the
values of the subgrade reaction k_h,
multiplied by the diameter of the
pier D, representing the elastic
part of the p-y curves, were
assumed equal to 1.5-2.5 times the
equivalent Young's moduli discussed
in chapter 7.

The ultimate horizontal pressures
were determined according to what
has been experienced by Bushan et
al.1979 and Sullivan et al.1979.
Limit tangential adhesions were
extimated according to conventional
methods applied for drilled shaft
in clay; values of about 20 kPa in
units A.1 and A.2II and 150-200 kPa
in units A.2I and B were assumed.
Such values were considered fully
mobilized for relative tangential
displacements equal to 1 cm.

8.3 Obtained results

In the worst case represented by
thicknesses of unstable materials
equal to 12 m (pier 1-4 and
abutment S1 of viaduct 6) allowable
horizontal displacements at ground
surface of 4 cm and rotations in
the range of $1-2 \cdot 10^{-3}$ radiants were
achieved, without the help of
active anchors, with limited total
embedded lengths of 30-32 m, of
which 18-20 m inside the stable
materials and 7-11 m inside unit B.
The appropiate choice of the
elastic characteristics of the resisting materials was in this respect resolutive.

9 PIER CONSTRUCTION

In order to limite swelling pheno-

mena and consequent deterioration of the soil properties, the piers execution were made according to the following modalities:
-exavation below the depth previously reached by the supporting steel sets not higher than 1.5 m;
-installation of wire netting against the vertical shaft of the excavated soil;
-installation of 60 cm spaced supporting steel sets, composed by 2 circular IPN 180 I-beams, crosswise connected;
-installation of wire netting against the sets and filling of the interspace between the wire nwttings with spritz-beton.

The excavations were always made by means of diggers and pneumatic hammers; soil profiles, revealed inside the piers, substantially confirmed the design assumptions related both to the depth of the "bedrock" and to the presence of water flow inside the body of the slopes.

Once the design depths was reached, the piers were immediately filled with concrete, steel reinforced in the upper part, where necessary.

Because of the uncertainties linked with the evaluations of the thrusts acting on the supporting sets during exavations, one pier of viaduct 6 was instrumented.

Instrumentation consisted of the installation of four couples of strains gauges at three different levels (5.5, 8 and 14 m below ground level). The strain gauges were connected by means of cables to a data acquisition system. Typical obtained results are shown in fig.13.

The instrumentation results showed that:
-the stresses in steel rapidly increased after the set installation; the rapid increase of stresses was limited only to the first 4-5 days, after which the measured values remained pratically constant with time;
-the stresses had a slightly tendence to increase with depth; for this reason below 25 m depth it was decided to use IPN 200 I-beams, 50 cm spaced;

The total horizontal pressure (σ_h) acting on the sets was evaluated according to the following equation:

$$\sigma_h = \frac{A \cdot \sigma}{R \cdot I} \qquad (3)$$

being:
A=55.8 cm^2 =IPN 180 I-beams area
σ=measured stresses in steel
R=5 m =pier radius
I=60 cm =set spacing

In the hypotheses of hydrostatic pressure distribution behind the supporting sets and of water table close to the ground surface, the total horizontal pressures were slightly higher (1.3-1.5 times) than the ones evaluable with the method proposed by NAVFAC DM-7 1971 for earth supporting structures in cohesionless materials. The associated values of the earth pressure coefficient (K) resulted slightly lower than 1.

10 CONCLUSIONS

Pier foundations design in unstable slopes represent a difficult geotechnical problem, especially in presence of considerable thicknesses of unstable materials.

The choice of the more convenient solutions depends on many factors; those linked to geological and geotechnical aspects are certainly of primary importance.

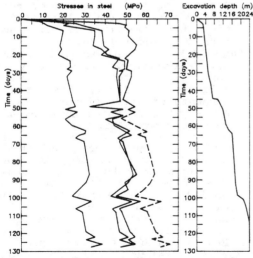

Fig. 13 — VIADUCT 6 — PIER 3 — INSTRUMENT SECTION N. 2 (8 m below G.L.)

In the described case the geological setting and the geomorphological studies were of capital importance in the phases of planning of suitable and resolutive site and laboratory investigations.

The monitoring of the slopes and the choice of appropiate elastic characteristics for the "bedrock" (based on reconaissances of the highly non linear clay behaviour and of the occurrence of small strains changes generally induced in the ground at working conditions), permitted to calibrate a confident pier foundations design, also in situations where the feasibility, without the help of additional supporting elements, might be put into discussion.

Topographic measurements of the heads of the piers, carried out for a relatively long period after construction, have till now confirmed the goodness of the assumptions made.

AKNOWLEDGEMENTS

The Writers thank the cooperation of Dr C.Longaretti and Dr O.Vacca (Studio Geotecnico Italiano srl) for the activities related to the geomorphological studies, the geotechnical characterization and the foundations design; the diligence of Mr F.Chini (OTR srl) and of the on site Del Favero SpA technicians during the phases of strain gauges installation and subsequent readings is also gratefully aknowledged.

REFERENCES

Atkinson J.H. and Bransby P.L. 1978. The mechanics of soils-An Introduction to Critical State Soil Mechanics. Univ. Series in Civil Eng., Mc Graw -Hill.

Bhushan K., Haley S.C. and Fong P.T. 1979. Lateral load tests on drilled piers. Journal of Geot. Eng. Div., ASCE, vol.105, n°GT8.

Bolton M.D. 1986. The strength and dilatancy of sands. Géotechnique 36, n°1.

Brinch Hansen J. 1961. The ultimate resistance of rigid piles against transversal forces. Bulletin n°12, Geoteknisk Institute, The Danish Geotechnical Institute.

Burland J.B. 1989. Small is beautiful-The stiffness of soils at small strains. IX L.Bjerrum Memorial Lecture, Can. Geot. Journal n°26.

Caquot A. and Kerisel J. 1948. Tables for the calculation of passive pressure, active pressure and bearing capacity of of foundations. Gautiers Villars, Paris.

Chiocchini U. and Cipriani N. 1986. Seismic microzoning to rebuild Caposele village destroyed by the November 23, 1980, earthquake (Irpinia, Campano-Lucano Apennine). Proc. of Int. Symp. on Eng. Geology Problems in Seismic Areas.

Chung R.M., Yokel F.Y. and Wechsler H. 1984. Pore pressure build-up in resonant column tests. Journal of Geot. Eng. Div., vol. 110, n°2.

Cotecchia F. 1989. A landslide in the Irpinian Unities of the Ofanto valley. R.I.G., Year XXIII, n°2 (in italian).

D'Elia B., Esu F., Pellegrino A., Pescatore T.S. 1985. Some effects on natural slope stability induced by the 1980 Italian earthquake. Proc. XI Int. Conf. on Soil Mech. Found. Eng., San Francisco.

Elson W.K. 1984. Design of laterally-loaded piles. CIRIA Report n°103.

Jardine R.J., Potts D.M., Fourie A.B. and Burland J.B. 1986. Studies of the influence of non-linear stress-strain characteristic in soil-structure interaction. Géotechnique 36.

Lambe T.W. and Whitman R.V. 1968. Soil Mechanics. John Wiley & Sons.

Marsland A. 1971. The shear strength of stiff fissured clays. B.R.S., Dept. of the Environment, Current Paper n°21.

Matsuda H. and Ohara S. 1989. Treshold strain of clay for pore pressure build-up. Earthquake Geot. Eng., XII Int. Conf. on Soil Mech. and Found. Eng., Rio de Janeiro.

Mononobe N. 1929. Earthquake-proof construction of masonry dams. Proc World Eng. Conf.,vol.9.

Mueller-Breslau 1906. Erddruck anf Stuetzmanern. Kroener.

NAVFAC DM-7 1971. Design Manual-Soil Mechanics, Foundations, and Earth Structures. Dept. of the Navy, Naval Facilities Engineering Command, Washington.

Ogniben L. 1969. Introductive scheme to the Calabro-Lucano geology. Men. Soc. Geol. Ital.,vol.8 (in italian).

Ogniben L. 1985. Report on the "conservative" geodynamic model of the Italian region. Edited by ENEA (in italian).

Okabe S. 1926. General theory of earth pressure. Journal Japanese Soc. of Civil Eng., vol.12, n°1.

Padfield C.J. and Mair R.J. 1984. Design of retaining walls embedded in stiff clay. CIRIA Report n°104.

Pellegrino A., Picarelli L. and Bilotta E. 1985. Geotechnical properties and slope stability in structurally complex clay soils-PartIII. Geotechnical Engineering in Italy, A.G.I. volume, ISSMFE Golden Jubilee.

Poulos H.G. and Davis E.H. 1974. Elastic solutions for soil and rock mechanics. John Wiley & Sons.

Reese L.C., Cox W.R. and Koop F.D. 1974. Analysis of laterally loaded piles in sand. OTC 2080, VI Annual Offshore Technology Conference, Houston.

Seed H.B. and Idriss I.M. 1970. Soil moduli and damping factors for dynamic response analysis. Rep. n°EERC 70-10, Un. of California, Berkeley.

Simpson B., Calabresi G., Sommer H. and Wallays M. 1979. Design parameters for stiff clays. Gen. Report, Proc VII Europ. Conf. on Soil Mech. Found. Eng., vol.5, Brighton.

Skempton A.W. 1977. Slope stability of cuttings in brown London clay. Special Lecture, Proc. IX Int. Conf. on Soil Mech. Found. Eng., vol.3, Tokyo.

Stroud M.A. 1988. Standard Penetration Test-Its application and interpretation - Part II. Penetration testing in UK, Proc Geot. Con. organized by ICE, Birmingham.

Stroud M.A. and Butler F.G. 1975. The Standard Penetration Test and the engineering properties of glacial materials. Proc Symp. Eng. Behaviour of Glacial Materials, Un. of Birmingham.

Sullivan W.R., Reese L.C., Fenske C.W. 1979. Unified method for analysis of laterally loaded piles in clay. Proc. of Num. Met. in Offshore Piling, ICE, London.

S2 Landslides and reservoirs
    Glissements de terrain et réservoirs

Landslides, Bell (ed.)© 1995 Balkema, Rotterdam, ISBN 90 5410 032 X

Keynote paper: Landslides and reservoirs

W. Riemer
Ehner, Luxemburg

ABSTRACT: A review of case histories reveals a wide spectrum of landslide/-reservoir interactions. Initial studies of mass movements in relation to reservoirs have to cover the catchment area, the reservoir, a downstream reach and adjacent valleys. Impacts of landslides on reservoirs are mostly undesirable but under certain conditions the reservoir may also eliminate slope stability problems. Remedial treatment of landslides should be adequate to reasonably exclude potentially catastrophic hazards. For other hazards a risk based approach is admissible. Difficulties are encountered in defining the time dependence of reservoir landslide hazards. The majority of slide events takes place early in the life of a project. Remedial measures for landslide risk management at reservoirs include conventional geotechnical stabilization but also adjustments in the design of project structures and project operation. The "brittleness" of slides introduces a particularly severe problem for the assessment of the landslide/-reservoir interaction. Because of this uncertainty, as well as because of uncertainty commonly encountered in geotechnical analysis, it appears mandatory that the "observational approach" is duly considered. The "observational approach" should be initiated with ample lead time and its perpetuation into the operation phase has to be assured. Deferring remedial action with reliance on the observational approach appears attractive from an economic point of view but geotechnical constraints and time dependence of landslide events favor early action.

1. INTRODUCTION

In the creation of reservoirs man essentially relies on favorable conditions in the morphological, hydrological and physical environment. The reservoir is vulnerable to the hazards implied by its environment. Moreover, the amounts of mass and energy stored above the land surface in the reservoir constitute a large hazard potential. Additionally, the impacts of the reservoir on its physical and biological environment may tend to augment the level of the hazard significantly and expose large areas to them.

Dams and reservoirs are built for their benefits which most prominently derive from agricultural irrigation and from socio-economically and environmentally acceptable power production. Especially for countries with scarce resources the construction of reservoirs is a strategy for survival. Under such conditions also negative impacts caused by the reservoirs may, within limits, be accepted provided the risks connected

with them can be adequately identified and quantified. Among the potential hazards linked to reservoirs the landslides occupy a prominent position. The Vajont catastrophe in 1963 sharply focussed the interest on this subject and still provides the most dramatic precedent associated with reservoirs (Fig. 8A). Since Vajont the theme landslides and reservoirs has featured on several international congresses. But, despite the important developments which took place in the methods of investigation and analysis this subject continues to offer a series of problems.

For the purpose of this landslide symposium it was attempted to develop an overview of the existing experience with landslides and reservoirs, of the spectrum of the underlying problems, the ramifications of associated hazards and risks and of the solutions which were adopted. However, it was not possible to establish a fully representative data base, because

1. many reservoir landslide incidents were treated as a routine with standard methods and no

need for publication was seen

2. negligible risks did not require treatment

3. pending issues in some cases still render publication inappropriate.

2. SCOPE AND DEFINITIONS

In relation to reservoirs it is appropriate to apply the term "landslide" in its widest sense to cover general "mass movements", including also falls, flows, avalanches and even subsidence. But emphasis is here given to the gravitational movements of soil and rock.

The majority of the existing reservoirs are located on a water course. The effects of landslides can propagate along the water course from the catchment area to the reservoir and from the reservoir to the downstream reach. Changes in groundwater regime caused by the reservoir may even affect slope stability in adjacent catchments. Therefore, the study of landslides in relation to reservoirs must notably extend beyond the confines of the reservoir itself.

Descriptive terms for landslides preferentially follow the definitions by Varnes (in Schuster et al. 1978). In this context it must, however, be pointed out that large landslides are normally complex, combining vertically and laterally components of different characteristics which can also change their geometry and properties in time.

Many landslides result from construction activities on access roads, spillways, etc. but are not unique to reservoirs and for this reason are not generally treated here. Reservoirs which serve the purpose of accumulating sediments rather than water will only marginally be mentioned. Also, the off-channel reservoirs are not generally included here because that would open another range of problems, which deserve separate treatment, as the catastrophic failures of the Baldwin Hills reservoirs in the USA and of the Lengede tailings pond in Germany show.

3. SPECIFIC ASPECTS OF RESERVOIR-LANDSLIDE INTERACTION

Most reservoirs are intended as permanent or semi-permanent features. Therefore, sustained static loads caused by the reservoir have to be considered. But also transient loads are involved which may exercise an even more important influence. Transient loads are imposed during first impounding, in the course of regular operation, in exceptional floods or droughts and by environmental seasonal variations.

Normally it is attempted to avoid landslides in the zone of influence of the reservoir but there are also situations where the mass movements are irrelevant or even beneficial for the purpose of the project. Although in the majority of the cases the reservoir tends to destabilize slopes and augment the natural hazards, the lake may on the other hand improve the stability of the slopes along its banks and it may help to mitigate or eliminate risks which were experienced prior to the creation of the reservoir.

In the following paragraphs the landslide/-reservoir interaction is treated in geographic order from upstream to downstream and subdividing static and transient, positive and negative effects.

3.1 Mass Movements in the Catchment Area

Mass movements in the catchment area of a reservoir can constitute the source of sediment supply or create run-off effects which are crucial for the design and operation of a reservoir. Particularly pronounced effects derive from landslide dams (e. g. Schuster 1986b) and glacier and/or moraine outbreaks (Hewitt 1982). Floods resulting from such events are notorious in the Himalayas, the Andes, in parts of China; there have also been several historical cases in the Alps. In recent years there has been remarkable success in controlled drainage of landslide dams, e. g. in Italy in 1987 (Govi 1988) and in USA (Schuster 1986). Nevertheless, for projects in remote areas of the Himalayas and of the Andes the effects of landslides in the catchment area have to be accepted as a natural phenomenon which must be considered in the project evaluation and design of a reservoir. Landslide related floods may influence the design for the following reasons:

1. Flood peaks may be higher than those extrapolated from normal run-off or from PMP. The 1945 landslide flood on the Mantaro river in Peru peaked 50 times higher than normal maximum flood (35000 m^3/s, Schuster 1986b).

2. Hydrographs of landslide floods may differ significantly from normal flood hydrographs, in the worst case rising nearly instantaneously to the peak and thus leaving little time for reaction (Krumdieck 1991).

3. Sediment concentration in a landslide flood does not conform with the normal sediment rating curves which are controlled by sediment supply.

4. Landslide floods may develop a specific transport mechanism (Krumdieck 1991) enabling them to carry exceptionally large blocks (Schuster 1986b, Valenzuela et al. 1991).

Sediments derived from landslide floods may be too coarse to be flushed from a reservoir and they may cause severe damage on the spilling and outlet structures. The Coca river in Ecuador provides an example for this type of risk (Moammar et al. 1982). Investigations for a storage dam were under way in March 1987 when an intensity MMVII earthquake mobilized semi-consolidated volcanic deposits, saturated after extreme precipitation. This resulted in debris flows which devastated the Coca valley. After this event the project of a storage dam was abandoned.

But reservoirs also help to contain the sediment transport hazard. Some reservoirs exclusively serve this purpose, as for instance the check dams built after the 1982/83 eruptions of the Galunggung in Indonesia which retained 14 hm³ of lahars, the check dams at the Pinatobo volcano in Luzon and the numerous check dams in the Alps (Eisbacher et al. 1984). The reservoirs of Mattmark and Mauvoisin in Switzerland, intended for hydropower, incidentally also eliminated the risk of outburst floods which were notorious in these valleys.

3.2   Mass Movements at the Reservoir

3.2.1 Effects of the Reservoir

1. Saturation may drastically change mechanical properties of the geological materials in contact with the reservoir. The most conspicuous examples are experienced with loess. According to Qian (1982) bank failures in loess caused the shores of the Sanmenxia reservoir in China to retreat up to 200m at initial impounding. At the Nurek reservoir the banks retreat at a rate of 4 to 10 m/a. Sergeev (1979) indicates that subsidence in loess may extend to a distance of up to 110 km from the reservoir shore. Although this phenomenon rarely produces dramatic events it may account for the largest mass movements along reservoir shores and affect the landuse of a vast perimeter. Jones et al. (1961) rank saturation as primary factor in the triggering of the extensive landslides

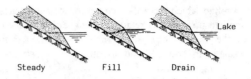

Translational Slide

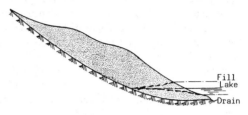

Rotational Slide

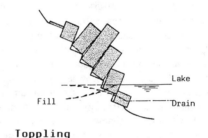

Toppling

Buckling

Figure 1 Sketches illustrating steady state and transient effects of reservoirs on landslides

which followed the creation of Roosevelt Lake on the Columbia River. Saturation may also be the main factor for superficial slides and the activation of frontal lobes of old slides which is commonly observed at the beginning of impounding (e. g. Karakaya Lake in Turkey). At the Polifiton reservoir in Greece in hydrothermally altered gneiss cracks appeared up to 100 m above the shoreline which gave the impression of subsidence and had no apparent relation to outward movement of the slope. They were tentatively attributed to changes in strength and to stress re-adjustments following saturation.

2. Gravity Loading by the rising lake frequently causes collapse in karst terrain, e. g. Keban in Turkey, Perdikkas in Greece and Röddinghausen in Germany. Earthquake shocks at Keban were attributed to the violent collapse of large caves in the limestone. At Perdikkas it was not possible to build a stable floor for the reservoir after the thin impervious cover had foundered into the karstic limestone when the reservoir filled. Originally intended for industrial water supply this reservoir now only serves for flood retention. At Röddinghausen an extensive grouting program eventually stabilized the reservoir. The load of the reservoir water also causes subtle deformations of the slopes which may decrease their stability.

3. Changes of effective Stress on a large scale, resulting from the reservoir load and from the infiltration and pressurizing of water in the underground and along the rim of the reservoir are factors which induce seismicity. There are more than 100 cases of seismic activity on record related to the filling and operation of reservoirs. Kariba, Koyna, Kremasta and Hsingfengkiang figure most prominently with strong shocks reaching up to $M_s=6,5$. In many other cases the seismic energy was released by temporary swarms of microseismic activity. The influence of the reservoir on potential seismic centers typically extends to a distance of 20 km from the shore (Meade 1983) but this figure certainly varies subject to geological features, geometry and dimensions of the reservoir. Keefer (1984) in a comprehensive evaluation of 40 major earthquakes and a large number of minor shocks detected an influence on slope stability starting above magnitude $M_s=4$, and affecting a perimeter stretching to a distance of 100 km from the rupture zone for $M_s=6,5$ earthquakes (as released at the Koyna reservoir). Thus, the filling of a reservoir can set free seismic energy stored in the crust which is capable to cause landslides. According to Keefer the phenomena which are to be expected primarily comprise rockfalls and disrupted soil slides whereas large scale landslides appear to be more inert to seismic activation. One cannot, however, discount chain reactions as represented by the 1970 Huascarán avalanche in Peru where the earthquake caused a comparative-

ly small ice and rockfall which then was followed by the collapse of the west facing flank of the mountain (Plafker et al. 1978). Methods for a better assessment of the seismic hazard during reservoir filling are developing and may help to reduce the resulting risks by an adjusted lake filling program.

4. Hydrostatic Uplift is most drastically demonstrated by peat floating up in a reservoir. Also pumice will float but normally the reaction of rocks and soils to the uplift created by the reservoir water is not so directly obvious. The effect of the uplift on slope stability depends on the failure mode of a slope, on the shape and location of an existing or potential failure plane and hydrogeological conditions (Fig. 1). For planar, translational slides with uniform shear strength the safety factor should increase if cohesion is available on the failure plane. The safety factor would remain unchanged if only friction is considered. In slides with an irregular or curved failure plane where movement involves rotation or internal deformation the effect of static uplift changes as a function of the elevation of the reservoir level. A "critical" elevation of the lake can be defined where the uplift pushes the safety factor to a value below 1. There may also be a "neutral" point or range in which the variation of the reservoir level does not change the safety factor. If the shape of the failure plane and the antecedent groundwater conditions are adequately defined the concept of the "influence line" offers a tool to probe the effect of impounding which have to be anticipated (Hutchinson 1977). At the Saint-Pont slide, France displacement monitoring has proven the stabilizing effect of the Sainte Croix reservoir. At the start of impounding the creep velocity of the slide dropped sharply from nearly 30 cm/a to only 8 cm/a (CFGB 1982). At the Yangtze Gorge reservoir the large Jipazi slide is expected to reach a minimum safety factor with the reservoir level at 160 m and to improve notably in stability at higher reservoir levels (Zhang Zhuoyuan et al. 1986). For buckling type failures the uplift would reduce the stabilizing moment and as a consequence also the stability factor as long as the water level is not high enough to cause buoyancy also in the driving slab of the slide. A toppling type failure will be adversely affected in its stability by toe submergence. The toppling failure would be indifferent to complete submergence or, if cohesion is available, would improve in stability. Generally, the seismic yield acceleration of slides is reduced by uplift.

5. Steady State Groundwater Flow frequently builds notable non-hydrostatic upward gradients in the valley floor which are unfavorable for slope stability. The reservoir may balance these gradients, eventually resulting in a new configuration of the groundwater table which is more favorable for the stability of the flanks.

6. Groundwater Transients occur during filling of the reservoir, as seasonal fluctuations in the aquifers or, as one of the most critical situations, during rapid drawdown of the reservoir. To determine the effect of the transient loading the hydrogeological parameters of the slide and the underlying slope and the normal groundwater conditions must be defined in adequate detail for hydrogeological modelling. A study carried out under the Clyde project in New Zealand for the Brewery slide illustrated the effects to be expected during impounding. Immediately after start of reservoir filling the infiltrating water creates a gradient directed into the slope which raises the safety factor. When the water table approaches its new equilibrium the safety factor asymptotically goes to its final value. At the Brewery slide the anticipated changes in the safety factor for one of the failure planes would range between +4% and -8%. The time to reach steady state depends on various factors, primarily on the permeability and the storage coefficient. Until steady state is achieved the displacement data have to be interpreted cautiously and in due consideration of changes in the groundwater table. If clay is involved in the landslide process the effects of pore water pressure must be taken into consideration. Uplift in the slide mass during raising of the reservoir level will result in a state of overconsolidation in the clay. Therefore, even old landslides which had arrived at residual strength can temporarily display a higher apparent strength. Rapid drawdown is normal reservoir operation practice in most reservoirs for pumped storage and many peaking power plants. Drawdown rates in pumped storage reservoirs are typically in the range of 10^{-3} to $5*10^{-2}$ cm/s. Reservoirs may also have to be emptied rapidly for flood control and maintenance purposes. The failure of a dam or landslide dam can cause extremely rapid draw-down. The effect of the drawdown on slope stability depends mainly on the permeability and the porosity of the rock or soil formation. In Japanese experience permeabilities of less than 10^{-3} to 10^{-4} cm/s can already prove critical for rapid reservoir operation (Nakamura 1985). In some of the most sensitive cases sliding was triggered by $2*10^{-4}$ cm/s drawdown rate (Nechranice reservoir, Rybar 1977) and $4*10^{-3}$ cm/s (Mingchukur landslide on the Akhangaran reservoir, Khasanov 1986). The destabilizing effect of rapid drawdown derives from unfavorable transient groundwater gradients and excess pore pressure.

7. External Erosion along reservoir shores may have the same detrimental effects which are observed on the sea shore where they have been studied in much detail, especially in Britain. Lukac (1982) found an exponential progression of mass movement due to wave abrasion induced instability of the kind:

$$Q[m^3/m] = a*t^b \quad \text{t in years, } 7 < a < 33$$
$$0,32 < b < 0,67$$

At the Möhne reservoir in Germany it took 50 years for wave erosion induced instability to develop to a state requiring remedial measures. On the other hand the impounding halts the erosion of the river in the valley floor which in many cases is a driving factor for slope destabilization. Without the erosion at the toe the slides can start to buttress themselves or are buttressed by sediments settling in the reservoir. It is, for instance, expected that the Malvaglia slide on the Blenio reservoir in Switzerland will stabilize with time (Baumer 1988). Reservoir sedimentation generally enhances the slope stability provided it does not block natural drainage and causes an increase in the non-hydrostatic groundwater gradients at the toe of the slopes.

8. Internal Erosion in the slopes as a consequence of lake level fluctuations is mainly observed in fine grained soils, e. g. silt and fine sand. But an interesting case was also reported from France where critical erosion occurred in dolomite, accentuated by dissolution of the rock (Avène reservoir, CFGB 1982).

9. Chemical effects on slope stability appear to be comparatively rare but they are certainly significant if formations with evaporites are concerned, e. g. at the Mont Cenis reservoir (CFGB 1982), at the Kama reservoir (Mikhailov 1982) and at the Sidi Salem reservoir in Tunesia (Mouelhi et al. 1982). Reservoir water is mostly quite soft and, therefore, may affect the hydration of clays by ion exchange and leach weak cementitious substance as for example in some sensitive clays.

10. Thermal effects may result from the circumstance that the reservoir water differs from the ambient ground- and groundwater-temperature

which prevailed prior to impounding. At the Atatürk damsite the groundwater temperature has dropped more than 10°C within 2 years after impounding. There may be a subtle influence on the shear strength of clay and certainly on the rate of pore pressure changes. The most drastic thermal effects will of course materialize if the reservoir is created in a permafrost region or in contact with a glacier. The Grimsel West reservoir in Switzerland will inundate the tongue of a glacier and it is expected that the reservoir will accelerate the calving and the eventual retreat of the glacier providing additional storage (Vischer 1991).

3.2.2 Effects of Slides on the Reservoir

Most of the effects are undesirable or even dangerous with few exceptions.

1. Loss of Storage. For most reservoirs the life storage provides the basis for the benefits of the project. The investment cost for the creation of life storage may be as low as a few cents for large reservoirs but is more commonly of the order of 1 $US/m³. Thus, there is a direct cost to every m³ of slide mass which ends up occupying life storage. Nevertheless, the Vajont and the Pontesei projects where particularly large portions of the reservoir were lost to slides still remained operational. If the slide mass settles in the dead storage the life storage may even increase. But, if the dead storage was allocated for sediments carried by the river the life span of the project will be reduced.

2. Operational Restrictions. To avoid activation of slides operational restrictions in respect of rate or range of drawdown had to be imposed for some reservoirs (e. g. Avène, CFGB 1982) or the freeboard had to be increased to contain impulse waves.

3. Landuse along Shoreline. Population will have to be relocated from potentially unstable areas near the reservoir (Lake Roosevelt, Jones 1961; Lake Karakaya in Turkey) and restrictions on the type of landuse may have to be imposed which have to be charged as a cost to the project.

4. Recreational Value. Landslides have destroyed marinas and other recreational facilities. Landslide generated waves endanger boating and swimming.

5. Infrastructure near the Shores. Along several reservoirs slides have seriously damaged roads and railroads or are demanding continued maintenance for them (e. g. Palagnedra reservoir, SwissCOLD

1982, Blenio reservoir, Baumer 1988).

6. Water Quality. Slides may temporarily or permanently increase the turbidity and the sediment concentration in the reservoir water, augmenting the operational costs and reducing the value of the water released from the reservoir. For the Sidi Salem project in Tunesia (Mouelhi 1982) the concern was that slope movements would continue to create new exposures of gypsum and by this effect raise the salinity of the water making it unsuitable for irrigation.

7. Landslide Dams. The Vajont reservoir was split in two parts by the Mount Toc slide (Fig. 8A). Also at the Couesque reservoir in France the need for construction of a diversion tunnel is envisaged in the case that a landslide dam starts to form. The risks associated with landslide dams are the obstruction to the reservoir operation, upstream flooding and, if the dam is breached in an uncontrolled way, severe downstream flooding. Even slow moving slides can built dams as for instance the Thistle slide in Utah with an average advance of 1 m/h (Kaliser et al. 1986).

8. Impulse Waves. Impulse waves caused by rapid slides or falls entering the reservoir have been experienced on several projects. The most serious event was the Vajont slide where the wave created damage on the opposite slope up to an elevation of 250 m above the original water line, oscillated over the shores in the upstream part of the reservoir and overtopped the dam with a height of 100 m and wiped out Longarone. Overtopping of the dam by impulse waves is reported from two other projects, Pontesei in Italy with a height of 13 m and Zhaxi in China (Fig. 8B, Tanyankuang slide) with a wave height of 21 m (Qi Xiaojun 1986). At the Three Gorge reservoir in China impulse waves up to 100 m high are predicted. For a number of projects the potential occurrence of impulse waves posed the criteria for operation rules of the reservoir and for the design of remedial measures. Impulse waves are a complex phenomenon. Analytical solutions were elaborated by Noda (1970). The impact of the landslide may generate three basic types of waves, oscillatory, solitary and bore, depending on the Froude number of the slide and the dimensional relation between reservoir and slide. The characteristics of the waves differ significantly in respect of propagation and height and there is uncertainty which type of wave will be generated, either because basic parameters as the Froude number cannot reliably be estimated or because the domains for the different

wave types cannot be sharply delimitated. Further investigations have improved the treatment of the hydraulic term of the problem (e. g. Slingerland et al. 1979, Huber 1980, Pugh 1982, Vischer 1986) and, in addition to conventional laboratory experiments also advanced numerical modeling is now feasible. The main difficulty for impulse wave predictions thus now resides with the physical input from the slide. Methods to estimate probable velocities of slides are still either simplistic or highly theoretical/-hypothetical. The relations developed by Scheidegger 1973 (Fig. 6) may serve as a rough guideline, using the difference between arc tan(f) and the slope angle to assess acceleration and velocity. Other than by water the impulse of a slide can also be transmitted by air to the environment. For instance with rapid snow avalanches the impact of the air blast on the reservoir has to be analyzed. The jobsite at the Pejerreyes reservoir in Chile was devastated by the air blast of a snow avalanche.

9. Effects on Structures. Surprisingly, neither the Vajont nor the Pontesei concrete dams were seriously damaged by the overtopping impulse wave. But, considering that overtopping is one of the more common causes of dam failures, obviously not all types of dams would have resisted equally well. Overtopping also can result as a consequence of the blockage of outlet or spillway structures by slides. Spillways requiring extensive cut slopes are frequently troubled by sliding (e. g. Sta. Rosa in Mexico, Marsal et al. 1979). Outlet, respectively intake structures are particularly vulnerable. The "slide no. 5" on the Tablachaca reservoir (Fig. 8D) endangers power intake, desanding and flushing facilities and the spillway gates (Morales 1984) and at the Amaluza dam a comparatively minor case of slope instability presents a serious problem for the power intake and the flushing gate (Fig 12B). The outlet structures may also be at hazard if a subaquatic slide causes the lake sediments to pile up against the gate and block it.

3.3 Downstream and Reservoir Rim Effects

Energy dissipation, changes in the run-off regime and in groundwater conditions can destabilize the slopes downstream of the reservoir and in adjacent valleys.

1. Scour by the spillway jet or at the bottom outlet is a notorious cause of slope stability

problems near the dam site. It is now common practice to take precautions against this effect in the design of the project but the extend of scour can not always be reliably predicted and remedial measures have to be developed once the effects start to materialize. Among several other dams this was the case at the Karakaya dam in Turkey, the Tablachaca dam in Peru and at Kariba.

2. Spillway Spray infiltrating into the rock was recognized as destabilizing factor at the Kariba dam (Lane 1970). Also at Karakaya sealing of the rock with shotcrete against infiltration of the spray was found helpful.

3. Downstream Degradation. The interception of sediments leads to downstream degradation and more active bank erosion affecting the stability of sensitive slopes.

4. Seepage at the dam site and through the reservoir rim causes hydraulic loads on the slopes which will be most detrimental in weak materials with permeable zones. A notoriously awkward feature is a buried river channel which on several projects caused slope stability problems and required remedial measures. Examples are the El Bosque reservoir in Mexico (Marsal et al. 1979), Piedra del Aguila in Argentina (Fig. 10E) and San Jacinto in Bolivia (Clarke et al. 1989). Extensive drainage to stabilize downstream slopes was required also under a variety of different geological settings at Lazaro Cardenas in Mexico, Polifiton in Greece, Atatürk in Turkey and several other sites. A minor failure on the reservoir rim at the Pinios Ilias reservoir in Greece occurred when the lake pressurized the gas contained in a sand layer. Once the infiltrating lake water had expelled the gas the higher viscosity of the water reduced the gradients to a safe level (incidentally, as at Arapuni in New Zealand the gas was found to consist essentially of nitrogen). A slide triggered by seepage to an adjacent valley is also reported from Cedar Falls in the US (ASCE 1975). If the reservoir rim is slender the slides on the downstream flank seriously endanger the entire project (e. g. Lajes in Brazil, Cabrera 1992). The Laguna Zapos which stores water for El Teniente mine in Chile is an extreme case where the entire flank of the reservoir shows distress and remains comparatively stable only thanks to the natural drainage afforded by the high permeability of the loosened rock.

5. Downstream Run-off Regime. Peak demand in power production and in water supply leads to rapid fluctuations of discharge from the reservoir. Especially if the banks of the river are of fine

grained and erodible material the groundwater transients occasioned by the fluctuations will result in slides.

6. Effects of Waterways. Seepage from canals and tunnels which convey the reservoir water for downstream use is a frequent cause of landslides. It is beyond the scope of the present text to deal with this topic.

7. Effects of Downstream Slides on Reservoirs. The most serious hazard is landslide damming of the river downstream of the reservoir. This obstructs operation of the reservoir and thus causes direct losses. It may lead to consequential damages if, for instance, the outlets cannot be operated and flood control cannot be exercised. The Restitución hydro-power plant on the Mantaro in Peru was equipped with bulkheads and high level emergency access to provide for the event that another huge landslide should inundate the valley.

4. HAZARD AND RISK ASSESSMENT

The summary review of the scenario of experienced and potential reservoir/-landslide interactions reveals a wide spectrum of hazards and risks implied by the association of these two elements. Several countries and institutions have issued regulations which stipulate procedures for design and operation of reservoirs, intended to keep the risk at a very low level, acceptable to the society (e. g. ENEL 1970, SwissCOLD 1982, CFGB 1973). Definitions of hazard and risk in association with landslides were spelled out for instance by Gruetter et al. (1982) and Varnes (1984). With:

Hazard H=probability of occurrence within a given time of a potentially damaging phenomenon

Vulnerability V=degree of loss to a given element resulting from the occurrence of the damaging phenomenon one obtains:

Risk=H*V*(number of elements at risk).

For the purpose of economical evaluation of a project or of a specific landslide case the "Risk" is a "cost".

The objective is to minimize the risk. This task can be approached from any one or any combination of the three factors in the above equation. The task is rarely as straightforward as the simple equation would tend to imply. A number of consecutive steps has to be negotiated:

1. The existence of a hazard has to be recognized.

2. A full inventory of hazards has to be made.

3. The hazard has to be quantified in its geological and geotechnical dimensions.

4. An event tree has to be conceived which helps to analyze the damage propagation; for instance: rapid drawdown -> landslide -> impulse wave -> overtopping of the dam -> failure of the dam -> downstream inundation.

5. Probabilities for the items in the event tree have to be defined and the likely magnitude of the consequences of an event has to be assessed.

6. The technical and economical viability of the risk reduction has to be studied and the remedial measures, if any, have to be justified as to their necessity and adequacy.

The following paragraphs are intended to point out some specific aspects of hazard and risk in relation to landslides and reservoirs.

4.1 Identification of the Hazard

Geological precedent or the evident susceptibility of the formations in the project area in many cases directly indicate the hazard. In other cases, as for instance the buried river channel at El Bosque (Marsal et al. 1979) the hazard is difficult to detect. Difficulties are also encountered with the identification of old landslides in hard rock formations. At Polifiton and Thissavros (Fig. 9) geologists had debated the existence of slides until eventually movements started. At the Tillo-Bego-Senketo slide on the Atatürk reservoir there had been a tentative to attribute the contorted morphology to karst which did not appear unreasonable in this specific environment. The presence of the slides at the Tablachaca reservoir was known when the construction of the dam started but they were not expected to create problems. Finally, the difficulty of predicting a slide is illustrated by Marsal's conclusion after analyzing the slide at the Sta. Rosa dam: "the calculations show that the rock should not have slid" (Marsal et al. 1979). Conversely also cases could be quoted where an important slide hazard was suspected but the ground eventually proved perfectly stable.

As pointed out above the search for landslide hazards has to comprise the catchment area, the zone of the reservoir itself, a part of the downstream reach of the river and occasionally also neighboring valleys. This is a formidable task. If it is approached rigidly, schematically and with a quest for comprehensive detail it is likely to prove prohibitive in time and cost for many projects. Rather, it is mandatory to proceed judiciously with stepwise refinement in scope and detail of the investigation.

The first step is to build an inventory and data base. Pertinent recommendations can be found e. g. in the texts by Schuster et al. (1978), Brunsden et al. (1984), Varnes (1984), etc. It is considered appropriate to mention a few methods which proved particularly useful in this context on several projects:

1. Geological and morphological evaluation of aerial photographs and imagery.

2. Morphological and morphometric analysis of topographic maps.

3. Field surveys covering general geology and engineering geology with special attention to lithology, structure and hydrogeological factors.

4. Inventory and classification of existing slides and their correlation with geological, morphological and hydrological factors.

The results of a surface reconnaissance may remain ambiguous and they rarely render accurate information on the depth and shape of the failure plane and on the driving forces. Especially if the destabilizing effect of toe inundation of a slide by the reservoir is to be analyzed the geometry of the toe and the hydrogeological conditions are of decisive importance. Subsurface investigations are then required. Because large areas may have to be investigated it is convenient to start with geophysical surveys. In order to obtain more detailed information on subsurface structure it may become necessary to take oriented drill cores or run television logs of the boreholes. The best insight into the subsurface geology is, of course, afforded by pits, shafts and adits. For most of the critical and complex slides at reservoirs the driving of exploratory adits proved indispensable.

However, even comprehensive and well performed subsurface explorations do not always provide fully conclusive information on the geometry and the mechanism of sliding. Therefore, the surface reconnaissance and the subsurface investigations have to be complemented by measurements from a monitoring system. Typical instrumentation systems comprise: displacement gauges on cracks, lines of sight, levelling trails, tilt meters, trilateration, gauges for run-off, piezometers and pore pressure gauges, subsurface displacement monitoring by inclinometers and extensometers, acoustic and microseismic monitoring. Significant improvements in the instru-

mentation have been achieved in the course of the past 30 years. This refers mainly to trilateration, borehole inclinometers and multi-port piezometers.

It is desirable to start monitoring one or preferably two years before filling of the reservoir. This will allow to study seasonal influences for instance on the groundwater regime and to detect slow creep.

4.2 Hazard Factors

The investigations compile sets of parameters which have to be weighted in two respects: the severity and the uncertainty associated with them. The weighting can be based on analytical evaluation, on local precedent or on general experience. The weighting of the parameters may have to be tailored to the potential risks of a specific project. The review of diverse case histories indicated a number of factors which appear of major importance in relation to reservoirs.

1. Ground Slope has to be evaluated in relation to the material. Some of the most voluminous mass movements along reservoirs occurred in rather flat ground in soils. Steep slopes introduce uncertainty, especially regarding first time failure and they provide conditions which permit acceleration of the slides.

2. Volume of Slides has to be weighted in relation to the volume of the reservoir and to the location of the slide relative to vulnerable structures. The 40 million m³ slidemass at the Bighorn reservoir in the USA remained almost unnoticed (Dupree et al. 1979) but for the small Martinie reservoir in France already 8000 m³ slide proved critical (Fig. 12 C, Couturier 1986). The larger the volume of an individual slide the more difficult it will be to define accurately its mechanism and its geometry. Thus, uncertainty tends to increase with volume. The volume parameter has also to be weighted for the density of mass flow and put in relation to the geometry of the reservoir at the front of the slide. Another distinction has to be made in respect of the coherence of a mass which enters a reservoir. Compact, coherent slides convert a larger proportion of their energy into waves than falls of disintegrated material (Vischer 1986).

3. Lithology. The materials most likely to generate mass movements at reservoirs appear to comprise loess, clay, especially sensitive clay, silt

and sand. This is illustrated by the Roosevelt (Jones et al.1961), the Fort Randall reservoir (Erskine 1973) and a number of projects in central Asia and China. In about 2/3 of the reported reservoir slides in hard rock clay is also involved in the form of interbeds, seams or gouge. A statistical evaluation of case histories for hazard ranking of rock types is difficult because it would have to consider the incidence of slides as well as the relative frequency of the rock types. Therefore, only an indicative rating can be given. First place, 30%, is occupied by schistose rocks (phyllite, slate, schist, gneiss) closely followed by stratified clastic sediments with 25%, mainly alternations of pervious and impervious strata with argillaceous beds. Slides in carbonate sediments are also quite frequent, 15%, and mainly affect alternations of limestone and marl. About 10% of the slides occurred in volcanic rocks. Still another category comprises ice and snow which is a common hazard in mountain regions although the lower specific weight and the fact that ice and snow will float to the surface and dampen the impact renders them somewhat less dangerous than rock slides.

4. Geometry of Slide. The first parameter to be considered is the location of the toe. If the toe is at or below the valley floor there will be a direct influence of the reservoir but there is also the possibility that the slide will be self-buttressing before growing into a major hazard (provided the buttress can be allowed to build without blocking the reservoir). With partially submerged slides the most important parameters are slide mechanism and toe geometry. With the toe flatter than the driving mass of the slide (e. g. Vajont) or curving upward the inundation of the toe will normally reduce the safety factor of the slide. Therefore, in the investigation program particular attention has to be directed to this part of the slide. Complex and curved failure planes are less likely to accelerate but there is also more uncertainty in their analysis.

5. Hydrogeological Conditions. High and uniform permeability is favorable for slope stability. Low, irregularly distributed and anisotropic permeability is a hazard and a factor of uncertainty under conditions of reservoir operation. The reservoir filling is less likely to destabilize the slope if the original groundwater levels are high.

6. Brittleness. This term is used here to denote processes resulting in substantial reduction of strength. It covers the transition from peak to residual strength in clay on the failure plane, the loss of strength due to saturation, loss of resisting

forces in the course of a progressive failure, loss of lateral confinement, changes in the shape or the location of the basal failure plane or of planes of internal deformation which decrease the resisting forces in the slide. Brittleness introduces uncertainty because the prediction of the residual resistance of the slide is difficult. Therefore, brittle slides, if they involve risks, require particularly careful treatment. The Vajont slide which accelerated in 70 seconds from 20 cm/d to at least 25 m/s is one of the most impressive examples of brittle failure. Slowly creeping old landslides are least likely to change their behavior drastically (except for the temporary effect of overconsolidation due to uplift), reactivation of dormant slides always involves a measure of brittleness and first time failures would mainly be expected to be brittle. Again, the distinction between hazard level and uncertainty must be pointed out. About 2/3 of the significant slide events on reservoirs occurred with old slides. On average, therefore, the hazard rating for old slides is higher than for first time failures. For first time failures the hazard is latent and its level is difficult to assess. Brittleness may finally intervene in the viscous state of a slide. As shown by Salt (1988) for slide gouge material from New Zealand the viscous resistance does not increase any more once the shear rate goes beyond a critical value. When the critical value is reached the excess forces will accelerate the slide. At that stage thermal effects may further precipitate the process.

10. Accessibility. Slides require monitoring and maintenance. If there is no access to carry out these works the preventive maintenance has to be taken out of the event tree and the full hazard level is to be applied for the project evaluation. Projects in remote areas should therefore be designed to be conservative and robust.

4.3 Hazard Assessment

The inventory of landslide related parameters in the study area has to be weighted and evaluated to translate it into a quantified assessment of the hazard, i. e. the probability of occurrence of a landslide event and of its capacity to cause damage.

Certain hazards entail extreme risks which the profession will not consider acceptable. In those cases a deterministic evaluation is preferred which is intended to reasonably exclude such risks. Public awareness, the increasing desire for safety and growing population density continue to lower the level of acceptability for such extreme risks, even if they are hypothetical.

The remaining risks can then be evaluated deterministically, statistical-deterministically or probabilistically. The large extent of the study area as well as the uncertainty attached to many parameters would tend to favour statistical and probabilistic evaluation. A probabilistic evaluation offers the advantage to be easily incorporated in the economic project evaluation. Reservoirs represent large investments and they are intended to produce benefits over an extended period of time. Typically a lifespan of 50 or more years is assumed and the costs accruing from hazards (risk) should not endanger the economic feasibility of the project.

The study of Jones et al. (1961) for Lake Roosevelt on the Columbia river provides a successful example of statistical evaluation. The study benefited from a uniform geotechnical setting and from a population of 300 sample slides which could be evaluated. An inventory of several parameters was made for the slides and the parameters were tested for their significance in relation to the slide phenomenon. Eventually only 4 parameters proved significant: the material, the original slope angle, groundwater conditions and submergence. Multiple regression with these parameters provided a simple equation to predict the locations and dimensions of landslides to be expected.

The statistical approach has since been applied successfully on other projects and modern methods of data processing have rendered its application still more convenient. However, the approach needs to be amended to include also the time coordinate. For instance, for the feasibility analysis of a regulating reservoir one should predict how much of its operating storage could be lost to slides and at what time during the life span of the project the loss will occur (e. g. Riemer et al. 1988).

If a repetitive process is involved as for instance with snow avalanches, glacier calving and in some cases also debris flows the problem can be solved by statistical determination of the recurrence interval.

Other landslides, however, are singular events and methods used to estimate their potential for failure are static in concept. This applies also to the probabilistic methods of stability analysis. The probabilistic methods use distribution functions for strength parameters which are time-invariant. Time can be introduced if a stochastic variable is involved in the driving forces or in the shear strength of the slide as e. g. meteorological factors and seismic acceleration.

Meteorological factors are known to control rockfall (Whalley 1984). For the Vajont slide Hendron et al. (1986) have found a precipitation of 700 mm in 30 days to be capable to trigger the slide also without submergence by the reservoir. The probability of high precipitation could be estimated from meteorological records and, in combination with the expected fluctuations of the reservoir, a probability of failure could have been predicted.

The Vajont slide was the worst reservoir-landslide disaster which is on record. It can be used in the definition of the extreme risk for reservoirs and landslides. ICOLD (1984) listed 35000 dams. The cumulated life time of these dams is of the order of 900000 dam-years. The annual probability of a Vajont magnitude landslide disaster then corresponds to $1,1*10^{-6}$. In a risk diagram (Fig. 2, modified from Whitman 1984) the Vajont event plots in the "accepted" range, at about 10% of the general risk associated with dams. It may be questioned how reliably defined this risk level is because most of the large dams are of comparatively recent construction. They may still produce problems which will call for a re-adjustment of the risk level. On the other hand there has been important progress in analytical procedures which should help to make dams and reservoirs safer.

For hazards of lesser magnitude only few data are available which help to define their probability, as for instance:
- large landslides in the Alps probability 1/1000
- large landslides on the Mantaro River in Peru probability 1/20 to 1/250 (Morales et al. 1984)
- impulse waves in Swiss lakes probability 1/10.

Even if such generalized probabilities are known they cannot be unreservedly applied to reservoirs. The landslide hazard at reservoirs is not random in time in all its components. At the Roosevelt lake 50% of the recorded landslides took place during the two years of first filling and another 10% at the first drawdown (Jones et al. 1961). A review of 60 published case histories generally confirms this time dependence of landslide incidence at reservoirs (Fig. 3). The most critical phase is the first filling of the reservoir. On average 85% of the slide events had started during construction and filling and less than 2 years after completion of the project. In Japanese experience 40% of the land-slides are triggered during first filling of the

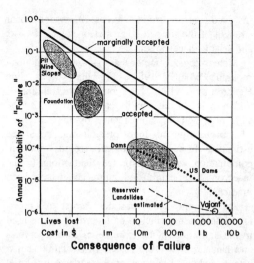

Figure 2 Risk and probability of failure (modified from Whitman 1984)

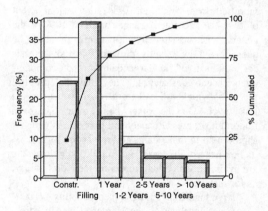

Figure 3 Slide-frequency at reservoirs in function of time

reservoir (Nakamura 1985). Therefore, the landslide hazard at reservoirs is subject to a pronounced recession with time.

A time dependent recession of mass movements due to bank failures and bank erosion was also stated by Lukác 1982 (see also above).

Additionally, processes of ageing or progressive failure should be perceived which will materialize after sustained operation of the reservoir. They could result for instance from ion exchange in clays, from leaching of cementitious substance or progressive wave abrasion.

According to the above deliberations the land-slide hazard on reservoirs contains four terms:

1. a random term (landslides triggered by extreme precipitation, snow avalanches, rock fall)

2. a term sharply receding with time (mainly slides triggered by construction activities, first submergence and initial reservoir operation)

3. a term slowly receding with time

4. a term gradually increasing with time (slow progressive failure).

For the time being it appears doubtful if useful generalized probability distributions can be obtained for all of the above defined four terms. But interesting approaches to the individual problems have been made and it is considered important to continue efforts in this direction. The probability distributions would greatly help in the process of making decisions. On the other hand the apparently "exact" formulation of a process and of decision criteria may generate an over-confident attitude. It is necessary to caution against such attitude because we have to cope with a changing environment. The reality of climatic change has eventually been recognized but it is unknown if the presently observed trends will persist (fig. 4). In such environment the prediction of hazard will always involve a considerable margin of uncertainty, even if comprehensive historic records should exist for a specific phenomenon. The ice avalanche which hit the job site of the Mattmark dam may serve as example. The Allalin glacier was known since the 6th century as a source of glacier outburst floods but no significant ice avalanche had ever been observed (fig. 5). Another case in point is the Cairnmuir slide above the Clyde reservoir. After remaining dormant for more than thousand years it reactivated during the last fifty years. A contributing factor to its reactivation is seen in the proliferation of rabbits which have degraded the vegetation, followed by an increase in infiltration capacity. The slide now reacts sensitively to intense precipitation.

5. RISK REDUCTION

5.1 General Considerations

When a landslide hazard at a reservoir is identified it must be decided:

1. if it is necessary to take action to reduce the risk,

2. to what extent the risk needs to be reduced and what means to apply to that purpose,

3. which is the appropriate timing for remedial action.

Preferably a deterministic approach should be taken in the treatment of extreme hazards to the reasonable elimination of unacceptable, catastrophic risks. The deterministic approach may have to be backed up by probabilistic evaluation or parametric studies which help to visualize the confidence levels. For the remaining hazards, not expected to be capable of catastrophic consequences, a risk based approach to the decision process would be attractive. For this approach resort can be taken to an optimization model which balances the risks against the cost of remedial measures. The concept proved successful in the design of mine slopes (e. g. Sperling et al. 1987). Risk based decisions are very simple in some cases. At the Furnas reservoir rockfall from a quartzite cliff caused impulse waves. With the excavation of only 15000 m³ the hazard was brought under control, the freeboard on the dam could be reduced and the increased operating range in the reservoir provided a benefit of 88 MW firm power (Szpilman 1976). But the consequences are not always so clearly defined. At the Amaluza dam (Fig. 12B) the cost quoted by the contractor for the anchoring of a slide near the intake appeared disproportionate to the dimensions of the slide. In hindsight the problems now experienced with that slide would have justified the costly investment into the preventive stabilization. This example shows that with the uncertainty involved in the hazard assessment it may be difficult to find a meaningful formulation for the objective functions for optimization. A particular problem, as mentioned above, is the time dependence of the reservoir landslide hazard, for which there are only rudimentary empirical data, not yet suitable for general purpose application. But, even if this complication which is introduced by the presence of the reservoir, is excluded, the risk based design for remedial measures on large landslide may run into difficulties. In this respect reference is made to the 17th Terzaghi lecture by Whitman (1984). Admitting a coefficient of variation of only 10% in the resisting forces of a landslide a factor of safety of 1,5 would correspond to a probability of failure of 1/1000. In large landslides uncertainty attaches to both driving and resisting forces. In a complex geological setting the volume of investigations required to prove coefficients of variations in the forces less than 10% could easily prove prohibitive. But, using the higher coefficients of variation the risk based design founded on probabilistic evaluation chances to end up in very costly remedial works.

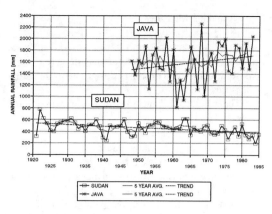

Figure 4 Climatic change displayed in trend of precipitation

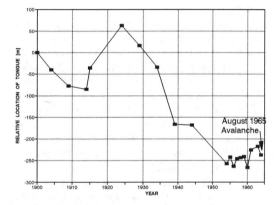

Figure 5 Retreat of the Allalin glacier before the event of ice avalanche

In conclusion, the risk based design offers a convenient tool and it would in general be desirable to promote its application. But if too many variables and too large uncertainties are involved, a simple parametric analysis of the problem may be more appropriate. Eventually, however, engineering judgement will still be required to decide on the design of remedial measures.

If a large landslide is to be partially submerged by a reservoir, it would be reasonable to concentrate the analysis on the effect of submergence. The remedial measures could than be conceived to offset the negative effects of inundation with a modest margin of safety (cf. Hoek 1991).

5.2 Remedial Measures

As long as uncertainty in hazard and risk cannot be excluded, preference should always be given to the most flexible and robust designs for remedial measures.

The routine of landslide treatment offers three fronts of attack:

1. avoid the problem
2. increase the resisting forces
3. reduce the driving forces.

The routine applies also to landslides at reservoirs. However, owing to the presence of the reservoir and the man-made structures pertaining to it there are more ramifications to the event and damage tree of a landslide in this particular situation which also widen the range of options for remedial measures. In the following some typical options are briefly discussed.

1. No Stabilization. The loss of dead and life storage due to landslides is accepted. The owner of the project obtains an easement on the land along the banks (e. g. Lake Roosevelt, USA), population from endangered areas is compensated for their land and resettled (e. g. Karakaya, Turkey), allowance is made for maintenance and occasional reconstruction of roads near the shores (e. g. Blenio, Switzerland). In the worst case it may be preferable to omit regulating storage adopting a run-of-river scheme instead.

2. Adjust the Design to Avoid Hazards. It has to be checked if the dam site can be shifted or the storage level can be changed to get away from the hazard areas. For instance the Kotmale dam site was removed from the vicinity of a slide and the reach of rockfall was mapped in relation to the reservoir (Fig. 10D, Doornkamp et al. 1982). Hazard areas can be diked off from the main reservoir (Koka, Ethiopia). Avoiding inundation of slides has the added advantage that potentially negative effects of submergence are avoided. On the other hand, there is also the possibility to raise the reservoir sufficiently above the neutral point of the slide or to submerge the slide completely. Check dams can be built to retain debris flows (extensively used in the Alps, cf. Eisbacher et al. 1984) or structures are used to divert avalanches (Ferden, Switzerland, Volkart 1974). In high mountain areas there may always remain a residual hazard from very large landslides which travel long distances and may even jump over drainage divides (fig. 6, Scheidegger 1973).

3. Adjust Design to Reduce Vulnerability. As the experience from Vajont and Pontesei shows, a concrete dam is less vulnerable to overtopping than an embankment dam. But embankment dams can also be designed to resist overtopping (e. g. Tunsbergdalsvatn, Norway, Lied et al. 1976). If the height of the impulse waves is modest an increase of the freeboard will exclude overtopping. On the Revelstoke dam in Canada the embankment on the left abutment is built 4,6 m higher than the main concrete dam on the river course (Forster 1986). An inexpensive way to gain freeboard on an embankment dam for emergencies is to place a parapet on the crest. Outlet structures should be dimensioned adequately to allow control of lake level as may be required in an emergency. Intake and outlet structures should be located such that they are not likely to be obstructed by sediments from slides. Additional facilities may be needed for flushing to clear the structures. Outlet structures may require special lining to resist abrasion from extreme sediment loads due to slide events. An interesting design was adopted for Agoyan in Ecuador. The reservoir is split in two parts. One part is reserved for clear water for the power intake, the other part at the main dam has large gates to pass lahars which may be produced by the Tungurahua volcano in the catchment area. Structures which are essential for the save operation of the reservoir should be well protected against damage from landslide events. If there is a tangible risk of landslide dams in the catchment

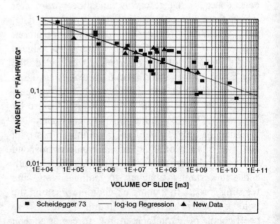

Figure 6 Tangent of "Fahrweg" of large slides as function of slide volume (modified from Scheidegger 1973)

area the spillway must be designed capable to handle also outbreak floods (see Code et al. 1986). Alternatively, if the location of the landslide or glacier dam is defined, a diversion tunnel can be driven through the valley flank (which was done e. g. for the Allalin glacier in Switzerland).

4. Adjust Reservoir Operation. If the hazard is seasonal, a temporary freeboard needs to be provided regularly. This applies for instance to snow avalanches but it was also considered appropriate for the large rock slide at Polifiton in Greece. At Polifiton the spillway gates were opened during the rainy season when the slide tended to accelerate. Anticipation of a dangerous movement may call for exceptional draining of the lake. In that case care must be exercised not to accelerate the movement by the effects of rapid drawdown. At the Gepatsch dam the filling of the reservoir was carefully adjusted to keep the Hochmaiß slide under control (Fig. 10C, Breth 1967). There are several reservoirs where a limitation on the rate of drawdown was imposed in order to avoid triggering of slides. If the effects of a slide can be contained with confidence rapid drawdown can be used intentionally to pull the slide down. If the slide is self-buttressing and if there is no risk of retrogression this is a very economic way to get rid of a slide. Groundwater conditions and the slide geometry may impose a limit also for the total range of drawdown, irrespective of the rate of drawdown.

5. Reduce Driving Forces on the Slides. If the slide mass constitutes a risk which cannot be avoided and which for economical, technical or political reasons is not considered acceptable it has to be treated. A basic step is to protect critical masses against destabilizing effects, for instance against erosion by surface run-off or by waves. Rip-rap with or without filters, gabbions or pavements are used for this purpose. Further steps are to reduce the driving forces and/or increase the resisting forces. The driving forces can be reduced by head unloading. A prerequisite is an adequate definition of the slide geometry and of the slide mechanism. For instance at Polifiton in Greece excavation started before a concluding investigation of the slide had been carried out and, therefore, remained ineffective. Sealing of the surface, surface drainage and grassing or reforestation help to control infiltration and the rapid building of destabilizing groundwater conditions inside and beneath the slide mass. Shotcrete (e. g. Lane 1970)

was applied to seal fractures in rock at Kariba, in other cases bituminous compounds and geotextiles were used. To drain water from the slide to the surface horizontal drain holes were drilled at the Amaluza dam in Ecuador and at Tablachaca in Peru. Other alternatives are drainage trenches and pumped wells. Particularly effective drainage is achieved with a system of tunnels and drain holes. The system proved successful at Revelstoke (Fig. 10C, Moore et al. 1982) and Mica in Canada, Polifiton in Greece, Tablachaca in Peru, Lazaro Cardenas in Mexico, Grand Maison in France, Schräh in Switzerland and it has also produced very favorable reactions on the Clyde slides in New Zealand (Fig. 11B). In practice a decisive advantage of drainage is its flexibility. The orientation of drain holes can be adjusted to conditions encountered during construction and additional holes can rapidly be drilled whenever required. Drain holes serve multiple purposes of investigation, stabilizing and monitoring. With the present experience it is conceivable that suitably located drainage tunnels might possibly have at least mitigated the Vajont disaster.

6. Increase Resistance of the Slide. Although in a majority of the reservoir slides clay is involved it was apparently not practical to treat the clay on a large scale. Only for the Lajes reservoir lime treatment of a slide on the reservoir rim is reported. Rather, mechanical strengthening and support of the slides is preferred. The most common measure is toe buttressing. The Mornos dam is an interesting sample in this category (Fig. 9B, Schetelig 1989) because the dam is placed against the slide and thus forms part of the toe buttress. At the Slide No. 5 on the Tablachaca reservoir (Morales et al. 1984) the buttress had to be placed on poorly consolidated lake sediments which required additional strengthening by concrete piles and vibroflotation gravel piles. Moreover, because of the spillway gates directly downstream and the desander and intake on the opposite bank there was not enough space to accommodate sufficient material in the toe buttress. Therefore, prestressed anchors and drainage systems had to be added. The slide has decelerated notably but creep is still observed. Also for the Rönkhausen pumped storage scheme prestressed anchors were selected for stabilization because of lack of space for the cheaper alternative of a toe buttress (Fig. 12A). Anchors were also installed for slope stabilization at Libby (Fig. 8E) and McKays dams in the USA,

at Chicoasen in Mexico and Amaluza in Ecuador. Anchors are comparatively expensive and their use is therefore normally limited to situations which require rapid support for comparatively small volumes of hard rock or which call for a flexible response where other means are not practicable. To stabilize the debris of an old landslide near the Sir dam in Turkey shafts were excavated and filled with concrete. The shafts eventually formed a retaining wall which protected the excavation for the dam.

5.3 Observational Approach

Rib (in Schuster et al. 1978) stated: "No one is capable nor is money available of studying in detail and guaranteeing the stability of all slopes for most large construction projects." If reservoirs are involved the task is additionally compounded in three respects:

1. the impacts of the reservoir on the slopes and slides which create a new, unprecedented situation,

2. the augmentation of risk by the presence of the reservoir and of the structures which create it,

3. a still inadequate definition of the time dependence of the hazard in the landslide-reservoir system.

Additionally, the geological setting of the reservoir may imply the extreme hazard of the exceptionally large landslides. Such hazard may be hypothetic but there are no viable means to eliminate it.

Under such conditions it must be recognized that uncertainty inevitably remains, irrespective of the most conscientious analysis, competent design and careful construction and that this uncertainty may correspond to an as yet poorly defined risk. At this point resort must be taken to the observational approach.

The observational approach proceeds from the concept that landslide events do not occur without warning. (Instantaneous movements triggered by earthquakes constitute an exception but also the capability to predict earthquakes continues to improve.)

With the premise of the warning phase preceding the landslide the method comprises three elements:

1. a monitoring system which is capable to detect, reliably and accurately, significant warning signs

2. a strategy of actions to be taken in the event of an impending dangerous landslide

3. an efficient organizational setup which reduces incoming data, evaluates the observations and decides which action to take when.

The 40 years since the concept of the observational approach emerged have seen an important development in technology which greatly contributes to the efficiency and the reliability of the method. This refers to the quality and capacity of the monitoring equipment, the ease of real time remote readout and instantaneous data reduction afforded by modern electronics. The decision maker who conducts the observational approach now gets better information and is left with more time for remedial action.

Meanwhile, the validity of the observational approach has been established by case histories from reservoirs worldwide. Some countries have institutionalized it by stipulating a minimum of monitoring and periodic evaluation (e. g. ENEL 1970). Nevertheless, the Vajont disaster leaves a question mark. The question was taken up by Belloni et al. (1987) who point out that modern instrumentation would have supplied a significantly better understanding of the problem. But, perhaps the most important conclusion is contained in the remark:

"The monitoring of the events leading to the Vajont slide was carried out after they had happened, rather than by trying to anticipate them."

Clearly, the observational approach must commence before first filling of the reservoir and it must pay particular attention to the first filling and early stages of operation. After that, at least with reference to the statistic of past experience, the hazard level recedes.

Important topics for the observational approach which are still under investigation and need to be pursued further refer primarily to the "brittleness" of the slides, regarding progressive limit equilibrium failure and particularly the viscous behavior. Many slides have continued creeping at fairly high but constant rates (fig. 7). At the Gepatsch dam a creep of 34 cm/d was rapidly brought under control but at Vajont the rate of only 20 cm/d preceded the terminal acceleration. Presently, generalized alarm criteria cannot be defined, they have to be selected in consideration of the specific conditions for each case and they have to be refined in the process of the observational approach.

5.4 Cost of Reservoir Landslides

Published case histories do not offer enough information to justify a comprehensive review but a few general aspects will be discussed.

1. Total Cost of a slide to the project of the reservoir includes investigation, construction of stabilizing works, monitoring, maintenance, costs of interruption or restriction of project operation, cost of land acquisition or easements, etc. From the few available data it would appear that in many cases the construction costs for stabilization are minor compared to the losses in operational benefits that would have been entailed by the respective landslide. Case histories which can be cited to this end are: 1. Revelstoke with an installed capacity of 2700 MW and a construction cost of 25 million $ for the stabilization of the Downie slide and 2. the Tablachaca slide where 45 million $ were expended to permit continued power production of the two plants which supply half of the electrical energy for Peru.

2. Cost Effectiveness of remedial measures, as far as geotechnical involvement is concerned, is controlled mainly by two factors: the selection of the type of treatment and the selection of the time for the treatment. Among the technical alternatives for the stabilization of a slide a drainage system offers a particularly attractive option which should be considered wherever appropriate. Adopting the observational approach it would appear that investment into remedial measures should be deferred as long as possible. In the economical evaluation of a project, for instance on the basis of net present worth, a delayed investment will be more favorable for the cost/benefit ratio and with more time for the study of the reaction of a slide it might become admissible to reduce the requirements for the remedial works. But there are geotechnical constraints which have to be considered when remedial action is postponed. A quotation from Rib (1978) summarizes the pertinent experience:

"The cost of preventing landslides is less than the cost of correcting them...".

The main reason behind this observation is the "brittleness" in the development of a landslide. A first-time slide and to a lesser degree also a dormant old slide have resisting strength which is above the residual strength. Therefore, when a slide starts to move it loses part of its strength and the posterior stabilization has to compensate for this loss. In the case of the reservoir-landslide system there is another strong argument against deferred remedial action. The majority of the slide events is experienced during filling of the reservoir and the initial phase of operation. Therefore, on average, little is gained by delaying the stabilization but there is the risk that operation of an already completed project has to be interrupted.

6. CONCLUSIONS

Landslide catastrophes on reservoirs are comparatively rare events but it is estimated that at least 1% of all reservoirs has to some extent been affected by mass movements. This refers to cost of construction, restrictions of operation, loss of production and consequential costs. Both, the frequency of landslide incidence as well as the extreme risk in the exceptional situation require routine investigations into landslides for every reservoir project.

The landslide study should commence with a rapid survey of regional conditions and should be progressively refined in scope and detail to concentrate on the essential features.

The reservoir does not always adversely impact the stability of a landslide. Under certain conditions the landslide remains indifferent to the reservoir or the reservoir may even stabilize it directly or indirectly.

The effects of the landslides on the reservoir are mostly undesirable but some reservoirs have gained live storage from mass movements.

There is a wide range of options for remedial action to defend against landslide problems. In

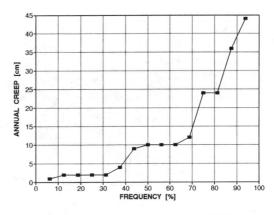

Figure 7 Observed steady creep rates of slides at reservoirs

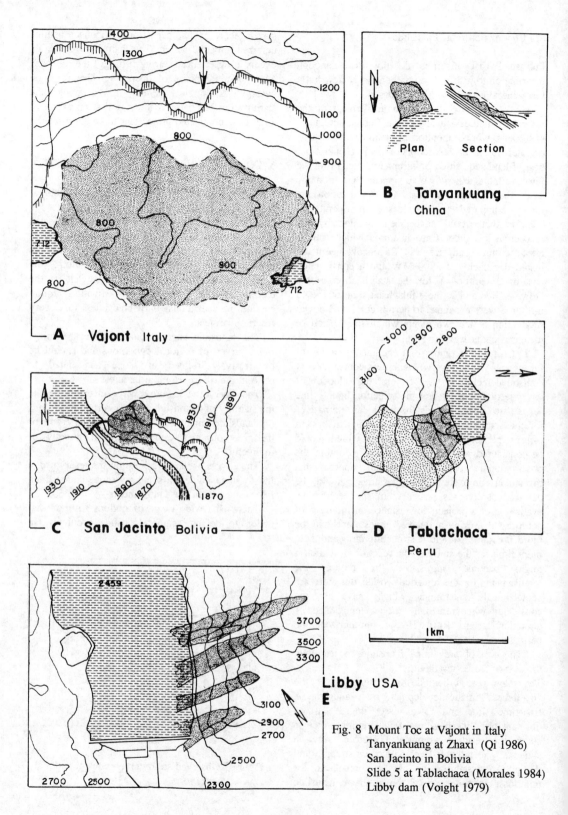

Fig. 8 Mount Toc at Vajont in Italy
Tanyankuang at Zhaxi (Qi 1986)
San Jacinto in Bolivia
Slide 5 at Tablachaca (Morales 1984)
Libby dam (Voight 1979)

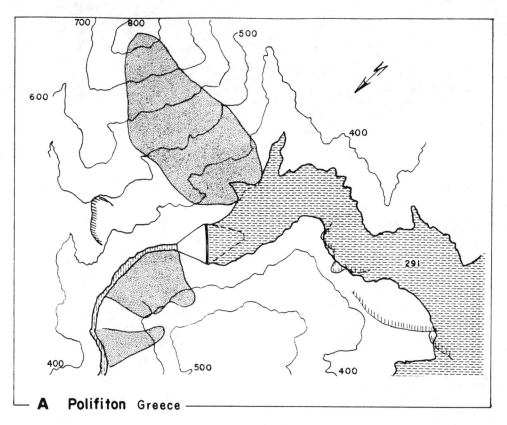

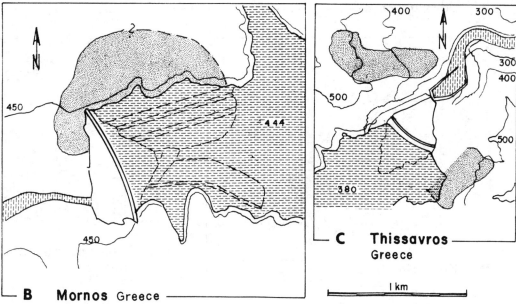

Fig. 9 Polifiton dam and reservoir in Greece
     Mornos dam (Schetelig 1989)
     Thissavros dam (Krapp et al. 1989)

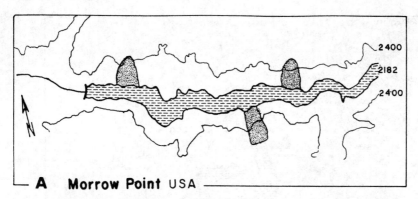

A Morrow Point USA

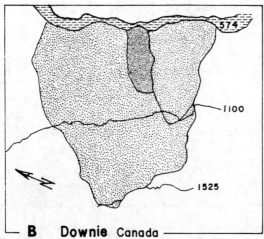

B Downie Canada

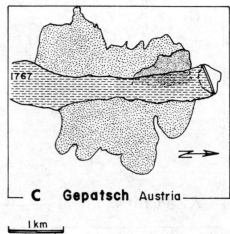

C Gepatsch Austria

1 km

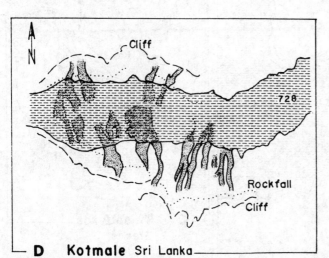

D Kotmale Sri Lanka

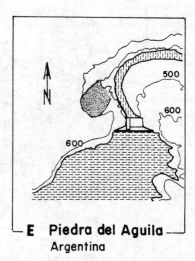

E Piedra del Aguila
Argentina

Fig.10 Morrow Point (Pugh et al. 1982)
    Downie slide at Revelstoke (Forster, 1986)
    Hochmaiß at Gepatsch (Lauffer 1967)
    Kotmale (Doornkamp et al. 1982)
    Paleocauce at Piedra del Aguila

1992

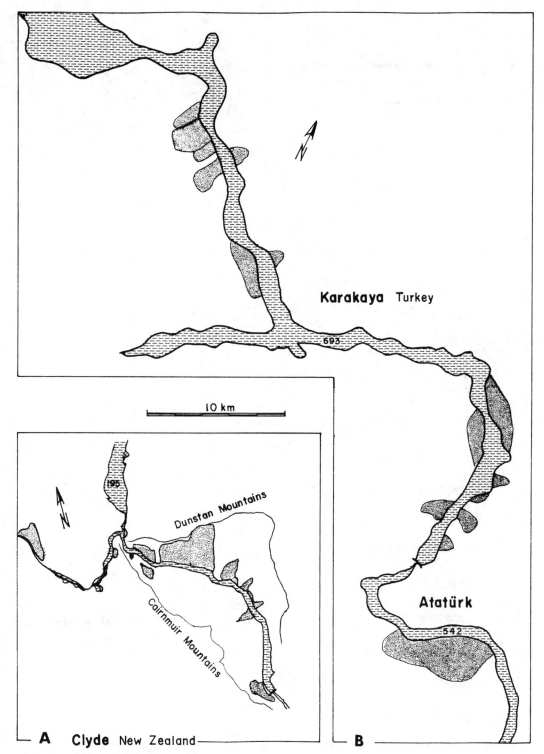

Fig.11 Karakaya and Atatürk reservoirs
　　　Clyde project (Gillon et al. 1992)

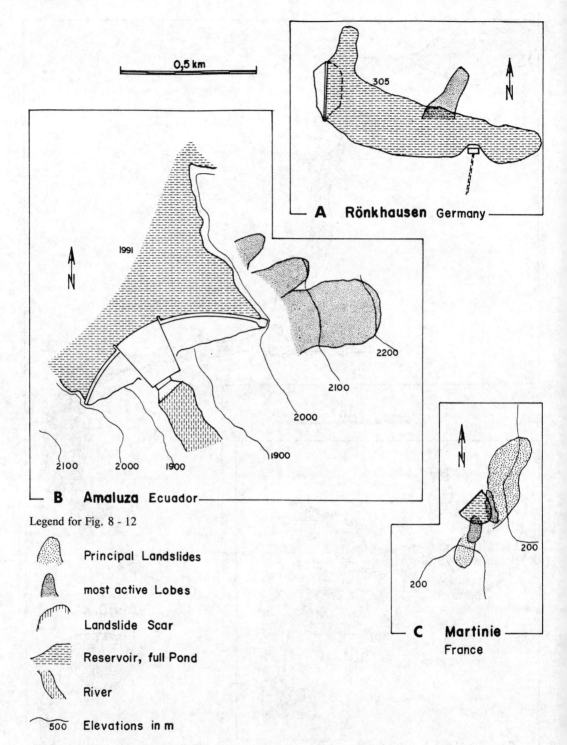

0,5 km

A Rönkhausen Germany

1991

2200
2100
2000
1900
2100   2000   1900

B   Amaluza Ecuador

C   Martinie
France

Legend for Fig. 8 - 12

Principal Landslides

most active Lobes

Landslide Scar

Reservoir, full Pond

River

500   Elevations in m

Fig. 12 Rönkhausen pumped storage (Heitfeld 68)
    Amaluza in Ecuador
    Martinie (Couturier 1986)

No.	Project	Country	Reference	Project Status	Year Incid.	Project Type	Dam Type	Height [m]	Lake Vol. [mio.m3]	Geology	Type of Slide	Location	Slide Vol. [mio.m3]	Slide Status	Effects	Remedial Type	Measures Cost mio$	Remark
1	Pd. del Aguila	Argentina		C. 91	ancient	HEPP	G	120	10000	Soil	SI	d/s Rim		part.act		Dr		first time failure, causes: scour, seepage, rain.
2	Parana Medio	Argentina	Morbidoni 86	Design		Mu.Purp.		25	25	Soils	SI,Fl,Fa	Res		stable		Resetl		
3	Gmünd	Austria	Horninger 67	C. 45	1963	HEPP	A,G	39	0.9	Schist	RSI	d/s	0.002	creep		Reconstr		creep max. 34 cm/d, actually 2 cm/a
4	Gepatsch	Austria	Lauffer 67	C. 65	1964	HEPP	Emb	153	250	Moraine	SSI	Res	20	creep		DrMon		
5	Bolgenach	Austria	Innerhofer 82	C. 78	ancient	HEPP	Emb	100	8.4	Sdst.Marl	Wedge	Abt	max. 0,05	stable	insign.	not req.		
6	San Jacinto	Bolivia		C. 89	1987	Mu.Purp.	A	47	55	Sdst.Slate	We.Top	Abt		stable	19 D	CBu		slide caused by seepage after raising dam
7	Lajes	Brasilia	Cabrera 92	C. 58	1958	Mu.Purp.	But	63	1052	DebSl	RFa	d/s		stable		DrGrt		excavate 15000m3 to permit reduction of freeboard
8	Furnas	Brasilia	Szpilman 76	C. 63	1969	HEPP	Emb	127	22950	Gneiss	RFa	Res	0.008	stable	Wave	Removal		3 slides caused by reservoir
9	Ivailovgrad	Bulgaria	Zanev 73	C. 64		HEPP	G	71	185	Quartzite		Res						sample for several dams
10	Stewartville	Canada	Peggs 82	C. 48	reptd.	HEPP	G	58	36	Till,Clay	SSI,SFI	d/s		stable	no	Ercon		
11	Mica	Canada	Moore 90	C. 72	ancient	HEPP	Emb	247	24700	Gneiss	Hydr.	Res	115			DrMon	13	
12	Amprior	Canada	Peggs 82	C. 76	ancient	HEPP	G+E	35	1.2	Till,Clay	SSI,SFI	Res	max. 0,9	stable	no	Ercon		
13	Revelstoke	Canada	Forster 86	C. 85	ancient	HEPP	G+E	153	5180	Schist	SSI	Res	1800	creep	no	DrMon	25	shore strip 200-300m wide affected
14	Various Projects	Central Asia	Mikhailov 82		contin.					Loess	Various	Res						mainly triggered by drawdown 1,5 m/d
15	Various Projects	Central Asia	Mikhailov 82		contin.					Permafrost		Res	total 1					reactivated slide
16	Pachkamar	Central Asia	Khasanov 86	C. 70	1972			65		Loess	Various	Res	up to 20					
17	Akhangaran	Central Asia	Khasanov 86	C. 81	1954			90			Various	Res						
18	Shorsu	Central Asia	Khasanov 86	C. 84	1984					Soil,Gyps	Various	Res						
19	Orava	Checoslov.	Lukác 82	C. 53		M.Purp.	G	41	346	Soil	WEr,SSI	Res			Sedimentn			
20	Velka Domasa	Checoslov.	Lukác 82	C. 66		M.Purp.	Emb	35	180	Soil	WEr,SSI	Res			Sedimentn			
21	Nechranice	Checoslov.	Rybar 77	C. 68	1969	M.Purp.	Emb	50	27	Sdst.Clst	RSI	Res		creep		Mon		
22	Sance	Checoslov.	Novosad 79	C. 71	ancient	M.Purp.	Emb	65	54	Flysch	RSI	Res	8	creep		Mon		
23	Liptovská Mara	Checoslov.	Lukác 82	C. 76		M.Purp.	Emb	53	360	Loess	WEr,SSI	Res			Sedimentn			
24	Nové Mlyny	Checoslov.	Ambroz 90	C. 89		M.Purp.	Emb	15	140	Loess	SSI	Res	2	creep		Riprap		creep <= 2 cm/a
25	El Cobre	Chile	Dobry 65	C. 84	1965	Tailings	Emb	35	10	Tailings	Liquef.	Res,Rim						
26	Lg. Sapos	Chile		C. 68	contin.	Wat.Sup.	Emb	15	3	Volcanic	Complex	Res,Rim		active	200 D			
27	Colbun,P.Norte	Chile	Valenzuela 91	C. 85	1985	HEPP	Emb	30	1490	Soils	Hydr.	Cat	4		30 D, hEcL	LoadDr		
28	Alfalfal	Chile		C. 92	1987	Mu.Purp.	Emb	16	0.6	Volcanic	RAv,DFI	Cat,Rim			3 D			
29	Cortaderal	Chile		Feasib.	1984	HEPP, RoR				Sed.Volc	SAv,GLOF	Cat						
30	Ertan	China				HEPP	Emb	16		Volcanic	SlideDams	Cat,Rim						
31	Yangtze Gorge	China	Zhuoyuan 86		1982					Sdst,Mudst	RSI	Res,Cat	15					Tanggudong slide dam breached with 53000m3/s
32	Sanmenxia	China	Qi Xiaojun 86	C. 60		Mu.Purp.	G	106	35400	Sdst.Slate	RSI	Res,Cat			Sedimentn	Reconstr.		Jipazi slide activated by 331mm rainstorm,
33	Zhaxi	China	Qi Xiaojun 86	C. 61		HEPP	But	104	3560	Loess	SSI	Res			overtopped			expected to stabilize when lake filled >160m
34	Guishi	China	Qi Xiaojun 86	C. 66		Mu.Purp.	G	43	595	Loess	RSI	Res	60Slds<1,4					
35	Fengtan	China	Qi Xiaojun 86	C. 80		Mu.Purp.	G	111	1550			d/s	1,7					1,5 billion sediment/2a, 10km waterlogging of banks
36	Three Gorge	China	Chen De Ji 86	Design		Mu.Purp.		237	815			d/s	33 Slds >1			GrtDr		slideplane dip 30-40deg,v=25m/s,wave 21m,40dead
37	Chivor	Columbia		C. 75	ancient	Ir	Emb	70	8	Diabase	Flow	d/s		part.act				Jiang Jia Po 18 mio. m3 active
38	Phievas	Cyprus		Design		Ir	AG	170	120	Plut.Sch	Various	Res,Abt	2,5	stable	Op.Loss	An		leakage from headrace
39	Amaluza	Ecuador		C. 83	1983	HEPP		43	0.76	Volcanic	Lahar	Cat	18	part.act		Flushing		
40	Agoyan	Ecuador		C. 87	1987	HEPP	G	160	120	Volc.Soils	DFI	Cat		potential				
41	Coca	Ecuador		Des.		HEPP	G	42	413	Schist	Various	Res,Abt	74	part.act				
42	Paute Mazar	Ecuador		Design		Mu.Purp.		113	1900	Volcanics	Subsid	Res			5D hEcL			
43	Koka	Ethiopia	Riemer 88	C. 60	none	HEPP		38	4,5	Gravel	SSI	Res		stable		Dikes		
44	Ferrières	France	Heitfeld 75	Design		HEPP	G			Gneiss	RSI	Res		creep		HUniTbu		moving bridge pylon
45	Sarrans	France	CFGB 82	C. 32	1953	HEPP	G	104	296	Gneiss	RSI	Res		stable		PilesAn		
46	Génissiat	France	CFGB 82	C. 48		HEPP	G	104	50	Soils	SSI	Res	15	creep		MonOper		if accelerated draw down lake creep 1 cm/a frontal lobes 2Mm3, faster with lake empty, slide dam possible.
47	Couesque	France	CFGB 82	C. 50	1953	P.Storage	A	64	56	Slidedebr.	DebSl	Res		creep		MonBypass		
48	Avène	France	CFGB 82	C. 62	1963	Ir	A	63	33,6	Dolomite	Erosion	Res				Oper		drawdown limited to 5 cm/h

No.	Project	Country	Reference	Project Status	Year Incid.	Project Type	Dam Type	Height [m]	Lake Vol. [mio. m3]	Geology	Type of Slide	Location	Slide Vol. [mio. m3]	Slide Status	Effects	Remedial Type	Measures Cost mio$	Remark
49	La Laye	France	CFGB 82	C. 65	1973	W.Supply	G	29	3,3	Marl,Lign	RSI	Res	0,2	creep		Oper		<2.25 m excess pressure, part of old slide
50	Mont Cenis	France	CFGB 82	C. 68		HEPP	Emb	120	332	Sch,Gyps	RSI	Res	15	creep		Mon		dissolution gypsum+anhydrite, 100000 to/a
51	Sainte Croix	France	CFGB 82	C. 74	1906	M.Purp.	A	95	780	Marl	RSI	Res		creep				creep 2 cm/month, triggered by flood 1906, lake filling has slowed movements.
52	Martinie	France	Couturier 87	C. 80		Ir	Emb	8	0,015	Lst,Marl	SSI	Res	0,008	stable		TrimTBuDr		
53	Montezic	France	Colombet 83	C. 82		P.Storage	Emb	33	0,57	Plut	RSI	Res		stable	prevented	AnMon		
54	Verney	France	CFGB 82	C. 84		P.Storage	Emb	42	15,6	Moraine	SSI	Res				TBuDrFr		850m galleries,1300m drainholes, < =12cm/a creep
55	Grand'Maison	France	Dubie 88	C. 85	1986	P.Storage	Emb	160	140	Plut,Lst	RSI	Res	<= 2			DrMon		
56	O.Röddinghausen	Germany	Heitfeld 79	abandon.	ancient	Tailings	Emb		12	Limestone	Subs.,RSI	Res				Grout		
57	Elta	Germany	Albiker 77	C. 68	1967	P.Storage	Emb	29	1,3	Limestone	Wedge	d/s Abt	2	stable	prevented	ExcAn		creep max. 1 cm/d
58	Rönkhausen	Germany	Heitfeld 68	shelved		W.Supply	Emb	53	40	Graywacke	RSI	Res+d/s	0,2	stable	insign.			
59	Hafenlohr	Germany		Constr.		HEPP	CDec	135		Sandstone	RSI	Res		creep				
60	Messokhora	Greece	Krapp 89	C. 62	1986	HEPP	Emb	180	10	Flysch	RSI	Res	> 5	stable	inoperatl	UnlDrMon		
61	Thissavros	Greece		C. 67	63	W.Supply	Emb	30	460	Gneiss	RSI	Res	7	stable	insign.			
62	Perdikas	Greece		C. 69	1972	Ir	Emb	53	765	Limestone	Subsid.	d/s				Mon		
63	Pinios Ilias	Greece	Wong 79	C. 74	1973	HEPP	Emb	98	2244	Claystone	Pneumatic	Res	507	dormant		none		local creep 20cm/a. Risk to flood Kremasta
64	Kastraki	Greece	Krapp 89	C. 79	ancient	HEPP	Emb	112	780	Flysch	Wedge	Res,d/s	max. 50	stable		UnlDrMon		max. 3cm/d creep
65	Politrion	Greece	Schetelig 89	abandoned		W.Supply	Emb	126	456	Gneiss	RSI	Res+Abt	25	active	Sedimentn	TBu		
66	Momos	Greece		Design		Ir	Emb	160		Marl,Tuff	Various	Res,Cat		active	Sedimentn			
67	Karangsambung	Indonesia		abandoned		W.Supply	A			Marl,Tuff	Various	Res,Cat					40	
68	Maung	Indonesia		Design			But		6		Various	Res,Cat						
69	Fahrranaz Pahlawi	Iran	Lane 70	C. 57	1968	W.Supply	A	93	6	Dol,Marl	RSI	Res	> 0,57	stable		Dr		at regulation basin, creep up to 2 cm/d
70	Pontesei	Italy	Jaeger 69	C. 60	1959	HEPP	A	55	1,4	Moraine	RSI	Res,d/s	> 4		overtopped	CBu		overtopped by wave >13m
71	Fusino II	Italy	Barioli 67	C. 61	none	P.Storage	A			Lst,Marl	RSI	Res		stable		TrimFilt		>100 m wave overtopped dam
72	Vajont	Italy	Jappelli 81	C. 65	contin.	HEPP	A	262	169	Lst,Marl	RSI,SSI	Res,d/s	240	creep	3000 D	Diversion		Creep < =0,3cm/d on rising lake level
73	Licodia Eubea	Italy				M.Purp.	G	66	22	Flysch		Res	6			Mon		
74	Shin Minawa	Japan	Mizukoshi 67	C. 49		HEPP	G		32	Graph.Sch	RSI	Res						
75	Nagasawa	Japan	Mizukoshi 67	C. 52		M.Purp.	Emb	72	16	Graph.Sch		Res						
76	Ishibuchi	Japan	Mizukoshi 67	C. 55		HEPP	Emb	53	92	Tuff,Sdst		Res						
77	Kamishiiba	Japan	Mizukoshi 67	C. 59	1958	HEPP	A	110	48	Grayw.Slt		Res						
78	Kanogawa	Japan	Mizukoshi 67	C. 57		M.Purp.	G	61	50	Slate,Sdst	RSI	Res	0,007		minor			First filling 20m movement, restarts at drawdown
79	Narugo	Japan	Mizukoshi 67	C. 57		HEPP	A	95	2,6	Talus		Res	0,2					
80	Yatogawa	Japan	Mizukoshi 67	C. 58	1960	HEPP	G	32	35	Shale		Res						c. 1cm/d on first filling, 0,4 cm/d on drawdown
81	Akiba	Japan	Mizukoshi 67	C. 62		M.Purp.	G	89	27	GreenSch		Res						
82	Futase	Japan	Mizukoshi 67	C. 63		M.Purp.	A	95	261	Phyllite		Res						
83	Hitotsuse	Japan	Fujita 77	C. 69	1969	M.Purp.	A	130	130	Sdst,Slate		Res						At first drawdown less than 2 mio m3 loss of storage expected. There are old landslide dams.
84	Shimokubo	Japan	Mizukoshi 67	C. 69		HEPP	A	129	123	Slate,Sdst		Res	>40 Slides	part.act				At first drawdown >40 slides triggered by drawdown 1,3 m/d seepage caused instability, 130+190m dr-gallery
85	Nagawado	Japan	Mizukoshi 67	C. 67		M.Purp.	A	155	13			Res						
86	Shingu	Japan	Fujita 77	C. 75	1976	M.Purp.	G	42	1750	Clay	SSI	Res	< 0,4	part.act		AnPil		
87	Andizhan	Kirghizia	Khasanov 86	C. 80	1982	Mu.Purp.	Emb	115	3580	Tuff		d/s,Res				GrtDr		
88	Lazaro Cardenas	Mexico	Hungsberg 76	C. 47	1947	Ir	Emb	68	224	Volcanic	SI	d/s Rim						
89	El Bosque	Mexico	Marsal 79	C. 54	1962	HEPP	Emb	114	400	Volcanic	SI	Res						
90	Santa Rosa	Mexico	Marsal 79	C. 63		HEPP	Emb		1680	Volcanic	SI	Res	0,4	stable		TBu		
91	Chicoasen	Mexico	Espinosa 82	C. 80	1985	HEPP	Emb	261	145	Lst,Shale		Res	6	stable		ExcAnDr		
92	Namche	Nepal	Schuster 86	Design		HEPP	G	65	100	Volcanics	GLOF	Cat			Flooding	Flushing		
93	Arun 3	Nepal				HEPP	G		350	Volcanics	GLOF	Cat						
94	Arapuni	New Zealand	Legget 62	C. 26		Mu.Purp.	A	64		Schist	Subsid.	Abt		stable	insign.			cracking and deformation creep 10 to max. 90 cm/a
95	Roxburgh	New Zealand	NZGS 90	C. 56		HEPP	G	78		Schist	Various	Res		creep				
96	Clyde	New Zealand	Gillon 92	Constr.	ancient	HEPP	G	102		Schist	Various	Res	up to >1000	part.act		DrTBuUnl	145	creep max. 9 cm/a

No.	Project	Country	Reference	Project Status	Year Incid.	Project Type	Dam Type	Height [m]	Lake Vol. [mio.m3]	Geology	Type of Slide	Location	Slide Vol. [mio.m3]	Slide Status	Effects	Remedial Measures Type	Measures Cost mio$	Remark
1	2	3	4	5	6	7	8	9	10	11	12	13	14	15	16	17	18	19
97	Tunsbergdalsvatn	Norway	Lied 76	C. 78		HEPP	Emb	42	1,1	Snow	Av,GLOF	Res	2,4	repeated	Wave	FrebCrpr		At 50m/s wave 5m expected. Damcrest and d/s toe protected against erosion.
98	Tablachaca	Peru	Morales 84	C. 72	1972	HEPP	AG	75	16	Phyllite	Complex	Res	9	part.act	Op.Loss	TBuDrAn	45	almost flooded by Mayunmarca slide. Creep 0,1cm/d, down from 2cm/d.
99	Tresna	Poland	Bujak 67	C. 67	1960	Mu.Purp.	Emb	38	100	Sdst,Shale	Wedge	Abt	0,1	stable				triggered by construction works
100	Rybinsk	Russia	Mikhailov 82	C. 41		Mu.Purp.	G	30	25400	Peat	Floating	Res						
101	Kama	Russia	Mikhailov 82	C. 54		Mu.Purp.	G+E	37	17500	Gypsum	CliffFall	Res						2 m/a shoreline recession, 45 m3/m/a up to 100m shore retreat
102	Novosibirsk	Russia	Savkin 1977	C. 57	contin.	Mu.Purp.	G+E	27	8800	Sediments	SSI	Res						
103	Bratsk	Russia	Trzhts. 78	C. 64	1959	Mu.Purp.	G+E	125	169000		RSI,SSI	Res	up to 3					many activated and new slides on filling up to 280m shore retreat on first filling
104	Krasnoyarsk	Russia	Savkin 1977	C. 67	1967	Mu.Purp.	G	124	73300		RSI,SSI	Res						
105	Montefurado	Spain		C. 54	1959	HEPP	G	42	1,1	Slate	Creep	Res						
106	Pas	Spain	Garcia 86	Design	1960	HEPP	Emb		175	Sdst,Shale	Rot	Res,Abt	up to > 50	active	Sediment	undecided		local damming, debris flows
107	Kotmale	Sri Lanka	Doornkamp 82	C. 85	ancient	HEPP	Emb	87		Metam,Soil	Various	Res,Abt	up to 4		3m wave,1D	ExcRedes		failures on draining of reservoir
108	Lake Davos	Switzerland	Huber 82	C. 24	1923	PowerInt.	Nat.Lake			Soil	SSI	Lake	0,9	creep				triggered by drawdown for construction of intake
109	Schräh	Switzerland	Swiss COLD 82	C. 24	1924	HEPP	G	112	147	Talus,Sch		Res		creep		DrOper		Drain galleries, drawdown rate <15cm/d, total
110	Palagnedra	Switzerland	Swiss COLD 82	C. 52	1924	HEPP	A	72	6		RSI	Res	0,6	creep	minor	Mon		<20m movement, creep max. >10cm/a creep max.4cm/a, damage to road and railroad. 20 mio$ losses in 1978 debris flow
111	Mauvoisin	Switzerland	Hanke 66	C. 57	ancient	HEPP	A	250	1822	Ice	IceAv,GLO	Res						Dam prevents GLOF at Giétro glacier.
112	Blenio	Switzerland	Baumer 88	C. 59		HEPP	A	92	4,6	Gneiss	Wedges	Res	>37	creep	minor	Oper		Waves of 2m at dam, volume 16000m3 expected creep c. 2cm/a, triggered by erosion, expected to stabilize
113	Sufers	Switzerland	Swiss COLD 82	C. 62	1966	HEPP	A	58	21	Talus	SSI	Res	0,15		none	Mon		
114	Gries	Switzerland	Swiss COLD 82	C. 65	1965	HEPP	G	60	19	Ice	Calving	Res			0,2 m wave			
115	Mattmark	Switzerland	Hanke 66	C. 67		HEPP	Emb	120	101	Ice	IceAv	d/s			88 D	Oper		dam prevents GLOFS at Allalin glacier, historic records since year 563
116	Santa Maria	Switzerland	Huber 82	C. 68	1975	HEPP	A	117	67	Snow	SnowAv	Res	0,15	creep	1,5 m wave			Waves <=5m, drawdown 2m required
117	Ferden	Switzerland	Volkart 74	C. 75		HEPP	A	67	1,9	Snow	SnowAv	Res	0,2			Oper		
118	Mapragg	Switzerland	Swiss COLD 82	C. 76		P.Storage	A	75	4,7			Res			9 m wave	TBu,Rr		
119	Grimsel West	Switzerland	Vischer 91	Design		HEPP	A	200		Ice	Calving	Res	1					Calving is expected to provide 50 hm3 storage 105 slides, bank recession <= 10m/a
120	Nurek	Tadjikistan	Khasanov 66	C. 80		HEPP	Emb	300	10500	Loess	Various	Res	24Slds<0,5	creep				Expected annual removal <=330000 to creep c=10 cm/a
121	Techi	Taiwan	Gallico 82	C. 74		HEPP	A	181	232		CliffFall	Res			insignif.			
122	Sidi Salem	Tunesia	Mouelhi 82	Constr.		Mu.Purp.	Emb	179	550	Marl,Gyps	RSI,MuFl	Res			Lduse,Dam	TBuDrMon		
123	Ataturk	Turkey		C. 88		HEPP	AG	173	48700	Lst,Shale	Various	Res	>200	active		Ercon		total 790 mio.m3 slides, 1/3 active
124	Karakaya	Turkey		C. 91		HEPP	G	120	9580	Heterog.	Various	Res,Abt	up to 240	part.act	Landuse	ResetlMon		
125	Sir	Turkey		C. 14	1918	Mu.Purp.	Emb	69	1120	Quartzite	Various	d/s Rim		stable		PilesAn		
126	Cedar Falls	USA	ASCE 75	C. 42	contin.	Mu.Purp.	G	168	116	Silt,Sand	SSI,SFI	Res,d/s			Lduse.wave			
127	F.D.Roosevelt	USA	Jones 61	C. 42		F.Co.	G		11795	Silt,Sand	SSI,SFI	Abt	>500 Slds			Resetl	7	d/s banks stabilized by pumped drainage, riprap 20mio$ losses in res. Waves 5 m.
128	Surry Mountain	USA	ASCE 1975	C. 42	1943	F.Co.	Emb	28	131	Soils	SSI,SFI	Res				ClearFbd		
129	Mud Mountain	USA	Galster 92	C. 48	1974	Mu.Purp.	Emb	130	5696	Sh,Chalk	Various	Res	0,05		4 m wave			
130	Fort Randall	USA	Erskine 73	C. 52		F.Co.	Emb	50	0,84		SSI,SFI	Abt						
131	Jemez	USA	ASCE 75	C. 53	1958	Mu.Purp.	G	42	752	Silt,Sand	SSI	Res						Similar to Lake Roosevelt
132	Chief Joseph	USA	Jones 61	C. 55	contin.	F.Co.	G	72	88	Silt,Sand	SSI,SFI	Res						
133	North Hartland	USA	ASCE 75	C. 61	1969	Mu.Purp.	Emb	58	366	Soils	SSI	Res				Cleanup		Hazard of landslide from Mt. Baker
134	Upper Baker	USA	Pugh 82	C. 62		HEPP	G	101	0,2		Volcanic	Res						
135	Olive Hills	USA	ASCE 75	C. 75	1963	W.Supply	G+E	43	1076	Volcanic		Res	Prevented					
136	Bighorn	USA	Dupree 79	C. 66	1966	Mu.Purp.	A	160	617	Shale		Res	40					total Vol. 10000
137	John Day	USA	ASCE 79	C. 68		Mu.Purp.	G+E	70			RSI	Res			no	none		
138	Morrow Point	USA	Pugh 82	C. 68		Mu.Purp.	A	143	144		RSI	Res	4,5	creep	no			slide A creep 44 cm/a

No.	Project	Country	Reference	Project Status	Year Incid.	Project Type	Dam Type	Height [m]	Lake Vol. [mio. m3]	Geology	Type of Slide	Location	Slide Vol. [mio. m3]	Slide Status	Effects	Remedial Type	Measures Cost mio$	Remark
1	2	3	4	5	6	7	8	9	10	11	12	13	14	15	16	17	18	19
139	Brookstail 3 N	USA	ASCE 75	C. 70	1971	W.Supply	Emb	15	0,3	Soil,Sdst	SSI	d/s Abt		stable		Dr		
140	Silver Jack	USA	ASCE 75	C. 71	1969	Mu.Purp.	Emb	53	17	Soils	SSI	d/s Abt	0,05	stable		TBuDr		
141	Libby	USA	Voight 79	C. 73	1971	Mu.Purp.	G	129	7165	Plutonic	Wedge	Res	2	stable		TBuAn		rapid failure after 0,5cm/a creep
142	McKays Point	USA		C. 89		HEPP	A	74	4	Argillite	RSI	Res	0,1	stable		TBuAnDr		>20% of reservoir lost to slides
143	M.F.Eel River	USA	Golze 77	Shelved														
144	Charvak	Usbekistan	Khasanov 86	C. 77	1970	Mu.Purp.	Emb	168	560	Loess,Sed	Various	Res	up to 50	active				new and reactivated slides, drawd. <= 3,5m/d, total displacement up to 45m/a
145	Kariba	Zamb/Zimbab.	Lane 70	C. 59	1962	HEPP	A	128	160368	Quartzite	RSI	d/s		creep		DrSeal		spillway spray destabilizes, remedial works continued into 1988

Explanations:
Col. 4: see List of References
Col. 5:   C. 91:completed 1991; constr.:under construction
Col. 6: year of major event stated; rept:repeated; contin:movements continue
Col. 7: HEPP:Hydro Power Plant; Mu.Purp.:Multi-Purpose Project; Ir:Irrigation; RoR:Run of River; P.Storage:Pumped Storage; Powerlnt: Intake for Hydropower
Col. 11: Dol:dolomite; Graph:graphitic; Grayw:graywacke; Gyps:gypsum; Lign:lignite; Lst:limestone; Mudst:mudstone; Metam:metamorphics; Plut:plutonic; Sdst:sandstone; Sed:sedimentary rocks; Sch:schist; Sh:shale; Volc:volcanics
Col. 12: Av:avalanche; RAv:rock avalanche; SAv:snow avalanche; Fa:fall; RFa:rockfall; Fl:flow; SFl:soil flow; MuFl:mud flow; GLOF:glacier outbreak flood; Hydr:hydraulic failure; Rot:rotational; Sl:slide;
    RSl:rock slide; SSl:soil slide; DebSl:debris slide; To:toppling; We:wedge; WEr:wave erosion;
Col. 13: Abt:abutment; Cat:catchment; d/s:downstream; Res:reservoir
Col. 14: volume of main feature if not indicated otherwise
    Col. 15: act:active; part: partially
Col. 16: D:dead; hEcl:high economic loss; insig:insignificant; Op:operational; Sediment:reservoir sedimentation
    Except for Vaiont the listed fatal casualties occurred during the design/construction phase
Col. 17: An:anchors; CBu:concrete buttress; Crpr:crest protection; Dr:drainage; Ercon:erosion control; Exc:excavate; Filt:filter; Frb:freeboard; Grt:grout; HUnl:head unlading; Mon:monitoring; Oper:operational precautions;
    Reconstr:reconstruct; Redes:redesign; Rr:riprap; Reset:resettlement of population; Seal:surface sealing; TBu:toe buttress; Unl:unload

addition to conventional geotechnical practice in the treatment of landslides there is also the possibility or the need to make adjustment in the design of the project structures or in its operation.

Landslides with potentially catastrophic consequences should be treated such that unacceptable risks are reasonably excluded. Treatment for features of lower risk levels can be scaled to the desired mitigation of the risk. A difficulty for the risk based approach derives from the time dependence not being defined for most types of landslide phenomena.

In the treatment of landslides at a reservoir consideration should always be given to the observational approach. Depending on the characteristics of the problem involved the implementation of the observational approach may range from periodic routine inspections to sophisticated concepts in instrumentation and organization.

Aspects of particular interest for continuing investigation relate to "brittleness" of slides, in their static as well as in their dynamic phase and the time dependence of hazards. About half of the reported slide events occurred very early in the life of a project. But there are also factors which may increase the hazards with the ageing of the reservoir and the slopes.

Although it may appear attractive to defer remedial action on a landslide the economical benefit of the delay is not always assured because failure is likely to occur soon and because repair may be more costly than prevention.

A reservoir with landslide risk must remain under the supervision of competent professionals. This is recognized in many countries and the pertinent procedures have been institutionalized. But there are also cases which display a disparity between construction and operation phase of a project. Funds are available for the construction of the project and highest competence in engineering design and construction supervision is demanded. But after commissioning of the project the operator may encounter serious difficulty in raising the funds for maintenance works and for hiring the qualified staff. The designer of a reservoir with landslide hazards must decide if this phase of the project is prone to play a decisive role and if the pertinent precautions should already be introduced into the design.

REFERENCES

Albiker, B. 1977. Stabilität einer Rutschscholle unter Einstau. Rock Mechanics 9: 173-181

Alean, J-C. 1984. Untersuchungen über Entstehungsbedingungen und Reichweiten von Eislawinen. Mitt. VAG ETH Zürich 64, 111 p.

Ambros, J. 1990. Landslide history versus efficiency of remedial measures. Proc. 6th Int. Congr. IAEG: 1315-1320

ASCE/USCOLD 1975. Lessons from dam incidents. 387 p.

Barioli, E. 1967. Mesures prise pour assurer la stabilité et le étancheité des rives du réservoir de Valgrosina. Proc. 9th Congr. ICOLD Q.32: 819-829

Baumer, A. 1988. Three rock slides in the southern Swiss Alps. Proc. 5th ISL: 1307-1311

Belloni, L.G., Stefani, R. 1987. The Vajont slide: instrumentation, past experience and the modern approach. In Dam Failures, Elsevier 1987: 423-444

Bertacchi, P., Fanelli, M., Maione, U. 1988. An overall approach to the emergency hydraulic problems from the natural dam and lake formed by the Val Pola Rockslide. Proc. 16th Congr. ICOLD C.32: 1439-1456

Breth, H. 1967. The dynamics of a landslide produced by filling a reservoir. Proc. 9th Congr. ICOLD Q.32: 37-45

Brotodihardjo, A.P.P. 1990. Cold lahar flows from the Galunggung volcano and their influence on life environments. Proc. 6th Congr. IAEG: 1517-1523

Brunsden, D., Prior, B. 1984. Slope instability. Wiley, 620 p.

Bujak, M., Glab, W., Morczewski, K., Wolski, W. 1967. Preventive measures against the rock slide at the Tresna dam site. Proc. 9th Congr. ICOLD Q.32: 1027-1036

Cabrera, J.G. 1992. Investigation of a land-slide on a crucial reservoir rim saddle. Proc. 6th Int. Symp. Landslides, Christchurch

Carrillo-Gil, A., Carillo-Delgado, E. 1988. Landslide risk in the Peruvian Andes. Proc. 5th ISL: 1137-1142

Cavers, D.S. 1981. Simple method to analyze buckling of rock slopes. Rock Mech.14: 87-104

Chang, S.C. 1984. Tsao Ling landslide and its effect on a reservoir project. Proc. 4th ISL: 469-473

Chen De Ji 1986. The geologic study of the Three Gorge Project in China. Proc. 5th Int. Congr. IAEG: 1067-1075

Chowdhury, R. 1978. Analysis of the Vajont slide - new approach. Rock Mech. 11: 29- 38

Civis 1967. Mattmark - ein Rückblick. Natur und Mensch 10, 3/4: 52-58, 5/6: 93-97

Clarke, I.D., Roose, K., Riemer, W. 1989. Protection of the pleistocene sediments along the rim of San Jacinto reservoir rim. Quaternary Engrng. Geology, Geol. Soc. Engrng. Geol. Special Publ. 7,:611-617

Code, J.A., Sirhindi, S. 1986. Engineering implications of the impoundment of the Indus River by an earthquake-induced landslide. In ASCE Geotech. Special Publ. 3: 97-110

Colombet, G., Glories, M. 1983. Confortement de talus rocheux de la carrière chenal de Montezíc. Proc. 5th Int. Congr. Rock Mech.: C25-29

Comité Francais des Grands Barrages 1973. Mesures prises en France pour faciliter la protection des populations a l´aval des barrages. Proc. 11th Congr. ICOLD Q.40: 661-681

Comité Francais des Grands Barrages 1982. Études et travaux réalisés en France en raison de l´instabilité de versants de retenue. Proc. 14th Congr. ICOLD Q.54: 563-589

Couturier, B. 1986. Problèmes de stabilité de versants et d´étanchéité de cuvette dans une retenue collinaire. Proc. 5th Int. Congr. IAEG: 1241-1244

Cruden,D.M. 1976. Major rock slides in the Rockies. Can. Geotech. J. 13: 8-20

D´Appolonia, E. 1990. Monitored decisions. J. Geotech. Engrg. Div. ASCE 116, 1:4-34

de Mello, V.F.B. 1986. General report: Engineering geological problems related to hydraulic and hydroelectric developments. Proc. 5th Int. Congr. IAEG: 2657-2682

Dobry, Y. 1965. Efectos del sísmo de Marzo de 1965 en los tranques de relaves de El Cobre. Revista IDIEM, 4, 2: 85-107

Doornkamp, J.C., Brunsden, D., Russel, J.R., Kulasinghe, A.N, Gottschalk, E.M.1982. A geomorphological approach to the assessment of reservoir slopes and sedimentation. Proc. 14th Congr. ICOLD Q.54: 163-174

Dubie, J.Y., Benefice, P., Guitton, C. 1988. Télétransmission de l´auscultation d´un glissement: retenue de Grand´Maison, glissement du Billan. Proc. 5th ISL: 399-404

Dubie, J.Y., Guitton, C. Poupart, M. 1991. Physical and numerical analysis of a landslide in a reservoir. Proc. 17th Congr. ICOLD C.17: 893-917

Dupree, H.K. Taucher, G.J., Voight, B. 1979. Bighorn reservoir slides, Montana, USA. In Voight, Rockslides and Avalanches, Vol. 2: 247-268

Eisbacher, G.H., Clague, J.J. 1984. Destructive mass movements in high mountains: hazard and management. Geol. Survey Canada Paper 84-16, 230 p.

ENEL 1970. Control of dams of ENEL, Acciaierie e Ferriere Lombarde Falck and Societa Montedison. Proc. 10th Congr. ICOLD Q.38: 243-277

Erskine, C.F. 1973. Landslides in the vicinity of Fort Randall Reservoir, South Dakota. Geol. Survey Prof. Paper 675, 65 p.

Espinosa, L., Bernal, C. 1982. Rock slope stability of Canyon walls at Chicoasen Damsite. Proc. 14th Congr. ICOLD Q.53: 425-447

Forster, J.W. 1986. Geological problems overcome at Revelstoke. Water Power and Dam Construction, July 1986, August 1986

Fujita, H. 1977. Influence of water level fluctuations in a reservoir on slope stability. Bull. IAEG 16: 170-173

Gallico, A., Yeh, L.S., Bergamini, M. 1982. Performance of the Techi Reservoir. Proc. 14th Congr. ICOLD Q.54: 591-600

Galster, R.W. 1992. Landslides near abutments of three dams in the Pacific Northwest, USA. Proc. 6th Int. Symp. Landslides, Christchurch

Ganser, O. 1973. Precautions for the protection of the population in the case of danger from storage schemes. Proc. 11th Congr. ICOLD Q.40: 705-715

García, A., Fernández, A. 1986. Slope instabilities and site selection for the Pas reservoir, Santander, Spain. Proc. 5th Int. Congr. IAEG: 1259-1265

Gilg, B. 1988. Glissement de terrain le long des retenues artificielles. Proc. 5th ISL: 1165-1168

Gillon, M.D., Hancox, G.T. 1992. Cromwell Gorge landslides - a general overview. Proc. 6th Int. Symp. Landslides, Christchurch

Golze, A. R. 1977. Handbook of dam engineering. Van Nostrand Reinhold, 793 p.

Govi, M. 1990. Special lecture: Past and recent mass movements in Italian Alps. Proc. 5th ISL Vol. 3: 1509-1514

Gruner, E. 1969: Vigilance over reservoirs. Water and Water Engrng, Sept. 1969: 369-373

Gruner, E. 1973. Classification of risk. Proc. 11th Congr. ICOLD Q.40: 55-68

Gruetter, F., Schnitter, N.J. 1982. Analytical risk assessment for dams. Proc. 14th Congr. ICOLD Q.52: 611-625

Habib, P. 1988. A propos de la vitesse terminale du glissement du Vaiont. Proc. 5th ISL: 1415

Hanke, H. 1966. Gletscherkatastrophen. Der Bergsteiger, 33, 6: 433-556

Heitfeld, K.H. 1975. Engineering geological problems of dams and reservoirs. Seminarbericht d. Deutschen UNESCO Kommission Nr. 29: 223-231

Heitfeld, K.H. 1978. Beispiele von Felsrutschungen im Nordteil des rechtsrheinischen Schiefergebirges. 3. Nat. Tagung Felsmechanik Aachen: 337-366

Heitfeld, K.H. 1984. Ingenieurgeologie im Talsperrenbau. In: Bender, Angewandte Geowissenschaften, Band 3: 479-494

Hendron, A.J., Patton, F.D. 1986. A geotechnical analysis of the behavior of the Vaiont slide. Civil Engineering Practice, Vol. 1, no. 2: 65-130

Herzog, H. 1973. Method for the evaluation of the significance of storage dams. Proc. 11th Congr. ICOLD Q.40: 159-164

Hewitt, K. 1982. Natural dams and outburst floods of the Karakoram Himalaya. Proc. Exeter Symp. IAHS Publ. 138: 259-269

Hoek, E. 199. When is a design in rock engineering acceptable? Müller Lecture, 7th Int. Congr. Rock Mech. Aachen 1991

Horninger, G., Kropatschek, H. 1967. The rock slides downstream from the Gmuend dam (Austria) and the measures to safeguard the dam. Proc. 9th Congr. ICOLD Q.34: 657-669

Huber, A. 1980. Schwalwellen als Folge von Felstürzen. Mitt. VAG ETH Zürich 74: 222 p.

Huber, A. 1982. Impulse waves in Swiss lakes as a result of rock avalanches and bank slides. Proc. 14th Congr. ICOLD Q.45: 455-476

Hungr, O., Morgan, G.C., Kellerhals, R. 1984. Quantitative analysis of debris torrent hazards for design of remedial measures. Canadian Geotech.J., 21: 663-677

Hungsberg, U. 1976. The use of drainage galleries for the stabilization of slopes at the Lazaro Cardenas Dam. Proc. 12th Congr. ICOLD Q.45: 1063-1073

Hutchinson, J.N. 1977. Assessment of the effectiveness of corrective measures in relation to geological conditions and types of slope movement. Bulletin IAEG 16: 131-155

Hutchinson, J.N. 1983. Methods of locating slip surfaces in landslides. Bull. Assoc. Eng. Geol XX, 3: 235-252

ICOLD 1984. World register of dams. 753 p.

Imrie, A.S., Moore, D.P., Enegren, E.G. 1992. Performance and maintenance of the drainage system at Downie slide. Proc. 6th Int. Symp. Landslides, Christchurch

INECEL 1987. Proyecto hidroeléctrico Coca-Codo Sinclair. Analisis de los efectos producidos por el sísmo del 5 de Marzo.

Innerhofer, G., Loacker, H. 1982. The stability of the rock rim of the Bolgenach reservoir. Proc. 14th Congr. ICOLD Q54: 83-92

Jaeger, C. 1969. The stability of partly immerged fissured rock, and the Vajont slide. Civ. Eng. Public Works Rev., December 1969: 1204-1207

Jappelli, R., Musso, A., 1981. Slope response to reservoir water level fluctuations. Proc. 10th Int. Congr. SMFE 1981: 437-442

Jones, F.O., Embody, D.R., Peterson, W.L. 1961. Landslides along the Columbia River Valley, Northeastern Washington. Geol. Survey Prof. Paper 367, 98 p.

Kaliser, B.N., Fleming, R.W. 1986. The landslide dam at Thistle, Utah. In ASCE Geotech. Special Publ. 3: 59-83

Keefer, D.K. 1984. Landslides caused by earthquakes. Geol. Soc. Am. Bull. v. 95: 406-421

Khasanov, A.S., Niyazov, R.A. 1986. Landslides on mountain water storage banks in Central Asia and the problem of their time dependent forecast. Proc. 5th Int. Congr. IAEG: 1267-1276

Krapp, L., Pantzartzis, P. 1989. Besondere ingenieurgeologische und hydrogeologische Aufgaben beim Talsperrenbau in N-Griechenland. Mitt. Ing.- u. Hydrogeol. RWTH Aachen, 32: 77-107

Krumdieck, M.A. 1991. Las crecidas repentinas como fenómenos de remoción en masa. Proc. 9th Panam. Conf. SMFE: 541-554

Lane, R.G.T. 1970. Major problems in the operation and maintenance of dams and reservoirs. Proc. 10th Congr. ICOLD Q.38: 329-347

Lauffer, H., Neuhauser, E., Schober, W. 1967. Uplift responsible for slope movements during the filling of the Gepatsch reservoir. Proc. 9th Congr. ICOLD Q.32: 669-693

Legget, R.F. 1962. Geology and Engineering. McGraw Hill

Lied, K., Palmstrom, A., Schieldrop, B., Torblaa, I. 1976. Dam Tunsbergdalsvatn. A dam subjected to waves generated by avalanches and to extreme floods from glacier lake. Proc. 12th Congr. ICOLD Q.47: 861-875

Lukác, M. 1982. Failure of reservoir banks caused by wave abrasion. Proc. 14th Congr. ICOLD Q.54: 1-9

Marsal, J., Resendiz, D. 1979. Presas de tierra y de enrocamiento. Limusa, Mexico, 546 p.

Meade, R.B. 1983. Reservoir-induced macro earthquakes: a reassessment. In Seismic Design of Embankments and Slopes, ASCE

Mikhailov, L.P., Pecherkin, I.A., Uspensky, S.M., Sokolnikov, U.N. 1982. Reservoir shores engineering-geological aspects. Proc. 14th Congr. ICOLD Q.54: 229-237

Mizukoshi, T., Tanka, H., Inouye, Y. 1967. A geologic investigation on the stability of reservoir Banks. Proc. 9th Congr. ICOLD Q.32: 47-65

Moammar, G., Costa, H., Salvador, J. 1982. Evolution of the Salado dam design. Proc. 14th Congr. ICOLD Q.53: 503-517

Moore, D.P., Imrie, A.S. 1982. Rock slope stabilization at Revelstoke dam. Proc. 14th Congr. ICOLD Q.53: 365-386

Moore, J.P. 1982. Effect of seepage on reservoir slope stability. Proc. 14th Congr. ICOLD Q.54: 341-351

Morales, B., Garga, V.K., Wright, R.S., Perez, J. 1984. The Tablachaca slide no. 5, Peru, and its stabilization. Proc. 4th ISL: 597-604

Morbidoni, N.P., Larangeira, D. 1986. Stability analysis on the left margin of the reservoir in the Paraná Medio Project - Chapeton closure dam. Proc. 5th Int. Congr. IAEG: 1281-1293

Morgenstern, N. 1963. Stability charts for earth slopes during rapid drawdown. Géotechnique 13: 121-131

Mouelhi, M., Vigier, G., Huynh, P., Rondot, E. 1982. Barrage de Sidi Salem. Problèmes d´instabilité des rives en relation avec la présence de gypse dans la cuvette. Proc. 14th Congr. ICOLD Q.54: 615-636

Müller, L. 1987. The Vajont catastrophe - a personal view. In Dam Failures, Elsevier: 423-444

Nagy, L. 1967. The effect of rapid water level lowering on the shores of storage reservoirs. Proc. 9th Congr. ICOLD Q.34: 887-893

Nakamura, H. 1985. Mechanism of reservoir-induced landslides. Proc. 4th Int. Conf. Field Workshop Landslides: 219-226

National Research Council 1985. Safety of dams. National Academy Press, 276 p.

Nieuwenhuis, J.D. 1991. The lifetime of a landslide. Balkema, 144 p.

Noda, E. 1970. Water waves generated by landslides. J. Waterways, Harbors Coastal Engrng. Div. ASCE, 96 WW4: 835-855

Noguera, G., Garcés, E. 1988. Colbun reservoir seepage. Proc. 16th Congr. ICOLD C.18: 1223-1235

Novosad, S. 1979. Establishing conditions of equilibrium of landslides in dam reservoirs by means of the geoacoustic method. Bull. IAEG 20: 138-144

Oboni, F., Egger, P. 1985. A probabilistic analysis of the Vajont slide. Proc. 5th Int. Conf. Numerical Methods in Geomechanics:1013-1018

Plafker, G., Ericksen, G.E. 1978. Nevados Huascarán Avalanches, Peru. In Voight, B. Rockslides and Avalanches, Vol. 1: 277-314

Peggs, D.R.D., Valliappan, P. 1982. The behavior of sensitive marine clay on reservoir slopes. Proc. 14th Congr. ICOLD Q54:131-149

Pugh, C.A., Harris, D.W. 1982. Prediction of landslide generated waves. Proc. 14th Congr. ICOLD Q.54: 283-315

Qi Xiaojun 1986. Evolution of geological environment with water resources and hydropower constructions. Proc. 5th Int. Congr. IAEG: 1161-1172

Qian, N. 1982. General report Q54. 14th Congr. ICOLD Q.54: 639-690

Rib, H.T., Liang, T. 1978. Recognition and identification. In Schuster and Krizek, Landslides: 34-80

Riemer, W. Ruppert, F.R., Locher, T.C., Nunez, I. 1988. Regional assessment of slide hazard. Proc. 5th ISL: 1223-1226

Rißler, P. 1981. Zur Sicherheitsdiskussion über Talsperrendämme. Wasserwirtschaft 71, 7+8: 200-205

Rybár, J. 1977. Prediction of slope failures on water reservoir banks. Bull. IAEG 16: 64-67

Salt, G. 1988. Landslide mobility and remedial measures. Proc. 5th ISL: 757-762

Santa Clara, J.M.A. 1991. The complex geology of Kariba´s right bank. Proc. 17th ICOLD Q.66: 229-245

Savkin, V.M. 1977. Shore dynamics of large reservoirs impounded behind hydroelectric plants in Siberia and experience with their engineering protection. Hydrotechnical Construction, 4: 374-377

Schetelig, K. 1989. Eine ausgedehnte Felssackungszone am Mornos-Damm in Griechenland. Mitt. Ing.- u. Hydrogeol. RWTH Aachen, 32: 13-39

Scheidegger, A.E. 1973. On the prediction of the reach and velocity of catastrophic landslides. Rock Mechanics 5: 231-236

Schuster, R.L., Krizek, R.J. 1978. Landslides analysis and control. National Research Council, Transportation Research Board, 234 p.

Schuster, R.L. 1986a. Landslide dams: processes, risk, and mitigation. ASCE Geotech. Special Publ. 3: 162p

Schuster, R.L., Costa, J.E. 1986b. Effect of landslide damming on hydroelectric projects. Proc. 5th Int. Congr. IAEG: 1295-1307

Schwab, H., Pircher, W. 1982. Monitoring and alarm equipment at the Finstertal and Gepatsch rockfill dams. Proc. 14th Congr. ICOLD Q52: 1047-1076

Sergeev, E.M. 1979. Hydrotechnical construction and protection of the geological environment. Bull. IAEG 20: 256-259

Slingerland, R.L., Voight, B. 1979. Occurrences, porperties, and predictive models of landslide generated water waves. In Voight, Rockslides and avalanches, B: 317-397

Skempton, A.W., Hutchinson, J.N., 1969. Stability of natural slopes and embankment foundations. Proc. 7th Int. Conf. SMFE: 291-335

Snow, D.T. 1964. Landslide of Cerro Condor-Sencca. In Engineering Case Histories No. 5, Geol. Soc. Am., 6 p.

Sperling, T., Freeze, R.A. 1987. A risk-cost benefit framework for the design of dewatering systems in open pit mines. Proc. 28th US Symp. Rock Mechanics: 999-1007

Swiss COLD 1982. General paper. Proc. 14th Congr. ICOLD G.P.-R.S.8: 889-958

Szpilman, A., Ren, C. 1976. The effect of landslide on Furnas reservoir. Proc. 12th Congr. ICOLD Q.47: 1109-1119

Tipping, E.D., Maragotto, C.H., Aisiks, E.G. 1982. Geological investigations for a buried river channel at Piedra del Aguila damsite. Proc. 14th Congr. ICOLD Q. 53: 489-502

Trollope, D.H. 1980. The Vaiont slope failure. Rock Mechanics 13: 71-88

Trzhtsinskii, Y.B., 1978. Landslides along the Angara reservoirs. Bull. IAEG 17: 42-43

Valenzuela, L., Varela, J. 1991. El Alfalfal rock fall and debris flow in Chilean Andes mountains. Proc. 9th Panam. Conf. SMFE: 357-373

VanDine, D.F. 1985. Debris flows and debris torrents in the Southern Canadian Cordillera. Can. Geotech. J. 22: 44-68

Varnes, D.J. 1984. Landslide hazard zonation: a review of principles and practice. UNESCO Natural Hazards 3, 63 p.

Vedris, E. 1988. Special lecture: evaluation of risks associated with slope instability. Proc. 5th ISL: 1491-1498

Vischer, D.L. 1986. Rockfall induced waves in reservoirs. Water Power and Dam Construction, Sept. 1986: 45-48

Vischer, D., Funk, M., Müller, D. 1991. Interaction between a reservoir and a partially submerged glacier: problems during the design stage. Proc. 17th Congr. ICOLD Q.64: 113-135

Voight, B. 1979. Rockslides and avalanches. Elsevier, Vol. A, 833 p., Vol. B, 850 p.

Volkart, P. 1974. Modellversuche über die durch Lawinen verursachten Wellenbewegungen im Ausgleichsbecken Ferden im Lötschental. Wasser und Energiewirtschaft 8/9: 2-8

Wang, G.X. 1988. Landslides along the bank of the Yangtze Gorges reservoir in China. Proc. 5th ISL: 1259-1261

Wiegel, R.L., Noda, E.K., Kuba, E.M., Gee, D.M,, Torenberg, G.F. 1970. Water waves generated by landslides in reservoirs. J. Waterways Harbors Div. ASCE, WW2: 307-333

Wilson, S.D., Marsal, R.J. 1979. Current trend in the design and construction of embankmanet dams. ASCE, 125 p.

Whalley, W.B. 1984. Rockfalls. In Brunsden and Prior, Slope Instability: 217-256

Whitman, R.V. 1984. Evaluating calculated risk in geotechnical engineering. J. Geotech. Engrg. Div. ASCE, GT110, 2: 145-188

Wilson, S.D., Mikkelsen, P.E 1978. Field instrumentation. In Schuster and Krizek, Landslides: 112-138

Wong, T.H., Gilmore, J.J. 1979. Static stability of Kastraki reservoir slopes. Proc. 20th US Symp. Rock Mech.: 337-345

Zanev, Z., Dobrev, D. 1973. Changes in some local conditions due to the construction of dams. Proc. 11th Congr. ICOLD Q.40: 165-173

Zhang, Z., Li, Y., Liu, H., Yang, S. 1986. A typical case history of landslide revival induced by pore water pressure - mechanism and stability analysis of Jipazi landslide, in the projected Yangtze Gorge reservoir. Proc. 5th Int. Congr. IAEG: 1309-1316

ACKNOWLEDGEMENT

The author is indebted to organizations and individuals for material and information contributed for review, and wants to mention specifically: BC Hydro, ELECTRICORP-Clyde Power Project, Electro-Watt Engineering, Estudio Lombardi, SEO, TIWAG, H. Arthur, J. W. Hilf, K. W. John, A. Krumdieck.

Landslides, Bell (ed.) © 1995 Balkema, Rotterdam, ISBN 90 5410 032 X

Theme report

Alan S. Imrie
BC Hydro, Vancouver, B.C., Canada

1 INTRODUCTION

Theme S2, Landslides and Reservoirs, contains six papers in addition to the Theme Address and these are tabulated in Table 1. Four of the papers (1,3,4,6) deal with man-made reservoirs associated with hydroelectric projects while two papers (2,5) discuss natural reservoirs formed behind landslide dams. One of the latter reservoirs (5) has actually been used for power generation.

The imposition of a reservoir can have a major impact upon the stability of adjacent soil and rock slopes. Formation of a lake inundates geological strata and structures and can reduce their strength directly or the lake can provide a more severe ground water regime in the slopes above the reservoir level. These destabilizing effects have the potential of initiating movements ranging from local sloughing at the toe, to creep over a wide area, or ultimately to catastrophic, rapid movement of a large mass. The stability impacts not only occur during the initial lake filling and the associated adjustment period (say 2-3 years), but can last for the life of a project, especially if the reservoir operating level suddenly changes, such as for a major drawdown.

Another critical aspect of a reservoir is that not only does the impounding tend to destabilize flooded or adjacent slopes, but if mass movement were to occur rapidly into the reservoir, a hazardous wave may result. This wave could impact locally on an inhabitated area or it could travel up and down the reservoir even impacting on the dam itself and communities downstream, as tragically demonstrated by the Vaiont slide in 1963.

For a reservoir formed behind a landslide dam, it is necessary to evaluate the stability of the natural barrier which often tends to be a temporary structure and may be breached. Uncontrolled release of the reservoir behind such a barrier could also have a catastrophic result to downstream habitation.

In general the surrounding slopes of the reservoirs discussed under Theme S2 have not had major stabilization works applied in the manner described under both Theme G3 (Stabilization and Remedial Measures) and Theme S6 (Lake Dunstan Landslides) of this Symposium; however, two papers (3, 4) do discuss significant abutment treatment near the dams or barriers containing the reservoir.

The purpose of this Panel Report is not only to summarize the main points of each paper, but also to discuss common ideas and new directions, and finally to propose some points for discussion during the Panel Session.

2 SUMMARY OF MAIN POINTS OF PAPERS IN THEME S2

Anagnosti notes that newly formed reservoirs tend to create slope instability either by direct inundation, by wave erosion, or by rapid drawdown which accompanies certain reservoir operating procedures. It is necessary to carry out investigations prior to impounding to evaluate the risk of landsliding. Such investigations include geologic mapping, obtaining subsurface geology and geotechnical properties (piezometric pressures, shear strength), and documenting existing slides.

A key point is that for small reservoirs (a shoreline length of 5 to 10 km) it is not onerous time or cost wise for an agency to make a detailed stability investigation of all the slopes adjacent to the reser-

TABLE 1

LIST OF PAPERS IN
THEME S2 - LANDSLIDES AND RESERVOIRS

	Author	Title	Comment
1.	Anagnosti, P.	Probabilistic versus deterministic approach in hazard assessment of landslides along man-made reservoirs.	Proposed complementary use of probabilistic techniques in assessing reservoir slope stability around large reservoirs.
2.	Asanza, M.; Plaza-Nieto, G.; Schuster, R.L.; Yepes, H.; Ribadeneira, S.	Landslide blockage of the Pisque River, northern Ecuador	Landslide dam which failed after 24 days. Mitigative measures controlled impact of flood waters.
3.	Cabrera, J.G.	Investigation of a landslide on a crucial reservoir rim saddle	Raised reservoir caused seepage through saddle and triggered a debris slide.
4.	Galster, R.W.	Landslides near abutments of three dams in the Pacific Northwest, USA	Impact of various geologic conditions on dams and reservoir engineering.
5.	Riley, P.B.; Read, S.A.L.	Lake Waikaremoana - present day stability of landslide barrier	Dam safety analysis of prehistoric landslide dam and reservoir used for hydroelectric power generation.
6.	Vitanage, P.W.	Landslide Around Kotmale, Mousakelle reservoirs	Slope stability impacts of 2 reservoirs during normal operation and drawdown.

voir. However for larger reservoirs (say 25 - 50 km long or more), it is not always possible to complete the same detailed investigations.

In the former instance the stability evaluation is normally deterministic, that is for a typical suspect slope, to determine the respective driving and restraining forces, and thus calculate its individual factor of safety. Anagnosti suggests that there is a role for the probabilistic approach to be used for slope evaluation in large reservoirs. The approach is based on cataloguing existing landslides, providing a distribution function of the failed slope angles and comparing to a function describing suspect areas that have not failed. Attention must be paid not only to submergence but also to drawdown effects. A deterministic calcu-

lation is still used for a typical slope but the result is applied probabilistically to the remaining areas of interest. From the example given in the paper, it is implied that this approach is best used in areas of uniform geology (eg. surficial materials) rather than areas of complex geology where throughgoing structures may be the key to stability (eg. an outward dipping fault zone).

Asanza et al. have described a landslide blockage of the Pisque River (Fig. 1) in northern Ecuador and in a short presentation at the Symposium following this Panel Report, they will describe the role of the constructed spillway on mitigating the hazard posed by failure of the natural dam and release of its reservoir. In their paper the following conclusions were drawn:

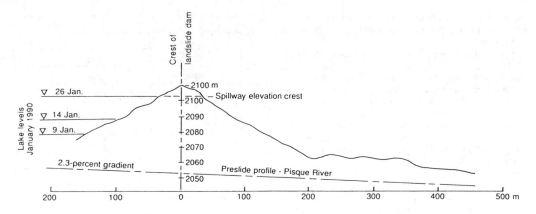

Fig. 1 Cross profile through the landslide dam, showing progressively higher lake levels during January 1990. (Asanza et al.)

a) Steep natural slopes can be de-stabilized by human activity (in this case irrigation).

b) In deep narrow valleys landslides can form high barriers and significant impoundments, and because of the barrier's often temporary nature, these reservoirs pose a significant risk to downstream communities of flooding.

c) For such natural dams, timely installation of mitigative works is required.

Cabrera provides an interesting case study which happened over 30 years ago but it still provides a key lesson involving the imposition of a hydroelectric reservoir on natural materials. In 1958 the existing Lajes Reservoir in the State of Rio de Janeiro, Brazil was to be raised 28 m by increasing the height of the main dam and adding several saddle dams around the reservoir rim. When the raising had reached 21 m, seepage adjacent to clay seams in gneissic bedrock caused piping and eventually initiated a debris flow on the downstream side of one of the saddle dams threatening catastrophic release of the reservoir. Mitigative measures were immediately implemented after lowering the reservoir, including providing an upstream clay blanket, grout curtain and drainage of the saturated zone. In addition the debris flow area, which remained active during any subsequent heavy rains because of montmorillonitic soils, was treated successfully with the addition of lime to increase its strength and reduce its plasticity. The important lesson from this study is to carefully evaluate the permeability and stability of the foundation materials forming the containment around the reservoir rim.

Galster describes three case studies involving landslides near the abutments of hydroelectric dams in the Pacific Northwest of the United States. One slide of about 46,000 m^3 and which was forecast occurred near the left abutment of Libby Dam during the construction phase before the reservoir was filled. It was a classical wedge failure (bounded by bedding plane faults and open joints) formed when excavation for a roadway along the reservoir resulted in daylighting of the key geologic structures, which initiated creep movements and ultimately resulted in failure. The failure area had been well instrumented with extensometers (Fig. 2) and provides an excellent example of the benefits of monitoring. Remedial measures (rock anchors, drainage and toe berms) prevented any further problems.

At Mud Mountain Dam the reservoir had operated without incident for about 26 years before a 54,000 m^3 soil slide moved rapidly into the reservoir just upstream of the left abutment after heavy rains. The slide caused a 4 m wave runup on the dam. At Chief Joseph Dam a large (3.8 x $10^6 m^3$) existing bedrock and soil slide was inundated by the reservoir behind the dam and was monitored very carefully during reservoir raising of 3 m in 1980. Between 1976 and 1981 the slide had moved about 0.3 m. After raising, and between 1981 and 1988, the slide moved about 2 m. All movements were slow and monitoring continues.

These case studies portray a variety of geologic conditions that often must be considered in development of hydroelectric or flood control projects and also empha-

2007

size the role of the proper monitoring in evaluating slope movements around reservoirs.

Riley and Read describe the present day stability of a huge (2.2 x 10⁹ m³) landslide dam which was formed 2200 years ago and impounded Lake Waikaremoana here in New Zealand. The ability of this barrier to survive for such a long period is most fascinating since many of these barriers last only briefly and it is therefore suggested that the authors give a short presentation on this aspect to the Symposium delegates. In their conclusions as well as describing the reasons for the barrier's longevity, the role of monitoring in assuring safety to the downstream public and lake users is emphasized as a key aspect. It is also of great interest to note that some enterprising hydroelectric engineers have used this natural dam and reservoir for energy generating purposes.

The final paper in this session by Vitanage describes the impacts of two hydroelectric reservoirs, Kotmale and Mousakelle in Sri Lanka, that were impounded in the mid 1980's and late 1960's respectively, resulting in instabilities that have been encountered during normal operation and drawdown. Landslides have damaged a number of buildings around the reservoirs. Studies have focused on finding whether there is a relationship between creep movements along structural lineaments (possibly triggered by microseismic activity) and the hydrogeological changes (water table, pore pressure) imposed by the reservoirs.

3 DISCUSSION OF COMMON THEMES AND SUGGESTION OF NEW DIRECTIONS

The presence of a reservoir can have a dramatic effect on ground stability, including the barrier which forms the reservoir, whether it be man-made or natural. A cautious approach must be taken by the involved scientists to ensure that the reservoir does not: (1) trigger major instability in the slopes around it or (2) cause a catastrophic release of water through the barrier containing it.

The assessment of reservoir impact on slopes involves the study of the complex interaction of many geologic and geotechnical factors. As a result the assessment of slope stability cannot result in definitive answers. Normally thorough engineering evaluations are undertaken for the reservoir containment structure (dam or dyke) which covers a relatively small area but it is much more difficult to carry out

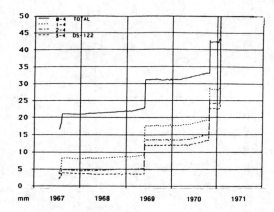

Fig. 2. Extensometer Record (Galster).

the same evaluation for the reservoir rim which may cover hundreds of kilometres as noted by Anagnosti.

Landslides are similar to other materials in that we have to know material properties, geometry and loads before we can make reliable predictions of behaviour. Geological engineers routinely make deformation predictions about complex rockslides with only limited information. One of the best methods to evaluate suspect areas is to make use of monitoring programs as indicated by Galster.

Monitoring programs can range from simple crack measurements during periodic visual inspections, to sophisticated monitoring systems continuously recording movement, piezometric pressures, hydrometeorlogical conditions and seismic activity. The results are used: (1) to check the interpretation of surface and subsurface geology; (2) to investigate movement mechanisms; (3) to correlate movements with other factors such as weather, water pressures, earthquakes or reservoir operation; and (4) to detect changes in slope behaviour (Moore and Enegren, 1992).

4 PROPOSED DISCUSSION TOPICS FOR PANEL SESSION, THEME S2

In addition to the topic of longevity of landslide dams as introduced by Asanza et al. and Riley and Read in their short presentations to the Symposium, the following two questions are suggested as discussion topics for the Panel Session.

1. Rock and soil slides are complex natural hazards analogous in some ways to floods and earthquakes, but generally slides are less predictable because of the lack of data that are normally available

and because there are no robust statistical methods of prediction.

Although the probability of failure cannot be defined accurately, a case could be made for bracketing a range of probabilities by: (1) appraising recent movement within the slide area of interest; (2) assessing past landslides which have occurred on similar features in the area; and (3) comparing case histories of stable and unstable slopes throughout the world.

Based on the above factors, and in particular the approach of Anagnosti described in this theme, what is the role of the probabilistic approach in slide assessment around reservoirs?

2. Monitoring is an important factor in the prediction of landslides. It is considered that large rockslides will undergo a period of increasing movement and deterioration prior to a large rapid movement. This behaviour is well illustrated by Vaiont in Italy and the Xintan slide in China (Wang et al. 1988) both of which exhibited increasing movements, small slides and crack openings long before the final catastrophic events. With a predictive monitoring system in place, it is suggested that changes in slope behaviour can be detected and steps can be taken to limit damage or to install preventative measures (Moore et al, 1991).

With the experience at Vaiont and Xintan and the description in this Theme given by Galster, what is the predictability of landslides adjacent to reservoirs (or elsewhere!) using monitoring systems?

5 REFERENCES

Moore, D.P. and Enegren, E.G. 1992. Perspectives from rockslide hazard assessments along B.C. Hydro's reservoirs. *Geological Association of Canada Special Paper on Landslide Hazards in the Cordillera* (in press).

Moore, D.P., Imrie, A.S. and Baker, D.G. 1991. Rockslide risk reduction using monitoring. *Proc. 3rd Canadian Dam Safety Conference,* Whistler.

Wang, L.S. et al. 1988. On the mechanism of starting, sliding and braking of Xintan landslide in Yangtze Gorge. *Proc. 5th Int. Symp. on Landslides,* Lausanne.

Landslides, Bell (ed.) © 1995 Balkema, Rotterdam, ISBN 90 5410 032 X

The working party on China's water resources related landslide inventory

Zuyu Chen

Institute of Water Conservancy and Hydroelectric Power Research, People's Republic of China

1.INTRODUCTION:

Landslides have caused serious damage of human lives and property in China's water resources development. The landslides that are related to water resources development are generally divided into three categories in our professions:

--Reservoir impounding induced landslides;

--Landslide dams;

--Excavation induced landslides.

Tables 1,2,3 give some typical cases of the three sorts of landslides. Some are very notifying such as the landslide of a 1,000 meter high slope; the avalanche with a velocity of 40m/s; the 380 meter high landslide dam; the dam overtopping due to landslide induced waves, etc.

During the coming ten years, China will construct a number of large scale water reclamation and hydropower projects in which serious landslide problems exist. Take the well known Three George Hydropower Project as an example, in the reservoir flooded area, there are a great number of potentially unstable slopes including 128 in the volume of $0.1-1.0 \times 10^6 m^3$; 32 in the volume of $1.0-10 \times 10^6 m^3$; 32 in the volume of over $10 \times 10^6 m^3$, in which 4 exceed $100 \times 10^6 m^3$.

In an effort to confront this challenging problem, a research program entitled *Study on the Stability of High Rock Slopes: Analysis and Stabilization* has been established, which is a sub-program of China's national key research program entitled *Study on High Dam Construction*, that will be undertaken in its 8th Five Year Plan period(1990-1995). With the financial support of the program, and in response to the call of the Working Party on World Landslide Inventory(WP/WLI, Cruden and Brown, 1992), the Working Party on China's Water Resources Related Landslide Inventory (abbreviated to WP/CWRLI) has been created by our community.

2.ORGANIZATION, POLICY, AFFILIATION OF WP/CWRLI:

WP/CWRLI is an organization jointly supervised by the Institute of Water

Table 1. Selected examples of reservoir unstable slopes

No.	Nature[1]	Project	Distance to dam (KM)	Dimension[2]				Horizontal Displacement[3]	
				L (M)	W (M)	H (M)	V ($10^6 M^3$)	rate (mm/d)	magnitude (M)
1	LS	Zaxi	1.55	160-280	30 35	140	1.65		
2	GM	Wujiang du	630	200	50-70	240	2.15	1.35	0.35
3	GM	Lubuge	0.5-2.1	600	20-30	90-150	2.6-2.7	119	3.2

Table 1. Selected examples of reservoir unstable slopes (continued)

No.	Geology	Hazards
1	Slate with clay seams, planar slide along the bedding plane and the weak seams.	On March 7th 1961, the unstable rock mass dumped into the reservoir, creating a 21 meter high surge that overtopped the concrete dam crest, killing at least 70 people working on the spillway.
2	Limestone, shale, mudstone and coal seams. The reversely dipping bedding planes created a toppling slope failure.	A total amount of $0.44 \times 10^6 M^3$ rock mass has been removed by directional blasting to stabilize the slope. No substantial displacement was measured since then.
3	Quaternary talus underlain by weathered sandstone and shales.	Almost immediately after reservoir impounding two earthquakes, measured 2.9 and 3.2 in magnitude on Richst scale, took place, followed by cracks and settlement in large area. Careful monitoring has been undertaken. No hazards were reported.

Note:1.Nature: LS=landslide; GM=large ground movement.

2.Dimension: L=length measured in the direction perpendicular to that of slide; T=average thickness measured vertically; H=maximum vertical height; V=volume.

3.Displacement: the maximum horizontal displacement rate and magnitude.

Table 2. Selected examples of excavation induced landslides

No.	Nature	Project, Location	Capacity (MW)	Dimension				Horizontal Displacement	
				L (M)	T (M)	H (M)	V (10^6 M^3)	rate (mm/d)	magnitude (mm)
1	SL	Tiansheng-qiao II, Sluice gate	447.5	63	10	30	0.007		
2	GM	Tiansheng-qiao II, Power plant	630	200	30	80	1.40	8.2	
3	LS	Manwan, left abutment	1,500	110	15-20	100	0.10		

Table 2. Selected examples of excavation induced landslides (continued)

No.	Geology	Hazards
1	Quarternary tulus covered by the excavation waste of rock fragments and gravels. Landslide happened righ after a layer of black silty clay was daylighted.	About 48 people were killed.
2	Shales and mudstone with a thick layer of clay interclation gently dipping in 10°, which formed as the rupture surface.	Workmen on the slope were evacuated. Extensive reinforcement including unloading, anchoring, drainage gallery, etc., were installed.
3	Rhyolite with heavily wethered and densely spaced joints. Landslide took place along one of the joint set including a small fault.	The on-line schedule of the power plant was delayed for one year. Large amount of prestressed cables were installed.

Note: The definition of Nature, Dimension are as same as those of the Table 1.

Table 3. Selected examples of river bank landslides

No.	Date (M/D/Y)	Location, River	Geology	Dimension			
				L (M)	T (M)	H (M)	V 10^6 M^3
1	6/8/1967	Tanggudong, Yalongjiang	Residual cohesionless soil	1,900	25	1,000	68
2	2/7/1943	Chana, Yellow River	Lacustrine stiff clay and sandstone.			430	127
3	6/13/1991	Nanyu, Bailong-jiang	Talus, Phylite slate	640	30-40	200	6.4
4	3/7/1982	Shalashan, Bashahe	Quarternary clay and losses underlain by red mudstone susceptible to softening upon to wetting.	700-1,100	70	330	40.0

Table 3. Selected examples of river bank landslides (continued)

| No. | Landslide dam | | Hazards |
	Height (M)	Storage ($10^6 M^3$)	
1	175-335	680	The river closed for 9 days. The breaching of the natural dam created a flood of 5,300m³/s and raised the water level of the river channel by 39 meters. Several small villages were evacuated.
2	unrecorded	unrecorded	The Yellow River was closed for a short period of time. The huge volume of rock mass burried three villages with a velocity of 40m/s, killing more than 100 people.
3	20	4	About 1400 houses and a part of highway were damaged.
4			Three villages were destroyed, 237 people died. A nearby reservoir was destroyed.

Note: The definition of dimension is as same as that of table 1.

Conservancy and Hydroelectric Power Research (IWHR) and the Institute of Planning, Designing for Water Resources and Hydroelectric Power. They are respectively China's leading institutions in scientific research and engineering consulting concerned with water resources development.

The main work to be undertaken by WP/CWRLI in the coming four years will be:

1.To implement a nation wide investigation and documentation of the landslides and engineering rock slopes as related to water resources development.

2. To create a computer database for these landslides and slopes;

3. To develop a rock mass classification system that will be uniquely used in our professional society to document these landslides and engineering slopes.

It can been seen that WP/CWRLI is an organization that is concerned with both stable and unstable slopes since the group is primarily engineering orientated. However the majority of the inventory will still be covered by unstable slopes which provide a large volume of information for collaboration and communication between WP/WLI and WP/CWRLI.

The members of WP/CWRLI come from the design and research institutions of water resources development, both of central and provincial level. They are responsible for providing information and in the meanwhile enjoy the full right to share the database created by WP/CWRLI.

3. THE FORMAT OF THE INVENTORY TABLES

WP/CWRLI believes that the success of the campaign depends to large extent on the completeness of the collection of all the necessary information to be investigated. It is therefore required that all the documentation be undertaken in accordance with a unified methodology and format. It is believed that the best way of assuring the usefulness of the recorded information is to provide a series of standardized forms that enable everyone to describe a landslide or a slope by the same format. A booklet entitled *Forms for Documenting Water Resources Related Slopes* has recently been issued, which includes eight parts as follows:

A. Fundamental information;

B. Geology;

C. Geotechnical properties of the slope material;

D. Design of the slope;

E. Construction of the slope;

F. Instrumentation and monitoring;

G. Landslide descriptions;

H. References.

To render the information computer processible, all the items are numbered. Efforts have been made to enable the writers to use 'Yes' or 'No' as much as possible while documenting. Where literal description is necessary, a key word column is provided to highlight the contents.

The Bieniawski's RMR (1989) rock classification system and Romana's updating (SMR, 1991) will be currently adopted by WP/CWRLI to rate rock mass quality. With the accumulation of the information, a report concerning the validity and feasibility of the these systems, as well as the recommendation of their modification and updating will be given by WP/CWRLI.

Reference

Bieniawski, Z. T., 1979, The Geomechanics Classification in Rock Engineering Applications, *Proc. 4th Int. Cong, Rock Mech., ISRM,* Montreux, Vol. 2, pp. 41-48.

Romana, M.,1985, New adjustment Ratings for Application of Bieniawski Classification to Slopes. *Proc. Int. Symp. Rock Mech. Excav. Min. Clv. Works, ISRM,* Mexico City. pp. 59-68.

Cruden, D. and Brown, W. Pregress towards the world landslide inventory, *Prec. Symp. 6th Int. Conf. Landslides,* pp. 59-64, Christchurch, Balkema.

Landslides, Bell (ed.) © 1995 Balkema, Rotterdam, ISBN 90 5410 032 X

The landslide phenomenon in the reservoir of Pasha-Kolla Dam in Northern Iran

A. Farajollahi
Mahab Ghodss Consulting Engineers, Tehran, Iran

S. D. Boghossian
Tehran University, Iran

ABSTRACT: This is an outline of a stability analysis study made to determine the landslide probability in the right flank slopes of the Pasha-Kolla dam reservoir. The slope consists of marlstone with a covering layer of residual soils formed by weathering. The stability of both marlstone and soil was determined. Stability analysis revealed that, in spite of the relative stability of the slope, water level fluctuations have a destabilizing effect. This destabilizing may lead to the occurrence of landslides and slope failures during earthquakes which may damage both the reservoir and the dam.

1 INTRODUCTION

This paper presents the results of the 1989 study of the Pasha-Kolla dam. Because Pasha-Kolla is an earthdam overflowing water may cause dam failure. Thus it is very important to determine the probability of slope failure.

2 ENGINEERING GEOLOGY STUDY OF LANDSLIDE PROBABILITIES

The Pasha-Kolla dam is located on the Babol River on the northern side of the Eastern Alborz Mountains (Longitude 52 40 E - 52 50 E, Latitude 36 10 N - 36 20 N)(Fig.1).

2.1 GEOLOGY OF THE AREA

The geology of the area consists of Cretaceous marls and lime-marls approximately 0.5 - 1m thick.

Geophysical studies have shown the Alborz Mountains to be rootless and thus isostatic uplift continues. The range is active and mobile with long, active faults in the area enhancing the probability of earthquakes.

2.2 CLIMATIC CONDITIONS

The annual rainfall is 820 - 830mm, promoting active weathering. The rock bodies and residual

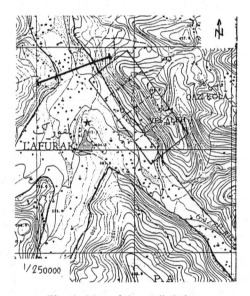

Fig. 1. Map of the studied slope.

soils have high moisture contents. Average winter temperatures of below zero mean ground freezing is inevitable.

2.3 SLOPE DESCRIPTION

The right slope above the reservoir is the least stable. It is 1000m in length, strikes N324 and dips at between 21 and 41°. The slope can be divided

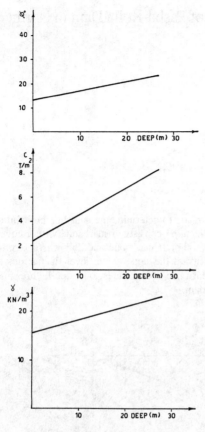

Fig.2- the relations between depth and
C , °, and ℽ.

Table 1-Values of C, Q°, and ℽ for zone a and
the fresh bed rock for block 1.

Geo. param eter Lith	ℽ KN/m³	C T/m²	Q°	ℽ Sat KN/m³
Zone a of the res.soil	1.54	2.4	14	1.96
Fresh bed rock	2.29	8.2	25	immeasurable

Average thickness of weathered zone is 27 m.

Table 2-Values of C, Q°, and ℽ for zone a and
the fresh bed rock for block 2.

Geo. param eter Lith	ℽ KN/m³	C T/m²	Q°	ℽ Sat KN/m³
Zone a of the res.soil	1.61	2.6	13	2.11
Fresh bed rock	2.32	8.3	23	immeasurable

Average thickness of weathered zone is 24 m.

Table 3-Values of C, Q°, and ℽ for zone a and
the fresh bed rock for block 3.

Geo. param eter Lith	ℽ KN/m³	C T/m²	Q	ℽ Sat KN/m³
Zone a of the res.soil	1.62	2.0	2 7	2.05
Fresh bed rock	2.34	8.8	25	immeasurable

Average thickness of weathered zone is 28 m.

into four morphologically separate blocks, each 250m in length.

The first block comprises the northern most section of the slope and dips at 33° in the lower and 26° in the upper part. The second block occurs to the south of the first and dips at 16° in the lower and 32° in the upper part. The third block is south of the second and dips at 28° in the lower and 35° in the upper part while the fourth block is the southern most and dips at 40° throughout.

2.4 THE INVESTIGATION OF INSTABILITIES

Instability studies were carried out in both rock and residual soil bodies in order to predict landslide movement. The following steps were taken to predict landslide activity:

1) a study of the marl discontinuities was undertaken, looking at the joint systems and their characteristics,

2) A geotechnical study of the marlstone and the covering weathered layer and the relationship between the two was undertaken using the RMR geomechanical classification.

The only significant factor affecting the value of C and ϕ was the compressive strength of the intact marlstone along the slopes, since the discontinuities were uniform. Because the compressive strength values were equal along the slope, the values of C and ϕ can be considered constant. The stability analysis of the rock body indicated a safety factor of 1.3 for slopes below 40°, thus there is little danger of landsliding occurring in the rock body.

Studies of the residual soil body stability were carried out next. Fig. 2 shows the relationship between depth and C, ϕ and gamma values.

It is evident from Fig. 2 that the relationship between depth and C, ϕ and gamma is linear. Following coring the weathered marl layer can be divided into the following zones:

Zone a; being the outermost weathered layer,

comprising silt and clay derived from the marls whose calcium carbonate has been removed by intense weathering. The layer is brown and has a high moisture content.

Zone b; comprises grey-brown highly weathered marl, with calcium carbonate traces.

Zone c; a brown-grey moderately weathered zone with marl debris with hackley fracture surfaces.

Zone d; slightly weathered marlstone.

The unit weights of the weathered material are higher than that of the intact marlstone due to the formers increased permeability and water content.

The linear relationship between depth, C, ϕ and gamma can be used to infer values at any depth in the weathered zones or in the intact marlstone. Measured values for C, ϕ and gamma for Zone a and fresh bedrock of each of the four blocks are given in tables 1-4.

The bedding plane of the weathered zones strikes east-west and dips at $32°$ N. As such the topographic slope is nearly perpendicular to the bedding plane and is in the direction of lithological transition.

Since the thickness of the weathered layer was negligible considering the total surface area, the plane of failure was regarded as being parallel to the slope surface and a planar analysis was chosen.

Isopleth mapping of landslide safety factors was the next step in this study. As geotechnical parameters could be calculated for any depth based on C, ϕ and gamma values for the outermost weathered layer and the bedrock. Thus it was possible to calculate the safety factor for any depth, considering the position of drill holes on the slope.By converting the data into landslide safety factor isopleth maps it was then possible to study sliding surfaces under different conditions. It was found that no safety factors below 1.1 exists and thus the weathered mass was stable under ordinary conditions.

CONCLUSIONS

In spite of the relative stability of this area under normal conditions, there exists a potential for sliding during earthquakes, especially in the weathered marl layer.

After the reservoir filled, stability conditions changed. While the upper slope remained as before, the lower slope, now under water, was experiencing new physical and mechanical conditions. After immersion, the lower slope unit weight increased while C and ϕ values decreased. The stability regime of the lower slope will be seriously affected

Table 4-Values of C, ϕ, and γ for zone a and the fresh bed rock for block4.

Geo. param eter Lith.	γ KN/m^3	C T/m^2	$\phi°$	γ Sat KN/m^3
Zone a of the res.soil	1.59	2.5	13	2.00
Fresh bed rock	2.30	8.3	24	immeasurable

Average thickness of weathered zone is 27 m.

by water level fluctuations. Weathered marls can absorb water due to their high clay content and can retain this water even after water levels have receded. The increased water content will further increase the unit weight and will cause small slides which will reduce the heeled resistance of the upper slope. The resulting lessened stability of the entire slope will now be more susceptible to the effects of earthquakes. If the slope (approximately 3000,000 cubic meters) collapsed into the reservoir it would create waves several meters high which could be very destructive to both the dam and the reservoir.

REFERENCES

Ayres, D.J. 1961. The treatment of unstable slopes and railway track formations. Journal Soc. of Eng.

Beer, F.E. 1969. Experimental data concerning clay slopes. Int. Conf. SMFE, Vol.2

Bishop, D.N. and Stevens, M.E. 1964. Landslides on logged areas in southeast Alaska.

Bowles, E.J. 1977. Foundation analysis and design.

Bowles, E.J. 1979. Physical and geotechnical properties of soils.

Campbell, R.H. 1975. Soil slips, debris flows and rainstorms in Santa Monica Mountains and vicinity, Southern California. U.S. Geol. Survey.

Hans. F. Winterkorn and Hsai-Yang. 1975. Handbook of foundation engineering.

Holmes, A. 1966. Principles of physical geology. Thomas Nelson.

S3 Open-pit mine slopes
    Pentes avec mines à ciel ouvert

Landslides, Bell (ed.) © 1995 Balkema, Rotterdam, ISBN 90 5410 032 X

Theme report

D.N.Jennings
Works Consultancy Services Ltd, Hamilton, New Zealand

ABSTRACT:
This report presents a brief overview of the ten papers allocated to the Open Pit Mine Slopes session (Theme S3) in this 6th International Symposium on Landslides.

1 INTRODUCTION

The objectives of this report are:

- to classify the range of topics
- to summarise the main points
- to identify common directions
- to highlight provocative/challenging elements

The papers are all generally in the form of case histories except for one paper which presents a parametric study of the effects of slope inundation. In preparing this overview the papers have been broadly grouped as follows:

- mine slope failures (2 papers)
- stability assessment (4 papers)
- mine slope deformation (3 papers)
- influence of underground mining on slope stability (1 paper)

2 MINE SLOPE FAILURES

2.1 Mules G.J. (1992) : Landslide features reflecting valley-wall rebound, Kaiya River, Porgera, Papua New Guinea.

Mules describes the geology and geomorphological features of Porgera area of Papua New Guinea where the Porgera Gold Mine is located adjacent to the Kaiya River. The terrain is of high relief with mountainous escarpments and dip slope valleys between elevations of 2000 - 4000 m. The

relatively low strength heavily over consolidated marine sediments (mudstone) have been tectonically deformed and are subject to regional uplift and high rates of erosion (46-72 mm/yr).

Mass wasting is extensive in the Kaiya River Valley with surficial failures including debris flows, debris slides, mudflows, rock falls, rock slides, debris avalanches and soil creep and slumping being common. Evidence of deep seated movement exists in the form of tension gulls or depressions, double crests, ridge top swamps, graben features back sloping and stepped ridge profiles, over steepened riverbanks, valley floor distortion of bedding and relocated bedrocked blocks. A suggested failure mechanism for the valley is shown in Figure 1.

To facilitate the mine development a major cut slope 70 m high was required. While the initial design required a 3H:1V batter, a trial slope of 1.5H:1V was incorporated in the initial excavation but it failed within two weeks. Failures involved rebound along steeply dipping pre-existing shear surfaces.

Mules discusses the complex aspects contributing to the failure mechanisms including high lateral stresses, pre-existing shear planes, rockiness deterioration (desegregation and softening), progressive failure, valley bulging, deep seated creep, sagging and possibly toppling.

Documentation of the geological features and developing an understanding of the geomorphology

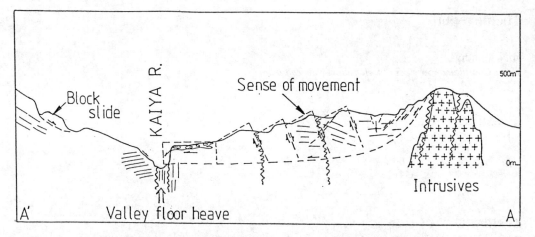

Figure 1 Suggested Valley Failure Mechanism (from Mules (1992))

and the geological processes is essential if effective engineering solutions are to be developed.

The area is obviously fascinating and challenging. This reporter looks forward to reading a paper on the slope engineering design in the future.

2.2 Read J.R.L. and Maconochie A.P. (1992) : The Vancouver Ridge Landslide, Ok Tedi Mine, Papua New Guinea.

The Vancouver Ridge Landslide at the Ok Tedi Mine occurred in August 1989 and included some 170 million tonnes of rock. Read and Maconochie describe the physical dimensions (1400 m long, 750 m wide and 600 m high) and geological features of Vancouver Ridge and the chronology and mechanism of the landslide.

The failure mechanism for the Vancouver Ridge landslide has been identified as an outward movement along the (15 to 25 m thick) Taranaki Thrust which dips at 23° to the north east towards to confluence of the Sulphide and Upper Sulphide Creeks.

Landslide mobilisation was observed in the weeks and days leading up to the failure - features included the development of cracks across Vancouver Ridge, spilling of rock from batters on the western side of the ridge and movement of colluvium from the eastern side of the ridge. No data has been presented on movement/displacement

observations or the interpretation/prediction of the failure event.

In an area of recognised active landsliding (80 landslides larger than one million cubic metres in the last 8000 years) it would be interesting to know how these hazards have been evaluated and the risks for the mining operation assessed.

3 STABILITY ASSESSMENT

3.1 Cojean J. and Fleurrision J.A. (1992) : Etude des instabilites de talus dues a la mise en eau de carrieres ou mines a ceil ouvert. (Study of slope instabilities due to the impoundment of quarries or open pit mines).

Cojean and Fleurisson (in French) present a parametric stability analysis of slopes in quarries or open pit mines which are progressively submerged by water at the end of the mining operation. Four slope profiles and five different potential failure surfaces are considered. The failure surfaces include planar, circular and three complex surfaces. Slope heights considered range from 50m to 700m. Strength parameters (effective?) in the study are c=0 to 200 kPa and $\phi = 10°$ to 40°.

Analyses have been undertaken for the various slope profiles as the quarry/mine fills with water. The effects of drawdown are considered. The analysis results are presented in the form of stability charts (Figure 2) which are provided for various slope/failure surface combinations.

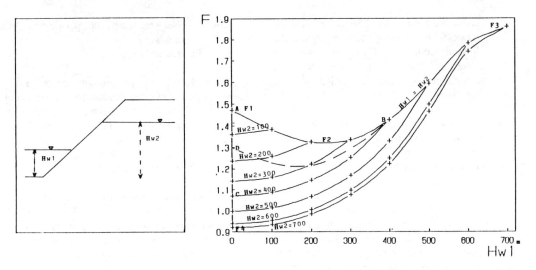

Figure 2 Example Stability Chart (from Cojean and Fleurisson (1992))

From Figure 2 four different situations are identified:

F_1 = F for a dry slope and the quarry empty

F_2 = minimum F as the quarry fills with water

F_3 = F for slope entirely submerged by water in quarry

F_4 = F after filling and complete rapid drawdown of quarry to empty

The authors describe the application of the results to the filling of an old quarry in the northern part of the Massif Central, France.

The authors conclude that the shape of the surfaces (convexity) is significant in the assessment of stability and the geological structures need to be identified deterministically. Cohesion is not found to be significant (c<500 kPa) for slopes several hundred metres high.

This paper illustrates the importance of considering long term stability of mine slopes where the abandoned mine is left to fill with water. A critical pool situation will develop during lake filling.

The stability charts presented in the paper need to be supplemented with physical information to define parameters such as slope geometry, slope height and slope piezometric profiles. It is not clear to this Reporter what the piezometric profile in the slope is.

3.2 List B.R. (1992) : Stability analysis techniques through the monitoring of Syncrude's open pit oil sand slopes.

List describes Syncrude Canada Ltd's open pit mining operation, pit slope failure mechanisms and the monitoring activities undertaken to maintain effective production.

The open pit mine within the Athabasca Oil Sand deposit in north eastern Alberta, Canada, produces an average of 160,000 m³ of ore daily. Draglines excavate the oil sand from the top of a 50 m high bench.

The oil sand deposit is contained within the McMurray Formation, the dominant geological unit. This formation is complex and variable and includes various sedimentary units associated with successive deposition in fluvial, estuarine and marine environments.

Three primary mechanisms for slope instability have been identified.

a) Block Slides - typically these are associated with high pore pressures in clay beds which dip (typically 10°) toward the pit.

b) Gas Ex Solution Induced Slope Deformation - slope failure develops as the gases in the oil sands come out of solution with slope unloading (excavation). Failures have resulted in slope crest regression of more than 20 m.

c) Translational Failures - more than 250 translational failures have been experienced over the past four years. Evaluation of failures indicates that the marine sediments are at residual strength when mined.

Monitoring of slope deformation using inclinometers is the primary means of managing production with minimal risk. Hundreds of inclinometers installed annually are read with frequencies varying from minutes to months. Monitoring facilitates the optimisation between production and safety.

Slope displacement criteria have been developed as follows:

- displacements of 0.45 mm/hr indicate a block slide is imminent (three hours or more) and a dragline is walked off

- mining bench crack widths form the basis for protection from gas ex solution failure

- extensive deformation studies have enabled protection from translational failures to be achieved by visual inspection.

The paper is an excellent example of an observational approach (complemented with extensive data evaluation) being utilised to ensure a safe production operation and effective management of slope stability hazards.

3.3 Kavvadas M., Marinos P. and Anagnostopoulos A. (1992) : Stability of open pit lignite slopes in Ptolemais, Greece.

Kavvadas, Marinos and Anagnostopoulos analyse the failures experienced in a lignite open pit mine at Ptolemais, Greece. Both slumping and wedge type failures are experienced. The mine geology comprises.

0 -15m   Quartenary sediments

15-65m   Neogene sands, silts, plastic clays and muds
65-145m   Lignite beds

An evaluation of the strength parameters of lignite has been presented including a comparison of results for different lignites. Strength parameters for the Ptolmais lignite were found to be in the range

$\phi' = 35°$ $c' = 150\text{-}600$ kPa (peak)
$\phi' = 30°$ $c' = 0$ (residual)

Stability analysis results for a composite wedge type analysis have been presented (Figure 3) which include the presence of water filled joints (tension cracks). The charts are for one case involving a 4:1 slope and a bedding plane dipping at 10°.

The depth of the water filled crack forming the rear of the wedge is shown to be a significant feature for highwall instability. Observations in the open pit are consistent with this.

The residual strength results appear to be rather high for lignite materials ($\phi' = 30°$). The authors have not achieved good definition of peak strength in the low stress range. Triaxial extension tests would assist in this area. This Reporter is concerned about the relationship between the different cases shown in the stability charts although this has not been evaluated.

3.4 Jermy C.A. (1992) : An assessment of the slope stability of some opencast coal mines in South Africa.

Jermy describes major coal deposits in South Africa. The paper discusses the influence of structural features, such as joints, fractures, faults and bedding planes, on the stability of rock slopes. The collection of structural geological data is described.

The stability of two mine slopes (Durha Colliery and Eikeboom Colliery) is assessed using established procedures. Planar wedge failures in the mine slopes were associated with variations in the orientation of a major joint set with respect to the highwall.

The author has not discussed the significance of variations in structural features although the small high wall failures are attributed to changes in the

Table 1: Representative properties

Mine Zone	γ(kN/m³)	w(%)	LL	PL	A
Allori	18.2	36	56	22	0.76
Castelnuovo	17.6	40	84	40	1.32

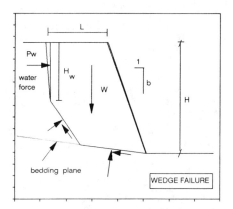

Figure 3   Wedge Failure Geometry (from Kavvados et al)

oreintation of major joints.   Stereo nets are presented. It would be interesting to consider how statistical variations in the structural data relate to the observed slope instability.

4   MINE SLOPE DEFORMATION

4.1 Bertuccioli P., Distefano D., Fesu F. and Federico G. (1992) : Initial Deformation of high cuts in overconsolidated clays.

Bertuccioli, Distefano, Fesu and Federico have presented observations of the deformations associated with high cuts in overconsolidated clays. The observations are related to the Santa Barbara Mine near Arezzo (central Italy). Deformation has occurred by slippage along discontinuities such as bedding planes, faults and contact surfaces between different soils.

The properties of the Santa Barbara clay are shown in Table 1 for two mine zones. Strength parameters are

$\phi' = 23\text{-}25°,$ $c' = 127\text{-}197$ kPa (peak)
$\phi' = 0\text{-}11°$ $c' = 0$ (residual)

In the two sites considered it was found that displacements developed as soon as excavation started at Site 2 while at Site 1 deformation developed relatively slowly and uniformly. At Site 2 inclinometer profiles are discontinuous and slow displacements occur along slip surfaces at various depths. Inclinometers failed in shearing soon after installation.

It was noted that at Site 1 the clay has few joints where as at Site 2 the clay is fissured and jointed. A resistance envelope study of the two sites illustrated that mobilized shear stress was near to but below the fully softened envelope.

The authors conclude that structural features in the clay mass influence deformation when the stress levels are well below the failure envelope. Where soil is highly fractured these features facilitate deformation by relative slippage of blocks and where discontinuities are widely spaced the deformation is controlled by the properties of the intact soil.

It is not clear from the paper what the nature of the joints in the clay is - are they persistent joint families or the randomly orientated fissures associated with heavily over-consolidated clays? The authors do not discuss the significance of the deformation with respect to stress relief or slope failure mechanisms.

4.2 Karatcholov P.G. and Stoeva P.Ch. (1992) : Investigation sur la deformation profonde des terrils dans les mines a ciel ouvert.

The paper by Karatcholov and Stoeva (in French) considers the stability of the large open cast mine at Maritza in eastern Bulgaria. Pit excavation and the deformation of overburden is accompanied by continuous processes of deformation. Through laboratory and insitu studies three characteristic zones in the overburden heap of differing strengths have been identified. The influence of heap formation on deformation has been considered.

Conclusions from the paper are:

- the disturbed tertiary clay in the overburden heaps are characterised by high porosity, high water content and a colloidal structure which determine the resistance of excavation and the deformation behaviour.

- the presence under the spoil heaps of plastic clay has an unfavourable influence and leads to deformation by means of an active/passive wedge mechanism

- the materials exhibit thixotropic properties

4.3 Mandzic E.H. and Zigic I. (1992) : Mobility of openpit unstable slope and influential factors.

Mandzic and Zigic describe the "Smreka" iron open pit mine in Yugoslavia. Mining has continued with slope moving at 10-50 mm/day for the last 13 years. Movement of some 10 million m^3 of rock occurs as regressive-progressive-regressive cyclic displacement. Mine slopes have a slope angle of 34° and are up to 360 m high. Two main factors influence the continuous deformation:

- discontinuities, weathering, relaxation and the phreatic water level in the slope

- artificial activities associated with mining, blasting and seismic effects.

The measurement of slope movements using survey points distributed over the slope is described. Typical movement profiles are presented and the influence of rainfall is illustrated. Because the moving slope indicates the classical factor of safety is less than 1.00 the authors propose a "coefficient of safety" (KS) where

$$KS = \frac{critical\ velocity}{over\ average\ velocity\ over\ the\ last\ period\ of\ measurement}$$

Using the method proposed by Fukuzono the critical movement velocity for the onset of progressive failure of the slope has been assessed as 50 mm/day. Table 2 presents the classification adopted for the coefficient of safety.

Table 2: Classification of rock slope activity and coefficient of safety regarding movement velocity (from Manzic and Zigic (1992))

```
1.Very small movement   v = 1 to 5 mm/day
Normal working process is possible on the
rock slope.
Blasting on the slope is normal.
Influence of water is very small.
KS = 10 to 50
2.Small movement        v = 5 to 10 mm/day
Normal working process is possible on the
rock slope.
Blasting must be decresed by 30 %.
Influence of water is increasing.
KS = 5 to 10
3.Middle movement       v =10 to 30 mm/day
Working on the slope is on the border zones
Blasting must be decreased for once a week.
Phreatic water level is increased.
KS = 5 to 1.7
4.High movement         v =30 to 40 mm/day
Working on the slope is outside the zone
of movement.
Blasting must be stopped apsolutely.
Phreatic water level is increased.
KS = 1.7 to 1.25
5.Very high movement    v =40 to 50 mm/day
Working on the slope is stopped.
Workers and machinery must be moved from
slope.
Phreatic water level is very high.
KS = 1.25 to 1.00
6.Critical              v > 50 mm/day
Calculated by Fukuzono method.
KS < 1,00
```

5 INFLUENCE OF UNDERGROUND MINING ON SLOPE STABILITY

5.1 Siddle H.J, Jones D.B., Whittaker B.N., Reddish D.J. (1992) : The influence of mining on hillslope stability.

The influence of underground mining activities on slope stability is often addressed where areas previously mined are exploited by open cast mining. Siddle, Jones, Whittaker and Reddish describe the results of recent research into the effects of mining on hillslope stability with reference to the landslide at East Pentwyn in north-east Gwent, UK. The landslide occurred in January 1954 but aerial photographs indicate slope distress had developed several years before the failure.

Both physical and analytical modelling have been undertaken to evaluate the effects of mining. Physical modelling was carried out using the large subsidence rating model at Nottingham University.

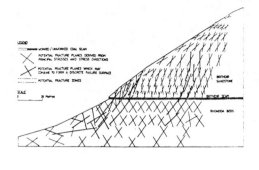

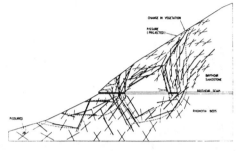

Figure 4   Directions of potential fracture planes before and after mining the Brithdir Seam (from Siddle et al (1992)).

The models demonstrated:

- high angle fracturing over the edge of coal pillars

- fracturing of floor rocks beneath the pillars

- outward displacement of the roof rocks about the pillar nearest the outcrop

- vertical displacement beneath the pillars indicating punching

Numerical modelling using finite element models showed similar displacement results to the physical model.

The modelling indicated that remnant pillars significantly effect slope stresses. Major stress concentrates remain about pillar locations after mining increasing pre-mining stresses by several hundred percent. The modelling enabled the identification of potential fracture planes before and after mining the Brithdir Seam (Figure 4).

The authors conclude that an affective model of

the 1954 landslide is developed. In particular they note the significance of remnant pillars within more extensive workings and their influence on fracture development.

The paper demonstrates the effectiveness of analysis in understanding the development of distress in a complex slope. The paper comments on the use of conventional emperical methods for the assessment of mine subsidence and their limitations particularly associated with steep slopes. This paper demonstrates there is a need for a balance between emperical and analytical methods.

6   DISCUSSION

The papers presented cover a range of topics. All of the papers are based on case histories and as such they reflect aspects of current practice. Aspects which deserve consideration in the design of mine slopes identified by the papers include:

- In all geotechnical tasks it is important to collect data and evaluate geological features in order to understand the problem

- Slope inundation due to lakefilling is a factor which needs to be considered paricularly where old pits are to be subsequently abandonded

- Deformation mechanisms in slopes are often complex

- Monitoring and observation can provide an effective an effective approach to managing slope stability hazards

- Underground workings can significantly effect slope stability.

7   REFERENCES

Bertuccioli P., Distefano D., Fesu F. and Federico G. (1992) : Initial Deformation of high cuts in overconsolidated clays. Proc 6th International Symposium on Landslides, Christchurch, NZ. February 1992.

Cojean J. and Fleurrision J.A. (1992) : Etude des instabilites de talus dues a la mise en eau de carrieres ou mines a ceil ouvert. (Study of slope instabilities due to the impoundment of quarries or open pit mines). Proc 6th International Symposium on Landslides, Christchurch, NZ. February 1992.

Jermy C.A. (1992) : An assessment of the slope stability of some opencast coal mines in South Africa. Proc 6th International Symposium on Landslides, Christchurch, NZ. February 1992.

Karatcholov P.G. and Stoeva P.Ch. (1992) : Investigation sur la deformation profonde des terrils dans les mines a ciel ouvert. Proc 6th International Symposium on Landslides, Christchurch, NZ. February 1992.

Kavvadas M., Marinos P. and Anagnostopoulos A. (1992) : Stability of open pit lignite slopes in Ptolemais, Greece. Proc 6th International Symposium on Landslides, Christchurch, NZ. February 1992.

List B.R. (1992) : Stability analysis techniques through the monitoring of Syncrude's open pit oil sand slopes. Proc 6th International Symposium on Landslides, Christchurch, NZ. February 1992.

Mandzic E.H. and Zigic I. (1992) : Mobility of openpit unstable slope and influential factors. Proc 6th International Symposium on Landslides, Christchurch, NZ. February 1992.

Mules G.J. (1992) : Landslide features reflecting valley-wall rebound, Kaiya River, Porgera, Papua New Guinea. Proc 6th International Symposium on Landslides, Christchurch, NZ. February 1992.

Read J.R.L. and Maconochie A.P. (1992) : The Vancouver Ridge Landslide, Ok Tedi Mine, Papua New Guinea. Proc 6th International Symposium on Landslides, Christchurch, NZ. February 1992.

Siddle H.J, Jones D.B., Whittaker B.N., Reddish D.J. (1992) : The influence of mining on hillslope stability. Proc 6th International Symposium on Landslides, Christchurch, NZ. February 1992.

S4 Slope instability in tropical areas
Instabilité des pentes en zones tropicales

Landslides, Bell (ed.) © 1995 Balkema, Rotterdam, ISBN 90 5410 032 X

Keynote paper: Slope instability in tropical areas

E.W. Brand
Civil Engineering Department, Hong Kong Government, Hong Kong

ABSTRACT: This paper presents an overview of the salient features of landslides in tropical areas, based heavily on the author's Hong Kong experience and his previous publications. It reviews rainfall/landslide relationships in high permeability materials, and suggests that rainfall intensity is the main agent of failure, with antecedent rainfall being of little account. An assessment is made of stability analysis in terms of its five 'components', with the conclusion that the state-of-the-art is generally poor for slopes in residual profiles. In addition, summary information is given about a recent study undertaken in Hong Kong to examine experimentally and theoretically the effects of boulders on the mass shear strength of residual materials and colluvium.

INTRODUCTION

This paper attempts to present a distillation of the essence of geotechnical engineering as related to slope stability under tropical conditions. The contents inevitably represent a very personal view of the topic, based largely as they are on a collection of the author's own perceptions and experiences over a period of more than twenty years in tropical countries. The paper contains very little that is new, virtually all the ideas presented and the information given having already been published elsewhere. In particular, the author draws heavily on three of his earlier state-of-the-art publications (Brand 1982, 1985a, 1985b).

There is nothing about slope instability in tropical areas which is especially different from that in non-tropical areas. The phenomenon of slope failure occurs in much the same way the world over, and the fundamental causes do not differ greatly with geographic location. The same methods of assessment, analysis and design apply, and similar remedial measures are applicable. The special feature of tropical areas, however, is that they are climatically both hot and wet, which together result in an abundance of slope failures. The climate conditions over geological time have resulted in deep weathering of the parent rocks to leave relatively weak materials on steep slopes, whilst the high rainfall is invariably the main agency of instability.

The fairly recent landslide sourcebook by Brabb & Harrod (1989) provides descriptions of landsliding in several tropical countries and regions. The author's contribution to this (Brand, 1989) reviews the whole of Southeast Asia (Figure 1), which can be regarded as a 'classic' tropical landslide region. Apart from experiencing annual rainfalls as high as 5 500mm, with intensities that can exceed 150mm per hour, much of the region is in the Pacific earthquake belt (Figure 2). The author's 1989 assessments of the significance of landslides and the state-of-the-art of landslide mitigation for the Southeast Asian countries are shown in Table 1.

Despite the presence of the Pacific earthquake belt, and of active volcanicity in Indonesia and parts of the Philippines, there is no doubt that rainfall is by far the main landslide-triggering agency. Indeed, the author believes that seismicity is often overemphasized worldwide as a direct cause of slope failures. This paper will concentrate on rain-induced slope failures of the kind so prevalent in steep terrain composed of deeply weathered rocks.

TROPICAL RESIDUAL MATERIALS

Characteristics

The warm wet climate in tropical regions produces materials which are the products of insitu weathering of rocks and which are commonly referred to as 'residual' materials, the degree of weathering and the extent to which the original structure of the rock mass is destroyed varying with depth from the ground surface. This process gives rise to weathering profiles which contain material 'grades' from fresh rock to completely weathered material,

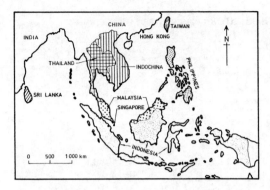

Figure 1. Southeast Asian countries reviewed by Brand (1989) for landslides

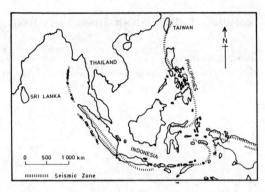

Figure 2. Southeast Asian countries in relation to the Pacific earthquake belt

the latter usually being described by geotechnical engineers as 'soil'. For engineering purposes, it is difficult to separate the several component parts of the weathering profile, and the whole profile is therefore best treated as a single entity.

The earth materials which comprise tropically weathered profiles are sometimes categorized simply as 'laterite', 'saprolite' and 'rock'. For the purposes of analysis and design, the engineering behaviour of both laterite and saprolite is usually considered to be governed by the principles of soil mechanics.

Residual materials have received very little attention from the geotechnical engineering community. This neglect, and the resulting paucity of relevant literature, stems largely from the fact that the very nature of residual material makes the application of soil mechanics principles problematical and renders the materials difficult to model for the purposes of engineering analysis and design. The main characteristics of residual materials which are responsible for this are :

(a) they are generally very heterogeneous, which makes them difficult to sample and test,

(b) they are nearly always unsaturated, which poses considerable difficulties for shear strength measurements, and

(c) they invariably have high permeabilities, which makes them subject to rapid changes in material properties because of external hydraulic influences.

As far as slope stability is concerned, the engineering behaviour of residual materials can be characterized very simply by :

(a) the importance of the complete weathering profile,

(b) the extreme difficulties of quantification, and

(c) the dominant role of rainfall.

From the point of view of analysis and design, residual materials undoubtedly represent the 'difficult' end of the broad spectrum of engineering earth materials. Engineering problems in residual materials and colluvium (see below) are not usually amenable to the principles of soil mechanics or rock mechanics alone, but must be combined with the appropriate elements of geology, geomorphology and hydrology. Geotechnical engineering in residual materials therefore spans the narrowly separated fields of soil mechanics, rock mechanics and engineering geology, and the engineering geological approach is generally the most satisfactory one for these materials.

Literature on Residual Materials

There is only a small amount of published literature on the engineering properties and behaviour of residual materials. As a starting point, reference should be made to the proceedings of the Specialty Session on Lateritic Soils held in Mexico City (Moh, 1969). Also of importance are the proceedings of the conferences on residual soils held in Hawaii (ASCE, 1982), Brazil (ABMS, 1985a) and Singapore (Balkema, 1988).

The proceedings of the twelve International Conferences on Soil Mechanics and Foundation Engineering each contain some papers directly relevant to residual 'soils', as do the proceedings of the many ISSMFE Regional Conferences held to-date; the session on slope stability in residual soils at the Fourth Panamerican Conference is particularly noteworthy (ASCE, 1971). To a lesser extent, there are also specifically relevant papers in the proceedings of the six Congresses of the International Association of Engineering Geology and the six Congresses of the International Society for Rock Mechanics.

Table 1.   Landslide significance and mitigation in Southeast Asian countries (Brand, 1989)

Country	Area (sq. km)	Population * (millions)	Population Density * (people/sq. km)	Per Capita GDP * (US$)	Volume of Relevant Literature	Human & Economic Significance of Landslides	State-of-the-Art of Landslide Mitigation
Hong Kong	1 033	6	5 421	6 268	Very High	Very High	Very High
Indochina	752 164	71	95	144	Nil ?	Low ?	Low ?
Cambodia	181 035	7	40	108			
Laos	236 798	4	17	105			
Vietnam	334 331	60	179	151			
Indonesia	1 919 263	163	85	529	Moderate (Lahars)	Very High	Moderate (Lahars)
Java	132 164	98	742		Very Low (Others)		Low (Others)
Sumatra	473 960	29	61				
Remainder	1 313 139	35	27				
Malaysia	330 669	16	47	2 002	Moderate	Moderate	Moderate
West	131 235	12	91				
East	199 434	4	20				
Philippines	299 765	54	180	603	Very Low	Moderate	Low
Singapore	580	3	4 414	6 827	Moderate	Low	Moderate
Sri Lanka	65 610	16	241	372	Low	High	Low
Taiwan	35 980	19	531	3 097	High	High	Moderate
Thailand	513 517	51	100	752	Low	Moderate	Low

* 1985 figures obtained from "The World in Figures", Economist Publications, London, 1987 edition.

A few key publications address themselves to the general philosophy and approach to engineering in residual materials. These include the papers on slope stability in residual soils by Vargas (1967), Patton & Hendron (1974), Morgenstern & de Matos (1975), Blight (1977, 1988), Oyagi (1984) and Massey & Pang (1988), as well as the state-of-the-art reports by Deere & Patton (1971) and Brand (1985b). In addition, there are the broad review papers by De Mello (1972) and Brand (1982, 1985a), together with the important reports produced by the ISSMFE Technical Committee on Tropical Soils (ABMS, 1985b; Brand & Phillipson, 1985). Lastly, slope stability in tropical areas features, to some degree, in the proceedings of all six International Symposia on Landslides, but this Sixth Symposium in Christchurch is the first that has devoted a whole session to tropical slopes. In this context should be mentioned the proceedings of the Southamerican Symposium on Landslides (Columbian Geotechnical Society, 1989).

Colluvium

Large deposits of colluvium often exist in conjunction with residual materials, parti-cularly as colluvial fans on footslopes of hillsides. Colluvium is material derived from the weathering of any parent rock which has been transported downhill by the agencies of gravity and water. It can range in general composition from a collection of matrixless boulders at one extreme to a fine slopewash material at the other. It poses many of the same general characteristics as residual material, particularly in the context of engineering behaviour. Because it is commonly found as slope cover over weathered rock profiles, it is sometimes difficult to distinguish between colluvium and the insitu material, particularly if only drillhole samples are available for examination. For geotechnical engineering purposes, colluvium can therefore be grouped with residual materials.

Colluvium features hardly at all in the technical literature, even though it is a fairly common engineering material. Few attempts appear to have been made to devise an engineering classification system, nor to investigate its wide range of material properties. Although colluvium features fairly prominently in some published descriptions of mass movements, there are only a few papers which describe engineering designs in this material, the most useful of which are probably

Table 2. Material grade classification system used by the Geotechnical Control Office (1988)

DESCRIPTIVE TERM	GRADE SYMBOL	GENERAL CHARACTERISTICS
Residual Soil	VI	Original rock texture completely destroyed Can be crumbled by hand and finger pressure
Completely Decomposed	V	Original rock texture preserved Can be crumbled by hand and finger Easily indented by point of geological pick Slakes when immersed in water Completely discoloured compared with fresh
Highly Decomposed	IV	Can be broken by hand into smaller pieces Makes dull sound when struck by hammer Not easily indented by point of pick Does not slake when immersed in water Completely discoloured compared with fresh
Moderately Decomposed	III	Cannot usually be broken by hand but easily broken by geological hammer Makes dull or slight ringing sound when struck by hammer Completely stained throughout
Slightly Decomposed	II	Not broken easily by geological hammer Makes ringing sound when struck by hammer Fresh rock colours generally retained but stained near joint surfaces
Fresh	I	Not broken easily by geological hammer Makes ringing sound when struck by hammer No discolouration

Table 3. Mass weathering zone classification system used by the Geotechnical Control Office (1988)

ZONE DESCRIPTION		ZONE SYMBOL	ZONE CHARACTERISTICS
Residual Soil		RS	Residual soil derived from insitu weathering; mass structure and material texture/fabric completely destroyed : 100% soil
Partially Weathered Rock	0/30% Rock	PW 0/30	Less than 30% rock Soil retains original mass structure and material texture/fabric (i.e. saprolite) Rock content does not affect shear behaviour of mass, but relict discontinuities in soil may Rock content may be significant for investigation and construction
	30/50% Rock	PW 30/50	30% to 50% rock Both rock content and relict discontinuities may affect shear behaviour of mass
	50/90% Rock	PW 50/90	50% to 90% rock Interlocked structure
	90/100% Rock	PW 90/100	Greater than 90% rock Small amount of the material converted to soil along discontinuities
Unweathered Rock		UW	100% rock May show slight discolouration along discontinuities

those by D'Appolonia et al (1966) and Murray & Olsen (1988).

IMPORTANCE OF WEATHERING PROFILE

The accurate logging of weathering profiles is fundamental to successful design and construction in residual profiles. These often contain a whole range of materials from an engineering point of view from 'soil' to 'rock'. The weathering profile is therefore of great importance for the stability of slopes, because it usually controls :

(a) the potential failure surface, and therefore the 'mode' of failure for analysis and design, and

(b) the groundwater hydrology, and therefore the critical pore pressure distribution in the slope.

There is no universally accepted system for describing and classifying the component parts of a weathering profile. Classification in terms of weathering 'zones' and weathering 'grades' is essential for engineering design, and there have been several major attempts

to provide a satisfactory description and clssification system for engineering purposes. Deere & Patton (1971) gave a valuable comparative summary of the classification systems available at that time, but there have been many systems proposed since.

Any weathering description and classification system must be suitable for the particular geological conditions and engineering purpose to which it is applied. In Hong Kong, where site formation and slope stability are the main geotechnical engineering problems, the Geotechnical Engineering Office (formerly Geotechnical Control Office) has adopted a system for the granites and volcanic rocks in which a profile is logged according to the six rock 'material grades' given in Table 2 and the six 'mass weathering zones' described in Table 3 (Geotechnical Control Office, 1988).

For the purposes of geotechnical analysis and design, the following should be noted :

(a) grades I to III material are usually treated as 'rock' and grades IV to VI material as 'soil', and

(b) the engineering behaviour of weathering

zones UW and PW 90/100 is considered to be governed by the principles of rock mechanics, whereas the other four zones are governed largely by the principles of soil mechanics.

There exists no engineering description and classification system for colluvium, although one is badly needed. Attempts have been made in Hong Kong and elsewhere to provide a framework for such a system, but these are entirely descriptive in character and require a great deal of further development.

SITE INVESTIGATIONS

For projects in deeply weathered rock profiles, site investigations must generally be more extensive and more expensive than for more homogeneous earth materials. The emphasis must be on the engineering geological approach, for which the following are the main components :

(a) execution of adequate surface and subsurface exploration to establish the site engineering geology, to define the 'engineering' materials involved, and to retrieve good quality samples for laboratory testing,

(b) identification of especially significant geological, geotechnical and hydrological features,

(c) study of existing local and other 'case histories' of similar projects and of sites with similar geology,

(d) routine insitu and laboratory measurement of selected engineering properties of materials to establish lower bound values, and comparison of these with generalized parameters developed from existing data, and

(e) continuous reappraisal of the site investigation results throughout the period of construction.

The importance of the study of good case histories cannot be overemphasized. Particularly important also is the continuous reappraisal process, since site investigations carried out in residual terrain before construction commences can in most cases only be regarded as 'initial' investigations, and the geotechnical engineer must be prepared to modify his design as excavations reveal much fuller subsurface information than was available prior to the construction phase. This is the basis of the 'observational method' of design expounded by Peck (1969).

Whereas the use of generalized material properties is not unsatisfactory where the materials are such that these can be established within a sufficiently sensible range, this is

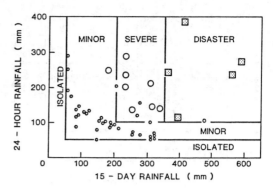

Figure 3. Erroneous rainfall/landslide correlation established by Lumb (1975) for Hong Kong

not so for residual materials, because of their variability and inhomogeneity. Generalized parameters for residual materials are difficult to obtain, and insufficient basic research has been carried out on most residual materials to provide even approximate values. This is an unsatisfactory situation which confronts everyone who works with residual materials, and it enhances the importance of the non-analytical methods of design described below.

Drilling and sampling in residual profiles can be difficult because of the vertical variability encountered, particularly since it is essential to obtain a high recovery for profile description purposes. A comprehensive international review of the methods used worldwide for the sampling and testing of residual soils (Brand & Phillipson, 1985) revealed that the sampling practice in most countries leaves much to be desired, heavy reliance often being placed on methods developed for sedimentary soils. In this context, the drilling and sampling methods used for good quality ground investigations in the weathered granite and volcanic profiles of Hong Kong may be of interest elsewhere (Brand & Phillipson, 1984); triple-tube rotary core barrels are used for sampling, and air-foam is sometimes employed as the flushing medium.

SLOPE INSTABILITY ASSESSMENT

The stability analysis of a slope is related directly to the prediction of the conditions under which the slope could fail. Apart from the application of sound judgement and experience alone, there are four basic methods available for the prediction of rain-induced failures in deeply weathered slopes. These are :

(a) correlations between slope failures and pattern of rainfall,

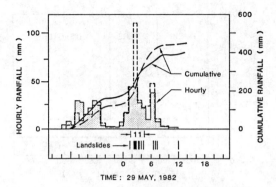

Figure 4. Example of landslide coincidence with time of occurrence of maximum hourly rainfall intensity in Hong Kong (Brand et al, 1984)

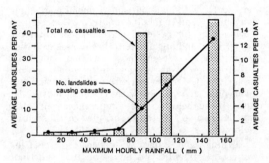

Figure 5. Correlation of landslides and casualties with hourly rainfall intensity in Hong Kong during the period 1963-1983 (Brand et al, 1984)

(b) terrain evaluation, mainly on the basis of geomorphological mapping,

(c) semi-empirical (or modified precedent) approach, which is based on an examination of the geomorphology and geology of stable and unstable slopes, and

(d) analytical methods, usually in the form of limit equilibrium analysis.

The first of these two approaches can be considered to be directly related, since they apply to the stability of a land area in one particular location. The last three methods can be regarded as methods of analysis and design; all three have been used extensively in residual materials. The terrain evaluation and semi-empirical methods are closely related, in that both are based on an explicit assumption that the stability characteristics of a slope can be assessed on the basis of observations of the performance of others with similar characteristics.

Whereas the vast majority of cut slopes in residual materials were not designed on the basis of rigorous soil mechanics methods, the soil mechanics approach is being increasingly adopted even in this difficult material. This is certainly true in Hong Kong, where design practice is governed largely by the Geotechnical Manual for Slopes (Geotechnical Control Office, 1984). The degree of safety of a slope can only be quantified on the basis of an analytical method, whereas methods (a), (b) and (c) provide no such quantification.

The four methods of slope stability assessment outlined in (a) to (d) above have been dealt with in some detail previously by the author (Brand, 1985b). Only rainfall/landslide correlations and limit equilibrium analysis will be considered further in this paper.

RAINFALL/LANDSLIDE CORRELATIONS

There have been many attempts over the years to establish relationships between rainfall and landslides, particularly in tropical areas where rain-induced landslides are commonly of such importance. Such correlations aid our understanding of landslide processes. For a particular geographical location, a reliable correlation also provides a broad basis for predicting widespread slope failures, which enables the establishment of a warning system for those whose lives and property might be endangered.

Published Correlations

Simple direct rainfall/failure correlations have been made for locations in Brazil (Barata, 1969; Guidicini & Iwasa, 1977; Campos & Menzes, 1992), Italy (Rossetti & Ottone, 1979), Japan (Onodera et al, 1974; Fukuoka, 1980), Sri Lanka (Bhandari et al, 1992) and the United States (Campbell, 1975; Nilsen et al, 1976; Pierson et al, 1992). More sophisticated correlation attempts have been undertaken in Brazil (Wolle & Hachich, 1989), New Zealand (Crozier, 1969, 1986; Eyles et al, 1978; Eyles, 1979; Crozier & Eyles, 1980), Hong Kong (Lumb, 1975, 1979; Brand et al, 1984), Japan (Aboshi 1979; Yagi et al, 1990), Korea (Kim et al, 1992) and China (Zhang & Sheng, 1992).

The more sophisticated attempts to establish rainfall/landslide correlations have generally assumed there to be a direct relationship between the occurrence of landslides and the quantity of rainfall, in terms of the 'short-term' rainfall immediately prior to the landsliding and the 'long-term' rainfall over some extended period before the landsliding. The short-term rainfall has typically been taken to be the 'one-day' or '24-hour' rainfall, and the long-term (antecedent) rainfall period has been suggested as being anything from a few days to one month or more. The first published correlation of this kind was probably that of Lumb (1975) for Hong Kong, in which landslide events were divided into 'minor', 'severe' and 'disastrous' on the basis of the daily (24-hour) rainfall and the 15-day antecedent rainfall (Figure 3).

In several parts of the world, the well-known approach used by Lumb (1975) has been used as a model by which to establish correlations between rainfall and landslides, including those reported by Bhandari et al (1992) and Kim et al (1992) in papers to this Symposium. It would seem logically and scientifically sustainable that antecedent rainfall is an important parameter in such correlations, because of considerations of moisture deficiency and the reduction in this caused by a long period of rainfall prior to some 'critical' short-term rainfall which brings about slope failure. The author believes, however, that far too much emphasis has been placed on the role of antecedent rainfall and, indeed, that antecedent rainfall is not an important factor in most tropical areas where soil permeabilities are relatively high. This opinion is based on the rainfall/landslide correlations undertaken a few years ago in Hong Kong (Brand et al, 1984; Brand, 1985b), and on an examination of the published correlations from elsewhere.

Hong Kong Correlations

The main features of the Hong Kong correlations are worth summarizing here, because this is probably the most thorough rainfall/landslide correlation study carried out anywhere. A system of about 50 automatic raingauges linked to a central micro-computer has allowed storm events to be monitored continuously in Hong Kong for a number of years. In some areas, the density of automatic raingauges exceeds one per square kilometre. In addition, excellent data exists on the times at which failures have occurred. Full details of these studies and the results have already been published elsewhere (Brand et al, 1984; Brand, 1985b), but the main conclusions are worth repeating, viz :

(a) The large majority of landslides are induced by localized short-duration rainfalls of high intensity, and these landslides take place at about the same time as the peak hourly rainfall.

(b) Antecedent rainfall of any duration is <u>not</u> a significant factor in landslide occurrence.

(c) A rainfall intensity of about 70mm/hour appears to be the threshold value above which landslides occur. The number of

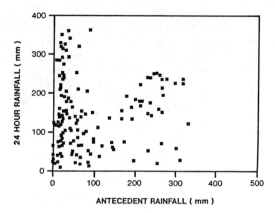

Figure 6. Lack of correlation between Hong Kong landslides during 1963-1983 and 9-day antecedent rainfall

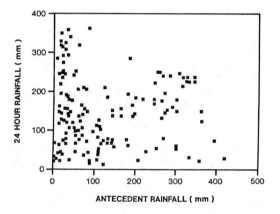

Figure 7. Lack of correlation between Hong Kong landslides during 1963-1983 and 14-day antecedent rainfall

landslides and the severity of the consequences increase dramatically as the hourly intensity increases above this level.

(d) The 24-hour rainfall usually reflects short-duration rainfalls of high intensity, and can therefore be used as an indicator of the likelihood of landslides. A 24-hour rainfall of less than 100mm is very unlikely to result in a major landslide event.

The importance of rainfall intensity in Hong Kong is illustrated vividly in Figures 4 & 5. Conclusion (b), that antecedent rainfall is <u>not</u> an important parameter for landsliding in Hong Kong, was reached after thorough analyses of the effects of antecedent rainfall periods that ranged from one day to 30 days. Figures 6 & 7 show the correlations of landslides with antecedent periods of 9 days

2037

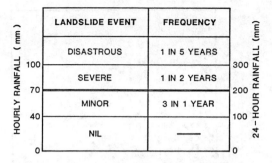

Figure 8. Historical relationship between rainfall intensity and landslide events in Hong Kong (Brand, 1985b)

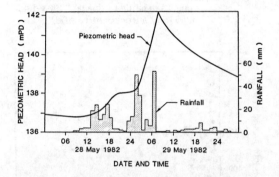

Figure 9. Example of a rapid piezometric pressure change in a Hong Kong slope, probably typical of high permeability slopes elsewhere

and 14 days. A weak correlation was found to exist only with a four-day antecedent period for very minor landslide events. In contrast, the very strong correlation between hourly intensity and landsliding leaves no doubt whatever that short-term intensity is the dominant parameter in Hong Kong. The practical significance of this conclusion for Hong Kong is summarized in Figure 8.

The author believes that short-term rainfall intensity is probably also the dominant landslide-controlling parameter in other parts of the world where soil permeabilities are high. With decreasing soil permeability, it is to be expected that antecedent rainfall will become increasingly important, and publications by Wieczorek (1987) and Addison (1987) lend support to this hypothesis. Also of interest in this regard is the theme lecture to this Symposium by Hutchinson (1992).

In many tropical areas, the dominance of short-term rainfall intensity as a landslide parameter is, the author believes, related mainly to rapid rises in piezometric pressure

of the kind illustrated in Figure 9. This increase in pore pressure in a Hong Kong slope of about six metres in four hours cannot be explained by the surface infiltration of rainwater alone, but must be largely the result of rapid rises in the groundwater table fed mainly by natural 'pipes' or 'tunnels' caused by subsurface erosion (Jones, 1981; Pierson, 1983). The importance of these in Hong Kong has been outlined by Brand et al (1986), and the mechanism of rapid groundwater rises has been discussed by Premchitt et al (1985). The author suggests that those seeking explanations for specific rain-induced slope failures in tropical areas would do well to investigate the possible existence of pipes and tunnels which, from Hong Kong experience, can be up to a metre in diameter.

Shortcomings in Other Correlations

In the author's opinion, virtually all rainfall/landslide correlations published to-date are unreliable, in varying degrees, in finding that antecedent rainfall is an important parameter. This is largely because the published rainfall/landslide correlations in high permeability soils were generally based on rainfall and/or landslide data of inadequate quality. More specifically, the published studies have suffered from one or more of the following deficiencies :

(a) The density of available raingauges has been inadequate, the precise rainfall at or near the sites of the landslides being unknown. The importance of this deficiency is illustrated in Figure 10, which shows an example of the enormous rainfall variations over short distances in Hong Kong. Such variations undoubtedly exist elsewhere.

(b) The raingauges have rarely been of the continuous-reading type, and they have not therefore measured short-term rainfall intensities.

(c) The most commonly used rainfall period for correlation purposes has been 24 hours, but this has invariably been chosen in an arbitrary manner (usually midnight to midnight) rather than as the 'rolling' (antecedent) 24-hour period.

(d) The precise times at which landslides occurred have not usually been known.

The well-known rainfall/landslide study by Crozier & Eyles (1980) in New Zealand, for example, seems to have suffered from all these deficiencies, since it was based on only three raingauges in an area of about 1 350 sq.km, and arbitrary 'daily' rainfalls were used for correlation purposes.

It is particularly noteworthy that virtually all published rainfall/landslide correlations have been based on the same 'daily' or '24-hour' rainfall, with the 24-hour period chosen in an arbitrary manner (typically midnight to midnight) as far as the rainfall pattern is concerned. If 24-hour duration rainfall is a significant parameter, then this will logically be the rainfall in the 24-hour period immediately prior to the occurrence of the landsliding. The published correlations would all be changed significantly if this 'correction' was made, which in practical terms means that landsliding should be correlated with the maximum rainfall in any 24-hour period.

It is unfortunate that the paper by Lumb (1975) is still often quoted as evidence of the importance of antecedent rainfall. Largely because Lumb used a midnight-to-midnight '24-hour' rainfall, his supposed rainfall/landslide correlation for Hong Kong (Figure 3) is <u>not correct</u>. The detailed study carried out more recently and described by Brand et al (1984) was based on much better rainfall and landslide information than was available for Lumb's earlier study, and the results totally supersede his findings.

SLOPE STABILITY ANALYSIS

Principles

Slope stability analysis is usually based on the limit equilibrium approach, for which the equilibrium of a sliding mass is examined (Figure 11). The degree of stability is quantified in terms of a 'factor of safety', F, which is most commonly defined as the ratio between the average shear resistance, S_a, and the average shear stress, τ_a, along the most critical slip surface, i.e. $F = S_a/\tau_a$.

The determination of the F-value for a particular slope therefore requires :

(a) prediction of the correct mode of failure (i.e. selection of the critical slip surface),

(b) prediction of the distribution of shear stress over the critical slip surface, and

(c) prediction of the distribution of shear resistance over the critical slip surface.

The distribution of shear stress along the critical slip surface is dependent on the loading and the method of analysis employed. The shear resistance along the slip surface is governed by the effective shear strengths of the materials of which the slope is composed and the normal effective stress distribution on the slip surface. The normal effective stress distribution is, in turn, a function of the pore pressure distribution at failure and the method of analysis. The five components of stability prediction are therefore :

(a) mode of failure,

(b) loading,

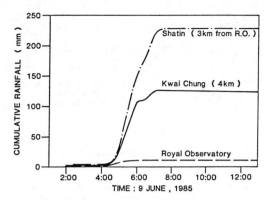

Figure 10. Example of large variations in rainfall over short distances in Hong Kong, which is probably characteristic of tropical rainfall

(c) method of analysis,

(d) shear strength, and

(e) pore pressure distribution at failure.

Mode of Failure

The selection of a suitable mode (mechanism) of failure is crucial to accurate stability predictions in all soil types. It is particularly crucial for slopes in residual profiles, where the potential failure surface is often governed by geological detail.

Whereas the pre-failure geometry of a slope is easily defined, it is often difficult to decide upon the critical potential failure surface for stability analysis or design. Occasionally, even a post-failure surface cannot readily be determined because of the multiple-failure nature of some landslides in residual materials.

As a general rule, shear strength in a residual profile increases with depth, and slope failures can therefore be expected to occur on relatively shallow slip surfaces. These surfaces are largely controlled by the weathering profile. Failures most frequently occur along surfaces dictated largely by relict joints or by boundaries between weathering zones (where clear boundaries exist). By the very nature of residual weathering profiles, non-circular failure surface are the most common, and these are often almost planar over a major proportion of their length.

Geological detail is often crucial to the location of the critical slip surface. A complex weathering profile can rarely be adequately investigated at the pre-failure stage. In those few instances where the profile is determined in considerable detail by extensive surface and subsurface investigations, the true geological situation usually does not lend itself

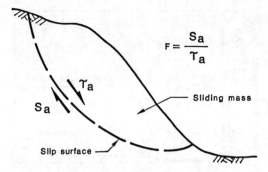

Figure 11. Analysis of slopes on the basis of the limit equilibrium approach

$$F = \frac{S_a}{\tau_a}$$

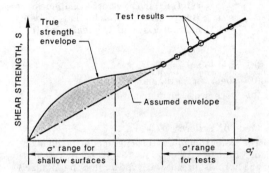

Figure 12. Possible underestimate of shear strengths in correct stress range by use of laboratory test data obtained in higher stress range

to being properly modelled for the purposes of meaningful engineering analysis.

The importance of soil fabric to the mode of slope failure cannot be over-emphasized. In particular, relict joints, especially when slickensided, can be instrumental in the initiation of a failure, and their presence can appreciably reduce the mass strength of the soil. Some examples of the importance of geological detail to slope failure modes in residual soils have been given by Deere & Patton (1971), Patton & Hendron (1974), Hencher et al (1984) and Brand (1985b).

Methods of Analysis

The state-of-the-art with respect to methods of stability anlaysis is undoubtedly very good, eventhough it is well-known that the available methods often produce slightly different theoretical F-values because of the different assumptions made in their formulation.

For residual soil slopes, methods of stability analysis which employ circular slip surfaces are usually inappropriate, and methods which

apply to any shape of surface must be used. In this category are the well-known methods by Janbu (1954, 1973) and Morgenstern & Price (1965). Although Morgenstern & Price made more satisfactory assumptions than Janbu, the latter's method is much easier to programme and requires much smaller computer capacity. For these reasons, it is more widely used for routine stability calculations.

Shear Strengths

Despite the obvious objections to the measurement of shear strengths on residual materials by means of laboratory tests, these still comprise the most satisfactory means of establishing the likely range of shear strengths on the softer materials (grades V & VI). The effects of corestones and other large-sized particles, however, cannot be readily determined, and there is no doubt that laboratory strength tests carried out on the 'matrix' material of residual soils and colluvium will usually underestimate the shear strength of the insitu material because of the neglect of the boulder content. The effect of boulders on shear strengths is considered later in this paper.

Residual materials and colluvium are invariably unsaturated and are of relatively high permeability. Stability computations must therefore always be made in terms of effective stresses; analysis on the basis of undrained strengths has no relevance.

Because of their high permeabilities (usually 10^{-4} to 10^{-6} m/sec.), rainwater infiltrates with ease into most residual soils and colluvium, and it is thought likely that saturation conditions will be approached at shallow depths in the field during the life of a slope. It is therefore generally felt to be appropriate to measure strength parameters on the basis of shear tests carried out on saturated soil specimens. Although this may often be a more severe condition than that experienced by the soil insitu, it remains the only certain means by which a 'base' shear strength envelope can be established.

The triaxial test is the most widely used method for shear strength measurement on residual 'soils' (Brand & Phillipson, 1985). Test specimens should be as large as possible, full-diameter lengths of drillhole sample being ideal for routine work, and they are usually saturated by the application of a sufficiently high back pressure prior to shear. Either drained tests (CD) or consolidated undrained tests with pore pressure measurement ($\overline{CU}$) can be used, but the latter are much to be preferred, because they are quicker and provide much more information about the stress-strain behaviour of the material.

Cell pressures used for triaxial testing must relate to the correct insitu stress range if the measured strengths are to be meaningful. Critical slip surfaces in residual slopes are most commonly shallow, and the effective stresses on these are therefore low (typically 30 to 200 kPa). At such low effective cell pressures, however, triaxial tests are difficult to control satisfactorily, and this stress range is not recommended for routine use. There is some evidence to suggest that the strength envelopes for some residual soils are curved at low effective stresses, and that the straightline projection of strengths measured at high stresses underestimates the strengths in the low stress range (Figure 12). It is almost certain that this is part of the explanation for why many stable residual soil slopes have theoretical factors of safety of less than 1.0 (Brand, 1985b).

As discussed below, failures of residual slopes are invariably caused solely by pore pressure increases. In its simplest form, the mechanism of failure is therefore that the soil suction (i.e. negative pore pressure) decreases (i.e. pore pressure increases), thereby reducing the shear strength of the earth material, as dictated by the effective stress principle. Failure occurs when the average shear resistance on the critical slip surface has decreased to the value of the average shear stress. Rain-induced slope failures therefore take place under conditions of almost constant total stress and increasing pore pressure (Brand, 1981).

The triaxial test is most commonly conducted by increasing the axial stress, σ_1, to failure while the cell pressure, σ_3, is kept constant. This is carried out either as a drained test ($\Delta u = 0$) or as an undrained test (or constant water content test) with pore pressure measured throughout the shearing process. In contrast, for rain-induced failure, σ_1 and σ_3 are almost constant, and the pore pressure increases to failure. The comparative stress paths for the common triaxial tests and for the real field situation are shown in Figure 13, in which the quantities p' & q are defined as p' = $(\sigma'_1+\sigma'_3)/2$ and q = $(\sigma_1-\sigma_3)/2$.

Figure 13 illustrates the fact that stress paths commonly followed in the triaxial test are quite different from the one which pertains in the field, and that the stress range over which triaxial tests are usually conducted are generally not appropriate to the field stress conditions. Where the strength envelope of a soil is markedly curved, this disparity will result in an appreciable underestimate of the correct shear strength for stability assessment (Figure 12). It is possible to follow the correct stress path in the laboratory simply by decreasing the cell pressure, but this does not simulate the correct mechanism of failure

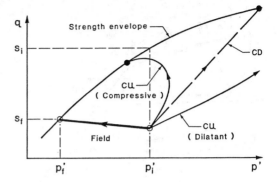

Figure 13. **Stress path for rain-induced slope failure compared with the stress paths for the common triaxial tests (Brand, 1981)**

Figure 14. **Insitu direct shear test equipment used successfully on grades IV, V & VI materials in Hong Kong (Brand et al, 1983b)**

in the field. This can only be done by means of a constant load test in which the pore pressure is increased from an initial negative value until failure occurs. Tests of this kind have been carried out by the Geotechnical Engineering Office in Hong Kong for some time, but the experimental difficulties are considerable, and this test procedure cannot be recommended for routine use.

Direct shear (shear box) testing is deservedly becoming more commonly used for the strength assessment of residual materials. Apart from its relative simplicity, this method has obvious merits over the triaxial test for strength measurements along relict joints, because correctly orientated specimens can be prepared. The drawbacks are that there is no certain method of saturation, and pore pressures cannot be measured during shear. Despite its lack of theoretical 'purity', however, the direct

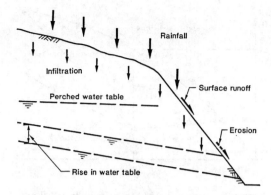

Figure 15. Multiple effects of rainfall on a slope of high permeability

shear test is seen by some as providing a means of obtaining shear strength data under conditions which model those in the field more closely than does the traixial test. It is also readily adaptable for insitu measurements of shear strength on relatively large masses of material. The equipment shown in Figure 14 has been used successfully for testing 300 x 300 mm soil and soft rock specimens in Hong Kong (Brand et al, 1983; Brand, 1988).

Relict joints and other such discontinuities are worthy of special mention, because they frequently play a major part in slope instability (St John et al, 1969; Hencher et al, 1984; Cowland & Carbray, 1988; Irfan & Woods, 1988). These are often slickensided and sometimes coated with thin deposits of very weak material. There have been some reported failures of quite flat slopes attributable to extremely low relict joint strengths. One such well-documented Hong Kong case, which was of major economic significance, led to an extensive investigation into the shear strength of jointed sedimentary volcanic material (Koo, 1982a, 1982b). Persistent joint surfaces were coated with a thin black-brown deposit believed to be the precipitation of iron and manganese oxide products that filled the joints in the parent rock during the course of weathering. The results of laboratory direct shear and triaxial tests showed that the effective strength parameters on the joints were much lower than those measured on the intact material.

Pore Pressures

For a residual slope in residual material, the prediction of the pore pressure distribution is by far the most critical factor for stability analysis. This is particularly so since most failures in residual slopes are caused by rainfall.

The hydrological effects of rainfall on a permeable slope are depicted in Figure 15. Some of the water runs off the slope and may cause surface erosion if there is inadequate surface protection. Because of the high soil permeability, however, the majority of the water infiltrates. This causes the water table to rise, or it may cause a perched water table to be formed at some less permeable boundary, usually dictated by the weathering profile. Above the water table, the degree of saturation of the soil increases, and the soil suction (i.e. negative pore pressure) therefore decreases.

Failures in residual cut slopes are thought to be caused mostly by the 'wetting-up' process by which the soil suction (and hence the soil strength) is decreased, but there is some evidence to suggest that transient rises in groundwater tables are responsible for some rain-induced landslides (Premchitt et al, 1985). Rain-induced slope stability failures thus occur as a direct result of pore pressure increases, and pore pressure distribution is therefore the variable of most concern.

The pore pressure distribution in a residual profile is dependent on the pattern of rainfall and the hydrogeology of the slope. Pore pressure is therefore a variable which is independent of soil mechanics considerations, being imposed upon a slope by external influences. For this reason, it is extremely difficult to predict the appropriate pore pressures for slope design or stability assessment. This fact is vividly illustrated in Figure 9 presented earlier.

Some approximate methods exist for the prediction of pore pressure changes in slopes during rainfall infiltration. For analysis of slope stability in residual soils and colluvium, however, measured pore pressure data is much to be preferred to the application of uncertain

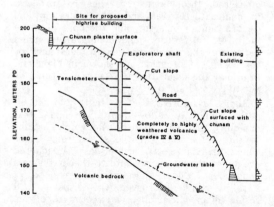

Figure 16. Cut slope in deeply weathered granite used for longterm measurements of soil suctions in Hong Kong (Sweeney, 1982; Sweeney & Robertson, 1982)

predictive methods. For such data to be meaningful, pore pressures must be monitored for a sufficiently long period of time, and piezometers must be installed at the appropriate depths at sufficient locations on the slope. In addition, an appropriate type of instrument must be used which can respond rapidly to pore pressure changes.

Because of the rapid changes in pore pressure that occur with rainfall in many residual soil slopes, the critical pore pressures are rarely obtained from normal piezometer measurements. In order to enable standpipe piezometers to record the maximum water levels attained during a given period, Halcrow 'buckets' (British Patent Hong Kong No. 1538487) have become popular in Hong Kong (Brand, 1985b). This system is cheap and simple, and it can be used to provide important design information when expensive automatic piezometer systems, of the type described below, cannot be justified.

During most of the year, suctions exist in many residual soil slopes. These suctions, which can be of very high magnitudes (Fredlund, 1991), are reduced dramatically by the process of water infiltration during rainfall. There are almost certainly situations where even the heaviest rainfall does not completely destroy the soil suctions, and these continue to contribute to the stability of a slope. Whereas slope design is commonly based on predictions of the maximum positive pore pressures that are likely to develop in the slope, no account is usually taken of negative pore pressures that might be sustained under the worst rainfall conditions. This neglect of the contribution of soil suction to the shear

Table 4. Author's summary assessment of the state-of-the-art of limit equilibrium stability analysis as applied to slopes in residual profiles

COMPONENT	SUB-COMPONENT	KNOWLEDGE
MECHANISM	—	GOOD – POOR
SHEAR STRESSES	LOADING	V.GOOD – GOOD
	ANALYSIS METHODS	V.GOOD – GOOD
SHEAR STRENGTHS	STRENGTH PROPERTIES	FAIR – POOR
	PORE PRESSURES	POOR – V.POOR

strength of the soil might be the main reason why stable residual soil slopes often have theoretical factors of safety of less than unity.

Programmes of laboratory testing carried out over many years on Hong Kong residual materials have proved that suction pressures act as 'modified' effective stresses, a matric suction of $u_a - u_w$ increasing the shear strength by $(u_a - u_w)\tan \phi^b$, where ϕ^b is the angle of internal friction with respect to matric suction (Fredlund, 1981; Ho & Fredlund, 1982; Fredlund & Rahardjo, 1985).

Much valuable data on soil suctions has been obtained on some Hong Kong slopes over several years. It has been found that the suctions on relatively shallow potential slip surfaces decrease everywhere close to zero during a wet season, and that suction cannot therefore normally be relied upon for slope stability purposes. However, some measurements (Figure 16) (Sweeney, 1982; Sweeney & Robertson, 1982) suggest that not all soil suction is destroyed by infiltration. The extent to which suctions play a significant part in slope stability in residual soil profiles elsewhere in the world clearly depends on the pattern of rainfall and the infiltration characteristics of the material concerned.

Reliable measurements of soil suctions can be obtained for shallow depths by means of simple agricultural tensiometers (Greenway et al, 1984). These instruments have suffered from the serious disadvantage, however, that they must be read manually, and critical readings during intensive rainfall are frequently unable to be obtained. As a result, some sophisticated automatic-reading equipment has been developed to measure both positive and negative pore pressures in Hong Kong conditions (Brand, 1988). This is based on similar equipment designed in the United Kingdom (Anderson & Burt, 1977; Burt, 1978).

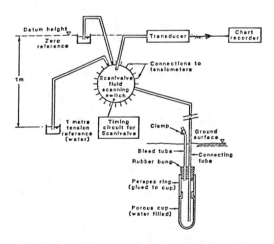

Figure 17. Automatic Scanivalve system used in Hong Kong for measurement of positive and negative pore pressures (based on Burt, 1978)

Small diameter tensiometer tips of high air-entry ceramic are each connected by twin water-filled flexible nylon tubes to a single transducer through a fluid scanning switch (Figure 17). The switch is analogous to an electrical wafer switch, in that it allows sequential measurement of a number of fluid pressure inputs. A rotating metal valve inside the water-filled wafer (Scanivalve) enables the pressure transducer to be connected to 24 different inputs in turn, two of which are used as reference reservoirs for calibration purposes. The measured pore pressures are recorded on a variety of types of strip-chart recorder, the pressure-sensitive paper type now being regarded as the most reliable. The system is powered by a twelve-volt battery, and all measuring and recording components are housed in a water-tight metal box. Although installation and maintenance costs are high, the systems have been found to be reliable for periods of up to three years.

State-of-the-art of Stability Analysis

Of the five 'components' of slope stability analysis listed earlier, detailed consideration has been given above to four of these. The fifth 'loading', requires no more attention than to say that the accurate determination of loading for slope stability analysis presents little difficulty, the dead weight of the sliding mass (Figure 11) being readily computed from the slope geometry and from data obtained from borehole samples.

Table 4 summarizes, in very broad terms, the Author's assessment of the overall state-of-the-art of slope stability analysis for slopes in residual profiles in terms of the five components of assessment. The determination of loads is straightforward. There are no reasons to believe that the available methods of analysis are anything but very good, as long as these are restricted to appropriate methods for the particular mechanism of failure. Our ability to select a correct mechanism of failure varies fairly widely depending on the geology and weathering profile for the slope. Because of the small amount of work that has been done to investigate the shear strength properties of residual earth materials, our knowledge of this component is only fair. However, by far the major difficulty with stability predictions in residual slopes is the poor state-of-the-art with respect to the prediction of pore pressure distribution at failure.

It is clear that the application of rigorous soil mechanics methods of slope stability analysis to residual profiles is extremely difficult. It is prudent to adopt this approach only in conjunction with a thorough engineering

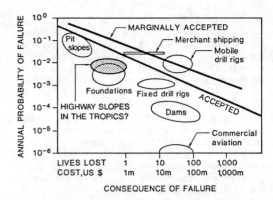

Figure 18. Probability and consequence of failure in the USA with 'accepted' and 'marginally accepted' risk levels (Whitman, 1984), compared with risk of failure of older highway cut slopes in Hong Kong

Table 5. Design factors of safety for cut slopes in Hong Kong (Geotechnical Control Office, 1984)

		RISK TO LIFE		
		High	Low	Negligible
ECONOMIC RISK	High	1.4	1.4	1.4
	Low	1.4	1.2	1.2
	Negligible	1.4	1.2	>1.0

geological assessment and with the liberal application of sound judgement. Analytical solutions alone cannot be relied upon, as succinctly stated by Peck (1975), thus :

" Analytical procedures have been developed for calculating the factor of safety of a slope under various conditions... The theories and their applications have met with some successes and more failures. Most of the failures are a consequence of oversimplification.

Even the most complex theories are necessarily oversimplifications of nature; they are general rather than specific. A theory that would take into account all the significant variables at a given location would be far too complex for use. Furthermore the parameters that must be evaluated to use the theory are not simple invariant quantities, few in number, but are instead complex, highly variable, and usually not constant with respect to time.

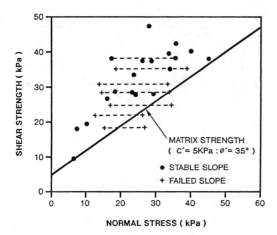

Figure 19. Minimum mass shear strengths of bouldery colluvium deduced from back-analyses of cut slopes in the Mid-levels, Hong Kong

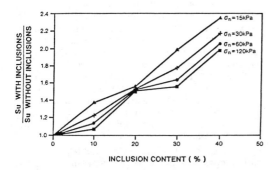

Figure 20. Results of laboratory direct shear tests carried out in Hong Kong to measure the effects of inclusions on shear strength

Frequently the range of variability of the parameters remains extraordinarily large even after completion of exploration and testing that costs as much as the project can possibly afford. "

FACTORS OF SAFETY

Because residual slopes commonly occur in steep natural terrain, it is often necessary to work to very low factors of safety in order to avoid the excessive costs of adopting higher values. In these circumstances, it is logical to relate design safety factors to the consequences of failure in terms of risk of casualties and risk of damage to property and services. Additionally, there is the public perception of acceptable levels of danger

and risk, with the inevitable political consequences.

Figure 18 shows 'accepted' and 'marginally accepted' risk levels for the USA as a function of probability and consequence of failure (Whitman, 1984). To this, the author has added a zone to represent some of the older highway cut slopes in Hong Kong, which are probably typical of many cut slopes in residual materials worldwide which were 'designed' largely on the basis of precedent without application of rigorous slope stability analyses.

Cut slope design practice in the dense urban environment of Hong Kong is governed by the Geotechnical Manual for Slopes (Geotechnical Control Office, 1984), which specifies acceptable factors of safety. Table 5 shows the recommended factors against loss of life and against economic risk for new soil and rock cut slopes. These values are for groundwater conditions resulting from a ten-year return period rainfall. There are three risk categories in each case – 'negligible', 'low' and 'high'. The risk-to-life category reflects the likelihood of loss of life in the event of failure, while the economic risk category reflects the likely magnitude of economic loss. Typical of high risk-to-life slopes are high cut slopes adjacent to occupied buildings. An example of a negligible risk-to-life slope is one which threatens only a lightly trafficked secondary road.

Hong Kong's Geotechnical Manual for Slopes also gives guidance on preventive and remedial works to existing slopes. When analysing an existing slope to determine the extent of any preventive (i.e. before failure) or remedial (i.e. after failure) works required, the performance history of that slope can be of considerable assistance to the designer. There is, for example, an opportunity to examine the geology of the slope more closely than for an undeveloped site, and to obtain more realistic information on groundwater. The Manual maintains that the designer is therefore able to adopt with confidence factors of safety for proposed preventive or remedial works that are slightly lower than those specified in Table 5 for new works, as long as rigorous geological and geotechnical investigations are conducted (which include a thorough examination of slope maintenance history, groundwater records, rainfall records and any slope monitoring records). These reduced factors of safety are 1.2, 1.1 and 1.0 for the high, low and negligible risk-to-life categories respectively.

EFFECT OF BOULDERS ON SLOPE STABILITY

Background

As discussed above, many slopes in residual

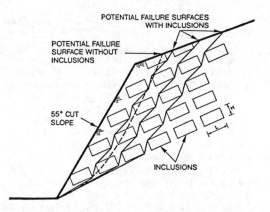

Figure 21. Theoretical model adopted to analyse the effects of inclusions on slope stability

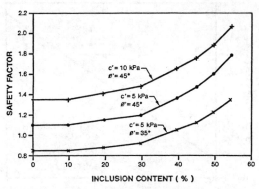

Figure 22. Typical results obtained from the theoretical model showing effects of inclusions on slope stability

materials remain stable even though stability analyses indicate that they should not be stable. In other words, theoretical factors of safety of stable slopes are often less than unity. The author can identify only one plausible reason for this in slopes for which the geology is well-defined (Table 4), namely, that the shear strength of the earth material is underestimated. This underestimate occurs because :

(a) laboratory strength measurements are carried out at stress levels which are too high (Figure 12),

(b) the critical pore pressures are over-estimated, most commonly in the form of the neglect of ever-present suctions, or

(c) the mass strength of the bouldery component of the residual profile is underestimated from shear strength tests conducted on the soil matrix alone.

Aspects (a) and (b) have been discussed above and at length in previous publications. In contrast, the mass strength of bouldery deposits has received very little attention. The small amount of published literature on the effects of inclusions on the shear strengths of soils nearly all relate to compacted fills, most commonly in connection with earth dam stability. The best known experimental studies are probably those by Holtz & Gibbs (1956), Holtz & Ellis (1961), Patwardhan et al (1970), Donaghe & Torrey (1979) and Rathee (1981).

Because of the importance of boulder content to slope stability in tropical areas, a study on this topic has been carried out recently in Hong Kong (Irfan & Tang, 1992) and the results of this will be of interest elsewhere.

Hong Kong Study

The Hong Kong study was aimed at assessing the contribution of boulders to the shear strength of colluvium and of the four Partially Weathered (PW) zones defined in Table 3 above. The scope of the study comprised a review of published data from similar studies, back analyses of some stable and failed Hong Kong cut slopes, triaxial tests and direct shear tests on compacted soil specimens containing inclusions, and the establishment and analysis of an appropriate theoretical slope model. The results of this work will be briefly summarized here.

A considerable number of stable and failed cut slopes were back-analysed in the Hong Kong study in an attempt to compare the mass shear strength of the bouldery slope materials with the shear strengths measured in laboratory tests on the matrix material. In all cases, it was assumed that the factor of safety was unity under the critical pore pressure conditions, which meant that the back-analysed mass shear strength was a lower bound value.

A typical set of results obtained from the back analyses is shown in Figure 19 for slopes in the Mid-levels area of Hong Kong, on which very detailed geotechnical investigations have been undertaken for many years. It will be seen that the deduced minimum shear strengths are generally appreciably above the effective strength envelope obtained from CU triaxial tests on the granitic soil matrix.

The results of some of the direct shear tests are shown in Figure 20. The compacted specimens consisted of crushed granitic aggregate (nominally 5 to 6.3 mm) embedded in a silty sand matrix with nominal particle size of 1.18 mm. The initial density was kept constant at 1.35 Mg/m³.

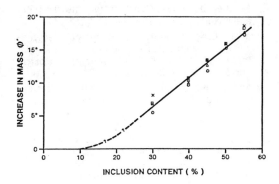

Figure 23. Summary of the theoretical model results showing effects of inclusions expressed as increases in mass angle of shearing resistance

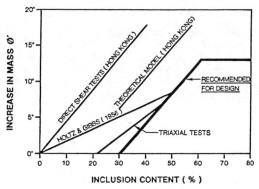

Figure 24. Summary of the Hong Kong study results, and the preliminary recommendation for mass strength increases for future slope analysis

A theoretical slope model was established to provide a direct quantitative assessment of the role of inclusions (boulders) on slope stability (Figure 21). The presence of regular inclusions causes the lengthening of potential failure surfaces, and the effect of this on the calculated factor of safety was computed by means of the Morgenstern & Price (1965) method of analysis. Very comprehensive analyses were conducted. The variables examined included inclusion size, shape, and spacing, as well as matrix strength and slope height. The slope angle was kept constant at 55°, which is fairly typical for Hong Kong cut slopes.

Typical sets of results obtained from the theoretical model for three soil matrix strengths are shown in Figure 22. From the analyses carried out, it was found that the normalized factor of safety (i.e. ratio of F with inclusions to F without inclusions) is, for all practical purposes, independent of the matrix strength. It was also found to be virtually independent of inclusion dimensions and inclusion shape.

The results from the theoretical model are summarized in Figure 23 in terms of increase in the mass angle of shearing resistance with inclusion content. These deduced ϕ' values are obviously very significant for stability assessments of slopes.

The Hong Kong study has culminated in a preliminary recommendation for mass shear strength increases to be used for slope stability analyses to take account of boulder content. This conservative recommendation is depicted in Figure 24 in comparison with the results from the direct shear tests, triaxial tests and theoretical model analyses; also shown are the results obtained by Holtz & Gibbs (1956). This recommendation could be of considerable

engineering and economic significance in Hong Kong and elsewhere.

CONCLUSIONS

The climatic conditions in tropical areas are largely responsible for the many slope instability problems that exist in steep terrain. Heavy rainfall is the main agent of failure, rapid infiltration of water occurring because of the high permeabilities of residual materials and colluvium. The heterogeneity of the residual profiles and the random nature of the rainfall mean that the application of theoretical methods of stability analysis is problematical. In particular, the determination of the mass shear strength and the prediction of the critical pore pressures are difficult, and these are the main reason why the overall state-of-the-art of stability analysis in tropical residual profiles must be judged as poor.

Correlations between the pattern of rainfall and the occurrence of slope failures are useful. The author is of the view, however, that the role of antecedent rainfall has been wrongly assessed in most such published correlations. The high quality Hong Kong data proves that short-term intensity is the only rainfall parameter of importance, and the author feels that this is also probably true in other geographical locations where the slope materials are of relatively high permeability.

The study carried out recently in Hong Kong to examine the effects of boulders on the stability of residual and colluvial slopes should be of interest elsewhere. It provides a good basis for correcting the measured shear strength of the matrix material of a bouldery slope to arrive at a mass angle of shearing resistance.

REFERENCES

ABMS (1985a). Proceedings of the First International Conference on Geomechanics in Tropical Lateritic and Saprolitic Soils, Brasilia, 4 vols, ABMS, Brazilian Soc. Soil Mech., Sao Paulo, 1758 p.

ABMS (1985b). Peculiarities of Geotechnical Behaviour of Tropical Lateritic and Saprolitic Soils. (ISSMFE Technical Committee on Tropical Soils, Progress Report 1982-1985). ABMS, Brazilian Soc. Soil Mech., Sao Paulo, 470 p.

Aboshi, H. (1979). Failure of granite slopes in Chugoku District under heavy rain: shearing strength at failure. Jour. Natural Disaster Science, (1), No. 1, 77-87.

Addison, K. (1987). Debris flow during intense rainfall in Snowdonia, North Wales: A preliminary survey. Earth Surface Processes and Landforms, (12), 561-566.

Anderson, M.G. & Burt, T.P. (1977). Automatic monitoring of soil moisture conditions in a hillslope spur and hollow. Jour. Hydrol., (33), 27-36.

ASCE (1971). Session 11: Slope Stability in Residual soils. Proc. 4th Panam. Conf. SMFE, San Juan, Puerto Rico, (2), 105-167, and (3), 125-146.

ASCE (1982). Proceedings of the ASCE Specialty Conference on Engineering and Construction in Tropical and Residual Soils, Honolulu, 746 p.

Balkema, A.A. (1988). Proceedings of the Second International Conference on Geomechanics in Tropical Soils, Singapore. A.A. Balkema, Rotterdam, 2 vols, 641 p.

Barata, F.E. (1969). Landslides in the tropical region of Rio de Janeiro. Proc. 7th Int. Conf. SMFE, Mexico City, (2), 507-516.

Bhandari, R.K., Senanayake, K.S. & Thayalan, N. (1992). Pitfalls in the prediction on landslide through rainfall data. Proc. 6th Int. Symp. Landslides, Christchurch, (2), 887-890.

Blight, G.E. (1977). Slopes and excavations in residual soils. Proc. 9th Int. Conf. SMFE, Tokyo, (3), 582-590.

Blight, G.E. (1988). Keynote paper: Construction in tropical soils. Proc. 2nd Int. Conf. Geomech. in Tropical Soils, Singapore, (2), 449-467.

Brabb, E.E. & Harrod, B.L. (Eds.)(1989). Landslides: Extent and Economic Significance. A.A. Balkema, Rotterdam, 396 p.

Brand, E.W. (1981). Some thoughts on rain-induced slope failures. Proc. 10th Int. Conf. SMFE, Stockholm, (3), 373-376.

Brand, E.W. (1982). Analysis and design in residual soils. Proc. ASCE Spec. Conf. Engineering and Construction in Tropical and Residual Soils, Honolulu, 89-143.

Brand, E.W. (1985a). Geotechnical engineering in tropical residual soils. Proc. 1st Int. Conf. Geomechanics in Tropical Lateritic and Saprolitic Soils, Brasilia, (3), 23-99.

Brand, E.W. (1985b). Predicting the failure of residual soil slopes. (Theme Lecture). Proc. 11th Int. Conf. SMFE, San Francisco, (5), 2541-2578.

Brand, E.W. (1988). Some aspects of field measurements for slopes in residual soils. Proc. 2nd Int. Symp. Field Measurements in Geomechanics, Kobe, Japan, (1), 531-545.

Brand, E.W. (1989). Occurrence and significance of landslides in Southeast Asia. Landslides: Extent and Economic Significance, ed. E.E. Brabb & B.L. Harrod, 303-324. A.A. Balkema, Rotterdam.

Brand, E.W., Dale, M.J. & Nash, J.M. (1986). Soil pipes and slope stability in Hong Kong. Qtrly Jour. Eng. Geol., (19), 301-303.

Brand, E.W. & Phillipson, H.B. (1984). Site investigation and geotechnical engineering practice in Hong Kong. Geotech. Eng., (15), 97-153.

Brand, E.W. & Phillipson, H.B. (Eds.)(1985). Sampling and Testing of Residual Soils - A Review of International Practice. Scorpion Press, Hong Kong, 194 p.

Brand, E.W., Phillipson, H.B., Borrie, G.W. & Clover, A.W. (1983). In-situ direct shear tests on Hong Kong residual soils. Proc. Int. Symp. Soil and Rock Investigations by In-situ Testing, Paris, (2), 13-17. (Discussion, (3), 55-56).

Brand, E.W., Premchitt, J. & Phillipson, H.B. (1984). Relationship between rainfall and landslides in Hong Kong. Proc. 4th Int. Symp. Landslides, Toronto, (1), 377-384.

Burt, T.P. (1978). An automatic fluid-scanning switch tensiometer system. British Geomorpholog. Res. Group., Tech. Bull. No. 21, 1-30.

Campbell, R.H. (1975). Soil slips, debris flows and rainstorms in Santa Monica mountains and vicinity, Southern California. US Geol. Survey, Prof. Paper 851, 55 p.

Campos, L.E.P. de & Menzes, M.S.S. (1992). A proposed procedure for slope stability analysis in tropical soils. Proc. 6th Int. Symp. Landslides, Christchurch, (2), 1351-1355.

Columbian Geotechnical Society (1989). Proceedings of the Southamerican Symposium on Landslides, Paipa, Columbia, 2 vols. Columbian Geotechnical Society, Bogota.

Cowland, J.W. & Carbray, A.M. (1988). Three cut slope failures on relict discontinuities in saprolitic soils. Proc. 2nd Int. Conf.

Geomech. in Tropical Soils, Singapore, (1), 253-258.

Crozier, M.J. (1969). Earthflow occurrence during high intensity rainfall in eastern Otago. Eng. Geol., (3), 325-334.

Crozier, M.J. (1986). Climate triggering of landslide episodes. Landslides: Causes, Consequences and Environment, by M.J. Crozier, 169-192. Crown Helm Pub., Auckland.

Crozier, M.J. & Eyles, R.J. (1980). Assessing the probability of rapid mass movement. Proc. 3rd Aus.-NZ Conf. Geomech., Wellington, (2), 47-51.

D'Appolonia, E.R., Alperstein, R. & D'Appolonia, D.J. (1966). Behavior of a colluvial slope. Proc. Conf. Stability and Performance of Slopes and Embankments, Berkeley, Cal., 489-515. (Reprinted Jour. Soil Mech. Found. Div., ASCE, (93), SM4, 447-493, 1967).

De Mello, V.F.B. (1972). Thoughts on soil engineering applicable to residual soils. Proc. 3rd S.E. Asian Conf. Soil Eng., Hong Kong, 5-43.

Deere, D.U. & Patton, F.D. (1971). Slope stability in residual soils. Proc. 4th Panam. Conf. SMFE, San Juan, Puerto Rico, (1), 87-170.

Donaghe, R.T. & Torrey, V.H. (1979). Scalping and replacement effects on strength parameters of earth-rock mixtures. Proc. 7th Europ. Conf. SMFE, Brighton, (2), 29-34.

Eyles, R.J. (1979). Slip-triggering rainfalls in Wellington City, New Zealand. New Zealand Jour. Sc., (22), 117-121.

Eyles, R.J., Crozier, M.J. & Wheeler, R.H. (1978). Landslips in Wellington City. New Zealand Geographer, (34), 2, 58-74.

Fredlund, D.G. (1981). The shear strength of unsaturated soils and its relationship to slope stability problems in Hong Kong. Hong Kong Engr, (9), 4, 37-45.

Fredlund, D.G. (1991). How negative can pore-water pressures get? Geotechnical News, September 1991, 44-46.

Fredlund, D.G. & Rahardjo, H. (1985). Theoretical context for understanding unsaturated residual soil behaviour. Proc. 1st Int. Conf. Geomechanics in Tropical Lateritic and Saprolitic Soils, Brasilia, (1), 295-306.

Fukuoka, M. (1980). Landslides associated with rainfall. Geotech. Eng., (11), 1-29.

Geotechical Control Office (1984). Geotechnical Manual for Slopes (2nd ed.). Geotech. Control Office, Hong Kong, 295 p.

Geotechnical Control Office (1988). Geoguide 3: Guide to Rock and Soil Descriptions. Geotechnical Control Office, Hong Kong, 189 p.

Greenway, D.R., Anderson, M.G. & Brian-Boys, K.C. (1984). Influence of vegetation on slope stability in Hong Kong. Proc. 4th Int. Symp. Landslides, Toronto, (1), 399-404.

Guidicini, G. & Iwasa, O.Y. (1977). Tentative correlation between rainfall and landslides in a humid tropical environment. Bull. IAEG, No. 16, 13-20.

Hencher, S.R., Massey, J.B. & Brand, E.W. (1984). Application of back analysis to some Hong Kong landslides. Proc. 4th Int. Symp. Landslides, Toronto, (1), 631-638.

Ho, D.Y.F. & Fredlund, D.G. (1982). Increase in strength due to suction for two Hong Kong soils. Proc. ASCE Spec. Conf. Engineering and Construction in Tropical and Residual Soils, Honolulu, 263-295.

Holtz, W.G. & Ellis, W. (1961). Triaxial shear characteristics of clayey gravel soils. Proc. 5th Int. Conf. SMFE, Paris, (1), 143-149.

Holtz, W.G. & Gibbs, H.J. (1956). Triaxial shear tests on pervious gravelly soil. Jour. Soil Mech. Fdn Div., ASCE, (82), 1-22.

Hutchinson, J.N. (1992). Landslide hazard assessment. (Theme Lecture). Proc. 6th Int. Symp. Landslides, Christchurch, (3), in press.

Irfan, T.Y. & Tang, K.Y. (1992). Effect of Coarse Fractions on the Shear Strength of Colluvium. Geotechnical Engineering Office, Hong Kong, Special Project Report (draft), 217 p. (Unpublished).

Irfan, T.Y. & Woods, N.W. (1988). The influence of relict discontinuities on slope stability in saprolitic soils. Proc. 2nd Int. Conf. Geomech. in Tropical Soils, Singapore, (1), 267-276.

Janbu, N. (1954). Application of composite slip surfaces for stability analysis. Proc. Europe. Conf. Stability of Earth Slopes, Stockholm, (3), 43-49.

Janbu, N. (1973). Slope stability computations. Embankment Dam Engineering (Casagrande Volume), ed. R.C. Hirschfield & S.J. Poulos, 47-107. Wiley, New York.

Jones, J.A.A. (1981). The Nature of Soil Piping - A Review of Research. Geo. Books, Norwich, UK, 315 p.

Kim, S.K., Hong, W.P. & Kim, Y.M. (1992). Prediction of rainfall-triggered landslides in Korea. Proc. 6th Int. Symp. Landslides, Christchurch, (2), 989-994.

Koo, Y.C. (1982a). Relict joints in completely decomposed volcanics in Hong Kong. Canad. Geotech. Jour., (19), 117-123.

Koo, Y.C. (1982b). The mass strength of jointed residual soils. Canad. Geotech. Jour., (19), 225-231.

Leroueil, S. & Tavanas, F. (1981). Pitfalls of

back analyses. Proc. 10th Int. Conf. SMFE, Stockholm, (1), 185-190. (Discussion, (4), 603-604).

Lumb, P. (1962). Effect of rain storms on slope stability. Proc. Symp. Hong Kong Soils, Hong Kong, 73-87.

Lumb, P. (1975). Slope failures in Hong Kong. Qtrly Jour. Eng. Geol., (8), 31-65.

Lumb, P. (1979). Statistics of natural disasters in Hong Kong, 1884-1976. Proc. 3rd Int. Conf. Applications of Statistics & Probability in Soil & Structural Engineering, Sydney, (1), 9-22.

Massey, J.B. & Pang, P.L.R. (1988). General report: Stability of slopes and excavations in tropical soils. Proc. 2nd Int. Conf. Geomech. in Tropical Soils, Singapore, (2), 573-579.

Moh, Z.C. (Ed.)(1969). Proceedings of the Specialty Session on Engineering Properties of Lateritic Soils, 7th Int. Conf. SMFE, Mexico City, 2 vols., 419 p.

Morgenstern, N.R. & de Matos, M. (1975). Stability of slopes in residual soils. Proc. 5th Panam. Conf. SMFE, Buenos Aires, (3), 369-384.

Morgenstern, N.R. & Price, V.E. (1965). The analysis of the stability of general slip surfaces. Geotechnique, (15), 79-83.

Murray, L.M. & Olsen, M.T. (1987). Colluvial slopes - A geotechnical and climatic study. Proc. 2nd Int. Conf. Geomech. in Tropical Soils, Singapore, (2), 573-579.

Nilsen, T., Taylor, F.A. & Brabb, E.E. (1976). Recent landslides in Alameda County, California (1940-71): An estimate of economic losses and correlations with slope, rainfall and ancient landslide deposits. US Geol. Survey, Bull. 1398, 24 p. & 1 drg.

Onodera, T., Yoshinaka, R. & Kazama, H. (1974). Slope failures caused by heavy rainfall in Japan. Proc. 2nd Int. Cong. IAEG, Sao Paulo, (2), V.II.1-V.II.10.

Oyagi, N. (1984). Landslides in weathered rocks and residual soils in Japan and surrounding areas. Proc. 4th Int. Symp. Landslides, Toronto, (3), 1-31.

Patton, F.D. & Hendron, A.J. (1974). General report on "Mass Movements". Proc. 2nd Int. Cong. IAEG, Sao Paulo, (2), V-GR.1-V.GR.57.

Patwardhan, A.S., Rao, J.S. & Gaidhane, R.B. (1970). Interlocking effects and shearing resistance of boulders and large size particles in a matrix of fines on the basis of large scale direct shear tests. Proc. 2nd SEAsian Conf. Soil Eng., Singapore, 265-273.

Peck, R.B. (1969). Advantages and limitations of the observational method in applied soil mechanics. Geotechnique, (19), 171-187.

Peck, R.B. (1975). General introduction to the session "Landslides and Their Prevention". Proc. 4th Guelph Symp. Geomorphology, Guelph, Ontario, 133-136.

Pierson, T.C. (1983). Soil pipes and slope stability. Qtrly Jour. Eng. Geol., (16), 1-11.

Pierson, T.C., Iverson, R.M. & Ellen, S.D. (1992). Spatial and temporal distribution of shallow landsliding during intense rainfall, southeastern Oahu, Hawaii. Proc. 6th Int. Symp. Landslides, Christchurch, (2), 1393-1398.

Premchitt, J., Brand, E.W. & Phillipson, H.B. (1985). Landslides caused by rapid groundwater changes. Groundwater in Engineering Geology, ed. J.C. Cripps et al, 87-94. Geol. Soc., London.

Rathee, R.K. (1981). Shear strength of granular soils and its prediction by modelling techniques. Jour. Inst. Engrs India, (62), 64-70.

Rossetti, R. & Ottone, C. (1979). Esame preliminaire delle condizioni pluviometriche dell'Oltrepo Pavese e dei valori critici delle precipitazioni in relazione ai fenomeni di dissesto franoso (Preliminary analysis of rainfall in Oltrepo Pavese and the critical rainfall values as related to landslide failures). Geologia Applicata e Idrogeologia, (14), 3, 83-99.

St John, B.J., Sowers, G.F. & Weaver, Ch. E. (1969). Slickensides in residual soils and their engineering significance. Proc. 7th Int. Conf. SMFE, Mexico City, (2), 591-597.

Sweeney, D.J. (1982). Some in situ soil suction measurements in Hong Kong's residual soil slopes. Proc. 7th S.E. Asian Geotech. Conf., Hong Kong, (1), 91-106. (Discussion, (2), 93-96).

Sweeney, D.J. & Robertson, P.K. (1982). Slope stability in residual soils in Hong Kong. Canad. Geotech. Jour., (19), 521-525.

Vargas, M. (1967). Design and construction of large cuttings in residual soils. Proc. 3rd Panam. Conf. SMFE, Caracas, (2), 243-254.

Whitman, R.V. (1984). Evaluating calculated risk in geotechnical engineering. (7th Terzaghi Lecture). Jour. Geotech. Eng., ASCE, (110), 145-188.

Wieczorek, G.F. (1987). Effect of rainfall intensity and duration on debris flows in central Santa Cruz Mountains, California. Debris Flows/Avalanches: Process, Recognition and Mitigation, ed. J.E. Costa & G.F. Wieczorek, 93-104. Geol. Soc. Amer., Boulder, Colorado.

Wolle, C. & Hachich, W. (1989). Rain-induced landslides in southeastern Brazil. Proc. 12th Int. Conf. SMFE, Rio de Janeiro, (3), 1639-1642.

Yagi, N., Yatabe, R. & Enoki, M. (1990). Prediction of slope failure based on amount of rainfall. Proc. Japan Soc. Civ. Engrs, No. 418, 65-73. (In Japanese).

Zhang, N. & Sheng, Z. (1992). Probability analysis of rain-related occurrence and revival of landslides in Yunyang - Fengjie area in East Sichuan. Proc. 6th Int. Symp. Landslides, Christchurch, (2), 1077-1083.

Landslides, Bell (ed.) © 1995 Balkema, Rotterdam, ISBN 90 5410 032 X

Theme report

L.D.Wesley

Department of Civil Engineering, University of Auckland, New Zealand

1. INTRODUCTION

The papers allocated to this session have been divided into the six categories shown in Table 1. This division is somewhat arbitrary but it provides a basis for identifying and discussing particular aspects of slope instability and landslide behaviour in tropical areas. It should perhaps be appreciated that while slope behaviour in tropical areas does involve a number of specific or unique features, it also involves factors common to other areas. As a consequence, some of the issues

Table 1

CATEGORY	PAPER NO	TITLE	AUTHORS
1.INFLUENCE OF HUMAN ACTIVITIES AND LAND USE.	1	Human factor in landslides and related phenomena.	Sinnatamby and Kandasammy
	2	Urban landslides as a consequence of old mining in Ouro Preto, Brazil	Sobreira
	3	Mechanism of natural slope instability in Cuaratingueta City, Brazil	Tsutiya and Macedo
	4	Increased debris flow activity due to vegetation change.	DeGraff
2.INFLUENCE OF RAINFALL AND PORE WATER PRESSURE.	5	Unprecedented landslides in granitic maintains of Southern Thailand	Phien-wej, Tan Zhibin and Zin Aung
	6	A proposed procedure for slope stability analysis in tropical soils	L.E.P. de Campos and Menezes
	7	Unsaturated colluvium over rock slide in a forested site in Rio de Janeiro, Brazil	T.M.P. de Campos, Andrade and Vargas
	8	Spatial and temporal distribution of shallow land-sliding during intense rainfall, south-eastern Oahu, Hawaii.	Pierson, Ellen and Iverson
3.BACK ANALYSIS STUDIES.	9	A statistical approach to cut slope instability problems in Peninsular Malaysia.	Othman, Hassan, Aziz and Anderson
	10	Study of slope stability in tropical regions utilising back analysis.	Queiroz and Gaioto
	11	Slope stability study along the Malaysian Kuala Lumpur Seremban expressway.	Amin, Taha, Salleh, and Ahmad
4.LANDSLIP HAZARD ZONING.	12	Elaboration of a map of stability of slopes in tropical regions through geotechnical mapping at low cost.	Albrecht, Vecchiato and Zuquette
	13	Photo interpretation for landslide hazard mapping.	Sithamparapillai and Thayalan
5.DESCRIPTIONS OF FAILURES.	14	A survey of slope failures along the Senawang - Air Keroh highway, Negri Sembilan/Melaka, Malaysia.	Tan
	15	Deep seated landslide of a granulite residual soil.	Magalhaes, Dias, Presa and Campos
6.REMEDIAL WORK	16	The structural control within a landslide in Rio de Janeiro.	Amaral and Porto

considered and discussed in this session may also be dealt with in other sessions. I have attempted in this report to try to focus on those aspects which tend to be the most characteristic of tropical areas. Comments will be made in turn on each of the categories listed in Table 1. For ease of reference I have numbered the papers 1 to 16 as indicated in the table.

2. INFLUENCE OF HUMAN ACTIVITIES AND LAND USE.

On first reading of all the 16 papers allocated to this session the topic which seemed to gain greatest emphasis was the influence which human activity has on slope stability. Although only four papers deal specifically with this topic, it is of indirect significance in a number of other papers. High rates of population growth in many tropical countries, together with the desire for rapid economic development is clearly putting enormous pressure on the environment in such countries, especially with respect to land use and land stability.

The specific factors mentioned in the four papers include the following:

(a) Deforestation
(b) Mining
(c) Inappropriate cultivation
(d) Discharge of sewage and wastewater into the ground
(e) Poor quality house construction

The first paper (by Sinnatamby and Kandasammy) presents a rather frightening picture of the effects of deforestation in particular areas of Sri Langka. Both erosion and mass landslide movement appear to be major problems.

The second paper relates specifically to uncontrolled residential settlement in an old mining area in Brazil, with the result that the population lives in a situation of constant landslide threat. The danger is aggravated by particular geological features of the site. The third paper, also from Brazil, deals with unplanned occupation of hill slopes in a particular part of Brazil, and further emphasises some of the points made in the second paper.

The fourth paper in this section describes a particular storm event and associated landslides in Southern Thailand in November 1988. Some 370 people were killed, 50,000 homes destroyed and extensive damage done to agricultural crops as a result of this storm. The author analyses the incidence of slides in both areas of forest cover and in areas of cultivation and concludes that the incidence of slips (or "debris flows") was more widespread in areas of cultivation than in areas of forest.

Although perhaps not specifically stated, it appears that the authors of these four papers agree that the effects of removal of vegetation cover is to allow increased infiltration into the ground with accompanying rise in pore water pressure, and it is this latter factor which is the predominant influence in initiating failure. This brings us to the second group of papers.

3. INFLUENCE OF RAINFALL AND PORE WATER PRESSURE

An interesting fact which emerged as I went through these papers was that papers No.4 and 5 deal with the same event (a severe storm in Southern Thailand) but have been prepared independently and actually come to different conclusions. Paper No.5 presents considerable statistical data which does not show any clear correlation between incidence of landslides and type of vegetation cover, and the authors conclude that "under this extremely heavy downpour -- the susceptibility to failures of the slopes was irrespective of vegetation cover".

Papers 6 and 7, both from Brazil, have some similarities in that they both raise the question of the pore pressure state in the slopes under consideration. Paper 6 deals specifically with partly saturated slopes, or slopes in which pore water tension (suction) plays a significant role in maintaining stability, and the ground water table remains permanently well below the zone in which instability could occur. For this situation the authors suggest that the stability analysis should be based on shear strength parameters from tests on samples at their natural water content, without any attempt to saturate these samples. Based on experience, the authors believe this procedure gives appropriate safety factors. Tests on these samples will generally give higher c and ϕ values than tests on saturated samples; these higher values reflect the influence of the pore water suction, and thus their use in analysis can be considered to take account of the suction which in the field contributes toward the stability of such slopes. Presumably, in the analysis, the pore pressure is take as zero, although this is not specifically stated in the paper. The alternative, of using strength parameters from saturated or soaked samples leads to low safety factors which are not compatible with actual field situations. In conjunction with the above method, the authors recommend that suitable surface drainage measures be adopted to prevent rainfall infiltration which would tend to invalidate the results of their analysis as it would alter the apparent effective shear strength parameters of the soil.

Paper 7 deals with quite a different situation to that analysed in paper 6 and the method proposed in paper 6 would clearly not be applicable to the slip described in paper 7. The latter paper deals with a shallow translational type of slip involving colluvium slipping on a hard rock surface. In this situation, positive pore water pressures build up in the colluvium, which during peak infiltration correspond approximately to seepage flow parallel to the surface, with the water table close to the surface. The use of shear strength parameters obtained from saturated samples is then appropriate and appears to lead to safety factors close to unity (i.e. in keeping with the field situation).

Paper No.8 is perhaps slightly out of place in this section as it has considerable similarity with papers 4 and 5. It describes intense landsliding in the island of Oahu in Hawaii caused by a sever rainstorm at the end of 1988. More than 400 landslides occurred, 80% of which developed into debris flows. Interesting points made in this paper are:

(a) The landsliding began within hours of the onset of the storm; in such situations time available for warning residents is very limited.

(b) The landslides were concentrated in the uppermost parts of the slopes; 75% occurred within 100m of ridge crests. No explanation was offered for this.

4.   BACK ANALYSIS STUDIES

The three papers here describe three totally different exercises in back analysis.

Paper No.9 presents a rather novel statistical approach which rejects height and slope inclination as controlling factors and in their place uses a number of empirical parameters related to the shape of the cutting and the hill slope above it. The method is applied to cut slopes in weathered granite in Malaysia where the upper part of the cuttings consists of residual soil and the lower part consists of granite in varying stages of weathering. The method has the advantage that it takes account of topographical features likely to influence the seepage situation in the slope (and thus the pore water pressures), which is undoubtedly a very important factor in determining stability. There is insufficient detail in the paper to be able to fully evaluate the method. It seems strange that slope height and inclination do not appear in the parameters; it is also surprising that the proportion of the cut in sound or reasonably sound rock is not taken account of.

Paper No.10 describes a more conventional approach to back analysis. This involves a procedure by which "average" values of the soil parameters c' and ϕ' are obtained. This procedure is not fully detailed but is based on earlier work by Hoek. Once the values of c' and ϕ' are obtained, these are used to obtain a family of curves of slope height versus slope angle for a range of values of factor of safety. These curves are of course valid only for the particular ground water condition
assumed in deriving them. The value of ϕ' obtained from the back analysis is only 19.6° which seems far too low for the relatively coarse grained soils involved. The value of ϕ' of 30.5° obtained from triaxial testing would appear to be much closer to the value expected for these materials.

Paper 11 utilises a three dimensional procedure together with an assumption of a purely cohesive soil (i.e. a total stress, $\phi = 0$ analysis). The method is applied to five particular slips which occurred in surprisingly low cuttings (heights ranged from 2.4m to 3.5m). The back analysis yields undrained strength values ranging from 3.3 to 7.3kPa. The reason for using an undrained analysis is not explained; it seems unlikely that cuts in residual soils would behave in an undrained manner.

5.   LANDSLIDE HAZARD MAPPING

Papers 12 and 13 describe techniques for hazard mapping in Brazil and Sri Langka respectively. The latter paper is of interest as it emphasises the use of a variety of photographic techniques in addition to conventional vertical stereopairs. In particular it mentions stereopairs taken obliquely and produced in colour. Such photos may provide much useful information not brought out by the conventional photos.

6.   DESCRIPTION OF FAILURES

Paper 14 gives an account of a study of a substantial number of slope failures along a particular highway in Malaysia. The conclusion from the study is that there is a distinct correlation between instability and material type; instability is mostly associated with graphitic schist. Paper 15 describes a reasonably detailed investigation of one particular landslide in a cut slope in a granulite residual soil. This investigation involved shear strength measurements as well as measurements of the ground water table, although data on the latter does not appear to be very comprehensive. The results of the analysis highlight the difficulties of fitting measured parameters to the actual situation.

7.   REMEDIAL WORK

The final paper (No.15) describes drainage measures to stabilise a major slip in Rio de Janeiro. The geology of the slip area is complicated by the presence of several dykes, which have an important influence on the slip behaviour and on the drainage measures designed to help stabilise the slip. Unfortunately, the diagrams accompanying this paper do not enable the reader to gain a clear picture of the landslide topography and geology, or of the remedial drainage measures.

S5 Landslides in Australasia

Glissements de terrain en Australasie

Landslides, Bell (ed.) © 1995 Balkema, Rotterdam, ISBN 90 5410 032 X

Keynote paper: Landslides in Australia

Robin Fell
School of Civil Engineering, University of New South Wales, Sydney, N.S.W., Australia

ABSTRACT: Landsliding in Australia is extensive and occurs mainly in areas underlain by tertiary basalt, tertiary and cretaceous sediments, interbedded sedimentary and coal measure formations. Typical landslide mechanisms and hydrogeological conditions in these environments are described. Maps showing the location of landslides, and a detailed bibliography are presented.

1 INTRODUCTION

In their letter of invitation to present this theme address, the Organising Committee left the exact scope and content of the paper open.

After some consideration I decided that I should cover:
• the location and extent of landsliding
• the geological, hydrological and geotechnical conditions which are conducive to landsliding, and typical mechanisms of landsliding
• a detailed bibliography.

The paper largely deals with landslides in soil and weathered rock and does not include instability in mining operations.

About 70 practitioners were approached, seeking published and unpublished information. More than 30 responded, and it is considered that the result is that most important information has been accessed. Clearly there will have been omissions, and the author apologises to those who may feel their efforts were not adequately recognised.

2 LOCATION OF LANDSLIDING, AND AVAILABLE REPORTS AND PAPERS

Australia is a large country, and it is necessary to divide it into smaller areas to describe the conditions. Figure 1 shows those areas which are largely determined by state boundaries, but modified to account for similar geological conditions in South East Queensland and Northern New South Wales.

Figures 2, 3, 4 and 5 show locations where landsliding has been observed.

Also shown on Figure 1 are those landslide locations which lie outside the areas in Figures 2, 3, 4 and 5. From this it will be apparent that most of the country is not affected by landsliding. There is good reason for this in that for much of the country:
• the topography is very flat
• the rainfall is very low (less than 250mm/year)
• the population is low, so landsliding would not be noted even if it was present.

Figures 2 to 5 show the occurrence of tertiary basalt. Also shown in Figure 3 is the Sydney Basin sediments, in Figure 4 the quaternary basalts, tertiary sediments and cretaceous sediments, and in Figure 5 tertiary fluvial deposits. These are included because of their importance to slope instability. The boundaries are drawn from the Bureau of Mineral Resources 1:2,500,000 maps, and are approximate Small deposits of basalt and tertiary alluvials do not show at this scale, and reference to more detailed maps is recommended.

There are also large areas of tertiary basalt in Northern and Central Queensland. Much of this area is relatively flat so there is little natural instability.

Tables 1 to 5 briefly describe the type of landsliding, geological conditions and reference

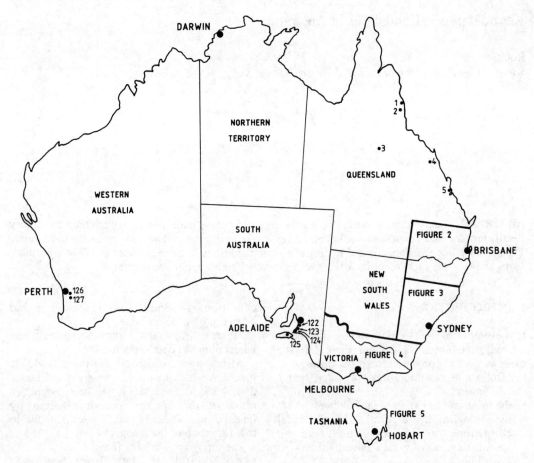

Figure 1. Map of Australia showing the boundaries of the landslide location maps in Figures 2, 3, 4 and 5.

from which more information can be obtained. The type of project affected by the landsliding is shown. For this table the following definitions of size have been adopted, based on the plan area of individual slides.

Small	$<1000m^2$
Medium	$>1000m^2$ and $<10,000m^2$
Large	$>10000m^2$ and $<50,000m^2$
Very large	$>500,000m^2$

Many of the landslides referred to are zoning studies which include a number of observed landslides. To allow description of this in Tables 1 to 5, the following terms have been used:

"a" landslide	—	when only one slide is being described
"several" landslides	—	less than 10 landslides
"many" landslides	—	10 to 100 landslides
"extensive" landslides	—	greater than 100 landslides

In most cases, reports or papers describing the landslide have been seen by the author. However, information for landslides 53 to 98 was taken from Joyce (1979) and Evans and Joyce (1974) supplemented with other reports for several of the landslides.

TABLE 1

LANDSLIDING IN AREAS NOT COVERED BY FIGURES 2 TO 5

Landslide Number	Project Description	Types of Landsliding	Geology	Reference(s)
1.	Cairns area Zoning Study	Debris slides, small medium	Granite and argillite colluvium	Hofman (1985)
2.	Palmerston Highway	Rotational and translational slides, small, medium	Basalt over granite	Simmons (1991)
3.	Flinders River Dam Site	Multiple rotational/ translational slide, very large	Cretaceous mudstone and sandstone	Eades and Eadie (1992)
4.	Eungella Range Highway	Translational, debris slide, small, medium	Granite	Litwinowicz Personal Comm.(1992)
5.	Yeppoon Bluff Roadway	Translational slide, medium	Mudstone	Litwinowicz (1989)
122.	Cowling Property, Old Norton Summit Road, Adelaide	Translational slide, medium	Metamorphic	Bell (1981) Bell and McInnes (1981)
123.	Aldinga Cliff Stability	Rotational slide, small to medium	Quaternary sediments	Proudman and Morris (1990)
124.	Sellicks Beach Coastal Landsliding	Rotational slide, large	Quaternary sediments	May and Bowman (1984)
125.	Kingscote, Kangaroo Island, Cliff Stability	Translational/ rotational, medium	Interbedded sedimentary overlain by quaternary sediments	Morris (1987)
126.	Canning Tunnel	Rotational slides and earthflows, small to medium	Granite and dolerite	Snowy Mountains Engineering Corporation (1972)
127.	Wungong Dam	Rotational slide, medium	Granite and dolerite	Lilly (1986), Fell, MacGregor and Stapledon (1992)

TABLE 2

LANDSLIDES SOUTH EAST QUEENSLAND AND NORTHERN NEW SOUTH WALES

Landslide Number	Project Description	Types of Landsliding	Geology	References(s)
6.	Rural housing, Gympie	Multiple rotational slump and flow medium to very large	Interbedded metamorphic (Amamoor Beds)	Findlay (1981)
7.	D'Aguilar Highway	Rotational/ translational slides, medium and large	Interbedded sedimentary/shale	Queensland Roads (1976)
8.	Buderim Town Zoning and Mooloolaba Road Sliding	Many rotational and translational slides, and earthflows	Basalt over tertiary sediments over interbedded sedimentary	Coffey & Partners (1981) (a), Robertson (1977) and Main Roads (1976)
9.	Mapleton – Maleny Plateau Zoning	Extensive rotational & multiple rotational slides – medium to very large and	Basalt and basalt colluvium, some over tertiary sediments	Willmott (1983)(a)
		Debris slides – small – medium	Basalt and interbedded metamorphic (Amamoor Beds)	
10.	Toowoomba City Eastern Slopes Zoning	Many rotational slides, medium to very large and Debris slides, small to medium	Tertiary basalt	Holmes (1981)

11.	Toowoomba Range Foothills Zoning	Many rotational slides, small to large and	Interbedded sedimentary (Marburg and Walloon)	Willmott (1984)(a)
		Debris slides, small to medium	as above and tertiary basalt.	Zahawi (1983)(a) and (b), (1981);
12.	Lockyer Valley Zoning	Extensive rotational and rotational/ translational slides, small to large	Interbedded sedimentary (Marburg and Walloon) lesser tertiary basalt	Zahawi & Trezise (1981)
13.	Rosewood–Marburg Zoning	Several rotational slides, small to medium, and debris slides, small.	Interbedded sedimentary (Marburg and Walloon)	Willmott (1987)
14.	Brisbane City Zoning	Several rotational slides, medium and large	Tertiary sediments	Hofman and Willmott (1984)
		Extensive rotational slides, small to medium, several large	Recent alluvial (banks Brisbane River)	Coffey and Hollingsworth (1975)
		Several/many rotational slides(?), small	Sedimentary (Tivoli) and metamorphic (Neranleigh Fernvale)	Findlay (1981) Findlay ((1988)
15.	Tamborine Mountain Zoning	Several complex rotational slides, large to very large; Many debris slides, small	Tertiary basalt Tertiary basalt and metamorphic (Neranleigh Fernvale)	Willmott (1981)
		Several rock falls	Tertiary basalt	
16.	Canungra–Beechmont–Numinbah Zoning	Many multiple rotational slides, large to very large and extensive rotational slides, small to medium; and many debris slides, small to medium	Tertiary basalt, tuff, tertiary basalt colluvium, over sedimentary	Willmott 1983(b)
17.	Raby Bay Canals	Several rotational slides, medium	Tertiary basalt and tuff	Hughes, Beal and Wright (1988–89), Coffey Partners Int.(1984–90) Fell (1988), Moon (1992)
18.	Tweed River Catchment Management Study	Extensive rotational and translational slides, small to medium	Tertiary basalt	Fell, MacGregor, McManus and Mostyn (1989)
		Many rotational/ translational slides, medium to large	Tertiary basalt	
		Several translational slides, large to very large	Interbedded sedimentary (Walloon)	
19.	Richmond River Catchment Management Study	Extensive rotational and translational slides small to medium	Tertiary basalt	MacGregor, McManus and Fell (1990)
		Several translational slides, large to very large	Interbedded sedimentary(Walloon, Grafton & Marburg)	
		Earthflow, large to very large	Tertiary basalt colluvium	
20.	Ballina Shire Zoning	Several rotational slides, medium to very large	Tertiary basalt	Coffey and Partners (1986)
21.	Brisbane – Sydney Railway instability – Kyogle to Queensland border	Several rotational and rotational/ translational slides, medium	Tertiary basalt	Arup Geotechnics 1989–90 and Coffey Partners International 1991–92
22.	Casino–Murwillumbah Railway, Haughtons Gap Tunnel	Rotational slide, medium	Tertiary basalt	Coffey Partners International (1990)(a)
23.	Casino–Murwillumbah Railway, Geotechnical Risk Assessment Review	Various	Various	Coffey Partners International (1991)(a)
24.	Grafton – Casino – Murwillumbah Railway Geotechnical Risk Assessment	Various	Various	Arup Geotechnics (1990)

TABLE 3

LANDSLIDING IN CENTRAL NEW SOUTH WALES

Landslide Number	Project Description	Types of Landsliding	Geology	Reference(s)
25.	North Coast Railway Maitland – Port Macquarie Geotechnical Risk Assessment	Various	Various	Coffey Partners International (1990)(b) Arup Geotechnics (1991)
26.	Northern Railway Ardglen	Translational/ rotational slide, medium	Tertiary basalt sediments contact	Coffey Partners International 1990(c)
27.	New England Highway, Devils Pinch	Rotational slide, medium	Interbedded of sedimentary (shale) below tertiary basalt cap	Dept. Main Roads NSW (1987)
28.	Muswellbrook to Ulan Railway, Assessment of Geotechnical problems	Various	Various	Coffey Partners International 1989(a) and (b)
29.	Landslides in the Newcastle – Lake Macquarie area review paper	Several translational and translational/ rotational slides, medium to large	Coal measures (Newcastle)	Fell, MacGregor and Sullivan (1989) and Fell Sullivan and MacGregor (1988)
30.	Speers Point Landslides	Several translational and translational/ rotational slides , medium to large	Coal measures (Newcastle)	Fell, Sullivan and Parker (1987), Chapman, Tsui and Fell (1987), Fell, Chapman and Maguire (1991), Fell (1987), Fell, Mostyn, O'Keeffe & Maguire (1988), Coffey and Partners (1983, 1984), Tsui (1987), Coffey Partners International (1990)e) and (f)
31.	Lake Macquarie City Council Area Zoning	Several translational and translational/ rotational slides, medium to large
Several small to medium rotational slides | Coal measures (Newcastle)

Interbedded sedimentary (Narrabeen) | Flentje (1991) and

Fell and Flentje (1991) |
32.	Carisbrooke Avenue	Translational slide, medium	Coal measures (Newcastle)	Forrester (1987)
33.	Gretley Colliery	Translational slide, medium	Coal measures (Newcastle)	Rigby and Carr (1987)
34.	Tickhole Tunnel Area	Two translational slides, large	Coal measures (Newcastle)	Fell, MacGregor and Sullivan (1989) Fell, Sullivan & MacGregor (1988) Coffey and Partners (1981)(b) Coffey Partners International (1990)(d)
35.	Sydney–Newcastle Freeway, Hue Hue Road, Wyong	Translational slide, large	Fine grained sedimentary (Patonga claystone)	Fell, MacGregor, Williams and Searle (1987)(a) and (b) Coffey & Partners (1983)
36.	Blackwell Mountain Woy Woy	Rotational slide, medium	Interbedded sedimentary (Narrabeen)	Coffey & Partners 1988(a)
37.	Pacific Highway Cheero Point	Rotational slide, medium	Fill and colluvium over sandstone (Hawkesbury)	Arup Geotechnics (1989)(a)
38.	Warringah Shire – urban landsliding	Many rotational slides, small to medium	Interbedded sedimentary (Narrabeen)	Burgess (1987) Walker et al. (1985)
39.	North West, West and South West Sydney Wianamatta Group Landslides	Many translational slides, medium to large	Fine grained sedimentary (Wianamatta)	Fell (1985), Jeffery (1987) DJ Douglas (1990) Soil Conservation Service of NSW (a,b,c and d) Jeffery & Katauskas (1990)

40.	Building Site / Quarry Artarmon	Translational/ rotational slide, medium	Fine grained sediamentary (Wianamatta)	Douglas et.al (1987)
41.	Brush Road, Eastwood	Translational/ rotational slide, medium	Fine grained sedimentary (Wianamatta)	Coffey & Partners (1988)(b)
42.	Razorback Range	Many rotational slides and earthflows, medium to very large	Fine grained sedimentary (Wianamatta)	Ferkh (1991), Fell (1985) Blong and Dunkerley (1976), Hazell & Kennedy 1957, Satardo (1957) Dent (1973) Longmac (1989), Main Roads (1929,1952,1971) Binns (1950), Dunkerley Dunkerley (1978),Dyer (1966)
43.	Prospect Creek bank instability	Several rotational slides, small	Alluvial over fine grained sedimentary	D.J. Douglas (1989)(b)
44.	Bells Line of Road Mt. Tomah	Rotational slide, medium	Tertiary basalt over fine grained sedimentary	Arup Geotechnics (1989)(b)
45.	Hume Highway Bendooley Hill	Several rotational slides, small to medium	Tertiary basalt over tertiary sediments over fine grained sedimentary	Golder Associates (1985)
46.	Hume Highway Cullarin Duration	Translational/ rotational slide, large	Tertiary sediments over metamorphics	Coffey & Partners (1985)
47.	South Coast Railway Line remedial works	Many rotational and translational/ rotational slides, medium to large	Fine grained sedimentary (Bald Hill claystone) and coal measures (Illawarra)	Longmac Associates (1989–91) Pitsas (1992)
48.	Lawrence Hargreaves Drive , Scarborough–Clifton	Several rotational slides, small to medium	Fine grained sedimentary (Wombarra claystone) and Interbedded sedimentary (Narrabeen)	Roads and Traffic Authority (1989)
49.	South Coast Railway Coledale Landslide	Rotational slide and debris flow	Coal measures (Illawarra)	Mostyn and Adler (1992)
50.	Coledale Area, Zoning	Many rotational and rotational/ translational slides, medium to large	Interbedded sedimentary (Wombarra claystone, Narrabeen Group) and coal measures (Illawarra)	Walker, Amaral and MacGregor (1987)
51.	Urban landslides in Woonona Heights and Fig Tree	Two translational slides, medium	Coal measures (Illawarra)	Coffey Partners International (1990)(g) and (1991)(b)
52.	Urban landslides in in Cordeaux Heights	Several rotational/ translational slides, medium to large	Coal measures (Illawarra)	Jeffery and Katauskas (1988)
53.	Shoalhaven Scheme Bendeela–Kangaroo Valley Area	Many rotational/ translational slides, large to very large	Interbedded sedimentary (Berry formation)	Snowy Mountains Engineering Corporation (1970–72)
54.	Maldon–Dombarton Railway upgrading, Wollongong	Several rotational and rotational/ translational slides, medium to large	Interbedded sedimentary (Narrabeen) and Coal measures (Illawarra)	Coffey & Partners (1984–85)
55.	Mt. Ousley Road, Wollongong	Several translational and translational/ rotational slides, medium to very large	Interbedded sedimentary (Narrabeen) and coal measures (Illawarra)	Hawkins (1991)

TABLE 4

LANDSLIDES IN VICTORIA AND SOUTHERN NEW SOUTH WALES

Landslide Number	Project Description	Types of Landsliding	Geology	Reference(s)
56.	Talbingo Dam Borrow area	Rotational slide, over volcanics	Tertiary basalt	Snowy Mountains Hydroelectric Authority (1967), Fell, MacGregor and Stapledon (1992)
57.	Tooma Dam	Translational, large	Granite	Fell, MacGregor and Stapledon (1992), Hunter (1982) Hunter and Hartwig (1962) Pinkerton & McConnell (1964)
58.	Lake Tarli Karng	Translational rock slide, large	Volcanic	Hills (1940) Spencer Jones (1971)
59.	Bogong and Dargo High Plains	Many "rock rivers" (Debris flow?)	Tertiary basalt	Talent (1965)
60.	Cobberas–Wombargo Area	Many "Rock Rivers" (Debris flow?)	Volcanic	Talent (1965)
61.	Parwan Valley Bacchus Marsh	Many rotational and translational slides, and earthflows, small to medium, and a large rotational slide	Tertiary sediments with quaternary basalt capping	Joyce and Evans (1976) Evans (1973) Forbes (1948)
62.	Dundas Tablelands, Casterton & Coleraine area	Many rotational slides and earthflows small to large	Tertiary sediments and volcanics	Joyce (1979)
63.	Valley of Moorabool River (West Branch) Bungal	Many rotational or translational slides and earthflows, small to medium (?)	Tertiary sediments and granite	Curry and Cox (1972) and Forbes (1948)
64.	Cathedral Ranges near Buxton	Earthflow	Sedimentary	Joyce (1979)
65.	Curdies River, Port Campbell	Many rotational slides and earthflows small to medium (?)	Tertiary sediments	Joyce (1979)
66.	Olivers Hill Frankston– Mt. Eliza	Several rotational and rotational/ translational slides, small to very large	Tertiary sediments	Golder Associates (1983,1990) Hills (1940) Joyce (1979)
67.	Melba Parade, Anglesea	Translational/ rotational slide and earthflow	Tertiary sediments	Neilsen et al.(1987) Joyce (1979) Neilsen (1974)
68	Lake Bullenmerri Camperdown	Rotational and translational slides and earthflow, small to large	Tertiary sediments capped by quaternary basalt	Evans (1973) Joyce and Evans (1976)
69.	Johanna River Lavers Hill, Otway Ranges	Not known	Cretaceous sediments	Medwell (1971) Joyce (1979)
70.	Coranderrk Creek Healesville	Not known	Sedimentary	Joyce (1979)
71	Cardinia Creek, Ferntree Gully	Rotational slide	Sedimentary	Joyce (1979) Downs (1963)
72.	Balcombe Bay, Mornington	Rotational(?)	Tertiary sediments	Joyce (1979)
73.	Eastern View	Many rotational/ translational slides and earthflows, medium to large (?)	Cretaceous sediments	Joyce and Evans (1976) Evans (1973) Spencer – Jones (1952)

74.	SEC No. 4 Power Station Kiewa	Not known	Metamorphic	Thomas (1946)
75.	Railway between Hamilton and Coleraine	Rotational	Tertiary sediments	Thomas (1946)
76.	Yallourn North Open Cut	Translational, large	Tertiary coal measures	Urie (1950)
77.	Mt. Dandenong Area (Mt. Kalorarma, Montrose)	Many rotational and rotational/ translational slides, small to very large, one very large, debris flow, affecting Montrose	Volcanics	Coffey Partners International (1991)(c), 1990(h) Coffey and Partners (1988)(c), Danvers– Powers (1982) McLennan (1987), Lundy Clarke (1975), Shire of Lillydale (1990) Moon, Olds and Wilson (1992) Fell (1992)
78.	Werribee Vale Bacchus Marsh	Many rotational slides and earthflows, small to medium (?)	Tertiary sediments	Evans (1973) Harding (1952)
79.	Barwon River near Forrest	Translational slide, Very large	Cretaceous sediments	Neilsen et al (1987), Joyce (1979), Knight, (1953)
80.	Road, Foster to Fish Creek	Not known	sedimentary	C.R.B. (1963) Joyce (1979)
87.	Chinamans Creek, Metung	? rotational, slide,	Tertiary sediments	Joyce (1979)
88.	4 km north–west of Colrossie	Not known, large	Tertiary sediments and basalt	CRB (1964) Joyce (1979)
83.	Warragul to Korumburra Road	Not known	Cretaceous sediments	CRB (1965) Joyce (1979)
84.	1.6 km north of Carrajung	Rotational slide, large (?)	Cretaceous sediments	CRB (1969)(a) Joyce (1979)
85.	1 km south of Carrajung	Not known	Cretaceous sediments	CRB 1969(B) Joyce (1979)
86.	Windy Point, Great Ocean Rd Lorne	Translational rock slide, medium	Cretaceous sediments	Williams and Muir (1972) Joyce and Evans (1976) Evans (1973) Neilsen et al (1987)
87.	Ben Cruachin	Earthflow	Sedimentary	Evans (1973)
88.	Freestone Creek	Translational slide and earthflow	Sedimentary	Evans (1973)
89.	Stoney Creek, Shoreham	Not known	Volcanics	Joyce (1979)
90.	Eglington Cutting Maroondah Highway Alexandra	Rockfall	Sedimentary	Joyce (1979)
91.	Glenelg River and Lower Crawford River Dartmoor area	Many rotational slides and earthflows	Tertiary sediments	Kenley (1971)
92.	East bank of Moorabool River, opposite Batesford quarry	Not known	Tertiary sediments	Joyce (1979)
93.	Coastline from San Remo to Kilcunda	Several (?) rotational slides	Cretaceous sediments	Edwards (1942)
		A medium rotational slide	Tertiary basalt over cretaceous sediments	Flintoff (1987)
94.	Greendale	Not known	Sedimentary	Joyce (1979)
95.	Hanlon Parade, Portland Joyce (1979)	Rotational slide, large	Cretaceous sediments overlain by quaternary basalt	Williams (1976)

96.	Mt. Leura Maar Crater, Camperdown	Rotational slide, and earthflow	Tertiary sediments	Joyce (1979)
97.	Wannon River, upstream from Morndal	Rotational slide and earthflow	Cretaceous sediments	Marker (1976)
98.	Minapre St, Lorne	Earthflow	Tertiary sediments	Joyce (1979)
99.	Scotts Creek Valley Heyetesbury	Translational slide, very large	Cretaceous sediments	Neilsen et al (1987)
100.	Princes Highway East Berwick Bypass	Rotational/ translational slide, medium	Tertiary on sedimentary	Flintoff (1987), Haustorfer & Flintoff (1985)
101.	Strzelecki Ranges overview	Many rotational and slides and earthflows Many rotational slides, debris flows, medium to very large	Tertiary basalt Tertiary and cretaceous sediments Cretaceous sediments	Brumley (1983)
102.	Ben Cairn Estate, Launching Place	Several rotational slides, medium to large	Volcanics	Golder Associates (1979) CRB (1969)(c) and (d)
103.	Howletts Road, Yallourn North	Translational slide, large	Cretaceous sediments overlain by tertiary sediments	McKinley and Raisbeck (1988) and McKinley and Pedler (1990)
104.	Terminal Station Richmond	Rotational slide, small	Tertiary basalt	McKinley and Pedler (1990)
105.	McKay Creek Power Station	Rotational slide, medium	Granite	McKinley and Pedler (1990)
106.	Bogong Creek Raceline	Rotational slide, medium (active) very large overall	Metamorphic (Mica schist and gneiss)	McKinley & Pedler (1990)
107.	Sugarloaf (Winnecke) dam	Translational slide, medium	Interbedded sedimentary	Stapledon and Casinader (1977), Casinader and Stapledon (1979); Casinader (1980); Fell, MacGregor and Stapledon (1992)
108.	Thomson Dam	Translational slide, large	Interbedded sedimentary	Marshall (1985) Fell, MacGregor and Stapledon (1992)
109.	Blackwood Avenue, Warburton	Translational slide, very large	Volcanics	Golder Associates (1982) CRB (1977)
110.	Plantes Hill, Mooroolbark Coffey and Partners	Several rotational slides, small (active) large overall	Tertiary basalt sediments over sedimentary	MMBW (1985–86) over tertiary (1987) MacGregor, Olds, and Fell (1990)
111.	Upper Yarra and Dandenong Ranges Zoning	Many rotational and translational slides, small to very large	Volcanics, tertiary basalt, sedimentary	Coffey and Partners (1988)(c) SCA/UYV and DRA (1981)
112.	Shire of Lillydale Zoning	Many rotational and translational slides, small to very large Debris flow, very large	Volcanics, tertiary basalt (sedimentary minor) Volcanics	Coffey Partners International (1990)(h) Shire of Lillydale (1990) Fell (1992) Lundy–Clarke (1975)
113.	Montrose Debris Flow Zoning	One historic debris flow (very large) and many potential debris flows, medium to very large	Volcanics	Coffey Partners International (1991)(c) Moon, Olds and Wilson (1992) Fell (1992)

TABLE 5

LANDSLIDING IN TASMANIA

Landslide Number	Project Description	Types of Landsliding	Geology	Reference(s)
114.	Tamar Valley Zoning	Several rotational/ translational slides, large and very large Many rotational/ translational slides, small to medium	Tertiary sediments interbedded with tertiary basalt	Knights and Matthews (1976), Sloane (1985), Telfer (1988), Stevenson (1971–84) Moore (1973–89), Sloane (1976–87) Knights (1972–77) Mines Dept Tasmania (A)
115.	Penguin Area	Rotational and translational slides, (?)	Tertiary basalt	Mines Dept. Tasmania (B)
116.	Burnie Area	Rotational and translational slides, (?)	Tertiary basalt	Mines Dept Tasmania (C)
117.	Devonport Area	Rotational and translational slides, (?)	Tertiary basalt	Mines Dept Tasmania (D)
118.	St. Helens Area	Rotational slides and earthflows, medium	Tertiary sediments	Sloane (1985) Mines Dept. Tasmania (E)
119.	Cooee Bypass	Translational and and rotational slides	Tertiary basalt and sediments over sedimentary	Baynes (1989)
120.	King River Bridge Lyell Highway	Rotational/ translational, large	Fluivo glacial	Paterson, Bayes and Bowling (1986)
121.	Bass Highway, Penguin to Howth	Rotational slides, medium	Tertiary basalt and sediments	Department of Main Roads Tasmania (1987–90)

In Figure 5, no attempt has been made to identify individual landslides or to list reports of areas investigated by the Engineering Geology and Groundwater Section of the Department of Mines, Tasmania. More than 100 reports have been prepared by them under the authorship of Stevenson (1971-84), Moore (1973-89), Weldon (1985-90), Sloane (1970-87), Moon (1979-84), Matthews (1971-89), Donaldson (1972-91), Knights (1972-77), Elmer (1971).

3 GEOLOGICAL, HYDROLOGICAL AND GEOTECHNICAL CONDITIONS CONDUCIVE TO LANDSLIDING

From Tables 1 to 5, it is apparent that there are particular geological conditions in which most of the landsliding in Australia occurs. These are:
• tertiary basalts
• tertiary and cretaceous sediments
• interbedded and fine grained sedimentary rocks
• coal measures

3.1 Tertiary basalts

In a series of reports on detailed zoning studies carried out at 1:25,000 scale by the Geological Survey of Queensland, Willmott (1981, 1983(a), 1983(b), 1984(a), and Holmes (1981) describe sliding in tertiary basalts in Southern Queensland. Willmott (1984(b)) summarises much of the information.

The basalts are volcanic lavas from the tertiary period, about 23 million years ago. The basalts in the Mapleton-Maleny and Buderim areas (#8 and #9) probably originated from volcanoes in the Glasshouse Mountains, and the others (and the basalts in the Tweed and Richmond Rivers) mainly from Mount Warning. These were multiple lava flows resulting in very thick accumulations over the

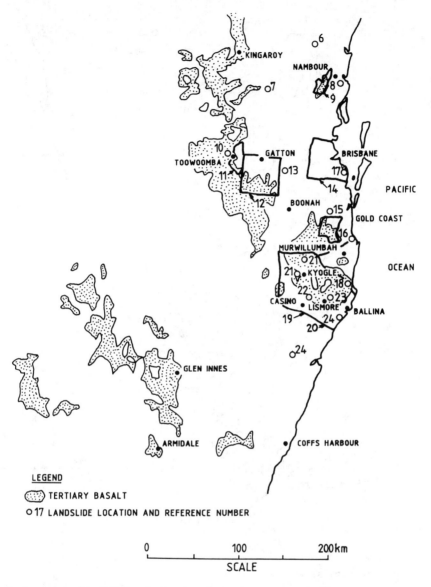

Figure 2. Location of landsliding in South East Queensland and Northern New South Wales.

buried landform, eg. 300m at Toowoomba (#10, Willmott 1984(a)) and 150m to 300m at Canungra-Beechmont-Numinbah (#16, Willmott 1983(b)). The layers of basalt have different characteristics leading to layers of higher and lower permeability, eg. thin layers would cool more rapidly leading to open fractures and high permeability. As shown in Figure 6, in some areas they are also interbedded with tuffs which, being low

permeability and low shear strength when weathered, promotes landsliding.

Willmott (1983(b)) reports that one such layer, the Hillview Rhyolite//Chinghee conglomerate, forms a distinct marker bed, consisting mainly of aglomeratic tuff, with pebbles and boulders of Rhyolite, and some fine grained sediments. This bed acts as a low permeability horizon, with high groundwater pressures above it, with many landslides.

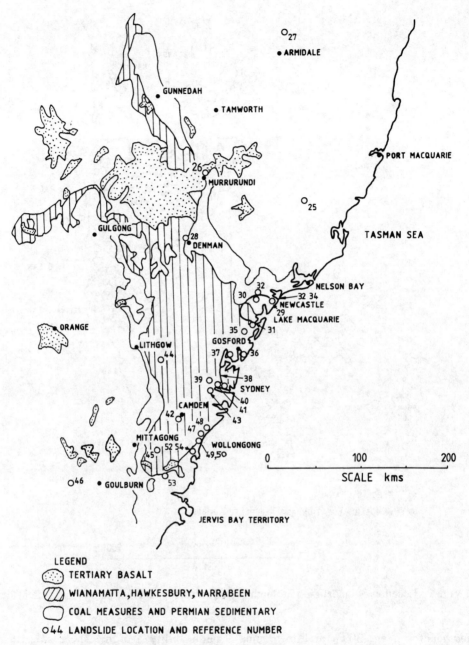

Figure 3. Location of landsliding in Central New South Wales.

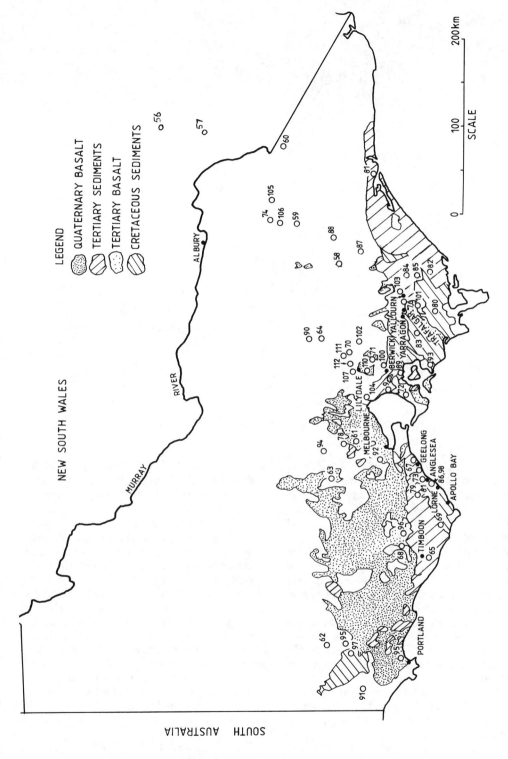

Figure 4. Location of landsliding in Victoria and Southern New South Wales.

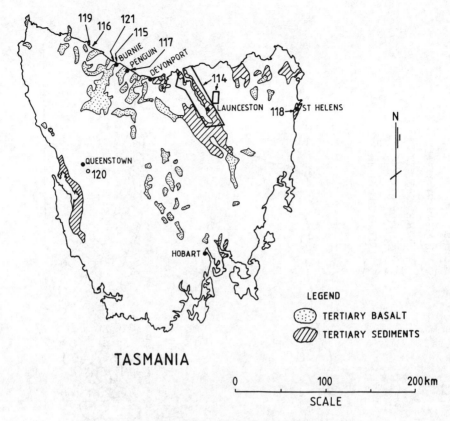

Figure 5. Location of landsliding in Tasmania.

As shown in Figure 7, in the Mapleton-Maleny area, tertiary sediments are also present, probably formed in lakes when the first basalt flows blocked streams. These are also often low permeability and strength, and promote landsliding.

The features shown on Figures 6 and 7 are interpretive and based on little detailed subsurface data. They are, however, largely consistent with the author's own experience. The important features are:

• there are large plateau areas above the slopes with deeply weathered, often laterititised basaltic soils. These are relatively permeable, and little instability occurs in these areas

• the "massive" basalt has joints which allow vertical flow of the groundwater, which emerges on the surface through more permeable layers

• the colluvial soils formed from basalt, and which overlie the basalt or other rocks

between the basalt, is lower permeability than the rock. This can lead to a buildup of pore pressure in the colluvium sufficient to cause landsliding

• tuff and tertiary sediments are often lower permeability and, when weathered, lower strength than the basalt, promoting deep rotational and translational landsliding.

Willmott (1984(b)) presents evidence that clearing of the trees has caused a raising of groundwater levels, and has led to relatively high levels of landslide activity in these areas.

Baynes (1989) has noted that layering of basalt, and interlayering of tuffs and tertiary sediments, has promoted landsliding at Cooee in North West Tasmania (#119). Figure 8 shows an interpretive section through the landsliding. As can be seen, the hydrological characteristics are similar to those noted by Willmott in Queensland.

Pells, reported in Coffey Partners International (1991-92); and Arup Geotechnics

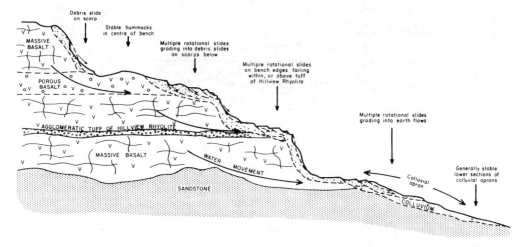

Figure 6. Interpretive hydrogeological conditions in landsliding in tertiary basalts in the Canungra-Beechmont-Numinbah area, Queensland (#16, Willmott, 1983(b)).

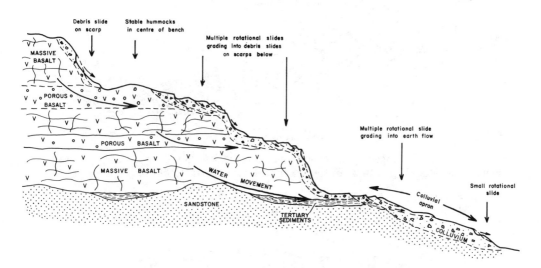

Figure 7. Interpretive hydrogeological conditions in landsliding in tertiary basalts in the Mapleton-Maleny area, Queensland (#9, Willmott, 1983(a)).

(1989-90) have noted similar features in landsliding in the tertiary basalts in Northern New South Wales (slides #18 in Table 2). Figures 9, 10 and 11 show hydrogeological conditions at a cutting failure and fill failures on the Brisbane-Sydney railway line, where extensive landsliding has occurred in cuts and fills requiring major remedial works.

In these cases, there is evidence that the basaltic colluvial soils have lower permeability than the underlying rock, and this causes build

up of pore pressures, which can lead to instability. The author's own experience in basaltic areas is consistent with these observations. Figure 12 shows typical conditions observed at Brunswick Heads, New South Wales, and in Mooroolbark, Victoria. In these areas the landsliding has occurred with relatively thin layers of basalt overlying low permeability sedimentary rocks. Groundwater flows through the open jointed fresh to slightly weathered basalt, and causes a build up of pore

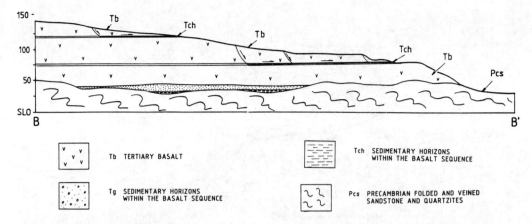

Figure 8. Interpretive hydrogeological conditions in landsliding in tertiary basalts at Cooee, Tasmania (#119, Baynes, 1989).

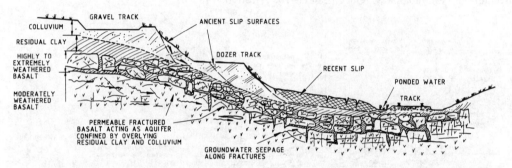

Figure 9. Interpretive hydrogeological conditions on a landslide in railway cutting near Naughtons Gap tunnel, New South Wales (#22, Coffey Partners International, 1990(a)).

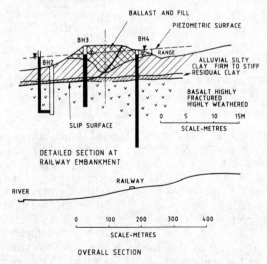

Figure 10. Recorded artesian piezometric pressures (in Brisbane-Sydney railway #21, Coffey Partners International, 1991-92).

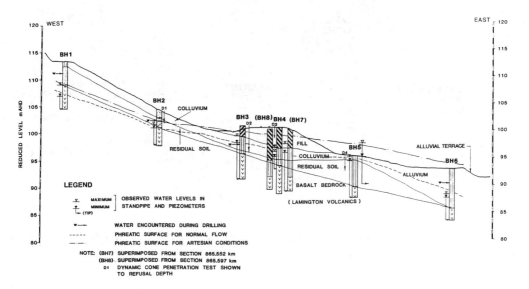

Figure 11. Section through landsliding on Brisbane-Sydney railway at Loadstone (#21, Arup Geotechnics, 1989-90).

pressure under the basaltic soil, and landsliding.

The other factor which makes basalt more susceptible to landsliding than most other geological conditions is the mineralogy of the clays (often montmorillonite is present as well as kaolinite), and the tendency of basaltic soils to be fissured.

In the Plantes Hill area, Mooroolbark (#110) the extremely weathered basalt had soil properties, but still contained joints, which gave the material the properties of a fissured clay. These were mapped in shafts, and it became apparent that in parts of the site the defects dipped out of the slope, were often polished and slickensided, but were not continuous. Details are given in MacGregor, Olds and Fell (1990). Extensive triaxial and direct shear testing led to the adoption of peak substance strength of c'=10 kPa,ϕ'=25°, residual strength c'=0,ϕ'=11°. The assessed mass strength in the zone where the defects were unfavourably oriented was c'=5kPa,ϕ'=18°, the average of peak and residual, assuming 50% continuity of defects at residual strength. Backanalysis of small failures at the site was inconclusive because piezometric conditions were not known accurately, but confirmed the design mass strength parameters were reasonable. It will be noted that this would generally be a lower strength than the fully softened (c'=0,ϕ'=25°) except at low stresses.

The author was also involved in the Raby Bay Canal project (#17), where failures of the canal banks occurred at several locations. This allowed backanalysis of two failures, one which occurred about 2 years after the canal was filled (they are constructed "in the dry" using conventional earthmoving equipment) and another in an oversteepened section during construction.

The basaltic and tuffaceous soils at the site are in places highly fissured where old land surfaces have been exposed, and fissuring is probably due to desiccation. Moon (1992) and Coffey Partners International (1990(a)). Hughes, Beal and Wright (1988-89) give details.

Testing of the most fissured clays indicated that the peak strength was in the range c'=3kPa to 11kPa,ϕ'=18° to 27° (the lower values probably being influenced by failure at least in part on fissures), the residual strength in the range c'_R=0 to 5kPa, ϕ'_R=7.5° to 11.5° (most ≈ c'_R=2kPa, ϕ'_R=8° to 10°. The backanalysis indicated a strength of c'=1kPa,ϕ'=13° was appropriate for first time slides, again much less than the fully softened strength of c'=0,ϕ'=18°. The quality of the knowledge of pore pressure in the backanalysis was not good, but adequate to strongly influence the decision to adopt the backanalysed values for design purposes.

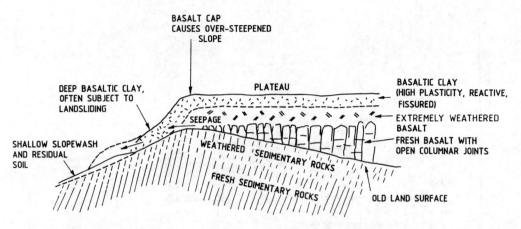

Figure 12. Typical hydrogeological conditions where basalt overlies sedimentary rocks (Fell, MacGregor and Stapledon, 1992).

Subsequent to the author's involvement, A. Moon from Coffey Partners International did extensive work in later parts of the canal development to relate the strengths to fissure orientation and continuity, and introduced the concept of a strength reduction factor. This is described in Moon (1992) and Coffey Partners International (1990(a)).

3.2 Tertiary sediments

The main landslide areas in tertiary sediments are in Victoria and Tasmania, but there are also instances of landsliding in tertiary sediments in Queensland and New South Wales. The extent of the larger tertiary deposits in Victoria and Tasmania is shown in Figures 4 and 5. As for basalts, it is important to recognise these are simplified boundaries, and other smaller areas of tertiary sediments occur.

The tertiary sediments are characterised by being relatively unconsolidated, and they behave as highly overconsolidated soils. They are often interbedding of clay rich and sand/gravel rich beds, with sliding occurring on the clays, but driven by pore pressures transmitted through the more permeable sand/gravel beds. Being overconsolidated, the clays are often fissured. One would suspect stress relief effects on stream downcutting through the sediments could lead to lower strength bedding surfaces. The tertiary sediments are often associated with tertiary basalts, and in many cases many have formed

when old landforms were blocked by basalt flows, allowing lakes to form.

The most detailed investigations of tertiary sediments have occurred in Tasmania, where extensive landsliding has occurred in the Tamar River Valley and St Helens. Knights and Matthews (1976) and Knights (1977) describe the geological conditions and types of landsliding which occur in the Tamar Valley (#114). Figure 13 shows sections through three of the landslides.

Knights and Matthews (1976) describe the tertiary sediments as lacustrine clay, silty clay, sandy clay and sand, with some lignite rich zones. They are overconsolidated but not lithified. The sediments are flat bedded or dip irregularly at low angles. They are interbedded with basalt towards the top of the 300m thick sequence. The basalt and cainozoic gravel terraces are more resistant to erosion forming steeper slopes. They also generally act as aquifers, although Sloane (1985) indicates the basalt was relatively low permeability at Beauty Point. The landslides form with a base on high plasticity clay, often with a sandy aquifer just above the clay. Shallow slides are common on the top 3m, superimposing on the main large to very large slides.

The clays often have a high montmorillonite content and plot well above the "A" line in the Casagrande plasticity chart. Backanalysis indicates residual strengths of the order of $c'=0, \phi'=12°$ to $16°$. Laboratory testing indicated slightly higher values with some residual effective cohesion. Piezometric levels

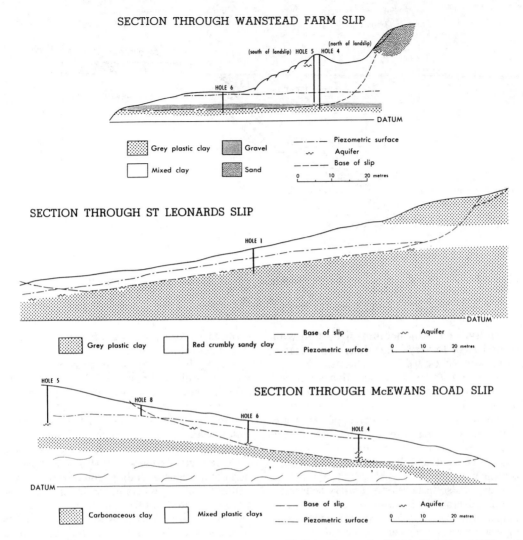

Figure 13. Section through landslides in the Tamar Valley, Tasmania (#114, Knights, 1977, and Knights and Matthews, 1976).

are rain controlled, with a lag between the onset of rain and peak values.

Sloane (1985) describes landsliding at St Helens and presents Figure 14.

The sediments are interbedded clays, sand, sand and gravel which vary laterally. Sliding is rotational, and individual slides are up to 100m wide along the sea front. Erosion by the sea is a contributtary factor, but landsliding is driven by pore pressures built up in periods of high rainfall.

As a result of the landslide activity in these areas, and in the basaltic areas of North West

Tasmania, zoning studies have been carried out and restrictions imposed on development. These are discussed in Telfer (1988) and Stevenson and Sloane (1980).

Landsliding in tertiary sediments in Victoria occurs in the Parwon Valley (#61, Joyce and Evans (1976), Evans (1973);, Dundas Tableland (#62, Joyce (1979(, Mooroobool River Valley (#63, Curry and Cox (1972)), Cathedral Ranges near Buxton (#65, Joyce (1979), Warnervale (#78, Evans (1973), Glenelg River (#96, Kenley (1971)). In all areas many rotational or translational slides and earthflows occur,

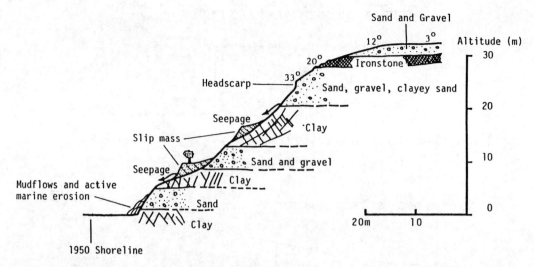

Figure 14. Section through landsliding at Chimney Heights, St Helens, Tasmania (#118, Sloane, 1985).

generally small to medium size, but with several large to very large slides are reported (at Parwon Valley and Dundas Tablelands).

At Parwon Valley, Joyce and Evans indicate that the sediments are fluviatile, horizontally bedded, and vary greatly in composition, including "incoherent boulder deposits, sands and clays, ferruginous sandstones and mudstones and almost pure limonite. They are poorly cemented and their resistance to erosion is weak".

There is also instability of tertiary sediments in the Olivers Hill, Frankston-Mt Eliza area of Port Phillip Bay (#66). Several rotational-translational landslides have occurred on a clay layer underlaying sandy clays and clayey sand. A progressive mechanism is thought to be occurring, with the slide surface at around sea level (Golder Associates 1983, 1990). The slides are large and affect residential areas.

Tertiary sediments have been an influence on construction of the Hume Highway west of Mittagong, at Bendooley Hill (#45) and Cullarin Range (#46). There was natural instability at both sites, and the highway was rerouted to avoid the tertiary sediments at Cullarin. At Bendooley Hill, tertiary basalt overlies the sediments. There has been post-construction instability of the cutting in the sediments despite adopting 3 horizontal to 1 vertical slopes. This appears to relate to low strength bedding

surfaces and/or stress relief effects. The area is still under investigation.

Sliding in tertiary sediments overlain by tertiary basalt occurs in Buderim, Queensland (#8). Figure 15(a) shows a plan of part of the area showing the high degree of activity. Figures 15(b) and (c) show some of the types of sliding which occurs in the area. The sediments are underlain by fine grained sedimentary rock.

Coffey and Partners (1981(a)) recognised a number of types of sliding, some being wholly within the basalt, others in basalt colluvium and tertiary sediments, others in tertiary sediments alone. Figure 15(b) shows sliding which is predominantly in basalt rock, residual soil or colluvium, with the toe in tertiary sediments. This sliding is usually rotational. Figure 15(c) shows sliding in the tertiary sediments overlain by a mantle of colluvium. These are predominantly translational slides, on slopes as flat as 7°.

Investigations of a slide in Buderim by the Main Roads Department (1976), Robertson (1977), showed that the tertiary sediments were horizontally bedded clays and sandy clays with sandy aquifers. Sliding occurred on clay layers beneath an aquifer in a similar way to the Tasmanian slides.

Landsliding in tertiary sediments has also been experienced in Oxley, a suburb of Brisbane (#14, Hofmann and Willmott (1984),

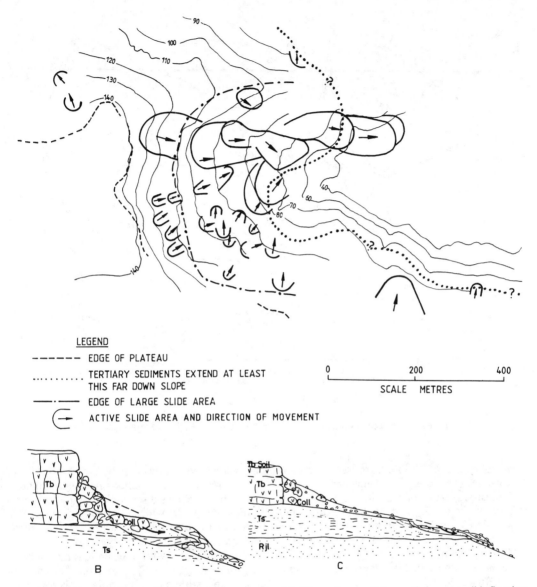

LEGEND

------- EDGE OF PLATEAU

.... TERTIARY SEDIMENTS EXTEND AT LEAST THIS FAR DOWN SLOPE

----·---- EDGE OF LARGE SLIDE AREA

⊂→ ACTIVE SLIDE AREA AND DIRECTION OF MOVEMENT

Figure 15. Landsliding at Buderim, Queensland: (a) Plan of one of the active areas, (b) Section through Type "A" sliding, (c) Section through Type "B" sliding (#8, Coffey and Partners, 1981(a).

Findlay (1981)). There is also extensive instability of the banks of the Brisbane River, with more than a hundred slides occurring after the 1974 floods of the river. These slides are in recent alluvial soils (#14, Coffey and Hollingsworth, 1975, Findlay, 1981).

3.3 Cretaceous Sediments

Extensive landsliding occurs in the cretaceous sediments of Western Victoria. Much of the larger sliding is translational, along bedding, but there is also extensive rotational sliding and earthflows. Neilson et al (1987) describe the sediments, and indicates that landslides occur in all units but are most frequent in the Otway and

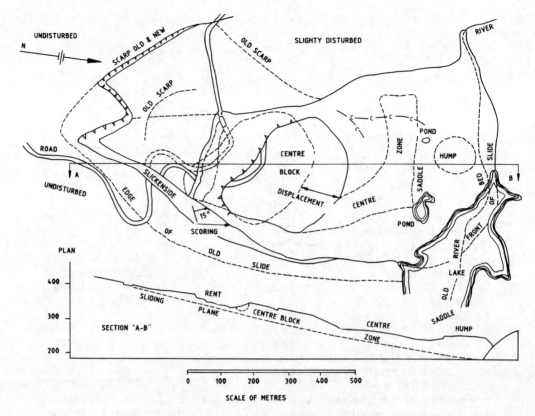

Figure 16. Plan and section of the Forrest landslide (#79, Neilson et al, 1987).

Heytesbury groups. Table 6 from Neilsen et al summarises the geology and landsliding.

Extensive rotational/translational landsliding at Eastern View (#73) is described in Joyce and Evans (1976), and translational sliding in rock at Windy Point (#36) by Joyce and Evans (1979). Williams and Muir (1972) describe remedial works carried out there on the Great Ocean Road.

Very large translational slides have occurred at Forrest (#79) and Scotts Creek (#99).

Figure 16 shows a plan and section of the Forrest landslide which occurred in June 1952, following record monthly and annual rainfall totals.

Neilson et al (1987) indicate that the slide involved 6 million m³ of Otway group sediments. Failures was along bedding which dips 12° to 20° towards the Barwon River. The river was blocked for a length of 400m by the slide which formed a natural dam 35m high. This breached in August 1953, releasing "a wall of water which was still 7m high 10km downstream." Fortunately damage was slight.

Figure 17 shows a section through the Scotts Creek landslide which Neilson et al (1987) describe as covering an area 15 hectares with 2 hectares being active recently.

Neilson et al (1987) indicate that failure is in gently dipping (5° to 7°) interbedded calcareous clays and silty clays, and minor clayey sand bands of the Gellibrand Marl, which is weathered to a depth of 9m. Failure occurred at or near the weathered unweathered contact. The author has not seen either of these landslides, but the experience would seem to indicate relatively low effective shear strengths on the bedding, possibly with the presence of bedding surface shears caused by differential movement on stiff and "soft" beds as the rocks are folded, or caused by stress relief as valleys are formed. One might question whether the Forrest slide is likely to reactivate and block the river again, with possibly worse consequences

Table 6. Lithology and relationship to landsliding — cretaceous sediments of western Victoria (adapted from Neilsen, Cooney and Dahlhaus, 1987).

Geological Group	Typical Failure Types	Slopes (°)	Lithology
Heytesbury	Earth flow	9.5.-16.5	Calcareous clay and silt
Nirranda	Translational	14	Calcareous clay and silt
Wangerrip	Rotational	19.5+	Sand and silt
Otway	Rotational/translational, Earth flow, topple	22-37	Lithic sandstone and siltstone

now there is more development downstream.

Landsliding in cretaceous clays overlain by quaternary basalts occurred in Hanlon Parade, Portland, in November 1974 (#95). Figure 18 shows a section through the landslide.

Williams (1976) describes the basalt as completely weathered and fissured. The Maretimo clay is a stiff fissured clay or silty clay. Tests show it to be montmorillonite with some fine quartz. Laboratory testing and backanalysis indicated low residual strengths for the basalt and Maretimo clay (c'=0,φ'=10°). Analysis indicated the pre-failure strength would have been between the assessed peak and residual values, and a progressive mechanism was postulated. The effect of fissuring on the mass behaviour may well have yielded a reduced strength without a progressive mechanism.

As shown in Figure 5, the cretaceous sediments also crop out in Eastern Victoria. McAuley and Raisbeck (1987) describe sliding at Howletts Road, Yallourn North (#103) in calcareous sediments overlain by tertiary sediments.

Brumley (1983) describes extensive landsliding in cretaceous sediments in the Strzelecki Ranges south of Yarragon, Trafalgar and Moe. The cretaceous sediments are overlain by tertiary sediments and tertiary basalts.

Brumley (1983) indicates that rotational (and, to a lesser extent, translational) slides and earthflows of medium to very large size occur extensively throughout all the geological units. Active small to medium slides develop on the layer features. Brumley includes maps showing the distribution of landslides and geology.

Translational sliding in cretaceous mudstones in North Queensland is described by Eades and Eadie (1992) (#3).

3.4 Interbedded and fine grained sedimentary rocks

Extensive rotational, translational and rotational/translational sliding occurs in areas underlain by interbedded fine and coarse grained sedimentary rocks (claystone, siltstone, shale, sandstone, conglomerate) and fine grained sedimentary rocks (claystone, shale, eg:

• Lockyer Valley, Queensland (#11,12,13). Zahawi (1983(a) and (b), Zahawi and Trezise (1981) and Willmott (1987) indicate 1000 landslides occurred in this area following heavy rains in 1974. Most of these slides occurred in rural areas. The slides occur in the Marburg and Walloon groups. Sliding in the Walloon is also evident in the Tweed River Valley (#18, Fell et al, 1989)

• Sydney Basin. Sliding occurs mainly in the Narrabeen group, within interbedded shale, siltstone and sandstone (Gosford, Terrigal, Warringah Shire, Port Hacking areas); within the relatively thick claystone or shale beds; eg. in the Bald Hill, Stanwell and Wombarra claystones in the Wollongong area, and Patonga claystone in the Wyong-Gosford area; and in the Wianamatta shale in the north west to south west Sydney area, and Razorback Range near Camden. Much of the sliding in the Sydney Basin has affected urban and infrastructure development (roads and railways).

There is not a great deal of detailed information on the Lockyer Valley landsliding, so the discussion on these landslides will be restricted to the Sydney Basin.

(a) Interbedded sedimentary — Gosford-Terrigal-Warringah Shire-Port Hacking

Landsliding in these areas has been investigated

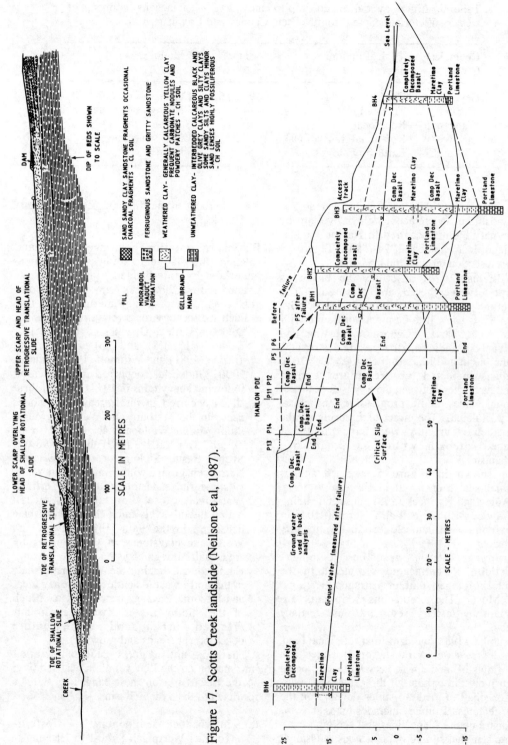

Figure 17. Scotts Creek landslide (Neilson et al, 1987).

Figure 18. Section through the landslide at Hanlon Parade, Portland (#95, Williams, 1976).

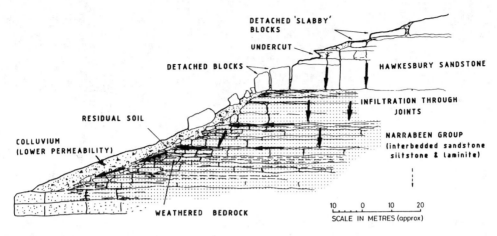

Figure 19. Typical hydrogeological conditions — interbedded sedimentary Narrabeen Group, Sydney Basin (adapted from Burgess (1987).

at hundreds of locations by consultants and government organisations. Burgess (1987) and Walker et al (1985) give overviews. Figure 19 shows typical conditions.

The colluvium often has a clayey matrix derived from weathering of the siltstone and laminate, and has lower permeability than the underlying weathered rock. Water infiltrates into the weathered rock, often from the overlying open jointed sandstone, and builds up pore pressure beneath the colluvium. There is often a slickensided shear surface at the contact of the colluvium and residual soil indicating existing landsliding and marginal instability in the natural state. Instability occurs during periods of heavy rain, where the colluvium is undercut or filled upon by house or road construction. Most sliding is small to medium rotational. It is unusual for the sliding to extend into the bedrock. Figure 20 shows a case south of Woy Woy where this did occur (#36, Coffey and Partners, 1988(a)). Sliding probably is a result of the presence of "bedding surface shear at the contact of the fine grained and coarser grained rock, and high water pressure in the subvertical joints."

(b) Fine grained sedimentary rock, Wyong-Gosford-Wollongong

Again, hundreds of sites have been investigated for housing and other development. Bowman (1972) provides an overview for the Wollongong area, but there has been a great deal of information gathered since then. Walker, Amaral and MacGregor (1987)

summarise conditions in the Coledale area north of Wollongong (#50).

Figure 21 shows the geological profile in the Coledale area (Walker et al, 1987).

Some of the most detailed investigations have been carried out on the South Coast Railway (#47, Longmac, 1989-91), Pitsis (1992) and Mostyn and Adler (1992)); the Maldon-Dombarton railway west of Wollongong (#54, Coffey and Partners 1984-85)); and Mt Ousley Road, Wollongong (#55, Hawkins, 1991-92, based on investigations by Roads and Traffic Authority).

Figure 22 is typical of the conditions which have caused extensive landsliding on the South Coast railway in the Stanwell Park area. This sliding has necessitated remedial works worth many millions of dollars.

Detailed site investigations, including use of inclinometers and survey to monitor movement vectors, show the sliding to be rotational-translational. In the case of sliding shown in Figure 22, part of the shear plane is within the claystone bedrock (Bald Hill) parallel to bedding, and may well be a bedding surface shear near the claystone/sandstone interface. Other slides in the area occur on the interface at colluvium-claystone, and probably represent remobilization of existing slide surfaces. Construction of the railway has overloaded the marginally stable slopes and, in some cases, altered surface and groundwater flow, leading to instability (eg. the Coledale slide, Mostyn and Adler (1992)).

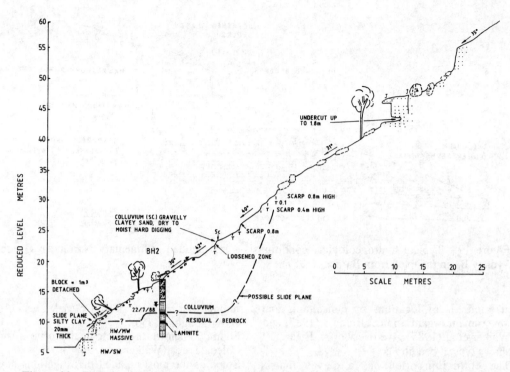

Figure 20. Sliding at Blackwell Mountain, Woy Woy (#36, Coffey and Partners, 1988(a)).

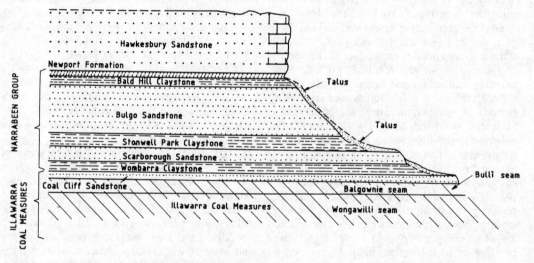

Figure 21. Geological profile in Coledale area, north of Wollongong.

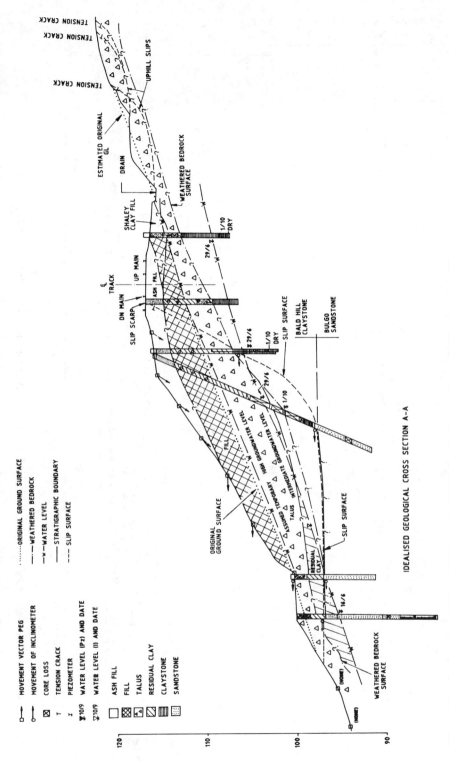

Figure 22. Section through landsliding at ch 48.75, South Coast railway (#49, courtesy of Longmac Associates and State Rail Authority of NSW, Longmac, 1989-91).

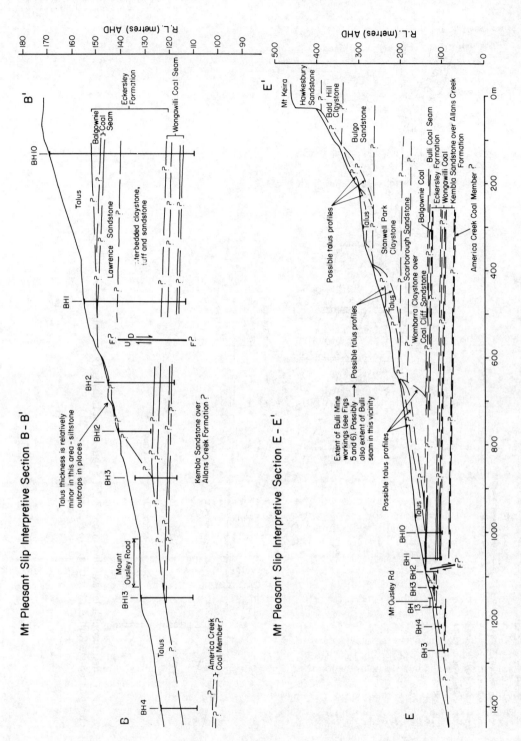

Figure 23. Interpretive sections through the Mt Ousley Road, Wollongong (#55, Hawkins, 1991).

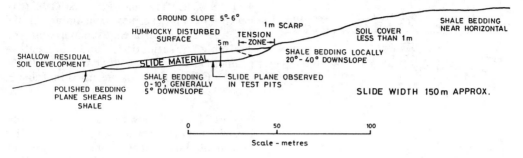

Figure 24. Section through typical landslide in Wianamatta shale, north west Sydney (Fell, 1985).

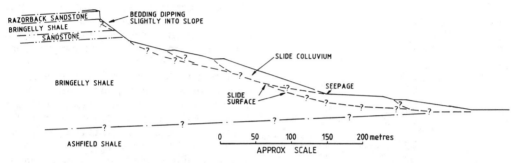

Figure 25. Interpretive section through landsliding in the Douglas Park-Menangle Road area, Razorback Range (adapted from Ferkh, 1991).

Figure 23 shows interpretive sections through sliding in the Mt Pleasant area on the Mt Ousley Road, Wollongong.

The Mt Ousley Road traverses the full geological profile from the Hawkesbury sandstone through the Narrabeen group and Illawarra Coal Measures. Large deep talus deposits cover much of the area.

These have probably formed by sliding on the Stanwell Park and Wombarra claystone, and on clay(stone) seams associated with coal seams in the Illawarra Coal Measures. Active landsliding has occurred in the downslope areas of the talus deposits, where the area has been influenced by the road and urban development.

Rockfalls are a part of the slope former weathering process in the Wollongong area, with the more resistant Hawkesbury sandstone being undercut by weathering of the claystones and shales of the Narrabeen formation. There has been recent active sliding in the Maldon-Dombarton railway area, and a very large rock

slide in the Burregorang Valley west of Camden, and other large rockfalls and slides which are described in Pells, Braybrooke, Kotze and Mong (1987). It is apparent that mine subsidence has had an influence on this activity.

Large rotational landslides have also been observed in the Bendeela-Kangaroo Valley area (#53, Snowy Mountains Engineering Corporation, 1970-72). These occur on siltstone of the Lower Berry formation.

There is no overview of landsliding in the Gosford-Wyong area, due in part to the lack of a detailed geological map of the area. Landsliding has occurred in the Patonga claystone in the Hue Hue Road cutting on the Newcastle-Sydney Freeway (#55). This is described in Fell, MacGregor, Williams and Searle (1987(a) and (b)) and Coffey and Partners (1983). Sliding occurred on low strength (c'=0,ϕ'=10° to 12°) bedding surface shears, and required remedial works worth several million dollars. The author has

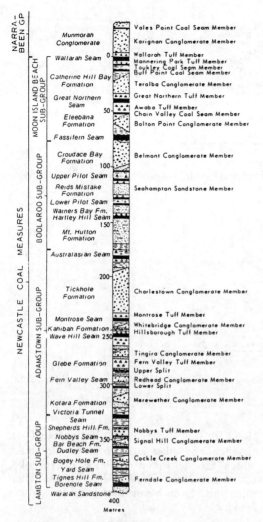

Figure 26. Stratigraphic subdivision of the Newcastle Coal Measures (reproduced from Herbert and Helby (1980)).

observed other natural sliding on the Patonga claystone within forestry reserves.

(c) Wianamatta shale — north east to south east Sydney and Razorback Range, near Camden

Again there are many sites which have been investigated for housing and development. An overview of sliding within the Wianamatta is given in Fell (1985) and Blong and Dunkerley (1976). Examples of sliding are given in Jeffery (1987), D.J. Douglas (1990), Douglas et al (1987), Coffey and Partners (1988(b)). Ferkh (1991) describes landsliding in the Razorback Range area, summarising the available information, and presenting geomorphological maps of some of the larger slides. Other references are listed in Table 3, slide #42.

Figures 24 and 25 show sections through typical landslides in the Wianamatta shale. Figure 24 is typical of those slides in the north west of Sydney (West Pennant Hills, Castle Hill, Glenhaven).

The landslides are naturally occurring and have almost certainly formed on bedding surface shears caused by local folding of the bedrock. The colluvial material is low permeability, and confines the water pressure in the relatively more permeable weathered rock beneath it.. Piezometric levels reach ground surface. Many of these slides are active in wet periods, the activity possibly being increased by removal of trees, and from the influence of urban development. They occur on quite gentle slopes (6°), indicating strengths on the slide plane of $c'=0, \phi'=12°$.

Figure 25 is typical of the larger landslides in the Razorback Range area. Work by Ferkh (1991) and by consultants such as Bruce Walker, Jeffery and Kautauskas, has shown that the model for these slides suggested in Fell (1985) is possibly incorrect. In that paper it was suggested that sliding was restricted to the upper part of the area, with flow being the dominant mechanism lower down the slope. The later work has shown that many are large to very large, deep rotational and rotational-translational slides, up to 500m from top to toe, 300m across, and 10m to 15m deep.. Earthflow does occur, but it is not the dominant mechanism intimated by the 1985 paper.

It can be seen that these landslides are quite different to those in Figure 24. Sliding is across the bedding, and on much steeper slopes. They have formed because of the more resistant sandstone beds on the top of the hill causes oversteepening of the slope. Many of the landslides are active, some moving several metres during 1990-91 which was a wet period. Assuming (as seems reasonable) that piezometric surface is at ground level, the backanalysed shear strength on the slide surface

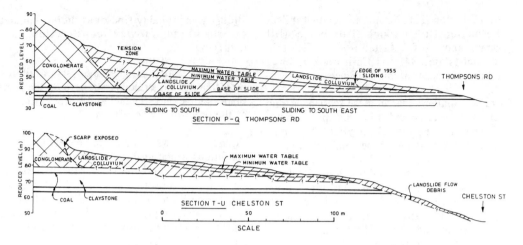

Figure 27. Sections through landsliding at Thompson Rd and Chelston St, Speers Point (#30, Fell, MacGregor and Sullivan, 1989).

is c'=0,φ'=20° to 24°, much higher than for the slides shown in Figure 24.

3.5 Coal measures

Coal measures occur extensively in Queensland, New South Wales, Victoria, and to a lesser extent Tasmania, South Australia and Western Australia.. The author is only aware of landsliding in the Sydney Basin, where the coal measures crop out as shown in Figure 3. In particular, landsliding is extensive in the Newcastle-Lake Macquarie and Wollongong areas.

The coal measures are characterised by the presence of several coal seams, interbedded with claystone, siltstone, sandstone and conglomerate. Much of the landsliding can be attributed to the presence of claystone, which is of tuffaceous origin (either as ash falling into water, or transported by water from surrounding areas), and weathered to a low strength, and often has low (effective) strength bedding surface shears present (Fell, Sullivan and MacGregor, 1989). These have formed by stress relief, or folding and faulting combined with the claystone. The coal seams often tend to act as aquifers. The coal seams often have thin (less than 50mm) clay(stone) bands in them, which are continuous and low (effective) strength. Much of the sliding is translational, along these low strength surfaces parallel to bedding which is usually near horizontal.

Landsliding in these areas has been investigated at hundreds of locations for housing and infrastructure development. It is convenient to consider the Newcastle-Lake Macquarie and Wollongong areas separately, although they are the north and south limits of the same geological strata.

(a) Newcastle-Lake Macquarie

Landsliding in this area is summarised in Fell, MacGregor and Sullivan (1989) and Fell, Sullivan and MacGregor (1988). Flentje (1991) and Fell and Flentje (1991) reviewed all geotechnical and geological information available in the Lake Macquarie City Council area, and produced maps at 1:4000 scale over most of the area showing the outcrop/subcrop of coal seams, and geotechnical zoning. It is understood that these will be available soon to consultants through Lake Macquarie City Council.

Figure 26 shows the stratigraphic subdivision of the Newcastle Coal Measures.

It is known that the majority of slope instability can be related to:
• the Booragul Tuff-Great Northern Seam-Awaba Tuff-Eleebana Formation--Fassifern Coal Seam. Instability at Speers Point (slide #30), Bareki Road (Fell, Sullivan and MacGregor, 1988) and Coal Point occur in this sequence

- Stockrington Tuff-Montrose Seam-Kahibah Formation-Hillsborough Tuff-Wave Hill Seam and Tuff. Instability at Tickhole Tunnel (slide #34) and Carisbrook Avenue (slide #32) occur in this sequence

There are smaller slides in these sequences, and other small to large slides in other parts of the coal measures, eg. the Gretley Colliery slide (#33) occurs on the top of the Nobbys Tuff.

Figure 27 shows a section through the Chelston St and Thompson Rd landslides, the former sliding in 1988 for the first time since 1950. Sliding is essentially translational, driven by water pressure since the bedding along a base of claystone is near horizontal.

Figure 28 shows sliding near the Tickhole Tunnel in Newcastle (Coffey Partners International, 1990(d)).

Landsliding at Tickhole is ancient, with recent activity, and involves a block slide along a claystone band in the Wave Hill seam. Bedding is near horizontal, and sliding is driven by water pressures in the subvertical joints in the conglomerate. An infilled tension zone up to 6m wide has formed (only a small portion of this movement has occurred in historic time). Similar translational sliding occurred on the opposite side of the railway line, and is described in Coffey and Partners (1981) and in Fell, Sullivan and MacGregor (1988). This

sliding was probably first time sliding initiated by road and railway construction, and quarrying.

Laboratory testing, and backanalysis of failures, indicates that the effective strength on the shear planes at Tickhole Tunnel and Speers Point are of the order of $c'=0,\phi'=10°$ to $12°$.

Figure 29 shows a model of the evolution of conditions which lead to the development of the Speers Point landslides. The Chelston St landslide represents a phase 3 slide, the Thompson Rd slide a phase 4 slide. While not in the same sequence, the Tickhole Tunnel slides are reasonably represented as phase 2 sliding.

(b)   Wollongong

Landsliding in this area is summarised by Bowman (1972), and parts of the area are covered in Walker, Amaral and MacGregor (1987) and Hawkins (1991). Detailed investigations of specific areas are detailed in Coffey and Partners (1984-85) for the Maldon-Dombarton Railway (#54), Coffey Partners International (1990(g)) and (1991(b)) in Woonona and Figtree (#51), Jeffery and Katauskas (1988) for Cordeaux Heights (#52), and Longmac (1989-91) for the South Coast Railway (#49)

Figure 23 shows the types of landsliding which occur:

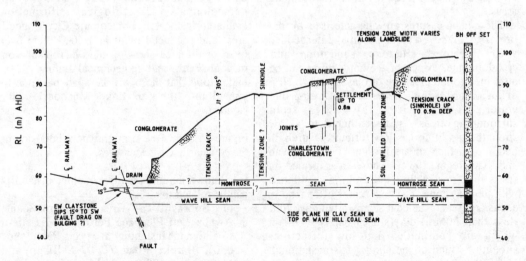

Figure 28. Landsliding near the Tickhole Tunnel, Newcastle (#34, Coffey Partners International, 1990(d)).

- translational sliding on low strength claystone layers in the coal measures
- rotational/translational sliding of colluvium/talus, largely on claystone/tuff beds.

As in Newcastle/Lake Macquarie, the coal seams often act as aquifers.

4 COMMENTS

In looking over all of the information gathered for this paper, there are some common features:

- most landslides are in soil and weathered rock, reflecting the deeply weathered profile

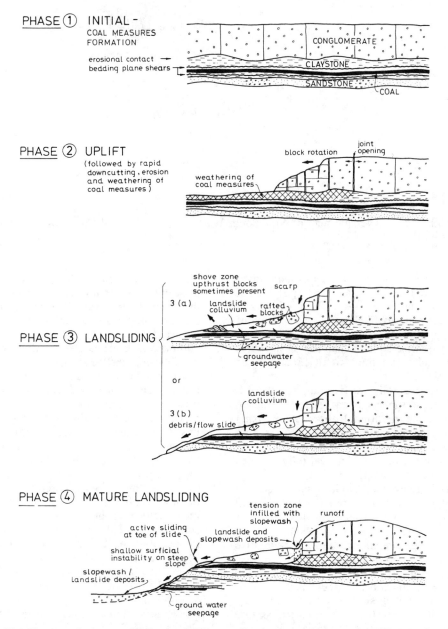

Figure 29. Slope evolution model for landsliding at Speers Point, Newcastle (Fell, MacGregor and Sullivan, 1989).

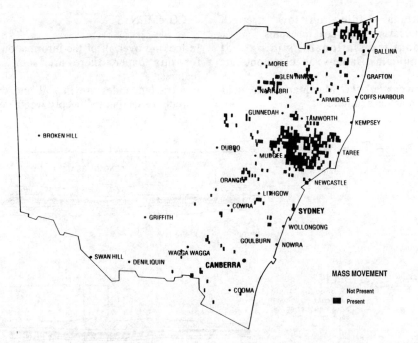

Figure 30. Mass movement map of New South Wales prepared by Soil Conservation Service of NSW (1990).

over much of the country. Most landsliding is restricted to a few geological environments

- the vast majority of sliding is reactivation of existing natural instability. There are very few first time slides which are of sufficient size and economic significance to be investigated or reported
- many of the soils are fissured, and there is evidence to suggest that at least in weathered basalts, the traditional approach of using fully softened strength to assess first time slides, may not be conservative, and strengths between residual and fully softened are more appropriate.
- similar geological environments lead to similar geotechnical and hydrogeological conditions. A knowledge of this can often lead one to the most economic way of stabilising landslides, and to recognise potentially unstable areas
- many of the sedimentary rock and tertiary sediments slides occur where there are low residual strength soils and rocks. In many cases, it is likely sliding has occurred because of the presence of bedding surface shears
- there is a lot of active sliding, which is

rainfall related, but the level of activity can be seen to have been increased by clearing of the material tree cover. This appears to have created conditions which are worse than have occurred for a long time preceding the coming of Europeans to the country, possibly since the Pleistocene

- much of the landsliding in Queensland, northern New South Wales and Victoria is in rural areas, and while there is some reduction in land productivity, and increase in sedimentation of streams, it does not have a very large direct cost, at least in the short term. It is interesting to note that a study of land degradation of New South Wales, carried out by the Soil Conservation Service of New South Wales (1988), indicated mass movement was present in 2.9% of the surface area of the state. Figure 30 shows the distribution. It will be seen that there is some relationship to the sliding shown in Figures 2 and 3, but the correlation is poor.

Figure 30 indicates the worst areas are in the Liverpool and Mount Royal ranges of the Upper Hunter Valley, in parts of the McLeay Valley, and the Richmond-Tweed river basins. The poor correlation probably relates to the

2092

absence of urban and infrastructure development in most of these areas. The Liverpool and Mount Royal Ranges are underlain by older sedimentary and metamorphic rocks. The area is relatively mountainous:

- Blong (1988) and Blong and Eyles (1989) discuss the economic significance of landsliding, detailing cases where there has been significant direct cost. Since preparing that paper, tens of millions of dollars have been spent by the State Rail Authority of NSW on the Sydney-Wollongong and Sydney-Brisbane lines. These, and remedial works on freeways at Hue Hue Road, and more recently near Mittagong and on Lawrence Hargreaves Drive north of Wollongong, tend to overwhelm the costs detailed by Blong and Eyles.

In any case, it is the author's view that the greatest costs are in the large areas of potential residential land which are effectively sterilised by being zoned as subject to the risk of landsliding. This includes large areas of basalt country in the hinterland of the Gold Coast, and areas of Newcastle-Lake Macquarie, Wollongong, Western Sydney, around the Dandenong Ranges of Melbourne, the Tamar Valley, and Bernie-Devonport areas of Tasmania.. As discussed in Fell (1992), the author is of the view that there is a need to introduce insurance cover against landsliding, as this would allow development on much of the marginal stability land, and thereby reduce much of the hidden cost of slope instability

5 ACKNOWLEDGEMENTS

The author would like to thank all those who were good enough to respond to the call for information,. Particular assistance in this was given by Fred Baynes, Ralph Rallings, Max Ervin, John Neilson, Jim Williams, Phillip Pells, Alan Moon, Patrick MacGregor, Bruce Walker, Jim Holden, Lloyd Matthews, Alex Litwinowicz, Ian Pedler, Don Raisbeck, Tony Phillips, John Morris and Peter Andrews.

6 REFERENCES

Arup Geotechnics (1988-89). Geotechnical reports to State Rail Authority of New South Wales on sliding at Loadstone, The Risks and Kyogle (unpublished).

Arup Geotechnics (1989(a)). Cheero Point landslip study, stage 2. Report for Roads and Traffic Authority of NSW (unpublished).

Arup Geotechnics (1990). Geotechnical problems, North Coast line

Arup Geotechnics (1989(b)). Slip on the Bells Line of Road, Mount Tomah. Report for Roads and Traffic Authority of NSW (unpublished).

Arup Geotechnics (1990). Geotechnical problems, North Coast line — Murwillumbah Branch line. Report for State Rail Authority of NSW (unpublished).

Arup Geotechnics (1991). Geotechnical problems, Telarah-Wauchope. Report for State Road Authority of NSW (unpublished).

Baynes, F.J. (1989). Cooee Bypass Geotechnical Assessment. HEC Enterprises Corporation Report for Department of Roads (unpublished).

Bell, D.H. (1981). Preliminary engineering geology assessment, landslide on Cowling property, Old Norton Summit Rioad, Adelaide. Report for owner (unpublished).

Bell, D.H. and McInnes, D.B. (1981). Preliminary geotechnical assessment, landslide on Cowling property, Old Norton Summit Road, Adelaide. Report for owner (unpublished).

Binns, C.S. (1950). Report on slips and landslides on State Highway No. 2, Hume Highway, Shire of Wollondilly between Camden and Picton at 26 July. DMR Report 39M.

Blong, R.J. (1988). Mass movement. Soil Conservation Service Golden Jubilee Conference, Leura, October.

Blong, R.J. and Dunkerley, D.I. (1976). Landslides in the Razorback area, New South Wales, Australia. Geografiska Annaler 58A.

Blong, R.J. and Eyles, G.O. (1989). Landslides: extent and economic significance in Australia, New Zealand and Papua New Guinea, in Landslides and Economic Significance, editors Brabb and Harrod., Balkema.

Bowman, H.N. (1972). Natural slope instability in the City of Greater Wollongong. Records Geological Survey of NSW, Dept. of Mines, Vol 14 Pt2, December.

Brumley, J. (1983). Slope stability in the Strzelecki Ranges, Victoria, in Collected

Case Studies in Engineering Geology, Hydrogeology and Environmental Geology, editors M.J. Knight, E.J. Minty, R.B. Smith, Geological Society of Australia.

Burgess, P.J. (1987). Urban slope stability in the Warringah Shire of Sydney, in Soil Slope Instability and Stabilization, editors B.F. Walker and R. Fell, Balkema.

Casinader, R.J. (1982). Systematic weak seams in dam foundations, in Balasubramaniam et al (eds). Geotechnical Problems and Practice of Dam Engineering, Balkema, 253-264.

Casinader, R.J. and Stapledon, D.H. (1979). The effect of geology on the treatment of the dam foundation interface at Sugarloaf Dam. 13th Int. Cong. on Large Dams, New Delhi, Q48, R.32, 591-619.

Chapman, T.G., Tsui, S. and Fell, R. (1987). A daily model for water-table variations on a hillslope. Symp. on Groundwater Recharge, July, Mandourah, W.A.

Coffey and Hollingsworth (1975). Bank stability study, Brisbane River, William Jolly Bridge to Bremer River junction (unpublished).

Coffey and Partners (1981(a). Landslip occurrence at Buderim Mountain. Report No B10031/AF (unpublished).

Coffey and Partners (1983). Lake Macquarie City Council slope stability and urban capability study, Speers Point, NSW. Report No 7089/1-AH, December.

Coffey and Partners (1984). Lake Macquarie City Council. Speers Point landslides. Installation of piezometers and assessment of stability. Report S7089/4-AC, December.

Coffey and Partners (1983). Freeway F3 — cutting at ch91km. Investigations of reasons for failure and design of stabilization works. Report for Dept. of Main Roads, NSW (unpublished).

Coffey and Partners (1984-85). Maldon-Dombarton Railway duplication and investigation and design of remedial works — several reports to Sinclair Knight and Partners for the State Rail Authority of NSW.

Coffey Partners International (1984-90). Raby Bay Development project reports (unpublished).

Coffey and Partners (1985). Geotechnical investigation, Cullarin Deviation Stage 2. Report for Dept. of Main Roads of NSW.

Coffey and Partners (1986). Slope instability, residential areas of Ballina Shire (unpublished report for Ballina Shire Council).

Coffey and Partners (1987). Landslip risk assessment, Mooroolbark. Report for Shire of Lillydale (unpublished).

Coffey and Partners (1987(b)). State Highway 23, Cardiff Road-Park Road stability investigation. Report to Dept. of Main Roads, NSW (unpublished).

Coffey and Partners (1988(a). Blackwall Mountain slope features, geotechnical assessment. Report for Gosford City Council (unpublished).

Coffey and Partners (1988(b). Brush Road landslide, Eastwood. Geotechnical investigation. Report for Ryde Municipal Council (unpublished).

Coffey and Partners (1988)(c)). Landslip and subsidence risk study, prepared for the Upper Yarra Valley and Dandenong Ranges Authority and reproduced by them as a Technical Report Series, Report No 27.

Coffey Partners International (1989(a). Muswellbrook to Ulan line. Assessment of geotechnical problems. Report for State Rail Authority of NSW (unpublished). Also reports on instability at ch 320km, 341km and 367km.

Coffey Partners International (1990). Geotechnical studies Stage 13, 14, 15 and part of stage 12, Raby Bay. Report for Sinclair Knight and Partners (unpublished).

Coffey Partners International (1990(a). Naughtons Gap Tunnel. Report for State Rail Authority of NSW (unpublished).

Coffey Partners International (1990(b)). Risk classification Wauchope to Telerah. Report for State Rail Authority of NSW (unpublished).

Coffey Partners International (1990(c)). Northern railway ch 362km, Ardglen. Geotechnical investigation. Report for State Rail Authority of NSW (unpublished).

Coffey Partners International (1990(d)). Sydney-Newcastle railway, slope instability near Tickhole Tunnel, Kotara. Geotechnical Report for State Rail Authority of NSW (unpublished).

Coffey and Partners (1990(e)). Assessment of land Lake Road, Cockle Creek stability assessment. Report for Pasminco Metals Sulphide (unpublished).

Coffey Partners International (1990(f)). Land capability study, Speers Point. Report for DeKagra Pty Ltd (unpublished).

Coffey Partners International (1990(g)). Arter Reserve landslide study, Figtree. Report for Wollongong City Council (unpublished).

Coffey Partners International (1990(h). Review of landslip risk with the Shire of Lillydale, prepared for the Shire of Lillydale. (unpublished)

Coffey Partners International (1991(a). Casino to Murwillumbah, geotechnical risk assessment review. Report for State Rail Authority of NSW (unpublished).

Coffey Partners International (1991(b)). Woonona Heights landslide study. Geotechnical investigation. Report for Wollongong City Council (unpublished).

Coffey Partners International (1991(c)). Shire of Lillydale. Study of the risk of debris flows and other landslips, Montrose, Victoria. Report for Lillydale Council (unpublished).

Coffey Partners International (1991-92). Geotechnical reports to State Rail Authority of NSW on sliding at The Spiral, Cougal, Loadstone, Kyogle (unpublished).

CRB (1963). Internal report, Traralgon District,, Country Roads Board, 8/1/1963.

CRB (1964). Internal report, Traralgon District, Country Roads Board, 5/10/1964.

CRB (1965). Internal report. Traralgon District, 25/1/1965.

CRB (1969(a). Internal report. Traralgon District, 20/1/69.

CRB (1969(b). Internal report. Traralgon District, 18/4/1969.

CRB (1977). Landslide affecting Blackwood Avenue, Warburton, Shire of Upper Yarra. Country Roads Board of Victoria, Report 750214.

CRB (1969(c)). Don Road landslip investigation. Country Roads Board (unpublished).

CRB (1969(d)). Landslip investigation on Don Road. Country Roads Board (unpublished).

Currey, D.T. (1952). Landslide on the East Barwon. Aqua, 3(11):18.

Currey, D.T. and Cox, F. (1972). The geology of Bungal Dam, Victoria. West Moorabool Water Board.

Danvers-Power, F. (1892). Notes on the late landslip in the Dandenong Ranges, Victoria. Report of the Fourth Meeting of the Australasian Association for the Advancement of Science, Hobart, 337-340.

Dent, B.B. (1973). The engineering geology of the Oaks-Picton district, New South Wales. University of NSW, BSc (App.Geol) Thesis (unpublished).

Department of Main Roads, NSW (1987). New England Highway. Treatment of slips on Devils Pinch, proposal for control of main landslide area. Report No G1665 (unpublished).

Department of Main Roads, Tasmania (1987-90). Reports on Bass Highway, Penguin to Howth — inland route. Geotechnical feasibility of MASG route (unpublished).

Donaldson, R.C. (1972-91). Reports on landsliding in Tasmania — Dept. of Mines, Tasmania (unpublished).

Douglas, D.J. (1989(a). Site investigation for proposed subdivision, Douglas Park. Report for Lean, Lackenby and Hayward (unpublished)..

Douglas, D.J. (1989(b). Landslips along Prospect Creek, Fairfield environs. Report for Dalland and Lucas (unpublished).

Douglas, D.J. (1990). Proposed subdivision, Buckleys Road, Winston Hills. Report for Bain Pty Ltd (unpublished).

Douglas, D.J., Maconochie, A.P. and McMahon, B.K. (1987). Slip failure in Ashfield shale at Artarmon, Sydney, in Soil Slope Instability and Stabilization, editors B.F. Walker and R. Fell, Balkema.

Downes, E.A. (1963). Speed limit 20. Pub. Aust. Railway Historical Society.

Dunkerley, D.L. (1972). The relation of slope form to mass movement near Picton, New South Wales. BA (Hons) Thesis, Macquarie University.

Dyer, D.S. (1966). The general geology of the Razorback district with special emphasis on the engineering geology of the landslides. BSc (Hons) Thesis, University of Sydney.

Eades, G.W. and Eadie, A. (1992). Flinders River damsite, case study of a feasibility investigation (to be published).

Edwards, A.B. (1942). The San Remo Peninsula. Proc. Royal Society of Victoria, 54, 59-78.

Elmer, S. (1971). Reports on landsliding in Tasmania — Dept. of Mines, Tasmania (unpublished).

Evans, R.S. (1973). Slope stability studies in Victoria with particular emphasis on the eastern Otway Ranges. BSc (Hons) Report, School of Geology, University of Melbourne (unpublished).

Evans, R.S. and Joyce, E.B. (1974). Landslides in Victoria, Australia. Vic. Nat., 91, 240-245.

Fell, R. (1985). Slope instability in the Wianamatta Group, in Pells, P.J.N., Engineering Geology of the Sydney Region, Balkema.

Fell, R. (1988). Raby Bay Canal Development. Unisearch Limited Report (unpublished).

Fell, R. (1992). Some landslide risk zoning schemes in use in Eastern Australia and their application. Sixth ISL, Christchurch, New Zealand, February.

Fell, R., MacGregor, J.P., Williams, J. and Searle, P. (1987(a)). A landslide in Patonga claystone on the Sydney-Newcastle Freeway, Geotechnique 37(3), 255-270.

Fell, R. MacGregor, J.P., Williams, J. and Searle, P. (1987(b)). Hue Hue Road landslide, Wyong, in Soil Slope Instability and Stabilization, editors B.F. Walker and R. Fell, Balkema.

Fell, R., Sullivan, T.D. and Parker, C. (1987). The Speers Point landslides, in Soil Slope Instability and Stabilization, editors B.F. Walker and R. Fell, Balkema.

Fell, R., Mostyn, G., O'Keeffe, L. and Maguire, P. (1988). Assessment of the probability of rain induced landsliding. Fifth Australia-New Zealand Conference on Geomechanics, Sydney.

Fell, R., Sullivan, T.D. and MacGregor, J.P. (1988). The influence of bedding plane shears on slope instability. Proc. Vth Intl. Symp on Landslides, ISSMFE, Lausanne, Vol 1, 129-134.

Fell, R., MacGregor, J.P. and Sullivan, T.D. (1989). Landslides in the Newcastle-Lake Macquarie area. Civil Engineering Transactions, Vol CE 31, No 1.

Fell, R., MacGregor, J.P., McManus, K.J. and Mostyn, G.R. (1989). Management of protected lands subject to mass movement in the Tweed River catchment. Unisearch

Fell, R., Sullivan, T.D. and MacGregor, J.P. (1988). The influence of bedding plane shears on slope instability. Proc. Vth Intl. Symp on Landslides, ISSMFE, Lausanne, Vol 1, 129-134.

Fell, R., MacGregor, J.P. and Sullivan, T.D. (1989). Landslides in the Newcastle-Lake Macquarie area. Civil Engineering Transactions, Vol CE 31, No 1.

Fell, R., MacGregor, J.P., McManus, K.J. and Mostyn, G.R. (1989). Management of protected lands subject to mass movement in the Tweed River catchment. Unisearch Limited Report for Soil Conservation Service of NSW.

Fell, R., Chapman, T.G. and Maguire, P. (1991). A model for prediction of piezometric levels in landslides, in Slope Instability Engineering, Institution of Civil Engineers, editor R.J. Chandler.

Fell, R. and Flentje, P. (1991). Lake Macquarie City Council, geotechnical zoning study. Unisearch Report A14082-01 for Lake Macquarie City Council (unpublished).

Fell, R., MacGregor, J.P. and Stapledon, D.H. (1992). Geotechnical engineering of embankment dams, Balkema (in press).

Ferkh, Z. (1991). Landsliding in the Razorback Range area. MAppSc Thesis, Dept. of Applied Geology, University of NSW.

Findlay, J.K. (1981). A review of landslide and slope stability hazards in Queensland. Geological Survey of Queensland, Record 1981/15.

Findlay, J.K. (1988). Landslide at Seventeen Mile Rocks Road, Oxley, April. Geological Survey of Queensland report.

Flentje, P. (1991). Geotechnical zoning study of Lake Macquarie City. MAppSc Thesis, Dept. of Applied Geology, University of NSW.

Flintoff, W.T. (1987). Recent landslips. RCA Seminar, 15th November. Road Construction Authority.

Forbes, I.G. (1948). Erosion on the Melton Reservoir catchment. State Rivers and Water Supply Commission, Victoria.

Forrester, K. (1987). The Carisbrooke Avenue landslide, in Soil Slope Instability and Stabilization, editors B.F. Walker and R. Fell, Balkema.

Golder Associates (1979). Ben Cairn Estate stability conditions. Report to Ministry for Planning (unpublished).

Golder Associates (1982). Geotechnical investigation landslip, Blackwood Avenue, Warburton. Report for Shire of Upper Yarra (unpublished).

Golder Associates (1983). Stability assessment, Ballar Creek, Mt Eliza. Report for Gutteridge Haskins and Davey (unpublished).

Golder Associates (1990). Foundation investigation, Brookwood Drive, Mt Eliza. Report for Bonacci Winward (unpublished).

Harding, A. (1952). Extensive landslide at Bacchus Marsh, Aqua 4(2):110.

Haustorfer, I.J. and Flintoff, W.T. (1985). Geotechnical influence of the design and construction of the bypass of Berwick. Australian Geomechanics News, No 9, June.

Hawkins, B. (1991). Slope instability in the Mount Ousley Road area, Wollongong. MEngSc Thesis, School of Civil Engineering, University of NSW.

Hazell, W.R. and Kennedy, G.A. (1967). Landslides on the Razorback Range. BE Thesis, University of NSW.

Hofmann, G.W. (1985). Slope stability and its constraints on closer settlement in the foothills of the Toowoomba range, Gatton Shire. Geological Survey of Queensland, Record 1984/76.

Hofmann, G.W. and Willmott, W.F. (1984). Landslide susceptibility of natural slopes in the City of Brisbane. Geological Survey of Queensland, Record 1984/10.

Holmes, K.H. (1981). Land stability of the eastern slopes of Toowoomba. Geological Survey of Queensland, Record 1981/2.

Hughes, Beal and Wright (1988-89). Raby Bay Development project reports (unpublished).

Hunter, J.R. (1982). Some examples of changes during construction, in A.S. Balasubramanian and J.S. Younger (eds) Geotechnical Problems and Practice in Dam Engineering, Balkema, 99-108.

Hunter, J.R. and Hartwig, W.P. (1962). The design and construction of the Cooma-Tumut project of the Snowy Mountains Scheme. Jour. Inst. Enggs. Aust., Vol 34, No 7-8, 163-185.

Jeffery, R.P. (1987). A case study of subsurface drains at Rogans Hill, in Soil Slope Instability and Stabilization, editors B.F. Walker and R. Fell, Balkema.

Jeffery and Katauskas (1988). Preliminary geotechnical and stability study, subdivision off Staffs Road, Cordeaux Heights. Report for Forbes, Rigby and Associates (unpublished).

Jeffery and Katauskas (1990). Construction of Grebert Road, Glenhaven. Report for K.R. Stubbs and Associates (unpublished).

Joyce, E.B. (1979). Landslide hazards in Victoria in Natural Hazards in Australia, editors R.L. Heathcote and B.G. Thom.

Australian Academy of Science, Canberra, 234-247.

Joyce, E.B. and Evans, R.S.. (1976). Some areas of landslide activity in Victoria, Australia. Proc. Royal Society of Victoria, 88, 95-108.

Knight, R.G. (1953). Landslide lake — East Barwon at Forrest: Breaching of natural dam. State Rivers and Water Supply Commission, Victoria. Investigations and Designs Branch Internal Report.

Knights, C.J. (1972-77). Reports on landsliding in Tasmania — Dept. of Mines, Tasmania (unpublished).

Knights, C.J. (1977). Piezometric measurements in tertiary lacustrine sediments in the Tamar Valley. Dept. of Mines, Tasmania, Report 1977/37 (unpublished).

Knights, C.J. and Matthews, W.L. (1976). A landslip study in tertiary sediments, Northern Tasmania. Bulletin Int. Assoc. Engineering Geology, No 14, 17-22.

Lilly, R.N. (1986). Wungong Dam landslide. Australian National Committee on Large Dams, Bulletin 74, 38-49.

Litwinowicz, A. (1989). Design of remedial works for instability at Yeppoon Bluff,. Queensland Transport Report R1671 (unpublished).

Longmac Associates (1989). Geotechnical assessment of the Picton slip, Old Hume Highway, Picton, for Roads and Traffic Authority of NSW. Longmac Ref. AGT5067.484, RTA Ref.. G1905.

Longmac Associates (1989-91). South Coast railway geotechnical investigation. Several reports covering the section from Helensburgh to Wollongong.

Lundy-Clarke, J. (1975). Mountains of Struggle. published by author. Copies in Shire of Lillydale libraries.

McDonald, M.J. (1972). Mass-wasting at Lal Lal. The Ballarat Institute of Advanced Education Studies Geological Society Newspaper, 22nd June.

MacGregor, J.P., McManus, K.J. and Fell, R. (1990). Management of protected lands subject to mass movement in the Richmond River catchment. Unisearch Limited Report for Soil Conservation Service of NSW.

MacGregor, J.P., Olds, R. and Fell, R. (1990). Landsliding in extremely weathered basalt, Plantes Hill, Victoria, in The Engineering Geology of Weak Rocks, 26th Annual Conference of the Engineering Group of the

Geological Society, University of Leeds, September.

McKinley, T. and Raisbeck, D. (1988). Monitoring an active landslide at Howletts Road, Yallourn North. Fifth Australia-New Zealand Conference on Geomechanics, Sydney.

McKinley, T. and Pedler, I. (1990). Examples of the treatment of local landslides. Australian Geomechanics Society student/graduate meeting, Monash University, July.

McLennan, J. (1987). Montrose settlement in the foothills — A History. Shire of Lillydale, Lilydale, Victoria.

Main Roads (1929). Razorback deviation. Main Roads, November, 36-40.

Main Roads (1952). Landslides on the Razorback Range and near Wollongong, 1949 and 1950. Main Roads, March, 77-83.

Main Roads (1971). Improvement to Hume Highway over Razorback Range and the problem of landslides. Main Roads, September, 4-7.

Main Roads Department (1970). Slip on Mooloolaba Road. Report R766.Marker, M.E. (1976). Soil erosion 1955 to 1974. A review of the incidence of soil erosion in the Dundas Tableland area of Western Victoria, Australia. Proc. Royal Society of Victoria, 88, 15-22.

Marshall, A.J. (1985). The stratigraphy and structure of the Thomson damsite and their influence on the foundation stability. South Australian Inst. Tech., MSc Thesis (unpublished).

Matthews, W.L. (1971-89). Reports on landsliding in Tasmania — Dept. of Mines, Tasmania (unpublished).

May, R.I. and Bowman, R.P. (1984). Coastal landslumping in Pleistocene sediments at Sellecks Beach, South Australia. Trans. R.Soc. South Australia 108(2), 85-94.

Mines Department of Tasmania (A). Several landslip zoning maps and reports, Tamar region.

Mines Department of Tasmania (B). Zoning of unstable land, Penguin area.

Mines Department of Tasmania (C). Zoning of unstable land, Burnie area.

Mines Department of Tasmania (D). Zoning of unstable land, Devonport area.

MMBW (1985-86). Plantes Hill tank. Several reports on engineering geology, soils and remedial works. Melbourne and Metropolitan Board of Works (unpublished).

Moon, A.T. (1979-84). Reports on landsliding in Tasmania — Dept. of Mines, Tasmania (unpublished).

Moon, A.T. (1992). Stability analysis in stiff fissured clay at Raby Bay, Queensland. Sixth ISL, Christchurch, New Zealand, February.

Moon, A.T., Olds, R.J. and Wilson, R.A. (1992). Debris flow risk zoning at Montrose, Victoria. Sixth ISL, Christchurch, New Zealand, February.

Moore, W.R. (1973-89). Reports on landsliding in Tasmania — Dept. of Mines, Tasmania (unpublished).

Morris, L.J. (1983). An investigation of an occurrence of subsidence in cliff face materials at Kingscote, Kangaroo Island. Dept. of Mines and Energy, South Australia. Report No 83/86.

Morris, L.J. (1987). Kingscote Cliff stability study, Kangaroo Island. Dept. of Mines and Energy, South Australia. Report No 87/95.

Mostyn, G.R. and Adler, M.A. (1992). Design of the remedial works for the Coledale landslide. Sixth ISL, Christchurch, New Zealand, February.

Neilson, J.L. (1974). Observations on the Melba Parade Land Slip, Anglesea. Geological Survey of Victoria. Report 1974/27 (unpublished).

Neilson, J.L., Cooney, A.M. and Dahlhaus, P.G. (1987). ANZ Slide '87. Field workshop on landslides, Victorian Section. August 4-5.

Paterson, S.J., Baynes, F.J. and Bowling, A.J. (1986). King River Bridge, Lyell Highway. Hydro-Electric Commission Report (unpublished).

Pells, P.J.N., Braybrooke, J.C., Kotze, G.P. and Mong, J. (1987). Cliff line collapse associated with mining activities, in Soil Slope Instability and Stabilization, editors B.F. Walker and R. Fell, Balkema.

Pinkerton, I.L. and McConnell, A.D. (1964). Behaviour of Tooma Dam. Trans. 8th Congress on Large Dams, R20, Q29, 351-375.

Pitsis, S. (1992). Slope instability in the Stanwell Park-Scarborough area. MEngSc Thesis, School of Civil Engineering, University of NSW (in preparation).

Proudman, A.R. and Morris, L.J. (1990). Cliff stability, Aldunga, South Australia. Dept. of Mines and Energy, South Australia. Report No 90/52.

Queensland Roads (1976). Construction of the Blackbutt Range section of the D'Aquilar Highway, Kilcoy-Yarraman. Queensland Roads, Vol 15, No 30. Main Roads Dept., Queensland.

Roads and Traffic Authority of NSW (1989). Several internal reports on investigations and remedial measures for repairing landslides on Lawrence Hargreaves Drive, Scarborough-Clifton.

Robertson, N.F. (1977). Stabilization of land slopes. Local Government Engineers Conference, November.

SCA/UYV and DRA (1981). Land constraints to residential development in the Upper Yarra Valley and Dandenong Ranges region, including landscape living policy areas and urban policy areas. Upper Yarra Valley and Dandenong Ranges Authority.

Shire of Lillydale (1990). Development in areas of possible slope instability. Resident Information Guide.

Simmons, J.V. (1991). Influences of geological detail on the time dependent instability of a cut slope in residual soils. Proc. Third Int. Conf. on Tropical and Residual Soils, Lesotho.

Sloane, D.J. (1985). Landslide zoning at Beauty Point and St Helens, Tasmania. 12th Int. Conf. and Field Workshop on Landslides, Tokyo, August. The Japan Landslide Society.

Snowy Mountains Engineering Corporation (1970-72). Shoalhaven Shire, Bendeela-Kangaroo Valley area. Reports on geology for Metropolitan Water Sewerage and Drainage Board.

Snowy Mountains Engineering Corporation (1972). Conning Tunnel, Report on Geology. Report for Metropolitan Water Supply, Sewerage and Drainage Board, Perth, Western Australia (unpublished).

Soil Conservation Service of NSW(a). Urban Capability, West Pennant Hills study area.

Soil Conservation Service of NSW (1988). Land Degradation Survey.

Soil Conservation Service of NSW(b). Urban Capability, Third Stage Glenhaven release area.

Soil Conservation Service of NSW(c). A land use and erosion study of the outer Sydney region.

Soil Conservation Service of NSW(d). Reconnaissance urban capability survey, Elderslie-Narellan.

Spencer-Jones, D. (1952). Landslides on the Ocean Road near Eastern View. Unpublished report, Mines Dept., No 2.

Spencer-Jones, D. (1971). Unpublished report, Mines Dept., No 5.

Stapledon, D.H. and Casinader, R.J. (1977). Dispersive soils at Sugarloaf Dam site, near Melbourne, Australia. Dispersive clays, related piping and erosion in Geotechnical Projects, ASTM STP623, American Society for Testing and Materials.

Stevenson, P.C. (1971-84). Reports on landsliding in Tasmania — Dept. of Mines, Tasmania.

Stevenson, P.C. and Sloane, D.J. (1980). The evolution of a risk zoning system for landslide areas in Tasmania, Australia. Third Australia-New Zealand Conference on Geomechanics.

Sutarto, N.R. (1975). Landslides in Picton and Douglas Park area. MAppSc Thesis, University of NSW.

Talent, J.A. (1965). Geomorphic forms and processes in the Highlands of Eastern Victoria. Proc. Royal Society of Victoria, 78, 119-135.

Telfer, A.L. (1988). Landslides and land use planning. Geological Survey Bulletin 63, Tasmania, Dept. of Mines.

Thomas, D.E. (1946). Landslip near State Electricity Commission's No 4 Power Station Area, Kiewa. Mines Dept, No 14 (unpublished report).

Tsui, S.W.H. (1987). Prediction of piezometric levels in landslides. MEngSc Project Report, School of Civil Engineering, University of NSW.

Urie, R.L. (1950). Yallourn North Open Cut. Investigations into the cause of the landslide adjacent to the camp area. State Electricity Commission of Victoria, Internal Report.

Walker, B., Dale, M., Fell, R., Jeffery, R., Leventhal, A., McMahon, M., Mostyn, G. and Phillips, A. (1985). Geotechnical risk associated with "hillside development". Australian Geomechanics, No 10.

Walker, B.F., Amaral, B. and MacGregor, J.P. (1987). Slope instability in the Coledale area of the Illawarra Escarpment, in Soil Slope Instability and Stabilization, editors B.F. Walker and R. Fell, Balkema.

Weldon, B.D. (1985-90). Reports on landsliding in Tasmania — Dept. of Mines, Tasmania (unpublished).

Williams, A.F. (1976). Landslip at Hanlon

Parade, Town of Portland, geotechnical investigation (post failure). Foundation Investigations Section, Materials Research Division, Country Road Board, Report No 608075, with appendices by J.L. Neilson and H.W. Pander.

Williams, A.F. and Muir, A.G. (1972). The stabilization of a large moving rock slide with cable-anchors. Proc. Third Southeast Asian Conference on Soil Engineering, 179-187.

Willmott, W.F. (1981). Slope stability and its constraints on closer settlement on Tamborine Mountain, southeast Queensland. Geological Survey of Queensland, Record 1981/14.

Willmott, W.F. (1983(a)). Slope stability and its constraints on closer settlement on the Mapleton-Maleny plateau, southeast Queensland, Geological Survey of Queensland, Record 1983/9.

Willmott, W.F. (1983(b)). Slope stability and its constraints on closer settlement in the Canungra-Beechmont-Numinbah area, southeast Queensland. Geological Survey of Queensland, Record 1983/64.

Willmott, W.F. (1984(a)). Slope stability and its constraints on closer settlement in the foothills of the Toowoomba range, Gatton Shire. Geological Survey of Queensland, Record 1984/44.

Willmott, W.F. (1984(b)). Forest clearing and landslides on the basalt plateaux of southeast Queensland. Queensland Agricultural Journal, January-February, 15-20.

Willmott, W.F. (1987). Slope stability and its constraints on building in the Rosewood-Marburg area. Geological Survey of Queensland, Record 1987/4.

Zahawi, Z. (1981). Landslides in the Lockyer Valley. Queensland Government Mining Journal, 82(961), 529-539.

Zahawi, Z. (1983(a)). Potential landslide areas in the Lockyer Valley. Geological Survey of Queensland, Record 1983/56.

Zahawi, Z. (1983(b)). Mount Sugarloaf landslide, investigation of recent movement. Geological Survey of Queensland, Record 1983/14.

Zahawi, Z. and Trezise, D.L. (1981). Lockyer Valley landslide investigations. Geological Survey of Queensland, Record 1981/20.

Landslides, Bell (ed.) © 1995 Balkema, Rotterdam, ISBN 90 5410 032 X

Keynote paper: Landslides in New Zealand

Warwick M. Prebble
Geology Department, University of Auckland, New Zealand

ABSTRACT:

Landslides in New Zealand vary in size from small earth flows, regolith slides and spalling failures to large, complex, deep-seated slides and toppling masses. High intensity rainfall, widespread and frequent slope movement and a high rate of erosion are characteristic of the New Zealand landmass. This is a consequence of a very active tectono-geomorphic environment and a humid climate.

Oblique convergence in the plate boundary results in thrusting, shearing, earthquakes and rapid uplift. Subduction produces volcanism in the North Island.

In the axial mountain ranges closely fractured rock dilates upon uplift, creating rock mass bulges with large scale toppling, sliding and debris avalanches.

Clay seams and crush zones, probably produced by thrusting and flexural slip, act as basal ruptures for deep seated slides in young soft, weak rock. Similar foliation - parallel defects promote large slides in schist. Collapse of andesite volcanoes generates debris avalanches and lahars. Ultrasensitive clayey palaesols in ignimbrites fail as slide-flows.

Significant engineering and environmental problems result from landsliding, throughout New Zealand.

1. INTRODUCTION

Landslides in New Zealand are widespread and varied.

Slope movements of different size and mechanism are found throughout the country. They range in size from small, shallow seated earth flows and regolith slides to very large, deep seated, complex landslides and toppling failures.

Size and mechanism are not an index of the economic and social impact of these slope movements. The shallow, small, but very numerous earth flows and regolith slides on the young mudrocks of the East Coast of the North Island produced devastating soil loss, pasture damage and flood plain aggradation during recent cyclonic storms.

By way of comparison the several huge, deep seated landslide complexes in the chlorite schist rock masses of Otago, have led to the construction of toebuttressing

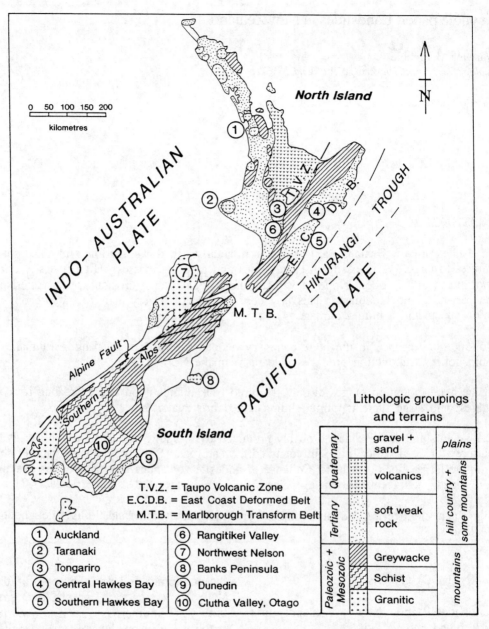

Fig. 1. Major tectonic features, main lithologic groupings and geotechnical terrains of New Zealand.

and slope drainage on a massive scale in order to reduce the risk of slope failure into the reservoir of the Clyde Dam.

In suburban Auckland, slabbing and spalling of weak mudrock in coastal cliffs has prompted a variety of retaining measures. Although small and shallow these failures are numerous and widespread.

Block slides in weak mudrock have

forced a small suburban area to be abandoned in Dunedin. Similar movements in similar rocks in the central North Island resulted in special design and construction procedures for large engineering works.

Although landslides in New Zealand are a major geologic hazard, they also contribute in a very positive sense to the landscape and earth resources of the country. For instance sector collapse of andesitic volcanoes has provided extensive ring plains in Taranaki and the central North Island. Large rock slides and rock avalanches have impounded beautiful lakes in the high country of both islands.

Nearly all of the recognisable landslides are late Pleistocene or Holocene in age. However the deposits of much older landslides are recognised in the Miocene rocks of both islands. Some examples are: huge rafts and blocks of limestone in conglomerates in Marlborough, ring plain breccias on Auckland's west coast and the very extensive allochtohonous chaos breccias of the Northland Peninsula.

This paper gives a brief description of the main types of slope movement, in relation to the geology, and provides an outline of several examples.

2. THE TECTONIC SETTING

The very reasons for New Zealand's existence as an isolated but mountainous micro continent ensure that slope movement is frequent, widespread and severe. The New Zealand landmass is elevated, steep, deeply dissected, well-watered, highly deformed and very active. This is a result of the on-going oblique convergence of the Pacific and Indo-Australian plates in this corner of the south-west Pacific Ocean (Fig. 1).

The high rate of mass movement and erosion results in serious soil loss (Eyles, 1983; Crozier, 1986).

Westward subduction beneath the North Island produces mainly rhyolitic volcanism, but also andesites and basalt. Migration of the plate boundary has left a series of Neogene volcanic arcs. From West to East these are, one which has subsided off the West Coast of the North Island, the Coromandel - eastern Northland arc and the present active Tonga-Kermadec arc. The latter arc extends on land as the Taupo Volcanic Zone. Rifting, caldera subsidence, geothermal activity, hydrothermal alteration and most recent eruptions are concentrated into this zone. However, outside it, to the west, lie the potentially active volcanic zones of Taranaki, Auckland and Kerikeri (Northland).

The Taupo Volcanic Zone (T.V.Z.) and the three western zones referred to above are Quaternary. Pyroclastic deposits and air-fall tephra are widespread, especially from sources within the T.V.Z. Andesitic volcanoes of Taranaki and the southern T.V.Z. have extensive ring plains, built up largely from lahars and sector-collapse debris avalanches. Rifting within the T.V.Z. has produced a belt of active normal faulting and shallow seismicity.

In the southern North Island and northern South Island oblique convergence in the plate boundary results in a major transform belt and a wide zone of crustal shearing. Thrusting, seismicity and uplift occur in this zone. Active faulting, much of it as major dextral faults with a reverse component, and low angle compressive thrusts, are numerous and widespread throughout the East Coast and southern North Island. (Sporli, 1980; Prebble, 1980; Pettinga, 1982; Walcott, 1987; Berryman, 1988).

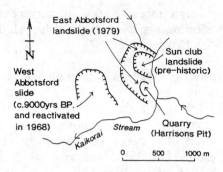

Fig. 2. Map of Kaikorai Stream and Abbotsford landslides, Dunedin, South Island. (After Coombs and Norris, 1981).

Most of the South Island is dominated by the Southern Alps and other mountain ranges. Greatest uplift occurs in these ranges, especially parallel to and just several km east of the Alpine Fault (Wellman, 1979). This is a major dextral fault, with locally complex subdivision into shear and thrust sectors. Overall dextral shift of 480 km, probably in late Neogene time is postulated on this fault which runs the length of the western side of the South Island. Uplift rates of 1 cm yr^{-1} are documented from modern geodetic data. Shearing, thrusting and uplift is widespread over a wide belt of the South Island. It is broadest in Otago, in the south, and the Marlborough transform and Nelson region in the north.

The result of the shearing, compression and rapid uplift is a very steep, high terrain formed of very closely fractured and crushed rock. Most of the uplifted undermass is either greywacke sandstone, indurated mudstone or chlorite schist. Granitic rocks in Nelson and Fiordland are also fractured and sheared. However, the very closely fractured tabular rock masses with extensive shear zones, crush zones and gouge seams typically belong to the greywacke and schist mountains. Geotechnically similar, but less abundant rock masses are found in Cretaceous to Palaeogene limestones and shales.

Seismicity within the East Coast, the transform and the alps is frequent and severe.

High intensity rain is common along the western front of the Alps. It is less frequent, along the east coast.

Around the rapidly rising spine of Mesozoic greywacke, schists and granitic rock masses, soft weak sedimentary rock is being uplifted, much of it rapidly and in recent geologic time. The soft weak sedimentary rock is most widespread in the North Island, where it is a thick cover over much of the west, centre, and north, as well as throughout the East Coast Deformed Belt. Tertiary rocks within this belt, especially Palaeogene strata, are intensely crushed and sheared, with broad zones of mélange gouge. (Pettinga, 1980, Prebble, 1987).

Within this active tectono-geomorphic setting, there are geotechnically distinct terrains. Each shows a range of slope movements from small earthflows, and larger regolith slides to very large complex deep-seated failures. However, each terrain is also dominated to some extent by one or a few types of slope movement which are major geotechnical hazards.

Five terrains are recognised (Fig. 1):-

3. SOFT, WEAK ROCK TERRAIN

Mudstone, shale and sandstone dominate this terrain which is widespread throughout the North Island and is restricted to narrow belts around the edges of the mountain ranges in the South Island. (Fig. 1).

The rocks are very weak to weak (1 to several MPa unconfined compressive strength), porous, wet and bedded.

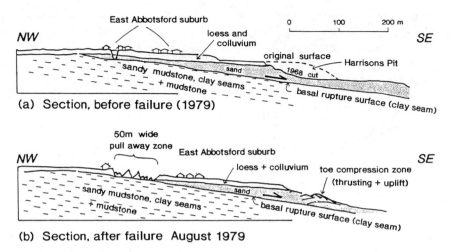

Fig. 3. Sections (before and after failure) through the East Abbotsford (1979) landslide. (After Coombs and Norris, 1981)

Stronger limestones are conspicuous in some regions (Hawkes Bay, Marlborough and western North Island). However, the limestones are interbedded or interleaved with the weak mudrocks which tend to influence the slope failure mode.

Continuous bedding-parallel clay seams and crush zones provide basal rupture surfaces for deep seated *block slides and rock slides*. Steeply dipping and intersecting tectonic joints and faults provide side shears and pull away zones. The clay seams and crush zones are of tectonic and sedimentary origin (Thompson 1981, Pettinga 1987, Prebble 1990, Wylie 1989). Current uplift rates of 5 to 8 mm yr[-1] on actively growing folds are associated with extensive areas of sliding on clay seams (Thompson 1981). In other areas deep weathering (HW to 11 m depth) has contributed to a continuous zone of *block sliding* and *regolith slide/debris flows* on soft weak mudrocks of south Auckland (Prebble 1990, Kermode 1991) (Fig. 7, "southern landslide zone").

The Abbotsford (August 1979) block slide (Coombs and Norris 1981) slid on a 25

mm thick smectite-rich clay seam for 50m downslope and opened up a large pull away graben within the East Abbotsford suburb (Fig. 2 and Fig. 3). A toe compression zone developed at the foot of the slope, where there was a pre-historic block slide (Sun club landslide) and a quarry (Harrisons Pit). An area of roughly 1/2 km^2 was involved in the slide.

Block slides, rock slides, debris slides and debris flows are widespread in the weak rock terrain. These large *complex deep seated failures* show multistage development over the last approximately 10,000 years (Stout 1977, Thompson 1981, Prebble 1987). Areas involved range up to several km^2 (Fig. 4) and volumes from a few million m^3 through tens of millions of m^3 to 1/10 km^3. Large wedge slides are also described (Pettinga 1987). Earthquake triggering of 1850 landslides, many in weak rock, is discussed by Pearce and O'Loughlin (1985).

Short displacement block slides, which have moved only several m, on a basal rupture at a few tens of m depth are

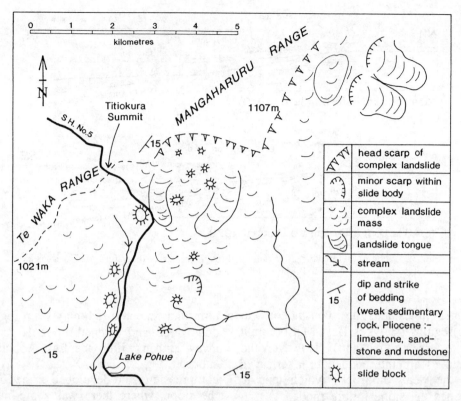

Fig. 4. Map of landslide complex on the dip slope of the Te Waka Mangaharuru Range, Central Hawkes Bay, North Island.

recognised by Thompson (1981).

Proto block slides in which the movement has been only a few cm are recognised on the banks of the Rangitikei River (Fig. 5 and Fig. 6). Tension cracks in the sandstone, above a near horizontal clay seam define the incipient pull away zones and potential basal rupture surface respectively (Fig. 5). Downstream of the proto block slide, a small hillock and a trough in the terrace surface above the Rangitikei River (Fig. 6) indicate where a block slide has already occurred and another is developing. *Incipient block sliding (a proto - block slide)* was also recognised in the hills of south Auckland, within the "southern landslide zone" (Fig. 7). A 2cm thick clay seam was uncovered in a water supply tunnel and is

postulated as the basal rupture surface for the proto block slide. Injection of the clay up into tension cracks above the seam, but not below is recorded by Wylie (1989) and suggests incipient block sliding. A section through the proto-blockslide is shown in Fig. 8.

Spalling and slabbing are ubiquitous in the soft weak rock, most noticeably in areas of more rapid uplift. Retreat of coastal cliffs in weak rock around Auckland has necessitated widespread retaining and rock mass reinforcement. (See for example N.Z. Engineering, 1989). The rate of spalling in the Rangitikei Valley (Central North Island) has been estimated at 1m per decade for one particular site, but may well exceed this elsewhere. These rates were of

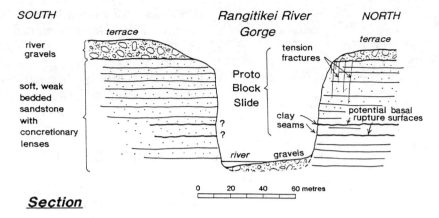

Section

Fig. 5. Section through a proto block slide, Rangitikei Valley, North Island.

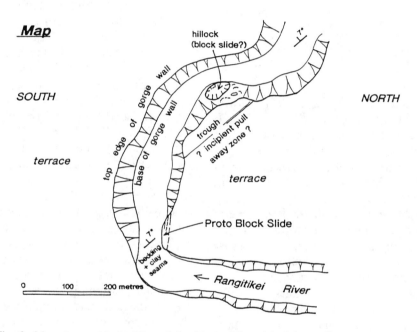

Fig. 6. Map of proto block slide, existing block slide and incipient pull away zone trough, Rangitikei Valley, North Island.

importance to the design criteria for piles of viaduct abutments in this region. The spalling occurs on all slopes irrespective of aspect, height, gradient or water content, although these factors may well vary the rate. The spalling is considered to be an unloading feature. De-stressing, as a result of rapid uplift and erosion, results in slope-parallel tensile failure of the intact rock. Changes in the microfabric of the rock accompany the uplift and spalling, and are described by Huppert (1988).

The clay-rich regolith of soft, weak rock is susceptible to shallow, sliding, especially after high intensity rainfall. Recent cyclone damage in the East Coast

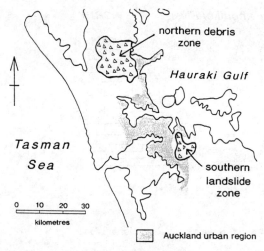

Fig. 7. Map of Auckland region showing location of southern landslide zone (Holocene) and northern debris zone (Miocene, but Holocene reactivation). (After Kermode, 1991).

was caused by ubiquitous *earth flows, debris flows and regolith slides* in the top few metres of regolith. Some slopes lost approximately one third of their soil cover. Severe *gullying and sheet erosion* resulted. Damage was also from *downslope deposition* of debris and *flood plain sedimentation* with mud for several km down the valleys. Deforested slopes were severely affected compared to those in full forest or conservation planting. The distribution of mudflows and landslips (shallow) on mudrocks in the eastern to southern North Island is discussed by Crozier, Eyles, Marx, McConchie and Owen (1980) and again by Crozier (1986).

4. VOLCANIC TERRAIN

Landslides are found in the andesitic massifs and rhyolite (ignimbritic) plateaus of the North Island (Prebble 1986). Neall (1982) has mapped the ring plain deposits of pre-historic debris avalanches from Taranaki volcano and more recently Latter

(1987) and Palmer and Neall (1989) have mapped similar deposits around Ruapehu Volcano. Lithofacies organisation of unconfined wet-avalanche flows from these volcanoes is discussed in some detail by Palmert, Alloway and Neall (1991).

4.1 IGNIMBRITE PLATEAUS

The volcanic terrain of the central to northern North Island is dominated by a thick sequence of rhyolitic ignimbrites which vary from welded, but fractured rock to thick porous deposits of low density pumice soil masses containing clay-rich palaeosols. Several tens of ignimbritic eruptions in the last 2 million years have left a complex and widespread pile of rhyolitic debris and rock masses, up to a few km thick in the Taupo Volcanic Zone (Fig. 1). To the West and East of the T.V.Z. rift, high standing ignimbritic plateaus extend on to the greywacke ranges and the weak mudrock hill country. Repeated eruptions, each followed by weathering, erosion and redeposition of part of the ignimbrite has resulted in a series of complex and superposed buried topographies. Each has considerable relief and clay-rich palaesols which were developed on the ignimbritic materials. Originally coarsely porous and glassy, these materials were weathered to finely porous, clayey, wet deposits. They show extreme sensitivity and could be described as quick-clays. Between the clays there are permeable, less weathered sands, gravels and rock masses. Some tephra deposits were reworked and redeposited to loess. (Prebble 1984 and 1986).

The result is now a series of ribbon aquifers perched in deposits which readily pipe and tunnel-gully, on top of ultra sensitive clays which can fail by *block*

2108

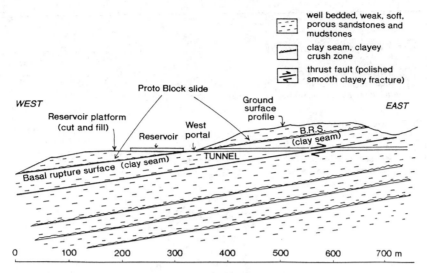

Fig. 8. Section through proto block slide, clay seams and crush zones in water supply tunnel, southern landslide zone, Auckland. (After Wylie, 1989).

sliding and rapid slide-flows.

The most recent demonstration of *rapid blockslide - flow* failure of these deposits was at Ruahihi adjacent to the northern T.V.Z. (Fig. 1). Water from a recently filled canal (Hatrick et al. 1982) and possibly also in trapped ribbon aquifers beneath the canal buttress fill, promoted a block slide on sensitive clays in the in-situ weathered ignimbrites beneath the canal. The movement was rapid and catastrophic, removing a 200 m long section of canal, an entire buttress fill and creating a 500 m long washout.

Instability and excessive deflections in ignimbrites, after filling of a headrace canal, was also recorded at Arapuni (Natusch 1984), west of the T.V.Z. Piping and tunnelling of tephric loess contributed to the undermining and rapid collapse of a leaking headrace canal at Whaeo (Jones et al. 1983), east of the T.V.Z.

4.2 ANDESITIC MASSIFS

Active andesitic cones in the Taranaki and Taupo Volcanic Zones have a history of *massive collapse* and huge *debris avalanches, debris flows and mudflows* (Palmer, Alloway and Neall, 1991).

Sector collapse, similar to that seen at Mt St Helens in 1980 (Schuster 1983) is inferred to have occurred several times at Taranaki and Ruapheu. The debris flow deposits of the Murimoto Formation, and Whangaehu Fan are largely considered to be the result of massive debris avalanches caused by sector collapse (Palmer and Neall 1989, Latter 1987). Similar deposits are mapped on the northern ring plain of Tongariro Volcano (Fig. 9).

Near Tokaanu, hydrothermally weakened and altered ground has collapsed twice in the last 150 years, producing clayey debris avalanches which overwhelmed local villages. Several more prehistoric deposits are found in the area also (Fig. 9).

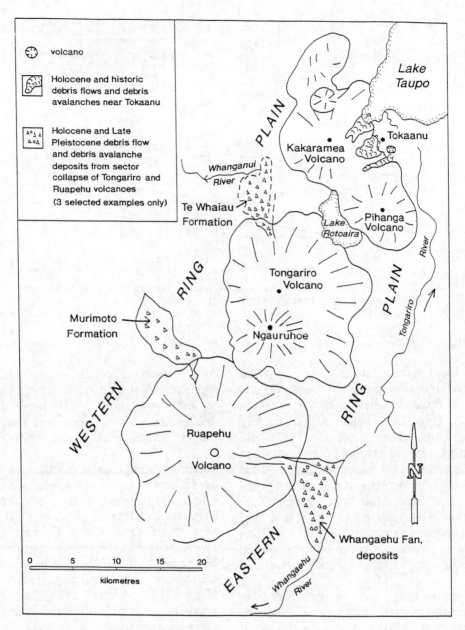

Fig. 9. Debris avalanche and debris flow deposits, Tongariro Volcanic District, southern Taupo Volcanic Zone, North Island.

5 SCHIST TERRAIN

The chlorite schist of Otago is noted for *huge complex landslides* on foliation parallel slopes. Large landslides are also evident on foliation disjunctive slopes, but these are usually localised rock falls, toppling failures and regolith slides (involving glacially and periglacially derived debris).

Multiple rupture surfaces on foliation

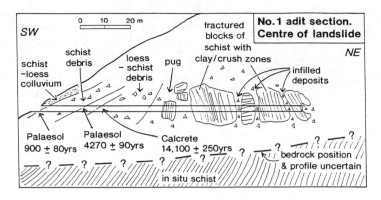

Fig. 10. Section through centre of K9 landslide complex, Kawarau River gorge, Clutha Valley, Otago: Showing multiple palaeosols and multistage landslide development. (After Bell, 1976(a)).

parallel shears and crush zones, with stepwise connections along intersecting defects, have generated rock slides and debris slides. Areas involved vary from less than 1 to several km². Multistage development over the late Pleistocene and Holocene is indicated by dates from palaeosols (Fig. 10) (Bell, 1976a), Bell (1976a). The sequential development of these large landslide complexes is paralleled by their equally complex ground water conditions (Williams, 1990).

Active creep has been demonstrated on the slide masses in the Kawarau Gorge of the Clutha River catchment (Fig. 11). The Cromwell Gorge landslides are also moving, at a few tens of millimetres per year (Williams 1990). The size of these landslides ranges from 7.5 to 1200 hectares in area, and from 3 to 1200 million m³ in volume. The largest (Nine Mile Creek landslide) has a slope length of 4000m and a river frontage of 3900m (Williams 1990).

Remedial measures (toe buttressing and gravity drainage) are necessary on all the major Cromwell Gorge landslides (Williams 1990).

In the Kawarau and Cromwell Gorges a large proportion of the landscape is occupied by landslide complexes as indicated by the mapping of Bell (1976) and Turnbull (1987) (Fig. 11). As Bell pointed out, rock mass anisotropy and mass movement have exerted a major influence on geomorphic development and slope stability in the chlorite schists of the Clutha Valley region.

A similar control by foliation parallel shears on block slides in chlorite schist of eastern Central Otago is described by Paterson, Hancox, Thompson and Thomson (1984).

6. GREYWACKE TERRAIN

The axial greywacke mountains of both islands, but most notably the Southern Alps, produce large *rock slide avalanches* (Whitehouse 1983). 46 rock avalanches of Holocene age, with volumes from 1 to 500 million m³ are reported by Whitehouse (1983). He points out the very significant sediment yield that is probably derived from these slope movements.

Debris avalanching is also a dominant slope process (Prebble 1987, Whitehouse

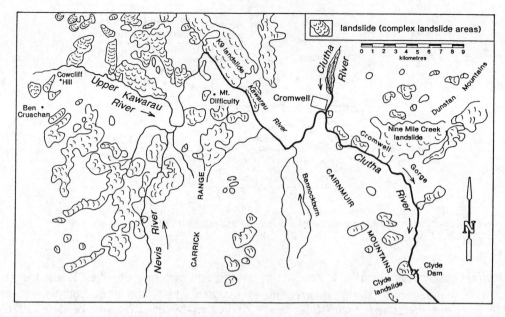

Fig. 11. Map of Kawarau River and Cromwell Gorge, Clutha River, Otago: showing distribution of major landslide complexes (After Turnbull, 1987).

1988 and this study).

Uplift rates of 10 to 12 mm yr^{-1} in the central Southern Alps are recorded by Wellman (1979), Whitehouse (1988), and Basher, Tonkin and McSaveney (1988). A maximum of 17 mm yr^{-1} is suggested by Wellman (1979) for one part of the Alps.

The steady-state balance between uplift and erosion, proposed for the Southern Alps by Adams (1985) cannot be substantiated (Williams 1988). The work of Whitehouse (1988), and Basher et al. (1988) indicate a crude balance is possible, but the data are as yet very imprecise. The uplift and denudation rates in the Southern Alps are amongst the highest anywhere on Earth. *Debris avalanching*, with associated *rock fall* and *mass creep* are cited by the above authors as common slope movement mechanism.

The role of toppling is probably very underestimated.

The rapid uplift of closely fractured rock and the equally swift denudation and dissection of the rock mass into high, steep slopes will lead to a rapid release of the compressive and shear stresses generated in the plate boundary.

Dilation and slip of fractures will follow, especially in closely fractured rock. This leads to a type of slope movement referred to as *rock mass bulging* (Prebble, 1987).

The style and mechanism of slope failure in rock mass bulging will be determined by the relative dominance of defects and their angle and direction of dip. Both *sliding* and *toppling* are observed in greywacke slopes (Fig. 12), and both give rise to *rock debris avalanches* and *screes*.

Deep seated toppling is probably widespread in greywackes. Examples are seen near Mt Cook (Fig. 12) and in north-east Marlborough (Prebble 1987). Dilational block toppling affects anaclinal (scarp) slopes and dilational flexural toppling and block- flexural toppling affect dip slopes (cataclinal underdip slopes). The latter failure mode is almost ubiquitous as *rock mass bulges* in limestones of the subsidiary mountain ranges of Marlborough.

Rock mass bulging, especially in *toppling mode,* produces uphill facing scarps, ridge rents, ridge top trenches and ridge cracking. *Rock mass bulging* may be an alternative mechanism to the gravity - faulting postulated by Beck (1968), and may also help to explain the features noted by Tabor (1971) in the north-west U.S.A., Zischinsky (1966) and Nemcok (1972) in the European mountains, and Plafker (1967) in Alaska.

Collapse of rock mass bulges, when the mass behaves like soil in the geotechnical sense rather than a true rock mass, gives rise to rock debris avalanches. The December 1991 rock avalanche from the summit of Mt Cook (Aoraki) may well be the latest example.

Recent glacier retreat over much of the Southern Alps has probably encouraged rock mass bulging through removal of valley ice and the consequent loss of lateral support. Further, acceleration of uplift about 134 Ka ago, referred to by Williams (1988), may also increase susceptibility to rock mass bulging in slopes.

Regolith slides and *debris flows* in highly weathered greywacke are widespread in lower altitude and warmer regions. Basal ruptures for these slope movements is the boundary between rock and residual soil. High intensity rainstorms produced severe regolith sliding, debris flows and clayey debris avalanches in Marlborough in 1975 (Bell 1976b; Bowring, Cunliffe, MacKay and Wright, 1978). Similar events in South Auckland were investigated by Rogers and Selby (1980).

Muddy debris flows, avalanches and debris torrents were experienced in Wellington in 1974, 1976 and 1977, during high intensity rainfall (Taylor, Hawley and Riddolls, 1977; Lawrence, Depledge, Eyles, Oakley and Salinger, 1978).

Extremely high sedimentation rates were experienced during these events, as the debris flows and avalanches reached flood plains or the coastal fringe.

Debris flows from wide and persistent crush zones in greywacke and argillite are also a source of extreme aggradation, below the deeply eroding source areas (Pierson, 1980; Kenny, 1980).

7. EAST COAST DEFORMED BELT

The East Coast Deformed Belt (Fig. 1) of the North Island and north-east South Island is well-known for smectite-rich shales, mélange, thrust belts and crush zones (Prebble, 1980; Kenny 1984; Pettinga 1987; Gage and Black 1979; Pearce, Black and Nelson, 1981). These authors have described a range of slope movements in these rocks, from debris flows and earth flows to deeper seated rock slides. The degree of fracturing and crushing in the belt is very intense and occupies wide and continuous zones. As a result, much of the terrain is underlain by soil in the geotechnical sense, rather than rock. This is reflected in the almost ubiquitous presence of mass movement.

8. GRANITIC TERRAIN

Fiordland, Westland and north west Nelson (Fig. 1) differ to some extent from other mountainous regions in that

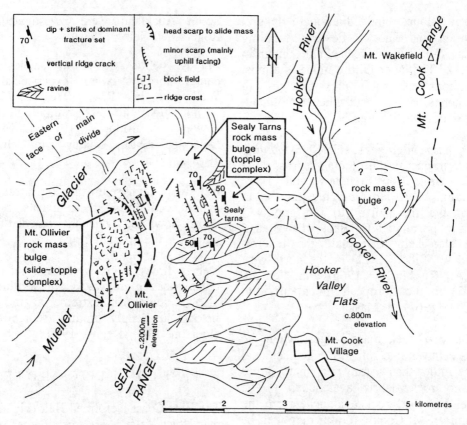

Fig. 12. Map of Mt Ollivier and Sealy Tarns Rock Mass bulges, in the Sedly Range, and a probable rock mass bulge in the Mt Cook Range, Southern Alps.

they are underlain by high grade, crystalline metamorphic rocks and igneous (granitic) intrusives. However, there are similarities in terms of height and steepness of slopes, recent glacial retreat and deglaciation and seismicity. Northwest Nelson especially is known for widespread, *rock slide avalanches* of the large 1929 and 1968 shallow earthquakes (Henderson, 1937; Lensen and Suggate 1968). *Landslide dammed lakes* and many large pre-historic rock slides are prevalent throughout the region (Adams 1980). Landsliding on Tertiary limestones, mudstones and sandstones are also very widespread and in many areas dominate over failures in granite (Pearce and O'Loughlin, 1985).

9. MORAINE AND LOESS

Glacial retreat in the Southern Alps in the last century alone has resulted in a marked increase in instability of lateral moraine walls. For instance, in the Mt Cook National Park "slumping" of the Tasman Glacier lateral moraine wall has closed part of the Ball Hut road to vehicles and seriously affected hut sites (Department of Lands and Survey, 1986). Reverse (uphill-facing) scarps on lateral moraine walls in the Hooker Valley and Tasman Valley may be attributable to dilation of the moraine walls as an alternative to true "slumping". The failure may therefore be a type of soil topple. Uphill facing scarps and an absence of slump-related features are evident also on the Mueller Glacier lateral moraine wall.

Glacially derived loess in the South Island and volcanigenic loess in the North Island are readily undermined by piping and tunnel-gully erosion. This encourages sliding of the loess deposits (Bell and Trangmar 1987, Prebble 1986).

10. ENGINEERING AND ENVIRONMENTAL PROBLEMS

Landslides, and slope movements in general, require engineering and land management throughout New Zealand.

Problems in soft weak rock terrain are many, and include stabilisation and support of coastal cliffs around Auckland's shoreline (NZ Engineering, 1989), relocation of roads and railways in central North Island, (Stout, 1977) including specific abutment designs, analysis for sliding stability of a water supply tunnel and reservoir complex in South Auckland and widespread soil conservation, erosion control, and river control measures.

Land use planning, in both urban and74 rural regions, is increasingly having to deal with landslides (Hawley and Luckman 1980; Prebble, 1983).

In the schist terrain of central Otago, vast toe buttressing and drainage systems are required to increase the stability of huge translational landslide complexes in the slopes of reservoir to be filled behind the Clyde Dam. Roads, tunnels and irrigation canals in Otago have also met problems of sliding in the schist (Bell 1976, Turnbull 1987).

11. SUMMARY

No region of New Zealand is free from landslides, which vary from numerous and widespread *regolith slides* and *earth flows* to *huge deep seated slide complexes*. *Clay seams, crush zones* and clay coated *fractures* of tectonic origin form continuous basal rupture surfaces for large landslides. A tectonic setting of high compression, shear and uplift is coupled with very weak and porous wet rocks and highly fractured and crushed greywacke and schist. This makes the alpine regions and hill country highly susceptible to *large scale sliding and toppling*. In this active tectonic and geomorphic setting, *rock mass bulging* is considered to be widespread and is caused mainly by deep seated toppling. In its severest form, *overturning of tabular rock* is caused by dilational flexural -block toppling in *cataclinal underdip slopes*. Volcanic terrain is characterised by huge debris avalanches from collapse of andesitic cones, and rapid slide-flow failures in ultrasensitive weathered rhyolitic ignimbrites.

A steady state between denudation (slope movement plus steam transport) and uplift is both postulated and disputed.

The rapid tectonic rate and the frequency and scale of slope movement is increasingly recognised and impinges on every region of New Zealand. Landslide investigations will continue to increase and diversify in geotechnical, environmental and planning projects.

Landslide complex in Neogene weak rock, on the dip slope of the Mangaharuru Range, central Hawkes Bay. Viewed from Titiokura summit looking East across several km^2 slide mass and the 3km long head scarp.

Ponui wedge slide in Neogene weak rock, Southern Hawkes Bay. Rupture surfaces of the slide are a clay gouge seam dipping from right down to left and sets of intersecting tectonic joints which form the 1km long side scarp at upper left.

Omatane bridge, and short displacement block slide in weak sandstone, Rangitikei Valley. Three slide blocks and two secondary pull-away chasms between the slide blocks, to the left of the bridge abutment. Bedding, and basal rupture clayseam (at river level) dip 12^0 to the right.

Rock slide, in limestone, flint and shale, Marlborough. (East Coast Deformed Belt), South Island. Slide body is 1.4km across. Swamp ponded by toe at left dates at c 4000yrs B.P.

Spalling in very weak sandstone, Rangitikei River gorge, North Island. Bedding is nearly horizontal, with concretions. The spalled slab is approximately 10m high, 6m wide and 1.5m thick.

A 2km long regolith slide in schist debris, near Cow Cliff Hill, Upper Kawarau Valley, central Otago. Head scarp of large slide at upper right.

Topple/rock slide complex of Mt Ollivier rock mass bulge, Sealy Range, Mt Cook National Park. Open ridge cracks, uphill-facing scarps, fracture dilation and block sliding are found in this complex landslide. Upper Meuller Glacier behind, to the South-West.

Toppling is closely fractured and crushed greywacke of the Sealy Tarns rock mass bulge, Mt Cook National Park, South Island. Deep gullying cuts through the highly dilated and loosened rock mass bulge. Exposed face is a few hundred m high. Hooker Valley 700m below. View is to the South.

12 REFERENCES

Adams, J.E. 1980. Contemporary uplift and erosion of the Southern Alps, New Zealand. *Geological Society of America Bulletin* 11, (2):1-114.

Adams, J.E. 1985. Large-scale tectonic geomorphology of the Southern Alps, New Zealand. In: M. Morisawa and J.T. Hack (eds): *Tectonic Geomorphology* (Allen and Unwin) Boston: 105-128.

Basher, L.R., P.J. Tonkin and M.J. McSaveney 1988. Geomorphic history of a rapidly uplifting area on a compressional plate boundary: Cropp River, New Zealand. In: P.W. Williams (ed.). The Geomorphology of Plate Boundaries and Active Continental Margins. *Zeitschrift Für Geomorphologie.* Supplementband 69: 117-132.

Beck, A.C. 1968. Gravity faulting as a mechanism of topographic adjustment. *New Zealand Journal of Geology and Geophysics* 11(1):191-199.

Bell, D.H. 1976a. Slope evolution and slope stability, Kawarau Valley, Central Otago, New Zealand. *Bulletin of the International Association of Engineering Geology* 14:5-16.

Bell, D.H. 1976b. High intensity rainstorms and geological hazards: Cyclone Allison, March, 1975, Kaikoura, New Zealand. *Bulletin of the International Association of Engineering Geology* 14:189-200.

Bell, D.H. and B.B. Trangmar 1987. Regoloth materials and erosion processes on the Port Hills, Christchurch, New Zealand. *Proceedings of the Fifth International Field Workshop on Landslides Christhchurch.*

Berryman, K. 1988. Tectonic geomorphology at a plate boundary: a transect across Hawkes Bay, New Zealand. In: P.W. Williams (ed.). The Geomorphology of Plate Boundaries and Active Continental Margins. *Zeitschrift Für Geomorphologie.* Supplementband 69: 69-86.

Bowring, L.D., J.J. Cunliffe, D.A. Mackay and A.F. Wright 1978. East Coast Survey. A study of catchment and stream condition with recommendations. *Marlborough Catchment and Regional Water Board, Blenheim, New Zealand.*

Coombs, D.S. and R.J. Norris, 1981. The East Abbotsford, Dunedin, New Zealand Landslide of August 8, 1979. An interim Report. *Bulletin de liaison de Laboratoire des Ponts et Chaussées. Special X.* 27-34 (Proceedings of 26th International Geological Congress, Paris 1980).

Crozier, M.J., R.J. Eyles, S.L. Marx, J.A. McConchie and R.C. Owen 1980. Distribution of landslips in the Wairarapa Hill Country. *New Zealand Journal of Geology and Geophysics.* 23: 575-586.

Crozier, M.J. 1986. Landslides: causes, consequences and environment. *Croom Helm* 252p.

Department of Lands and Survey, 1986. The Story of Mount Cook National Park. 6th edition.

Eyles, G.O. 1983. The distribution and severity of prsent soil erosion in New Zealand. *New Zealand Geographer* 39(1):12-28.

Gage, M. and R.D. Black 1979. Slope stability and geological investigations at Mangatu State Forest. *Forest Research Institute New Zealand Forest Service Technical Paper* No. 66.

Hatrick, A.V., A. Howarth, J.H.H. Galloway and G. Ramsay, 1982. Report of the Committee to Inquire in to the Failure of the Ruahihi Canal. *Ministry of Works and Development,* Wellington, N.Z.

Hawley, J.G. and P.G.Luckmann, 1980. The Geomechanics of Soil Conservation. *Proceedings of the Technical Groups of the New Zealand Institution of Engineers.* 6(1)G:2-53 to 2-60 (Proceedings of 3rd Austalia-New Zealand Conference on Geomechanics 1980).

Henderson, J. 1937. West Nelson Earthquake of 1929. *New Zealand Journal of Science and Technology* 19:66-144.

Huppert, F. 1988. Influence of microfabic on geomechanical behaviour of Tertiary fine-grained sedimentary rocks, from Central North Island, New Zealand. *Bulletin of the International Association of Engineering Geology,* 38: 83-93.

Jones, O.T., N.H.H. Galloway, A. Howarth and G. Ramsay, 1983. Report of Committee to Inquire into the Canal failure on the Wheao Power Scheme. *Ministry of Works and Development.* Wellington, N.Z.
Kenny, J.A. 1980. Geology of the Ihungia catchment Raukumara Peninsula. *University of Auckland MSc thesis in Geology* 151p.

Kenny, J.A. 1984. Stratigraphy, sedimentology and structure of the Ihungia decollement, Raukumara Peninsula, North Island, New Zealand. *New Zealand Journal of Geology and Geophysics* 27:1-19.

Kermode, L.O. 1991. Whangaparaoa - Auckland. Info map 290 Sheet R10/11 1:100,000. New Zealand Land Inventory, Rock Types. Department of Survey and Land Information, Wellington, N.Z.

Lawrence, J.H., D.R. Depledge, R.J. Eyles, D.J. Oakley and M.J. Salinger 1978. Landslip and flooding hazards in Eastbourne Borough - A Guide to Planning. *Ministry of Works and Development, Water and Soil Division,* Wellington, N.Z.

Lensen, G.J. and R.P. Suggate 1968. Inangahua Earthquake - Preliminary account of the Geology. In: Preliminary Reports on the Inangahua Earthquake New Zealand, May 1968. *Department of Scientific and Industrial Research Bulletin* 193:17-36.

Natusch, G.G. 1984. Arapuni in retrospect. Transactions of the Institution of Professional Engineers New Zealand ii(1, C.E.):1-12.

Nemcok, A. 1972. Gravitational slope deformation in high mountains. *Proceedings of the 24th International Geological Congress* 13, 132-141.

New Zealand Engineering, 1989. Clifftop stabilisation - solutions for anxious property owners. *New Zealand Engineering,* August 1st, 1989:11-13.

Palmer, B., B.V. Alloway, and V.E. Neall 1991. Volcanic-debris-avalanche deposits in New Zealand -Lithofacies organisation in unconfined, wet-avalanche flows. In: Sedimentation in Volcanic Settings. SEPM Special Publication No 45: 89-98.

Paterson, B.R., G.T. Hancox, R. Thomson and B.N. Thompson, 1984. Investigations in hard rock terrain. In: Engineering for Dams and Canals. Philosophy and methods of investigation used in New Zealand. *The Institution of Professional Engineers, New Zealand. Proceedings of Technical Groups,* 9, 4(G): 5.1-5.20.

Pearce, A.J., R.D. Black and C.S. Nelson 1981. Lithologic and weathering influences on slope form and process, eastern Raukumara Range, New Zealand. *International Association of Hydrological Sciences Publication* 132; 95-122.

Pearce, A.J. and C.L. O'Loughlin 1985. Landsliding during a M7.7 earthquake. Influence of geology and topography. *Geology* 13; 855-858.

Pettinga, J.R. 1982. Upper Cenozoic structural history, coastal southern Hawkes Bay, New Zealand. *New Zealand Journal of Geology and Geophysics,* 25; 149-191.

Pettinga, J.R. 1987. Ponui Landslide: a deep-seated wedge failure in Tertiary weak-rock flysch, Southern Hawke's Bay, New Zealand. *New Zealand Journal of Geology and Geophysics,* 30:415-430.

Pierson, T.C. 1980. Erosion and depositon by debris flows at Mt Thomas, North Canterbury, New Zealand. *Earth Surface Processes,* 5:227-247.

Plafker, G. 1967. Surface faults on Montague Island associated with the 1964 Alaska Earthquake. *Geological Survey Professional Paper* 543-G. United States Department of the Interior.

Prebble, W.M. 1980. Late Cainozoic sedimentation and tectonics of the East Coast Deformed Belt in Marlborough, New Zealand. In: Sedimentation in oblique-slip mobile zones (Ed by P.F. Ballance and H.G. Reading). *Special Publication of the International Association of Sedimentologists,* 4. 217-218.

Prebble, W.M. 1983. The Influence of Lithology, structure and slope upon erosion mechanisms, particularly Mass Movements. In: D.J. Moore (Ed.). Catchment management for optimum use of land and water resourcse: Documents

from an ESCAP seminar. Part 2 - New Zealand Contributions *Water and Soil Miscellaneous Publication* 46, 263-282.

Prebble, W.M. 1984. Investigations in an Active Volcanic Terrain. In: Engineering for Dams and Canals: Philosophy and methods of investigation used in New Zealand. *The Institution of Professional Engineers, New Zealand. Proceedings of Technical Groups* 9, 4(G):7.1-7.15.

Prebble, W.M. 1986. Geotechnical problems in the Taupo Volcanic Zone. In J.G. Gregory and W.A. Watters (Eds). Volcanic hazards assessment in New Zealand. *New Zealand Geological Survey Record 10:* 65-80.

Prebble, W.M. 1987. Slope movements in limestones and shales, North-East Marlborough, New Zealand. *PhD thesis in Geology, The University of Auckland.*

Prebble, W.M. 1990. Manukau Sanitary Landfill. Report on the Geology of the site north of Caldwells and Sandstone Road, near Whitford *Auckland Uniservices Ltd.*

Rogers, N.W. and M.J. Selby 1980. Mechanisms of shallow translational landsliding during summer rainstorms: North Island, New Zealand. *Geografiska Annaler* 62(A)1-2:11-21.

Schuster, R.L. 1983. Engineering Aspects of the 1980 Mount St Helens Eruptions. *Bulletin of the Association of Engineering Geologists* XX, 2:125-143.

Sporli, K.B. 1980. New Zealand and oblique slip margins. Tectonic development up to and during the Cainozoic. In: Sedimentation in oblique-slip mobile zones (Ed. by P.F. Ballance and H.G. Reading) *Special Publication of the International Association of*

Sedimentologists 4, 147-170.

Stout, M.L. 1977. Utiku landslide, North Island, New Zealand. Geological Society of America Reviews in Engineering Geology, III:171-184.

Tabor, R.W. 1971. Origin of ridge-top depressions by large-scale creep in the Olympic Mountains, Washington. *Geological Society of America Bulletin,* 82:1811-1822.

Taylor, D.K., J.G. Hawley and B.W. Riddolls, 1977. Slope Stability in Urban Development. *New Zealand Department of Scientific and Industrial Research Information Series* 122.

Thompson, R.C. 1981. Landsliding in Cenozoic softrocks of the Taihape-Mangaweka area, North Island, New Zealand. *Bulletin de liaison de Laboratoire des Ponts et Chaussées. Special X 93-100.* (Proceedings of the 26th International Geological Congress, Paris, 1980.

Turnbull, I.M. 1987. Sheet S133 - Cromwell. *Geological Map of New Zealand 1:63,360* Map and notes. Department of Scientific and Industrial Research, Wellington, New Zealand.

Walcott, R.I. 1987. Geodetic strain and the deformational history of the North Island of New Zealand during the Cainozoic. *Philosophical Transactions of the Royal Society London* A 321:163-

Wellman, H.W. 1979. An uplift map for the South Island of New Zealand, and a model for Uplift of the Southern Alps. In: R.I. Walcott and M.M. Cresswell (eds). The Origin of the Southern Alps. *The Royal Society of New Zealand Bulletin* 18:13-20.

Whitehouse, I.E. 1983. Distribution of large rock avalanche deposits in the central Southern Alps, New Zealand. *New Zealand Journal of Geology and Geophysics,* 26: 271-279.

Whitehouse, I.E. 1988. Geomorphology of the central Southern Alps, New Zealand: the interaction of plate collision and atmospheric circulation. In: P.W. Williams (ed.). The Geomorphology of Plate Boundaries and Active Continental Margins. *Zeitschrift Für Geomorphologie* Supplementband 69, 105-116.

Williams, M.J. 1990. Clyde Power Project - Cromwell Gorge landslides. *New Zealand Engineering* September 1, 1990: 11-13.

Williams, P.W. 1988. The Geomorphology of Plate Boundaries and Active Continental Margins. (Introduction) *Zeitschrift Für Geomorphologie* Supplementband 69: V-IX.

Wylie, C.A. 1989. Engineering Geology of a Soft, WEak Rock Tunnel: A.R.A. Redoubt Road No 2 Inlet Tunnel. *MSc thesis in Geology, The University of Auckland.*

Zischinsky, U. 1966. On the deformation of high slopes. Proceedings of the 1st Conference of the International Society of Rock Mechanics, Lisbon 2:179-185.

Landslides, Bell (ed.) © 1995 Balkema, Rotterdam, ISBN 90 5410 032 X

Theme report

A.T. Moon

Coffey Partners International Pty. Ltd, Brisbane, Qld, Australia

ABSTRACT: The seven papers included in this session are reviewed under the following headings: Landslide dammed lakes; Landslides and rural land management; and other topics. Questions are presented on the two main topics for the purpose of encouraging discussion.

1 INTRODUCTION

The seven papers included in this session are those papers originating from Australia or New Zealand which have not been included in other general or specialist themes.   A rather miscellaneous variety of topics has been the result of this method of selection. However, for the purpose of this review the following grouping of papers has been adopted:-

* Landslide dammed lakes (2 papers)
* Landslides and rural land management (2 papers)
* Other topics (3 papers)

Within each group the papers are discussed in alphabetical order of their authors. Following a review of the papers, specific questions relating to the two major topics are presented for the purpose of encouraging discussion.

2 LANDSLIDE DAMMED LAKES

2.1 *Landslide-dammed lakes in New Zealand - preliminary studies on their distribution, causes and effects*

Perrin and Hancox describe and discuss about 80 lakes and former lakes which have been formed by the damming of rivers by large landslides.   The initial purpose of the study was to contribute to a world-wide compendium of such lakes but further work is planned on the age, failure mechanisms and hazards associated with specific landslides.   The volumes of water dammed by the landslides are rarely greater than the volumes of the landslides themselves.

The most striking aspect of the study is the association of large landslide dams with large earthquakes and the authors even suggest that landslide dams may be used as an indication of palaeoseismicity.

2.2 *Lake Waikaremoana barrier - a large landslide dam in New Zealand*

A specific study of the largest existing landslide dammed lake in New Zealand has been carried out by Read, Beetham and Riley. The Lake Waikaremoana dam is over 500m high and occupies an area of about 17km². The dam was formed by a two phase landslide thought to be triggered by a large nearby earthquake about 2200 years ago. The first phase is assumed to have been a rock avalanche and this was followed a short time later by a large block slide along a bedding plane in weak mudstone with an apparent dip of 7°.

The collapse of the left bank of the river valley initiated the rock avalanche. For the block side to occur a shear surface at residual strength parallel to bedding would have been required. Such a surface could have resulted from flexural slip in the tilted and probably gently folded rocks in the area. Hutchinson (1988) has pointed out that flexural sheared zones can be expected at very flat dips where thick incompetent beds occur. The lack of exposure of "clays at residual strength" during a "preliminary inspection" of a "possible failure zone" is no surprise. It is often difficult to identify sheared surfaces because of desiccation of near surface material or negative pore pressures caused by undrained unloading during excavation.

A companion paper by Riley and Read (in press) assessing the present day stability of the landslide barrier is included in this Symposium under another theme.

3 LANDSLIDES AND RURAL LAND MANAGEMENT

3.1 *Management of lands subject to mass movement*

Macgregor and McManus describe a study carried out for the Soil Conservation Service of New South Wales in the northeast of the State. The purpose of the study was to review the criteria for assessing land vulnerable to mass movement and develop guidelines on land management for implementation by Soil Conservation Service officers. Site conditions and the types of failure that occur in the study area are described and the relationship between slope stability and rock type is discussed. General discussion is provided on the factors contributing to mass movement with particular emphasis given to the influence of vegetation. Reference is made to the preparation of guidelines and field assessment check sheets and to zoning and risk levels.

The authors appear to have been involved in a worthwhile study which met its objectives. The inclusion of general discussion on mass movement and a review of the literature on the influence of vegetation would have been relevant and useful in the authors' report to their client. However, in a technical paper to a specialist conference it may have been better to summarise this material with appropriate references in order to devote more space to technical aspects particular to the project. The inclusion of the guidelines or an example of the application of the field assessment sheets may have been useful.

3.2 *Engineering geological assessment of slope instability for rural land-use, Hawkes Bay, New Zealand*

The two objectives of this well illustrated paper by Pettinga and Bell are to review the slope instability problems in the area studied and to outline "engineering geology guidelines for cost-effective farm management practices". Folded and faulted weak rocks subject to periodic uplift has led to the development of two characteristic terrain types. The relict (older) landscapes are higher with subdued and generally stable slopes while the rejuvenating landscapes consist of steep sided gullies with many landslides. While the relict terrain is suitable for pasture (and cropping in some areas) the rejuvenating terrain is more difficult and less suited to the traditional grassland farming. Reafforestation and total retirement of the land in some cases may be required to reduce erosion and consequent downstream flood aggradation.

The authors recommend that engineering geologists should be employed by local catchment authorities to ensure satisfactory input into farm management planning. In the study in New South Wales by MacGregor and McManus the input was by external consultants. Whether or not full time employment by the catchment authorities themselves is appropriate or cost-effective depends to some extent on the size, budgets, terms of reference, and existing expertise within the authority.

4 OTHER TOPICS

4.1 *Landslide damage on the East Coast Region arising from tropical cyclone Bola, March 1988*

McKelvey and Murton describe the investigation and repair of landslide damage to a highway. The damage was caused by high rainfall associated with a tropical cyclone. More than 60 individual sites requiring investigation and major repairs occurred on a 150 km length of highway underlain by weak folded and faulted sedimentary rocks. Brief descriptions of the major failures are given and some of the remedial options adopted are listed. Poor attention to surface drainage contributed to many of the landslides.

Only one diagram of a landslide is included in the paper and that diagram is not discussed in the text. The conclusion referring to structural changes in administration and possible inadequacy of future response is political rather than technical. For a specialist technical conference it would have been preferable to illustrate and discuss one or two examples of technically innovative remedial design at the expense of the lists and brief general descriptions that were included.

4.2 *Timing of relief and landslides in Central Otago, New Zealand*

McSaveney, Thomson, and Turnbull discuss the relationship between uplift, gorge erosion, and landslides in the active tectonic area of Central Otago. This paper could also have been included in Theme S6 - Lake Dunstan Landslides as it provides useful background information on the development of the Cromwell Gorge area. The rate of river incision and uplift over the past 500,000 years has been assessed on the basis of dated glacial and interglacial deposits.

In discussion of mass movement the authors refer to landslides, sagging (Hutchinson, 1988) and sagging metamorphic terrain. It is understood that most of the mass movement features of concern in the Cromwell Gorge would best be described as compound listric (spoon-shaped) slides using Hutchinson's terminology. It is presumed that the relationship between sagging and landslides and the nature of the Cromwell Gorge Landslides will be discussed further at the Pre-Symposium Field Seminar and during Theme 6 of this symposium.

4.3 *Weathering and strength loss at an earthflow site*

Trotter, Pinkney and Tod describe variations in geotechnical properties in two soil profiles overlying weak mudstone. One profile is taken within an active earthflow or mudslide (Hutchinson, 1988) while the other profile is taken in the adjacent stable slope. Five weathering zones are identified in each profile but the depth of weathering is greater at the earthflow site.

In describing the variations in physical properties within and between the soil profiles further comment on their relationship to the earthflow would have been useful. For example, while the greater depth of weathering may have predated the slope failure the higher moisture content and the lower bulk density and penetration resistance may be the result of the earthflow. Fuller descriptions of the profiles (preferably tabulated) including systematic descriptions of the defects would also have been useful in assessing the study. Similar studies overseas by Chandler and Apted (1988) and Hawkins, Lawrence and Privett (1988) have also addressed the clay mineralogy and effective shear strength of the various weathering zones and further work in this area would increase the value of the research.

5 DISCUSSION QUESTIONS

5.1 *Landslide dammed lakes*

The papers identify large earthquakes as the most likely trigger for large landslides which form dams and refer to further work on failure mechanisms and hazard evaluation. Another paper at this symposium discusses the seismic stability of landslides and reservoir slopes at the Cromwell Gorge (Gillon and Hancox in press).

General question

• What further work has been carried out on identifying failure mechanisms and evaluating hazards?

Specific questions

• Has there been work at Lochnagar and Polnoon to explain why they have failed so dramatically compared to the slower movement of other Central Otago landslides?

• Has there been work on more recent events in the north of the South Island or elsewhere where more information may be available? How many of these have been first time failures?

5.2 *Landslides and rural land management*

Both papers recognise the role of trees in reducing the risk of landslides and MacGregor and McManus comment on effective spacing. Blanket reafforestation of all land identified as being vulnerable to landslide (on a slope and geology criterion in New South Wales and the rejuvenating terrain in New Zealand) may be a drastic and uneconomic solution.

Question:

• Would it be possible to adopt a more targeted approach of spaced trees and open ground which allows a combination of pasture and forestry? Such an approach would involve an understanding of failure mechanisms and careful subzoning of the broad areas identified as vulnerable.

REFERENCES

Chandler, R.J., and Apted, J.P. 1988. The effect of weathering on the strength of London Clay. *Quarterly Journal of Engineering Geology*, London 21:59-68.

Gillon, M.D. and Hancox, G.T. in press. Cromwell Gorge Landslides - a general overview. *Proc. 6th Int. Symp. Landslides*, Christchurch 1992, Balkema.

Hawkins, A.B., Lawrence, M.S. and Privett, K.D. 1988. Implications of weathering on the engineering properties of the Fuller's Earth formation. *Geotechnique* 38(4): 517-532.

Hutchinson, J.N. 1988. General report: morphological and geotechnical parameters of landslides in relation to geology and hydrogeology. *Proc. 5th Int. Symp. Landslides*, Lausanne, 1, 3-35, Balkema.

Riley, P.B. and Read, S.A.L. in press. Lake Waikaremoana - present day stability of landslide barrier. *Proc. 6th Int. Symp. Landslides*, Christchurch 1992, Balkema.

Landslides, Bell (ed.) © 1995 Balkema, Rotterdam, ISBN 90 5410 032 X

Land stability in urban sites of the North Shore, Auckland, NZ

John S. Buckeridge
Engineering Department, Carrington Polytechnic, Mt Albert, Auckland, New Zealand

ABSTRACT: The lithologic character of Auckland's North Shore is predominantly sedimentary, with localized volcanic centres. Much of the steeper, previously undeveloped land is underlain by Waitemata Group sediments, which have a regolith of variable thickness, characterized by high clay content. Rapid population growth from the mid twentieth century has resulted in utilization of sites previously considered unusable.
Ground failure in these areas is seen to originate principally in two ways: natural, where slippage and slumping develops along pre-existing lithologic features/structures, and man made, where slippage of overburden develops on buried ground surfaces.
Two examples of slope failure are considered: a rotational slump along lithological boundaries, developed after extensive rainfall coupled with inadequate runoff disposal; and a translational block slide, developed during latter stages of subdivisional construction, following placement of stormwater reticulation parallel to land contours. The rôle of groundwater is critical in both cases: as a lubricant of slippage planes, effecting lower shear strength in clays, and as a transporter of leached cementing agents.

1 INTRODUCTION

1.1 *Demographic Pressures*

Although the present population of Auckland's North Shore is greater than 151,000, growth has been a rapid and relatively recent phenomenon. The first census, in 1858, gave the "non-Maori" population of the region as 969. A little over a century later, in 1959, the population had reached 49,000; in the following decade, this had more than doubled, reaching 102,000 by 1971. The key limiting factor to early growth was restricted access to Auckland City, the commercial centre of the region, but this was overcome following the opening of the Auckland Harbour Bridge in 1959.

The earliest development in the region, apart from pre-European fortified hill sites at North Head, was rural; however bungalows, primarily for holiday accommodation, soon appeared on the more level areas adjacent to beaches. Within two decades of the bridge being built, the nature of the population changed, with the North Shore becoming in large part a dormitory suburb for Auckland City. At this time, most of the flatter land had been built upon, and as the pressure for new sites increased, less favourable, steeper areas were subdivided.

This paper analyses two specific examples of ground failure on less favourable building situations within North Shore City, in both cases, study was carried out by the author after failure had occurred.

2 PHYSIOGRAPHY AND REGIONAL GEOLOGY

The North Shore is characterised by deeply eroded hills rising to about 100 metres above sea level, with cliffed headlands along the eastern seaboard reaching altitudes of more than 40 metres a.s.l. The two principal lithotypes are Holocene basalts and middle Cainozoic flysch deposits. Basalt comprises the scoria cones and flows around Devonport and Takapuna plus the phreato-magmatic craters at Tank Farm, Onepoto Basin and Lake Pupuke; weathered flysch however, is by far the predominant rock of the North Shore, it outcrops over most of the remaining part, forms the cliffed headlands and is the immediate basement. This flysch (designated regionally as the Waitemata Group) is made up of interbedded mudstone, siltstone and sandstone with occasional coarser, volcanogenic horizons. In many situations it is weathered to a depth of up to 10 metres, with a resultant clay/silt rich regolith.

Both the cases considered here lie within that part of the Waitemata Group mapped by Ballance (1976) as the Warkworth Subgroup. Further subdivision, primarily on the basis of bedding attitude, was undertaken by Schofield (1989), but structural relationships in the area are still unclear. Schofield recognised two formations, the Paremoremo Formation, characterised by interbedded conglomerates and steeper bedding (dip averaging about 30°SE), and the coastal East Coast Bays Formation, which is gently dipping, generally with attitudes of between 1° - 12°SE. The formations are separated by the northwest-trending East Coast Bays Fault. Both may be described as "volcanic poor" flysch, characteristically comprising non-calcareous light grey siltstone and poorly sorted, grey to greenish grey sandstone. Upon weathering, these rocks form products that include a very stiff kaolinitic clay above the water table, and firm to stiff montmorillonme below the water table (Kermode, 1986). Soil above the water table is characterised by an upper, thin veneer of organogenic loam overlying an elluvial zone of pale, cream-brown, banded, silty-arenaceous sediments from which much of the ferrugenous material had been leached. Below this, approximating the water table, ferrugenous-rich laminae develop in slightly coarser horizons. That swelling clays may be present, intercalcated with these horizons, is testified to by the difficult drilling conditions often encountered.

Land instability on the North Shore is a function of lithology, weathering and bedding attitude, exacerbated by a relatively high rainfall. Bedding plane failure is common in a number of locations, and is of particular concern near the coast. Further inland, failure is known to have occurred following inadequate compaction of fill in new subdivisions, and also on steeper slopes in accordance with the down-slope dip of the weathering profile (irrespective of bedding plane attitude).

3 THE EAST COAST ROAD SLUMP

The site on which the failure occurred lies on the coastal side of East Coast Road, Browns Bay, where it faces and slopes at about 15° to the north. The section was developed as part of a new subdivision in 1975 and involved limited amounts of cut and fill. In July 1975, a dwelling was erected in a cut near the top of the section and although some provision for interception of stormwater was made at the time, (including on site soakage pits for disposal of stormwater), this later proved inadequate, with water accumulating in, on and around the apron of the cut during heavy rainfall.

Slippage developed upon the site following a prolonged period of abnormally high rainfall in mid 1990 (Figure 1). The first significant land movement was recorded on July 28th, eleven days after a daily precipitation record of 32.3 mm was measured. Further heavy rainfall occurred on July 26th (11.4 mm) and on the day of failure, July 28th (24.0 mm), the latter being more than six times the daily expected average for July.

Although movement of the slump was for the most part gradual, it was punctuated by more rapid movement on days 1, 9 and 16 after its inception, finally reaching the situation depicted in Figure 2. The more rapid movement occurred penecontemporaneously with heavy rainfall (see Figure 1).

3.1 *Profile of the failure*

The slump developed as a medium sized rotational flow within the upper 2 metres of clayey silt rich regolith (Figure 3). The uppermost portion (head) was characterised by tension cracks and escarpments, the main escarpment being up to 1.85 metres in height. Blocks of soil retained some coherency in the

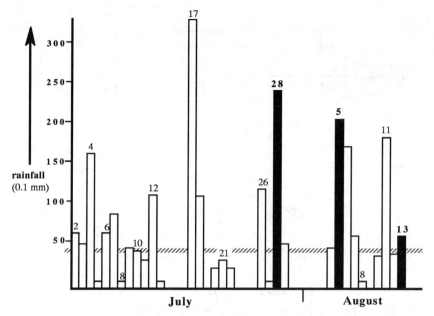

Figure 1: Rainfall data for the period from July 2nd to August 13th, 1990. Shaded bars give precipitation figures for days of significant movement on the site. The horizontal hatched line represents the average daily precipitation for the months of July-August. Note that units are given as 0.1mm. (Data courtesy of New Zealand Meteorological Office, Christchurch).

upper third of the slump, this being in part due to the restraint provided by what had been a 1 metre high log retaining wall (see Figure 2). The central and lower parts of the slump developed as debris flows, which resulted in a central depression and a distal bulging (Figure 4a,b). Ponding developed near the head in local "mini-grabens", and water was found to be flowing from these and from seepages around the perimeter of the slump on September 1, following a period of four rain-free days.

3.2 The failure plane

A series of boreholes were drilled with associated in situ shear strength testings in an attempt to intersect the failure plane. A typical borehole (Figure 3), near the head of the slump, showed undrained shear strengths in the mobilised material (clayey fine sands) at around 35 - 50 kPa; however a rapid increase in shear strength occurred at about 2 metres, with values greater than 130 kPa being recorded. The accompanying lithologic change at this point, from pale silty sands to medium brown sand with grey silt laminae, accompanied by bands of illuvial iron, defined the depth of the (now altered) water table. Sediment below this was logged as sand, but was found to be sensitive to remoulding, with the relic arenaceous structure quickly degenerating to release clay minerals. It is anticipated that the failure plane formed at this depth from minor slippage planes, which developed as surface water moved into the system, lowering shear strength and acting as a lubricant.

Failure is determined as having developed in two distinct phases: the first phase involved development of small slips, following critical hydrostatic loading of the fill and soil adjacent to, and immediately below the dwelling. The second phase followed the resultant rapid increase in hydraulic conductivity, which lead to increased pore pressure and the lubrication of lenses of swelling clays (smectite and/or montmorillonite) at the developing failure plane. The presence of swelling clay being confirmed by expansion in boreholes at 2 metres depth.

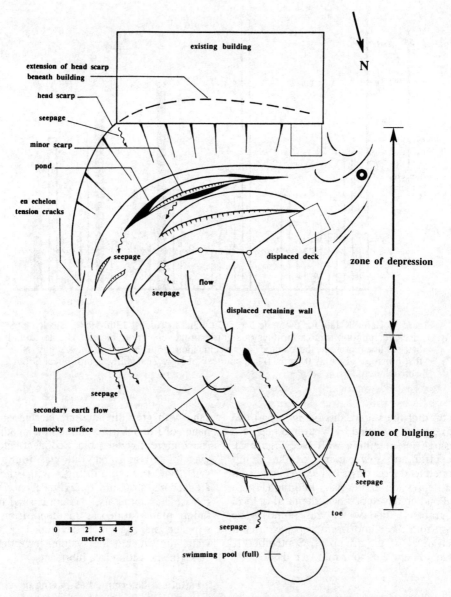

Figure 2: Plan view of the East Coast Road Slump on August 30, 1990, one month after the first major slippage was recorded. Greatest movement occurred near the centre of the slump, as denoted by the arrangement of the cracks. Dislocation of a 300mm diameter stormwater pipe situated 4 metres to the northwest of the building (denoted by a heavy round circle), will have contributed to increased groundwater infiltration. Seepage, shown here as sinuous arrows, occurred over much of the area, but only significant flows are recorded. The filled swimming pool at the northern end of the slump is considered likely to have been a key element in the earlier stabilisation of the lower slope.

3.3 Remedial work

The dwelling was evacuated soon after movement occurred on July 28th. On August 30th, following a geotechnical analysis, two options were identified for the future usage of the property: the first to remove and relocate the building at more suitable site, (there had been very minimal damage and no loss of structural integrity); the second being to underpin and to re-establish services to the exisiting building, and to stabilize the ground beneath the building. The second of the two options was chosen, and the following works were embarked upon in late 1990.

1   Underpining existing house: 300 mm timber poles were placed at 2.5 metre centres to 4.0 metres depth, exposed reinforcing (from the existing undermined foundation) was cleaned and treated. The existing house was jacked back to level as required, and dry packed between new corbel and existing footings.

2   Ground stabilization: the slumped material was removed and the site was re-landscaped with appropriately designed 300mm timber pole retaining walls; pressure grouting, via holes drilled in the basement slab, was carried out to establish firm contact with the substrate.

3   Establishment of services: down pipes have been connected to a new stormwater line established on the western boundary;   links were re-established to the sewer line on the northern boundary, and stormwater that previously collected and ponded on the southern (upper part) of the site, is piped via a silt trap directly into the Council stormwater line.

Costs: Tenders for the above remedial work ranged from \$116,000 to \$50,000, with a figure at the lower end of this range being agreed upon.

4 BROWNS BAY TRANSLATIONAL BLOCK SLIDE

The site on which this failure occurred lies immediately to the southwest of Browns Bay township, in a subdivisional development that

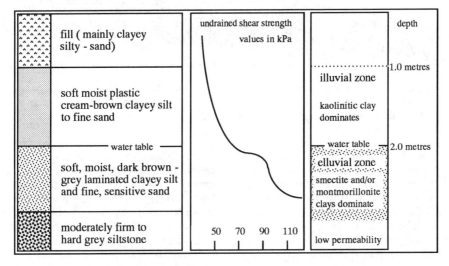

Figure 3: Soil profile (with regolith developed down to below 2 metres), as exposed at the East Coast Road Slump, showing lithological changes, strengths and properties. Undrained shear strengths were measured *in situ* with a hand held shear vane. Depth below surface in metres.

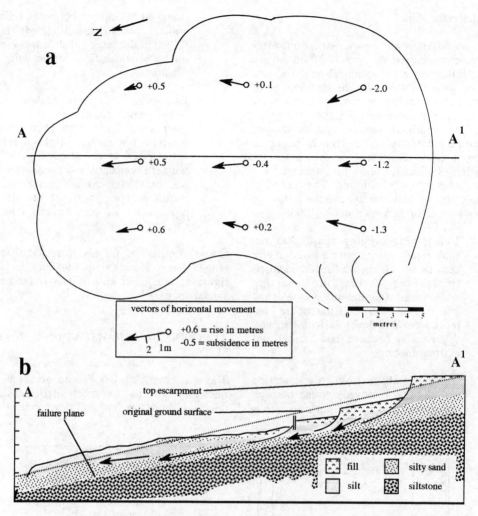

Figure 4: Schematic representation of the East Coast Bays Slump failure. **a** : Plan showing vectors of movement on the slump as observed on August 30, 1990. **b** : Cross-section A-A1, with vertical scale (left hand side) in metres.

started in mid 1985. The area lies on moderate to gently sloping ground, with an easterly aspect. In September 1985, after much of the initial services, including stormwater systems, footpaths and curbing had been emplaced, a major landslide developed, with a top escarpment, aligned with surface contours, of about 60 metres in length forming (Figure 5).

The failure is interpreted as a composite one, involving both translational slide (including some degree of underthrusting) and slump components. The slide appears to have developed within the weathering profile, with failure being precipitated by the infiltration of excess surface water to the upper surface of the less permeable horizon. Water entered via trench excavations carried out for stormwater services, where insufficiently compacted backfill acted as both a water trap and conduit, resulting in an increase of pore pressure and a lubrication of the lower, impermeable clayey horizon. The instability of this particular site was exacerbated by the emplacement of much of the stormwater system parallel to contours, with insufficient downslope linkage (Buckeridge, 1991).

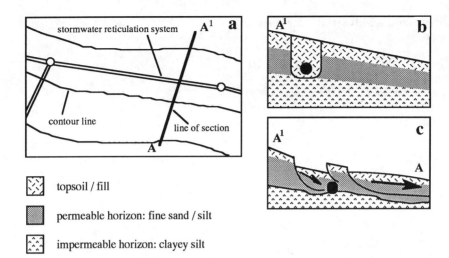

topsoil / fill

permeable horizon: fine sand / silt

impermeable horizon: clayey silt

Figure 5: Schematic representation of a translational block slide-slump failure at Browns Bay. **a** : Plan showing disposition of stormwater reticulation along contours, **b** : section A-A[1] before failure, where insufficiently compacted backfill has acted as a water trap and conduit, **c** : section A-A[1] after failure, with water having reached and lubricated the lower, impermeable clayey horizon via the fill. The emplacement of much of the stormwater system parallel to contours, with insufficient downslope linkage, contributed to the instability of this particular site.

5 CONCLUSIONS

The above failures are typical of landslides in the North Shore region. Key aspects of both are the movement of surface water to the potential failure plane with the resultant increase of pore pressure at this level and the presence and formation of swelling clays at or near the top of the water table. The chemistry of the groundwater at this level is generally acid, (containing H_2CO_3), leaching of soil alkali components occurs within this zone, resulting in poorer bond strengths and the formation of smectite and montmorillonite. In both the above cases, instability has been exacerbated by either poorly placed or poorly drained fill. The climate of the North Shore is also a contributary factor, as long dry periods occur, alternating with periods of intense rainfall. During the dry periods, extensive dessication cracks open up in the soil (shrinkage limits for most soils being about 20% water content), providing ready access for surface water to the lower horizons.

In both situations, instability problems would have been less likely to have developed if proper attention had been paid to drainage design and landfill placement.

ACKNOWLEDGEMENTS:

The author wishes to thank Peter Bone of Bone and Associates, Consulting Civil Engineers, Auckland, who carried out the structural engineering components of the East Coast Road Slump remedial work, and who critically read the manuscript.

REFERENCES:

Ballance, P.F., 1976. Stratigraphy and bibliography of the Waitemata Group of Auckland, New Zealand. *New Zealand Journal of Geology and Geophysics* 19(6): 897-932.

Buckeridge, J.S., 1991. Educational Response to a Changing Environment. *Proceedings of the New Zealand Institute of Surveyors Annual Conference* 1991 : New Horizons - Facing the Challenge. Auckland, October, 1991. Section 3: 1-5.

Kermode, L.O., 1986. Sheet N42/1, Hobsonville. *Geological Map of New Zealand 1:25,000.* Industrial Series. Department of

Scientific and Industrial Research, Wellington.

Schofield, J.C., 1989. Sheets Q10 & R10, Helensville and Whangaparaoa. *Geological Map of New Zealand* 1:50,000. Department of Scientific and Industrial Research, Wellington.

Landslides, Bell (ed.) © 1995 Balkema, Rotterdam, ISBN 90 5410 032 X

Engineering geology of schist landslides, Cromwell, New Zealand

D. F. Macfarlane
DSIR Geology & Geophysics, Dunedin, New Zealand
B. W. Riddolls
Riddolls Consultants Ltd, Queenstown, New Zealand

N. A. Crampton
Crampton Engineering Geological Ltd, Cromwell, New Zealand
M. R. Foley
Mark Foley and Associates, Nelson, New Zealand

ABSTRACT: Engineering geological work for the completion of the Clyde Power Project has mainly involved detailed assessment of the characteristics of landslide-affected schist terrain bordering the reservoir area in the Cromwell Gorge, immediately upstream from the Clyde Dam. Landslide development has been strongly influenced by various geological characteristics of the schist bedrock, with initial failure having been controlled to a large extent by foliation shears and low-angle faults, which commonly form persistent rock mass defects. Investigation of such defects in both bedrock and where they form slide surfaces was hampered by a high degree of geological variability. Geological modelling for assessment of the stability of reservoir slopes and landslide hazards has been aided by recognition of different categories of slope movement materials, distinguished on the basis of characteristics that reflect varying degrees of disruption and downslope displacement.

1 INTRODUCTION

The Clyde Power Project has involved the construction of a 102 m high concrete gravity dam near Clyde for Electricity Corporation of NZ Ltd. Upon full commissioning, it will impound a reservoir (Lake Dunstan) up to 62 m deep, extending through the Cromwell Gorge and upstream into the Upper Clutha Valley, and along the lower reaches of the Kawarau River. Through the Cromwell Gorge, nearly 40% of the shoreline will be bordered by large landslide areas (Figure 1), some of which currently exhibit creep movements (Gillon & Hancox, 1992).

This paper describes the principal engineering geological characteristics of schist terrain in the Cromwell Gorge, with particular reference to the nature and variability of geological materials and defects in relation to assessment of landslide extent and development. The results of this work were used in geological modelling for evaluation of the effects of lake-filling on stability, in hazard and risk assessments, and in determining the need for and design of remedial works.

2 GEOLOGICAL SETTING

The complex nature of the schist terrain may be attributed to the influence of several geological factors, both regional and local in extent. Those relevant in the understanding of many engineering geological characteristics are described as follows.

2.1 Geomorphology

The Cromwell Gorge is a broadly V-shaped antecedent valley, incised up to 1400 m below the adjacent mountain crests. The valley is more-or-less symmetrical in cross-profile, with average slopes of 22 to 24 degrees. However, a marked asymmetry of the lower slopes is evident above a steeper inner gorge incised into bedrock.

2.2 Stratigraphy

The main geological features of the project area are outlined by Gillon & Hancox (1992) and more detailed descriptions are provided by Turnbull (1981, 1987). The regional stratigraphy is dominated by lithologically variable schist which forms the basement rock. Metamorphism occurred under high strain, low thermal gradient conditions during the Jurassic to early Cretaceous periods. Regional uplift and erosion during the Cretaceous-early Tertiary period resulted in the development of an

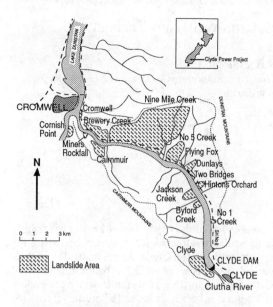

Figure 1. Location plan, Cromwell Gorge

extensive low relief surface (peneplain) over the entire region. During the mid Tertiary, terrestrial sediments were widely deposited but were subsequently stripped from rising mountain ranges.

Late Quaternary age (last 500,000 years) glacial moraines, interglacial lake sediments, outwash gravels and alluvial fans largely obscure the remaining Tertiary sediments and are well preserved throughout Central Otago, mainly in intermontane basins. Discontinuous remnants of fluvioglacial outwash terraces are preserved at several levels through the Cromwell Gorge.

2.3 Schist lithology

In the Cromwell Gorge area the dominant rock type is textural zone IV greyschist, with subordinate greenschist and metachert. The greyschist material grades lithologically between two main types (Turnbull 1981, 1987):
- mica-rich: pale greenish-grey to silvery grey, thinly and regularly laminated schist.
- quartz-rich: pale to mid-grey schist characterised by well developed quartz-albite segregation laminae.

The quartz-rich lithology predominates in the Gorge, except in the central part where, on the north side, mica-rich is more common. Planar mineral segregation laminae of variable thickness (≤ 5 mm) are typical of both lithotypes. Locally, quartz-rich schist exhibits either contorted, fre-

quently disrupted, segregation laminae, or has quartz occurring as rods or irregular masses.

2.4 Structure

Regionally, the structure of the gently-dipping Otago schists is dominated by successive generations of large recumbent folds. The Jurassic-early Cretaceous Rangitata Orogeny resulted in three recognisable phases of such folding in the schists, as well as considerable fault movement (Turnbull 1981, 1987). The subsequent Miocene-Recent Kaikoura Orogeny (which locally has occurred over the last 3-5 million years, and is still active) has further deformed the schists, by both folding and faulting, accompanied by regional uplift and down-warping to form extensive 'basin-and-range' topography (McSaveney et al 1992).

In the Cromwell Gorge schist foliation attitudes are very variable. Folding has warped the schist into a broad, open, westward-plunging fold (the Leaning Rock Antiform of Turnbull, 1981). Small (parasitic) drag folds occur adjacent to both low-angle faults and steeply dipping faults. A broad zone of steeply dipping to overturned schist has been recognised beneath the lower parts of several landslides along the left bank (eg. No. 5 Creek Slide, Nine Mile Downstream Slide), and is attributed to bedrock flow (Beetham, Smith et al 1992) rather than tectonic deformation.

Significant moderately steeply dipping faults include features along and parallel to the Cromwell Gorge (eg. in Brewery and Firewood Creeks), the Fish Creek Fault, the Mystery Fault Zone and the Cairnmuir Fault Zone (Gillon & Hancox, 1992). The latter, which crosses the valley about 1 km upstream from the dam, is known to have undergone late Quaternary displacement along parts of its length.

Two major low angle south-dipping faults have been recognised along the right bank of the gorge - a higher level fault (itself displaced) separating schists of distinctly different texture and a lower fault separating quartz-rich and mica-rich schist. Low angle faults have also been identified on the left bank during detailed investigation of the landslides (Gillon & Hancox, 1992).

3 ENGINEERING GEOLOGICAL DESCRIPTION OF SCHIST TERRAIN

3.1 Bedrock

In situ rock, widely affected by varying degrees of

stress relief, has been extensively investigated throughout the Cromwell Gorge. This work has provided much useful information for assessment of the main engineering geological controls on slope evolution and landslide formation, and has formed an important part of the evaluation of reservoir slope stability.

3.1.1 Rock Material Characteristics

Unweathered greyschist is typically light to dark grey where fresh, variation in colour normally being related to relative proportions of quartz/albite (light) and mica (dark) minerals. The rock is generally only slightly to moderately weathered in outcrop and within a few metres of the surface or, at depth localised adjacent to defects .

Uniaxial compressive strength varies markedly in relation to lithology, mica-rich lithologies tending to be of lower strength than quartz-rich. The more mica-rich schists are typically in the weak to moderately strong category (q_u 10 to 50 MPa) while schists dominated by quartz are strong (q_u 30 to 90 MPa). Massive and poorly laminated schists, and greenschists are very strong (q_u 100 to >250 MPa).

All lithologies display marked strength anisotropy, with strength normal to foliation 3 to 8 times greater than the strength parallel or moderately oblique to foliation. Strength anisotropy appears to be greatest in quartz-rich schist with well developed quartz/albite and mica segregation laminae.

The following peak effective strength parameters parallel to foliation have been determined for quartz-rich schists: $c' = 2$ MPa; $\emptyset' = 23$-32 degrees.

3.1.2 Rock Mass Defects

Rock mass defects are well exposed in outcrop and excavations throughout the project area, especially in Clyde Dam foundations, (Paterson et al 1983, Hatton et al 1987, 1991). They include weak seams (crushed and sheared zones and gouge), shattered zones, and joints.

Crushed and sheared zones typically have low shear strengths, commonly controlled by thin gouge seams consisting of moderately to highly plastic clay with persistent slickensided surfaces. Gouge seams include a clay-size fraction of at least 30%, mixed layer mica-smectite being the dominant clay mineral. They range from very thin slivers (0.5mm) to seams several tens of millimetres thick.

Figure 2 Typical subdued landslide topography, Cromwell Gorge

Both tectonic and stress relief mechanisms of defect formation are recognised. Tectonic defects include faults, consisting of crushed zones up to several tens of metres thick with large offsets; typically with distinctive footwall, hanging wall and internal gouge seams. Crushed, shattered, and incipiently fractured rock may persist for many metres away from the principal zone of deformation.

Crushed and sheared zones with smaller apparent fault offsets, and joints of similar attitude, typically form defect sets sub-parallel to the mapped major faults. Detailed mapping of the Clyde Dam foundations (and other exposures throughout the gorge) has revealed a number of defect sets, generally arranged orthogonally, but locally displaying a high degree of variability and structural complexity which constrains accurate modelling and projection of persistent defects. For example, even after surface exposure, core drilling, and a network of sub foundation drives, the absolute geometry of one of the most extensively investigated crushed zones at the Clyde Dam site remained uncertain beyond the immediate vicinity where it was physically defined (R. Thomson, pers. comm.)

Defects sub-parallel to foliation are common in the schist. These features, called "foliation shears", include crushed and sheared zones, resulting from both flexure in response to folding and direct shear displacement in response to faulting. Generally they are between 50 mm and 2 m thick, with very thin gouge seams present as either thin slivers located immediately adjacent to the wall rock, or as narrow seams within crushed material. Foliation shears are typically spaced 5 to 50 m apart and

2139

locally splinter, warp, terminate against other defects, or die out.

Stress relief features, formed in response to removal of overlying materials and valley downcutting, are recognised mainly by opening of defects, resulting in deterioration of rock mass quality. Suitably oriented pre-existing defects have accommodated often large differential movements. In addition, extensional joints, infill zones and conduits have developed as a result of gross rockmass dilation. Significant flexural deformation, attributed essentially to stress relief, is recognised in the weaker, mica-rich greyschists (Beetham, Smith et al, 1992). Crushed zones and joints along foliation are more closely spaced than in the undeformed rock, imparting a secondary fissility.

The field strength of typical defects is derived principally from two components: the residual friction angle and waviness effects. Residual strength is dependent on the presence of gouge seams, which are identified on surfaces with as little as 50 mm displacement, and their clay mineralogy. Typical values range from 8 to 14 degrees for samples of highly plastic clay. Waviness is almost always present and, depending on whether previous displacements have produced thick or thin crushed zones (containing the gouge), and on the nature of stress distributions during displacement, may contribute between 10 and 50% of the field strength (G. Salt, pers. comm.). Shear strengths of joints have not been evaluated in detail because most sets are steeply dipping and act as release surfaces rather than surfaces of sliding.

3.2 Landslide masses

3.2.1 Slide materials

Extensive deposits of schist-derived bouldery debris, and displaced schist masses have been mapped within the delimited slide areas shown in Figure 1. Their surface extent is often poorly defined (Figure 2) and well-developed scarps are rare, mainly because of the age of the features (Section 4.3). They also tend to be obscured by erosion and the accumulation of surficial materials, such as loess or colluvium. However the surface morphology of the schist debris is mostly hummocky, indicating that these deposits have undergone significant movement.

Schist debris is typically greyish-brown, and up to 80 m thick consisting of sub-angular blocks up to 20 m across, mostly randomly oriented in an unconsolidated, well graded, finer-grained matrix. Beneath the debris, defects commonly exhibit varying degrees of dilation or mechanical disruption resulting from slope movements along low-angle weak seams (eg. foliation shears, faults). These dilated rock mass conditions have been recorded both from direct investigations, and indicated from seismic refraction and reflection surveys (Bryant et al 1992).

Both within and at the base of the landslide masses, crushed zones have been proved in many surface excavations, tunnels, shafts and cored drillholes. These zones clearly comprise defects along which sliding has taken place, and contain either moderate to highly plastic clay gouge seams, derived from original tectonic crushed zones, or lower plasticity gouges thought to be due solely to gravitation processes.

The effective field strength of these defects has been determined using the resistance envelope method. Resultant frictional strengths have been estimated to lie in the range 21 - 29 degrees, with most slides of the order of 26 ± 2 degrees.

3.2.2 Geological controls on development

Slope movements in the Cromwell Gorge (and the nearby Kawarau Gorge; Bell, 1976, 1987) have been strongly influenced by the nature of the schist bedrock. Failure is favoured along pre-existing defects and is most common on slopes formed parallel or subparallel to foliation, persistent low angle faults, and weaker rock types such as mica-rich schist. Landslides are extensively developed where foliation generally dips into the valley, ie. along the left (north) side downstream of Brewery Creek Slide area to near Dunlays Slide. Defects are also recognised as controlling lateral margins, internal zonation, and headscarps to landslides.

Most of the initial movement appears likely to have been a result of stress relief along pre-existing defects within the rock mass. Where adversely oriented and critically located, ongoing movement is readily reactivated by other disturbing forces.

Low-angle defects along which stress relief displacement has occurred are considered to have little or no potential for on-going movement because the driving force causing stress relief displacement is necessarily dissipated during that displacement. However, such low angle defects at valley floor level may allow toe breakout of a slide across the rock fabric, and steeper features may be reactivated as slide failure surfaces if they are subsequently under-cut.

As well as forming failure zones, many defects also act as groundwater barriers (aquicludes or aquitards) below and within landslides. This commonly results in complex groundwater systems which include multiple perched aquifers within slides, and/or confined conditions both within and beneath slides (Macfarlane et al, 1992).

Confined groundwater systems locally exert significant uplift pressures which can have a major influence on landslide stability (Gillon & Hancox, 1992).

4 INTERPRETATION OF SLOPE MOVEMENTS

4.1 *Investigation difficulties*

Old, large landslides can be difficult to recognise (Patton & Hendron, 1974), and this is commonly the case in the Cromwell Gorge where most of the landslides are ancient features exhibiting subtle, subdued morphology. Because surface evidence of landslide boundaries is often poorly defined, mapping of foliation or lineation attitudes was important in helping to distinguish in situ outcrops from those affected by slope movements.*

While small diameter (HQ) core drilling has been used extensively to investigate sub-surface conditions (Gillon, et al 1992), the results obtained have rarely been unequivocal in terms of definition of slope movements. Particular difficulties have been experienced in the recognition and correlation of weak seams forming basal failure surfaces, even where exposed in drives or surface excavations. Not only are such seams often not recovered in drillcore but their projection and interpolation between investigation points is commonly constrained by the same high degree of geological variability, complexity and lack of continuity and marker horizons that has been demonstrated in the in situ schist. Similarly, while mechanically disrupted rock masses were readily distinguishable in excavations (eg. by open joints), recognition in drill core was more difficult, being based mainly on trends in fracture and weathering intensities. It follows that geological models developed for use in assessment of lake filling effects were inevitably a simplification of actual conditions, containing many uncertainties which had to be allowed for in numerical analyses.

4.2 *Modes of failure*

The complex structural controls on slide geometry have often made it difficult to identify failure mechanisms.

While several of the commonly described types of slide movements (Varnes, 1978) appear to have operated, translational failures controlled principally by pre-existing foliation shears or low-angle faults seem to predominate. Debris slides can also be assumed to approximate such surfaces in many places. It is likely that the debris has undergone significant movement (10's to 100's of metres), such that any asperities could have been reduced, and hollows infilled. Curvilinear failure surfaces may therefore be inferred locally within or at the base of the debris, especially in toe areas, implying an element of rotation within otherwise generally translational movement. Internal failure surfaces are also likely to be curvilinear in both upslope and cross slope directions.

For many of the slides some combination of structural control and adverse groundwater conditions has resulted in complex failure mechanisms, which may have varied through time to cause the development of the slides in their present form.

4.3 *Landslide evolution*

The slide areas delimited along the reservoir are considered to have developed as part of an ongoing process of landform evolution that has resulted in the formation of the Cromwell Gorge as the river has down-cut through the actively rising mountain blocks over at least the last 3-5 million years (McSaveney, et al 1992). Landslides have probably been present in the gorge since about 300 m of downcutting was reached. Incision of the rivers appears to have kept pace with uplift, and development of the very large landslides has probably been gradual.

During the long history of landsliding in the Cromwell Gorge, most of the larger slides would have gradually propagated to greater depths by ongoing erosion at their toes, and exploiting successively lower favourably inclined weaknesses in the rockmass. Deepening and/or enlargement will have occurred at irregular intervals with periods of greater activity as the river has oversteepened slopes or down-cut in response to episodic uplift,

---

Slope movement is used here in the sense of Varnes (1978), i.e. comprising falls, slides, spreads, and flows, (also widely known as *mass movement*).

Table 1 Terminology used in slope movement interpretation, Cromwell Gorge

Term	Mass Description	Surface Characteristics	Type of Movement	Degree of Displacement
Chaotic Debris	Gradation from large competent rock blocks to fine grained material (commonly intensely sheared and crushed schist) with gouge seams. Foliation attitudes of blocks normally highly variable. Slickensided downslope-dipping internal or basal failure zones.	Laterally impersistent breaks in slope (i.e. hummocky). Well developed slide scarps where active. Blocks on surface.	Rotational, translational, or complex slide.	10's to 100's of m.
Displaced Schist	Large competent rock blocks either in contact or separated by open and/or infilled joints or by zones of sheared/crushed material. Foliation either parallel or oblique to undisturbed rock, and adjacent blocks may be slightly rotated relative to one another. Slickensided near downslope-dipping internal or basal failure zones, typically sub-parallel to foliation or pre-existing rock defects (Also termed "Blocky Debris", where disruption is greatest).	Laterally persistent breaks in slope (i.e. broadly irregular). Forms outcrops locally.	Translational slide.	m's to 10's of m.
Basal Failure Zone	Crushed zones and gouge seams with some sheared and shattered schist of variable thickness. Fabric may be sub-parallel to boundaries, contorted, or totally disrupted. Gouge seams typically thin but persistent with slickensides, oriented downslope.	Outcrops rare.	Slide.	mm's to 100's of m.
Disturbed Schist	Sub-slide rock mass with partly open defects often infilled. Typically quartz-rich massive and laminated schist. (Termed "Relaxed Schist" in near-surface situations).	Relaxed schist forms prominent outcrops similar to undisturbed schist.	Stress relief processes.	mm's to m's.
Deformed Schist	Sub-slide rock mass, fissile with discrete sheared and crushed zones sub-parallel to flexurally deformed ("buckled") foliation, steepened to overturned locally. Typically mica-rich laminated schist.	Prominent outcrops with foliation dipping at moderate to high angles to undisturbed schist.	Bedrock flow, by stress relief and/or gravitational processes.	mm's to 10's of m.
Undisturbed Schist	*In situ* rock mass with closed defects.	Forms prominent outcrops.	None.	None.

climatic changes, variations in river sediment load and/or lowering of base level during glaciations. In contrast, the aggradation of gravels during warmer stadials during glaciations will have acted to reduce (or even halt) slide movements so that deep under-cutting and large-scale activity have been episodic over hundreds of thousands of years. Although there is no conclusive evidence, it is possible that large major earthquakes may have also influenced the early development of the landslides. However, geological evidence shows that there have been no significant large-scale rapid movements in the last 15-30000 years, even though many large earthquakes are inferred to have occurred during this time (Gillon & Hancox, 1992).

There is no evidence to suggest that any of the landslides have moved as single masses. Most can be zoned into areas with distinctly different surface morphology, degrees of slope deflation, and material types. Variation in the depth to slide base at several of the slides is indicative of steps across defects trending normal and parallel to slope direction, often approximately coincident with the boundaries of zones mapped at the surface. These characteristics strongly suggest that different parts of the slides have moved by differing amounts, at different depths and probably at different times. Slow creep movement has probably always been the dominant rate of movement.

4.4 Interpretive Categorisation of Materials

4.4.1 Basis of Terminology

Interpretation of the results of the extensive programme of subsurface investigations in the landslide-affected Cromwell Gorge was directed principally towards the development of the most likely geological models for use in numerical analysis assessment of the effects of lake filling. Due to frequent difficulties in recognising and correlating weak seams forming either internal or the deepest (basal) failure zones, these zones were commonly identified approximately by categorising and mapping various soil and rock mass characteristics attributable to varying degrees of disturbance inferred to have resulted from differing amounts of movement. Table 1 gives the terminology and characteristics for each category.

In summary, slide movement is inferred from the presence of "chaotic debris" and "displaced schist", bounded at depth by a "basal failure zone". "Disturbed" and "deformed schist" are categories of sub-slide slope movement, originating through rock creep and/or stress relief. Figure 3 illustrates general spatial relationships.

4.4.2 Recognition and distribution of materials

An example of interpretation of material distribution (part of Clyde Slide Area) is given in a cross-section, Figure 4, with emphasis on the typical evidence used for the definition of internal and basal failure zones. Additional examples (in Nine Mile Creek Slide Area) are described by Beetham

et al (1992). In most slide areas, recognition of the different slope movement material types was relatively straightforward. In some cases, however, if available data did not conclusively allow identification of a basal failure zone, it proved difficult to determine whether rock mass dilation was the result of landsliding or stress relief movement processes. Distinguishing between displaced and disturbed material categories was therefore somewhat arbitrary, requiring a conservative approach in selecting basal failure zones.

The presence and distribution of material categories both within and between landslide areas is typically variable. In the simpler landslides either chaotic debris or displaced schist are the only materials overlying the basal failure zone, whereas elsewhere, both are present.

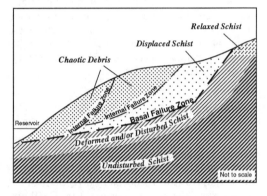

Figure 3 Schematic representation of general distribution of slope movement material categories, Cromwell Gorge

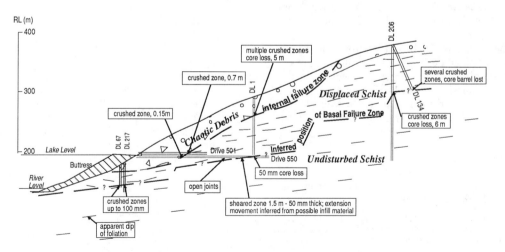

Figure 4 Typical data and interpretation of slope movement, Cromwell Gorge

The recognition of basal failure zones did not remove the need for recognition of defects in the undisturbed bedrock which could have the potential for "first time" landslide development.

5 CONCLUSIONS

Complex schist geology in the Cromwell Gorge has contributed to a range of controls on slope movement. As a result, subsurface definition of slide extent and geometry by conventional investigation methods has been constrained mainly by difficulties in correlating major defects, notably those forming basal failure zones. Geological models developed for use in hazard assessment and determination of lake filling effects were therefore necessarily based on a simplification of real conditions. Recognition of the inherent uncertainties in geological modelling was an important factor in engineering decisions on the need for and scope of remedial works carried out to offset possible adverse effects of lake-filling on reservoir slopes on Clyde Power Project.

ACKNOWLEDGEMENTS

The authors are part of a large team developing solutions for landslides around the shoreline of the Clyde Dam reservoir. The works are being carried out by staff from the Electricity Corporation of NZ Ltd, Works Consultancy Services Ltd, DSIR and specialist subconsultants. The contributions of specialist reviewers D. Stapledon and L. Richards and the Electricity Corporation Review Panel of J. Libby, D. Deere, W. Swiger and W. Reimer to our understanding of the landslides is gratefully acknowledged.

The paper has also benefitted from regular discussion with and criticism from fellow engineering geologists on Clyde Power Project.

The permission of the Electricity Corporation of NZ Ltd and Works Consultancy Services Ltd to publish this paper is also gratefully acknowledged.

REFERENCES

Bell, D.H. 1976: Slope evolution and slope stability, Kawarau Valley, Central Otago, New Zealand. *Bull Int Association Engineering Geologists* 14, 5-16.

Bell, D.H. 1987: The K9 landslide, Kawarau Valley, Central Otago. *Proc. Int. Symp. on Eng. Geol. Env. in Mountainous Areas*, Beijing, China.

Beetham, R.D., G. Smith, C.J. Newton, & D.N.Jennings, 1992: The engineering geology of the Nine Mile Creek schist landslide complex. *Proc. 6th International Symposium on Landslides*. Christchurch, NZ.

Beetham, R.D., K.E. Moody, D.A. Fergusson, D.N. Jennings, & P Waugh, 1992: Landslide development in schist by toe buckling. *Ibid.*

Bryant, J.M., T.C. Logan, D.E. Woodward, & R.D. Beetham, 1992: The use of seismic methods in defining landslide structure. *Ibid.*

Gillon, M.D., P.F. Foster, G.T. Proffitt, & A.P. Smits, 1992: Monitoring of the Cromwell Gorge landslides. *Ibid.*

Gillon, M.D., & G.T. Hancox, 1992: Cromwell Gorge landslides - a general overview. *Ibid.*

Gillon, M.D., B.N. Denton, & D.F. Macfarlane, 1992: Field investigation of the Cromwell Gorge landslides. *Ibid.*

Hatton, J.W., J.C. Black, & P.F. Foster, 1977: New Zealand's Clyde Power Station. *Water Power & Dam Construction* 39 (12) 15-20.

Hatton, J.W., P.F. Foster, & R. Thomson, 1991: The influence of foundation conditions on the design of Clyde Dam - *XVIIth Congress on Large Dams*, Vienna, 157-178.

Macfarlane, D.F., A. Pattle, & G. Salt, 1992: Nature and identification of Cromwell Gorge landslides groundwater systems. *Proc. 6th International Symposium on Landslides*. Christchurch, NZ.

McSaveney, M.J., R. Thomson, & I.M. Turnbull, 1992: Timing of relief and landslides in Central Otago, New Zealand. *Ibid.*

Paterson, B.R., G.T. Hancox, R. Thomson, & B.N. Thompson, 1983: Investigations in hard rock terrain in Symposium on Engineering for Dams and Canals *IPENZ Proc Tech Group*, 9, 4G

Patton, F.D. & A.J. Hendron, 1974: General report on "mass movements". *Proc. 2nd Int. Congr. IAEG*, Sao Paulo, 1, V-GR, 1-57.

Turnbull, I.M. 1981: Contortions in the schists of the Cromwell District, Central Otago, New Zealand. *NZ Journal of Geology and Geophysics* 24, 65-86.

Turnbull, I.M. 1987: Sheet S133 - Cromwell. *Geological Map of New Zealand, 1:63360*. Map and Notes. DSIR, Wellington.

Varnes, D.T., 1978: Slope movement types and processes *in Landslide Analysis and Control*. US Transportation Research Board, Special Report 176.

Landslides, Bell (ed.) © 1995 Balkema, Rotterdam, ISBN 90 5410 032 X

Landslides in tabular rock masses of an active convergent margin

Warwick M. Prebble
Geology Department, University of Auckland, New Zealand

ABSTRACT: Tabular limestones and shales form prominent hog back mountains in the southern part of the East Coast Deformed Belt, New Zealand. This belt is an on-land manifestation of Cenozoic convergence along the plate boundary and is a focus of continuing thrusting, shear, uplift, earthquakes and slope movement in the Upper Cretaceous and Cenozoic cover.

The limestones are strong, but well bedded and closely fractured. The interbedded shales are fissile, smectite-rich and weak.

Clay seams and crush zones are widespread. Many are parallel to bedding, others dip steeply across bedding and are very wide.

Rock slides are relatively few but large, covering areas up to 2 or 3 km ^{2}. The slide blocks are limestone, on basal rupture zones formed of bedding parallel crush zones and clay seams in shales.

Debris flows, earth flows and toppling are numerous and widespread, making up complex slope movements of several km^2 in area. Toppling in limestone produces masses of debris which give rise to stony debris flows in high intensity rainstorms, choking streams and causing severe aggradation. The shales, where crushed, are extremely weak and give rise to clayey debris flows and earth flows.

All the slope movements are probably Holocene, and many are active. Rapid tectonic uplift and stream downcutting, combined with the closely fractured and crushed rock, promote continuing slope movement.

Landslide mapping at scales of 1:5000 to 1:25,000 over an area of 300km^2 has defined elements and slope movement types.

This paper summarises some initial results.

1. METHODS - CHARACTERISATION OF ROCK TYPES AND GEOMORPHIC MAPPING.

Lithology of the intact rocks was characterised by strength, density, porosity and slaking properties combined with mineralogy and microstructure.

Defects were characterised by orientation, continuity, spacing, size (width), openness (aperture), relief, surface roughness, coatings and fillings.

A distinction was drawn between fractures (joints and cleavage), and zones (clay seams, crush zones, shear zones, gouge).

Geomorphic elements included ridge lines, streams, active faults, terraces, and slope movements.

Slope movements were mapped according to geometry (areal shape and internal features such as scarps, ridges, troughs, blocks, trenches), lithology, internal structure and relationship to adjacent rock masses (eg override zones, pull away zones, lateral shears, valleyward overturning, head scarps, rock debris interlocking).

From this information 3 main types of slope movement were defined (Fig. 1).

2. LITHOLOGIC UNITS - THE INTACT ROCKS.

Micritic limestones and smectite-rich shales dominate the slopes of the hog-back ranges in northeast Marlborough, at the southern end of the East Coast Deformed Belt, New Zealand (Fig. 2) (Spörli, 1980). Collectively, these units are known as the Amuri Limestone.

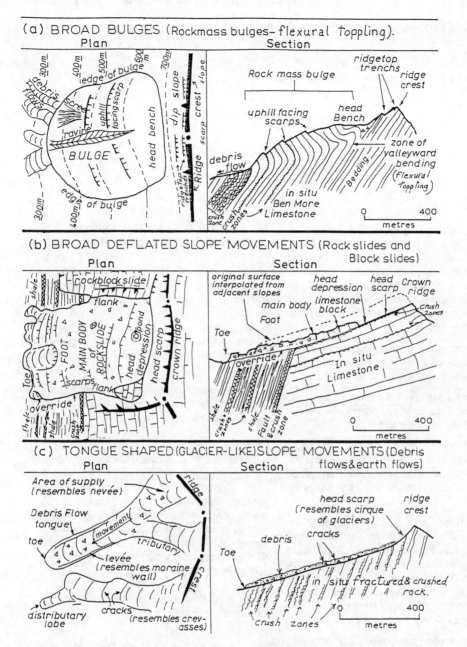

(a) BROAD BULGES (Rockmass bulges- flexural toppling).
Plan   Section

Rock mass bulge
ridgetop trenchs
ridge crest
edge of bulge
300m 400m 500m 600m 700m
debris flows
uphill facing scarp
ravine
BULGE
head bench
head Bench
uphill facing scarps
zone of valleyward bending (flexural toppling)
dip slope
Ridge scarp crest slope
ridgetop
debris flow
Bedding
in situ Ben More Limestone
crush Zone
crush zones
300m 400m
edge of bulge
0   400
metres

(b) BROAD DEFLATED SLOPE MOVEMENTS (Rock slides and Block slides)
Plan   Section

shale
Rock block slide
flank
MAIN BODY of ROCKSLIDE
pond
head depression
head scarp
Crown ridge
FOOT
scarps
flank
Toe
override
shale
crush zones
original surface interpolated from adjacent slopes
main body
Foot
Toe
head depression
head scarp
limestone block
Crown ridge
crush zones
override
shale
shale crush zones
Fault & crush zone
In situ Limestone
0   400
metres

(c) TONGUE SHAPED (GLACIER-LIKE) SLOPE MOVEMENTS (Debris flows & earth flows)
Plan   Section

Area of supply (resembles nevée)
ridge
Debris Flow tongue
toe
movement
tributary
levée (resembles moraine wall)
crest
distributary lobe
cracks (resembles crevasses)
head scarp (resembles cirque of glaciers)
ridge crest
debris
cracks
Toe
in situ fractured & crushed rock.
crush zones
0   400
metres

Fig 1   Plan & section views of
THE THREE MAIN TYPES OF SLOPE MOVEMENT
showing their typical features.

Siliceous shale (Whangai Formation), flint (Mead Hill Formation) and coarse clastic rocks are also widespread but do not feature prominently in the hog back ranges, and are usually absent from dip slopes.

2.1 Micritic limestones.

Limestones form several stratigraphically distinct units, separated by smectite shales. The limestones are siliceous (up to 50% SiO_2 as chalcedony and quartz), hard, dense, non porous and strong. They are very durable. Solution weathering and karst are absent. The lowermost limestone grades down into the Mead Hill flint.

2.2 Smectite shales.

The shales lie between the micritic limestones and usually contain regularly alternating interbeds of micritic limestone and fissile, micritic smectite-rich shale. Shale interbeds dominate over micritic interbeds, particularly in the upper shale units. The shales are fissile, very weak to extremely weak, slightly porous and slake readily.

Together the micritic limestones and smectite shales are a few hundred metres thick and Late Cretaceous to Oligocene in age.

3. Defects

The limestones are *tabular* rock masses, with highly continuous fractures and pressure-solution cleavage, very closely spaced, usually parallel to bedding. Occasionally, disjunctive pressure-solution cleavage dominates over bedding parallel cleavage. However, bedding parallel *fractures* truncate and dominate disjunctive defects.

The shales are *fissile* and *foliated*. Bedding parallel and very low angle *crush zones* (10^0 to 15^0 to bedding) are common and continuous in the shales. The low angle crush zones are thrust faults which give rise to considerable repetition and tectonic thickening of the shales and limestones. The crush zones vary from several cm to many tens of m in width. They are soft, wet and extremely weak.

Clay seams are laced through the crush zones. These seams are also wet, very soft and extremely weak.

Several large, wide crush zones of the main thrust faults can be mapped throughout the region. A few of these are more steeply dipping than the others and are at a higher angle to bedding. The main stream valleys have been carved out along these steeply dipping crush zones which consequently run along the toe of all the main dip slopes.

4. GROUNDWATER

No subsurface data exists for groundwater. Springs are widespread along the lower part of all dip slopes, immediately upslope of the wide, steeply dipping crush zones. This indicates that groundwater is 'ponded' to some extent in the lower part of the dip slopes.

5. TOPOGRAPHY AND TECTONIC SETTING

All the dip slopes are formed in limestone, with shales and other rocks exposed on the scarp slopes. Main stream valleys follow crush zones, tributaries following a combination of dominant and minor fracture sets. Rectilinear, fall-line ravines dissect some dip slopes, irrespective of defect orientation. The ravines are located in rock mass bulges (section 6).

The southern end of the East Coast Deformed Belt is a zone of dextral transform faulting, rapid uplift ($7mmyr^{-1}$), microseismicity and strong shallow earthquakes (Walcott 1978, Bibby, 1981). Faulting has produced continuous and wide zones of crushing. Thinner, more frequent, bedding-parallel crush zones are also produced by flexural slip folding.

Very wide (several 100's of m in places) belts of mélange are essentially clayey crush zones containing fractured blocks, lenses and lozenges of limestone. These belts behave as soil masses.

6. SLOPE MOVEMENTS

The most numerous slope movements are toppling, debris flows and earth flows. Rock slides are relatively rare. Complex combinations of failure, usually from debris flows, earth flows and toppling are numerous and widespread.

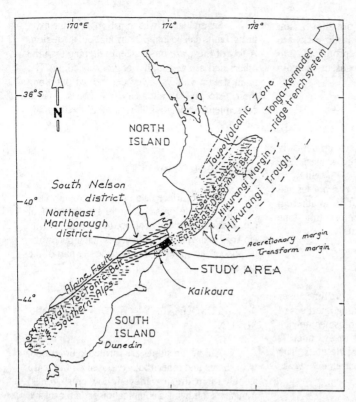

Figure 2   The New Zealand Mobile Belt, showing the
location of the study area.

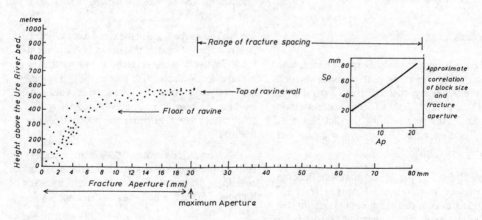

Relationships of slope height to fracture aperture, and of aperture to spacing,
Tarn rock mass bulge.

Figure 3

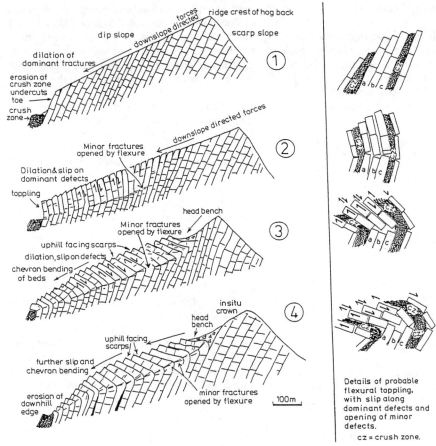

Figure 4 Scheme for development of rock mass bulge and valleyward bending over (Flexural toppling) Progression from ① to ④, commencing with undercutting of slope by erosion of extremely weak wide crush zone.

Details of probable flexural toppling, with slip along dominant defects and opening of minor defects.

cz = crush zone.

6.1 Toppling - rock mass bulging:

Whilst minor toppling (simple block toppling) is widespread on scarp slopes (anaclinal slopes) and leads to rock falls, the most common and widespread form of toppling is found on dip slopes. These are underdip cataclinal slopes (Cruden, 1988). The dip of the dominant defects (continuous bedding parallel fractures and crush zones - a tabular rock mass) is greater than the dip of the slope surface itself, and greater than 25°. The toppling involves *valleyward bending and overturning* of the rock mass, hinging at a depth of several tens of m in the slope.

Extreme *dilation* of fractures (Fig. 3) accompanies the toppling which was observed only in *extremely closely fractured tabular limestone.*

Dilation produces extreme loosening of the rock mass which then acts more like soil rather than rock, although as a toppling mass. Dilation of the rock mass produces a visible bulge of toppling rock usually a kilometre square. Fall-line ravines cut through the centre of some bulges, revealing the overturning down to the hinge line and exposing the loosened, dilated rock mass to erosion by debris flows.

Uphill facing scarps cut across the rock mass bulges, especially along crush zones and clay seams.

There are no head scarps and no overrides to the bulges, which have indistinct margins. A head bench, or indistinct head trench separates the rock mass bulge from undisturbed slope above it.

2149

6.2   Debris flows and earth flows.

Stony debris flows arise from rock mass bulges where the limestone is loosened to a state which enables it to behave more like gravel than fractured rock. Thick, extensive limestone gravel deposits downslope of the rock mass bulges indicate that large *rock debris avalanches* have also been generated.

High intensity rainstorms (Bowring, Cunliffe, Mackay and Wright, 1978) caused choking of stream channels and severe aggradation downstream of rock mass bulges. Valuable flatland pasture was buried and bridge clearance was seriously reduced.

Clayey debris flows and earth flows are very numerous and widespread. They arise from crush zones in the smectite shales and mélange.

The flows are tongue-shaped and have glacier-like morphology. They are at least 5 to 10m thick, 30 to 100m wide and 150m to 1km long.

6.3   Rock slides.

Only four rock slides and rock block slides were found and these are confined to dip slopes in the limestone where the dip of dominant defects is parallel to the slope and less than 25^0.

Bedding parallel clayey crush zones and clay seams form the basal rupture surfaces, in the underlying shales. Rock slides are a few to several tens of m thick and cover areas up to 2 or $3km^2$.

All the slope movements are probably Holocene (5000 years and less). Radiocarbon dates for one rock slide, terrace ages, absence of tephra known elsewhere in the region and comparative geomorphology indicate a Holocene age. Many are at least partially active.

7.   DISCUSSION.

Rapid tectonic uplift and stream downcutting have probably combined to release high compressive stresses in north-east Marlborough. The area coincides with a belt of locked subduction, crustal shortening, thrusting and the highest crustal compression and most rapid uplift in the East Coast Deformed Belt (Walcott, 1978; Bibby, 1981). It is suggested that rapid uplift and stress relief of the extremely fractured and crushed rock have promoted the fracture dilation and flexural toppling. Rock

mass bulging is the result. Although the toppling is probably initially flexural (Hoek and Bray, 1981), block flexural toppling and eventually, block toppling follow as the dilation and overturning in the rock mass bulges develop. The area is seismically active. Earthquakes probably facilitate the toppling.

The results of this study differ from other discussion of toppling (de Freitas and Watters, 1973; Hoek and Bray, 1981; Holmes and Jarvis, 1985; Cruden, 1988) in that:-
(i) The toppling involves rotation up to and through the vertical, so that overturning and valleyward bending of tabular rock takes place.
(ii) Extreme dilation and loosening of the rock mass assists this process and causes rock mass bulging (Fig. 4).

Only where the dip of dominant defects is less than 25^0 does sliding prevail over rock mass bulging. Hence nearly all the dip slopes are failing by toppling (rock mass bulging) and only a few by rock block sliding.

Mapping and characterisation of in-situ rock mass and slope movements, as described under 1.0 "Methods...", led to recognition of the rock mass bulging on slopes which were previously mapped as stable interfluves.

8.   References

Bibby, H.M., 1981. Geodetically determined strain across the southern end of the Tonga-Kermadec-Hikurangi subduction zone. Geophysical Journal of the Royal Astronomical Society 66, pp 513-533.

Bowring, L.D., Cunliffe, J.J., Mackay, D.M. and Wright, A.F., 1978. East Coast Survey. A study of catchment and stream condition with recommendations. Marlborough Catchment and Regional Water Board. Blenheim, New Zealand.

Cruden, D.M., 1988. Thresholds for catastrophic instabilities in sedimentary rock slopes, some examples from the Canadian Rockies. Zeitschrift für Geomorphologie Supplementband 67: 67-76.

de Freitas, M.H. and Watters, R.J., 1973. Some field examples of toppling failure. Géotechnique, 23, 4: 495-514.

Hoek, E. and Bray, J.W., 1981. Rock slope Engineering. Institute of Mining and Metallurgy, London.

Holmes, G. and Jarvis, J.J., 1985. Large-scale topplings within a sackung type deformation at Ben Attow, Scotland. Quarterly Journal of Engineering Geology. 18: 287-289.

Spörli, K.B., 1980. New Zealand oblique slip margins. Tectonic development up to and during the Cainozoic. In: Sedimentation in oblique-slip mobile zones. (Ed by P.B. Ballance and H.G. Reading). Special Publication of the International Association of Sedimentologists. 4: 147-170.

Walcott, R.I., 1978. Present tectonics and Late Cenozoic evolution of New Zealand. Geophysical Journal of the Royal Astronomical Society, 52: 137-164.

ACKNOWLEDGMENTS

The author thanks Margaret Walker for typing the text and Roy Harris for drafting the figures. Field work was supported by The University of Auckland Reserach Committee.

Landslides, Bell (ed.)© 1995 Balkema, Rotterdam, ISBN 90 5410 032 X

The Grave earthflow complex, South Westland, New Zealand

M. Eggers
Coffey Partners, New South Wales, Australia

D. H. Bell
University of Canterbury, Christchurch, New Zealand

ABSTRACT

The Grave Earthflow Complex is an ancient feature which was reactivated by construction of State Highway Six in 1963-65 through steep coastal hill country, South Westland, New Zealand. Engineering geological mapping and surface movement monitoring studies show the post-1965 movement is a complex rotational slide - earthflow failure. Near the toe a planar slide surface has developed in sheared grey mud derived from faulting in Cretaceous-Tertiary rocks. Movement rates in the range 1 to 4 metres per year were recorded, with the early part of the 15 month record identifying deceleration following unloading of toe support by benching of the road batter and the latter part showing some correlation with seasonal rainfall. Surface drainage and tree planting may provide effective remedial options to limit maintenance problems with the highway, but subsurface investigations are required to design and cost all potential remedial options. Potential for future slope failure in surrounding slopes is high, being governed by the high regional rainfall, wide distribution of weak fault zone materials in the Cretaceous-Tertiary rocks, and steep deeply incised coastal slopes. Future forest management activities on this landform type should be limited to low impact small scale timber harvesting systems.

1.0 INTRODUCTION

The Grave Earthflow Complex is located on a tributary of Grave Creek some 28km north of Haast township, South Westland, New Zealand (Figure 1). The failure complex has developed on steep coastal hill country in faulted Cretaceous-Tertiary bedrock. Evidence suggests that the complex was in a degraded, revegetated state between 1949 and 1963, with movement being reactivated between 1963 and 1965 by construction of State Highway Six. During rainstorm events the earthflow toe now encroaches on to the highway pavement surface, temporarily blocking the road right-of-way.

Engineering geological investigation of the Grave Earthflow was undertaken in 1985-86 to study the mechanism and causes of failure,

to preliminarily assess remedial options for maintenance of the highway. The methodology used involved engineering geological mapping, surface movement monitoring, and laboratory characterization of selected materials.

The investigation was also part of the government controlled South Westland Management Evaluation Program (SWMEP). That Programme was designed to provide key information on which to base decisions on the future management of the native forests in the region. Instability on Cretaceous-Tertiary hill country, such as the Grave Earthflow, presented significant implications to the future management of the native forests in South Westland. Study of the earthflow provided an opportunity to understand the existing slope movement processes

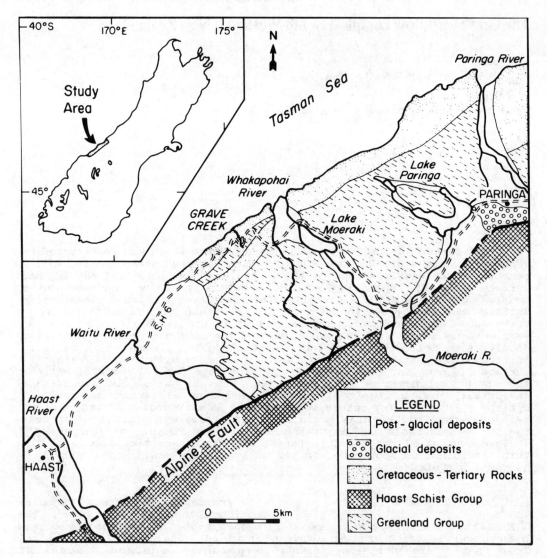

Figure 1: Location and Regional Geology

and their potential influence on future resource management activities.

2.0 GEOMORPHOLOGICAL & GEOLOGICAL SETTING

South Westland lies on a narrow coastal plain bounded to the east by the Alpine Fault at the foot of the Southern Alps (Figure 1).

The study area is situated on a remnant block of steep coastal hill country which escaped erosion by ice during the Pleistocene glacial advances.

In South Westland the Alpine Fault forms a marked topographical and geological boundary. West of the fault on the coastal plain the oldest rocks are interbedded greywacke and argillite of the Ordovician Greenland Group (Adams 1975). These basement rocks are unconformably overlain by a late Cretaceous and early Tertiary succession (Nathan 1977), part of which includes the sequence encountered at Grave Creek (Figure 2).

The main structural features of Grave Creek are two northeast striking faults which bound the Otumotu Formation and between which the Grave Earthflow occurs. The eastern Arnott Fault uplifts Greenland Group basement and the western fault, interpreted as a splinter from the Arnott Fault (Nathan 1977), uplifts a sliver of Lower Otumotu Formation against Tauperikaka Coal Measures and Upper Otumotu Formation. Northwest of this fault outcrops a sequence which essentially dips steeply to the west, extends from the upper Otumotu Formation through to the Arnott Basalt which outcrops in the seacliffs (Figure 2).

The climate of the region is dominated by heavy and prolonged orographic rainfall caused by the Southern Alps creating a barrier to the prevailing westerly airstream. Paringa, situated 25km north of

Grave Creek, receives a mean annual rainfall of 5441mm (N.Z. Meteorological Service 1973).

3.0   SLOPE MOVEMENT DESCRIPTION

3.1   History of Movement

The earliest evidence of movement comes from 1949 air photos which show a series of raised features on a broad spur above the present position of the earthflow. These appear to be degraded scarps which form part of a dormant, revegetated failure that indicates probable ancient movement of this section of the hillslope. Reactivation of movement was caused by construction of State Highway Six in 1963-65. Initial slide movement during this period was along a headscarp which has subsequently remained inactive, with current

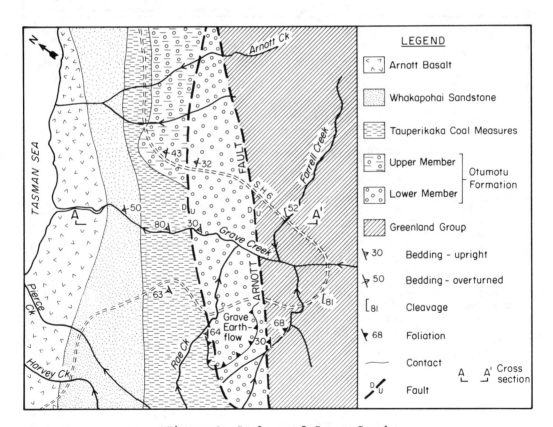

Figure 2: Geology of Grave Creek

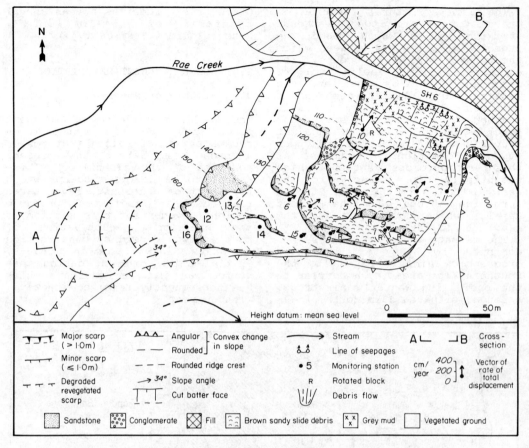

Figure 3: Summary Engineering Geology Plan of Earthflow Complex

movement being confined to the southeastern side of the earthflow complex (Figure 3). This active area has provided continuing maintenance problems with the highway.

In December 1984 the earthflow toe was recut and benched to create a catchment area for debris to collect adjacent to the road. This action significantly increased movement rates in the currently active portion, and possibly reactivated the upper part of the 1963-65 slide area near the headscarp.

3.2 Slope Morphology

The headscarp of the Grave Earthflow is located between 80 and 135m upslope of State Highway Six on a 27° slope, and forms a distorted square shape in plan view (Figures 3 & 4).

The hillslope on which the earthflow is located can be divided into the following three areas:

(a) the slope above the main headscarp extending to the ridgeline;

(b) the failure activated during road construction in 1963-65, which is outlined by the main headscarp system with a total area of approximately 1ha: and

(c) the currently active portion on the eastern flank outlined by a major scarp with an area of about 0.5ha.

The hillslope above the main headscarp contains a series of at least four degraded scarps which are semicircular in outline and surround areas of depressed ground. The degraded, revegetated nature of the

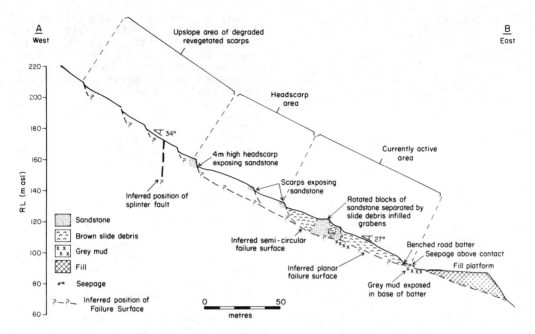

Figure 4: Slope Profile through Earthflow

features suggest that they are ancient, inactive (pre-1949) slope movements.

The failure area activated in 1963-65 is bounded by a 135m long headscarp up to 4.0m high, and by a shallow (<1m deep) graben structure. The area between this scarp system and the currently active portion contains a number of tension cracks and smaller shallow debris slide failures. The majority of this area is vegetated, with little bare ground.

The currently active system is confined to the eastern flank of the earthflow complex adjacent to the southeastern lateral shear zone (Figure 3). The head area contains a series of displaced and rotated sandstone blocks, while the main body is characterised by a series of transverse tension cracks and significant bare ground. The foot of the earthflow consists of a degraded road batter which is undergoing secondary failure by debris flow mechanisms and producing a debris lobe approximately 20m across that flows across the road during rainstorm events.

3.3   Earthflow Materials

Within the Grave Earthflow Complex Lower Otumotu Formation and a sandstone unit are exposed. The Lower Otumotu Formation occurs as a grey, massive cobble conglomerate of high and very high strength outcropping in major lateral scarps on both sides of the earthflow near the foot area.

The large rotated blocks in the earthflow head and body are a low strength sandstone, the origin of which is uncertain. The yellow-brown weathering of the sandstone is characteristic of the Tauperikaka Coal Measures which could not be in-situ in this position. The sandstone blocks may therefore form an ancient slope movement deposit originating west of the splinter fault upslope from the existing failure. Alternatively, the sandstone may be the upper member of the Otumotu Formation and would directly overlie the Lower Otumotu conglomerate.

Crushing and shearing of the Otumotu Formation by movement on the Arnott and splinter faults has

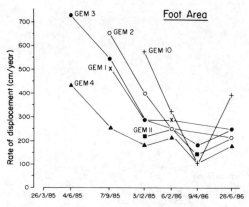

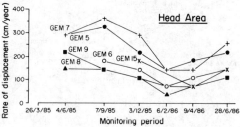

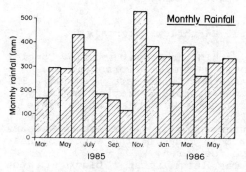

Figure 5: Rate of Movement and
Correlation with Rainfall

produced a grey clay of low
plasticity with bands of
carbonaceous material. This grey
clay outcrops at the toe of the
earthflow in the cut road batter,
and was also encountered in shallow
test pits excavated at the base of a
number of rotated blocks in the
failure body. The grey clay is
inferred to be the zone in which
basal sliding is taking place in the
main body and foot of the earthflow.

The majority of material which
makes up the earthflow is a
silty-gravelly sand with some clay
derived from disintegration of the
sandstone blocks by movement within
the earthflow. The clayey sand has

a high in-situ moisture content
which is close to its liquid limit.
Shallow test pits excavated below
the earthflow surface and exposures
in the road batter show a wet to
saturated zone about 0.5m thick
immediately above the clayey
sand:grey clay contact. X-ray
diffraction analysis showed the clay
mineralogy of both the grey clay and
clayey sand materials to be a
chlorite-illite-quartz assemblage
with no swelling clays identified.

4.0 SURFACE MOVEMENT MONITORING

4.1 Survey Network & Movement Data

Surface movement monitoring was
undertaken to establish the rate of
movement and to define any patterns
of movement which could help
determine the mechanism of failure.
A network comprising nine monitoring
stations positioned in the currently
active portion was initially set up
in March 1985. A further seven
stations were added to the system in
October 1985 to include the main
headscarp area, which was thought to
be experiencing a reactivation of
movement due to the development of
fresh tension cracks in this area
after March 1985. Seven surveys
were completed using EDM methods in
the period March 1985 to June 1986.

4.2 Rates of Movement

The greatest rate of movement was
recorded in the foot area of the
active portion, with a maixmum
recorded rate of 419cm/year. Rates
of movement decreased toward the
head region of the active portion,
with the lowest rate recorded being
108cm/year. Direction of movement
tended to follow the local slope
aspect, resulting in a change in
direction from east-northeast in the
head area to north-northeast at the
foot of the earthflow (Figure 3).
Beyond the active portion movement
occurred immediately above the
active scarp, and was separated from
the remaining area of the inactive
headscarp area by a fresh tension
crack system.

Changes in rate of movement over
the monitoring period (Figure 5)
shows a decline from an initial high
level in the foot area, while the

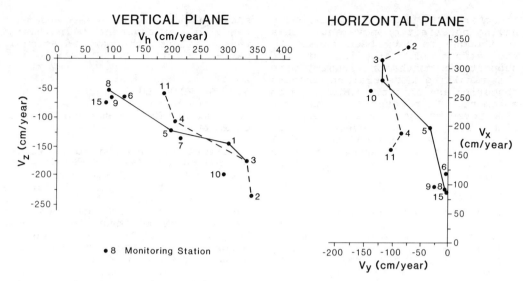

Figure 6: Hodographs of Movement

head area experienced a more subdued decline with some stations increasing in rate at the start of the period. This pattern of movement and the reactivation of movement in the headscarp area are inferred to be a response of the earthflow to unloading of toe support when the road batter was cut and benched in December 1984.

Removal of toe support would have caused an "instantaneous acceleration" in movement rates in the foot area. Monitoring recorded the subsequent deceleration of movement, and this obscured any other rate controls which may have been present. In the head region the increase in movement rates was delayed and reduced as the effect migrated upslope. Near the end of the monitoring period movement rates had returned to levels where the primary influence appears to be climatic.

4.3 Correlation with Rainfall

Correlation of rainfall with movement is poor over the majority of the monitoring period, with the best correlation occuring during the last three months of the record (Figure 5).

From April 1985 to April 1986 movement rates appear to be independent of rainfall, with a high

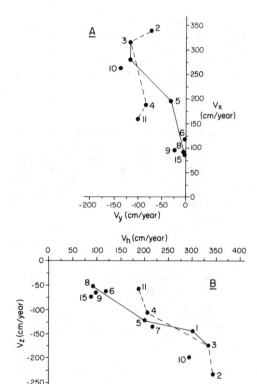

Fig 6·22: Hodographs of average displacement rates of monitoring stations on the Grave Earthflow, measured between 26/3/85 and 28/6/86
A = Horizontal Plane   B = Vertical Plane

2159

November 1985 rainfall apparently having no effect on movement. This is thought to have been caused by the effect of unloading of toe support six months previously, with the resulting deceleration process obscuring any climatic influences on movement. From about April to June 1986 movement shows some correlation with rainfall, and movement rates appear to have decreased to a level where climatic influences were once again the primary factor controlling movement.

4.4 Hodographs of Movement

Movement rates can be used to analyse the mechanism of movement by resolving the average movement rate vector into x, y and z components as:

$$V_x = x/t; \qquad V_y = y/t; \qquad V_z = z/t;$$

where t is the time interval between measurements (in this case the monitoring period). The horizontal projection of the average displacement vector (V_h) is determined from this data by:

$$V_h = (V_x^2 + V_y^2)^{1/2}.$$

Graphs showing the horizontal plane (V_x versus V_y) and the vertical plane (V_h versus V_z) displacement vectors are termed hodographs of the movement rates, and assume that displacement is continuous or steady creep (Ter-Stepanian & Ter-Stepanian 1971). Groups of geometrically related monitoring stations are separated on the graphs for interpretation (Figure 6), and the following features of the failure mechanism can be inferred from the hodographs :-

(a) Stations GEM 15, 8, 5, 3 and 1 are located along a line parallel to the earthflow axis. They plot as rectilinear hodographs with the foot area stations showing the greatest movement (GEM 3 and 1) and the head region stations showing the least. This is indicative of a retreating flow failure with the seat of movement located in the lower part of the slope.

(b) Stations GEM 2, 3, 4 and 11 are located across the earthflow normal to the axis. The hodograph shows a part rectilinear and part fan-like arrangement, indicating that failure is part-rotational, part-translational.

5.0 SLOPE FAILURE MODEL

5.1 Failure Mechanisms and Sequence of Movement

Failure is inferred to be taking place by a complex rotational slide-earthflow type of movement. Near the head of the complex, failure is by multiple rotational sliding in a sandstone unit which degrades into an earthflow movement in the main body and foot. The failure surface below the main body and foot is a planar zone of sliding in sheared grey clay derived from the Otumotu Formation, and it daylights in the cut road batter face. The combination of curved and planar elements on the failure surface gives the movement a part-rotational and part-translational character.

Movement may also be considered as multi-storied, with debris slides near the head and a debris flow at the foot developed on top of the main earthflow movement. At the time of monitoring, movement rates were retrogressive retreating upslope opposite to the direction of movement of material.

5.2 Causes of Failure

Factors contributing to failure can be separated into fundamental causes which combine to produce a slope at limiting equilibrium, and triggering events which initiate failure of the critical slope (Figure 7).

Two processes have contributed to the original (pre-1949) slope movements:-

(a) Reduction in shear strength of the rock mass by:

i) faulting of the lower Otumotu Formation producing a low strength grey clay material which forms part of the failure surface: and

ii) water infiltration during rainstorm events increasing positive pore pressures.

(b) Increase in shear stress in the slope by removal of underlying support as a result of

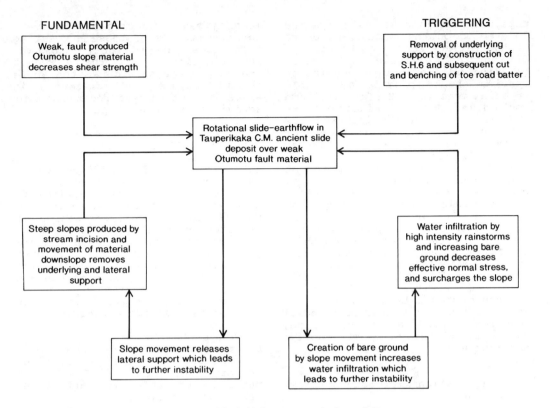

FUNDAMENTAL

Weak, fault produced Otumotu slope material decreases shear strength

TRIGGERING

Removal of underlying support by construction of S.H.6 and subsequent cut and benching of toe road batter

Rotational slide–earthflow in Tauperikaka C.M. ancient slide deposit over weak Otumotu fault material

Steep slopes produced by stream incision and movement of material downslope removes underlying and lateral support

Water infiltration by high intensity rainstorms and increasing bare ground decreases effective normal stress, and surcharges the slope

Slope movement releases lateral support which leads to further instability

Creation of bare ground by slope movement increases water infiltration which leads to further instability

Figure 7: Model of causes of slope failure

over-steepened slopes produced by deep stream incision.

Undercutting of the critical slope by construction of State Highway Six provided the triggering mechanism required to reactivate movement.

The Grave Earthflow slope failure model highlights three key elements of the coastal Cretaceous-Tertiary hill country considered to be relevant to slope stability in this landform type:

(a) major faulting which has produced low strength fault zone materials which form part of the underlying rock mass,

(b) high, steep slopes produced by deep stream incision due to uplift since the last glacial advance causing a drop in stream base level, and

(c) high regional rainfall.

6.0 REMEDIAL OPTIONS & IMPLICATIONS FOR FUTURE FOREST MANAGEMENT

6.1 Remedial Options for State Highway Six

Remediation of the earthflow to limit maintenance problems with State Highway Six can be carried out provided it is recognised that complete stabilisation of the earthflow may not be economically viable or necessary.

Potential remedial measures include:

(a) dewatering (including surface drainage),

(b) planting of trees,

(c) buttressing the toe,

(d) relocation of the highway, and

(e) maintaining the present situation with regular maintenance.

With little or no subsurface information it is difficult to design and cost potential remedial measures. However, based on the physical condition of the slope and inferences on the nature and depth of the failure zone a combination of surface drainage and close planting of appropriate vegetation species should provide acceptable remedial measures. Surface drainage would divertoverland rainwater flows and vegetation would increase the interception and retention of rainwater and eliminate groundwater by evapotranspiration.

To further investigate the cost effectiveness of potential remedial measures subsurface investigations by trenching or drilling and installation of piezometers and inclinometers is required. This would allow the nature and depth of the failure zone, groundwater levels and the material permeabilities to be assessed more accurately. Continued monitoring of surface movements should be carried out to assess the effectiveness of remedial measures adopted, and to detect any possible extension of the active movement area.

6.2 Implications for Future Forest Management

Evaluation of slope stability on forest lands can be undertaken at three levels of land management activities (Prellitz 1985):
(a) Resource Allocation,
(b) Project Planning, and
(c) Critical Site Stabilisation.
The aim of this study was to provide information at the Resource Allocation phase to give an overview of slope movement potential adequate for development planning. Studies of the Grave Earthflow complex found that the potential for coastal hill country in Cretaceous-Tertiary bedrock to undergo slope failure is high, and governed by the following factors:
a) high regional rainfall,
b) wide distribution of weak fault zone materials, and
c) steep, deeply incised coastal slopes.
Large scale timber harvesting operations such as clear-felling and selective logging involving extensive road development and deforestation would have a large impact on slope stability and cannot be endorsed. Small scale timber harvesting systems such as bush-mill logging may be applicable however, but the impacts of these harvesting techniques are largely unknown. Investigations at the Project Planning stage should use the above three factors as criteria in assessing future resource management activites on these slopes.

7.0 CONCLUSIONS

1. Investigation of the Grave Earthflow Complex has been carried out by engineering geological mapping, surface movement monitoring and laboratory characterisation of selected materials. Assessment of the mechanisms and causes of failure from this data has allowed preliminary evaluation of remedial options for highway maintenance and provided an understanding on the implications of slope instability for future resource management.

2. The earthflow complex is bounded by two faults which has produced a sheared grey clay derived from the Otumotu Formation. The fault materials probably form part of the basal zone of sliding.

3. Surface movement monitoring measured an initial high rate of movement, that subsequently declined during the monitoring period. This pattern of movement was considered to be the response of the earthflow to unloading of toe support by excavation and benching of the road batter three months prior to commencement of monitoring. The deceleration process initially obscured any climatic control on movement, however by the end of the monitoring period rainfall appeared to have an effect on movement rates.

4. Failure is taking place by a complex rotational slide-earthflow type of movement. Fundamental causes of failure are a reduction in shear strength of the rock mass by faulting in the Otumotu Formation, and water infiltration during rainstorm events increasing positive pore pressures. An increase in shear stress in the slope by removal of underlying support as a result of over-steepened slopes produced by deep stream incision is also a factor in the original slope factor.

5. Movement was reactivated by undercutting of the critical slope during construction of State Highway Six in 1963-65, and by further benching in late 1985-early 1986. As noted above, the deceleration phase of this later reactivation was recorded during the slope monitoring programme.

6. Remediation of the earthflow to limit maintenance problems with the highway may be achieved by surface drainage and close planting of vegetation. However, design and costing of remedial options requires more subsurface information, and a limited drilling and monitoring programme is outlined.

7. Studies of the Grave Earthflow Complex found that the potential for coastal hill country in Cretaceous-Tertiary bedrock to undergo slope failure is governed by the high regional rainfall, wide distribution of weak fault zone materials, and steep deeply incised coastal slopes. Implications for future forest management include the high impact potential on slope stability by large scale timber harvesting activities such as clear-felling and selective logging, although small scale harvesting systems may by more suitable and should be investigated by project planning studies.

REFERENCES

Adams, C.J.D. 1975. Discovery of Precambrian rocks in New Zealand: age relations of the Greenland Group and the Constant Gneiss, South Island. Earth & Planetary Science Letters 28:98-104.

Nathan, S. 1977. Cretaceous and lower Tertiary stratigraphy of the coastal strip between Buttress Point and Ship Creek, South Westland, New Zealand. N.Z. J. Geol. Geophy. 20(4): 615-54.

N.Z. Meteorological Service 1975. Rainfall normals for New Zealand 1941 to 1970. N.Z. Met. Serv. Misc. Publ. 145.

Prellitz, R.W. 1985. A complete three-level approach for analysing landslides on forest lands. Proc. Int. Symp. on Erosion, Debris Flow and Disaster Prevention, Tsukuba, Japan: 475-479.

Ter-Stepanian, G. & Ter-Stepanian, H. 1971. Analysis of landslides. Proc. 4th Conf. Soil Mech. Budapest: 499-504.

Discussions

G1   Monday 10th February 11am
S1   Monday 10th February 2pm
G2   Tuesday 11th February 9am
G1 + G2   Tuesday 11th February 11am
S4   Tuesday 11th February 4pm
G3   Thursday 13th February 9am
G5   Thursday 13th February 11am
S2   Thursday 13th February 2pm
G4 and World landslide inventory   Friday 14th February 11am
General discussion   Friday 14th February 2pm

Discussion session abbreviations: Q = Question
A = Answer
C = Comment

Landslides, Bell (ed.)© 1995 Balkema, Rotterdam, ISBN 90 5410 032 X

Discussions

MONDAY 10th FEBRUARY.
11am Technical Session G1

Q1: WHITTAKER

Do we know enough about the situation where slides occur without explanation? (ie. a situation where land is weakened and then a slide occurs much later).

A: BROWN

Lots of overseas reports to the inventory show new features, not observed for centuries.

New features developed due to :- glacial retreat, climatic change, landuse practices, earthquake shaking.

Q2: TORRANCE (to Brown)

For the Landslide Inventory work, are they getting information on mineralogy or just rock types?

A: BROWN

Just beginning. The National Academy of Science asked for this information but not capable of giving it as yet. Hopefully will be able to in the future.

Q3: TORRANCE (to Cruden)

Was there much information on mineralogy in the G1 session papers?

A: CRUDEN

In some cases, yes. Detailed study is encouraged.

Q4: CHOWDHURY

Could anyone give an idea of how to optimise the costs of investigation and geological model development?

A: MOON

Read the work of Dave Stapledon!

MONDAY 10th FEBRUARY.
2pm Technical Session S1.

C: McMAHON

There has never been a landslide caused by an earthquake recorded in a mine (compared with natural slopes). Is the difference the amount of cyclic loading that natural slopes have been subjected to? The explanation is the cumulative effect of a number of earthquakes. A slope subjected to 500 earthquakes is not necessarily stable but rather 500 closer to failure.

Q5: GILLON (to Crozier)

How was the correlation between landslides and earthquakes in the Taranaki region made?

A: M.CROZIER

First, the location of faults and their degree of activity was established for the region (see paper by B. Pillans, NZ J Geology and Geophysics, Vol 33, 1990). There are nine distinct faults within the region; all high-angle, normal and active. The distribution of all deep-seated landslides was mapped for three single-aged events. The landslides cluster around fault zones, with some of the largest slides having failed on the fault trace itself. Back analysis of slides employing worst case conditions (e.g. groundwater at surface) indicate static factors of safety above 1.0. Numerous large landslides belonging to the same age, and a liquifacted debris matrix also suggest earthquake triggering.

C: JACOBSON

The Newmark Analysis is a useful technique. It is important to distinguish whether a landslide block is in stable, unstable or neutral equilibrium.

C: ROLD

In 1970 the Rulison underground nuclear detonation experiment artificially generated an earthquake of 5.5 Richter Magnitude. Severe problems were expected from the several hundred or more landslides and two steep canyon highway and railroad rockfall areas, all within 60 miles of the detonation. Detonation was postponed from the spring to the early fall (for weather and other reasons). Groundwater conditions were better in the fall. Careful post-detonation investigation indicated only one small new landslide, no reactivation of the numerous landslides and negligible rock fall or topple of cliffs or balanced rocks.

C: SCHINDLER

Ch. Bonnard and F. Noverraz gave a paper and some additional oral information concerning the rockfall of Randa near Zermatt in Switzerland. Their report was written soon after the catastrophes of April 18th and May 9th 1991. Since then, many investigations and measurements have been made by the geologists involved. I was charged by the federal and cantonal authorities to evaluate all facts available now and to interview eye-witnesses, this with the aim to determine the reasons for the rockfalls and to make a forecast. Unfortunately, it is too early to give final results, but already now it is obvious the events of Randa were exceptional in many respects.

Very peculiar is the fact that the two big rockfalls with a total volume of 30×10^6 m^3 took many hours to break down. As a result, the debris accumulated in a very regular cone without any reverse slope. The extremely strong winds usually associated with big rockfalls were very weak here and caused no serious damage.

Some previous activation of the unstable slope was observed many years before the first catastrophe. The old forest at the bottom of this zone was destroyed. Previous to the second rockfall, a typical acceleration of the displacements could be observed. Both collapses started near the base of the unstable rock mass, mostly near springs. They were announced by outbreaks of dust or water under high pressure and were followed by bursting of the rock in a manner similar to an explosion. In addition, rock masses broke down along steep, intensively fractured zones. Seen as a whole, nearly no toppling happened and the first breakdowns undermined higher portions of the rock mass.

Preliminary conditions for the collapse are the intricate pattern of different systems of brittle tectonic deformation (schistocity and bedding planes are not critical) and the glacial erosion. When the ice support of the steep slopes definitively melted away some 10000 years ago, detension opened existing discontinuities and groundwater could start to circulate. It seems that this process slowly progressed to reach an uncommon degree today. Besides the very steep slopes, another important reason for this observation could be connected with the washing of fine detrital material by groundwater. In the paper published in the proceedings of this congress by A. Baumer and I, we point to the effects of cushions of sand and silt (clay is rare in these metamorphic crystalline series with fresh erosion) filling fissures within an unstable rock mass. At Dirinei, when combined with strong water supply, they allow a temporary gliding movement along rough, undulating joints inclined at an average of $43°$. At Randa, topography excludes long range gliding, but by filling open fissures and by dislocation of sand and silt in a very steep, shaky rock mass, stability may become critical during periods of strong water supply.

Another interesting aspect concerns the "Little Ice Age" ending here around 1880. Permafrost did not reach the later rockfall area, but seems to have partly blocked the supply of groundwater and sand and silt. The warming up of the climate could slowly have opened old circulation paths so the stability conditions degraded continuously. The breakdown happened during a period of massive inflow of groundwater when snow melted. On the other hand, an earthquake or sliding along planes of layered sediments can be excluded as having started the rockfalls. Investigations go on; well established results can be expected for summer 1992.

Q6: ANON.(to Morgenstern)
The softening model has 4 constants. What is the significance of the parameters and how are they determined? Models can lead to practical problems. What is the softening period of 100 years based on?

A: MORGENSTERN
The parameters reflect the curve of the failure envelope, but there is no simple answer on how they were derived. Based on experience, eg. in the U.K.

The process of softening involves dilation around blocks. Need experimental work on how to derive parameters.

C: JANBU
Assessment of soil stability has two major parts:
     1) assess subsurface
     2) numerical analysis - many methods from equilibrium point of view.

Q7: OLDS (to Morgenstern)
a) Is loosening just localised strength reduction? Quantify?
b) Does the use of factor of safety in deformation analysis lead to overprediction?

A: MORGENSTERN
a) Loosening minimises the joint strength (Canadian Geotech. Journal) and can quantify strain as a result of loosening. Quantify loosening in

fissures and joints as approximately equal to the reduction of strength and may be associated with softening.

b) Can get the factor of safety without deformation analysis by using finite element analysis and calculating where deformation reduces shear strength. Deformation is hard to predict as the factor of safety approaches 1.0. It is difficult to predict deformation due to the localisation of strength reduction.

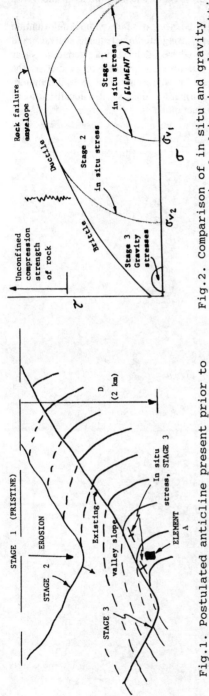

Fig.1. Postulated anticline present prior to valley formation, rather than creep of strata under gravity loadings.

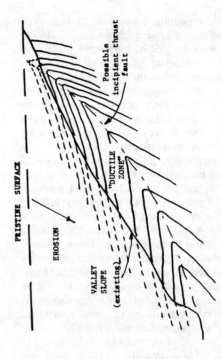

Fig.2. Comparison of in situ and gravity stresses involved in river downcutting.

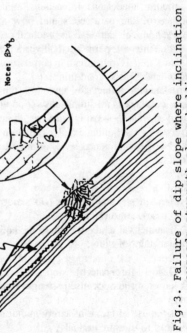

Fig.3. Failure of dip slope where inclination exceeds strength along bedding.

Fig.4. Postulated anticlinal structure present prior to valley formation.

C: JAMES

Several authors raised the topic of quasi-plastic deformation of rock slopes by gravity loadings. However, such an interpretation by no means represents a unique solution.

Figure 1 shows a sequence of events in river downcutting. At the outset, Stage 1, the stresses at an element A might be:

$\sigma_v = $ h, (40MPa, say); and $\sigma_h = 2 - 3$ x σ_v. A Mohr Circle can be drawn with $\sigma_h - \sigma_v$ as diameter and it falls well below the failure envelope of Figure 2, for this situation.

As the river cuts down to Stage 2, the vertical stress reduces to perhaps half the original value, while the horizontal stress at A remains much the same. The new Mohr Circle now cuts the failure envelope for the rock, and indicates that the rock mass is liable to failure mechanisms such as thrust faulting or some quasi-plastic deformations. This occurs at a level still well below the river, where the rock condition is still close to ductile.

At Stage 3, the vertical stress reduces to a small value and the horizontal stresses are reduced by cracking, etc. There is also some rotation of the stress directions, as shown. The Mohr Circle for a depth below the valley side of about 50m is shown. This Mohr Circle is quite small in comparison to the earlier in situ stress condition, and it lies well within the brittle range of behaviour of all rocks, except perhaps halite.

Thus the alternative interpretation of the "bending" of the strata on the valley side, as shown in Figure 1, would be that the geometry existed prior to the river valley.

Other valley side geometries were illustrated in the report by Dr. Cruden; some of these are repeated here, approximately, with comments.

Figure 3 shows the failure of a dip slope, which occurs in a situation where the dip slope, ß, exceeds the strength along bedding, Φ_b. Usually a slab type failure occurs at the outermost bed, and this could conceivably progress into the slope. Alternatively, at least some of the beds in this situation would be at or near residual strength, and this might allow a deeper slip to develop initially. This type of failure is amenable to simple analysis, since the bedding strength can be easily obtained, or estimated from inspection. The break-out strength of the rock toe can be deduced. The act of break-out is a strain softening mechanism, and this can lead to chaotic deformations in the toe area. Such slides do not develop by "toppling" failure within the slope, unless the slope angle exceeds the strength along the bedding by a substantial amount.

The situation in Figure 4 was also proposed as a case of creep deformation of a slope. The actual mechanism was not explained, but it would appear that all beds would need to turn over simultaneously. If not, they would loosen and become filled - especially when vertically orientated - with colluvium, etc. For this sort of creep to occur, the rock mass Φ would need to be less than the slope angle, which is unrealistic. If rock behaved in this way, it would not be possible to build dams.

C: FELL

I would like to caution against the generalisation implied by Dr. Brand's address that tropical soils should be considered as unsaturated for slope stability analysis. While this may be true for Hong Kong, it is certainly not true for many, if not most, high rainfall tropical areas. My experience in Papua New Guinea, Fiji, Tanzania and Northern Australia would confirm this. I would put forward the view that tropical soils are in fact not different to soils elsewhere and it would be better if we did not have sessions in conferences devoted to them. It would be better to differentiate on soil origin, eg. residual soils, lateritic soils, etc.

I would also caution against Dr. Brand's assertion that antecedent rainfall is not relevant to the occurrence of landsliding, and that 24 hour or less rainfalls are critical. His assertion may be correct for shallow landslides, but is not for larger, deeper slides.

Among others, monitoring of a 10 to 15m deep slide near Newcastle, Australia, showed that antecedent rainfall from 30 to 60 days was important. This is reported in Fell, Chapman and Maguire in the 1991 Isle of Wight Conference on Landslides.

C: WESLEY

I agree with Professer Fell's comment that residual or tropical soils are just as likely to be fully saturated as partly saturated. The fine grained residual soils of volcanic origin found in many parts of Southeast Asia, especially Indonesia, are generally fully saturated, apart from a very shallow zone at the ground surface. The water table may however be quite deep and above the water table the pore water pressure will generally be negative, that is it will be in a state of tension. This will clearly influence the effective stress and thus the stability of slopes. Compared to the zone of positive pore pressure, measurement of the pore water tension is very difficult and it is rarely taken account of in stability analysis.

C: BRAND

It is certainly true not all failures result from the loss of soil suction. There are without doubt situations where a positive pore pressure increase in the form of a rising groundwater table (or even a perched water table) is the main cause of failure. In fact, the direct correlation obtained in Hong Kong

between slope failures and short-term rainfall intensity might mean that groundwater movements are of more importance than suction decreases where the slope materials are of high permeability.

I certainly agree that there are no basic differences between tropical and sub-tropical landslide mechanisms. The vast majority of landslides throughout the world are caused by rainfall, even though the quantity of rainfall required to trigger failure varies greatly. I apologize if I appeared to have suggested otherwise in my theme lecture.

C: NIEUWENHUIS

Shallow landslides usually move seasonally and rarely move catastrophically. This type of landslide is uncommon in tropical areas. Shallow landslides depend on slow infiltration from rainfall and therefore rainfall intensity is not as important as rainfall duration in causing failure in this type of landslide.

C: BRAND

My theme lecture was obviously directed at areas with high rainfall. As you suggest, slow moving landslides are much less common in tropical areas than are catastrophic ones. There are only a few examples of slow moving landslides in Hong Kong - the vast majority of failures occur suddenly during high intensity rainfall, without prior warning.

It is worth mentioning that most failures in tropical areas, especially of cut slopes, are very shallow. As I said earlier, the Hong Kong correlations show that rainfall duration is not a significant parameter for landslide causation in our high permeability materials.

C: PIERSON

Soils in steep drainage basins on Oahu (Hawaii) are commonly clay rich (clay $> 80\%$) but highly aggregated. The largely sand-size, pellet like soil aggregates have considerable strength when dry or only slightly moist, but when saturated they typically swell and lose most of their shear strength. The clays are predominantly smectite/halloysite interlayers. Accompanying iron oxides and hydroxides are largely crystalline forms. Al-oxides are absent. Parent material is deeply weathered basalt.

Have others had experience with similar soils? I question how such soil aggregates can become so weak when wetted in a dish under the microscope but obviously retain enough strength not to fail under most circumstances at field sites that regularly receive heavy rainfall.

C: PREBBLE

Allophanes produce aggregates as sand size grains. The answer to clay behaviour in relation to landslides may lie in the clay microstructure and may require further research.

C: MEYER/BRAND

(Discussion of antecedent rainfall in relation to different types of landslides.)

Q8: CROZIER

The desire to pinpoint a single cause or trigger for landslides is an oversimplification.

Fiji examples indicate that critical factors change with the conditions. One rainstorm event triggered landslides in areas without adequate vegetation (eg. gardens), whereas another larger magnitude rainstorm event caused landsliding irrespective of vegetation cover.

Water above the shear plane is critical. This is usually the result of infiltration and the existing water table minus the losses of water flowing out of the slope. In respect to antecedent precipitation, do you (Brand) dismiss the water balance approach?

A: BRAND

The logic behind the water balance approach cannot be faulted, and I certainly cannot dismiss it as a sound basis for explaining rain-induced landslides. The Hong Kong correlation study, however, did not concern itself with mechanisms but merely examined the correlation between the occurrence of landslides and the pattern of rainfall. The results from this scientific piece of work show irrefutably that antecedent rainfall is of no importance in Hong Kong. As I said in my lecture, I think it likely that antecedent rainfall becomes increasingly important as soil permeability decreases.

C: ROLD

Antecedent Moisture and It's Relation to Slope Instability.

In Colorado, anecdotal data indicates that heavy fall snowfall before the freezing of the ground usually precedes heavy deep seated landslide activity the following spring and summer. Activity can be especially severe if an unusually high early fall snow season is followed by unusually high spring snowfall. Deep snows on unfrozen ground tend to thermally blanket the ground, thus preventing normal freezing and allowing melting snow to soak into the ground, raising deep soil moisture levels.

On the other hand shallow landslides and debris flows on steep slopes are caused by short term, high intensity thunderstorms.

C: HEARN

Following the discussion on the use of slope height/angle relationships as indicators of cut slope stability, I would like to make reference to a similar database collected by myself along an 80km highway in Nepal. The field data differentiated between soil slopes (most alluvial and colluvial) and phyllite rock slopes. Rock mass and soil mass classification together with slope drainage, landuse and stability data were collected and stability graphs were plotted accordingly. Reasonable relationships were obtained for soil slopes where relatively flat terrace surfaces were recorded above the cut slopes. No clear relationships were evident for the rock slope data and I suggest that this may reflect the influence of slope drainage from long steep slopes above the cut slope, as discussed by L.D.Wesley. There was sufficient scatter to make earthworks design recommendations inappropriate.

C: DeGRAFF

Dr.L.D.Wesley noted, in his panel report, a difference between conclusions stated in my contribution entitled, "Increased Debris Flow Activity Due to Vegetative Change", and the contribution from N.Phien-wej, T.Zhibin, and Z.Aung entitled, "Unprecedented Landslides in Granitic Mountains of Southern Thailand". This difference is the importance of vegetation as a factor in the landslide disaster which occurred in southern Thailand in November 1988.

The steep slopes present in the Khao Luang section of the Nakhon Si Thammarat Range are certainly an important factor contributing to landslide occurrence in this area. Similarly, the granitic bedrock weathers to a surficial material which also increases landslide-susceptibility for this same area. However, these factors exerted the same influence on historic landslide occurrence as they did on the November 1988 disaster. Slope inclination and bedrock type are examples of basic conditions influencing landslide susceptibility (Sharpe, 1938).

As Phien-wej et al note, the November 1988 event involved an unprecedented number of landslides over a large area. The reason for this extensive landslide activity must be a condition which has differed over time rather than a basic condition such as slope inclination or bedrock. Such time-variable factors are referred to as initiating conditions by Sharpe (1938).

The extreme rainfall event of more than 750 mm of rainfall within a 2-day period is the initiating factor identified by Phien-wej et al as responsible for the November 1988 landslide disaster. ESCAP

(1989) reported rainfall for the period 1951 to 1987 for comparison to the 1988 storm event. While the 1988 event was the largest recorded, some recent storm events were nearly as intense. For the recorded rainfall at Nakhon Si Thammarat (Station 552201), the 2-day rainfall for November 21-22, 1988 was 734 mm (ESCAP, 1989). This same station recorded 2-day rainfall amounts of 593 mm for December 12-13, 1984 and 687mm for January 4-5, 1975.

The specific rainfall amounts are probably an unreliable measure of rainfall in the Khao Luang section of the Nakhon Si Thammarat Range. No rainfall stations are located in the area where the landslides occurred. The station at Nakhon Si Thammarat is located about 15 km east of the mountain front at an elevation of less than 500 meters above sealevel. In contrast, the Nakhon Si Thammarat Range ranges from 500 meters to 1,835 meters at Khao Luang peak. As Dr.E.W.Brand noted in his theme presentation for slope stability in tropical areas, our ability to correlate rainfall with landslide occurrence is hampered by rainfall being recorded at stations significantly distant from the landslide locations. Rather than a specific rainfall-landslide correlation, we can only conclude that a storm of extreme magnitude affected the Nakhon Si Thammarat Range on November 1988. Past rainfall records indicate that storm events in 1984 and 1975 provided 2-day rainfall amounts which were of similar magnitudes. It is known these earlier storms triggered landslides in the Nakhon Si Thammarat Range. However, it seems earlier landslide occurrence was far less extensive than occurred in 1988.

My contribution to the proceedings indicated vegetation was the initiating condition which made the effect of the 1988 storm event much more severe than previous storms of a similar magnitude. The number of landslides in cultivated area was greater than for forested area based on my helicopter and ground-based field reconnaissance in early 1989. These observations are consistent with a later satellite imagery study of the Amphoe Phi Pun area. Rosenqvist et al (1990) used photogrammetric analysis of March 1988 and November 1988 imagery to determine that 1300 hectares of landslide scars present in 1988 were in cultivated areas compared to about 870 hectares of landslide scars found in forested areas. The cultivated area includes areas classified as bare ground, sparse vegetation generally associated with young rubber tree plantations, and rubber plantations. Rosenqvist et al (1990) also noted an obvious relationship between landcover and landslides with landslide frequency being highest for

bare ground and decreasing for landcover types with increasing vegetative cover. The tropical forest class had the lowest landslide occurrence.

Phien-wej et al compared the percent of landslide area present for different slope gradients in the Amphoe Phi Pun. This 150 square kilometer area was one of the most heavily damaged in the Nakhon Si Thammarat Range by the November 1988 storm event. Their comparison of landslides to vegetation is given in Figure 7.

The vegetation classes used are young rubber plantations (less than 5 years old), old rubber plantations, and natural tropical forest. From Figure 7, it is clear most of the landslides occurred on slopes less than 30 degrees irrespective of vegetative cover present. Landslide area under natural forest cover is shown as a greater percentage of the three slope classes within this 30 degree range for slope gradient.

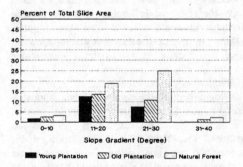

Fig. 7 Correlation of slide occurrence and vegetation type

If young plantations and old plantations are combined, which is comparable to the cultivated landcover category of Rosenqvist et al (1990), the percentage of total landslide area is greater for plantations than for natural forest cover for slope gradients of 0 to 10 degrees and 11 to 20 degrees and nearly equal for slope gradients of 21 to 30 degrees. This would suggest a greater correlation between landslide occurrence and rubber plantations more consistent with the conclusions drawn by Rosenqvist et al (1990) and my own field observations.

There is no indication of why rubber plantations were segregated into two classes. Rubber is a naturally shallow-rooted tree species. This tendency for a shallow root network is encouraged by silviculture practices. As a consequence, neither young or old rubber plantations offer a significant difference in root reinforcement to soil compared to natural forest vegetation. Likewise, the younger rubber plantations have very open, single-stored canopies. Old rubber plantations also have single-stored canopies. I observed canopy openings of 30 to 40 percent in old rubber plantations which does not seem significantly different from the openings in young rubber plantations. Therefore, there does not seem to be any slope stability effect from rubber tree plantations which would justify discriminating between young and old plantations.

The 1988 storm event was an extreme storm. However, it does not seem sufficiently greater than some previous storms in the area to create the extensive landslide disaster that occurred in southern Thailand. The other initiating condition which has changed significantly since the last comparably large storm in 1984 is the amount of rubber tree plantations present in the Nakhon Si Thammarat Range. Expansion of rubber trees into the natural forests began in the mid-1980's. Rosenqvist et al (1990) found nearly 4000 hectares of natural forest was converted to rubber plantations between 1984 and 1988.

Based on the discussion above, it would appear the contrary conclusions reported in my contribution and that of Phien-wej et al arose from the differences in representing vegetation data rather than an actual difference in causative factors for the landslide occurrence in 1988.

REFERENCES

ESCAP 1989. ESCAP technical assistance to the flood affected areas in southern Thailand. Bangkok: UN/ESCAP.

Rosenqvist,A., S.Mura, & S.Vilbulsreth 1990. Flood damage analysis in southern Thailand. Proc. 23rd Int. Symp. of Remote Sensing of Environment: 315-324. Ann Arbor, MI: ERIM.

Sharp,C.F.S. 1938. Landslides and related phenomena. New York: Columbia University Press.

C: RIEMER

Concerning long-term effectiveness of drainage. Methods similar to the RODREN drainage methods have been applied in Iran and Afghanistan for over 1000 years to collect groundwater (Qanat, respectively Karez systems).

Q9: HAWKINS

I understand that more than 92 reinforced earth embankments have failed out of 16000 mentioned by Schuster. How many have failed and by what mechanism?

A: SCHUSTER

Most have failed due to foundation collapse or due to structural corrosion. Others possibly have failed due to their being underdesigned.

Although I know of a few individual cases of failure of reinforced soil embankments, I have no information on the total number of such failures worldwide. In regard to mechanisms of failure: my feeling is that most reinforced soil failures have been due to inadequate foundation conditions and/or faulty design. With increasing long-term experience in design and construction of these walls, the percentage of failures of future embankments should decrease. A few reinforced soil embankments apparently have undergone distress caused by failure of structural elements due to corrosion of metals or chemical breakdown of geosynthetics. Long-term protection of metal and geosynthetic structural elements has been the subject of considerable world-wide research for the past few years. This research is leading to development of new chemically-resistant materials and material composites that will result in structural elements capable of increasing the long-term stability of future reinforced soil structures.

C: NIEUWENHUIS

Examples highlight the need for studies of both groundwater and slope movement in conjunction. An example is cited from Norway where the removal of groundwater caused settlement. Once started, long-term movement is very difficult to stop by controlling groundwater alone.

C: POPESCU

Commenting on the relative use of piles to control landslides in East and West countries - a problem raised by the Panel Report - it is to be said that the easiest to get technology prevailed in Eastern Europe during the last decade. Due to this reason, piles in rows or lattices have been used widely in Romania as remedial measures. Probably in some instances other stabilisation methods would be more cost effective.

However, the design of bored piles to control landslides has the merit of being flexible enough for changes during or subsequent to the construction of remedial measures. The number, distance between or embedded depth of piles may be easily modified according to the real ground conditions recognized during drilling operations.

Isolated restraining structures on piles are often combined with modification of slope geometry and drainage measures in order to reduce the lateral force on restraining structures.

C: FORRESTER

Mention has been made during this discussion session of the long-term deterioration of subsoil drainage systems. Within my experience, a major cause of deterioration of drains is ochre clogging. This is a subject that has been investigated for many years in agricultural work, but has been mostly ignored in civil engineering. A paper on the ochre problem, as applied to landslide control, has been submitted to this Symposium as a late contribution.

C: MOSTYN

The failure rates for reinforced earth walls are shown as about 1% and this should not be a cause of specific concern as this rate is not too different from those of most major types of civil engineering structures.

C: CHOWDHURY

Caution is required when recommending remedial or preventive measures in general terms. It is necessary to specify detailed conditions concerning an individual site before recommending a solution and retaining structures are more likely to be successful where the depth of potential movement or landsliding is relatively shallow. Drainage would have to be part of the solution if high pore water pressures are likely to be the cause of failure, even when retaining structures are used in implementing such a solution in it's totality. There is little information available in the literature about the earth pressures exerted on the structures retaining a moving earth mass. Such structures are likely to be subjected to dynamic forces.

The usual design procedures using earth pressure coefficients for static lateral loading are not necessarily appropriate. Research is needed to learn more about the magnitudes of earth pressure coefficients applicable to retaining structures for sloping earth masses subject to potential movement.

Drainage is a solution often adopted for deep-seated slides. It is used, in many cases, in combination with other measures. Retaining structures, on their own, have not often been employed successfully for deep-seated slides especially if high pore pressures are the likely cause of potential failure.

C: HUTCHINSON

Comment on stabilisation of rock slope by anchors, stabilisation of clay cut by trench drains, and overturning of wall across a slide by inferred passive pressures.

C: DUNNICLIFF

There have been a couple of recent innovations in the monitoring and instrumentation field - perhaps they will eventually be used instead of inclinometers! They are:
- time domain reflectivity
- fibre optics.

Brief information is given by Dunnicliff, 1988 (Geotechnical Instrumentation for Monitoring Field Performance. John Wiley and Sons Inc., New York, 577pp.)

C: MORGENSTERN

There is much improved instrumentation these days, such as that used where 500km was mined under highwall reaching temperatures of -40°. Also good instrumentation of a 17km periphery tailings dam which is moving.

Now moving towards the "smart borehole" with automated inclinometers, etc. Two new innovations are in monitoring localised strain and the automated EDM.

Q10: JACOBSON

In the monitoring at Clyde, has consideration been given to the possibility of stick-slip?

A: PROFFITT

Yes it has, and a number of holes have one tiltmeter.

C: FOSTER

The aspect of walkover survey of slides is important as well as instrumentation.

Q11: McMAHON

Talking about negative pore pressure, is the problem of air significant? If the air is from water, then it must be closer to atmospheric pressure than the water was. Obviously would be trouble if negative pore pressure and positive water pressure, but that is impossible.

A: DUNNICLIFF

When monitoring negative pore pressure, there will usually be both gas and water in the pores, with the gas pressure higher than the water pressure. In fine grained materials, the difference can be very significant, and we need to know which we are measuring. If we install a conventional vibrating wire or pneumatic piezometer, with a standard coarse filter, we will measure the pore gas pressure, whereas if we plan to make an effective stress stability analysis, we are more interested in the pore water pressure. Methods of measuring pore water pressure in soils where both gas and water are present, are discussed by Dunnicliff, 1988 (cited above).

Q12: McMAHON

There is instrumentation at Clyde to protect against disaster, but it basically assumes that the disaster will never happen. The monitoring centre would be wiped out in a Vaiont-type event and evacuation procedures, etc., are inadequate.

A: FOSTER

Have looked at the potential of waves if a rapid failure occurs. The Cromwell office would be okay as it is positioned 20m above the lake. Instruments can pick up movement of concern and there is good emergency preparedness. Inundation maps of Cromwell have been compiled, plus ones for Lake Hawea. Have linked the project to Civil Defence.

Now have ideas on what levels of readings on the instruments are unexpected or unacceptable and also how to handle them.

Q13: MIKKELSEN (to Cancelli)

Would like further details on the multiple extensometer.

A: CANCELLI

Concerning the multiple extensometer put in place at Camporaghena landslide, the drawing is very synthetic and can be misleading. Actually, it consists of 6 wires, individually connected to 6 mechanical anchors placed at different depth.

Q14: ANON.

Why has there not been more work done on negative pore pressure and unsaturated flow?

A: IVERSON

Significant advances in understanding and monitoring unsaturated flow have been made within the hydrology community. Not all of these have been appreciated by the geotechnical engineering community. Excellent sources of information include journals such as Water Resources Research and the Journal of the Soil Science Society of America.

C: IVERSON

I would like to point out the need for controlled, replicable experiments to better understand the performance of instrumentation and the reliability of monitoring and model predictions. The newly constructed U.S.Geological Survey debris-flow flume provides a unique opportunity for replicable experiments at nearly field scale.

C: MARTIN

There were various references to the importance of monitoring and performance assessment of drainage measures in both the G3 and G5 sessions this morning. I would like to briefly comment on our experience with assessing the effectiveness of horizontal drains in typical Hong Kong slopes.

The GEO began a review of horizontal drain performance in 1987. Information was gathered from about 35 sites where drains had been installed. The length of the data record (as at late 1991) varies from 5 to 17 years. Seven cases were found to be fairly well-documented and all the available information on construction techniques, ground conditions and drain performance was examined and evaluated in some detail.

The main conclusions from this review to date are:
(a) Horizontal drains are an effective means of lowering groundwater levels in slopes formed in saprolites and residual soils. A number of different drain types and installation techniques have been found suitable in Hong Kong for achieving significant reductions in piezometric levels. Drains varying in length from 10-35m (exceptionally 90m), typically spaced in several rows on the slope at fairly close horizontal centres (usually <20m), have resulted in drawdowns generally in the range 3-5m but locally up to 15m.
(b) Clogging does not appear to be a significant factor in reducing drain discharges over periods of 5-15 years in soils formed from insitu weathering of generally coarse-grained granites and tuffs.
(c) Flows from different drains at a site typically vary very widely. Often only a small percentage of the total number of drains account for most of the total discharge. The observational approach to design, combined with phased construction, is likely to be beneficial in many cases of drain installation in heterogeneous soils with complex near-surface hydrogeology, i.e. where possible individual new drain locations should be adjusted during construction by taking into account piezometric data and/or flows from existing drains.
(d) Monitoring frequencies can be reduced with time once the performance of the drainage system has been adequately assessed and found acceptable (say in the first two years of installation). The reduction in the frequency should be consistent with the risk of slope failure and the designer's confidence in the performance of the system. Regular maintenance, especially clearance of drain outlets and flushing of drains with markedly reduced flows, will help to ensure good long-term performance.

C: HANCOX

So far our preliminary work has been restricted to office studies, and in a few cases, brief aerial and ground reconnaissance. As indicated in our paper, we intend to continue our work by carrying out studies of selected landslide-dammed lakes (for example Green Lake, Tutira, Waikaremoana, Polnoon, Lochnagar and several 1929 lakes in NW Nelson) to get better data on landslide characteristics and slide models. This will help clarify our thinking on failure mechanisms.

As far as hazard is concerned, all lakes identified so far will be subjectively assessed and assigned a hazard rating. Special attention will be given to lakes where communities or structures downstream would be at risk if the landslide dam failed. The criteria that would need to be taken into account for hazard assessment are generally well known, and include landslide dam materials and volume, lake volume, catchment outflow, and dam overtopping potential during major floods.

Geological evidence of past overtopping and partial breaches (as shown by Fig 3 of our paper) would also be sought. The age of the dam would also need to be considered, but with some caution, since even dams that have survived hundreds or thousands of years (such as the Polnoon barrier, Fig 3) can be overtopped and breached. On the other hand, some historical dams have survived only a few years before failure (e.g. the Te Hoe landslide dam, formed during the 1931 Napier earthquake). The hazard assessment process needs to allow for these and other uncertainties. This will be attempted using appropriate criteria and weighting factors to take known characteristics and uncertainties into account.

As an adjunct to the hazard issue, it is interesting to note the longevity characteristics of landslide dams in New Zealand. Studies by Costa and Schuster (1988) suggest that 85-90% of the landslide-dam failures occurred within one year of formation. However, although they do not give specific statistics for the percentage of dams that have failed compared to the total number of dams formed, a failure rate of about 32% may be inferred (of the 225 natural dams studied, 73 had failed). A similar failure rate has been determined for landslide-dammed lakes in New Zealand. Data presented in our paper show that, of about 80 known landslide dams in New Zealand, 28% (22) have failed, and some 72% remain intact with existing lakes (Table 1a).

Some authors (for example Imrie, in press) believe that landslide dam barriers "last only briefly", and find it unusual for landslide dams, such as that at Lake Waikaremoana, to survive for thousands of years. However, our work has shown that many (about 72%) landslide dams in New Zealand, both historic and prehistoric, have survived for tens to thousands of years. Similar figures appear to be presented by Costa and Schuster (1988), although their paper is not explicit on this issue.

One interesting observation that emerged from our studies was the large number of new lakes and former lakes that were found to be impounded by landslide dams. Often the reason for a lake's existence remains undetermined until a comprehensive search of maps and aerial photos is undertaken. Generally, there are far more

Table 1a. Statistics on failure rate of prehistoric and historic landslide dams in New Zealand
(addendum to Perrin and Hancox, in press)

LANDSLIDE-DAMMED LAKES IN NEW ZEALAND			
PREHISTORIC LANDSLIDE DAMS [#52 = 65%]		HISTORIC LANDSLIDE DAMS [#28 = 35%]	
Existing Lakes (dam intact)	Former Lakes (dam failed)	Existing Lakes (dam intact)	Former Lakes (dam failed)
#38 – 73% prehist/lakes – 48% of total lakes	#14 – 27% prehist/lakes – 8% of total lakes	#20 – 71% hist lakes – 25% of total lakes	#8 – 29% of hist lakes – 10% of total lakes
Failure Rate: 27% of the prehistoric dams have failed		Failure Rate: 29% of the historic dams have failed	
Overall Failure Rate: Of about 80 known landslide dams some 22 (28%) have failed, 72% have existing lakes.			

landslide-dammed lakes than is believed, and many will have existed for hundreds or thousands of years.

C: RAHARDJO

Significant progress has been made in developing sophisticated and advanced technology for measuring positive pore-water pressures. However, we appear to be slow in advancing the technology for measuring negative pore-water pressures. Commercially available devices for measuring negative pore-water pressures have limitations in their applications to geotechnical problems. The ability to measure the negative pore-water pressures correctly will greatly assist us in understanding the relationship between the rainfall and infiltration rates as pertinent to the analyses of shallow landslides. Perhaps the panel members can comment on the slow response from the geotechnical community in addressing the problems of measuring the negative pore-water pressures and applying this theory to practical problems. For example, it is a well known and established fact that slopes often fail after prolonged periods of rain. This influx of water must pass through the zone of soil with negative pore-water pressures. As a result, the understanding of flow through the unsaturated zone becomes the key theory to understand the potential for catastrophic landslides associated with prolonged rainy periods.

A: DUNNICLIFF (to Rahardjo)

Most advances in this field have been made by pavement engineers. Negative pore water pressure cannot readily be measured using conventional diaphragm piezometers.

A: ANON (to Rahardjo)

Information on advances in the hydrological community can be found in the Water Resources Journal.

Q15: ANON

Confused about negative pore pressure in shallow or superficial failures. Negative pore pressure not so involved in deep failures.

A: MOSTYN

Generally many slopes in Australia can only be considered stable if some account is taken of negative pore pressures that must exist. Nevertheless it appears that the available instrumentation is not adequate to monitor high negative pore water pressures in the field. The effect of these negative pore water pressures generally is very small on failures deeper than, say, 10m.

C: GEE

The subject of landslides forming dams that create lakes has come up in many sessions of this conference. There seems to be a general feeling that these lakes are unusual and rare features. I ask the question "How are lakes normally created?" It appears to me that most large lakes are dammed either by glacial moraines or by landslides. Other ways of creating lakes are coastal spits etc., folding and warping tectonically, volcanic disturbance and glacial cirques - but these seem relatively rare compared to moraine and landslide dammed lakes. I suggest that landsliding is a comparatively normal way to create a lake. Does anyone have any comment on these thoughts?

C: VANDINE

Can I suggest a practical method for monitoring the affect of a reservoir on shoreline erosion, beach formation and slope stability. That is, taking video recordings at an appropriate oblique angle from a helicopter. These recordings should be taken before, at intervals during filling, and at intervals after filling. This technique has been successfully used filling. This technique has been successfully used for coastal shoreline stability and regression studies along the Gulf of Mexico and along the Arctic Coast of Canada.

C: ROLD

Problems of hazard assessment are not so much related to the state of art but to other factors. Consultants' work is limited by the client's desire. Academicians follow their research interests or fads of research funding. Government agencies are usually underfunded and forced to yield to political direction or following public interests of the moment. A major problem is obtaining the funding and authorisation for adequate hazard evaluation. Consultants must convince their clients of the need for deeper and more areally extensive investigations. Academicians need longer term funding, Government agencies must obtain greater funding and support in times when hazards appear to be low.

In addition, a major problem is a lack of basic geologic data and sufficiently detailed geological mapping. No matter how sophisticated the systems such as GIS, RIVET, SISYPPE may be, or how exact the geomechanical or engineering calculations may be, without adequate geologic and geohydrologic data the system will be of little value.

C: TAYLOR

Geotechnical Engineers and Engineering Geologists are engaged to make technical decisions about land stability - not economic ones.

In urban or sub-urban land development for residential subdivision particular problems arise and the technical advisers are often squeezed into accepting a responsibility which should be borne by others.

Prior to development of the land, all the pressure comes from the owner/developer to realise his asset but, in New Zealand at least, the decision to accept or decline the proposal has to be taken by the Local Body, which will commonly rely upon a report (on land stability) prepared by an Engineer engaged by the developer.

All too often development proposals relate to land which has not been assessed for stability before zoning for particular uses by the Local Body in which case there is already a strong predisposition to approve rather than reject the proposal.

After development subdivision and sale, the vulnerability factor in the equation

RISK = HAZARD X VULNERABILITY

has another dimension because the land is now the (perhaps the only real) asset of people of modest means with little capacity to investigate, pursue at law, or mitigate precautionary or remedial works if instability develops.

This calls for conservative decisions about land use. The Engineer/Geologist adviser has a duty not to convey a precision of calculation of certainty of evaluation out of step with all the variables of the prevailing ground conditions. He/she must convey the risks and uncertainties to the owners (or the Local Body) in terms which allow them to make their own decisions. Unfortunately, Engineers at least, often are not objective in their advice, and become the "fall guys".

A solidarity within the professional societies, at least as to the extent of site investigations appropriate to the residential land, would help Engineers and Geologists resist pressures from their clients.

Q16: FELL (to Brown & Cruden)

Would you please advise whether you are now accepting data on individual landslides for the World Inventory, and if so, in what format (presumably in the format recommended by the Landslide Working Party), and do you want historical data, or only data from now?

A: BROWN

All information will be accepted in all forms.

C: OLDS

There is a need for standards within geomechanics societies, and people must stand up to clients.

(Also comments on landslide hazard zoning in residential areas and some issues raised by a recent debris flow zoning study).

C: IVERSON

An important element of hazard assessment involves prediction of post-failure runout distances, paths, and velocities of debris flows and avalanches. Our ability to make such predictions is presently limited by a poor understanding of some of the physical processes operating in these rapid mass movements. Building a better understanding requires systematic hypothesis testing by way of replicable experiments. The U.S. Geological Survey has recently developed a debris flow flume facility for performing replicable experiments at almost field scale.

Q17: BONNARD (to Hutchinson)

How would you establish alarm criteria on the basis of movement records? Will displacements, velocities or accelerations be considered? Will these data be combined with piezometric data to set an alarm threshold? Can these questions be illustrated by the case of Dunstan reservoir monitoring programme? Can we use these data to produce local hazard maps?

A: HUTCHINSON

Ideally, piezometric data, displacements, velocities and accelerations would be considered. Where data was sufficient it may be feasible to set a piezometric level alarm threshold to warn of impending movement. Once movements are occuring, the level of these in relation to the elements at risk can be assessed. Also it is clear that accelerating movements are more hazardous than steady ones and those are worse than decelerating ones. Whether the slide is first time or on pre-existing shears is relevant. In the former case a Saito/Varnes/Fukuzono analysis of the tertiary creep phase may be useful. Such data form a very useful component of landslide hazard maps (but say little about run-out).

The question re Dunstan Reservior would be best answered by the local experts.

Q18: ANON (to Hutchinson)

What is the expected annual dollar cost of the monitoring programme at Clyde?

A: HUTCHINSON

Not sure!

C: CHEN ZUYU

The Water Resources Related Landslide Inventory in China has received information on reservoir impounding induced landslides, excavation induced landslides, and landslide dams.

C: ROLD (to Popescu)

Dr. Popescu's figure #1 shows three categories of stability; "stable, marginally stable, and unstable". I would suggest a very important classification "potentially unstable". Many areas are standing stable at present by whatever criteria one might use. However, the geologic environment is such that any experienced engineering geologist or geomorphologist would be certain that there's a high probability that serious or significant movement will occur.

C: POPESCU

In our suggested method for reporting landslide causes, slopes are visualised in one of the three following stages: stable, marginally stable and actively unstable. The three stability stages provide a useful framework for understanding the causes of landslides and classifying these into two groups on the basis of their function:

1) preparatory causes which make the slope susceptible to movement without actually initiating it and thereby tending to place the slope in a marginally stable state.

2) triggering causes which initiate movement. These causes shift the slope from a marginally stable state to an actively unstable state.

A particular cause may perform either or both functions, depending on it's degree of activity and the margin of stability. From this point of view, potentially unstable slopes are a particular case of marginally stable slopes.

C: HEARN

In 1983 I carried out landslide hazard mapping studies for route alignment purposes in east Nepal. I found the system based on terrain units and mapped terrain factors to be far more successful in explaining the distribution of landslides than the approach based on a grid system.

Nine years later I have been fortunate enough to test the terrain unit hazard mapping scheme with landslide data from recent air photographs. I have not yet completed the analysis but it would seem that the predicting ability of the technique is better in some catchments than in others. The reasons for this have yet to be ascertained.

C: HEARN

I think that the need to validate and calibrate hazard and terrain mapping and classification with real life landslide events is an important one. Since submission of my hazard mapping report to Ok Tedi Mining Ltd. an erosion event leading to the temporary loss of a power pylon has occurred. This event took place in a high hazard zone according to my mapping scheme. There has also been a recommendation for remedial action. This outcome provides some support to the classification scheme adopted, although it is early days yet.

C: FORRESTER

During the last few years, Roads and Traffic Authority, New South Wales (RTA) have made use of acoustic emission (AE) monitoring in certain slope stability problems. In appropriate circumstances, it has proved to be particularly useful for assessing the effectiveness of landslide control measures, as given by comparing sets of readings taken before and after completion of the work. AE monitoring can also indicate whether any parts of the landslide remain unstable.

An example of how useful the method can be may be seen from a landslide project near Newcastle, New South Wales. A cross section showed four separate blocks of moving material (Fig 1): two blocks towards the top of the slope, a graben block and a toe block.

Stability calculations showed that the safety factor of the whole system would be significantly below 1.5 with drainage alone, assuming a conservative lowering in piezometric levels. The need for some kind of additional restraining system was therefore anticipated. Because of the mass of moving material involved, such a system would have to be large, probably involving prestressed ground anchors. AE monitoring before placing horizontal drains, and afterwards following heavy rain, showed event count rates (per minute) of, respectively, 14.1 and 2.8 for the two head blocks, 3.5 and 0.2 for the graben block, and 5.2 and 6.1 for the toe block. The head blocks and the graben blocks were therefore considered to be stable, with the toe block alone still in doubt. To stabilise this, a simpler and much cheaper structure, such as a cribwall, would only be needed. In the six years since the project was finished, a period that has included an earthquake and several very wet periods, particular attention has been given to the toe block, but so far the maximum count rate subsequently recorded there has been 2.4 events/minute, and no movement has been noted. Consequently, no further control work has been considered necessary up to the present time, although monitoring and observation will continue. This landslide was originally described in the following reference:

Forrester,K. 1987. The Carisbrooke Avenue landslide. In: Walker,B.F. & Fell,R. (eds) 1987. Soil Slope Instability and Stabilisation. Sydney: 337-346.
Rotterdam: Balkema.

RTA use a Slope Indicator Co MS-2 for this work. It has hydrophone transducers, 22 mm dia. The UPVC tubing used in RTA's standpipe piezometers has been increased from 20 to 25 mm bore, so that the piezometers can be used as AE monitoring locations, as well as for water level measurements.

RTA have recently published a document on AE entitled: Materials Guide No.2. Acoustic Emission.

C: HUTCHINSON (to Forrester)

I believe that A.E. tends to be irregular rather than monotonically increasing towards failure - and that even periods of quiescence can occur on the way to failure.

I therefore feel that it is unsatisfactory to rely on A.E. measurements taken only once and that a continuing series of such measurements are needed to form a sound basis for judgements about stability.

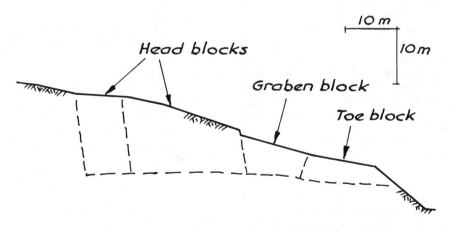

Fig 1. Carisbrooke Avenue landslide. Typical cross section.

Q19: BONNARD

Comment on landslide instability mapping. Has work been done on assessment of landslide runout zones?

C: BELL

Comment on the importance of runout zone definition.

A: ROLD

Colorado Geological Survey has done considerable work to determine runout of snow avalanches and rockfall. Snow avalanche work built on formulae developed by the Swiss (Frutiger and others). Rockfall was developed with the Colorado Department of Transportation. In each, the models have utilized considerable field and empirical data.

References for snow are: Mears, Art; Bulletin 38 (now revised as Bulletin 49) Colorado Geol.Survey, and Mears, Art; Special Publ.#7, Colorado Geol.Survey. Rockfall work is in press by Rick Andrew and Bob Barrett of Colorado Geol.Survey and Colorado Dept. of Transportation respectively. Field evidence from mudflow and debris flows show runouts several times greater than would be indicated by numerical calculation.

Q20: EVANS

a) What studies have been done to assess the effects of a catastrophic failure at Clyde, what methods were used and what results were obtained?

b) Is there any evidence of past catastrophic sliding in the Cromwell Gorge, and if not, what is the justification for the remedial works?

A: HEER

I can only answer in a general sense as follows: Studies have been undertaken to establish the magnitude of overtopping waves generated by the rapid movement of the Clyde, Jackson Creek and Nine Mile Slides. These studies were based on estimates of slide volumes and velocities and took into account the shape of the reservoir in assessing the extent to which waves would be attenuated. From memory, the Clyde Slide could generate an overtopping wave of 25m, Jackson Creek, 5m and Nine Mile Creek 5m.

To my knowledge there is no evidence in the Cromwell gorge of past catastrophic failures.

A: STAPLEDON (to Evans (b))

No evidence has been found in the Cromwell Gorge area of rapid catastrophic landslides which have occurred in the past. However, even relatively slow movements could be extremely serious to the project if they resulted, for instance, in blocking of the reservoir.

That such movements could occur if no stabilizing works were constructed was illustrated by the behaviour of Jackson Creek Slide during preparation of the foundation for the stabilizing fill. Excavation of only 4000m^3 of loess near the toe of this 5 million cubic metre slide, caused the slide to accelerate from 10-20 mm/year to 8 mm/day.

A: HUTCHINSON

Catastrophic failures are commonly associated with:
- very high relief (rock falls and sturzstroms)
- brittleness in shear:
  eg. peak -> residual (first time slides)
  eg. liquefaction (flow slides)
  eg. negative rate effects on residual shear (Vaiont gouge)
- also, sudden access of water to loose regoliths (as in volcanic and non-volcanic debris flows).

C: MORGENSTERN

A variety of materials are responsible. For waste dump material need a geological precedent or model to guide the investigation and find analogues.

C: CHOWDHURY

We are lacking models for catastrophic landslides. Need a geotechnical model of progressive failure including how failure develops and post-failure behaviour. Monitoring and observation are good, but finite element models are not great.

Q21: McMAHON

Can the risks at Clyde be considered extremely low when there are obvious model uncertainties?

A: GRAHAM

The risk of catastrophic failure during lakefilling is considered extremely low based on geological precedent arguments that show satisfactory performance for a minimum of 16000 years and the fact that engineering works have been designed to offset lakefilling effects. Your comment is correct, there are uncertainties in the models; the approach to managing these uncertainties is to observe and assess landslide response to lakefilling and to carry out additional engineering works to correct any undesirable and unexpected response. The monitoring system is the key to managing lakefilling. The system is robust, has redundancy built in, and key instruments are telemetered to the office.

Q22: CHOWDHURY (to Graham)

Were any numerical calculations of probabilities done at any stage of the Clyde Power Project, and was any quantitative risk modelling done?

A: GRAHAM

No quantitative probabilistic analysis was carried out because of the difficulty in obtaining necessary input data for realistic probabilistic analysis.

Q23: CHOWDHURY

How often are probabilistic calculations carried out in North America?

A: MORGENSTERN

The role of probabilistic analysis is about zero. Probabilistic analysis always seeks out the weakest link.

C: McMAHON

In Australia, pit slopes are designed on the safety factor and probabilistic analysis. About 50% of those designed from safety factor fail, but those with a probabilistic basis are lower risk.

C: VALENZUELA

Following comments on the difficulties in assessing hazard of catastrophic landslides (Q19: Evans), a case history was shown on an attempt to assess such a hazard in the Chilean Andes Mountains for a mining company.

C: IVERSON

On the basis of previous discussion in this meeting, I formed the impression that considerable disagreement exists about the role of groundwater in landslides. In contrast, I personally believe that the fundamentals of groundwater flow in slopes are understood quite well. Data are not always adequate, of course, but the physical processes themselves are not mysterious. Is there disagreement on this subject?

C: BONNARD

There are difficulties in resolving the amount of rainfall versus movement.

C: PATTON

There is a gap between hydrogeological knowledge and the application of this to landslides.

C: BRAND

It seems that not enough is known about groundwater.

Q24: PATTON (to Hutchinson)

In Figure 8 of your paper you have presented results of a slow and fast ring shear test made on a sample of material from the Vaiont slide slip surface. You conclude that the effects shown (both higher and lower shear strengths than the slow residual strength), "if representative, is sufficient to explain the high speed of the Vaiont slide without the need to involve slip surface heating".

Did you measure pore pressures and temperature of the failure surface in such tests, and if not, how did you reach your conclusion?

A:HUTCHINSON

Measurements of temperature and pore-water pressure were made in the Vaiont rapid ring shear test, and were found to be insignificant. We believe, therefore, that a mechanism of strength loss independant of the slip surface heating mechansim does exist.

Landslides, Bell (ed.) © 1995 Balkema, Rotterdam, ISBN 90 5410 032 X

Errata

Volume 2, pages 899-904

A methodological approach to landslide hazard assessment: A case history by L.Cascini, S.Critelli, S.Di Nocera & G.Gullà

p. 903 Left column. The figure below replaces Fig.6:

p. 903 Right column, lines 8 to 11:

ranges from 0.5 to 4 ha. These landslides involve 13% of the rocks of class VI, 8% of the rocks of class V, 8% of the rocks of class IV and, finally, 6% of the rocks of class III.

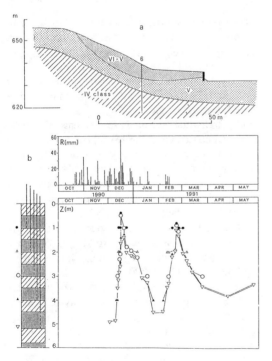

Fig. 6 "Vigne" landslide; a) Cross-section; b) Depth of piezometric levels from ground surface

Volume 2, pages 1031-1036

Rock-avalanche risk at Platì, Southern Italy by
P.G. Nicoletti
Text of subheading 4 must be read as follows:

4 GEOMORPHIC AND STRUCTURAL CONTROLS

Local morphology would play a key role, so that an attempt will be made to determine its effects. Nicoletti and Sorriso-Valvo (in press) have identified three basic ways in which local morphology can control the shape and mobility of rock avalanches. They can be briefly described as (Figure 4): A) Channeling, which favours mobility; B) Spreading on a free surface, which is less favourable to mobility and, C) High-angle (> 60°-70°) impact against an opposite slope, which hampers it.

Unfortunately, in the case under examination, it is difficult to establish whether A or C would occur. In particular, it is difficult to establish what the angle of impact between the moving mass and the opposite slope would be in the initial stage of the event. Should the angle be less than 60°-70° we can expect a situation similar to A, otherwise we can expect situation C.

At the foot of the Sackung, the fiumara di Platì runs through a narrow valley which strikes 210° (downstream) and has a 14% slope (A-B, Figures 1 and 2). The segment which follows strikes 185° and has a 13.8% slope, the final segment strikes 141° and has a 11.7% slope. The slope facing the Sackung, which would work as a deflection surface, strikes 190°-195°.

The movements that have already taken place within the Sackung mass and along its crown were examined from aerial photos and seem to have occured with directions between 130° and 155°, averaging 145° (141° in the crown area). These directions coincide only occasionally with the direction of maximum dip of the slope (about 130°), but this is not surprising as phenomena of the kind we are dealing with are often controlled more by structural factors than by surficial morphology (Dramis and others 1985). The upper margins of the scars left by these masses tend to follow the main fracture set visible in the aerial photos (Figure 5), which strikes between 040° and 070°.

Field measurements taken on fractures and schistosity resulted in the picture given in Figure 6, whose most interesting features are perhaps the remarkable concentration of poles in the NW sector and the secondary concentration in the NE sector. These are consistent with both the above mentioned directions of movement and the main fracture sets observed in the aerial photos.

The average attitude of schistosity is around 115°/42°, and the dip is greater than the slope, but in the NW sector of the equal-area projection of Figure 6 there are also several fractures dipping with angles both higher and, in a few cases, lower than 42°. Some of these fractures have dip-angles smaller than 22°, i.e. they dip out of the slope.

The poles forming the secondary concentration (NE sector) are in good agreement with the fracture set striking between 110° and 160°, also detected from aerial photos (Figure 5). This fracture set and its remarkable dip (55° to 88°) could well be critical factors as far as direction of movement is concerned, as they are probably responsible for the tendency to move with direction 140°-145°.

Data presently available do not permit the estimation of the stability of the rock mass. It can however be noted that the observed distribution of fracture and schistosity planes 1) represents a necessary condition for the instability of the rock mass and, 2) could cause the formation of a fairly curvilinear slip surface.

From the whole set of these elements it may be conjectured that the collapse of the Sackung would occur with a movement striking about 140°-145°, and therefore at an angle of about 50° to the opposite slope. Such an angle seems too small to seriously hinder motion, so that channelling would probably take place.

After the first deflection, the debris should move along the valley segment striking 185°, then swerve anticlockwise to enter the final segment striking 141°. About 700 m before Platì, because of a change in the underlying terrane, the valley suddenly widens. The minor sinuosity of path, as well as its final widening, could lead to a certain braking effect, but how much influence it would have, is impossible to state.

A further complication arises from the narrowness of the initial segment of the valley, which might provoke clogging. Moreover, since the motion would start as a rockslide, the initial speed should not be particularly high. On the other hand, if the failure occurs in seismic conditions, as seems more likely, the vibration could help to keep the mass in motion.

Finally, since in the Aspromonte Massif the weather is severe and tectonic uplifting is active, the possibility of a failure due to the effects of long-term, unidirectional processes like weathering and uplift should not be totally disregarded. As Terzaghi (1950) noted, these were the causes of the disaster of Goldau (Switzerland, 1806).

Volume 2, pages 1265-1270

Initial deformations of high cuts in overconsolidated jointed clays by P. Bertuccioli,
D. Distefano, F. Esu & G. Federico

Page	Column	Row	Errata	Corrigenda
1266	right	12th from top	85	75
1267	right	8th from bottom	250	140

Page 1268: Fig. 5c should be substituted with the one pasted below:

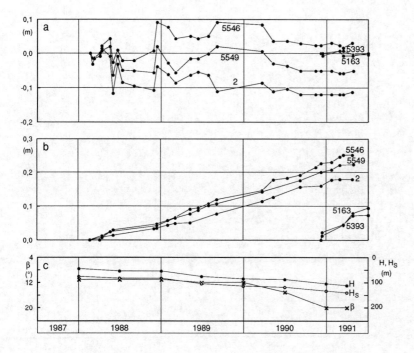

Landslides, Bell (ed.) © 1995 Balkema, Rotterdam, ISBN 90 5410 032 X

Author index*